CANCER BIOLOGY

Cancer Biology

THIRD EDITION

RAYMOND W. RUDDON, M.D., Ph.D.

New York Oxford
Oxford University Press
1995

Oxford University Press

Oxford New York
Athens Auckland Bangkok Bombay
Calcutta Cape Town Dar es Salaam Delhi
Florence Hong Kong Istanbul Karachi
Kuala Lumpur Madras Madrid Melbourne
Mexico City Nairobi Paris Singapore
Taipei Tokyo Toronto

and associated companies in
Berlin Ibadan

Published by Oxford University Press, Inc.,
198 Madison Avenue, New York, New York 10016

Library of Congress Cataloging-in-Publication Data
Ruddon, Raymond W., 1936–
Cancer biology / by Raymond W. Ruddon.—3rd ed.
p. cm.
Includes bibliographical references and index.
ISBN 0-19-509690-8 (cloth).—ISBN 0-19- 509691-6 (pbk.)
1. Cancer. 2. Carcinogenesis. 3. Molecular biology. I. Title.
[DNLM: 1. Cell Transformation, Neoplastic. 2. Neoplasms—etiology. QZ 202 R914c 1995]
RC261.R85 1995
(616.99'407—dc20)
DNLM/DLC
for Library of Congress 94-35524

9 8 7 6

Printed in the United States of America
on acid-free paper

Preface

The goals of the third edition of *Cancer Biology* are similar to those of the first two editions, namely, to provide a historical perspective on key developments in cancer research and an intellectual framework for understanding new developments in cancer research. This is indeed a formidable task and requires an integration of knowledge from several fields of research, including epidemiology, pathology, genetics, virology, immunology, molecular biology, and cell biology. A potential disadvantage of a single-author text in cancer biology is that no one person can be an expert in all these fields. An advantage, however, is that diverse knowledge can be digested, integrated, and presented through the eyes of one "cancer biology watcher" in a way that students of the field can understand without necessarily having advanced expertise in a given area.

I have tried to approach the writing of this book as if I were a student coming into the field and wanted answers to the following questions: What were the early discoveries that led to development of a particular field of cancer research? What are the bedrock facts that have stood the test of time? What is the current state-of-the-art? Where is the field going in the future? I have also tried to show the connection between basic findings in the laboratory and clinical cancer medicine and to provide an extensive bibliography for each area of research.

The book has been extensively expanded and reorganized since the last edition. There are also new chapters on genetic alterations in cancer cells, tumor suppressor genes, growth factors and signal transduction mechanism, cell cycle regulation and apoptosis, tumor immunology, gene-targeted therapy for cancer, and cancer prevention.

I would like to thank the many investigators who have allowed me to use data from their own research to illustrate key points. I would also like to thank the numerous colleagues who have read the book and used it in their teaching. Their comments and the feedback from their students have helped me think about ways to improve the presentation. I am greatly indebted to Joan Amato for her patient and diligent work in preparation of the text and illustrations. Finally, I want to thank Bill Pratt, a longtime friend and colleague at the University of Michigan, who has understood what not all do—why one does crazy things like write books.

Omaha R.W.R.
January 1995

Contents

CANCER BIOLOGY

1

Characteristics of Human Cancer

BASIC FACTS ABOUT CANCER

Cancer is a complex family of diseases, and carcinogenesis—the turning of a normal cell into a cancer cell—is a complex, multistep process. Clinically, cancer is a large group of diseases, perhaps a hundred or more, that vary in age of onset, rate of growth, state of cellular differentiation, diagnostic detectability, invasiveness, metastatic potential, response to treatment, and prognosis. In terms of molecular and cell biology, however, cancer may represent a relatively small number of diseases caused by similar molecular defects in cell function and resulting from similar alterations to a cell's genes. Ultimately, cancer is a disease of abnormal gene expression. This altered gene expression occurs through a number of mechanisms, including direct insults to DNA (such as gene mutations, translocations, or amplifications) and abnormal gene transcription or translation. A major thrust of this textbook is to describe these mechanisms and show how they can alter cellular function in a way that leads to cancer.

Cancer is a leading cause of death in the western world. In the United States and a number of European countries, cancer is the second leading killer after cardiovascular disease. Over 1 million new cases of cancer occur in the United States each year, not including basal cell and squamous cell skin cancers, which add another 700,000 cases annually. These skin cancers are seldom fatal, do not usually metastasize, and are curable with appropriate treatment, so they are usually considered separately. Melanoma, on the other hand, is a type of skin cancer that is more dangerous and can be fatal, so it is considered with the others. The highest mortality rates are seen with lung, colorectal, breast, and prostate cancers. Over 500,000 people die each year in the United States from these and other malignancies (see Chap. 2).

In many cases, the causes of cancer are not clearly defined, but both external, factors (e.g., environmental chemicals and radiation) and internal ones (e.g., immune system defects, genetic predisposition) play a role (see Chaps. 3 and 6). Clearly, cigarette smoking is a major causal factor. Many such factors may act together to initiate (set off the initial genetic malfunction) and promote (stimulate growth of initiated cells) carcinogenesis. Often 10 to 20 years may pass before an initiated neoplastic cell grows into a clinically detectable tumor.

Cancer can occur at any time, but it is usually considered a disease of aging. On average, the diagnosis of the most common types of cancer comes at about age 67. Although cancer is relatively rare in children, it is still a leading cause of death in those between ages 1 and 14. In this age group, leukemia is the most common cause of death, but other cancers—such as osteosarcoma, lymphoma, Wilms' tumor (a kidney cancer), and lymphoma—also occur.

Over 8 million Americans alive today have had some type of cancer. Of these, about half are considered cured. About one in three people now living in the United States will eventually develop some type of cancer.

For reasons that are not entirely clear, there has been a steady rise in U.S. death rates from cancer during the past 50 years. The age-adjusted cancer mortality rate was 143 per 100,000 population in 1930; by 1950, it was 157 per 100,000; and by 1990, it was 174 per 100,000.[1] The major increase has been in deaths due to lung cancer. Thus, cigarette smoking is highly suspect. In addition, pollution, diet, and other life-style changes may have contributed to the increase in cancer mortality (Chap. 2).

The good news is that more and more people are now being cured of their cancers. In the 1940s, for example, only 1 in 4 persons diagnosed with cancer lived at least 5 years after treatment; in the 1990s, that figure rose to 40%. When normal life expectancy is factored into this calculation, the relative 5-year survival rate is about 53% for all cancers taken together.[1] Thus, the gain in number of survivors (from 1 in 3 to 4 in 10) means that almost 85,000 people are alive now who would have died from their disease within 5 years had they been living in the 1940s. This progress is due to better techniques of diagnosis and treatment, many of which have come about from our increasing knowledge of cancer cell biology.

THE HALLMARKS OF MALIGNANT DISEASES

Malignant neoplasms or cancers have several distinguishing features that enable the pathologist or experimental cancer biologist to characterize them as abnormal. The most common types of human neoplasms derive from epithelium—that is, the cells covering internal or external surfaces of the body. These cells have a supportive stroma of blood vessels and connective tissue. Malignant neoplasms may resemble normal tissues, at least in the early phases of their growth and development. Neoplastic cells can develop in any tissue of the body that contains cells capable of cell division. Though they may grow fast or slowly, their growth rate frequently exceeds that of the surrounding normal tissue. This is not an invariant property, however, because the rate of cell renewal in a number of normal tissues (e.g., gastrointestinal tract epithelium, bone marrow, and hair follicles) is as rapid as that of a rapidly growing tumor.

The term *neoplasm,* meaning "new growth," is often used interchangeably with the term *tumor* to signify a cancerous growth. It is important to keep in mind, however, that tumors are of two basic types: benign and malignant. The ability to distinguish between benign and malignant tumors is crucial in determining the appropriate treatment and prognosis of a patient who has a tumor. The following are features that differentiate a malignant tumor from a benign tumor:

1. Malignant tumors invade and destroy adjacent normal tissue; benign tumors grow by expansion, are usually encapsulated, and do not invade surrounding tissue. Benign tumors may, however, push aside normal tissue and become life-threatening if they press on nerves or blood vessels or if they secrete biologically active substances, such as hormones that alter normal homeostatic mechanisms.
2. Malignant tumors metastasize through lymphatic channels or blood vessels to lymph nodes and other tissues in the body. Benign tumors remain localized and do not metastasize.
3. Malignant tumor cells tend to be "anaplastic" or less well differentiated than normal cells of the tissue in which they arise. Benign tumors usually resemble normal tissue more closely than malignant tumors do.

At first, some malignant neoplastic cells resemble the normal tissue in which they arise, both structurally and functionally. Later, as the malignancy progresses, invades surrounding tissues, and metastasizes, the malignant cells may look less like the normal cell of origin. The development of a less well differentiated malignant cell in a population of differentiated normal cells is sometimes called *dedifferentiation.* This term is probably a misnomer for the process, because it implies that a differentiated cell goes backward in its development after carcinogenic insult. It is more likely that the anaplastic malignant cell type arises from the progeny of a tissue *stem cell* (one that still has a capacity for renewal and is not yet fully differentiated) that has been blocked or diverted on its way to becoming a fully differentiated cell.

Examples of neoplasms that maintain some differentiation include islet cell tumors of the pancreas that still make insulin, colonic adenocarcinoma cells that form glandlike epithelial structures and secrete mucin, and breast carcinomas that make abortive attempts to form structures resembling mammary gland ducts. Hormone-producing tumors, however, do not respond to feedback controls regulating normal tissue growth or to negative physiologic feedback regulating hormonal secretion. For example, an islet cell tumor may continue to secrete insulin in the face of extreme hypoglycemia, and an ectopic adrenocorticotropic hormone (ACTH)-producing lung carcinoma may continue to produce ACTH even though circulating levels of adrenocortical steroids are sufficient to cause Cushing's syndrome (see Chap. 12). Many malignant neoplasms, particularly the more rapidly growing and invasive ones, only vaguely resemble their normal counterpart tissue structurally and functionally. They are thus said to be *undifferentiated* or *poorly differentiated*.

4. Malignant tumors usually, but not always, grow more rapidly than benign tumors. Once they reach a clinically detectable stage, malignant tumors generally show evidence of significant growth, with involvement of surrounding tissue, over weeks or months, whereas benign tumors often grow slowly over several years.
5. Malignant neoplasms continue to grow even in the face of starvation of the host; they press on and invade surrounding tissues, often interrupting vital functions; they metastasize to vital organs—for example, brain, spine, and bone marrow—compromising their functions; and they invade blood vessels, causing bleeding. The most common effects on the patient are cachexia (extreme body wasting), hemorrhage, and infection. About 50% of terminal patients die from infection (see Chap. 12).

Differential diagnosis of cancer from a benign tumor or nonneoplastic disease usually involves obtaining a tissue specimen by biopsy, surgical excision, or exfoliative cytology. The last of these is an examination of cells obtained from swabbings, washings, or secretions of a tissue suspected to harbor cancer; the "Pap test" involves such an examination.

CLASSIFICATION OF HUMAN CANCERS

The classification of human tumors by tissue type is given in Table 1-1. Although the terminology applied to neoplasms can be confusing for a number of reasons, certain generalizations can be made. The suffix *oma*, applied by itself to a tissue type, usually indicates a benign tumor. Some malignant neoplasms, however, may be designated by the *oma* suffix alone; these include lymphoma, melanoma, and thymoma. Rarely, the *oma* suffix is used to describe a nonneoplastic condition such as granuloma, which is often not a true tumor but a mass of granulation tissue resulting from chronic inflammation or abscess. Malignant tumors are indicated by the terms *carcinoma* (epithelial in origin) or *sarcoma* (mesenchymal in origin) preceded by the histologic type and followed by the tissue of origin. Examples of these include adenocarcinoma of the breast, squamous cell carcinoma of the lung, basal cell carcinoma of skin, and leiomyosarcoma of the uterus. Most human malignancies arise from epithelial tissue. Those arising from stratified squamous epithelium are designated *squamous cell carcinomas,* whereas those emanating from glandular epithelium are termed *adenocarcinomas.* When a malignant tumor no longer resembles the tissue of origin, it may be called *anaplastic* or *undifferentiated.* If a tumor is metastatic from another tissue, it is designated, for example, an adenocarcinoma of the colon metastatic to liver. Some tumors arise from pluripotential primitive cell types and may contain several tissue elements. These include mixed mesenchymal tumors of the uterus, which contain carinomatous and sarcomatous elements, and teratocarcinomas of the ovary, which may contain bone, cartilage, muscle, and glandular epithelium.

Neoplasms of the hematopoietic system usu-

Table 1-1 Classification of Human Tumors by Tissue Type

Tissue of Origin	Benign	Malignant
Epithelium		
Surface epithelium (nonglandular)	Papilloma	Carcinoma (squamous cell, epidermoid, transitional cell)
Glandular epithelium	Adenoma	Adenocarcinoma
Basal layer of epidermis	—	Basal cell carcinoma
Trophoblasts of placental villi	Hydatidiform mole	Choriocarcinoma
Connective tissue		
Fibrous tissue	Fibroma	Fibrosarcoma
Cartilage	Chondroma	Chondrosarcoma
Bone	Osteoma	Osteosarcoma
Smooth muscle	Leiomyoma	Leiomyosarcoma
Striated muscle	Rhabdomyoma	Rhabdomyosarcoma
Fat	Lipoma	Liposarcoma
Endothelial tissue and its derivatives		
Blood vessels	Hemangioma	Hemangiosarcoma
Lymph vessels	Lymphangioma	Lymphangiosarcoma
Bone marrow		
Granulocytes	—	Myelocytic leukemia
Erythrocyes	Polycythemia vera	Erythrocytic leukemia
Lymphocytes	Infectious mononucleosis	Lymphocytic leukemia
Plasma cells	—	Multiple myeloma
Monocytes	—	Monocytic leukemia
Endothelial lining	—	Ewing's sarcoma
Lymphoid tissue		Non-Hodgkin's malignant lymphomas Lymphocytic type Histiocytic type Undifferentiated, pleiomorphic type Undifferentiated, Burkitt type
Thymus	—	Thymoma
Neural tissue and its derivatives		
Glial tissue	"Benign" gliomas (some ependymomas and oligodendrogliomas are considered nonmalignant)	Glioblastoma multiforme, medulloblastoma, astrocytoma, ependymoma, oligodendroglioma
Meninges	Meningioma	Meningeal sarcoma
Neuronal cells	Ganglioneuroma	Neuroblastoma
Nerve sheath	Neurilemmoma	Neurilemmal sarcoma (schwannoma)
Nerve sheath	Neurofibroma	Neurofibrosarcoma
Melanocytes	Pigmented nevus (mole)	Malignant melanoma
Adrenal medulla	Pheochromocytoma	Malignant pheochromocytoma
Retina	—	Retinoblastoma
Specialized nerve endings	Carcinoid tumors	Carcinoid tumors
Mixed tumors derived from more than one cell type		
Embryonic kidney	—	Nephroblastoma (Wilms' tumor)
Gonadal tissue	Teratoma	Teratocarcinoma
Gonadal tissue	—	Embryonal carcinoma with choriocarcinoma

ally have no benign counterparts. Hence the terms *leukemia* and *lymphoma* always refer to a malignant disease and have cell-type designations, such as acute or chronic myelogenous leukemia, Hodgkin's or non-Hodgkin's lymphoma, and so on. Similarly, the term *melanoma* always refers to a malignant neoplasm derived from melanocytes.

MACROSCOPIC AND MICROSCOPIC FEATURES OF NEOPLASMS

The pathologist can gain valuable insights into the nature of a neoplasm by careful examination of the overall appearance of a surgical specimen. Often, by integrating clinical findings with the macroscopic characteristics of a tumor, a tentative differential diagnosis can be reached. Also, notation of whether the tumor is encapsulated, has extended through tissue borders, or has reached the margins of the excision provides important diagnostic information.

The anatomic site of a neoplasm is important for several reasons. The site of the tumor dictates several things about the clinical course of the tumor, including (1) the likelihood and route of metastatic spread, (2) the effects of the tumor on body functions, and (3) the type of treatment that can be employed. It is also important to determine whether the observed tumor mass is the primary site (i.e., tissue of origin) of the tumor or a metastasis. A primary epidermoid carcinoma of the lung, for example, would be treated differently and have a different prognosis than an embryonal carcinoma of the testis metastatic to the lung. It is not always easy to determine the primary site of a neoplasm, particularly if the tumor cells are undifferentiated. The first signs of a metastatic tumor may be a mass in the lung noted on x-ray or a spontaneous fracture of a vertebra that had been invaded by cancer cells. Because the lungs and bones are frequent sites of metastases for a variety of tumors, the origin of the primary tumor may not be readily evident. This is a very difficult clinical situation because, to cure the patient or to produce long-term remission, the oncologist must be able to find and remove or destroy the primary tumor so as to prevent its continued growth and metastasis. If histologic examination does not reveal the source of the primary tumor or if other diagnostic techniques fail to reveal other tumor masses, the clinician has to treat blindly; thus he or she might not choose the best mode of therapy.

Another consideration is the accessibility of a tumor. If a tumor is surgically inaccessible or too close to vital organs to allow complete resection, surgical removal is impossible. For example, a cancer of the common bile duct or head of the pancreas is often inoperable by the time it is diagnosed because these tumors invade and attach themselves to vital structures early, thus preventing curative resection. Similarly, if administered anticancer drugs cannot easily reach the tumor site, as is the case with tumors growing in the pleural cavity or in the brain, these agents might not be able to penetrate in sufficient quantities to kill the tumor cells.

The site of the primary tumor also frequently determines the mode of metastatic spread and the target organs for metastasis. In addition to spreading locally, cancers metastasize via lymphatic channels or blood vessels. For example, carcinomas of the lung most frequently metastasize to regional lymph nodes, pleura, diaphragm, liver, bone, kidneys, adrenals, brain, thyroid, and spleen. Carcinomas of the colon metastasize to regional lymph nodes and, by local extension, they ulcerate and obstruct the gastrointestinal tract. The most common site of distant metastasis of colonic carcinomas is the liver, via the portal vein, which receives much of the venous return from the colon and flows to the liver. Breast carcinomas most frequently spread to axillary lymph nodes, the opposite breast through lymphatic channels, lungs, pleura, liver, bone, adrenals, brain, and spleen.

Some tissues are more common sites of metastasis than others. Because of their abundant blood and lymphatic supply as well as their function as "filters" in the circulatory system, the lungs and liver are the most common sites of metastasis from tumors occurring in visceral organs. Metastasis is usually the single most important criterion determining the patient's prognosis. In breast carcinoma, for example, the 5-year survival rate for patients with localized disease and no evidence of axillary lymph node involvement is about 85%; but when more than four axillary nodes are involved, the 5-year survival is about 30%, on average.[2]

The anatomic site of a tumor will also determine its effect on vital functions. A lymphoma growing in the mediastinum may press on major blood vessels to produce the superior vena caval syndrome, manifested by edema of the neck and face; distention of veins of the neck, chest, and upper extremities; headache; dizziness; and fainting spells. Even a small tumor growing in the brain can produce such dramatic central nervous system effects as localized weakness, sensory loss, aphasia, or epilepticlike seizures. A lung tumor growing close to a major bronchus will produce airway obstruction earlier than one growing in the periphery of the lung. A colonic carcinoma may invade surrounding muscle layers of the colon and constrict the lumen, causing intestinal obstruction. A frequent symptom of prostatic cancer is inability to urinate normally.

The cytologic criteria that enable the pathologist to confirm the diagnosis or at least to suspect that cancer is present (thus indicating the need for further diagnostic tests) are as follows:

1. The morphology of cancer cells is usually different from and more variable than that of their counterpart normal cells from the same tissue. Cancer cells are also more variable in size and shape (Fig. 1-1).
2. The nucleus of a cancer cell is often larger and the chromatin more apparent ("hyperchromatic") than the nucleus of a normal cell; the nuclear/cytoplasmic ratio is often higher; and the cancer cell nucleus contains prominent, large nuceoli (Fig. 1-1).

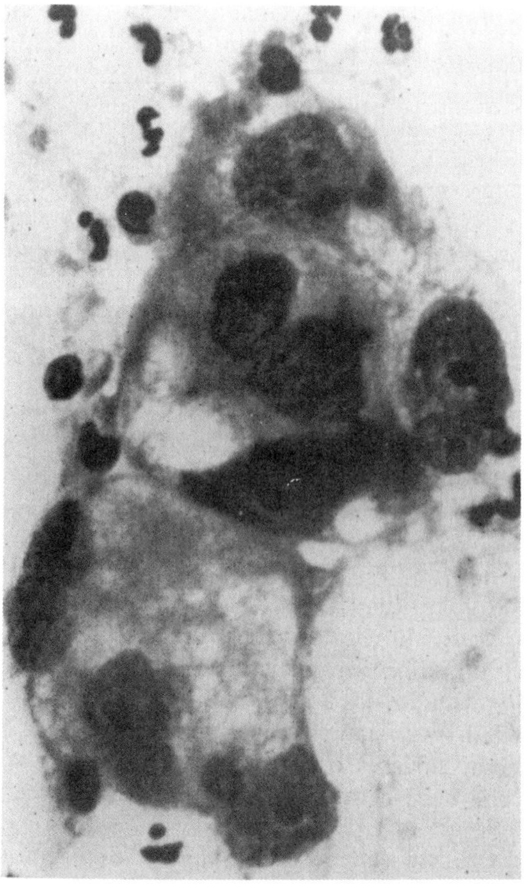

Figure 1-1 Malignant cells in vaginal secretion prepared by Papanicolaou technique. Note abnormal size and shape of cells, density of chromatin in nuclei, and prominent nucleoli. Numerous polymorphonuclear leukocytes are also present (X650). (From Warren.[3])

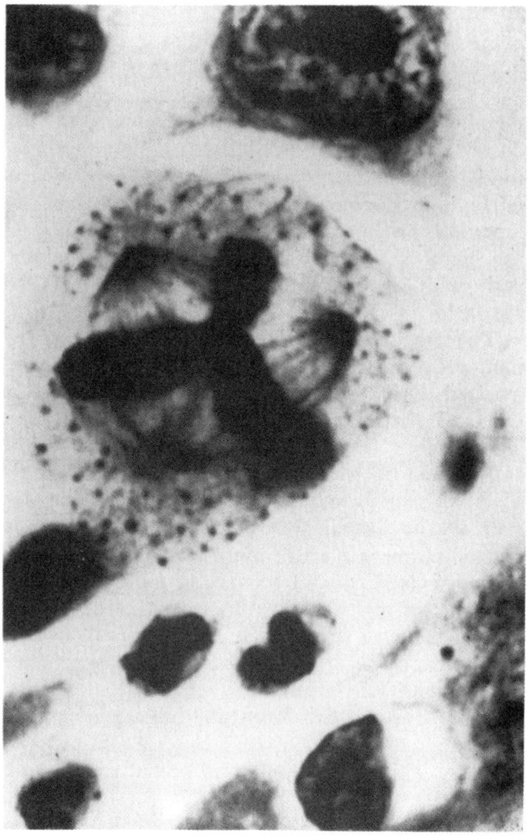

Figure 1-2 Abnormal tripolar mitosis occurring in malignant melanoma. (Phosphotungstic acid-hematoxylin stain; X2350). (From Warren.[3])

3. The number of cells undergoing mitosis is usually greater in a population of cancer cells than in a normal tissue population. Twenty or more mitotic figures per 1000 cells would not be an uncommon finding in cancerous tissue, whereas less than 1 per 1000 is usual for benign tumors or normal tissue.[3] This number, of course, would be higher in normal tissues that have a high growth rate, such as bone marrow and crypt cells of the gastrointestinal mucosa.
4. Abnormal mitosis and "giant cells," with large, pleomorphic (variable in size and shape) or multiple nuclei, are much more common in malignant tissue than in normal tissue (Fig. 1-2).
5. Obvious evidence of invasion of normal tissue by a neoplasm may be seen (Fig. 1-3), indicating that the tumor has already become invasive and may have metastasized.

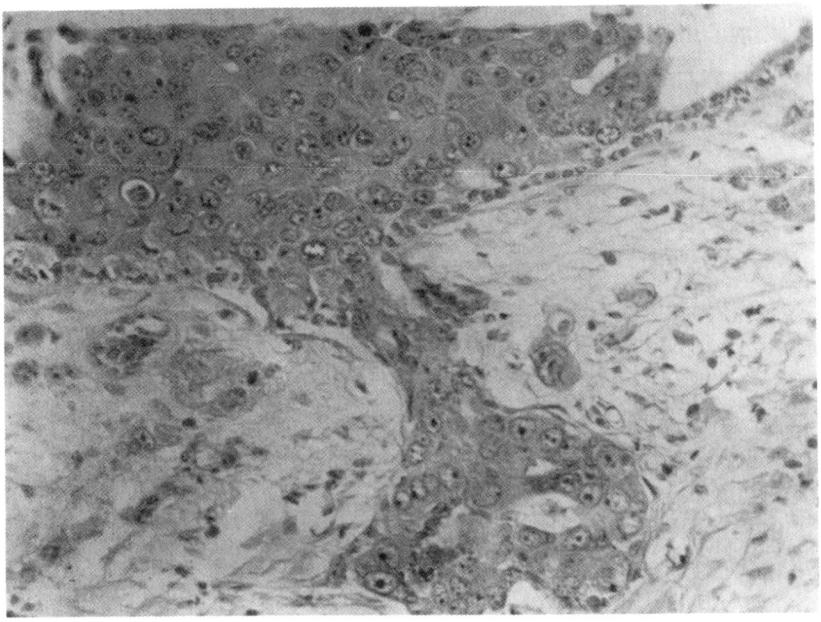

Figure 1-3 Mammary carcinoma invading surrounding breast tissue in a 32-year-old woman (X400). (From Warren.[3])

The histologic classification of neoplasms determines the growth patterns, propensity to metastasize, type of treatment, and thus the prognosis. Several examples will serve to illustrate this.

Lung carcinomas fall into three general categories: (1) epidermoid or squamous cell, (2) undifferentiated (small or large cell), and (3) adenocarcinomas.

Epidermoid carcinomas (Fig. 1-4A) tend to grow more slowly and remain localized longer than the other cell types. They more often grow in larger bronchi, producing symptoms of airway obstruction and thus bringing these patients to the physician earlier than the other forms of lung cancer. The treatment of choice for this tumor is surgery. Early diagnosis and detection of localized disease, followed by resection, can produce 5-year survival rates approaching 50%.[3] Epidermoid carcinomas may be detectable as distinct "coin" lesions that appear round and smooth on chest x-ray films. They may cause complete bronchial obstruction, leading to atelectasis (collapsed areas of the lung), which is detectable by x-ray study and may assist in locating the primary tumor when it is not evident as a discrete density in the lung. Histologically, this tumor varies widely, ranging from sheets of stratified squamous epithelium to less differentiated forms that are difficult to distinguish from an undifferentiated large-cell carcinoma.[4]

Undifferentiated small-cell bronchogenic carcinomas (Fig. 1-5) may occur in the smaller peripheral bronchi. They have a high growth fraction, are rapidly invasive, and about 70% will have distant metastases by the time of diagnosis.[5] The treatment of choice here is chemotherapy, and the life expectancy is frequently less than a year. Small-cell lung carcinomas have some unusual characteristics. They frequently have the properties of an APUD (*a*mine *p*recursor *u*ptake and *d*ecarboxylation) cell type,[6] which is characteristic of cells of neuroendocrine origin. These tumors produce a number of markers that support this concept. Thus, patients who have small-cell carcinomas may have abnormal blood levels of the hormones serotonin, ACTH, antidiuretic hormone (ADH), or calcitonin, which appear to be ectopically produced by these tumors. A second unusual feature of this malignant disease is the change in histologic cell type during the course of the disease after chemotherapy. At autopsy, the pathologic diagnosis frequently is "squamous cell" carcinoma rather

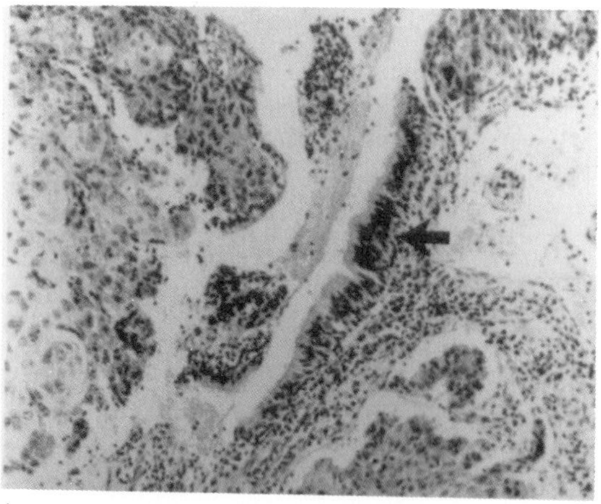

A

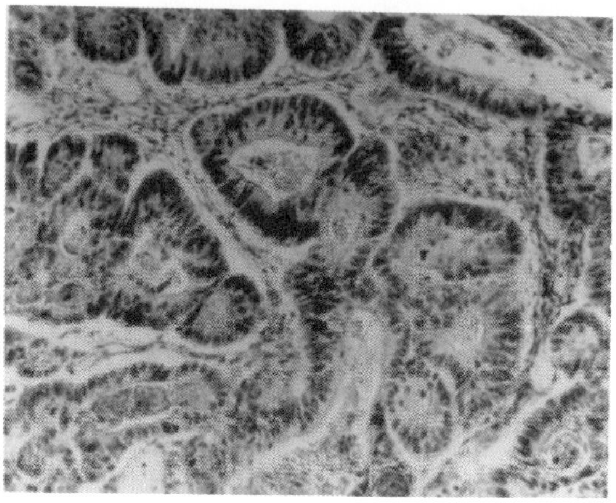

B

Figure 1-4 (A) Invasive epidermoid carcinoma of bronchus showing transition from columnar epithelium (*arrow*) to neoplastic squamous cell carcinoma above it. Note that part of the tumor projects into the bronchial lumen. (B) Well-differentiated bronchogenic adenocarcinoma. (From Gunn.[4])

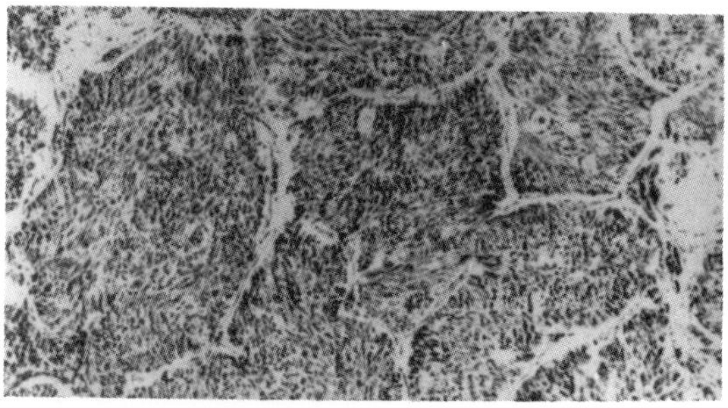

A

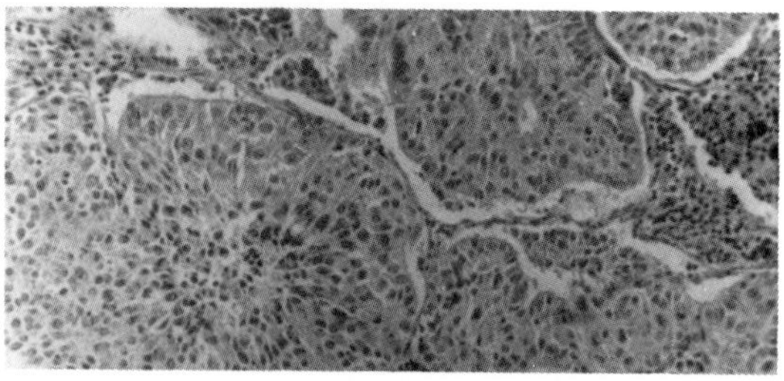

B

Figure 1-5 Examples of undifferentiated lung carcinomas. (A) Small-cell ("oat cell") carcinoma. (B) Anaplastic large-cell carcinoma. (From Gunn.[4])

than small-cell carcinoma. This may be the result of the destruction of the predominant cell type of the original tumor and the proliferation of a minor population of drug-resistant cells.

In contrast with the other two major groups of bronchogenic carcinomas, bronchial adenocarcinomas (Fig. 1-4B) most often arise in the periphery of the lung and frequently do not produce signs of pulmonary involvement until late in the disease. Unlike epidermoid and small-cell carcinomas, this form of lung cancer is not closely associated with smoking.[3] These patients are treated by surgery but generally have a poor prognosis. Histologic patterns vary from a well-differentiated mucin-secreting columnar epithelium formed into alveoluslike structures to poorly differentiated cells arranged in irregular sheets.[4]

The leukemias provide another example of the importance of histologic identification of tumor cell type to the diagnosis, treatment, and prognosis of cancer. These are a group of diseases that arise from the hematopoietic (blood cell-forming) stem cells of the bone marrow. The basic defect in the malignant cell type is one of faulty differentiation and uncontrolled proliferation. This causes abnormal cells in the marrow to accumulate, crowding out the normal blood-forming cell types and spilling immature cell types into the peripheral blood. Normally, only mature blood cells appear in the peripheral blood. Although the term *leukemia* usually refers to white blood cell malignancies, a rare human hematopoietic malignancy arising from red blood cell precursors is *erythroleukemia* (Di Guglielmo's disease).

Leukemic cells may infiltrate the lymph nodes, spleen, liver, and other tissues, greatly enlarging these organs. Certain infectious diseases may mimic the leukemias in certain respects. For example, in infectious mononucleosis, there may be a peripheral leukocytosis, many circulating immature lymphocytes, enlarged lymph nodes, hepatomegaly, and splenomegaly. There are specific laboratory tests for this disease, however, and the patient usually recovers within a month. The more common forms of leukemia are of four types: acute myelocytic leukemia (AML), chronic myelocytic leukemia (CML), acute lymphocytic leukemia (ALL), and chronic lymphocytic leukemia (CLL).

GRADE AND STAGE OF NEOPLASMS

Histologic Grade of Malignancy

The histologic grading of malignancy is based on the degree of differentiation of a cancer and on an estimate of the growth rate as indicated by the mitotic index. It was once generally believed that less differentiated tumors were more aggressive and more metastatic than more differentiated tumors. It is now appreciated that this is an oversimplification and, in fact, not a very accurate way to assess the degree of malignancy for all kinds of tumors. However, for certain epithelial tumors, such as carcinomas of the cervix, uterine endometrium, colon, and thyroid, histologic grading is a fairly accurate index of malignancy and prognosis. In the case of epidermoid carcinomas, for example, in which keratinization occurs, keratin production provides a relatively simple way to determine the degree of differentiation. Based on this criterion and others like it, tumors have been classified as grade I (75% to 100% differentiation), grade II (50% to 75%), grade III (25% to 50%), and grade IV (0% to 25%).[3] More recent methods of malignancy grading also take into consideration mitotic activity, amount of infiltration into surrounding tissue, and amount of stromal tissue in or around the tumor. The chief value of grading is that it provides a general guide to prognosis for certain cancers and an indicator of the effectiveness of various therapeutic approaches.

Tumor Staging

Although the classification of tumors based on the preceding descriptive criteria helps the oncologist determine the malignant potential of a tumor, judge its probable course, and determine the patient's prognosis, a method of discovering the extent of disease on a clinical basis and a universal language to provide standardized criteria among physicians are needed. Attempts to develop an international language for describing the extent of disease have been carried out by two major agencies—the Union Internationale Contre le Cancer (UICC) and the American Joint Committee for Cancer Staging and End Results Reporting (AJCCS). Some of the objectives of the classification system developed by these groups are (1) to aid oncologists in planning treatment, (2) to provide categories for estimating prognosis and evaluating results of treatment, and (3) to facilitate exchange of information.[7] Both the UICC and AJCCS schemes utilize the TNM classification system, in which T categories define the primary tumor, N the involvement of regional lymph nodes, and M the presence or absence of metastases. The definition of extent of malignant disease by these categories is termed staging. Staging defines the extent of tumor growth and progression at one point in time; four different methods are involved:

1. Clinical staging: estimation of disease progression based on physical examination, clinical laboratory tests, x-ray films, and endoscopic examination.
2. Radiographic staging: evaluation of progression based on sophisticated radiography—for example, computed tomography (CT), arteriography, lymphangiography, and radioisotope scanning.
3. Surgical staging: direct exploration of the extent of the disease by surgical procedures.
4. Pathologic staging: use of biopsy procedures to determine the degree of spread, depth of invasion, and involvement of lymph nodes.

These methods of staging are not used interchangeably, and their use depends on agreed-upon procedures for each type of cancer. For example, operative findings are used to stage certain types of cancer (e.g., ovarian carcinomas) and lymphangiography is required to stage Hodgkin's disease. Although this means that different staging methods are used to stage different tumors, each method is generally agreed on

Table 1-2 Specific Criteria Related to T Categories

	T_1	T_2	T_3	T_4
Depth of invasion				
Solid organs	Confined	Organ capsule, muscle	Bone, cartilage	Viscera
Hollow organs	Submucosa	Muscularis	Serosa	
Mobility	Mobile	Partial mobility	Fixed	Fixed and destructive
Neighboring structures	Not invaded	Adjacent	Surrounding	Viscera
Diameter (cm)	<2	2 to 4–5	5–10	>10

Source: Adapted from Rubin.[7]

by oncologists, thus allowing a comparison of data from different clinical centers. Once a tumor is clinically staged, it is not usually changed for that patient; however, as more information becomes available following a more extensive workup, such as a biopsy or surgical exploration, this information is, of course, taken into consideration in determining treatment and estimating prognosis. Staging provides a useful way to estimate at the outset what a patient's clinical course and initial treatment should be. The actual course of the disease indicates its true extent. As more is learned about the natural history of cancers and as more sophisticated diagnostic techniques become available, the criteria for staging will likely change and staging should become more accurate.

It is important to remember that staging does not mean that any given cancer has a predictable, ineluctable progression. Although some tumors may progress in a stepwise fashion from a small primary tumor to a larger primary tumor and then spread to regional nodes and distant sites (i.e., progression from stage I to stage IV), others may spread to regional nodes or have distant metastases while the primary tumor is microscopic and clinically undetectable. Thus, staging is somewhat arbitrary, and its effectiveness is really based on whether or not it can be used as a standard to select treatment and to predict the course of disease.

Although the exact criteria used vary with each organ site, the staging categories listed below represent a useful generalization[8]:

Stage I ($T_1N_0M_0$): Primary tumor is limited to the organ of origin. There is no evidence of nodal or vascular spread. The tumor can usually be removed by surgical resection. Long-term survival is from 70% to 90%.

Stage II ($T_2N_1M_0$): Primary tumor has spread into surrounding tissue and lymph nodes immediately draining the area of the tumor ("first station" lymph nodes). The tumor is operable but it may not be completely resectable because of local spread. Survival is 45% to 55%.

Stage III ($T_3N_2M_0$): Primary tumor is large, with fixation to deeper structures. First-station lymph nodes are involved; they may be more than 3 cm in diameter and fixed to underlying tissues. The tumor is not usually resectable, and part of the tumor mass is left behind. Survival is 15% to 25%.

Stage IV ($T_4N_3M_+$): Extensive primary tumor (may be more than 10 cm in diameter) is present. It has invaded underlying or surrounding tissues. Extensive lymph node involvement has occurred, and there is evidence of distant metastases beyond the tissue of origin of the primary tumor. Survival is under 5%.

A detailed classification of the TNM system as proposed by Rubin[7] is shown in Tables 1-2, 1-3, and 1-4: T_1 defines a primary tumor that is confined to the organ of origin and usually no larger than 2 cm in diameter (it may be spreading into superficial tissue layers, but it is not deeply invading and is freely movable by palpation); T_2 represents a localized tumor, 2 to 5 cm in diameter, that is invading adjacent tissues or structures and, in the case of "hollow" organs, the tumor extends into the lumen but occupies less than half the lumen circumference (it is partially movable by palpation); T_3 tumors are usually 5 to 10 cm in diameter, invading through the visceral wall and serosal layers into adjacent structures and occluding more than one half of the lumen circumference in the case of hollow organs (they are fixed to adjacent or underlying structures and are not movable by palpation); T_4 tumors are large, greater than 10 cm in diame-

Table 1-3 Specific Criteria Related to N Categories

Nodal Metastases	N_1	N_2	N_3	N_4
Number	Solitary	Multiple	Multiple	Multiple
Size (cm)	<2–3	>3	>5	>10
Mobility	Mobile	Partially matted, muscle invasion	Fixed to vessels, bone, skin	Fixed and destructive

Source: Adapted from Rubin.[7]

ter, extend into adjacent organs, and often produce a fistula. They are destructive to adjacent tissues such as bones, blood vessels, and nerves; in hollow organs, complete occlusion of the lumen can occur to cause symptoms of obstruction.

The criteria for establishing lymph node involvement (N categories) are based on size, firmness, amount of invasion, mobility, number of nodes involved, and distribution of nodes involved (i.e., ipsilateral, contralateral, distant involvement): N_0 indicates that there is no evidence of lymph node involvement; N_1 indicates that there are palpable lymph nodes with tumor involvement, but they are usually small (2 to 3 cm in diameter) and mobile; N_2 indicates that there are firm, hard, partially movable nodes (3 to 5 cm in diameter), partially invasive, and they may feel as if they were matted together; N_3 indicates that there are large lymph nodes (over 5 cm in diameter) with complete fixation and invasion into adjacent tissues; N_4 indicates extensive nodal involvement of contralateral and distant nodes.

The criteria applied to metastases (M categories) are as follows: M_0, no evidence of metastasis; M_1, isolated metastasis in one other organ; M_2, multiple metastases confined to one organ, with minimal functional impairment; M_3, multiple organs involved with no to moderate functional impairment; M_4, multiple organ involvement with moderate to severe functional impairment. Occasionally a subscript is used to indicate the site of metastasis, such as M_p, M_h, M_o for pulmonary, hepatic, and osseous metastases, respectively.

As mentioned earlier, clinical staging is not an exact science, but it is a useful way to estimate the course a malignant disease will take in a given patient. One of the biggest imponderables is the presence of small metastatic foci in lymph nodes or other organs. These are often undetectable by current diagnostic procedures, and their presence means that a potentially fatal event has already occurred, even though the involvement appears to be minimal (i.e., stage I or II disease). Ultimately, of course, the true pathologic stage of the disease becomes apparent as the tumor grows, becomes more accessible, or metastasizes to other sites. Thus, the accuracy of predictions based on clinical staging can be evaluated. The knowledge gained over a number of years has provided some confidence in clinical staging. How fast a tumor will grow and metastasize, how aggressive it will be, and what its ultimate mortality rate will be can now be predicted with a fair degree of accuracy for certain tumors. For example, a basal cell carcinoma of the skin, although it can be locally invasive and destructive, is not likely ever to metastasize to distant structures, and because it is on the body surface, it is readily accessible for diagnosis. Thus, it is unlikely that stage I disease will progress to further stages. On the other hand, bron-

Table 1-4 Specific Criteria Related to M Categories[a]

Distant Metastases	M_1	M_2	M_3	M_4
Number of metastases	1	>1	Multiple	Multiple
Number of organs affected	1	1	Mutliple	Multiple
Impairments	0	Minimal	Minimal to moderate	Moderate to severe

[a]M categories may be modified to show viscera involved by lettered subscript, as pulmonary (M_p), hepatic (M_h), osseous (M_o), skin (M_s, brain (M_b). M_+ is sometimes used to indicate microscopic evidence of metastases, confimred by pathological examination.

Source: Adapted from Rubin.[7]

chogenic carcinoma does progress from localized disease to local nodal metastasis, distal nodal metastasis, and metastasis to other organs. Therefore, in this case, a high percentage of patients will ultimately progress to stage IV disease. In certain types of lung cancer, unfortunately, this has already occurred by the time the disease is diagnosed. Thus, 70% of patients with small cell carcinoma of the lung will have distant metastases by the time of diagnosis. Another example of a progressive tumor is colonic carcinoma, which initially occurs in the epithelial layer lining the inside of the colon. The tumor then invades the smooth muscle layer beneath the epithelium and eventually spreads through smooth muscle and the outer serosal surface, leading to invasion and destruction of surrounding tissues, fistual formation with adjacent organs, and distant metastasis, almost always ultimately involving the liver. Thus, in colonic cancer, also there is an inexorable progression from the early stages to advanced stages of disease unless, or course, the disease is diagnosed early enough to allow complete surgical removal before metastases have occurred.

It is clear that early diagnosis is the sine qua non of effective cancer treatment. In the past there have been three established types: surgery, radiation, and chemotherapy with drugs. These treatment modalities have been used together, frequently leading to a higher number of long-term remissions and cures than were attainable by a single method. Localized disease is usually best treated by surgical removal or irradiation. However, disseminated disease is frequently not amenable to these types of treatment. Chemotherapy has been the most feasible method of treating multifocal disease. The realization that clinically undetectable micrometastasis may already have occurred by the time of diagnosis of many cancers has led to the use of so-called *adjuvant therapy*, utilizing anticancer drugs together with surgery or irradiation. This has produced some encouraging results, for example, in breast cancer.[9]

The aggressiveness of the malignant disease dictates how radical a treatment should be. Most cancers are treated as though they were at least one stage beyond their clinically defined stage.[8] Very aggressive cancers that are known to metastasize early—for example, choriocarinoma of the uterus and small-cell carcinoma of the lung—are treated as if they were already metastatic, even though they may appear to be localized at the time of diagnosis.

Usually the survival rates of cancer patients are expressed in terms of 5 years. This fact in itself indicates how gloomy the long-range survival of the most common cancers is at the present time. The most important reason for the incurability of these common solid tumors—that is, cancers of the lung, breast, colon-rectum, uterus, and prostate—is that they are not diagnosed in time to be eliminated completely. It is estimated that in the United States more than 50% of all cancers have metastasized by the time of diagnosis.[10] Most cancers are curable if they are detected early enough. Thus, the importance of improving diagnostic techniques is evident, and major research efforts have been launched to develop better methods to do this. In addition, of course, vigorous efforts to identify and eliminate causative factors in our environment should continue.

SPECIAL DIAGNOSTIC PROCEDURES

In a small percentage of cases (about 10%), routine diagnostic procedures may not produce a firm diagnosis, either because certain tumor cells are histologically similar or because the cells are so poorly differentiated that their tissue of origin cannot be determined. Often such tumors may be metastatic from an unknown primary site. In such cases, differential diagnosis may be extremely difficult.

Additional diagnostic procedures may then be helpful in designating the tissue type or source of the unknown primary tumor. These procedures include electron microscopy, immunohistochemistry, cytogenetics, probes for abnormal gene expression, and levels of various tumor markers in the patient's serum or urine. Many of these techniques are also research tools that help the experimental oncologist to determine the causes of certain types of neoplasms or to learn more about the fundamental biology of the neoplastic process.

Electron Microscopy

Electron microscopy (EM) procedures are carried out on specially prepared, fixed, and embedded tissue sections impregnated with heavy metals, allowing for differential absorption of a focused electron beam that is passed through the specimen onto a photographic plate. A "picture" of the intracellular organelles, magnified several thousandfold (up to 200,000 times or more), is thus obtained. Certain cellular features that distinguish one cancer cell type from another may then be seen. Features that have value in the differential diagnosis of cancer include cytoplasmic tonofibrils that identify squamous differentiation in epithelial tumors, neurosecretory granules that indicate a tumor of neuroendocrine origin, and cytoplasmic mucin granules that help to distinguish between a malignant mesothelioma and a poorly differentiated adenocarcinoma.[11]

Immunohistochemistry

This approach uses antibodies directed against tumor-associated antigens (usually cell surface proteins, glycoproteins, mucins, or glycolipids) in a reaction procedure that allows differential staining of a given cell type. A commonly used procedure is to overlay a tissue section with an antibody directed to a tumor cell antigen, wash out the unbound antibody, and then react the tissue with a second antibody that recognizes the first antibody. For example, if the primary antibody is a monoclonal antibody from a mouse, the second antibody could be a rabbit antimouse antibody. The second antibody is tagged with biotin, which provides a reactant for an avidin-horseradish peroxidase complex, producing a colored oxidation-reduction product. This product can then be seen in the light microscope. Sometimes the second antibody is tagged with a fluorescent compound that can be seen in a flu-

Table 1-5 Selected Antigenic Moieties of Diagnostic Value in Practical Immunohistochemistry

Antigen	Predominant Distribution	Diagnostic Use
Cytokeratin	Epithelial cells	Distinction between lymphoma or melanoma and carcinoma
Epithelial membrane antigen	Epithelial cells	Distinction between melanoma and carcinoma
Leukocyte common antigen	Leukocytes	Distinction between lymphoma, carcinoma, and melanoma
Desmin	Myogenous cells	Identification of myogenic sarcomas
Muscle-specific actin	Myogenous cells	Identification of myogenic sarcomas
Thyroglobulin	Thyroid follicular cells	Identification of certain thyroid carcinomas
Prostate-specific antigen	Prostatic epithelium	Identification of metastatic prostatic carcinomas
Calcitonin	Parafollicular thyroid epithelium	Distinction of medullary thyroid carcinoma from other thyroid tumors
Carcinoembryonic antigen	Endodermally derived epithelium	Identification of certain carcinomas; distinction of mesothelioma and adenocarcinoma
Placental alkaline phosphatase	Placental tissue and germ cell tumors	Screening identification of possible germ cell and trophoblastic tumors
Alpha-fetoprotein	Neoplastic hepatic tissue and selected germ cell tumors	Identification of possible hepatocellular carcinoma, embryonal carcinoma, or endodermal sinus tumor
Beta-human chorionic gonadotropin	Placental tissue; trophoblastic and germ cell tumors	Identification of possible trophoblastic or germ cell tumors
CA-125	Mullerian epithelium	Identification of possible female genital tract carcinomas
CA-19-9	Alimentary tract epithelium	Identification of gastrointestinal or pancreatic carcinomas
Gross cystic disease fluid protein-15	Pathologic mammary epithelium	Identification of metastatic breast carcinomas
HMB-45	Melanocytic cells	Identification of melanomas
Chromogranin-A	Neuroendocrine cells	Identification of neuroendocrine carcinomas
Synaptophysin	Neuroendocrine cells	Identification of neuroendocrine carcinomas and neuroectodermal tumors

Source: Pfeiffer and Wick.[11]

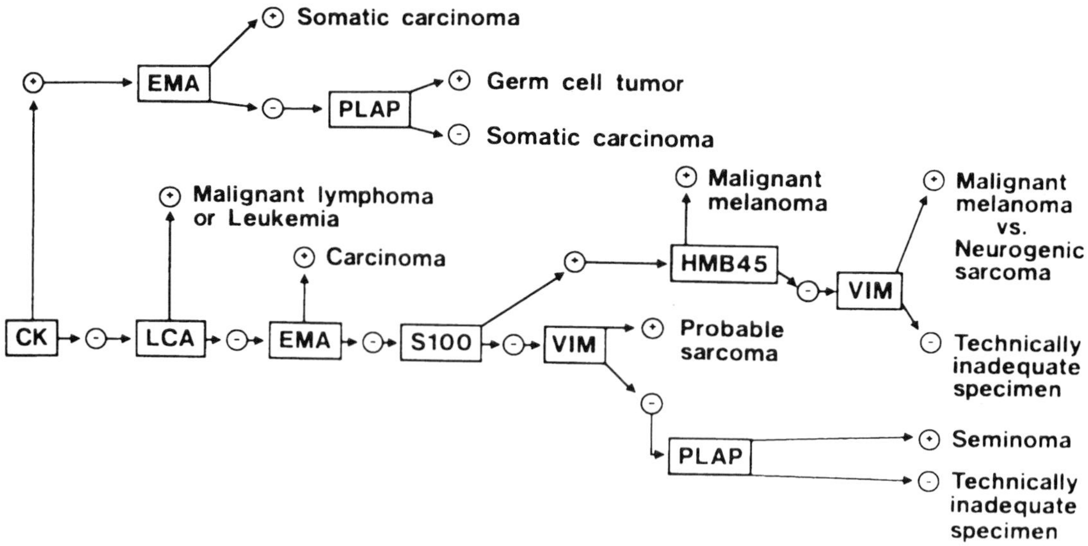

Figure 1-6 Immunohistochemical diagnosis of large-cell malignancies. (From Pfeiffer and Wick.[11])

orescence-detecting microscope. Table 1-5 provides a list of common tumor-associated factors that can be used in differential diagnosis.

Several of these tumor-derived "markers" are also shed into the peripheral blood and can be detected in the blood or urine by radioimmunoassay or other immunologic techniques. These tumor markers include human chorionic gonadotropin (hCG), calcitonin, alpha-fetoprotein, (AFP), carcinoembryonic antigen (CEA), prostate-specific antigen (PSA), and CA-125. An example of a strategy for using immunohistochemistry as a way of diagnosing differentially among cancers of a large cell–type histology is shown in Figure 1-6.

Cytogenetics

Cytogenetic analysis looks at a "spread" of the individual chromosomes from a population of cells. Since a number of cancers have distinctive chromosomal abnormalities, this procedure has some usefulness in determining the type of cancer cell present in the body. The classic example is the Philadelphia chromosome seen in chronic myelogenous leukemia (CML). This distinctive chromosomal abnormality results from the translocation of pieces of chromosome 22 to chromsome 9. More than 90% of CML patients have this abnormality in their leukemic cells, and those who do not have a less favorable prognosis.[12] Thus, some clinically important information can be gained by chromosomal analysis in CML patients. A number of other translocations and deletions have been observed in cancer cells (Chap. 3), and although these have not necessarily been helpful in differential diagnosis, they have been extremely useful in providing clues to the cancer disease process. For example, the translocation between chromosomes 8 and 14 in Burkitt's lymphoma led to an explanation for the activation of the *myc* oncogene in that disease. The finding of homogeneously staining regions (HSRs) and double minute chromosomal fragments (DMs) in cancer cells led to the discovery of some gene amplifications in cancer cells.

Probes for Abnormal Gene Expression

Molecular biology has contributed greatly to our understanding of cancer cell biology. One of the best examples of this is the discovery of oncogenes and tumor suppressor genes (Chaps. 7 and 8). Current molecular biological techniques, in some instances, allow for the detection of mutated oncogenes or tumor suppressor genes in cancer cells. These mutations can activate oncogenes or inactivate tumor suppressor genes and disrupt important cell regulatory functions. Moreover, the DNA blotting technique developed by E. M. Southern can be used to detect gene deletions or rearrangements in cells. A similar "Northern" blotting analysis—in which RNA is extracted from cells, separated by gel electrophoresis, blotted onto a nitrocellulose filter, and probed with a radioactive cDNA probe—is used to detect and quantitate the level of an mRNA gene transcript. The newer polymerase chain reaction (PCR) technique can be used to detect minute levels of nucleic acids from very small numbers of cells (the number of cancer cells available from a biopsy frequently limits detection of DNA or RNA by other techniques). The PCR technique amplifies small pieces of nucleic acid thousands of times so that a low level of a given gene or gene transcript can be detected. Some examples of the way these techniques can be used are given below.

Southern blot analysis has been utilized to genotype lymphoid malignancies, on the basis of their immunoglobulin expression, with probes to immunoglobulin heavy and light chains and T-cell receptor.[13] Southern blot analysis and PCR techniques have been used to detect chromosomal translocations (t) between chromosomes 14 and 18 [t(14;18)] involving the *bcl*-2 gene in B-cell lymphoma and to detect the *bcr-abl* fusion gene transcript in CML. The PCR technique is so sensitive that it can find previously undetectable numbers of CML cells in patients in apparent clinical remission.[14]

A modified PCR reaction reveals human papillomavirus (HPV) types 16 and 18 in cervical tissues.[15] Since HPV infection has been implicated in the causation of cervical carcinoma, this is a valuable procedure to detect patients who are at risk for this disease.

Molecular biological techniques have also been used to detect amplification of the N-*myc* oncogene in neuroblastoma,[16] amplification of the Her-2/*neu* oncogene in breast cancer,[17] and mutations of the *ras* oncogene in colonic cancer.[18] Since these changes in oncogene expression are often correlated with patient prognosis, their detection provides an important adjunct to cancer diagnosis and treatment.

Tumor Markers

A tumor marker is a biological signal that a tumor is present in the body. These markers are biochemical compounds produced and shed by tumor cells into the bloodstream and usually detected by radioimmunoassay (RIA) or enzyme-linked immunoadsorbent assay (ELISA) in serum, plasma, urine, or other body fluids. Tumor markers fall into several categories: tumor-associated oncofetal proteins (e.g., AFP, CEA), mucins (e.g., Du-PAN-2), hormones (e.g., hCG, calcitonin, ACTH, parathyroid hormone), enzymes (e.g., prostatic acid phosphatase, placental alkaline phosphatase), tumor-associated proteins (e.g., prostate-specific antigen, CA-125 ovarian antigen). These tumor-derived products are often overproduced or inappropriately produced (e.g., ACTH by certain lung cancers) by cancer cells as compared to normal tissues. Unfortunately, they are not usually specific for cancer and are sometimes elevated in nonmalignant proliferative or inflammatory conditions such as inflammatory bowel disease. Nevertheless, they are useful adjuncts to diagnosis and can aid in following the results of therapy. For example, a young man with a testicular mass and detectable levels of hCG or AFP in his blood almost certainly has testicular cancer. A person with colon cancer and very high CEA levels most likely has liver metastases. Furthermore, when hCG levels become undetectable and remain so after treatment in a man with testicular cancer or a woman with choriocarcinoma, there is a high probability of cure.

REFERENCES

1. *Cancer Facts and Figures—1994.* Atlanta: American Cancer Society, Inc., 1994.
2. I. C. Henderson and G. P. Canellos: Cancer of the breast—The past decade. *N Engl J Med* 302:17, 1980.
3. S. Warren: Neoplasms, in W.A.D. Anderson, ed.: *Pathology.* St. Louis: C. V. Mosby Co., 1961, pp. 441–80.
4. F. G. Gunn: Lung, in W.A.D. Anderson, ed.: *Pathology.* St. Louis: C. V. Mosby Co., 1961, pp. 664–704.
5. P. A. Bunn, Jr., M. H. Cohen, D. C. Ihde, B. E. Fossieck, Jr., M. J. Matthews, and J. P. Minna: Advances in small cell bronchogenic carcinoma. *Cancer Treat Rep* 61:333, 1977.
6. A.G.E. Pearse: Common cytochemical and ultrastructural characteristics of cells producing polypeptide hormones (the APUD series) and their relevance to thyroid and ultimobranchial C cells

and calcitonin. *Proc R Soc Lond [Ser B]* 170:71, 1968.
7. P. Rubin: A unified classification of cancers: An oncotaxonomy with symbols. *Cancer* 31:963, 1973.
8. P. Rubin: Statement of the clinical oncologic problem, in P. Rubin, ed.: *Clinical Oncology.* Rochester, NY: American Cancer Society, 1974, pp. 1–25.
9. G. Bonnadonna, A. Rossi, P. Valagussa, A. Banfi, and U. Veronesi: The CMF program for operable breast cancer with positive axillary nodes: Updated analysis on the disease-free interval, site of relapse and drug tolerance. *Cancer* 39:2904, 1977.
10. V. T. DeVita, Jr., R. C. Young, and G. P. Canellos: Combination versus single agent chemotherapy: A review of the basis for selection of drug treatment of cancer. *Cancer* 35:98, 1975.
11. J. D. Pfeiffer and M. R. Wick: The pathologic evaluation of neoplastic diseases, in A. J. Holleb, D. J. Fink, and G. P. Murphy eds.: *Clinical Oncology.* Atlanta: American Cancer Society, Inc., 1991, pp. 7–24.
12. C. M. Morris, A. E. Reeve, P. H. Fitzgerald, P. E. Hollings, M.E.J. Beard, and D. C. Heaton: Genomic diversity correlates with clinical variation in Ph1-negative chronic myeloid leukaemia. *Nature* 320: 281, 1986.
13. D. Kamat, M. J. Laszewski, J. D. Kemp, J. A. Goeken, C. T. Lutz, C. E. Platz, and F. R. Dick: The diagnostic utility of immunophenotyping and immunogenotyping in the pathologic evaluation of lymphoid proliferations. *Mod Pathol* 3:105, 1990.
14. M.-S. Lee, K.-S. Chang, E. J. Freireich, H. M. Kantarjian, M. Talpaz, J. M. Trujillo, and S. A. Stass: Detection of minimal residual *bcr-abl* transcripts by a modified polymerase chain reaction. *Blood* 72:893, 1988.
15. D. K. Shibata, N. Arnheim, and W. J. Martin: Detection of human papillomavirus in paraffin-embedded tissue using the polymerase chain reaction. *J Exp Med* 167:225, 1988.
16. R. C. Seeger, G. M. Brodeur, H. Sather, A. Dalton, S. E. Siegel, K. Y. Wong, and D. Hammond: Association of multiple copies of the N-*myc* oncogene with rapid progression of neuroblastomas. *N Engl J Med* 313:111, 1985.
17. D. J. Slamon, G. M. Clark, S. G. Wong, W. J. Levin, A. Ullrich, and W. L. McGuire: Human breast cancer: Correlation of relapse and survival with amplification of the HER-2/*neu* oncogene. *Science* 235: 177, 1987.
18. E. R. Fearon and B. Vogelstein: A genetic model for colorectal tumorigenesis. *Cell* 61:759, 1990.

2

The Epidemiology of Human Cancer

TRENDS IN CANCER INCIDENCE AND MORTALITY

Long-range trends in the incidence of various cancers in different populations provide clues to the causes of cancer. Because of the long latency period (up to 20 to 30 years; see Chap. 11) between the first exposure to carcinogenic agents and the appearance of clinically detectable cancer, current trends probably reflect an exposure that began decades earlier. Another major factor affecting the overall incidence of cancer is the change in the average age of the population. The average age at the time of diagnosis (averaged for all tumor sites) is 67,[1] and as a higher proportion of the population reaches age 60 and above, the incidence of cancer will go up as a result of this demographic factor alone. Moreover, with the long-term downward trends in other causes of death—primarily infectious and cardiovascular diseases—more people live to an age when the risk of developing cancer becomes high. It is projected, for example, that about 25% of males and 20% of females born in 1985 in the United States will eventually die of cancer.[2] This is up from about 18% for males and 16% for females born in 1975. If current trends continue, about 33% of Americans now living will develop some form of cancer.[3]

The age-adjusted cancer death rates from 1930 to 1990 for selected cancer types are shown in Figure 2-1. Interestingly, if one excludes respiratory tract and skin cancer, the two types increasing most dramatically in recent years, the incidence of cancer in males and females has remained relatively flat since the 1950s.[3] However, there has been a disturbing increase in the incidence rates of breast cancer (about 2% a year) since 1980.[3] Part but not all of this is due to increased detection by mammography. Neither have the overall mortality rates changed markedly since the 1950s, although the cure rates for childhood leukemia, Burkitt's lymphoma, Hodgkin's disease, certain pediatric solid tumors (e.g., Wilms' tumor, Ewing's sarcoma, and retinoblastoma), testicular cancers, and choriocarcinoma have improved considerably since then. These successes are not reflected in the overall mortality rates because they involve relatively rare cancers compared with the "big three": lung, breast, and colorectal cancer. Indeed, these three themselves account for about 50% of all cancer deaths (Fig. 2-2).[4]

Overall, there does not appear to be an "epidemic" increase in mortality from any type of cancer other than respiratory tract and melanoma, and there are several encouraging trends in younger age groups. For example, in the under-65 age group, there was a decrease in mortality rates in the 1973 to 1989 reporting period for breast (4.6%), brain (8.1%), colorectal (18.3%), ovarian (26%), stomach (28.5%), uri-

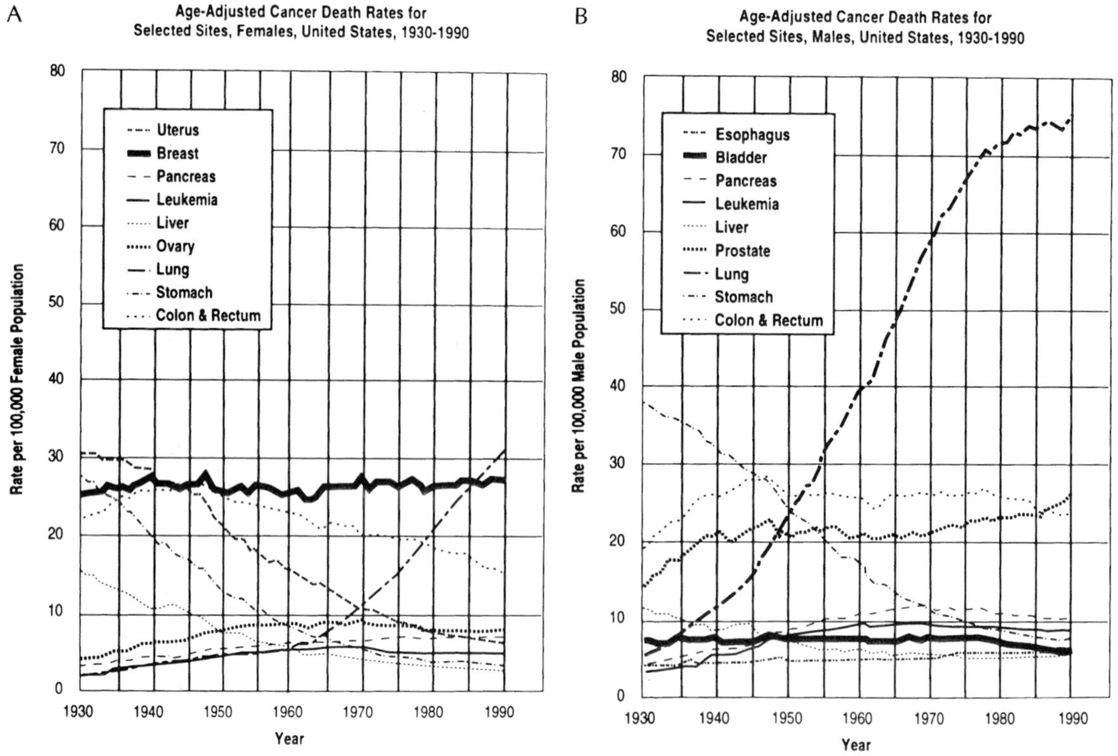

Figure 2-1 Age-adjusted cancer death rates for selected sites, United States, 1930 to 1990. (A) Females. (B) Males. Age-adjusted to the 1970 U.S. standard population. (From Boring et al.[4]).

nary bladder (29.6%), cervical (40.8%), testicular (64.2%), and uterine corpus (37.5%) cancers as well as for Hodgkin's disease (55.3%) and leukemia (16.6%).[1] These changes appear to represent primarily improvements in early diagnosis and treatment; for some of these cancers, however, the incidence rate between 1973 and 1989 has also decreased for the under-65 age group. These cancers include Hodgkin's disease (3.1%) as well as colorectal (3.9%), ovarian (3.9)%, stomach (12.5%), cervical (32.4%), and uterine corpus (45%) cancers. The overall trends in cancer incidence and mortality from 1973 to 1989 for all ages and sexes is shown in Fig. 2-3.

There are, however, some disturbing trends in the other direction. Chief among these is the epidemic increase in lung cancer in women.[1,3] In 1985, lung cancer accounted for as many deaths as breast cancer in women in the United States, and by 1986 it had overtaken breast cancer as the primary cause of cancer mortality in women (Fig. 2-2). The overall mortality for lung cancer is still much higher in males, but the rate of increase in females has been higher than that in males for the past several years. This is almost certainly because of the relatively later development of smoking as a social habit among women (Fig. 2-4).[5]

Other cancers for which mortality rates are increasing are brain cancer (over age 65 only), melanoma, non-Hodgkin's lymphoma, multiple myeloma, lung cancer (males, mostly over 65), leukemia (over 65 only), and cancer of the pancreas (over age 65 only).

Another way to examine trends in cancer mortality is to look at 5-year survival rates for various cancers, which are an index of how well diagnosis and treatment have progressed over time. Table 2-1 illustrates the percent change in incidence, mortality, and 5-year survival rates from the early 1950s to the late 1980s. While marked increases in 5-year survival rates for melanoma, prostate cancer, testicular cancer, bladder cancer, Hodgkin's disease, leukemia, and childhood cancers have occurred, more modest improvements have been observed in most other can-

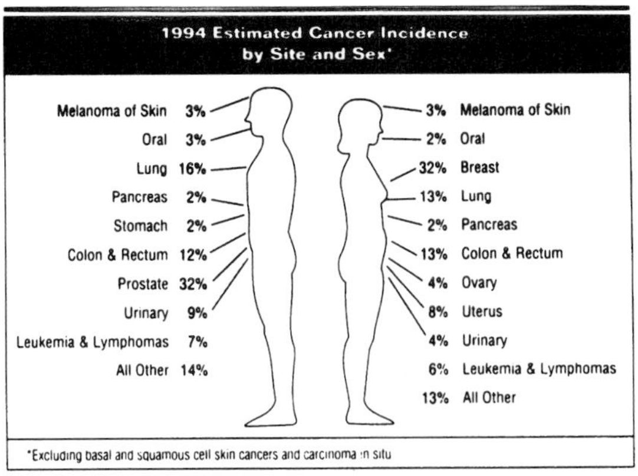

A

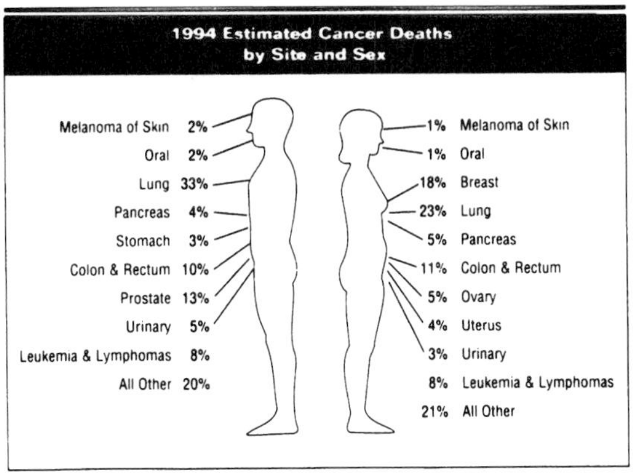

B

Figure 2-2 1994 estimated cancer (A) incidence and (B) deaths by site and sex, excluding basal and squamous cell skin cancers and carcinoma in situ. (From Boring et al.[4])

cers. Some cancers remain extremely difficult to diagnose early and treat successfully.

The latter include cancers of the stomach, lung, pancreas, liver, and esophagus. These all have 5-year survival rates well under 20%.

There is a difference between the sexes and the races in mortality and relative survival rates (Fig. 2-5). The cancer incidence rates are highest for black males, followed in order by white males and then black and white females, who have about the same rate. Mortality occurs in a similar order: black males > white males > black females > white females. Five-year survival rates, however, are higher for both white males and females than for black males and females. The reasons for these racial and sexual differences in incidence and mortality may reflect differences in environmental exposure to carcinogens, diet, smoking habits, and other lifestyle differences. The lower survival rates for minority groups, however, probably also reflect later diagnosis, which in turn reflects poorer access to and utilization of health care providers.[3] Blacks have significantly higher incidence and mortality rates than whites for cancers of the esophagus, uterine cervix, stomach, liver, prostate, and larynx and for multiple myeloma. Whites have higher incidence rates for melanoma, other skin cancers, and cancer of the breast, uterine corpus, ovary, testis, urinary bladder, and brain as well as lymphoma.[1]

The cost of cancer, in both personal and economic terms, is enormous. The National Cancer Institute estimated that in 1990 the total cost of cancer in the United States was $104 billion: $35 billion for direct medical costs, $12 billion for the cost of lost productivity, and $57 billion for terminal care.[3] The cost of diagnostic tests—including mammograms, Pap smears, and screens for colorectal cancer—adds another $3 to $4 billion. The average number of years of life lost per person dying of cancer is second only to the average number for accidents. The average number of years of life lost per person for different cancers is shown in Figure 2-6. Some basic facts about selected individuals cancers are given below.

Lung Cancer

About 172,000 new cases of lung cancer were diagnosed in the United States in 1994. The incidence rate for men was 80 per 100,000 in 1990, down from a high of 87 per 100,000 in 1984.[1,3] The incidence rate for women, however, continues to increase and reached 41 per 100,000 in 1990. Falling rates among men are due largely to a decline in the incidence for the 35- to 54-year age group. An estimated 153,000 deaths occurred in 1994 due to lung cancer.

Cancers of the lung and bronchus represent 90% of all respiratory tract malignancies. The remainder are mostly neoplasms of the larynx (7%) and nasal cavity (2.5%). Squamous cell carcinomas account for about 35% of all lung cancers.[6] This histologic type is closely correlated with smoking and is the most common type of lung cancer in men. These tumors display varying levels of differentiation, from well-differen-

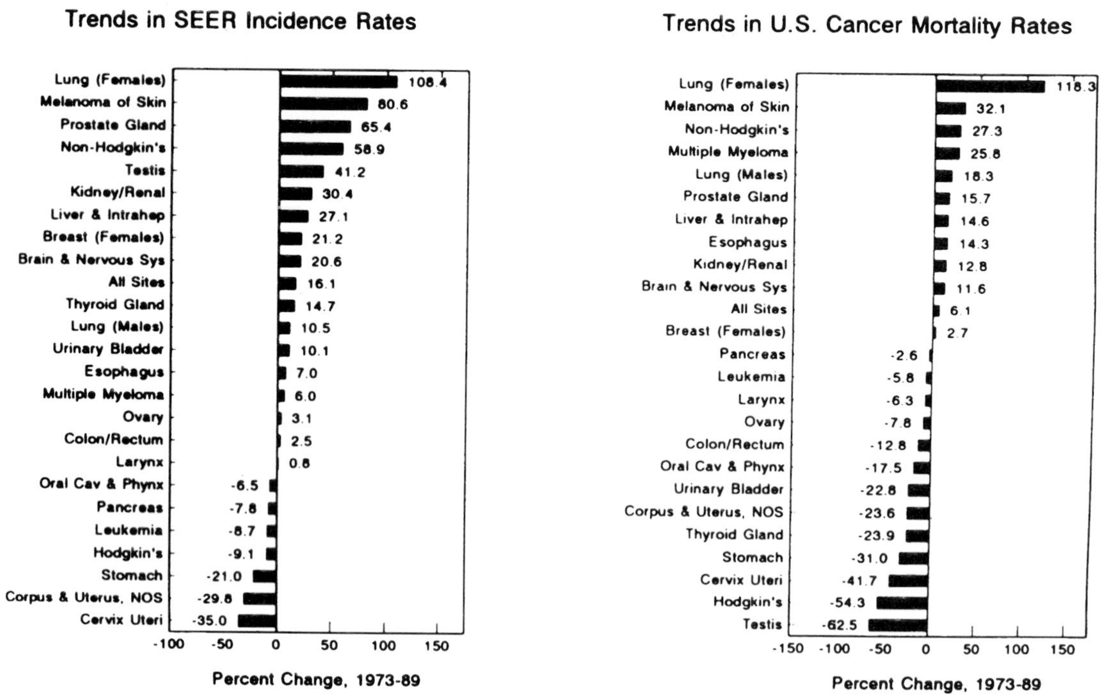

Figure 2-3 Trends in SEER incidence and U.S. mortality rates by primary cancer site, 1973 to 1989. (From Hankey et al.[1])

tiated keratinized squamous epithelium to highly undifferentiated (anaplastic) cells.

Adenocarcinomas of the lung have become more common and now account for almost 35% of pulmonary malignancies.[6] The growth rate of adenocarcinomas is faster than that of squamous cell carcinomas, and they often metastasize to the brain. Lung adenocarcinomas vary from differentiated-looking glandular epithelium to more highly mitotic anaplastic cells. Bronchoalveolar carcinomas are a subtype of lung adenocarcinomas that account for about 5% of all lung cancers. This type of cancer is treatable by surgery if it is localized to a solitary lesion, but it may be multicentric, with additional lesions showing up later.

Large cell undifferentiated carcinomas account for about 15% of lung cancer. Histologically, these tumors contain large anaplastic cells and frequent mitotic cells; they lack squamous or glandular cell characteristics. They tend to metastasize early and have a poor prognosis.

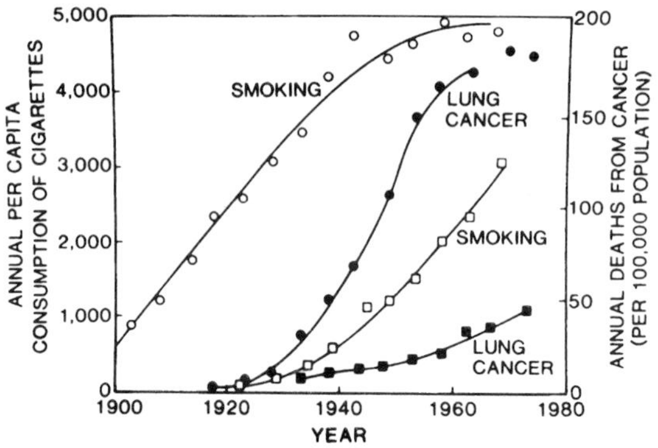

Figure 2-4 Trends in smoking prevalence and lung cancer in British males and females. The data for this chart are from England and Wales. In men, smoking (○) began to increase at the beginning of the 20th century, but the corresponding trend in deaths from lung cancer (●) did not begin until after 1920. In women, smoking (□) began later, and the increase in lung cancer deaths in women (■) has appeared only recently. (From Loeb et al.[5])

Table 2-1 Summary of Changes in Cancer Incidence and Mortality, 1950–89 and 5-Year Relative Survival Rates, 1950–88 (by Primary Cancer Site, Males and Females)

	All Races		Whites					
			Percent Change 1950–89				5-Year Relative Survival Rates (Percent)	
			Incidence		U.S. Mortality			
Primary Site	Estimated Cancer Cases in 1989[a]	Actual Cancer Deaths in 1989	Total	EAPC[b]	Total	EAPC	1950–54	1983–88
Stomach	20,000	14,185	−73.5	−3.0	−76.0	−3.8	12.0	16.0
Colon and rectum	151,000	57,023	10.0	0.4	−25.6	−0.7	42.0	58.4
Larynx	12,300	3,727	62.4	1.2	−10.1	−0.2	52.0	67.2
Lung and bronchus	155,000	137,013	263.8	3.3	245.2	3.4	6.0	13.3
Males	101,000	88,973	217.5	2.7	211.5	3.1	5.0	11.8
Females	54,000	48,040	530.9	5.2	503.0	5.6	9.0	16.0
Melanomas of skin	27,000	6,161	321.0	4.2	152.4	2.4	49.0	82.6
Breast (females)	142,000	42,836	52.5	1.2	4.7	0.1	60.0	79.3
Cervix uteri	13,000	4,487	−76.0	−3.7	−73.9	−3.9	59.0	68.0
Corpus and uterus	34,000	5,850	−6.8	−0.5	−63.5	−2.2	72.0	84.2
Ovary	20,000	12,256	8.2	0.1	−0.2	−0.2	30.0	39.3
Prostate	103,000	30,519	108.8	2.0	14.8	0.2	43.0	77.6
Testis	5,700	392	115.0	2.1	−66.4	−2.8	57.0	92.7
Bladder	47,100	10,121	55.7	1.2	−35.6	−1.2	18.7	79.1
Kidney and renal pelvis	23,100	9,638	109.4	1.9	28.0	0.5	34.0	55.2
Hodgkin's disease	7,400	1,721	29.2	0.4	−65.5	−3.1	30.0	77.7
Non-Hodgkin's lymphomas	32,800	18,064	158.6	2.8	108.7	1.6	28.0	51.9
Leukemias	27,300	18,406	7.8	0.2	−2.1	−0.3	10.0	37.6
Childhood cancers	6,600	1,768	9.8	0.4	−61.1	−2.6	30.0	65.1
All sites excluding lung and bronchus	855,000	359,117	29.9	0.6	−19.4	−0.5	41.4	57.0
All sites	1,010,000	496,130	44.3	0.9	3.2	0.2	39.1	53.5

[a]American Cancer Society (ACS): *Cancer Facts and Figures 1989*—excludes basal and squamous cell skin and in situ carcinomas except bladder.

[b]The EAPC is the estimated annual percent change over the time interval.

Source: From Hankey.[1]

Small cell lung cancers (SCLC) make up most of the remainder of pulmonary malignant neoplasms (10%). These are vicious forms of lung cancer because they grow rapidly, metastasize early, and usually have a very poor prognosis. Symptoms often occur late in the course of the disease; thus it is frequently well advanced by the time of diagnosis. It is also associated with smoking history. The SCLC tend to produce a variety of neuroendocrine substances—such as ACTH, ADH, calcitonin, and neurospecific enolase—that often cause symptoms related to the biological activity of the hormonal substance. These effects constitute a class of symptoms called the *paraneoplastic syndrome* (see Chap. 12). About 10% of all patients with SCLC suffer from this syndrome.[7]

Therapy for lung cancer differs depending on the type of tumor. Surgery is the treatment of choice for non-SCLC, but chemotherapy is the primary treatment for SCLC. Because SCLC has a high growth fraction (i.e., a high percentage of dividing cells), it is a tumor type that responds to anticancer drugs. About 70% of patients will have an objective response; of these, about 30% will have a complete remission.[6] In some situations, other treatment modalities may be of benefit—for example, "neoadjuvant" therapy (pretreatment with chemotherapy followed by surgery) or radiation therapy with or following surgery or chemotherapy. In any case, overall survival rates for lung cancer are not good. The average 5-year survival rate for all patients and all stages of disease is only 13%.[8] The survival rate is 46% if disease is detected when it is still localized, but only 16% of lung cancers are discovered that early.[3]

Colorectal Cancer

New cases of colonic cancer were expected to reach 107,000 in 1994 and to cause over 49,000

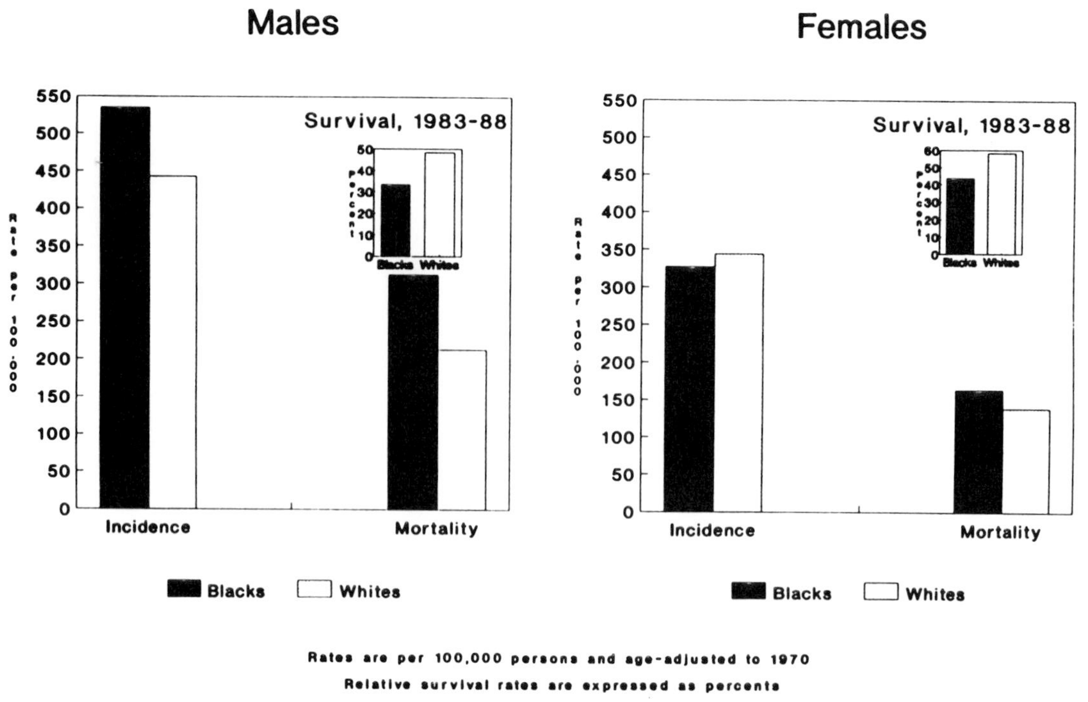

Figure 2-5 Incidence and mortality rates (1985 to 1989), relative survival rates (1983 to 1988), black and white americans. Rates are per 100,000 persons and age-adjusted to 1970. Relative survival rates are expressed as percents. (From Hankey et al.[1])

deaths.[3] The incidence and mortality figures for rectal cancer were 42,000 and 7,000, respectively. The mortality rates for colonic cancer may be somewhat overestimated and those for rectal cancer underestimated because some death certificates state "colon cancer" when the listed hospital diagnosis was cancer of the rectum or rectosigmoid.[9]

Colorectal cancer increases with age and has increased significantly in white males, black males, and black females between 1973 and 1989. Since 1985, however, there have been declines for white males and females as well as a small decline for black females.[9] Moreover, mortality rates have declined for white males and females and for black females but increased for black males in the 1985 to 1989 reporting period. This is most likely due to earlier diagnosis, since more than one-third of colorectal cancer patients are now diagnosed while the cancer is still localized. Five-year survival rates are about 90% when the cancer is detected while still localized; they fall off rapidly with spread of the disease. Survival rates for patients with distant metastases are less than 7%.[3]

A number of predisposing factors have been identified for colorectal cancer.[10] These include familial adenomatous polyposis, chronic ulcerative colitis, and the "cancer family syndrome."[11] There is also a correlation with high-fat, low-fiber diets, although the cause-effect relationship between diet and colorectal cancer is not clearly understood.

Familial polyposis is an inherited autosomal dominant trait that leads to the production of multiple adenomatous polyps in the colon. These polyps have a tendency to undergo malignant transformation by age 30. In familial polyposis, an associated gene mutation has been identified. In one study, mutations of the af-

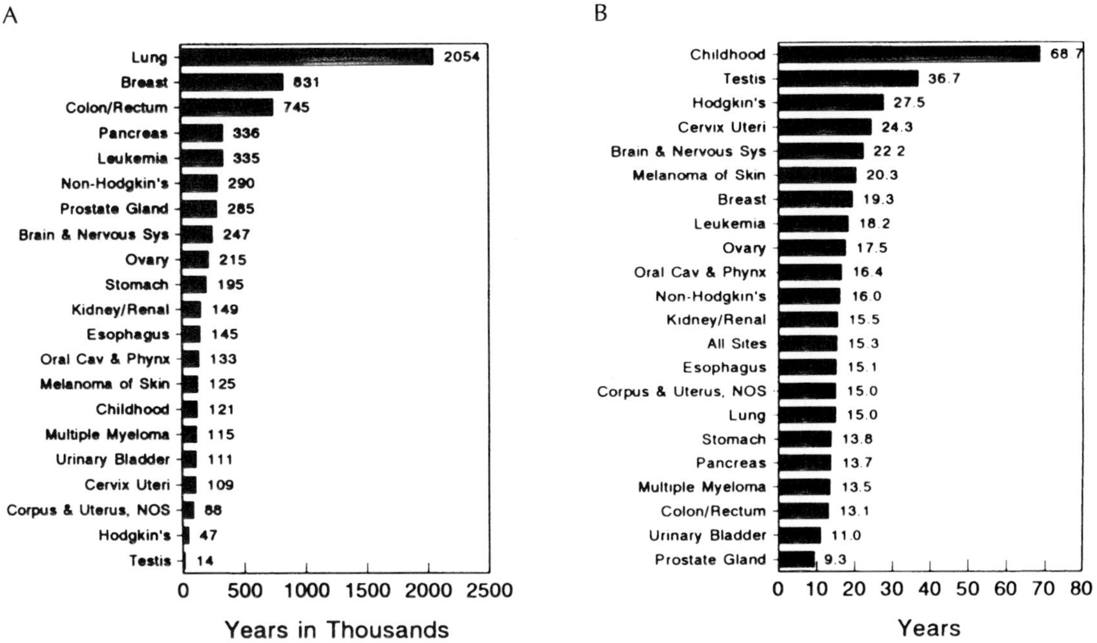

Figure 2-6 (A) Person-years of life lost due to cancer, all races both sexes, 1989, and (B) Average years of life lost per person dying of cancer, all races, both sexes, 1989. (From Hankey et al.[1])

fected gene, called APC, were found in the earliest tumors that were analyzed, including adenomas 0.5 cm in diameter, and the frequency of mutations remained constant as benign adenomas progressed to adenocarcinomas.[12] These data indicate that mutations of the APC gene, thought to be a tumor suppressor gene (see Chap. 8), play a major role in initiation of colonic cancer. Other genetic changes such as mutations to the *ras* oncogene and in the p53 tumor suppressor gene occur later in tumor progression than alteration of the APC gene (Chap. 3).

Treatment of colonic cancer may be by surgery, radiation therapy, chemotherapy, or a combination of treatment modalities, depending on the stage and location of the tumor.[10] Surgery with or without radiation therapy is the most effective treatment, but survival is closely correlated with early detection of localized disease. Chemotherapy by itself is minimally effective in this disease, but combinations of 5-fluorouracil (5-FU) and immune system stimulators such as levamisole or 5-FU plus leucovorin have shown some promise.

Pancreatic Cancer

About 28,000 new cases of pancreatic cancer occur each year in the United States.[3] An estimated 25,000 deaths occurred in 1994. Although pancreatic cancer accounts for about 2.5% of all annual diagnoses of cancer, it is responsible for 5% of all cancer deaths.[13] It is a particularly virulent form of cancer and is usually diagnosed late in the disease, when distant metastases have developed. The overall 5-year survival rate is only 3%. Pancreatic cancer has no clear early warning signs or symptoms and is usually "silent" until the disease is well advanced. No effective therapeutic approach has been developed, although surgery, radiation, and chemotherapy have all been tried—probably because the disease is already so far advanced by the time treatment usually begins.[14]

Risk factors include age (most cases are diagnosed between ages 65 and 79), smoking, sex, and possibly diets high in fat.[3] The disease is 30% more common in men and 60% more frequent in black Americans than in white Americans.[3] Incidence and mortality rates have been fairly constant since the 1970s except for the small increased incidence among black women.

Breast Cancer

An estimated 182,000 new cases of breast cancer occurred in 1994.[3] Incidence rates for breast cancer have increased about 2% per year since 1980; however, the incidence rate has leveled off at about 108 per 100,000 since 1989.[15] The increase in incidence rate from 1980 to 1987 appears to reflect, to a large extent, mammography screening and earlier diagnosis. The evidence for this includes the fact that the increased incidence occurred mainly for smaller tumors and at earlier stages of disease. Thus, one would expect mortality rates to start to fall by the middle of the next decade. For patients diagnosed in the 1970s, the 5-year survival rates changed very little. Those patients diagnosed in the 1980s have had survival rates only about 4% better than those diagnosed in the 1970s.[15] Since the overall mortality rate for breast cancer has not changed dramatically in the past several years (Fig. 2-1), the implication is that so far, earlier diagnosis has not had an impact. Part of the reason for this may be that breast cancer can recur late, in some cases well after 5 years. Therefore, although the 5-year survival is going up, overall survival has not yet been affected. Five-year survival rates are about 92% for localized breast cancer; they drop to about 18% when there are distant metastases.

Risk factors for breast cancer include age over 50, early menarche and late menopause, first childbirth after age 30, nulliparity, family history of breast cancer, obesity, possibly dietary factors such as a high-fat diet, and high doses of radiation to the chest before age 35.[16,17] However, a majority of women who develop breast cancer do not have a known risk factor, which indicates that the causes may be multiple and are not yet clearly understood.

There is a clear correlation to family history of breast cancer.[18] Significant increases in risk are associated with a history of breast cancer in a young sister or mother or with a family history of bilateral disease. The risk of developing breast cancer by age 65 approaches 50% for women with a strong family history of breast cancer (e.g., multiple family members with breast cancer). Two genes involved in genetic susceptibility to breast cancer (BRCA1 and BRCA2) have been identified (see Chap. 3).

A type of autosomal dominant inherited condition that predisposes to breast cancer at an early age is the Li-Fraumeni syndrome.[19] Members of these families may develop breast cancer in their late 20s or early 30s. Multiple other kinds of cancer also occur in these families, including sarcomas, brain tumors, leukemia, and adrenocortical tumors.

Treatment of breast cancer includes surgery, radiation, chemotherapy, and hormonal modification. Localized breast cancer can often be treated by partial mastectomy (lumpectomy) alone or with radiation therapy. Total mastectomy may be still the treatment of choice for more extensive but localized disease, but breast-conserving surgery has been an increasing trend. Such surgery, for example, constituted 31% of surgical procedures in 1983 and 44% in 1988[13]; however, the percentage of patients receiving the two types of treatment varies in different parts of the United States.[20] Current data show equivalent survival for patients treated with conservative surgery plus radiation therapy and patients treated with total mastectomy for localized disease.[21] Chemotherapy is recommended for patients with metastatic spread to regional lymph nodes. Use of the antiestrogen drug tamoxifen has proved useful in combination with chemotherapy for postmenopausal women with positive lymph node involvement who have tumors containing estrogen (ER+) or progesterone (PR+) receptors.[22] Taxmoxifen is also undergoing clinical trials as a chemopreventive agent in disease-free women who are at high risk for developing breast cancer (see Chap. 15).

Ovarian Cancer

New cases of ovarian cancer are expected to reach an annual count of 24,000 in 1994.[3] This accounts for 4% of all cancers among women. An estimated 13,600 deaths occurred in 1994 due to ovarian cancer, more deaths than those caused by any other cancer of the female reproductive system. A reason for the high mortality rate is that there are no obvious symptoms until late in the disease. Thus, it is often not diagnosed until advanced stages are reached. Although a serum tumor marker (the glycoprotein CA-125) is elevated in about 50% of women with early-stage disease, it is not specific or sensitive enough to be useful in screening women for cancer.[23]

Risk factors include age over 60, multiparity, history of breast cancer, and certain genetic predispositions. Epithelial ovarian cancer is far more common in industrialized nations (with the exception of Japan); and in the United States it is much more common in Caucasian women of northern European descent, suggesting that lifestyle and dietary factors play a role in causation.[3,24] Several studies have indicated that there is a diminished risk in women who have used oral contraceptives.[23]

Treatment usually involves surgical removal of both ovaries, the uterus, and fallopian tubes. In advanced disease, an attempt is made to "debulk" the tumor by removing all intraabdominal disease. In a young woman of childbearing age with early-stage disease localized to one ovary, it may be possible to remove only the affected ovary.

Chemotherapy for advanced disease has met with only limited success. The new drug taxol, derived from the bark of the Pacific yew tree, has shown some promise in advanced disease refractory to other drugs.[23]

The overall 5-year survival rate for ovarian cancer is 41%; it is 88% for those diagnosed with localized disease and 17% for patients with metastatic disease.[3] Unfortunately, only about 23% of all cases are detected at the localized stage.

Uterine Cancer

Of the estimated 70,000 new cases of cancer of the uterine cervix diagnosed in 1994, some 15,000 were invasive cervical carcinomas and 55,000 were carcinomas in situ.[3] Cancers of the uterine endometrial epithelium accounted for about 31,000 new cases. It is estimated that 4,600 deaths from cervical cancer and 5,900 from endometrial cancer occurred in 1994. The death rate from uterine cancer has decreased more than 70% over the past 40 years, primarily due to earlier diagnosis and treatment, in which the Pap test has played an important part.

Risk factors for cervical cancer include early age of first sexual intercourse, multiplicity of partners, cigarette smoking, and infection with certain sexually transmitted diseases, particularly papillomavirus (see "Infectious Etiology of Cancer," below, and Chap. 6). For endometrial cancer, risk factors, include family history, a history of infertility due to failure to ovulate, prolonged estrogen therapy, and obesity.[3,24] Some 75% of endometrial cancers occur after age 50 and only 4% occur before the age of 40. Peak age for diagnosis of endometrial cancer is 58 to 60, whereas the peak age for diagnosis of cervical cancer is age 48 to 50.[24]

Cancer of the uterine body is usually treated by hysterectomy with or without irradiation for grade II or III neoplasms. Treatment with progestins may be effective in premenopausal women with grade I disease and as an adjuvant to surgery in more advanced disease. Chemotherapy is used for metastatic disease and appears to be most effective in poorly differentiated tumors that are not hormone-dependent.[24]

Cervical cancers diagnosed at the in situ stage may be treated by local surgery, cryotherapy, or electrocoagulation. More advanced disease is treated by hysterectomy or radiation therapy.

Overall survival varies greatly with the stage of the disease. Five-year survival for patients treated for cervical carcinoma in situ is virtually 100%. For endometrial cancer, 5-year survival for patients diagnosed early is 93%. These survival rates drop off dramatically if disease is metastatic at the time of diagnosis.

Urinary Bladder Cancer

Bladder carcinoma is the most common malignant neoplasm of the urinary tract. An estimated 51,200 new cases occurred in 1994: 38,000 in men and 13,200 in women.[3] It is the fourth most common cancer in men and the ninth most common in women. In 1994, some 10,600 deaths due to bladder cancer were expected to occur. Mortality rates for bladder cancer have been decreasing since 1973, most likely due to the increasing number of cases diagnosed early and because of improved treatment.[25]

The disease is most common in the 50- to 70-year age group. Risk factors include smoking, exposure to certain chemicals, and parasitic infection. The discovery that bladder cancer occurred in industrial workers exposed to high levels of aniline dyes is one of the classic stories in chemical carcinogenesis (see Chap. 6). Chronic infection with schistosomal parasites has been associated with bladder cancer in parts of the world where schistosomiasis is indigenous (see "Infectious Etiology of Cancer," below).

Treatment of localized disease may be by endoscopic resection and fulguration.[25] Total cystectomy is indicated when the tumor is invasive and presents as a grade III or IV lesion. Radiation with surgery and/or chemotherapy has been used for advanced disease. Instillation of chemotherapeutic agents into the urinary bladder has been used with some success for superficial bladder cancer, and this treatment has been shown to reduce recurrence.[26]

Five-year survival for patients with superficial disease is 90%. For regional disease, it is 45%; and for disease with distant metastases, it is 9%.[3]

Prostate Cancer

Annual new cases of prostate cancer are expected to reach 200,000 in the United States in 1994; 38,000 mortalities are expected.[3] Prostate cancer is a disease of aging: it is relatively rare among men under age 55 and reaches an incidence rate of 1000 per 100,000 men over age 75.[27] Black males in the United States have the highest incidence rate in the world for reasons that are not known. Overall incidence rates for all ages are 101 per 100,000 black males versus 70 per 100,000 white males in the United States. Over the 1973 to 1989 reporting period, the age-adjusted incidence rate rose 3% per year, and it has been rising more rapidly in whites than blacks. This is due to a number of factors: the increasing average age of the male population, increased reporting due to more transurethral resections, and more screening tests being carried out.[27] However, not all of the rise in incidence can be due to increased detection, because mortality rates also have been increasing, particularly for black males.

Detection methods for prostate cancer include digital rectal examination, use of a serum tumor marker called prostate-specific antigen (PSA), and ultrasonography. Prostate-specific antigen is a prostate-associated glycoprotein that is shed into the bloodstream and detected by immunologic techniques. Serum levels are elevated both in benign prostatic hypertrophy and in prostatic cancer, but the levels are higher in cancer. It has been suggested that PSA testing be used as a screening procedure in men over 50, but its lack of specificity is a problem. In combination with digital exam and ultrasound, PSA increases the ability to detect cancer of the prostate.[28]

Treatment of prostate cancer includes surgical removal of the prostate, irradiation for locally invasive tumors, and hormonal therapy. Hormonal treatments traditionally have employed estrogens such as diethylstilbestrol (DES) or removal of androgenic hormones by removing the testes. More recently, the nonsteroidal antiandrogen flutamide administered with luteinizing hormone releasing hormone (LHRH) agonists have been shown to be equivalent to DES or orchiectomy.[26]

Over the past 30 years, 5-year survival rates have improved from 50% to 78%.[3] Some 60% of all prostate cancers are detected while they are still localized; the survival rate for these patients is 88%, although there is a significant difference between whites (78%) and blacks (63%).[27]

Leukemia

The leukemias are a group of malignant diseases that involve the blood-forming lineage of cells. In leukemia, there is a block in cellular differentiation (see Chap. 5), so that immature cells pile up in the bone marrow and spill into the peripheral blood. Depending on the cell lineage involved and the degree of cellular differentiation, this disease is classified as acute or chronic myelogenous leukemia (AML and CML) or acute or chronic lymphocytic leukemia (ALL and CLL). Rarer forms include monocytic, basophilic, eosinophilic, and erythroid leukemia. Acute myelogenous leukemia is also called acute nonlymphocytic leukemia (ANLL), particularly in adults, and consists of a group of leukemias characterized by uncontrolled proliferation of primitive hematopoietic stem cells in which myeloid, monocytic, erythroid, or megakaryocytic cell types may predominate.[29]

New cases (28,600 per year) are about equally divided between acute and chronic leukemias.[3] Leukemia strikes more adults (over 26,000 annually) than children (2,600 per year), and ALL accounts for 80% of the leukemias of childhood.[29]

Leukemias are associated with certain risk factors, environmental and genetic. The ANLL form has been linked to exposure to radiation (e.g., radiologists, Japanese atomic bomb survivors), toxic chemicals (e.g., benzene), previous treatment with mutagenic drugs (e.g., melphalan and other anticancer alkylating agents), and genetic predispositions, (e.g., Down's syndrome). The predominant form of leukemia in children under 15 is ALL, with a 4:1 incidence ratio over AML, whereas in adults it is the reverse.[30] Peak age for ALL is 3 to 4 years. Siblings of affected children have a fourfold higher risk of developing leukemia in the first decade of life, suggesting a strong genetic predisposition. Some autosomal recessive genetic disorders associated with chromosomal instability such as Fanconi's anemia, Bloom's syndrome, and ataxia telangiectasia are also associated with a higher risk of leukemia[30] (see below). Viral infection is associated with adult T-cell lymphocytic leukemia and adult hairy cell leukemia (see Chap. 6), but virus-induced leukemias have not been clearly identified in children.[30]

Several distinct chromosomal translocations that are known to activate certain oncogenes have been observed in various forms of leukemia. These include the *bcr-abl* chimeric gene in CML and activation of the *myc* and *bcl*-2 genes in B-cell lymphocytic leukemias (see Chap. 3).

Treatment of leukemia is one of the success stories in cancer chemotherapy. In the 1950s, childhood ALL, for example, was virtually 100% fatal. Now, with newer drugs and drug regimens, 95% of children with ALL attain a complete remission, and long-term survival is achieved in about 73% of patients.[3] Survival has also improved for CLL, but it remains dismally low for patients with AML or CML (20% to 25%).[29,31]

Skin Cancer

Nonmelanoma skin cancer is by far the most common cancer in the United States. Over 700,000 new cases a year are diagnosed. Because most forms of skin cancer are slow-growing, only locally invasive, do not readily metastasize, and—if treated appropriately—can be cured, skin cancer is usually not considered in the calculations of incidence, mortality, and survival rates for cancer. However, because it is so common and in most situations preventable, it is an important disease to consider.

The two most common types of nonmelanoma skin cancer are basal cell and squamous cell carcinomas. Basal cell carcinoma (BCC) accounts for most of the cases.[32] It is a malignant neoplasm of the basal cells of the epidermis and occurs predominantly on sun-damaged skin. In this case, the inciting carcinogen is clearly known: ultraviolet (UV) irradiation. It rarely affects dark-skinned people. Because of the sunbathing craze among younger people, BCC is now seen in individuals still in their 20s, and the incidence rate increases with age and amount of sun exposure. Depletion of the ozone layer in the atmosphere, which filters out some of the sun's UV rays, has been implicated in the increasing incidence of BCC.[32]

Squamous cell carcinoma (SCC) is a malignant neoplasm of the keratinizing cells of the epidermis. It has some propensity to metastasize to regional lymph nodes and distant sites if it is not treated early. Squamous cell carcinoma is the second most common skin cancer in humans. It usually occurs later in life than BCC and is more common in men than in women. As for BCC, extensive exposure to UV irradiation from sunlight is the most important causal factor in SCC.

Treatment of BCC or SCC involves surgical removal (e.g., Mohs micrographic surgery with histologic verification of complete removal of tumor) or destruction of the lesion by a variety of methods (e.g., electrodesiccation, cryosurgery, or laser therapy).[32] Overall cure rates for BCC and SCC are high (>90%) if the lesion is detected and treated early, though there may be recurrences requiring retreatment.

Malignant melanoma of the skin, however, is a different beast. It is a more aggressive form of cancer and, once it becomes invasive, can metastasize rapidly and widely. The incidence of melanoma has been rising about 4% per year in the United States and currently there are about 32,000 new cases and over 7000 deaths annually due to melanoma.[3] This malignant disease is also associated to some extent with excessive exposure to sunlight and is more common in fair-skinned individuals. Surgery is the treatment of choice for localized disease. Radiotherapy is used for palliative treatment of central nervous system (CNS) or bone metastasis. Chemotherapy is also used for metastatic disease, but the response rate is only 15% to 20%.[33]

Brain Cancer

An estimated 16,900 new cases of primary brain and other CNS tumors (e.g., meningeal and spinal cord tumor) occurred in 1992.[34] The number of deaths due to all CNS tumors in 1992 was expected to reach 11,800. Gliomas are the most common type and may be classified as benign astrocytomas, anaplastic astrocytomas, and glioblastomas, which represent a continuum from fairly benign to highly malignant tumors.[35] The most common malignant type is glioblastoma. The overall 5-year survival rate for brain cancers is about 25%.[34]

Brain cancer is one of those cancers for which the incidence rate is rising, particularly for those 65 years of age and older. In the over-65 age group, the mortality rate has been increasing about 3% per year since 1973.[34] The reasons for this are not clear but appear to relate in part to better diagnostic techniques such as computed tomography (CT), nuclear magnetic resonance imaging (MRI), and positron emission tomography (PET). However, mortality rates began to rise before such techniques were widely available and have continued to rise since they were introduced. This suggests that some environmental factors may be involved. Brain cancer mortality rates for whites are 75% higher than for blacks.

GEOGRAPHIC DISTRIBUTION OF HUMAN CANCERS

Worldwide Variation in Cancer Incidence

The incidence of malignant neoplastic disease varies greatly in different parts of the world. This reflects all the variables affecting the way people in different areas live, such as their diet, individual behavioral characteristics, social customs, exposure to environmental agents, exposure to endemic parasitic infections, and average life expectancy. The data in Table 2-2, which illus-

Table 2-2 Worldwide Variation in the Incidence of Common Cancers, with the Range of Variation Expressed for Ages 35–64

Type of Cancer	High-Incidence Area	Low-Incidence Area	Range of Variation
Skin	Australia (Queensland)	India (Bombay)	>200
Buccal cavity	India	Denmark	>25
Nasopharynx	Singapore[a]	England	40
Bronchus	England	Nigeria	35
Esophagus	Iran	Nigeria	300
Stomach	Japan	Uganda	25
Liver	Mozambique	Norway	70
Colon	U.S.[b] (Connecticut)	Nigeria	10
Rectum	Denmark	Nigeria	20
Pancreas	New Zealand[c]	Uganda	5
Breast	U.S.[b] (Connecticut)	Uganda	5
Uterine cervix	Colombia	Israel	15
Uterine corpus	U.S.[b] (Connecticut)	Japan	10
Ovary	Denmark	Japan	8
Bladder	U.S.[b] (Connecticut)	Japan	4
Prostate	U.S.[d]	Japan	30
Penis	Uganda	Israel	300

[a]Chinese.

[b]The U.S. data are taken from the Connecticut Tumor Registry because it is the oldest continued cancer registry based on a defined population in this country.

[c]Maori.

[d]Blacks.

Source: Data from Doll.[36]

Table 2-3 Standardized Mortality Ratios for Selected Cancer Sites

Cancer Site[a]	Native Japanese	Nisei	General U.S. Population
Stomach	100	38	17
Colon	100	288	489
Lung	100	166	316
Breast	100	136	591

[a]Data for stomach, colon, and lung are for males only; breast data for females only.
Source: Data from Haenszel and Kurihara.[40]

trate the range of variation in cancer incidence, are expressed for similar age groups in the different countries in order to correct for the age variable. Several examples will serve to point out some of the environmental or behavioral factors that may be involved.

In 1961, two unusual outbreaks of disease in animals were observed: hepatic necrosis was noted in domestic poultry in England and a number of hatchery-bred rainbow trout in the northwestern United States died of hepatomas. Both of these outbreaks turned out to have been caused by feed contamination with a mold called *Aspergillus flavus,* which produces a chemical, aflatoxin, now known to be an extremely potent liver carcinogen. Aflatoxin is also a suspected etiologic agent in certain countries with a high endemic rate of liver cancer; in China, Kenya, Swaziland, and Mozambique, the incidence of liver cancer is proportional to the amount of aflatoxin in the diet.[37,38] Liver cancer in high-incidence areas of the world is also closely associated with hepatitis B virus (HBV) infection, which may be a cocarcinogen with aflatoxin. Some studies suggest that HBV infection is the more important causal factor.[39]

In Kashmir, an old custom of carrying a basket of hot coals called "kangri" under the clothes next to the abdomen to keep warm during the bitter winters led to a high incidence of skin cancers in the areas of skin continually exposed to this thermal trauma. The chewing of betel nut in India has been associated with a high frequency of oral cancers. In the United States, the incidence of lung cancer is much higher in cigarette smokers. Exposure to industrial chemicals is known to be associated with some cancers, such as bladder cancer in aniline dye workers. In Egypt and other countries, endemic infection with schistosomal parasites is common. The eggs of these parasites become encysted in the urinary bladder of the host, and this phenomenon is thought to be related to the high incidence of bladder cancer in these countries.

Ugandans, Nigerians, and South African blacks are at low risk to develop cancers of the larynx, lung, stomach, colon, rectum, kidney, and brain but at high risk for primary hepatocellular carcinoma. The Japanese have less malignant melanoma, Hodgkin's disease, and cancers of the prostate, corpus uteri, ovary, and breast but are at high risk for stomach cancer. The highest rates of esophageal cancer occur in northeastern Iran, central Asia, and northern China (Honan, Hopei, and Shansi provinces). Nasopharyngeal carcinoma is a relatively rare neoplasm in North America and Europe but is common in China and Southeast Asia.

The genetic composition of a given population can also affect the susceptibility to develop cancer, but genetic background does not appear to play the major role in the causation of most human cancers. That human cancer is more environmental than genetic in origin is demonstrated by cancer incidence rates in populations that emigrate from one environment to another. In the example shown in Table 2-3, the children of Japanese emigrants to the United States have cancer incidence rates approaching those of the U.S. population rather than those of the Japanese population in Japan.

In the industrialized parts of the world, cancer causes one-fifth of all deaths each year. In a study of 15 industrialized countries, 5% of all deaths due to cancer occurred in people less than age 45, while 60% occurred in those 65 years of age and older.[41] Different trends in cancer mortality rates were observed in six geographic areas: the United States, eastern Europe, western Europe, East Asia, Oceania, and Nordic countries. In the period from 1969 through 1986, recorded cancer mortality in people age 45 and over in these six regions increased for lung, breast, and prostate cancer and declined for stomach cancer. The rate of increased mortality due to lung cancer in men ages 45 to 54 has slowed or reversed in all six areas except eastern Europe and East Asia, reflecting smoking habits to a large extent. There have been rapid increases in mortality rates for several forms of cancer in people in the 45- to 84-year age group. Figures 2-7, 2-8, and 2-9 show the mortality rates and percent change in mortality rates for stomach, lung, and breast cancer, respectively, for males and females 45 to 84 years of age from 1969 to 1986.

Stomach cancer mortality rates declined in all six geographic areas but remain much higher in eastern Europe and East Asia than in other areas. This trend most likely reflects improvements in food storage and handling, such as reductions in toxic methods of food preservation (smoking and pickling), increased use of refrigeration, and decreased salt consumption.[41]

Lung cancer deaths have been increasing in most areas of the industrialized world, particularly in males in East Asia and in women (Fig. 2-8). Most of these increases are related to increased cigarette smoking, but they may also reflect, to a lesser extent, changes in air pollution, diet, and occupational factors as well as, in certain parts of East Asia, the use of smoky, hydrocarbon-emitting fuels for domestic cooking and heating.[41]

Breast cancer mortality is increasing in all regions and most age groups (Fig. 2-9) except for younger women in the United States and Nordic countries and women between ages 75 and 84 in Oceania.[41] The lowest mortality rates are seen in East Asia, although the rate there appears to be increasing.

Cancer Mortality by State in the United States

Clues to the origins of human cancer often come from clustering of certain cancers in a defined locality or in populations living or working in

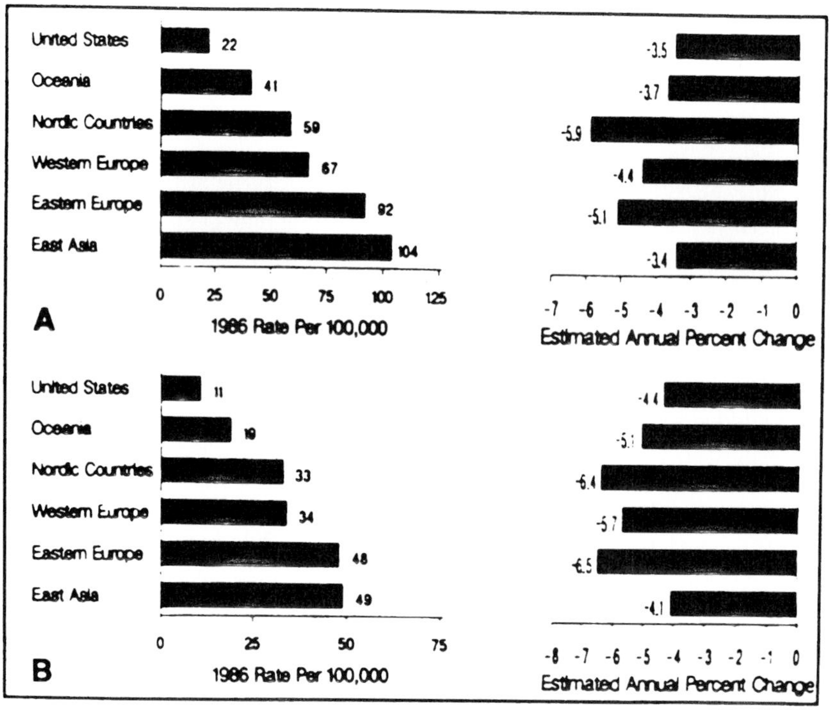

Figure 2-7 Stomach cancer in males (A) aged 45 to 84 years and females (B) aged 45 to 84 years. Mortality data provided by WHO (1969–1986). (From Hoel et al.[41])

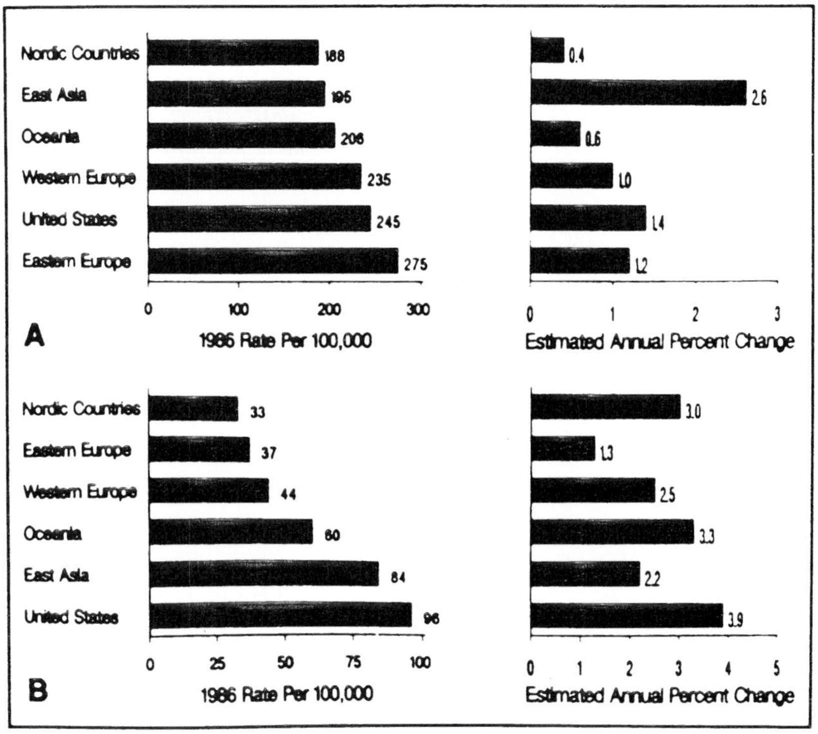

Figure 2-8 Lung cancer in males (A) aged 45 to 84 years and females (B) aged 45 to 84 years. Mortality data provided by WHO (1969–1986). (From Hoel et al.[41])

similar environments. Discovery of cancer clustering has led to the identification of environmental factors involved in increased cancer risk. This is most striking when particular tumors (often rather rare ones) are observed to be increasing in a population with a low background of risk for that tumor type. An example of this was the observation of an unusual clustering of hepatic angiosarcomas, a tumor that is very rare in the general population, in U.S. workers exposed to high levels of vinyl chloride. One way to identify clusters of more common cancers is to examine

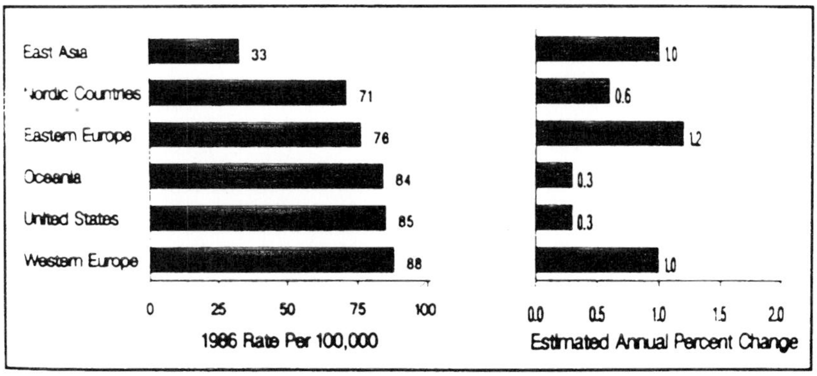

Figure 2-9 Breast cancer in females aged 45 to 84 years. Mortality data provided by WHO (1969–1986). (From Hoel et al.[41])

the geographic distribution of cancer systematically, comparing incidence and mortality rates between regions. When this is done by state in the United States, some interesting results emerge (Table 2-4).[42]

The highest mortality rates for breast cancer are seen in the District of Columbia and the states of Delaware, New Jersey, Rhode Island, and New York. Low-incidence states are those with a more rural population. These differences may reflect dietary and life-style differences. Urban women may be more likely to delay childbearing and eat a less well balanced diet or one high in fat, for example.

Colorectal cancer also tends to be higher in the northeastern states. The patterns for males and females are almost identical. High colonic cancer mortality rates are associated with urbanization, higher income, and certain ethnic groups (e.g., Irish, Germans, and Czechs).[43] Colonic cancer has been linked in some studies to dietary practices, especially to consumption of meat (beef in particular) and a high-fat, low-fiber diet. As yet, none of these factors has been clearly

Table 2-4 Average Annual Age-Adjusted Cancer Mortality Rates by State, All races, 1985–89

State	Rate	SE[a]	Rank	PD[b]
Breast, females				
Total, United States	27.4	0.06		
Five high states				
District of Columbia	33.3	1.31	(01)	21.5
Delaware	33.0	1.34	(02)	20.4
New Jersey	32.1	0.36	(03)	17.2
Rhode Island	31.8	1.01	(04)	16.1
New York	31.4	0.24	(05)	14.6
Five low states				
Arkansas	22.8	0.56	(47)	−16.8
Idaho	22.6	0.93	(48)	−17.5
Texas	22.6	0.23	(49)	−17.5
Utah	22.4	0.80	(50)	−18.2
Hawaii	18.5	0.83	(51)	−32.5
Colon and rectum, males and females				
Total, United States	19.8	0.04		
Five high states				
District of Columbia	25.6	0.84	(01)	29.3
New Jersey	23.9	0.22	(02)	20.7
Rhode Island	23.5	0.60	(03)	18.7
New York	23.1	0.14	(04)	16.7
Delaware	23.1	0.81	(05)	16.7
Five low states				
Idaho	15.8	0.54	(47)	−20.2
Texas	15.7	0.14	(48)	−20.7
New Mexico	15.6	0.46	(49)	−21.2
Wyoming	15.3	0.82	(50)	−22.7
Utah	13.7	0.45	(51)	−30.8
Lung and bronchus, males and females				
Total United States	48.0	0.06		
Five high states				
Kentucky	61.8	0.56	(01)	28.8
Nevada	59.4	1.07	(02)	23.8
West Virginia	56.9	0.71	(03)	18.5
Louisiana	56.8	0.52	(04)	18.3
Arkansas	55.6	0.63	(05)	15.8
Five low states				
Colorado	34.3	0.48	(47)	−28.5
North Dakota	33.8	0.96	(48)	−29.6
Hawaii	33.7	0.80	(49)	−29.8
New Mexico	33.2	0.68	(50)	−30.8
Utah	20.6	0.56	(51)	−57.1

[a]SE = standard error of the rate.

[b]PD = percent difference between state rate and total U.S. rate.

Source: Adapted from Hankey and Kosary.[42]

established as a cause of colonic cancer. Colorectal cancer increased significantly among white males, black males, and black females between 1973 and 1989, although for the most recent 5-year reporting period (1985 to 1989), a decline in incidence for white males and white females was observed.[9]

Lung cancer mortality shows a very different distribution from that of breast and colon cancer (Table 2-4). It is highest in Kentucky, Nevada, West Virginia, Louisiana, and Arkansas. Clusters of high mortality due to lung cancer appear around large metropolitan areas,[43] where air pollution may add to the risk of cigarette smoking.

Mortality among white males is elevated in counties with mining, paper, chemical, petroleum, and shipbuilding industries; however, the increases in mortality associated with the presence of such industries is 1 to 2 deaths per 100,000 population compared to increases of 10 to 20 deaths per 100,000 for urban in comparison with rural counties.

ROLE OF VARIOUS FACTORS IN THE DEVELOPMENT OF CANCER

Table 2-5 illustrates the agents or circumstances for which there is good evidence of carcinogenicity in humans. These are generally divided into three categories: occupational, medical, and social. The data are obtained from the International Agency for Research on Cancer (IARC), based on studies of a working group that periodically accumulates and evaluates the evidence for human carcinogenesis. Although the information presented in Table 2-5 may suggest that most human cancer is caused by occupational or medical exposures, this is not the case. It is simply that the cause of the bulk of human cancers is unknown, and only after certain discrete, well-documented exposures can cancer causation be laid at the feet of a particular agent. Sometimes rare cancers, such as the hepatic angiosarcomas mentioned above, call attention to certain agents. Most epidemiologists ascribe only about 2% to 6% of human cancers to occupational exposures.[44–46] The "best estimates"[44] of the proportion of cancer deaths due to various factors are as follows: tobacco = 25% to 40%, alcohol = 2% to 4%, diet = 10% to 70% (some authors include diet under life-style and attribute 30% to 60% to it),[45] reproductive patterns and sexual behavior = 1% to 13%, drugs and medical procedures = 0.5% to 3%, geophysical factors (mostly sunlight) = 2% to 10%, and infection (e.g., parasites, transmissible viruses) = 1% to 10%. Surprisingly, some authors ascribe only about 1% to 5% of cancer deaths to pollution.[44] There is considerable debate about the role of environmental exposure to air and water pollutants in cancer causation; this is discussed in greater detail below. Obviously, the ranges cited for each of the potential causes are very broad in some cases, reflecting the degree of uncertainty about the whole business of ascribing definitive causes at the current state of our knowledge.

Cigarette Smoking

As noted above, epidemiologists have attributed as many as 25% to 40% of all cancer deaths to tobacco use, primarily cigarette smoking. In 1992, for example, there were approximately 149,000 deaths due to lung cancer in the United States (93,000 men and 56,000 women), making up about 34% of all deaths from cancer in men and 22% of those in women.[4] Although the most direct correlation is between cigarette smoking and lung cancer, tobacco use has also been implicated in cancers of the mouth, pharynx, larynx, esophagus, urinary bladder, pancreas, and kidney. Smoking of pipes or cigars has been implicated in the occurrence of cancers of the mouth, pharynx, larynx, and esophagus, but this form of tobacco use is generally considered much less dangerous because the smoke is usually not inhaled. A number of studies have also suggested a correlation between "passive smoking"—that is, exposure to smoke and others' cigarettes in the home or workplace—and lung cancer.[47] An enormous amount of research on the relationship of tobacco smoking and cancer has been carried out in recent years. The vast majority of these studies indict cigarette smoking as a major cause of lung cancer. Epidemiologic studies, autopsy reports, and experimental animal data reviewed in the original U.S. Surgeon General's Report of 1964 and in subsequent U.S. Department of Health, Education, and Welfare

Table 2-5 Established Human Carcinogenic Agents and Circumstances

Agent or Circumstance	Type of Exposure[a] Occupational	Medical	Social	Site of Cancer
Aflatoxin			+	Liver
Alcoholic drinks			+	Mouth, pharynx, larynx, esophagus, liver
Alkylating agents				
Cyclophosphamide		+		Bladder
Melphalan		+		Marrow
Aromatic amines				
4-Aminodiphenyl	+			Bladder
Benzidine	+			Bladder
2-Naphthylamine	+			Bladder
Arsenic[b]	+	+		Skin, lung
Asbestos	+			Lung, pleura, peritoneum
Benzene	+			Marrow
Bis(chloromethyl) ether	+			Lung
Busulfan		+		Marrow
Cadmium[b]	+			Prostate
Chewing (betel, tobacco, lime)			+	Mouth
Chromium[b]	+			Lung
Chlornaphazine		+		Bladder
Furniture manufacturer (hardwood)	+			Nasal sinuses
Immunosuppressive drugs		+		Recticuloendothelial system
Ionizing radiations[c]	+	+		Marrow and probably all other sites
Isopropyl alcohol manufacture	+			Nasal sinuses
Leather goods manufacture	+			Nasal sinuses
Mustard gas	+			Larynx lung
Nickel[b]	+			Nasal sinuses, lung
Estrogens				
Unopposed		+		Endometrium
Transplacental (DES)		+		Vagina
Overnutrition (causing obesity)			+	Endometrium, gallbladder
Phenacetin		+		Kidney (pelvis)
Polycyclic hydrocarbons	+	+		Skin, scrotum, lung
Reproductive history:				
Late age at 1st pregnancy			+	Breast
Zero or low parity			+	Ovary
Parasites				
Schistosoma haematobium			+	Bladder
Chlonorchis sinensis			+	Liver (cholangioma)
Sexual promiscuity			+	Cervix uteri
Steriods				
Anabolic (oxymetholone)		+		Liver
Contraceptives		+		Liver (hamartoma)
Tobacco smoking			+	Month, pharynx, larynx, lung, esophagus, bladder
UV light	+		+	Skin, lip
Vinyl chloride	+			Liver (angiosarcoma)
Virus (hepatitis B)			+	Liver (hepatoma)

[a]A plus sign indicates that evidence of carcinogenicity was obtained.
[b]Certain compounds or oxidation states only.
[c]For example, from x-rays, thorium, Thorotrast, some underground mining, and other occupations.
Source: From Doll and Peo.[44]

reports strongly support a causal relationship. The data can be summarized as follows:[48]

1. A strong relationship between cigarette smoking and lung cancer mortality in men has been demonstrated in numerous prospective and retrospective studies, with risks for all smokers as a group ranging from 11 to 22 times those of nonsmokers.[49]
2. A dose-response relationship between cigarette consumption and risk of development of lung cancer for both men and women has been demonstrated in numerous studies, with risk much higher for men and women who are heavy smokers than for those of nonsmokers. Light smokers have an intermediate risk.[49,50]
3. Mortality from lung cancer directly attributable to cigarette smoking is increased in the presence of "urbanization" and such occupational hazards as uranium mining and exposure to asbestos.
4. Cessation of smoking results in lowered risk or mortality from lung cancer in comparison with continuation of smoking.[49,50]
5. Results from autopsy studies show that changes in the bronchial mucosa that are thought to precede development of bronchogenic carcinoma are more common in smokers than in nonsmokers, and there is a dose-response relationship for these changes.
6. Chronic inhalation of cigarette smoke or the intratracheal instillation of various fractions of tobacco smoke produce lung cancer in such experimental animals as dogs and hamsters.
7. Cell culture studies show that various constituents found in tobacco and cigarette smoke condensate produce malignant transformation of cells.
8. Numerous complete carcinogens and cocarcinogens (tumor promotors) have been isolated from cigarette smoke condensation.

Numerous mutagens and carcinogens have been identified in the particulate or vapor phases of tobacco smoke; these include benzo[*a*]pyrene, dibenza[*a*]anthracene, nickel, cadmium, radioactive polonium (^{210}Po), hydrazine, urethane, formaldehyde, nitrogen oxides, and nitrosodiethylamine.[51] Moreover, mutagenic activity is 5- to 10-fold higher in the urine of smokers than that of nonsmokers.[52] An increased incidence of chromosomal abnormalities has also been observed in smokers' peripheral blood lymphocytes compared with those of nonsmokers.[53,54]

There is a worldwide epidemic under way in lung cancer that reflects the increasing amount of tobacco use everywhere. Over 660,000 new cases of lung cancer are reported yearly by the World Health Organization (WHO).[55] A study by the WHO of cancer mortality trends between 1960 and 1980 in 28 industrialized countries showed that lung cancer mortality increased by 76% in men and 135% for women.[55] Mortality rates from lung cancer in women have increased more than 100% in Japan, Norway, Poland, Sweden, and the United Kingdom; more than 200% in Australia, Denmark, and New Zealand; and over 300% in Canada and the United States. For the United States, it is estimated that 171,000 of the 514,000 cancer deaths in 1991 were "excess," in that they were due to tobacco-related cancers that might have been avoided if the victims had never smoked.[49]

A WHO survey in the late 1980s also showed that 33% of males about age 15 in "developed" countries smoke cigarettes and that 50% of these teenagers smoke in developing countries. The rates for females are lower, but they are rising rapidly. It is estimated that the incidence of lung cancer in developing countries will be the same as that in developed nations in 40 years.[55] If this is the case, by the year 2000 there will be 1 million new lung cancer cases per year worldwide. By 2025, if this trend continues, there will be more than 3.5 million new cases annually. And this is just for lung cancer. When one considers the other tobacco-related cancers, the number of preventable new cancer cases could be more than double that estimated for lung cancer alone.[49] Clearly, unless smoking habits and other life-style and pollution factors change, the future looks bleak for attempts to prevent cancer.

An encouraging trend, however, has been noted in the United States, which could bode well for other nations if they reduce tobacco consumption. Lung cancer rates in men under age 54 started to decline in the 1980s, and there has been a leveling off of mortality rates in men age 55 to 64.[50] In the oldest age groups, however, a steady increase in lung cancer mortality rates continued into the late 1980s, reflecting a cohort of individuals who started smoking in high numbers at an early age. If these trends continue, the lung cancer mortality rate in men should begin to show a real decline by the year 2000.[8] The numbers for women are less encouraging, since they started smoking late in the century as a group, and the 20-year lag in lung cancer is still catching up with them (Fig. 2-4). Thus, lung cancer mortality rates in women are expected to climb for the remainder of the century and begin to decline after that.[56]

Alcohol

Alcohol is thought to "interact" with smoking in the causation of certain cancers, particularly oral and esophageal cancers. Alcohol appears to be synergistic with tobacco in causing cancers of the mouth, pharynx, larynx, and esophagus but not of the lung.[57,58] In the case of liver cancer, there is good evidence that alcohol consumption sufficient to cause cirrhosis of the liver increases the incidence of liver cancers, perhaps secondary to the chronic damage to the liver caused by alcohol abuse. Pure alcohol is not by itself carcinogenic in animals and may exert its carcinogenic effect secondarily to tissue damage, as in the case of hepatic cirrhosis, or by facilitating uptake of carcinogens by exposed tissues, as may be the case for oral and esophageal cancer.[44]

A small positive association between alcohol and breast cancer risk has been seen in some but not all epidemiologic studies.[59] In some studies, even moderate consumption (< three drinks per week) was associated with increased risk. However, considering all the data, a clear relationship between alcohol consumption and breast cancer has not been established.[59]

Diet

There is suggestive evidence that the diet is associated with several types of human cancer. Cancers of the stomach, colon, pancreas, breast, ovary, uterine endometrium, and prostate appear to fall into this category.[60,61] This hypothesis stems largely from the observed incidence of cancers in various parts of the world. Stomach cancer is more prevalent in Japan, western South America, and parts of northern and eastern Europe than in the United States. On the other hand, cancers of the colon, pancreas, breast, ovary, endometrium, and prostate are more common in the United States, western Europe, Australia, and New Zealand than in other parts of the world.

A number of hypotheses relating to dietary factors have been postulated to explain this variability in cancer incidence among countries. The consumption of smoked or pickled fish and meat has been implicated in stomach cancer. Prior to refrigeration, large amounts of nitrate and salt were used as food preservatives in the United States. Since that time, there has been a sizable decrease in the intake of nitrate and a dramatic decrease in the incidence of stomach cancer in the United States. But in areas of the world where consumption of nitrate is still high, stomach cancer is more prevalent.

Based on epidemiologic data, a high dietary intake of fat, protein, and beef as well as a dietary deficiency of fiber have been implicated in colon cancer. In the typical U.S. diet, 40% to 45% of calories come from fat; but in Japan, where the incidence of colon cancer is much lower, only 15% to 20% of dietary calories are derived from fat (Table 2-6). Much research has been directed to pinpointing the carcinogenic factors associated with high-fat diet. To date, however, such factors have not been clearly identified. In some studies, a correlation between fecal excretion of bile acids and colonic cancer has been noted.[62] There is evidence that bile acids may be tumor promotors rather than initiating agents (see Chap. 6). In animals, the bacteriologic flora of

Table 2-6 Comparison of Daily Dietary Components (grams or milligrams) in Japan and Hawaii

Nutrient	Japan	Hawaii
Calories	2132	2274
Total protein (g)	76	94
Animal protein (g)	40	71
Vegetable protein (g)	37	24
Total fat (g)	36	85
Principally saturated (g)	16	59
Principally unsaturated (g)	21	26
Total carbohydrate (g)	335	260
Simple (g)	61	92
Complex (g)	278	169
Alcohol (g)	28	13
Cholesterol (mg)	457	545

Source: Data from Weisburger et al.[60]

the gastrointestinal tract appear to be involved in the conversion of bile acids to tumor-promoting agents.[60] The incidence of breast cancer is also higher in countries where there is a high dietary intake of saturated fat (the United States and the Scandinavian countries) than in countries with low saturated fat intake (Japan). Furthermore, as noted above (Table 2-3), the incidence of breast cancer in descendants of Japanese emigrants to the United States is higher than in native Japanese, which suggests a dietary role.

The role of dietary fat in causing human cancer is far from clear, however. For example, a review of world data found no clear relationship between dietary fat and gastrointestinal or prostate cancer.[63] Studies in the United Kingdom, Sweden, and France found no relation between fat intake and breast or colorectal cancer (reviewed in Ref. 61). A study of 90,000 American nurses showed no correlation between levels of dietary fat and breast cancer.[64] However, an American diet "low in fat" is still higher in fat than a typical Japanese diet, for example, and a difference in risk may have been observed between cohorts if a lower cutoff for fat intake was used. Also, some studies in Israel and Canada have found a positive correlation between dietary fat and breast cancer (reviewed in Ref. 61).

Experimental studies in animals are also somewhat confounding. High-fat diets enhance formation of tumors in the colon, pancreas, and breast, and fats can have tumor-promoting activity; but caloric restriction appears to be at least as important and perhaps more so than the percentage of fat in the diet.[61]

Epidemiologic studies in humans with low and high rates of colorectal cancer support the hypothesis that colorectal cancer risk is increased by a diet that is high in fat, high in protein, and low in fiber.[65–67] Evidence from the combined analysis of 13 case control studies provides substantive evidence that intake of fiber-rich foods is inversely related to risk of cancers of the colon and rectum.[66] Moreover, data from a study of 7284 male health professionals, 170 of whom developed colorectal adenomas, showed that a diet high in fat and low in fiber increased the risk of developing adenomas.[67] These data suggest that dietary factors play a very early role in the malignant transformation process, since adenomatous polyps occur approximately 10 years before the clinical diagnosis of colorectal cancer.

Dietary "fiber," defined as plant cell wall material not hydrolyzed by human intestinal tract enzymes, is regarded by some epidemiologists as helpful in preventing colonic cancer. Some studies show an inverse relationship between dietary fiber content and mortality from colonic cancer.[61] Although these studies require further documentation, it is reasonable to assume that a dietary fiber content that keeps the stools soft and bulky and increases the rate of transit of contents through the gastrointestinal (GI) tract could decrease the concentration (per bulk) of carcinogens in the stool and reduce the time of exposure of the GI mucosa to carcinogens.

In addition to dietary fat and fiber, other constituents of the diet may play a key role in carcinogenesis within the GI tract and also at other target sites. Certain directly acting mutagenic and carcinogenic substances exist in the diet of various populations.[68] In addition to the aflatoxins mentioned above, certain other mycotoxins produced from molds are mutagenic in the Ames test (see below). One of these, sterigmatocystin, is a potent carcinogen in animal studies. Other carcinogens that may be present in the diet include safrole in sassafras and black pepper, hydrazine-type compounds in the false morel *Gyromitra esculenta*, pyrrolizidine alkaloids in certain plants (e.g., *Senecio*) used to make herbal teas, and gossypol in cottonseed oil, which is widely used as a cooking oil in certain countries such as Egypt. In Japan, some people eat bracken fern regularly, and those who do have a threefold greater risk of developing cancer of the esophagus.[69]

The cooking of foods may also generate potentially carcinogenic substances. Benzo[*a*]pyrene and other polycyclic hydrocarbons can be produced by pyrolysis when meat is broiled or smoked or when food is fried in fats that have been used repeatedly.[44] Moreover, Sugimura and Nagao[70] have shown that mutagenic products are generated from pyrolysis of amino acids or caramelization of sugars during heating of foods. This probably means excessive heating and may not be a significant problem at usual cooking temperatures.

Agents added to food to preserve it or to give

it color, flavor, sweetness, or a certain consistency are consumed in the diet in large amounts. Most of these have been screened for mutagenic potential and some have been removed from the market because they were carcinogenic, even if weakly so, in animals. Saccharin is an example of such a product. Feeding large amounts of saccharin to rats produces bladder cancer. Furthermore, feeding saccharin to rats that received direct administration of the powerful carcinogen *N*-methyl-N-nitrosourea into the urinary bladder increases the production of bladder cancers. This has led to the conclusion that saccharin "promotes" more than "initiates" the carcinogenic process. Yet the doses necessary to do this are extremely large and may not be relevant to human exposure patterns. Moreover, it has now been shown that the bladder carcinogenicity of saccharin is most likely caused by the formation of protein-saccharin-silicate crystals that is peculiar to the rat.[71] These precipitates cause a chronic inflammatory response that leads to a focal regenerative hyperplasia. If this is sustained over the lifetime of the animal, as happens in chronic feeding studies, bladder cancer can result. It is highly unlikely that the usual doses of saccharin ingested by humans would do this, because human urine contains considerably less protein, has a lower pH, and contains less of the salts likely to form these crystals as compared with the rat. Furthermore, there is no epidemiologic evidence for increased risk of cancer in humans from ingestion of saccharin. Therein lies the rub. How do you regulate the use of a weak carcinogen that may or may not produce cancer after 20 to 30 years of low-dose exposure or that may cause cancer in animals exposed to doses that humans are highly unlikely to ever experience? This question is central to the whole issue of regulation of exposure to potentially carcinogenic substances and is extremely difficult to answer.

The addition of nitrites or nitrates (nitrate can be reduced to nitrate in the GI tract) to foods, especially meat, as preservatives is another controversial subject. If one considers the total consumption of nitrite in foods, most of it comes from natural sources; only about 10% is attributable to nitrite added as a preservative.[72] However, this added amount may still have some significance because it may reach the GI tract in "packets" of high concentration.[44] Because nitrites may combine with secondary amines to form nitrosamines in the GI tract, it seems reasonable to minimize their addition to foods.

Sexual Development, Reproductive Patterns, and Sexual Behavior

The duration of hormonal exposure appears to play a role in the susceptibility of women to breast cancer. The carcinogenic effects of hormones were first demonstrated in animals. In 1932, Lacassagne reported the induction of mammary carcinomas in mice injected repeatedly with an ovarian extract containing estrogen. Later, he also showed that the synthetic estrogen diethylstilbestrol produced mammary tumors in susceptible strains of mice.[73] Furthermore, ovariectomized mice and rats have a decreased frequency of breast cancer, whereas rodents subjected to increased levels of estrogen, progesterone, and prolactin have an increased frequency of breast cancer, although the timing of exposure to individual hormones appears to be crucial.[74,75] Similarly in humans, a role of hormones in the development of breast cancer has been deduced from the known risk factors associated with the disease. These factors include early age of menarche, delayed age of first pregnancy, and delayed menopause, suggesting longer duration of exposure to hormonal stimulation as an etiologic agent in breast cancer. Studies on blood levels and urinary excretion patterns of hormones in breast cancer patients or women at risk to develop breast cancer have yielded conflicting results. However, in a study of women whose mothers had bilateral breast cancer and who thus had a high familial risk of also developing the disease, it was concluded that this increased risk was associated with elevated plasma levels of prolactin, progesterone, and estrogen.[76]

In men, late descent of the testes is associated with an increased susceptibility to testicular cancer, for reasons that are not clear, though this may reflect some faulty differentiation response in the testicular tissue. This cancer usually occurs in younger men (average age at diagnosis is 32).

Reproductive patterns are related to cancers of the uterine endometrium, ovary, and breast, all three of which are less common in women who have had children—especially those who did so early in their reproductive lives—than in women who have not had children.

Cancer of the uterine cervix is associated with early and frequent sexual contact with a variety of partners. As indicated above, there is some evidence for an association of a transmissible virus, human papillomavirus, with cervical cancer (see Chap. 6).

Industrial Chemicals and Occupational Cancers

The chemicals and industrial processes that have a known or suspected etiologic role in the development of cancer are listed in Table 2-5. About 2% to 4% of all cancer deaths are attributed to occupational hazards.[44,77] Of those agents listed as carcinogenic for humans, a number were identified because of the close association between an abnormal clustering of certain cancers and exposure to an industrial chemical or process. For example, epidemiologic studies of workers occupationally exposed to industrial levels of 4-aminobiphenyl show that they have a higher incidence of bladder cancer.[78] Occupational exposure to asbestos fibers results in a higher incidence of lung cancer, mesotheliomas, GI tract cancers, and laryngeal cancers.[79] As mentioned earlier, cigarette smoking and occupational exposure to asbestos act synergistically to increase the incidence of lung cancer. Several epidemiologic studies have shown increased frequency of leukemia in workers exposed to benzene.[79–81] Two studies of workers exposed to bis(chloromethyl) ether have indicated an increased risk of lung cancer, primarily SCCL.[79] There is also an increased risk of lung cancer among workers in chromium industries. Occupational exposure to 2-naphthylamine has long been known to be associated with urinary bladder cancer.[79] Since the time of Percival Pott and his study of chimney sweeps (see Chap. 6), coal soot has been known as a cause of skin cancer. Since that time, occupational exposure to soot, coal tar, pitch, coal fumes, and some crude shale and cutting oils has been shown to be associated with cancers of the skin, lung, bladder, and gastrointestinal tract. The carcinogenicity of these latter agents is probably related to their content of polycyclic aromatic hydrocarbons (PAH).

The highest levels of human exposure to PAH occur in industrial processes involving the use of coal tar and pitch and in the production of coke from coke ovens.[82] Epidemiologic data indicate that ambient coal tar and pitch in iron and steel foundries contain carcinogenic substances and may lead to an increased incidence of lung cancer, particularly in smokers who work in such environments. A study by van Schooten et al.[82] showed that coke-oven workers were exposed to substantial concentrations of PAH in the air, including benzo[*a*]pyrene and pyrene. A total of 47% had detectable levels of PAH-DNA adducts in their white blood cells, compared to 30% of control subjects who worked in another part of the plant. In both groups, smokers had significantly higher levels of PAH-adducts than nonsmokers. Since the carcinogenic mechanism of PAH involves metabolic activation and alteration of DNA function (see Chap. 6), these data suggest that exposure to various carcinogens and/or susceptibility to their DNA-damaging effects could be monitored by measuring DNA-adduct formation in peripheral white blood cells or perhaps, if sensitive and specific enough assays could be developed, in urine. It should be noted that such measurements would be subject to individual variations relating to variation in daily exposure levels, genetic differences in metabolic activation of PAH and DNA repair mechanisms, smoking habits, amount of air pollution with PAH in the place of residence, and amount of PAH in the drinking water and diet.

Herbicides

Herbicides are a heterogeneous class of chemicals widely used in agriculture, forestry, and gardening to kill undesirable weeds and foliage. Although agricultural workers and workers in the plants that manufacture herbicides are exposed to the highest concentrations, the entire population is probably exposed to some level of herbicide contamination, albeit low. This could come about from residual contamination of foodstuffs, runoff into groundwater used for drinking supplies, or airborne contamination in areas of heavy spraying.

Herbicides that have been used commercially include the phenoxy compounds 2,4-dichlorophenoxyacetic acid (2,4-D) and 2,4,5-trichlorophenoxyacetic acid (2,4,5-T; also known as Agent Orange), triazines, amides, benzoics, carbamates, trifluralin, and uracils. 2,4,5-T and to some extent 2,4-D preparations were contaminated with dioxins and furans, particularly prior to 1975, when government manufacturing restrictions limited the amount of allowable contamination. One contaminant, 2,3,7,8-tetrachlorodibenzo-*p*-dioxin (TCDD), is a potent mutagen and a powerful carcinogen in animal studies. Prior to 1975, TCDD concentrations of 1 part per million (ppm) were observed in commercial phenoxy herbicide preparations, whereas current levels are below 0.1 ppm.[83]

A study of workers applying phenoxy herbicides has shown detectable blood levels of TCDD.[83] Because TCDD is stored in fat tissue, its half-life in the body may be as long as 7 years. Calendar period of exposure and intensity of use of 2,4,5-T were determinants of serum levels of TCDD. These serum levels were also associated with intensity of exposure to 2,4-D, but this was confined to individuals exposed before 1975. Some workers, based on the assumed half-life of 7 years, would have had serum TCDD levels up to 329 parts per trillion (ppt); a maximum of 26 ppt has been reported in the general population. In a Vietnam veteran heavily exposed to Agent Orange, serum levels of 1530 ppt were estimated to have been reached during the time of his peak exposure. Adipose tissue concentrations of 540 ppt have been shown to cause thyroid cancers in animals.[83]

There is considerable controversy over the long-range health effects of exposure to 2,4,5-T, but it is clear that some individuals heavily exposed to this substance before 1975 would most likely have achieved body concentrations known to be carcinogenic in some animals.[83] As yet, however, there is no conclusive evidence for TCDD carcinogenicity in humans.[84] The manufacture of 2,4,5-T has been discontinued in most western countries, and its use has been banned in the United States since 1983.

The association of herbicide use with cancer in humans has recently been reviewed.[85] Studies from several countries and states in the United States provide significant evidence supporting a relationship between non-Hodgkin's lymphoma in farmers and exposure to phenoxy herbicides. Several studies have also indicated a relationship to increased risk of soft tissue sarcomas with exposure to phenoxy herbicides. Although these herbicides have also been implicated in some studies as increasing the risk of cancers of the colon, lung, nasal passages, prostate, and ovary as well as leukemia and multiple myeloma, there have been too few definitive studies to demonstrate an exposure-risk relationship.

Another dilemma is that experimental animal studies do not convincingly demonstrate the carcinogenicity of 2,4-D and 2,4,5-T.[86] This makes it difficult to develop a consistent public policy on the use of such substances, particularly when most regulatory decisions are made on carcinogenicity testing in animals. This is the reverse of the saccharin situation noted above, in that the epidemiologic data are positive and the experimental animal data are not. Nevertheless, it seems only prudent to monitor exposure to herbicides and other occupationally related chemicals carefully and to promote minimal exposure safety practices, particularly for farmers and other workers who may experience high exposure. In the United States, the National Cancer Institute, in collaboration with the Environmental Protection Agency, is undertaking a long-term cohort study of pesticide-exposed farmers.

Air and Water Pollutants

Evidence that potential carcinogens in the air or water might cause cancer is based on several assumptions as well as on epidemiologic data. One of the assumptions is that there is a linear, nonthreshold dose-response relationship between the given dose of carcinogen and the number of cases of cancer. This assumption is based primarily on dose-response studies in experimental animals. Such a dose-response relationship carries the implication that there is no such thing as a safe level of exposure to a carcinogen. This may or may not be correct. From the pharmacologic evaluation of a very broad spectrum of drugs, one can conclude that there are dosage levels below which no biochemical or physiologic effects are seen (excluding, perhaps, such effects as allergic reactions to drugs). In the case of chemical carcinogens, however, one could argue that it would take only a "single hit" on a key nucleotide in a DNA molecule to produce a malignant change in a cell. From what is known about the ability of mammalian cells to "repair" damage to their DNA, the likelihood that a single hit could produce a cancer cell is remote except perhaps in individuals who have a genetic DNA repair defect (see below). Taking the nonthreshold approach to evaluation of exposure to environmental agents is, of course, the most conservative policy; it tends to predict the largest response (i.e., the largest number of cancers) for any given level of low-dose exposure. Since the possible consequences of exposure to carcinogens in the general environment are so enormous, a number of investigators think that it is appropriate to do this. Although this approach seems reasonable to environmentalists, currently only limited evidence supports it. Evidence from air pollution studies, for example, indicates that estimating cancer risk by the extrapolation of dose-response relationships may be an oversimplification of the problem. Large metropolitan areas have a substantially higher level of atmospheric carcinogens—such as benzo[*a*]pyrene, resulting from combustion of fossil fuels—than rural areas, yet some studies[87] show that nonsmokers in urban areas do not have a significantly higher risk of lung cancer than rural nonsmokers. However, urban *smokers* do have a significantly higher incidence of lung cancer than comparably heavy smokers in rural areas. These observations and others, such as the potentiation of lung cancer in uranium miners[88] and asbestos workers[89,90] who smoke, support the idea that a *combination* of urban air pollution and smoking is the most carcinogenic.

Numerous potential carcinogens have been found in air and water, particularly in areas near or downstream from large industrial complexes. For example, nitrosamines, a class of chemicals that are among the most potent carcinogens known from experimental animal studies, are present in the environment, albeit usually at very low concentrations. Fine et al.[91] reported detectable levels of one of these, *N*-nitrosodimethylamine (NDMA), in the air around a chemical plant manufacturing dimethyl hydrazine, for which NDMA is a precursor. Typical NDMA levels were between 6 and 36 $\mu g/m^3$ of air on the factory site, about 1 $\mu g/m^3$ in the residential area adjacent to the factory, and about 0.1 $\mu g/m^3$ 2 miles away. Typical daily human exposures were 39, 10, and 0.3 μg, respectively, in these areas. Also, NDMA was found in an estuary near the plant, in the effluent from a neighboring sewage treatment facility, and in the soil immediately adjacent to the factory. These sorts of data show how relatively small industrial emissions of noxious chemicals can accumulate. In the case of nitrosamines, such amounts could be potentially carcinogenic over time.

In addition to industrial sources, domestic sewage treatment plant effluents may contain carcinogenic substances that can find their way into drinking water supplies. More than 50 chlorinated hydrocarbons have been identified in domestic sewage effluents.[92] This same study estimated that over 1000 tons of chlorinated organic compounds are discharged by sewage treatment plants into the nation's waterways annually. Chlorinated hydrocarbons result from the chlorination of water heavily polluted with organic chemicals.[93] Some of these chlorinated compounds are known to be carcinogenic in animals.

Although discharges from industrial and municipal waste treatment plants may be continuous sources of pollution, spills resulting from industrial operations, transportation accidents, or dumping of chemical wastes on or near bodies of water can contribute significant levels of hazardous substances to public water supplies.

The Environmental Protection Agency and other groups have undertaken studies of several large metropolitan areas to evaluate the level of contamination of public drinking water supplies and to assess the carcinogenic risk associated with this contamination. In a survey of 80 cities, a number of potentially dangerous trihalomethanes—including chloroform, bromodichloromethane, dibromochloromethane, and bromoform—where detected.[94] Chloroform, a known carcinogen in animals, was found in the drinking water of 80 cities. Carbon tetrachloride, also a known carcinogen, was found in the drinking water of 10 cities. In one survey,[95] 325 organic chemicals were identified in the drinking water of various cities. Only about 10% of these have been adequately tested for carcinogenicity. Among the known or suspected carcinogens identified in drinking water are benzene, bis(chloromethyl) ether, carbon tetrachloride, chloroform, Dieldrin, polychlorinated biphenyls, 1,1,2-trichloroethylene, and vinyl chloride. Thus, it is evident that the general public is exposed to a wide variety of environmental chemical carcinogens. Since there is a 20- to 30-year latent period between exposure to certain carcinogenic agents and the development of clinically detectable cancer, it will probably take several decades to fully evaluate the impact of our contaminated environment. It should be pointed out, however, that the expected correlations between exposure to a given carcinogen in the drinking water and the type of cancer expected to result from such exposure have not been established. For example, even though chloroform, a hepatocarcinogen, is the predominant organic contaminant in the drinking water of certain communities in Louisiana that take their water from the Mississippi, there is no increased mortality due to hepatic cancers in those communities.[96]

Nevertheless, a considerable debate over the role of environmental pollutants in human cancer continues. Based on studies of cancer incidence in various regions in Africa, Higginson and Oettlé[97] provided some definitive data on the impact of environmental factors in the causation of human cancer. The work of Higginson and his colleagues has generally been credited with establishing the fact that about two-thirds of all human cancers have an environmental cause and thus, theoretically at least, are preventable. This has led many people to believe that the environmental agents responsible for cancer are chemicals that we inhale or ingest. However, as Higginson himself has reiterated,[98,99] what he meant by "the environment" is the total milieu in which people live, including cultural habits, diet, exposure to various infectious agents, average age of menarche, number of children a woman bears, age of menopause—in short, the cultural as well as the chemical environment.

Although we have seen that clear correlations between excess occupational exposure to carcinogenic chemicals and some cancers can be made, the contribution of these occupationally related cancers to the total incidence of cancer in industrialized nations is small. Furthermore, although urban–rural differences in cancer incidence have been reported in several countries, these differences tend to disappear when homogeneous populations with a similar life-style—for example, the Mormons—are studied.[100] And, although in England and Wales certain occupations have been associated with a different risk of cancer from that of the general population, nearly 90% of such variation is eliminated if individual groups of similar social class and habits are compared.[101] Other inconsistencies also occur.[98] For example, bladder cancer is linked to certain chemical and allied industries in the United States, but no clear industrial association has been found in Japan. Prostate cancer is higher in blacks than in whites living in the same counties in the United States, and both black and white males in the United States have a higher frequency of prostate cancer than men in the industrialized United Kingdom and Japan.

Thus, the overall distribution patterns of cancer observed in North America and western Europe, with high frequencies of lung, colon, breast, and uterine cancer, suggest some common factors in the environment of these regions in comparison with regions in Africa in which there is a much lower incidence of these malignant diseases. It is, however, unjustified, at present at least, to link these differences in incidence directly to recent food additives or chemical pollutants. Life-style differences appear to play a large role in the causation of these and other cancers.[98] For example, the varying incidence of cancer of the breast, ovary, and uterus can be related at least partly to differences in average age at onset of menarche, sexual behavior, and reproductive patterns among different population groups. Taken together, all the data accumulated to date suggest that cancer distribution patterns represent a variety of differences in life-style, with exposure to chemical pollutants in the ambient environment of industrialized societies contributing to some but as yet unclear percentage of the total number of cancer deaths.

Radiation

Ultraviolet Light

It has been known for a long time that exposure to UV or ionizing irradiation can cause cancer in humans. The association between skin cancer and exposure to sunlight was observed more than 100 years ago, and in 1907 William Dubreuilh, a French dermatologist, reported epidemiologic evidence implicating sunlight as a cause of skin cancer, supporting the earlier observations.[102] In 1928, George Findlay, a British pathologist, experimentally verified this by inducing skin cancer in mice exposed to UV radiation.[103]

Ultraviolet radiation is a low-energy emission and does not penetrate deeply. Hence, the skin absorbs most of the radiation and is the primary carcinogenic target. Because nonmelanomatous skin cancer is the most easily detectable and curable human cancer, the fact that it is also the most common human cancer and one of the few in which an etiologic agent is clearly identifiable is often overlooked. The fear of skin cancer, however, is apparently not sufficient to prevent people from overexposing themselves to this carcinogenic agent.

The evidence for the association of skin cancer and UV radiation is compelling and can be summarized as follows:[104]

1. Skin cancer occurs primarily on exposed areas, that is, the head, neck, arms, hands, and legs in women as well as the upper torso in men.
2. Skin cancer is relatively rare in dark-skinned races, in whom skin pigment filters out ultraviolet radiation, whereas it is common in fair-skinned people.
3. The incidence of skin cancer and the amount of exposure to sunlight are related.
4. Skin cancer frequency and the intensity of solar radiation are related. Going toward the equator, the prevalence of skin cancer in Caucasians increases in proportion to the intensity of ultraviolet radiation.
5. Skin cancer can be induced in laboratory animals by repeated exposure to UV radiation.
6. The inability to repair DNA damaged by UV radiation is associated with skin cancer. Thus, individuals with xeroderma pigmentosum, an inherited disease with a DNA-repair defect (see below), almost always develop skin cancer.

Both malignant melanoma and nonmelanoma skin cancers are associated with exposure to UV radiation, although the dose-response curve is less steep for melanoma.[104] The most common types of skin cancer are basal cell carcinoma, which may be locally invasive but is almost never metastatic; squamous cell carcinoma, which is more aggressive than basal cell carcinoma, invades locally, and may metastasize; and melanoma, which is less common than the other forms of skin cancer but is often highly malignant and rapidly metastatic, with an average 5-year survival rate of 70% to 80%.

Ionizing Radiation

The carcinogenic effects of ionizing radiation were discovered from studies of pioneer radiation workers who were occupationally exposed, individuals who were exposed to diagnostic or therapeutic radiation, and atomic bomb survivors. Malignant epitheliomas of the skin were observed in the earliest experimenters with x-rays and radium within a few years after their discovery in 1895 and 1898, respectively. By 1914, a total of 104 case reports of radiation cancers had been noted and analyzed.[105] In 1944, the role of ionizing radiation in the increased incidence of leukemia among radiologists was recognized.[106]

The type of neoplasm produced in individuals exposed to ionizing radiation depends on a number of factors, including dose of radiation, age at time of exposure, and sex of the individual. Within 25 to 30 years after whole body or trunk irradiation, there is an increased incidence of leukemia and cancers of the breast, thyroid, lung, stomach, salivary gland, other gastrointestinal organs, and lymphoid tissues (Table 2-7). Other malignant neoplasms have been observed in tissues that were locally exposed to high doses of radiation. A number of unfortunate but striking examples are available. During the 1920s, in a factory in Orange, New Jersey, watch dials were painted with radium and mesothorium to make them luminescent. To get a fine tip on the brushes used to paint the dials, the workers wetted the brush tips on their tongues, leaving a deposit of radioactive material on the tongue. Approximately 800 young women were exposed to the radioactive materials in this manner. Radium is a radioisotope that becomes deposited in bone, and several years later, a high incidence of osteogenic sarcoma became evident in these

Table 2-7 Risk Estimates for Cancers Occurring with Increased Frequency in Irradiated Populations

	Cancer Mortality or Incidence/10^6 Person-Years/Rad[a]					
			Therapeutic Irradiation		Diagnostic Irradiation	
Cancer Type	Atomic Bomb Survivors	Occupationally Exposed Workers	Adults	Children	Prenatal	Postnatal
All cancers	3–4	1–5	4–8	>2	0–50[b]	—[c]
Leukemia	0.5–1	0.2–1	1	1	0–25[b]	0–3
Breast (female)	1–3	—	(6)	—	—	(8)
Thyroid	(1–4)	—	—	(2–9)[b]	—	—
Lung	0.6–1	0.5–1	1	—	—	—
Stomach	0.05	—	0.5	—	—	—
Other GI tract	0.5	—	0.3	—	—	—
Bone	—	0.1	0.1–0.8	0.9	—	—

[a]Numbers represent mortaility rates except for those in parentheses, which are incidence rates.

[b]Values apply to first 10 years after birth.

[c]Where there is a dash, the risk is not readily quantifiable.

Source: Data from Upton.[107]

workers.[108] Another radioisotope that becomes deposited in internal organs is thorium, which was used in the preparation of a radiocontrast solution called Thorotrast, once used for diagnostic purposes. The overall incidence of malignant diseases in patients who received this material has been found to be twice the expected incidence, with liver tumors and leukemias being about sixfold higher than expected.[109] Another example of local irradiation producing cancer a number of years later is the observation of thyroid cancers in individuals who had been irradiated over the neck during childhood either for a so-called enlarged thymus gland or for hypertrophied tonsils and adenoids. An 83-fold increased risk for thyroid cancer has been noted in these cases.[110]

The period between irradiation and the appearance of cancer depends, to some extent, on the age at irradiation.[107] Juvenile tumors and leukemias associated with prenatal irradiation become evident in the first 2 to 3 years after birth, with a peak incidence at 5 years of age. The latency period following postnatal irradiation, however, is 5 to 10 years for leukemia and more than 20 years for most solid tumors. The increased incidence of leukemia and solid tumors also appears to be higher after prenatal than after postnatal irradiation. In general, the data suggest that the relative risk for cancers other than leukemia decreases with increasing age at the time of irradiation.

In the case of breast cancer, for example, women who received a radiation dose before age 40 have an increased risk of developing breast cancer.[59] After age 40, radiation has a small effect on breast cancer risk. Women who were below age 10 at the time of exposure have an increased risk that does not become apparent until they reach the age at which breast cancer usually occurs.[59] This increased risk persists for at least 35 years and may remain throughout life. Current evidence indicates that a very small number, probably less than 1%, of breast cancer cases result from diagnostic radiography.[59] Certainly the small exposure to radiation currently used in mammography should not discourage women from availing themselves of this potentially lifesaving procedure.

Radon

Radon, a radioactive gas, is ubiquitous in earth's atmosphere. It is formed from the radioactive decay of radium 236. Radium is found in substantial but varying amounts in soil and rocks and ends up in some building materials. Various parts of the country have varying amounts, as do certain localities within a small geographic area. There is extensive epidemiologic evidence that exposure to high levels of radon produces bronchogenic carcinoma (reviewed in Ref. 111). Most of this evidence comes from studies of workers involved in deep mining of uranium and other ores. Because this epidemiologic evidence is quite compelling and because radon is so widespread, the possibility that large numbers of people might develop lung cancer due to such exposure has produced a "radon scare" in the United States not unlike the "asbestos scare."

An increased incidence of lung cancer in deep-well miners was observed in uranium and other ore miners in eastern Germany and western Czechoslovakia more than 60 years ago. Exposure to radon among these miners was very high, approaching 3000 picocuries (pCi) per liter of air. In the early 1950s, an increase of lung cancer was noted in uranium miners in Colorado. Later, an increased rate of lung cancer was also noted in miners working in iron, zinc, tin, and fluorspar mines. In these mines, radon levels were also high. Although these miners were also exposed to other potentially carcinogenic dusts, the common feature was exposure to radon. The excess number of lung cancer deaths in these miners (compared to nonminers) ranges from 0.3% to 13% and varies depending on the ambient air concentration.[111] This risk goes up in more than an additive manner for individuals who are also smokers.

Monitoring of homes began in a rather haphazard fashion in the 1980s. Nevertheless, some regions with high indoor levels were found, including the Reading Prong geological region, extending from Pennsylvania to New York. Estimates of risk for lung cancer from radon exposure in residences is based on extrapolations from miner risk data. This may or may not be realistic. There is some evidence both ways. In a case control study of 400 women with lung cancer performed by the New Jersey Department of Health, an increased risk was found for exposure levels of 2 pCi/L of air, but the results were not statistically significant.[112] A study done in China, in which median household radon levels ranged from 2.3 to 4 pCi/L of room air, showed no positive associations between radon levels and lung cancer,[113] suggesting that projections of lung cancer risk from surveys of miners exposed to high radon levels are overestimates.

Estimates of increased risk due to radon residential exposure vary widely. For an average lifetime exposure to 1 pCi/L, estimates vary from 5000 to 20,000 excess lung cancer deaths per year in the United States.[111] These estimates uphold the conservative tradition of radiation protection. To put this in some perspective, a lifetime exposure to 4 pCi/L is estimated to cause a 1% increase in lung cancer, whereas the risk of smoking cigarettes increases the risk of lung cancer at least 10-fold over that of nonsmokers. Most homes in the United States have indoor radon levels below 2 pCi/L, and these levels are usually found in basements.[114]

Drugs

Of the cancer-associated chemicals listed in Table 2-5, some have been used as therapeutic or diagnostic agents in medical practice. Among these, the anticancer drugs in particular have been implicated as risk factors. A number of these (e.g., cyclophosphamide, melphalan, and busulfan) are alkylating agents, known to interact with DNA in a manner similar to that of known chemical mutagens and carcinogens (see Chap. 6).

Second cancers arising later in life from the effects of treatment of childhood cancers are a particular concern. It is estimated that from 3% to 12% of children treated for cancer will develop a new cancer within 20 years from the time of first diagnosis.[115] This is a 10-fold higher risk than that of age-matched controls. Exposure to therapeutic radiation and anticancer drugs—such as the alkylating agents nitrogen mustard, cyclophosphamide, procarbazine, and nitrosoureas—are known risk factors for second cancers. About 25% of those who develop second cancers are known to have some genetic susceptibility, such as Li-Fraumeni syndrome, retinoblastoma, neurofibromatosis, or a sibling with cancer[115] (see below). The most common malignant familial condition predisposing to second neoplasm is retinoblastoma; in patients with this condition, osteosarcomas and soft tissue sarcomas are most common.

The risk of second cancers in children surviving ALL, however, is relatively low. In a study of 9720 children treated for ALL from 1972 to 1988, 43 second cancers were seen.[116] Second cancers may occur within 5 years from first diagnosis. These are usually ANLL, CML, or non-Hodgkin's lymphoma. Secondary solid tumors, the most common of which are brain tumors, may be seen 5 to 15 years later. Brain tumors are most often seen in patients who received cranial irradiation. Most other secondary solid cancers in this patient population also appear to result from irradiation. Of those not treated with irradiation, most will have received alkylating agents and a number appear to have some genetic predisposition.

A higher incidence of lymphomas has been seen in patients who have received organ transplants, for which they have been treated with immunosuppressive drugs, some of which are also used in cancer chemotherapy. The most widely used immunosuppressive drug, cyclosporine, which has revolutionized the organ transplant field, has also been noted to cause lymphoproliferative disease, including lymphomas. This was particularly a problem in the earlier clinical trials when higher doses were being employed in combination with high doses of corticosteroids and antithymocyte globulin.[117] More recently, the incidence of lymphoproliferative disease has been reduced to about 1% in transplant patients treated with lower doses of cyclosporine.

Other drugs are also suspected of causing cancer in humans. For example, 10 cases of liver cell tumors have been reported in patients with blood disorders treated for long periods with the androgenic steroid oxymetholone.[79] Several studies indicate that chronic abuse of analgesics containing phenacetin leads to papillary necrosis of the kidney. It has been suggested that this is related to the subsequent development of transitional cell carcinoma of the renal pelvis in a number of these cases.[79]

Hormones

As noted earlier, the risk factors associated with breast cancer include age at menarche, age at the time of the first full-term pregnancy, and age at menopause. This strongly suggests a role for estrogens and prolactin in breast cancer. Production of these hormones increases near menarche and starts to decrease in the perimenopausal period. Prolactin levels have been reported to decrease in women after the first full-term pregnancy,[118] and this may provide some protective effect. Some studies indicate that the rate of cell proliferation is greater in nulliparous women than in parous women (reviewed in Ref. 118); this may reflect the lower hormonal levels in the latter group. Moreover, mitotic activity of breast epithelium varies during the menstrual cycle and peaks during the luteal phase, suggesting that progesterone also has a role in regulating the mitotic rate in breast tissue. Presumably, these hormones could act as promoters for cells initiated by some carcinogens, and the duration of exposure to these hormones could then increase the risk in a woman who has a propensity to develop breast cancer. That other risk factors are also involved is evident from comparative data from U.S. and Japanese women. In these two groups, data on the first birth, nulliparity, and age at menopause show that the lower rate in Japanese women is not accounted for by these factors.[118] The remaining difference may be related to dietary fat and total body weight, both of which are, on the average, higher in U.S. women. The breast cancer risk associated with body weight is thought to operate through the increased levels of conversion of adrenal androgens to estrogen and lower levels of sex-hormone–binding globulin in obese women.[59,119]

Estrogens have been used extensively in the treatment of postmenopausal symptoms and for the prevention of osteoporosis. There is a clear association between use of "unopposed" estrogen therapy (i.e., without progestins) and increased risk of endometrial cancer. Some evidence suggests that cyclical treatment of postmenopausal women with lower doses of estrogen followed by progestin does not carry this increased risk (see Chap. 15).

The role of hormone replacement therapy as a risk factor in breast cancer is controversial.[59,120] Over 30 epidemiologic studies of this issue have been carried out in the past 20 years.[59,120] Taken together, the data show that the risks associated with short-term use (i.e., $\leq$ 9 years) of estrogen replacement therapy are minimal. After 15 to 20 years of continual use, a relative risk of 1.5- to 2.0-fold over controls may be reached. The type of estrogen used and whether or not progestins are added to the replacement regimen may alter the risk of breast cancer. Additional data are needed to determine the relative risk of 15 or more years of estrogen-replacement therapy and the effect of adding progestins to the regimen.

Many epidemiologic studies have shown that oral contraceptives do not significantly affect the risk of breast cancer. However, a small increase has been reported in some studies in certain groups of women. These include women who have used contraceptives for several years before age 25 and/or before the first full-term pregnancy; women who continue to use oral contraceptives at age 45 and older; women with a history of benign breast disorders; multiparous, premenopausal women with early menarche; and women with a family history of breast cancer (reviewed in Ref. 59).

Several studies have reported that use of diethylstilbestrol (DES) during pregnancy is associated with an overall risk of about 1.5-fold for developing breast cancer.[59] The well-documented appearance of vaginal adenocarcinomas in women whose mothers had been treated with DES in early pregnancy with the intent of preventing abortion is another example of hormonally induced neoplasm.[121]

Although exogenous hormone use may be associated with some increased risk of breast cancer, that risk is relatively small compared to the major established risk factors (Table 2-8).

Infection

Cancer is not an "infectious disease" in the usual sense of the term. Doctors, nurses, and spouses who come into close contact with cancer patients do not have a higher risk of developing cancer than the rest of the population.[44] However, there is now known to be a clear association between infection with certain types of viruses and neoplastic disease. Such infection probably acts in concert with other carcinogenic agents or processes. This is discussed in detail in Chap. 6. Suffice it to say here that an association between infection with Epstein-Barr virus and Burkitt's lymphoma, hepatitis B virus and liver cancer, human T-cell lymphotropic virus (HTLV) and leukemia, and human papillomavirus and cervical cancer are examples of this linkage. A number of neoplasms have also been associated with HIV infection in AIDS patients. These include Kaposi's sarcoma and non-Hodgkin's lymphoma, primarily, but CNS lymphomas and Hodgkin's disease are also seen in patients with AIDS.[122]

Table 2-8 Established Risk Factors for Breast Cancer in Females

Risk Factor	High-Risk Group	Low-Risk Group	Mangitude of Differential[a]
Age	Old	Young	♦♦♦
Country of birth	North America Northern Europe	Asia, Africa	♦♦♦
Socioeconomic status	High	Low	♦♦
Marital status	Never married	Ever married	♦
Place of residence	Urban	Rural	♦
Place of residence	Northern US	Southern US	♦
Race ≥ 45 years	White	Black	♦
<40 years	Black	White	♦
Nulliparity	Yes	No	♦
Age at first full-term pregnancy	≥30 years	<20 years	♦♦
Oophorectomy premenopausally	No	Yes	♦♦
Age at menopause	Late	Early	♦
Age at menarche	Early	Late	♦
Weight, postmenopausal women	Heavy	Thin	♦
History of cancer in one breast	Yes	No	♦♦
History of benign proliferative lesion	Yes	No	♦♦
Any first-degree relative with history of breast cancer	Yes	No	♦♦
Mother and sister with history of breast cancer	Yes	No	♦♦♦
History of primary cancer in endometrium or ovary	Yes	No	♦
Mammographic parenchymal patterns	Dysplastic parenchyma	Normal parenchyma	♦♦
Radiation to chest	Large doses	Minimal exposure	♦♦

[a]♦♦♦ denotes relative risk of greater than 4.0; ♦♦ denotes relative risk of 2.0 to 4.0; ♦ denotes relative risk of 1.1 to 1.9.

Source: From Kelsey and Gammon.[59]

Infection with certain parasites also seems to be able to initiate a cascade of events culminating in malignant neoplastic disease in certain populations. The high incidence of bladder cancer in patients whose urinary bladders are infected with the schistosomal parasite indigenous to Egypt and other parts of Africa as well as the occurrence of a type of liver cancer (cholangiosarcoma) in patients with clonorchiasis, a parasitic infection of the liver common in parts of China, are examples of this association.

Genetic Factors in Cancer Causation

A number of inherited traits can be related to the causation of cancer. A few cancers have a definite inheritance, whereas others may arise in individuals with a genetic defect that makes them more susceptible to potentially carcinogenic agents.

Inherited Cancers

These neoplasms represent a small fraction, perhaps 1% to 2%, of total cancers.[122] A high percentage of certain tumors, however, is genetically determined. For example, dominant genetic inheritance accounts for about 40% of retinoblastomas and 20% to 40% of Wilms' tumors (embryonal renal tumors) and neuroblastomas. Familial multiple polyposis of the large bowel is another example of a disease that is transmitted as a Mendelian-dominant trait, with about an 80% penetrance rate. Cancer of the large bowel will eventually occur in nearly 100% of untreated patients with familial multiple polyposis. These patients also have a predisposition to develop a variety of other neoplasms, particularly subcutaneous tumors and osteomas.

The probability that an individual carrying the retinoblastoma gene will develop a tumor is about 95%, and an average of three to four tumors occur in such a gene carrier. A child born without the gene has only 1 chance in 30,000 of developing retinoblastoma. This amounts to about a 100,000-fold increased risk in the gene carrier group (assuming an average of 3 tumors per gene carrier with an incidence of 95 per 100 compared to 1 tumor per 30,000 in the general population). Although the presence of the retinoblastoma gene virtually ensures that the carrier will develop such a tumor, oncogenesis at the cellular level must be a rare event, because only three to four tumors, on the average, develop in a retinal cell population of several millions (this assumes that the cancer arises from the progeny of one or a small number of precursor cells, which appears to be the case for most cancers; see Chap. 4). Thus, the genetically dominant inherited mutation is not in itself sufficient to ensure that a retinal cell bearing the gene will become a cancer cell.

A model to explain these sorts of data is the two-stage model of Knudson, which hypothesizes that two mutational events must occur for oncogenesis to happen.[123,124] In the case of the genetically dominant inherited gene for cancer, one mutational event is already present in the germ-line cell (inherited from one of the parent's germ cells) and a second, somatic mutation must occur for the tumor to be expressed. Nongenetically linked tumors would arise by two somatic mutations, both of which would presumably occur after fertilization. The idea that somatic mutations are involved in oncogenesis is supported by the observations that many oncogenic agents (i.e., chemicals and radiation) interact with cellular DNA and are able to induce mutational events, and that many cancers appear to arise from a single clone of cells (see Chap. 4). This suggests that some mutational events can produce a single, abnormal, malignantly transformed cell, which is the precursor of a malignant tumor.

The two-step model predicts that in the instance of colonic cancer, for example, preneoplastic lesions present as polyps in the germ-line–inherited familial disease would be of multicellular origin—that is, from a variety of cells with a "built-in" first mutation already present—whereas the preneoplastic lesions present in patients with the nonfamilial form of colonic cancer would be clonal—that is, originating from a single mutated cell type. In both the hereditary and nonhereditary types of colonic cancer, however, a second mutation (a rare event at the cellular level, as indicated above) is necessary to produce a cancer cell; therefore, the malignant tumors arising from the second event would be expected to be of clonal origin in both situations (Fig. 2-10).

The evidence that the two-mutation model is

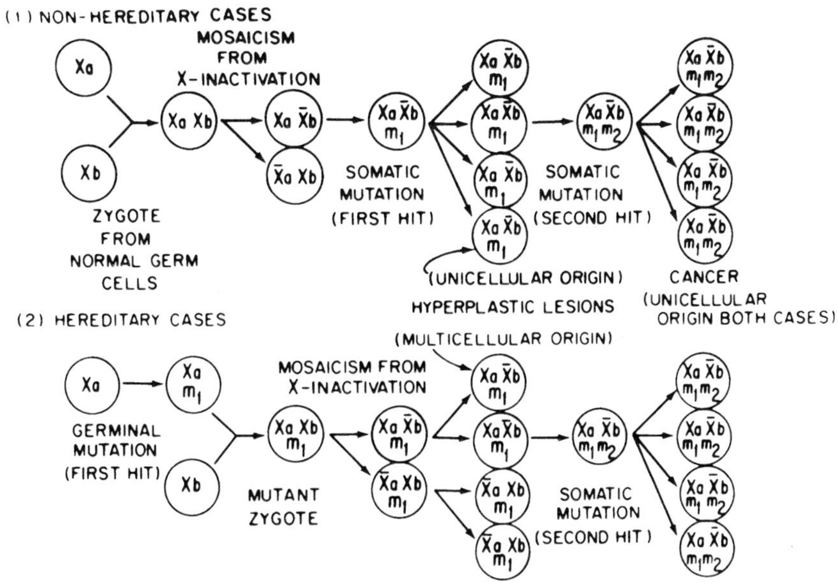

Figure 2-10 The use of information on females heterozygous at an X-linked gene to elucidate the cellular origin of cancer: The singly mutant intermediate hyperplastic lesions are differently constituted in hereditary (2) and nonhereditary (1) cases. Xa denotes an X-chromosome expressing a genetic allele *a*; Xb. X-chromosome expressing allele *b*; XaXb, cell in which Xa and Xb chromosomes are both active; $Xa\overline{Xb}$, cell in which Xb chromosome is inactivated; $\overline{Xa}Xb$, Xa chromosome inactivated; m_1 first cancer mutation; and m_2, second mutation. The sites of m_1 and m_2 are not specified except that they occur on autosomes, not X-chromosomes. If no second event occurs, a hyperplastic nodule of mutant cells remains. If the first hit is a somatic mutation (i.e., a nonhereditary case), the hyperplastic lesion is of unicellular origin. If the first hit is a germinal mutation (i.e., a hereditary case), the hyperplastic lesion is of multicellular origin. If a second hit occurs, a monoclonal cancer (nonhereditary or hereditary) occurs. (From Moolgavkar and Knudson.[124])

correct is primarily indirect and is based on observations of increased susceptibility of genetically predisposed persons to oncogenic agents. For example, children with the genetically inherited form of retinoblastoma have a greatly increased risk of developing osteogenic sarcoma after receiving irradiation.[124] These patients already have a germ-line mutation, and exposure to a second mutagenic agent, radiation, may produce the second mutation necessary for malignant transformation.

About 50 forms of hereditary cancers have been reported.[17] For example, breast cancer for many years has been considered to have a familial association. Similar associations have been noted for ovarian cancer. Genetic studies have also associated the occurrence of breast cancer with a variety of other tumors in the same families.[125] These include associations between breast cancer and ovarian cancer, gastrointestinal tract cancer, soft tissue sarcoma, brain tumors, or leukemia. In a study of 12 pedigrees that had a clustering of breast and ovarian cancer among female relatives, the data suggested that a genetic factor was transmitted from affected mothers to half their daughters and, in some families, father-to-daughter transmission appeared to occur.[125] These observations suggested the possibility of X-chromosome linkage in the transmission of breast and ovarian cancer in these families. In a number of these family-associated cancers, specific chromosomal abnormalities or genetic mutations have been observed. This will be discussed in detail in Chapter 3. The following are a few examples.

Retinoblastoma has been one of the best studied of the familial cancers. In this cancer, the long arm of chromosome 13 has a deletion in band 14 (del 13q14),[126] resulting in loss of the normal retinoblastoma gene (*RB*1) allele in cells with an already mutated *RB*1 allele. The first allele has suffered a mutational event in the germ line (all patients with bilateral disease) or in a somatic retinal cell (most patients with unilateral disease).[127] It is now known that the *RB* locus codes for a tumor suppressor gene product and that loss of this function fosters the neoplastic transformation of the cells (see Chap. 3).

In 1964, Miller et al.[128] first described the association between Wilms' tumor (a renal cancer in children) and the congenital defects aniridia and hemihypertrophy, indicating a hereditary transmission of this neoplasm. Cytogenetic studies revealed a deletion of the short arm of chromosome 11 in band 13 (del 11p13),[129] suggesting the presence of a tumor suppressor gene at this locus.

The Li-Fraumeni syndrome—noted earlier in this chapter, which results in the appearance of sarcomas, breast cancer, brain cancer, leukemia, and other tumors in afflicted family members—has been shown to be associated with the loss of part of chromosome 17 bearing a tumor suppressor gene called p53 (see Chap. 8).

Inherited Conditions That Predispose to Susceptibility to Carcinogens

Other genetically inherited conditions carry a predisposition to cancer involving the interaction of genetic and environmental factors. These include xeroderma pigmentosum (characterized by extreme sensitivity to sunlight, an incidence of skin cancer approaching 100%, and, frequently, neurologic deficiencies), ataxia telangiectasia (progressive cerebellar ataxia, dilated blood vessels, immune deficiencies), and Fanconi's anemia (hematologic deficiency leading to anemia and hemorrhage with an increased risk for leukemia). These rare autosomal recessive genetic disorders have in common an inability to repair DNA damaged by such potentially carcinogenic substances as ultraviolet radiation, ionizing radiation, or chemical carcinogens (for mechanisms of DNA repair, and repair defects, see Chap. 6). As noted previously, almost 100% of patients with xeroderma pigmentosum (homozygotes) will get skin cancer, but in addition, heterozygous carriers of the XP gene also have an increased risk of skin cancer.[130]

Another potential genetic polymorphism in the human population that could contribute to a genetic predisposition to develop cancer involves individual differences in the way people metabolize "foreign" chemicals (xenobiotics), particularly those that are potentially carcinogenic. Several such pharmacogenetic polymorphisms have been noted in the human population (Table 2-9).[131] Several of these polymorphisms relate to the enzymes involved in metabolic activation of chemical carcinogens—for example, metabolism of arylamines and polycyclic aromatic hydrocarbons (PAHs).

Occupational exposure of dye industry workers to arylamines has been known to be associated with increased evidence of urinary bladder cancer.[132] Based on epidemiologic studies and investigations of animal models, it is now clear that individuals with poor ability to acetylate arylamines (slow acetylator phenotype) have a highly increased risk of developing bladder cancer compared to rapid acetylators,[131] probably because rapid acetylators can detoxify aromatic amines to less carcinogenic metabolites.

Induction of the metabolism of polycyclic aromatic hydrocarbons is involved in their activation to ultimate carcinogens (see Chap. 6). This metabolic system, called arylhydrocarbon hydroxylase (AHH), is also subject to genetic variation in two ways. The *Ah* genetic locus codes for a receptor system involved in the ability of PAHs to induce the enzyme system (cytochrome P450IA1) that causes their own metabolism and that of other xenobiotics. The Ah receptor binds PAHs and transduces a signal to the nucleus of the cell that results in increased transcription of the gene coding for P450IA1. The *Ah* locus is inherited as an autosomal recessive trait in inbred strains of mice.[131] In humans, a significant correlation between AHH inducibility and enhanced risk of lung cancer in cigarette smokers has been noted.[133] Genetic polymorphism has also been observed in cytochrome P450IA1, which is responsible for metabolism of benzo[*a*]pyrene and other PAHs in cigarette smoke.[134] Individuals with a specific genetic alteration, seen by family pedigree analysis and restriction endonuclease mapping of the P450IA1 gene locus, were at higher risk to develop squamous cell carcinoma of the lung and to be more susceptible to develop cancer at lower exposure doses of cigarette smoke than people without this genotype.[134]

Table 2-9 Classification of Human Pharmacogenetic Disorders

Less enzyme or defective protein
Succinylcholine apnea
Acetylation polymorphism
Isoniazid-induced neurotoxicity
Drug-induced lupus erythematosus
Phenytoin-isoniazid interaction
Isoniazid-induced hepatitis
Arylamine-induced bladder cancer
Increaced susceptibility to drug-induced hemolysis
Glucose-6-phosphate dehydrogenase deficiency
Other defects in glutathione formation or use
Hemoglobinopathies
Hereditary methemoglobinemia
Hypoxanthine-guanine phosphoribosyltransferase (HPRT)-deficient gout
P450 monooxygenase polymorphisms
Debrisoquine 4-hydroxylase deficiency
Vitamin D–dependent rickets type I
C21-Hydroxylase polymorphism
Enzymes of methyl conjugation
Hyperbilirubinemia
Crigler-Najjar syndrome type II
Gilbert's disease
Fish-Odor syndrome
Increased resistance to drugs
Inability to taste phenylthiourea
Coumarin resistance
Possibility of (or proven) defective receptor
Steroid hormone resistance
Cystic fibrosis
Trisomy 21
Dysautonomia
Leprechaunism
Defective absorption
Juvenile pernicious anemia
Folate absorption-conversion
Increased metabolism
Succinylcholine resistance
Atypical liver alcohol dehydrogenase
Atypical aldehyde dehydrogenase
Change in drug response due to enzyme induction
The porphyrias
The *Ah* locus
Abnormal drug distribution
Thyroxine (hyperthyroidism or hypothyroidism)
Iron (hemochromatosis)
Copper (Wilson's disease)
Disorders of unknown etiology
Corticosteroid-induced glaucoma
Malignant hyperthermia associated with general anesthesia
Halothane-induced hepatitis
Chloramphenicol-induced aplastic anemia
Phenytoin-induced gingival hyperplasia
Thromboembolic complications caused by anovulatory agents

Source: From Nebert and Weber.[131]

AVOIDABILITY OF CANCER

If, as a number of cancer epidemiologists contend, "life-style" accounts for about 80% of all malignant cancers, then presumably the same proportion of cancers should be avoidable. To be more specific, about 30% of all cancers are thought to be related to smoking, 3% to alcohol consumption, 35% to diet, 7% to sexual and reproductive patterns, and another 5% to occupational hazards and industrial products.[44] Moreover, about 1% are estimated to be related to drugs and medical procedures (primarily x-rays), and 3% to geophysical factors (mostly exposure to sunlight). Thus, if these estimates are correct, about 84% of all cancers should be avoidable. The "ideal" man, then, should not smoke or drink; should eat a diet low in fat, rich in fiber and yellow vegetables; should protect himself from hazardous chemicals in the workplace and home; should minimize drug taking and avoid unneeded x-rays; and should avoid excessive exposure to sunlight. The "ideal" woman should do all this and, in addition, have at least one child early in her reproductive life and avoid multiple sex partners. Assuming that we cannot totally avoid the pollution in our environment, which the epidemiologists tell us accounts for no more than 2% of cancers, exposure to certain infectious agents (10% of cancers), and certain other unknown factors including genetic determinants (~4%), we should be able to decrease our cancer mortality rate by about 84% by simple, direct actions as individuals. This conclusion is almost certainly overly sanguine but appears to be worth the experiment.

Evidence to support the idea that most cancers are avoidable comes from different sorts of data: (1) differences in the incidence of cancer in various areas of the world (see Table 2-2); (2) differences in the incidence rates for various types of cancer between residents of a country and those of the same ethnic group who have emigrated to another country (see Table 2-3); (3) variations over time in the incidence of cancers within a given society or community (see Fig. 2-1); and (4) the identification of specific causes of cancer and preventive measures resulting therefrom (e.g., aniline dyes in bladder cancer, vinyl chloride in hepatic angiosarcomas, asbestos in mesotheliomas, etc.).

Risk Assessment

As discussed more fully in Chapter 6, most chemical carcinogens are also mutagens. Because mutagenicity is much easier to measure experimentally than carcinogenicity, many of the tests used to assess the carcinogenic risk potential of substances in our environment are based on mutagenicity assays. Exposure of human beings to mutagens occurs from chemicals in our diet, water, and air; from products that we use as cosmetics and drugs; and from cigarette smoking. As noted before, several mutagenic substances have been identified in cigarette smoke. Mutagens are also found among the natural products contained in foods, such as products elaborated by molds (e.g., aflatoxin) or by edible plants that synthesize a variety of toxins, presumably to ward off insects, as well as among synthetic chemicals such as pesticides, industrial pollutants, and weed killers.[68,135] In short, we live in a sea of mutagens and carcinogens. Identification of potentially mutagenic substances in our environment is a major public health and political issue. Of the myriad of potential mutagens and carcinogens in our diet, only a few have been studied in detail. In addition, more than 65,000 synthetic chemicals are produced in the United States, and about 1000 new chemicals are introduced each year.[135] Only a small number of these were examined for mutagenic and carcinogenic potential before being marketed. Obviously, this is a major epidemiologic problem and one that has a major economic impact on private industry as well as on the consumer and taxpayer. What is needed are accurate, rapid, and economically feasible tests to predict the mutagenic and carcinogenic potential of the numerous chemicals in our environment. In practice, however, it has not been possible to develop a "perfect" short-term test. A number of false positives and false negatives result from using these tests.

More than 100 short-term tests for mutagenicity and carcinogenicity have been developed. Some of the most widely used are bacterial mutagenesis, mutagenesis in cell culture systems, direct measurement of damage to DNA or chromosomes in exposed cells, and malignant transformation of cell cultures.[136] The Ames test, among the most popular of the short-term tests, was developed by Bruce Ames and his colleagues.[137] The basis of this assay is the ability of a chemical agent to induce a genetic reversion of a series of *Salmonella typhimurium* tester strains, which contain either a base substitution or a frameshift mutation, from histidine requiring (his^-) to histidine nonrequiring (his^+). These strains have been specially developed for this assay by selecting clones that have a decreased cell surface barrier to uptake of chemicals and a decreased excision repair system. Other advantages of this system are the small genome of the bacteria (4×10^6 base pairs), the large number of cells that can be exposed per culture dish (about 10^9), and the positive selection of the mutated organisms (i.e., only the *mutated* organisms will grow under the test conditions). This system has great sensitivity: only about 1 in 1000 to 1/10,000 of the mutated bacteria have to be detected to give a positive test, and nanogram amounts of a potent mutagen can be detected as a positive.[137] Both base substitution and frameshift mutagens can be detected, and, using the appropriate tester strains, the type of mutagen can be deduced, because frameshift mutagens usually revert only frameshift mutations of the tester strains and not base substitution mutations, and vice versa. Because many mutagens must be metabolized to be active, a liver homogenate fraction containing microsomes is usually added to the incubation to provide the drug metabolizing enzymes.

The potential of various chemical agents to mutagenize mammalian cells has also been used as a short-term test. Frequently, mutation at the hypoxanthine-guanine phosphoribosyltransferase (HGPRT) locus is used as a marker; the endpoint of the assay is loss of sensitivity to purine antimetabolites that must be activated by HGPRT to be effective, thus leading to the selection of $HGPRT^-$ clones. Cultured fibroblast cell lines such as Chinese hamster V79 or ovary cells (CHO) are frequently used in this way.

Agents that damage DNA can often be detected by examining an index of genotoxicity, such as unscheduled DNA synthesis, sister chromatid exchange, or chromosome breakage in cultured cells exposed to the agents in question.

Carcinogenic potential has been estimated by the ability of chemicals to "transform" smooth, well-organized monolayers of normal diploid fibroblasts into cells that grow piled up on one another (transformed foci) or into a cell type that can grow suspended in soft agar (normal fibroblasts do not usually grow on soft agar). Sometimes the putative malignant cells are then injected into immunosuppressed or immunodeficient ("nude") mice to further demonstrate that they are malignant. All of these estimates of carcinogenic potential are fraught with danger in that a significant number of false negatives or false positives can occur.

No single short-term test is foolproof; however, if definitive evidence of genotoxicity has been obtained in more than one test, a chemical is highly suspect. An agent that is found to be mutagenic, DNA damaging, and a chromosome breaker is almost certain to be carcinogenic also.[136] Final proof of mutagenicity and carcinogenicity involves the chronic exposure of whole animals to the test chemical. Although the short-term in vitro tests have several advantages, a number of important components, such as absorption, pharmacokinetics, tissue distribution, metabolism, age or sex effects, and species specificity, cannot be duplicated in vitro. Tests in whole animals take a long time and are, unfortunately, very expensive.

There has been considerable debate about how to extrapolate dose-response data for carcinogenesis in animal studies to estimates of human risk. According to one school of thought, there is no real "safe" level of exposure to a carcinogen. This is based on the idea that, theoretically, a single "hit" (e.g., an alkylation of a key base in a regulatory gene) would be enough to produce a mutation that could lead to a cancer. This is sometimes called the "no-threshold" concept. This sort of thinking is evident in the Delaney Clause of the federal Food, Drug, and Cosmetic Act, which specifies that any substance demonstrated to be carcinogenic in either animal bioassay or human study be banned as a food additive from all food products. Ideally, this is the best approach; however, it is extremely difficult to eliminate all carcinogens from the environment, and it is not cost-effective to try to do so. Moreover, realistic human exposures may be much lower than those used to produce cancers in animals.

How, then, do you estimate the danger of low-dose exposures? And more important, how do you estimate the risk of low-dose exposure over a lifetime? These are extremely difficult questions to answer, but in practical terms, as long as an individual's DNA-repair enzymes are working (see Chap. 6), there probably is some low level of exposure below which DNA lesions can be removed efficiently without permanent damage. The low level of mutation in human genes seems to argue in support of this conclusion.[138] If one could, in fact, measure the amount of DNA adduct formation (i.e., the amount of DNA bases bound to carcinogen) after exposure to various doses of carcinogen, one could probably get a much better estimate of the risks involved in exposure to various amounts of carcinogenic agents.[139] A shift in the dose-response curve with low doses of carcinogen could occur for several reasons, all of which make linear extrapolations of dose-response data from animal studies tenuous. Some of these reasons relate to differences in metabolism, distribution, and overall pharmacokinetics between species.

REFERENCES

1. B. F. Hankey, L. A. Gloeckler-Ries, B. A. Miller, and C. L. Kosary: Overview, in *Cancer Statistics Review 1973–1989*. NIH, Pub. No. 92-2789, I.1–17, 1992.
2. H. S. Seidman, M. H. Mushinski, S. K. Gelb, and E. Silverberg: Probabilities of eventually developing or dying of cancer—United States, 1985. *CA* 35:36, 1985.
3. *Cancer Facts and Figures—1994*. Atlanta: American Cancer Society, Inc., 1994.
4. C. C. Boring, T. S. Squires T. Tong, and S. Montgomery: Cancer statistics, 1994. *CA* 44:7, 1994.
5. L. A. Loeb, V. L. Ernster, K. E. Warner, J. Abbotts, and J. Laszlo: Smoking and lung cancer: An overview. *Cancer Res* 44:5940, 1984.
6. L. P. Faber: Lung cancer, in A. I. Holleb, D. J. Fink, and G. P. Murphy, eds.: *Clinical Oncology*. Atlanta: American Cancer Society, Inc., 1991, pp. 194–212.
7. P. A. Bunn Jr. and J. D. Minna: Paraneoplastic syndromes in V. T. Devita, S. Hellman, and S. A. Rosenberg, eds.: *Principles and Practice of Oncology*. Philadelphia: J. B. Lippincott, 1985, pp. 1797–1842.
8. L. A. Kessler: Lung and bronchus, in *Cancer Statistics Review 1973–1989*. NIH, Pub. No. 92-2789, XV.1–13, 1992.

9. L. A. Gloeckler Ries: Colon and rectum, in *Cancer Statistics Review 1973–1989.* NIH, Pub. No. 92-2789, VI.1–2, 1992.
10. R. W. Beart: Colorectal cancer, in A. I. Holleb, D. J. Fink, and G. P. Murphy, eds.: *Clinical Oncology.* Atlanta: American Cancer Society, Inc., 1991, pp. 213–218.
11. W. A. Albano, H. T. Lynch, J. A. Recabaren, C. H. Organ, J. A. Mailliard, L. E. Black, K. L. Follett, and J. Lynch: Familial cancer in an oncology clinic. *Cancer* 47:2113, 1981.
12. S. M. Powell, N. Zilz, Y. Bazer-Barclay, T. M. Bryan, S. R. Hamilton, S. N. Thibodeau, B. Vogelstein, and K. W. Kinzler: APC mutations occur early during colorectal tumorigenesis. *Nature* 359:235 1992.
13. B. A. Miller: Pancreas, in *Cancer Statistics Review 1973–1989.* NIH, Pub. No. 92-2789, XXI.1–9, 1992.
14. R. M. Beazley and I. Cohn: Tumors of the pancreas, gallbladder, and extrahepatic ducts, in A. I. Holleb, D. J. Fink, and G. P. Murphy, eds.: *Cancer Oncology.* Atlanta: American Cancer Society, Inc., 1991, pp. 219–236.
15. B. F. Hankey: Breast, in *Cancer Statistics Review 1973–1989.* NIH, Pub. No. 92-2789, IV.1–20, 1992.
16. J. L. Kelsey and G. S. Berkowitz: Breast cancer epidemiology. *Cancer Res* 48:5617, 1988.
17. E. F. Scanlon: Breast cancer, in A. I. Holleb, D. J. Fink, and G. P. Murphy, eds.: *Clinical Oncology.* Atlanta: American Cancer Society, Inc., 1991, pp. 177–193.
18. B. Newman, M. A. Sustin, M. Lee, and M.-C. King: Inheritance of human breast cancer: Evidence for autosomal dominant transmission in high-risk families. *Proc Natl Acad Sci USA* 85:3044, 1988.
19. F. P. Li: Cancer families: Human models of susceptibility to neoplasia. The Richard and Hinda Rosenthal Foundation Award Lecture. *Cancer Res* 48:5381, 1988.
20. R. T. Osteen, G. D. Steele, H. R. Menck, and D. P. Winchester: Regional differences in surgical management of breast cancer. *CA* 42:39, 1992.
21. S. M. Pierce and J. R. Harris: The role of radiation therapy in the management of primary breast cancer. *CA* 41:85, 1991.
22. B. Fisher: A biological perspective of breast cancer: Contributions of the National Surgical Adjuvant Breast and Bowel Project Clinical Trials. *CA* 41:97, 1991.
23. C. D. Runowicz: Advances in the screening and treatment of ovarian cancer. *CA* 42:327, 1992.
24. S. B. Gusberg and C. D. Runowicz: Gynecologic cancer, in A. I. Holleb, D. J. Fink, and G. P. Murphy, eds.: *Clinical Oncology.* Atlanta: American Cancer Society, Inc., 497, 1991, pp. 481–497.
25. B. F. Hankey: Urinary bladder, in *Cancer Statistics Review 1973–1989.* NIH, Pub. No. 92-2789, XXVI. 1–9, 1992.
26. I. N. Frank, S. D. Graham, and W. L. Nabors: Urologic and male genital cancer, in A. I. Holleb, D. J. Fink, and G. P. Murphy, eds.: *Clinical Oncology.* Atlanta: American Cancer Society, Inc., 1991, pp. 271–289.
27. B. A. Miller and A. L. Potosky: Prostate, in *Cancer Statistic Review 1973–1989.* NIH, Pub. No. 92-2789, XXII.1–8, 1992.
28. P. J. Littrup, F. Lee, and C. Mettlin: Prostate cancer screening current trends and future implications. *CA* 42:198, 1992.
29. A. J. Mitus and D. S. Rosenthal: Adult leukemias, in A. I. Holleb, D. J. Fink, and G. P. Murphy, eds.: *Clinical Oncology.* Atlanta: American Cancer Society, Inc., 1991, pp. 410–432.
30. C.-H. Pui and G. K. Rivera: Childhood leukemias, in A. I. Holleb, D. J. Fink, and G. P. Murphy, eds.: *Clinical Oncology.* Atlanta: American Cancer Society, Inc., 1991, pp. 433–452.
31. B. A. Miller: Leukemias, in *Cancer Statistics Review 1973–1989.* NIH, Pub. No. 92-2789, XIII.1–15, 1992.
32. R. J. Friedman, D. S. Rigel, D. S. Berson, and J. Rivers: Skin cancer: Basal cell and squamous cell carcinoma, A. I. Holleb, D. J. Fink, and G. P. Murphy, eds.: in *Clinicial Oncology.* Atlanta: American Cancer Society, Inc., 1991, pp. 290–305.
33. S. E. Singletary and C. M. Balch: Malignant melanoma, in A. I. Holleb, D. J. Fink, and G. P. Murphy, eds.: *Clinical Oncology,* Atlanta: American Cancer Society, Inc., 1991, pp. 263–270.
34. B. F. Hankey: Brain and other nervous system, in *Cancer Statistic Review 1973–1989.* NIH, Pub. No. 92-2789, III.1–12, 1992.
35. J. Ransohoff, M. Koslow, and P. R. Cooper: Cancer of the central nervous system and pituitary, in A. I. Holleb, D. J. Fink, and G. P. Murphy, eds.: *Clinical Oncology,* American Cancer Society, Inc., 1991, pp. 329–337.
36. R. Doll: Introduction, in *Origins of Human Cancer.* H. H. Hiatt, J. R. Watson, and J. A. Winsten, eds.: Cold Spring Harbor, NY: Cold Spring Harbor Laboratory, 1977, pp. 1–12.
37. H. Autrup, T. Seremet, J. Wakhisi, and A. Wasunna: Aflatoxin exposure measured by urinary excretion of aflatoxin B_1-guanine adduct and hepatitis B virus infection in areas with different liver cancer incidence in Kenya. *Cancer Res* 47:3430, 1987.
38. J. D. Groopman, Z. Jiaqi, P. R. Donahue, A. Pikul, Z. Lisheng, C. Jun-shi, and G. N. Wogan: Molecular dosimetry of urinary aflatoxin-DNA adducts in people living in Guangxi autonomous region, People's Republic of China. *Cancer Res* 52:45, 1992.
39. T. C. Campbell, J. Chen, C. Liu, J. Li, and B. Parpia: Nonassociation of aflatoxin with primary liver cancer in a cross-sectional ecological survey

in the People's Republic of China. *Cancer Res* 50:6882, 1990.
40. W. Haenszel and M. Kurihara: Studies of Japanese migrants: I. Mortality from cancer and other diseases among Japanese in the United States. *JNCI* 40:43, 1968.
41. D. G. Hoel, D. L. Davis, A. B. Miller, E. J. Sondik, and A. J. Swerdlow: Trends in cancer mortality in 15 industrialized countries, 1969–1986. *JNCI* 84:313, 1992.
42. B. F. Hankey and C. L. Kosary: U.S. cancer mortality rates by state, in *Cancer Statistic Review 1973–1989.* NIH, Pub. No. 92-2789, XXIX.1–22, 1992.
43. W. J. Blot, T. J. Mason, R. Hoover, and J. F. Fraumeni, Jr.: Cancer by country: Etiologic implications, H. H. Hiatt, J. R. Watson, and J. A. Winsten, eds.: *Origins of Human Cancer.* Cold Spring Harbor, NY: Cold Spring Harbor Laboratory, 1977, pp. 21–32.
44. R. Doll and R. Peto: The causes of cancer, *JNCI* 66:1191, 1981.
45. J. Higginson and C. S. Muir: Environmental carcinogenesis: Misconceptions and limitations to cancer control. *JNCI* 63:1291, 1979.
46. E. L. Wynder and G. B. Gori: Contribution of the environment to cancer incidence: An epidemiologic exercise. *JNCI* 58:825, 1977.
47. H. G. Stockwell, A. L. Goldman, G. H. Lyman, C. I. Noss, A. W. Armstrong, P. A. Pinkham, E. C. Candelora, and M. R. Brusa: Environmental tobacco smoke and lung cancer risk in nonsmoking women. *JNCI* 84:1417, 1992.
48. *The Health Consequences of Smoking.* Washington, DC: U.S. Department of Health, Education, and Welfare, HEW Publication No. (CDC) 74-8704, 1974, pp. 35–37.
49. D. R. Shopland, H. J. Eyre, and T. F. Pechacek: Smoking-attributable cancer mortaility in 1991: Is lung cancer now the leading cause of death among smokers in the United States? *JNCI* 83:1142, 1991.
50. L. Garfinkel and E. Silverberg: Lung cancer and smoking trends in the United States over the past 25 years. *CA* 41:137, 1991.
51. U.S. Public Health Service: *The Health Consequences of Smoking—Cancer: A Report of the Surgeon General.* Rockville, MD: U.S. Department of Health and Human Services, Office on Smoking and Health, 1982.
52. R. M. Putzrath, D. Langley, and E. Eisenstadt: Analysis of mutagenic activity in cigarette smokers' urine by high performance liquid chromatography. *Mutation Res* 85:97, 1981.
53. G. Obe, H. J. Voght, S. Madle, A. Fahning, and W. D. Heller: Double blind study on the effect of cigarette smoking on the chromosomes of human peripheral blood lymphocytes in vivo. *Mutation Res* 92:309, 1982.
54. H. J. Vijayalaxmi and H. J. Evans: In vivo and in vitro effects of cigarette smoke on chromosomal damage and sister-chromatid exchange in human peripheral blood lymphocytes. *Mutation Res* 92:321, 1982.
55. J. Stjernswärd: Battle against tobacco. *JNCI* 81:1524, 1989.
56. S. S. Devesa, W. J. Blot, and J. F. Fraumeni: Declining lung cancer rates among young men and women in the United States: A cohort analysis. *JNCI* 81:1568, 1989.
57. A. J. Tuyns, G. Péquinot, and O. M. Jensen: Les cancers de l'oesophage en Ille-et-Villaine en fonction des niveaux de consommation d'alcool et de tabac: Des risques qui se multiplient. *Bull Cancer (Paris)* 64:45, 1977.
58. B. Herity, M. Moriarity, G. J. Bourke, and L. Daly: A case-control study of head and neck cancer in the Republic of Ireland. *Br J Cancer* 43:177, 1981.
59. J. L. Kelsey and M. D. Gammon: The epidemiology of breast cancer. *CA* 41:146, 1991.
60. J. H. Weisburger, L. A. Cohen, and E. L. Wynder: On the etiology and metabolic epidemiology of the main human cancers. In H. H. Hiatt, J. R. Watson, and J. A. Winsten, eds.: *Origins of Human Cancer.* Cold Spring Harbor, NY: Cold Spring Harbor Laboratory, 1977, pp. 567–602.
61. D. Kritchevsky: Diet and cancer, in A. I. Holleb, D. J. Fink, and G. P. Murphy, eds.: *Clinical Oncology.* Atlanta: American Cancer Society, Inc., 1991, pp. 125–132.
62. B. S. Reddy and E. L. Wynder: Metabolic epidemiology of colon cancer: Fecal bile acids and neutral sterols in colon cancer patients and patients with adenomatous polyps. *Cancer* 39:2533, 1977.
63. R. MacLennan: Fat intake and cancer of the gastrointestinal tract and prostate. *Med Oncol Tumor Pharmacother* 2:137, 1985.
64. W. C. Willett, M. J. Stampfer, and G. A. Colditz, B. A. Rosner, C. H. Hennekens, and F. E. Speizer: Dietary fat and risk of breast cancer. *N Engl J. Med* 316:22, 1987.
65. K. S. Yeung, G. E. McKeown-Eyssen, G. F. Li, E. Glazer, K. Hay, P. Child, V. Gurgin, S. L. Zhu, J. Baptista, M. Aloe, D. Mee, V. Jazmaji, D. F. Austin, C. C. Li, and W. R. Bruce: Comparisons of diet and biochemical characteristics of stool and urine between Chinese populations with low and high colorectal cancer rates. *JNCI* 83:46, 1991.
66. G. R. Howe, E. Benito, R. Castelleto, J. Cornee, J. Estève, R. P. Gallagher, J. M. Iscovich, J. Deng-ao, R. Kaaks, G. A. Kune, S. Kine, K. A. L'Abbe, H. P. Lee, M. Lee, A. B. Miller, R. K. Peters, J. D. Potter, E. Riboli, M. L. Slattery, D. Trichopoulos, A. Tuyns, A. Tzonou, A. S. Whittemore, A. H. Wu-Williams, and Z. Shu: Dietary intake of fiber and decreased risk of cancers of the colon and rectum: Evidence from the combined analysis of 13 case-control stuides. *JNCI* 84: 1887, 1992.

67. E. Giovannucci, M. J. Stampfer, G. Colditz, E. B. Rimm, and W. C. Willett: Relationship of diet to risk of colorectal adenoma in men. *JNCI* 84:91, 1992.
68. B. N. Ames: Dietary carcinogens and anticarcinogens. *Science* 221:1256, 1983.
69. T. Hirayama: Diet and cancer. *Nutr Cancer* 1:67, 1979.
70. T. Sugimura and M. Nagao: Mutagenic factors in cooked foods. *CRC Crit Rev Toxicol* 6:189, 1979.
71. S. M. Cohen, M. Cano, R. A. Earl, S. D. Carson, and E. M. Garland: A proposed role for silicates and protein in the proliferative effects of saccharin on the male rat urothelium. *Carcinogenesis* 12:1551, 1991.
72. P. Shubik: Food additives, contaminants, and cancer. *Prev Med* 9:197, 1980.
73. A. Lacassagne: Apparition d'adénocarcinomes mammaires chez des souris mâles, traiteés par une substance oestrogéne synthétique. *C R Soc Biol (Paris)* 129:641, 1938.
74. J. E. Bruni and D. G. Monetmurro: Effect of pregnancy, lactation, and pituitary isografts on the genesis of spontaneous mammary gland tumors in the mouse. *Cancer Res* 31:1903, 1971.
75. C. J. Bradley, G. S. Kledzik, and J. Meites: Prolactin and estrogen dependency of rat mammary cancers at early and late stages of development. *Cancer Res* 36:319, 1976.
76. B. E. Henderson, M. C. Pike, V. R. Gerkins, and J. T. Casagrande: The hormonal basis of breast cancer: Elevated plasma levels of estrogen, prolactin, and progesterone, in H. H. Hiatt, J. R. Watson, and J. A. Winsten, eds.: *Origins of Human Cancer.* Cold Spring Harbor, NY: Cold Spring Harbor Laboratory, 1977, pp. 77–86.
77. J. H. Weisburger and C. L. Horn: The causes of cancer, in A. I. Holleb, D. J. Fink, and G. P. Murphy, eds.: *Clinical Oncology.* Atlanta: American Cancer Society, Inc., ed. 1991, pp. 80–98.
78. M. R. Melamed: Diagnostic cytology of urinary tract carcinoma. *Cancer* 8:287, 1972.
79. International Agency for Research on Cancer, Report of an IARC Working Group: An evaluation of chemicals and industrial processes associated with cancer in humans based on human and animal data: IARC Monographs Vols. 1 to 20. *Cancer Res* 40:1, 1980.
80. M. Aksoy, E. Erdem, and G. Dincol: Leukemia in shoe-workers exposed chronically to benzene. *Blood* 44:837, 1974.
81. P. F. Infante, J. K. Wagoner, R. A. Rinsky, and R. J. Young: Leukaemia in benzene workers. *Lancet* 2:76, 1977.
82. F. J. van Schooten, F. E. van Leeuwen, M.J.X. Hillebrand, M. E. deRijke, A.A.M. Hart, H. G. van Veen, S. Oosterink, and E. Kriek: Determination of benzo[a]pyrene diol epoxide-DNA adducts in white blood cell DNA from coke-oven workers: The impact of smoking. *JNCI* 82:927, 1990.
83. E. S. Johnson, W. Parsons, C. R. Weinberg, D. L. Shore, J. Mathews, D. G. Patterson, Jr., and L. L. Needham: Current serum levels of 2,3,7,8-tetrachlorodibenzo-p-dioxin in phenoxy acid herbicide applicators and characterization of historical levels. *JNCI* 84:1648, 1992.
84. E. S. Johnson: Human exposure to 2,3,7,8-TCDD and risk of cancer. *Crit Rev Toxicol* 21:451, 1992.
85. H. I. Morrison, K. Wilkins, R. Semenciw, Y. Mao, and D. Wigle: Herbicides and cancer. *JNCI* 84:1866, 1992.
86. A. Blair: Herbicides and non-Hodgkin's lymphoma: New evidence from a study of Saskatchewan farmers. *JNCI* 82:544, 1990.
87. W. Haenszel, D. B. Loveland, and M. G. Sirken: Lung-cancer mortality as related to residence and smoking histories: I. White males. *JNCI* 28:947, 1962.
88. V. E. Archer, J. D. Gillam, and J. K. Wagoner: Respiratory disease mortality among uranium workers. *Ann NY Acad Sci* 271:280, 1976.
89. W. J. Nicolson: Asbestos-The TLV approach. *Ann NY Acad Sci* 271:152, 1976.
90. E. C. Hammond, I. J. Selikoff, and H. Seidman: Asbestos exposure, cigarette smoking, and death rates. *Ann NY Acad Sci* 330:473, 1979.
91. D. H. Fine, D. P. Rounbehler, N. M. Belcher, and S. S. Epstein: N-nitroso compounds: Detection in ambient air. *Science* 192:1328, 1976.
92. R. L. Jolley: *Chlorination Effects on Organic Constituents in Effluents from Domestic Sanitary Sewage Treatment Plants.* Publication No. 565. Oak Ridge, TN: Environmental Science Division, Oak Ridge National Laboratory, 1973.
93. J. J. Rook: Formation of haloforms during chlorination of natural waters. *J. Soc Water Treat Exam* 23:234, 1974.
94. Environmental Protection Agency (EPA): *Preliminary Assessment of Suspected Carcinogens in Drinking Water.* Report to Congress. Washington, DC: EPA, 1975.
95. G. A. Junk and S. E. Stanley: *Organics in Drinking Water:* Part I. *Listing of Identified Chemicals.* Springfield, VA: National Technical Information Services, 1975.
96. T. A. DeRouen and J. E. Diem: Relationships between cancer mortality in Louisiana drinking water source and other possible causative agents, in H. H. Hiatt, J. R. Watson, and J. A. Winsten, eds.: *Origins of Human Cancer.* Cold Spring Harbor, NY: Cold Spring Harbor Laboratory, 1977, pp. 331–345.
97. J. Higginson and A. G. Oettlé: Cancer incidence in the Bantu and "Cape Colored" races of South Africa: Report of a cancer survey in the Transvaal (1953–55). *JNCI* 24:589, 1960.
98. J. Higginson and C. S. Muir: Environmental carcinogenesis: Misconceptions and limitations to cancer control. *JNCI* 63:1291, 1979.
99. J. Higginson: Perspectives and future developments in research on environmental carcinogen-

esis, in A. C. Griffin and C. R. Shaw, eds: *Carcinogens: Identification and Mechanism of Action*, New York: Raven Press, 1979, pp. 187–208.

100. J. L. Lyon, M. R. Klauber, J. W. Gardner, M.P.H. Melville, and C. R. Smart: Cancer incidence in Mormons and non-Mormons in Utah, 1966–1970. *N Engl J Med* 294:129, 1976.
101. A. J. Fox and A. M. Adelstein: Occupational mortality: Work or way of life? *J. Epidemiol Commun Health* 32:73, 1978.
102. W. Dubreuilh: Epithéliomatose d'origine solaire. *Ann Dermatol Syphiliq* 8:387, 1907.
103. G. M. Findlay: Ultra-violet light and skin cancer. *Lancet* 2:1070, 1928.
104. E. L. Scott and M. L. Straf: Ultraviolet radiation as a cause of cancer, in H. H. Hiatt, J. R. Watson, and J. A. Winsten, eds.: *Origins of Human Cancer.* Cold Spring Harbor, NY: Cold Spring Harbor Laboratory, 1977, pp. 529–546.
105. S. Feygin: *Du Cancer Radiologique.* Paris: Rousset, 1915.
106. H. C. March: Leukemia in radiologists. *Radiology* 43:275, 1944.
107. A. C. Upton: Radiation effects, in H. H. Hiatt, J. R. Watson, and J. A. Winsten, eds.: *Origins of Human Cancer*, Cold Spring Harbor, NY: Cold Spring Harbor Laboratory, 1977, pp. 477–500.
108. H. S. Martland and R. E. Humphries: Osteogenic sarcoma in dial painters using luminous paint. *Arch Pathol* 7:406, 1929.
109. J. S. Horta, L. C. Da Motta, and M. H. Tavares: Thorium dioxide effects in man: Epidemiological, clinical, and pathological studies. *Environ Res* 8:131 1974.
110. L. H. Hempelmann, W. J. Hall, M. Phillips, R. A. Cooper, and W. R. Ames: Neoplasms in persons treated with x-rays in infancy: Fourth survey in 20 years. *JNCI* 55:519, 1975.
111. N. H. Harley and J. H. Harley: Potential lung cancer risk from indoor radon exposure. *CA* 40:265, 1990.
112. J. Schoenberg and J. Klotz: *A Case-Control Study of Radon and Lung Cancer among New Jersey Women.* New Jersey State Department of Health Technical Report, Phase I. Trenton, NJ: New Jersey Department of Health, 1989.
113. W. J. Blot, Z.-Y. Xu, J. D. Boice, Jr., D.-Z. Zhao, B. J. Stone, J. Sun, L.-B. Jing, and J. F. Fraumeni, Jr.: Indoor radon and lung cancer in China. *JNCI* 82:1025, 1990.
114. A. V. Nero, M. B. Schwehr, W. W. Nazaroff, and K. L. Revzan: Distribution of airborne radon-222 concentrations in US homes. *Science* 234:992, 1986.
115. C. A. DeLaat and B. C. Lampkin: Long-term survivors of childhood cancer: Evaluation and identification of sequelae of treatment. *CA* 42:263, 1992.
116. J. P. Neglia, A. T. Meadows, L. L. Robison, T. H. Kim, W. A. Newton, F. B. Ruyman, H. N. Sather, and G. D. Hammond: Second neoplasms after acute lymphoblastic leukemia in childhood. *N Engl J Med* 325:1330, 1991.
117. P. E. Oyer, E. B. Stinson, S. W. Jamieson, S. A. Hunt, M. Billingham, W. Scott, C. P. Bieber, B. A. Reitz, and N. E. Shamway: Cyclosporin-A in cardiac allografting: A preliminary experience. *Transplant Proc* 15:1247, 1983.
118. M. C. Pike, M. D. Krailo, B. E. Henderson, J. T. Casagrande, and D. G. Hoel: "Hormonal" risk factors, "breast tissue age," and the age-incidence of breast cancer. *Nature* 303:767, 1983.
119. P. C. MacDonald, C. D. Edman, D. L. Hemsell, J. C. Porta, and P. K. Siiteri: Effect of obesity on conversion of plasma androstenedione to estrone in postmenopausal women with and without endometrial cancer. *Am J Obstet Gynecol* 130: 448, 1978.
120. B. S. Hulka: Hormone-replacement therapy and the risk of breast cancer. *CA* 40:289, 1990.
121. A. L. Herbst, P. Cole, T. Colton, R. E. Scully, and S. J. Robboy: Age-incidence and risk of DES-related clear cell adenocarcinoma of the vagina and cervix. *Am J Obstet Gynecol* 128:43, 1976.
122. B. Safai, B. Diaz, and J. Schwartz: Malignant neoplasms associated with human immunodeficiency virus infection. *CA* 42:74, 1992.
123. A. G. Knudson, Jr.: Genetic predisposition to cancer, in H. H. Hiatt, J. R. Watson, and J. A. Winsten, eds.: *Origins of Human Cancer.* Cold Spring Harbor, NY: Cold Spring Harbor Laboratory, 1977, pp. 45–52.
124. S. Moolgavkar and A. G. Knudson, Jr.: Mutation and cancer: A model for human carcinogenesis. *JNCI* 66:1037, 1981.
125. H. T. Lynch, R. E. Harris, H. A. Guirgis, K. Maloney, L. L. Carmody, and J. F. Lynch: Familial association of breast/ovarian carcinoma. *Cancer* 41:1543, 1978.
126. J. J. Yunis and N. Ramsay: Retinoblastoma and subband deletion of chromsome 13. *Am J Dis Child* 132:161, 1978.
127. X. Zhu, J. M. Dunn, R. A. Phillips, A. D. Goodard, K. E. Paton, A Becker, and B. G. Gallie: Preferential germline mutation of the paternal allele in retinoblastoma. *Nature* 340:312, 1989.
128. R. W. Miller, J. F. Fraumeni, Jr., and M. D. Manning: Association of Wilms' tumor with aniridia, hemiphypertrophy, and other congenital malformations. *N Engl. J Med* 270:922, 1964.
129. V. van Heyningear, P. A. Boyd, A. Seawright, J. M. Fletcher, J. A. Fantes, K. E. Buckton, et al: Molecular analysis of chromosome 11 deletions in aniridia-Wilms' tumor syndrome. *Proc Natl Acad Sci USA* 82:8592 (1985).
130. M. Swift and C. Chase: Cancer in families with xeroderma pigmentosum. *JNCI* 62:1415, 1979.
131. D. W. Nebert and W. W. Weber: Pharmacogenetics, in W. B. Pratt and P. Taylor, eds.: *Principles of Drug Action: The Basis of Pharmacology.* New York: Churchill Livingstone, 1990, pp. 469–531.

132. R.A.M. Case, M. E. Hosker, D. B. McDonald, and J. T. Pearson: Tumours of the urinary bladder in workmen engaged in the manufacture and use of certain dyestuff intermediates in the British chemical industry. *Br J Ind Med* 11:75, 1954.
133. R. E. Kouri, C. E. McKinney, D. J. Slomiany, D. R. Snodgrass, N. P. Wray, and T. L. McLemore: Positive coorelation between high aryl hydrocarbon hydroxylase activity and primary lung cancer as analyzed in cryopreserved lymphocytes. *Cancer Res* 42:5030, 1982.
134. K. Nakachi, K. Imai, S-i. Hayashi, J. Watanabe, and K. Kawajiri: Genetic susceptibility to squamous cell carcinoma of the lung in relation to cigarette smoking dose. *Cancer Res* 51:5177, 1991.
135. B. N. Ames: Identifying environmental chemcials causing mutations and cancer. *Science* 204:587, 1979.
136. J. H. Weisburger and G. M. Williams: Carcinogen testing: Current problems and new approaches. *Science* 214:401, 1981.
137. B. N. Ames, W. E. Durston, E. Yamasaki, and F. D. Lee: Carcinogens are mutagens: A simple test system combining liver homogenates for activation and bacteria for detection. *Proc Natl Acad Sci USA* 70:2281, 1973.
138. R. D. Kuick, J. V. Neel, J. R. Strahler, E.H.Y. Chu, R. Bargal, D. A. Fox, and S. M. Hanash: Similarity of spontaneous germinal and in vitro somatic cell mutation rates in humans: Implications for carcinogenesis and for the role of exogenous factors in "spontaneous" germinal mutagenesis. *Proc Natl Acad Sci USA* 89:7036, 1992.
139. D. G. Hoel, N. L. Kaplan, and M. W. Anderson: Implication of nonlinear kinetics on risk estimation in carcinogenesis. *Science* 219:1032, 1983.

3

Genetic Alterations in Cancer Cells

CHROMOSOMAL ABNORMALITIES

In 1914, Boveri[1] formulated the somatic mutation hypothesis of the origin of cancer. He thought that the origin of the cancer cell was due to a "wrongly combined chromosome complex" occurring in a somatic cell (rather than a germ cell) and that this caused abnormal cell proliferation. He believed further that this defect was passed on to all cellular descendants of the original cancer cell. He also thought that a single abnormal chromosome combination could account for the malignant character of a cancer cell. It is now well established that some human cancers have a familial distribution (see Chap. 2) and that certain chromosomal rearrangements are associated with human malignant neoplasia.

One of the first chromosomal abnormalities definitively associated with human cancer was the so-called Philadelphia chromosome (Ph^1), described by Nowell and Hungerford[2] in patients with chronic myelocytic leukemia (CML). The Ph^1 chromosome was at first thought to result from a deletion of part of the long arm of chromosome 22 in the leukemic cells. Rowley[3] later found that the lesion was really a translocation of a piece of chromosome 22 to chromosome 9. More than 90% of patients with CML have the Ph^1 chromosome in their leukemic cells, and the presence of this chromosome is perhaps the strongest argument that chromosomal aberrations are causally related to cancer. This hypothesis is strengthened by the fact that the Ph^1 chromosome is an acquired characteristic of leukemic cells; only the CML-affected individual in a set of identical twins has the Ph^1 chromosome in his or her bone marrow cells.[4] The development of chromosomal banding techniques provided a great advance in the identification of chromosome rearrangements in cancer cells. The interaction of certain alkylating fluorochromes (e.g., quinacrine mustard) and histochemical stains (e.g., Giemsa stain) with specific regions of chromosomes produces "bands" along the chromosomes that can be used to fingerprint each chromosome pair.[5,6] A number of definitive assignments of chromosomal changes in various cancers have now been made (see below).

The average chromosome band observed with standard banding techniques contains 5×10^6 nucleotide pairs, and deletions or duplications of 2×10^6 nucleotide pairs or less would be difficult to detect.[7] Since the gene size needed to code for an average protein with a molecular weight (MW) of 50,000 is about 1200 nucleotide pairs, approximately 1000 genes could be duplicated or deleted without being detected. Furthermore, it is likely that the entire gene would not have to be duplicated or deleted for its function to be altered. For example, the alteration of gene *cis*-regulatory elements or transcription termination sequences could have profound effects on gene function. Thus, the chromosome derangements detected in cancer cells by banding underestimate the actual number of cancers that have altered gene function. Other, more sensitive methods for detecting genetic alterations are discussed below.

TYPES OF GENETIC ALTERATIONS IN CANCER CELLS

Cancer is essentially a genetic disease in that all cancer cells have some alteration of gene expression. These genetic alterations include chromosomal translocations, of which the Ph^1 chromosome in CML is an example, inversions, deletions, amplifications, point mutations, and duplications or losses of whole chromosomes (aneuploidy) (Fig. 3-1).[8] Most of the current information about genetic alterations in cancer has

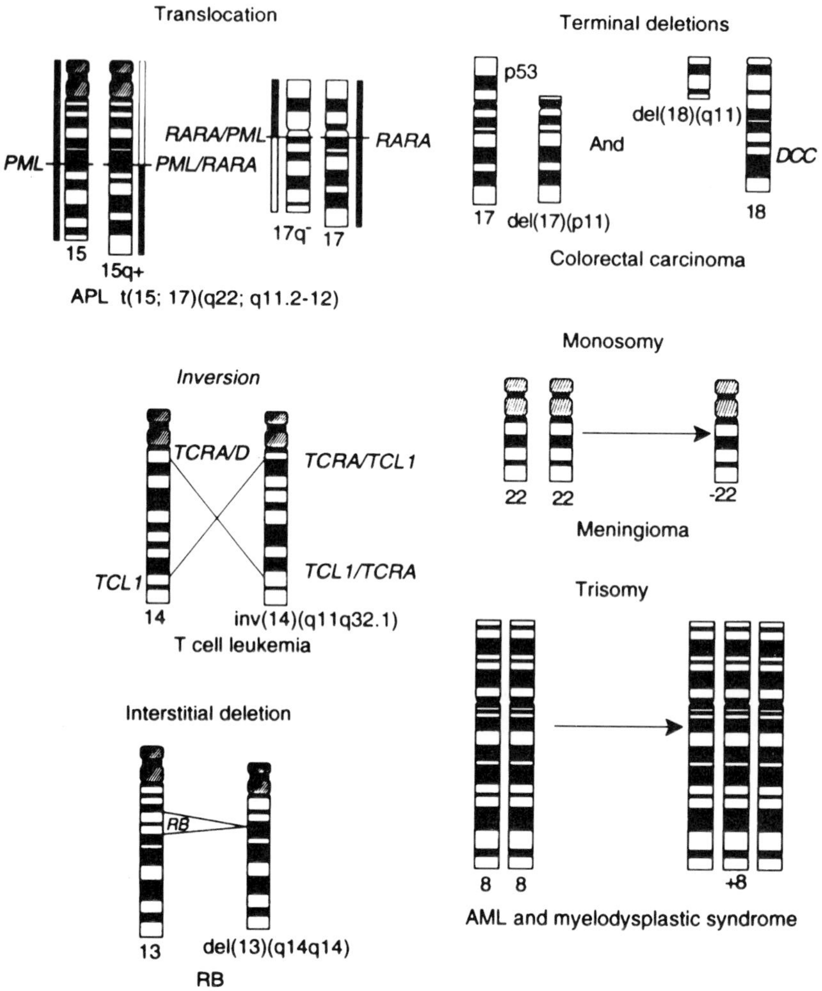

Figure 3-1 Schematic representation of chromosomal aberrations observed in tumors. Shown are the t(15;17)(q22;q11.2-12) seen in APL; the inv(14)(q11q32.1) observed in T-cell leukemia; the del(13)(q14q14) associated with RB; the terminal deletions of chromosomes 17p and 18q seen in colorectal carcinoma; monosomy 22 associated with meningioma; and trisomy 8 seen in AML and myelodysplastic syndrome. (Reprinted with permission from Solomon et al., Chromosome aberrations and cancer. *Science* 254:1153. Copyright 1991 by the American Association for the Advancement of Science.)

come from studies of leukemias and lymphomas[8,9] because it is easier to obtain relatively pure single-cell dispersions of populations of these types of cells from peripheral blood or bone marrow samples than it is to obtain them from pure cell populations from solid tumors such as colon, lung, or breast. Nevertheless, a significant amount of information has been obtained about the genetic alterations in solid tumors, and some of these are described in subsequent sections of this chapter.

Translocations and Inversions

Reciprocal translocations are typical of leukemias and lymphomas. More than 100 commonly occurring translocations have been observed.[8] The fact that many of these occur consistently in certain specific cancer types argues strongly that they are involved in a key way in generating the malignant phenotype.

As noted above, the first constant translocation observed was the reciprocal translocation between the long arm (called q) of chromosome 9, band 34 (for the location of the band; each arm, the short arm p above the centromere and long arm q below the centromere, is divided numerically) and band 11 of the q arm of chromosome 22. The shorthand used by cytogeneticists to describe this is t(9; 22)(q34; q11).

Later it became apparent that the t(9; 22) translocation in CML involved a breakpoint near the Abelson *(abl)* protooncogene. Indeed, as it turned out, this was just one of many such translocations involving protooncogenes in leukemia and lymphoma (Table 3-1). In fact, the common involvement of protooncogenes in these breakpoints is strong evidence for the involvement of these genes in the malignant process of leukemia and lymphoma. The first translocation junction involving a protooncogene to be analyzed was actually the t(8; 14) translocation seen in Burkitt's lymphoma[10,11] (Fig. 3-2). This translocation results in the translocation of the *myc* cellular

Table 3-1 Molecularly Characterized Neoplastic Rearrangements

Part	Disease	Rearrangement	Gene	Protein Type
A	BL	t(8; 14)(q24; q32) t(2; 8)(p11; q24) t(8; 22)(q24; q11)	*MYC*	HLH domain
	B-CLL	t(11; 14)(q13; q32)	*BCL1 (PRAD1?)*	PRAD1 is a G1 cyclin
	Follicular lymphoma	t(14; 18)(q32; q21)	*BCL2*	Inner mitochondrial membrane
	B-CLL	t(14; 19)(q32; q13)	*BCL3*	CDC10 motif
	Pre-B ALL	t(5; 14)(q31; q32)	*IL-3*	Growth factor
B	T-ALL	t(8; 14)(q24; q11)	*MYC*	HLH domain
	T-ALL	t(7; 19)(q35; p13)	*LYL1*	HLH domain
	T-ALL	t(1; 14)(p32; q11)	*TCL5 (TAL1, SCL)*	HLH domain
	T-ALL	t(11; 14)(p15; q11)	*RBNT1*	LIM domain
	T-ALL	t(11; 14)(p13; q11)	*RBNT2*	LIM domain
	T-ALL	t(7; 9)(q35; q34)	*TAN1 (TCL3)*	Notch homolog
	T-ALL	t(10; 14)(q24; q11)	*HOX11 (TCL3)*	Homeodomain
C	Parathyroid adenoma	inv(11)(p15; q13)	*PTH* deregs *PRAD1*	PRAD1 is a G1 cyclin
	B-CLL	t(8; 12)(q24; q22)	*BTG1* deregs *MYC*	MYC has an HLH
D	CML, B-ALL	t(9; 22)(q34; q11)	*BCR-ABL*	BCR, GAP for $p21^{ras}$ ABL, tyrosine kinase
	APL	t(15; 17)(q22; q11.2-12)	*PML-RARA*	PML, Zn finger RARA, Zn finger
	AML-M2, AML-M4	t(6; 9)(p23; q34)	*DEK-CAN*	DEK, nuclear CAN, cytoplasmic
	Pre-B ALL	t(1; 19)(q23; p13)	*E2A-PBX*	E2A, HLH PBX, homeodomain
	NHL	ins(2; 2)(p13; p11.2-14)	*REL-NRG*	REL, NF-κB family NRG, no homology

Key: **A,** Oncogenes juxtaposed to Ig loci; **B,** oncogenes juxtaposed to TCR; **C,** oncogenes juxtaposed to other loci; **D,** fusion oncoproteins; inv, inversion; deregs, deregulates; AML-M2, acute myeloblastic leukemia; AML-M4, acute monomyelocytic leukemia; HLH, helix-loop-helix.

Source: Reprinted with permission from Solomon et al., Chromosome aberrations and cancer. *Science* 254:1153. Copyright 1991 by the American Association for the Advancement of Science.

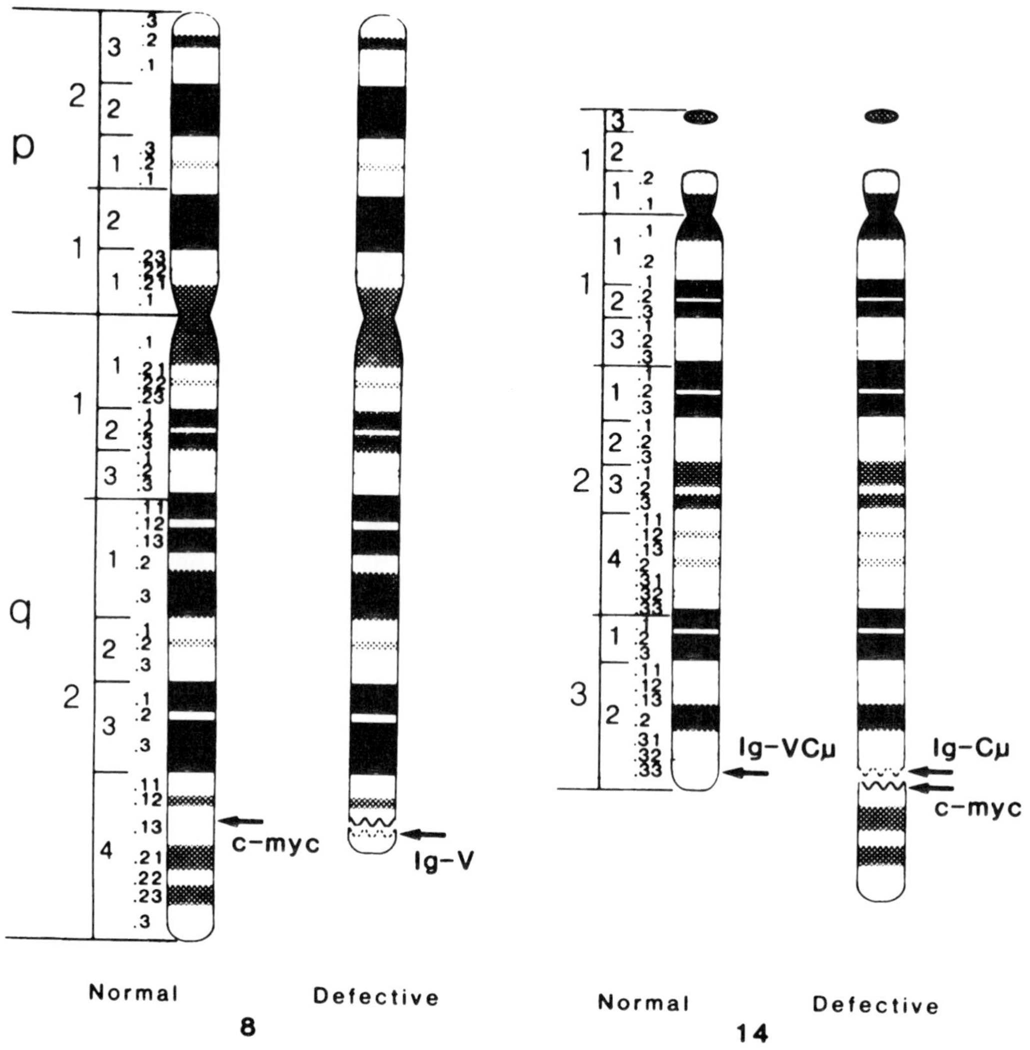

Figure 3-2 Location of c-*myc* oncogene and heavy-chain immunoglobulin variable (V) and constant μ (C_μ) genes on normal and defective chromosomes 8 and 14 in Burkitt's lymphoma, represented at the 1200-Giesma band stage. The defective chromosome 8 loses the c-*myc* and gains V genes. The defective chromosome 14 gains c-*myc* from chromosome 8, becoming contiguous or near to C_μ. Arrows point to the normal and rearranged location of these genes. Broken ends of defective chromosomes indicate breakpoint sites. (From Yunis.[22])

protooncogene from chromosome 8 to chromosome 14 near the immunoglobulin heavy chain Cμ (Ig-Cμ) gene, resulting in the activation of the *myc* gene.

The genes involved in the breakpoint junction of CML were the next to be identified. In 1982, Hagemeijer et al.[13] showed that the c-*abl* gene was translocated from chromosome 9q to the Ph[1] chromosome (chromosome 22). This was identified because of its homology to the viral oncogene v-*abl* isolated from a mouse pre-B-cell leukemia (see Chap. 6). Using a probe derived from the v-*abl* gene, Heisterkamp et al.[14] identified—by chromosomal "walking" (see below) across the translocation junction—sequences that were derived from chromosome 22, thus proving a reciprocal translocation event. The breakpoints on 22 in 17 out of 17 CML patients examined occurred within a 5.8-kilobase segment, which they called the breakpoint cluster region or *bcr*. The breakpoints in *abl* on chromosome 9 occur at variables sites, but always in introns. As a result of the translocation, the *abl* gene piece containing exon II through to its 3′ terminus is moved to the midpoint of the *bcr* gene, which encodes a GTPase activating protein (GAP), forming a fusion gene, which codes for a chimeric Bcr-Abl protein. This protein has high tyrosine kinase activity, a signal translocation mechanism often deregulated in cancer cells (see Chap. 7).

A conundrum arose when similar t(9; 22) translocations were found in a significant number of patients with adult acute lymphocytic leukemia (ALL). Since ALL is a very different disease than CML, it was difficult to reconcile this with a cause-effect relationship for these two diseases. However, it was later found that the Bcr-Abl fusion protein from CML cells results from a somewhat different breakpoint than that seen in most ALL patients. The CML fusion protein is 210 kilodaltons (kDa) in size, while that seen in ALL is 190 kDa.[15] Both have tyrosine kinase activity, but the Bcr-Abl fusion protein from ALL has higher activity, which may relate to the fact that ALL is a more aggressive disease.[16]

This conundrum will arise again and again because similar genetic changes occur in very different kinds of cancer (see below). For example, mutations in the *ras* protooncogene occur in several different types of cancer, as do mutations or deletions of the p53 tumor suppressor gene. The explanation seems to be that different patterns of gene alteration can produce common phenotypic changes in cells that lead to misregulated cell proliferation, invasion, and metastasis. Another possibility is that similar patterns of altered gene expression can lead to different endpoints in different cell types. Evidence for the former comes from the multiplicity of genetic changes seen in human cancer cells (Tables 3-1 and 3-2). Evidence for the latter comes from the observation of common genetic translocations in CML and B-cell ALL and in Burkitt's lymphoma and T-cell ALL, for example. In any case, "all roads lead to Rome," in the sense that cancer cells share several common features.

Another well-defined translocation involves the *bcl*1 gene, originally defined by its rearrangement with the immunoglobin heavy chain locus (IgH) in B-cell chronic lymphocytic leukemia (B-CLL), diffuse B-cell lymphoma, and multiple myeloma.[17] *bcl*2 is another oncogene identified by translocation, in this case by the t(14; 18) translocation observed in follicular lymphoma.[18] *bcl*2 is involved in regulation of lymphocyte proliferation and differentiation and acts to prolong cell survival by blocking programmed cell death (apoptosis).[19]

Table 3-2 Translocations in Solid Tumors

Tumor	Translocation
Breast adenocarcinoma	t(1)(q21-23)
Glioma	t(19)(q13)
Ewing's sarcoma	t(11; 22)(q24; q12)
Leiomyoma (uterus)	t(12; 14)(q13-15; q23-24)
Lipoma	t(3; 12)(q27-28; q13-15)
	t(6)(p22-23)
	t(12)(q13-15)
Liposarcoma (myxoid)	t(12; 16)(q13; p11)
Melanoma	t(1)(q11-q12)
	t(1; 6)(q11-12; q15-21)
	t(1; 19)(q12; p13)
	t(6)(p11-q11)
	t(7)(q11)
Myxoid chondrosarcoma	t(9; 22)(q22; q11.2)
Malignant histiocytosis	t(2; 5)(p23; q35)
Ovarian adenocarinoma	t(6; 14)(q21; q24)
Pleomorphic adenoma	t(3; 8)(p21; q12)
	t(9; 12)(p13-22; q13-15)
	t(12)(q13-15)
Renal cell carcinoma	t(3; 8)(p21; q24)
Rhabdomyosarcoma (alveolar)	t(2; 13)(q35-37; q14)
Synovial sarcoma	t(X; 18)(p11; q11)

Source: Reprinted with permission from Solomon et al., Chromosome aberrations and cancer. *Science* 254:1153. Copyright 1991 by the American Association for the Advancement of Science.

Genes encoding transcriptional regulatory factors are frequently involved in translocation breakpoints seen in hematologic malignancies. For example, two related helix-loop-helix (HLH)–type transcriptional regulators LYL1 and TCL5 are rearranged in T-ALL.[8] Myc is an HLH protein that is translocated and deregulated in both B- and T-cell neoplasms. The TCL3 locus identified in the t(10; 14) of some T-ALLs codes for a homeobox protein, HOX11, also a transcriptional regulator (see below).

The multiplicity of translocation events in various cancers strongly suggests that they have a causal relationship in inducing the cancer phenotype. However, some of them may occur as secondary events in the evolution of more aggressive phenotypic changes. As discussed in more detail in Chapter 6, the inherent genetic instability of malignant cells leads to further karyotypic abnormalities as the disease progresses, reflecting additional genetic alterations that increase growth potential. Evidence that malignant transformation of cells does not usually result from single translocation events comes from patients with ataxia telangiectasia, who have an increased likelihood of developing leukemia. These patients may have T lymphocytes with a translocation present for several years before leukemia develops.[20] Similarly, some patients with benign follicular hyperplasia have *bcl*2 gene rearrangements.[21] Thus, additional genetic mutation events must occur to trigger the development of the full-blown malignant phenotype.

Chromosomal Deletions

The history of the study of chromosomal abnormalities in cancer closely parallels advances in chromosomal banding techniques as well as the development of molecular biology. Before the discovery of banding techniques, only one cancer, CML, had been clearly associated with a consistent chromosomal defect, the Ph[1] chromosome. This was reported in 1960.[2] After the introduction of banding techniques in 1970,[5] more than 30 neoplastic conditions were shown to have consistent chromosomal anomalies. Before 1970, chromosomal defects were generally thought to be secondary or late changes in neoplasia, but by the early 1980s, it was rather widely believed that they are found in most cancers and that each cancer is associated with a specific chromosomal lesion.[22] We are now in a third phase of evolution in this field, the driving force of which is the application of the techniques of molecular biology—e.g., gene cloning, in situ hybridization, restriction endonuclease mapping of gene sequences, and polymerase chain reaction (PCR) analysis of gene transcription. The use of these techniques has led to the conclusion that a given chromosomal abnormality may be associated with a variety of neoplasms and that a given oncogene can be activated in a variety of human cancers.

Certain general statements can be made about the kinds of chromosomal abnormalities seen. The most common defects observed in solid tumors are deletions in specific gene sequences, sometimes observed as loss of a part of a banding region or the loss of heterozygosity of a specific genetic allele (Fig. 3-3).[23] Reciprocal translocations between two chromosomes are also observed; however, in the case of solid tumors, as contrasted with leukemia and lymphoma, few of these translocation breakpoints have been cloned. Gene amplifications are sometimes observed as homogeneously staining regions on chromosome banding patterns or as small chromosomelike fragments in cells called double-minute chromosomes. (Gene amplification is

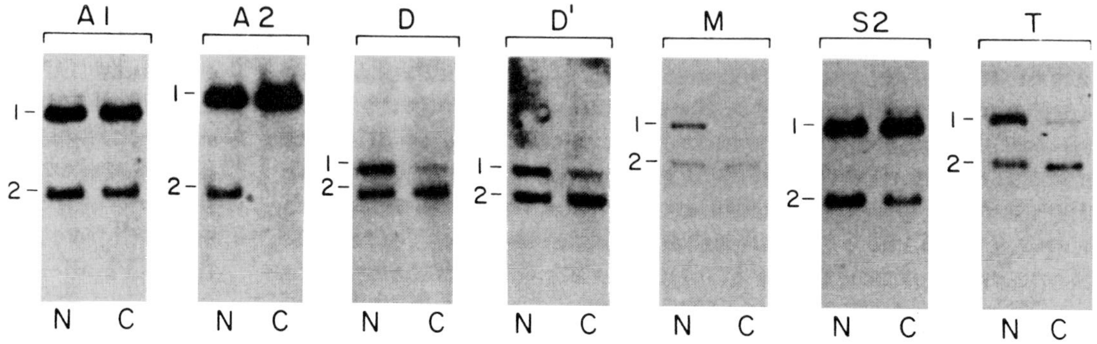

Figure 3-3 Representative Southern hybridizations of 11p gene probes to normal tissue (*N*) and transitional cell carcinoma of the urinary bladder (*C*) DNA. Patient designations are shown above the blots, and the numbers shown on the left indicate the polymorphic alleles (designated 1 and 2). Blots from patients A1, A2, M, S2, and T are *Msp*I restriction endonuclease digests of DNA hybridized to the c-Ha-*ras*-1 gene probe. DNA from patient D (and sample D′) was digested with *Hin*fI and hybridized to the insulin gene probe. Note the preservation of heterozygosity at the c-Ha-*ras*-1 locus in tumor DNA of patient A1, while tumor DNA from patients A2, D, M, S2, and T reveal loss, or marked reduction in the intensity, of one allele. In addition, note the increased intensity of the retained allele in patients A2 and D. D′ represents DNA extracted from frozen sections of patient D. (From Fearon et al.[23])

discussed in more detail below.) Single base substitutions or point mutations also occur in a variety of cancers (see below). As in the case of the translocations discussed above, many of the genetic changes seen in solid tumors result in activation of a cellular oncogene. In tumors with genetic deletions, a tumor suppressor gene may be lost (see Chap. 8).

Deletion of genetic material in a cancer cell suggests loss of function that regulates cell proliferation or differentiation. More than twenty human solid tumors have been shown to have some type of chromosomal deletion (Table 3-3), and the number is growing as more tumors are analyzed by molecular techniques. Some chromosome deletions appear to be most common in certain tumor types. These include deletion del(13)(q14q14) seen in retinoblastoma, which results in loss of the *Rb* tumor suppressor gene, the 11p13 deletion in Wilms' tumor, and the deletion of the *DCC* (deleted in colon cancer) gene in colonic carcinoma (see below). Deletions in the long arm of chromosome 5 (del 5q) are seen in a number of hematologic diseases, including acute nonlymphocytic leukemia and chronic myeloproliferative disorders. These deletions commonly involve the 5q21-31 region, which contains genes encoding growth factors and growth factor receptors involved in myeloid cell differentiation.[24]

Table 3-3 Deletion and Loss of Heterozygosity in Solid Tumors

Tumor	Chromosomal Deletion in Tumor	Allele Loss
	Cloned	
Retinoblastoma	13q14	13q
Colorectal carinoma	17p 18q	5q; 17p; 18q
Wilms' tumor	11p13	11p
	Noted	
Bladder adenocarcinoma	1q21-23 Monosomy 9	9q; 11p; 17
Breast adenocarcinoma	1p11-13 3p11-13 3q11-13	1p; 1q; 3p; 11p; 13q; 16q; 17p; 17q; 18q
Glioma	1p32-36 6p15-q27 7q22-q34 8p21-23 9p24-p13	17
Leiomyosarcoma (intestine)	1p12-12	NT[a]
Leiomyoma (uterus)	6p21 7q21-31	NT
Lipoma	13q12-13	NT
Lung adenocarcinoma	3p13-23	3p; 13q; 17p
Lung small cell carcinoma	3p13-23	3p; 13q; 17p
Mesothelioma	3p21-25	NT
Mesothelioma (pleura)	1p11-13	NT
Malignant fibrous histiocytoma	1q11	NT
Melanoma	1p11-22 6q11-27	1p
Meningioma	Monosomy 22 22q12-13	22q12-qter
Neuroblastoma	1p32-36	1p
Ovarian adenocarcinoma	3p13-21 6q15-23	3p; 6q; 11p; 17q
Prostatic adenocarcinoma	7q22 10q24	10; 16
Renal cell carcinoma	3p13-21	3p
Uterine adenocarcinoma	1q21-23	3p

[a]NT, not tested.

Source: Reprinted with permission from Solomon et al., Chromosome aberrations and cancer. *Science* 254:1153. Copyright 1991 by the American Association for the Advancement of Science.

Other chromosome deletions are observed in multiple kinds of cancer. These include deletions in the chromosome 3p13-23 region in small cell carcinoma and adenocarcinoma of the lung, renal cell carcinoma, and ovarian adenocarcinoma; deletion in the 1p32-36 region in neuroblastoma and glioma; 1p11-22 deletions in melanoma, breast adenocarcinoma, intestinal leiomyosarcoma, mesothelioma, and malignant fibrous histiocytoma (Table 3-3). The 1q21-23 region is often subject to deletions in uterine and bladder adenocarcinomas and is involved in translocations in breast adenocarcinomas. Deletions of 6q11-27 have been reported in melanoma, glioma, and ovarian carcinoma. Portions of the 7q21-34 region are lost in uterine leiomyoma, prostate carcinoma, glioma, and acute myeloid leukemia. The p53 tumor suppressor gene-containing region of chromosome 17p is deleted or mutated in a wide vareity of human cancers (see Chap. 8).

The fact that there is such commonality among cancer cell types in the loss of chromosomal material strongly suggests that these regions contain genes coding for regulatory factors involved in cell proliferation and/or differentiation of a wide variety of cell types. As discussed in more detail in Chapters 8, 9, and 10, many of these regions contain genes involved in cell-cycle regulation through interaction with factors called cyclins or in signal transduction pathways that regulate response elements of particular growth regulatory genes.

Induction of the malignant neoplastic process is thought to involve at least two genetic "hits," as described in Chapter 2. In the case of genetically predisposed tumors, the first genetic alteration may be inherited through the germ line, while the second alteration occurs after birth. In genetically predisposed cells, the remaining single normal allele may be sufficient to maintain normal growth regulation, and a second deletion or mutation is required to inactivate the remaining normal allele. In the case of a tumor suppressor gene, both alleles are then, in effect, lost or inactivated.

Loss of heterozygosity at a genetic locus has frequently been the mechanism for detecting deletion or mutation of genes involved in cancer causation. This is detectable by a technique called restriction fragment length polymorphism (RFLP) (see below).

Gene Amplification

Gene amplification can occur at different levels. A gene that is usually present in a single copy in normal cells may be duplicated or may undergo a small increase in copy number. This is ordinarily difficult to detect by molecular or cytogenetic techniques. A second level of amplification may involve 10- to 100-fold or so increases in copies of a genetic locus containing certain genes (the amplified region is sometimes called an *amplicon*). This level of amplification may be manifest by homogeneously staining regions (HSRs) or extrachromosomal double-minute chromosomes (DMs). A third level of amplification can result from the duplication of whole chromosomes, leading to a trisomy of an individual chromosome or, in some cases, to increases in a number of chromosomes (polyploidy).

Amplifications of genes observed in human cancers include amplification of the N-*myc* gene in stage III and IV neuroblastoma, of the epidermal growth factor receptor–regulated gene *her-2/neu* in advanced breast and ovarian carcinomas, and of the *int*-2, *hst*1, and *prad*1 oncogenes in breast and squamous cell carcinomas and in melanoma (see Chap. 7).

Trisomy of chromosome 8 has been observed in acute myelogenous leukemia (AML), acute lymphocytic leukemia (ALL), and myeloproliferative disease. Trisomy of chromosome 9 has been seen in myeloproliferative disorders and of chromosome 12 in malignant lymphoma and lymphoproliferative disorders. Some malignant lymphomas have a trisomy of 3, and trisomy of 7 has been found in some carcinomas and neurogenic tumors (reviewed in Ref. 8).

Aneuploidy

The genetic instability manifest during tumor progression is characterized by a variety of aberrations in the genome, including point mutations; gene deletions, rearrangements, and amplifications; chromosome translocations; and abnormal chromosome number, known as aneuploidy. Although the more subtle changes in the genome—namely, point mutations, gene deletions, and gene rearrangements—may be associated with initiation of the malignant transformation process, gross changes in the number of chromosomes usually occur as tumors progress in malignancy. As noted earlier, certain chromosomal deletions, translocations, and trisomies are characteristically associated with a particular form of cancer; these are called *nonrandom chromosomal alterations*. Changes in cell ploidy, however, are associated with a variety of tumor types in their advanced stages and may be random in the sense that no definitive pattern of chromosome number is associated with a given tumor type. In advanced cancers, both random and nonrandom chromosomal alterations may be found. These continuing genomic changes bring about tumor heterogeneity and the natural selection of more highly invasive and metastatic cancers (see Chap. 11). Thus, tumor progression, in a sense, may be viewed as a highly accelerated evolutionary process.[25]

Evidence for changes in ploidy during tumor progression comes from both human and experimental animal cancers. For example, Frankfurt et al.[26] examined chromosomal ploidy in 45 human prostate carcinomas by staining DNA with a fluorochrome and scanning cells for DNA content by flow cytometry. They found that localized tumors which had not metastasized had a much lower incidence of aneuploidy than more poorly differentiated tumors and tumors that had spread beyond the pelvis. In nearly two-thirds of patients with aneuploid tumors, pelvic or distant metastases were found. In general, the frequency of aneuploidy increased with progressive stages of the disease. Human urinary bladder carcinomas demonstrate the same association between degree of malignancy and degree of aneuploidy: most aneuploid bladder tumors have a high histologic grade and are invasive.[27,28] Primary human breast cancers are mostly diploid, as determined by flow cytometry and karyotype analysis, whereas cells taken from metastatic sites are often aneuploid.[29] "Dedifferentiation," as evidenced by loss of estrogen receptors, and poor prognosis have also been associated with aneuploidy. However, the diploid breast carcinoma cells had the ability to invade human amnion basement membrane in culture,[29] suggesting that they had achieved an invasive phenotype before they became aneuploid.

In a series of Dunning rat prostatic tumors of different stages in malignant progression, Wake et al.[30] found that the original parent tumor had a normal karyotype, while cytogenetic analysis of the Dunning sublines of later progression indicated a correlation between degree of aneuploidy and more advanced malignant phenotype. During the spontaneous evolution of Chinese hamster cells in culture to highly tumorigenic cells, there is a multistep progression of karyotypic changes, as determined by Giemsa banding and flow cytometry.[31] Four stages of neoplastic progression were identified: trisomy of chromosome 5, a change in banding pattern of chromosome 8, an insertion in chromosome $3(3q^{+})$, and a trisomy of chromosome 8. Trisomy of chromosome 5 preceded the acquisition of the cells' ability to be tumorigenic in nude mice.

Disomy

An unusual type of inheritance pattern has been observed in patients with a genetically determined large-fetus syndrome called Beckwith-Wiedemann syndrome (BWS). In these patients there is a propensity to develop malignant neoplasms, particularly Wilms' tumor of the kidney, but hepatoblastomas and rhabdomyosarcomas also occur. These patients have a uniparental paternal disomy for the 11p15.5 region of chromosome 15 and a loss of the maternal allele of this locus in the tumors that they develop.[32] In an analogous fetal overgrowth syndrome in the mouse, a region of chromosome 7 homologous to human chromosome 11p15.5 also contains a paternal disomy.[33] Interestingly, this locus contains the gene for insulinlike growth factor 2 (IGF-2), and an increased level of IGF-2 mRNA is seen in the tumors of BWS patients. Since the maternal allele of this locus is lost in these tumors, it suggests that overexpression of growth-promoting genes (IGF-2) and loss of a tumor suppressor function on the maternal chromosome locus 11p15 combine to cause the malignant tumors in these individuals.

Genomic Imprinting

Genomic imprinting, which is the selective repression of expression of a genetic allele from either the maternal or paternal gene set, is a known phenomenon in *Drosophila* and mice, but until recently was not observed in humans. It is now known to be the likely cause of loss of regulation of some genes in human cancer. As noted above, for example, in patients with Beckwith-Weidemann syndrome, who have a high propensity to develop Wilms' tumors of the kidney, there is loss of heterozygosity at the chromosome 11p15 locus, which contains two genes—H19 and insulinlike growth factor II (IGF-2)—that are genetically imprinted in the mouse. Paternal uniparental expression of IGF-2 and maternal monoallelic expression of H19 occur in the mouse. Both of these genes have the same type of monoallelic expression in humans.[33a] In contrast, about 70% of Wilms' tumors without loss of heterozygosity at this locus have biallelic-expression of H19 and/or IGF-2, suggesting loss of imprinting of these genes in the tumors. It is likely that more examples of loss of genomic imprinting will be found in human cancers. The potential mechanisms for genomic imprinting are discussed in Chapter 5.

Trinucleotide Expansion

Repeated DNA sequences are widely interspersed throughout the human genome. These interspersed repeats are frequently close to or even within structural genes. And while structural genes in general have a low mutation rate (e.g., about 1 amino acid out of 400 per 200,000 years),[34] repeated sequences have much higher mutation frequencies.[35] Since they are usually in noncoding regions of the genome, these mutations are tolerated by the organism and indeed may even be beneficial by allowing genetic recombination and alternate splicing events to produce new gene arrangements that help an organism adapt to new environments.

The interspersed repeated DNA sequences can undergo a unique form of mutation—namely, variation in copy number. This form of mutation, sometimes called "dynamic mutation,"[35] results from an increase in copy number of repeated trinucleotide sequences, hence the term *trinucleotide expansion*. This form of mutation has now been linked to a number of genetic diseases, including the fragile X syndrome, in which a CGG trinucleotide is amplified[36,37]; myotonic dystrophy and spinal bulbar muscular atrophy, in which the amplified repeat is trinucleotide CAG (reviewed in Ref. 35); and Huntington's chorea, in which the amplified repeat is also CAG.[38] While normal individuals may have 6 to 60 copies of these repeats, unaffected transmitting individuals may have 60 to 200 copies, and severely affected persons may have more than 1000 copies of a trinucleotide repeat. An unusual feature of this type of mutation is that the copy number increases with succeeding generations, explaining the phenomenon of "genetic anticipation," in which asymptomatic carriers in earlier generations pass on the mutant chromosome to their offspring such that, in successive generations, the repeat length and the severity of the disease increase.

Sequencing studies have revealed that the trinucleotide repeats occur in the 3′ untranslated regions of certain genes. In the case of the CAG repeat in myotonic dystrophy, for example, the repeat occurs near a region with a cyclic AMP–dependent protein kinase–like sequence. It is not apparent why these amplified trinucleotide repeats in uncoded regions near a gene would so dramatically affect function, but the data suggest that these sequences have some regulatory action. In the case of the fragile X syndrome, the amplification blocks transcription of a gene called *FMR*-1.[39] If one scans the human gene bank, more than 30 sequences with 5 or more copies of trinucleotide repeats can be found. For example, at least 10 human genes contain p(CCG)n repeats of 5 or more copies (Table 3-4).

While no such mutations have as yet been reported in human cancer, it seems likely that similar genetic changes will be identified in individuals with a susceptibility to develop cancer, especially since some protooncogenes and other regulatory genes contain such trinucleotide repeats (Table 3-4).

Point Mutations

Point mutations that lead to single base changes in a DNA sequence are discussed in more detail in Chapters 6 and 7, regarding mechanisms involved in chemical carcinogenesis and in activation of protooncogenes. Suffice it to say here that reaction of DNA with carcinogenic chemicals can lead to formation of base adducts that can cause base mispairing during DNA replication or loss of an adducted nucleic acid base, producing an abasic site in the DNA chain. Such abasic sites may then be filled with an inappropriate base during DNA repair or replication, leading to a point mutation. If this mutation is

Table 3-4 $(CCG)_n$ Repeats in Human Genes

Gene or Encoded Protein	Gene Bank Symbol	Copy Number	Location
znf6 (zinc finger transcription factor)	HUMZNF	8, 3, 3	5′ Untranslated region
CENP-8 (centromere autoantigen)	HUMCENPB	5	5′ Untranslated region
c-cbl (proto-oncogene)	HUMCCBL	11	5′ Untranslated region
Small subunit of calcium-activated neutral protease	HUMCANPO2	10,6	Coding region (*N* terminal)
CAMIII (calmodulin)	HUMCAM3X1	6	5′ Untranslated region
BCR (breakpoint cluster region)	HUMBCRD	7	5′ Untranslated region
Ferritin H chain	HUMFERH	5	5′ Untranslated region
Transcription elongation factor SII	HUMTEFSII	7	5′ Untranslated region
Early growth response 2 protein	HUMEGR2A	5	Coding region (central)
Androgen receptor	HUMAR	17	Coding region (central)

Source: From Richards and Sutherland.[35]

in a regulatory element of a gene, loss or alteration of regulation of gene expression can occur. If the mutation is in a coding region of a gene, an altered protein may be formed.

Microsatellite Instability

DNA sequences termed *microsatellites* are 1- to 6-nucleotide motifs randomly repeated numerous times in the human genome. For example, about 100,000 (CA)n dinucleotide repeats are found scattered throughout the human genome, and many of these exhibit genetic polymorphisms in the length of the repeats. In colorectal cancer, a number of studies have shown differences in the repeat (CA)n length between tumor and normal DNA from colon specimens from the same patient (reviewed in Ref. 39a). This microsatellite instability (MSI) correlated with tumor location in the ascending colon, with increased patient survival, and inversely, with loss of heterozygosity for chromosomes 5q, 17p, and 18q. Microsatellite instability has been found in both "sporadic" and familial colorectal cancers. Similarly, MSI has now been reported in breast cancer,[39a] small cell lung cancer,[39b] non–small cell lung cancer,[39c] urinary bladder cancer,[39d] and gastric cancer.[39e] Thus, the data suggest that MSI is a common genetic alteration in human cancer.

Mismatch DNA Repair Defects

The frequent occurrence of MSI and other genetic alterations suggests a generalized defect in human cancer. For instance, the genome of patients with hereditary nonpolyposis colorectal cancer (HNPCC) syndrome contains frequent alterations within (CA)n and other simple repeated sequences. This syndrome, which affects as many as 1 of 200 individuals in the western world, predisposes affected persons to cancers of the colon, endometrium, ovary, and other organs, often before age 50.[39f] Such alterations as those seen in the instability of (CA)n sequences indicate a DNA replication error called the RER phenotype. Tumor cells of the RER type display a biochemical defect in mismatch DNA repair analogous to a similar defect in bacteria and yeast (see Chap. 6). A gene involving this DNA repair defect has been found and maps to chromosome 2p. This gene, called hMSH2 (human *mut*S homologue 2), is homologous to the bacterial gene *mut*S, which is responsible for strand-specific mismatch repair.[39g,39h]

MAPPING THE HUMAN GENOME

The National Institutes of Health in collaboration with the Department of Energy has mounted an effort to map the entire human genome, the Human Genome Project. This is a monumental effort somewhat akin to the Apollo Project for putting a man on the moon. There are an estimated 3000 megabases (3×10^{15}) of DNA sequence in the human genome, containing 50,000 to 100,000 genes. So far, only about 2500 genes have been mapped, but the pace of discovery is picking up as new molecular techniques become available.

Historically, the first really powerful technique for mapping the location of genes in the human genome was the use of somatic cell hybrids. With this technique, cells of two different species—mouse and human, for example—one of which is usually a permanent line bearing drug-resistance markers [e.g., cells lacking hypoxanthine-guanine phosphoribosyltransferase ($HGPRT^-$)], are treated with an agent that induces cell membrane fusion such as Sendai virus or polyethylene glycol to form hybrid cells (hybridomas). The nuclei of these hybridomas also fuse to form a single nucleus containing all of the chromosomes from the parental cells. Usually, one of the parental cells is a primary-type cell that will not proliferate for extended periods of time in culture. The other cell type has an enzyme deletion, such as $HGPRT^-$, that will not allow it to grow in a selection medium such as HAT (hypoxanthine + aminopterin + thymidine). The $HGPRT^-$ cells cannot survive the block to purine synthesis induced by aminopterin unless they are fused to a cell containing HGPRT. This allows them to bypass the aminopterin-induced block of purine synthesis by using the purine salvage pathway. Thus, only the fused cells derived from both parental cell types survive. As the hybrid cells proliferate and are passaged in culture, chromosomes of one parental type (usually the human chromosomes in a $HGPRT^-$ mouse/primary human cell fusion) are lost, while the complete set of chromosomes from the other parent cell is retained. The loss of chromosomes is random and, with continual passage, a variety of hybridomas bearing only certain human chromosomes evolve. Because human chromosomes are clearly distinguishable from mouse chromosomes by banding techniques, the cells bearing specific human chromosomes are readily identifiable. The panel of cells is then examined for specific human phenotypic markers and, by a process of elimination, the assignment of the enzyme or marker to a specific chromosome can be deduced. For example, if only hybrid cells bearing chromosome 1 contain the human adenylate kinase-2 isoenzyme, the gene for this enzyme must be located on chromosome 1.

A second technique for chromosomal localization of genes is in situ hybridization. This technique directly localizes genes by using a radioactively labeled cDNA probe to hybridize with a "spread" of metaphase chromosomes (see Fig. 3-4).[40] This allows precise localization of the banding region of the chromosome containing the gene, but the technique sometimes suffers from the limitations of low sensitivity and interference of the hybridization solutions with binding of the chromosomal banding reagent.[41]

In the Southern blotting technique, genomic DNA isolated from somatic cell hybrids containing various human chromosomes is digested with restriction endonucleases to produce specific DNA fragments, which are then separated according to size by agarose gel electrophoresis. The DNA on the agarose gels is then denatured, blotted onto a nitrocellulose filter, and hybridized with radioactive DNA probes. Since the flanking and intervening sequences of even highly conserved structural genes have usually diverged to the point that similar genes of different species vary in their pattern of digestion by restriction endonucleases, identification of human genes can be made against the background of the other parent cell type's genome.

Restriction endonuclease mapping can be used to determine the linkage of various genes as well as their chromosomal location. Genetic variation at the nucleotide sequence–specific "cutting points" of the restriction enzymes produces variability in the pattern of restriction fragments seen by Southern blot analysis (Fig. 3-5).[42] The variation in restriction fragment size has been called restriction fragment length polymorphism (RFLP). Such RFLPs are inherited by Mendelian genetics and can be found using probes for specific genes, as shown in Figure 3-5 for the c-*ras* protooncogene, or using a "library" of DNA probes developed to cloned regions of the total human genome. As more and more probes for specific human genes are developed and their RFLPs are determined, it will be possible to find the location and linkage of most human genes and, more exciting yet, discover which specific genes (and hence which gene products) are involved in human genetic diseases and cancer.

Once a region of genetic inheritance related to a disease state is identified by RFLP, its chromosomal location can be discerned by its linkage to other genes known to be on a specific chromosome. Once a chromosomal region is identified, DNA probes for specific gene sequences that map to the same region are used to "walk" down the chromosome to find a genetic sequence that contains an altered restriction en-

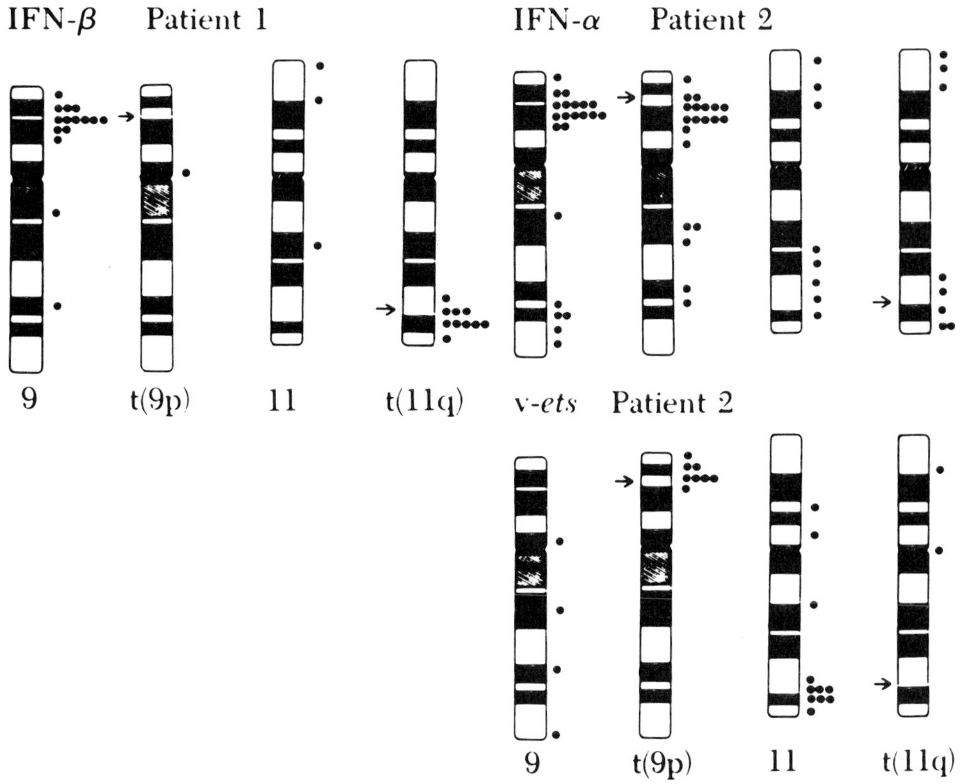

Figure 3-4 Example of the use of in situ hybridization to map chromosomal location of genes. In this example, the distribution of hybridization sites for the radioactively labeled cDNA probes of interferon -α(IFN-α) and -β(IFN-β) and of the oncogene v-*ets* on normal chromosomes 9 and 11 as well as on the translocated derivatives t(9p) and t(11q) are shown. Metaphase cells were obtained from two acute monocytic leukemia patients with t(9:11) (p22:q23) translocations. The arrows identify the breakpoint junctions on the rearranged chromosomes. (From Diaz et al.[15a])

donuclease clip site. The chromosome walk method is used to align or link pieces of DNA by consecutive hybridizations to probes that bind to one end of a cloned piece of DNA. These probes are used to identify an overlapping piece of DNA, i.e., the next gene sequence in line. This region, then, may be isolated and sequenced to identify the mutation.

The mapping of RFLP is also used to detect loss of heterozygosity of certain genes in cancer. Loss of heterozygosity of a given gene is a common occurrence in human cancer and has led to the discovery of a number of putative tumor suppressor genes (see Chap. 8).

A "first-generation" physical map of the human genome has been obtained by cloning human chromosome segments in yeast, called the yeast artificial chromosome (YAC) library.[42a] As many as 33,000 clones are contained in the YAC library; these YACs contain an average length of 0.9 megabases and cover the equivalent of 10 haploid genomes. The library has been screened with more than 2000 genetic markers, and the relative distributions on human chromosomes have been determined for many of them.

Another approach is to "map genes backward." This method starts by isolating mRNA from cells and making the mirror-image complementary DNAs (cDNAs). These cDNA sequences are entered into gene data banks and compared with known sequences from other genes from any source from bacteria to human. Since there is often significant sequence homology, at least for crucial control genes that are conserved during evolution, this often gives a clue about the function of a human DNA sequence. These cDNAs can also be used as probes to identify the chromosomal location of

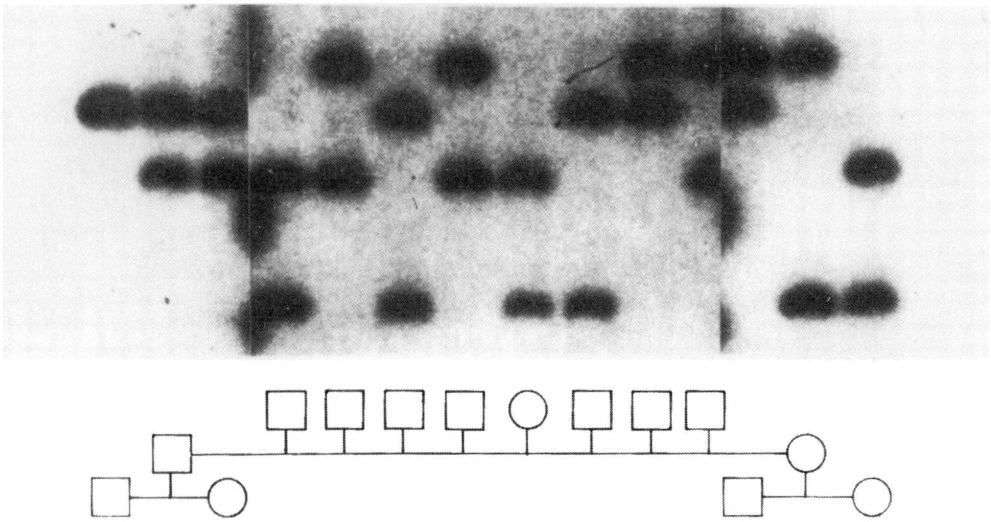

Figure 3-5 Restriction-length polymorphism in a human *ras* gene. The DNA from each member of the three-generation family was digested with a restriction enzyme and then probed with radioactive, cloned Harvey-*ras* DNA. The pattern of fragments detected by the probe is displayed for each individual under his or her position in the pedigree (females are represented by circles and males by squares). A close inspection of the patterns shows that the fragments are inherited in a Mendelian fashion. (From White et al.[42])

such sequences. In this way, one of the important gene defects in colon cancer, the hMLH1 mismatch repair gene, was identified (see below).

GENE DEREPRESSION IN CANCER CELLS

The fact that many malignant neoplasms produce polypeptides, oligosaccharides, and lipids that are inappropriate for the cell types of their tissue of origin indicates a derangement in the flow of genetic information in the transformed malignant cell. This could occur by means of an alteration of gene expression, resulting from gene amplification, rearrangement, translocation, or point mutations, as discussed above. Any of these mechanisms could result in the so-called derepression of genes normally present in cells but not at all or only minimally expressed in normal adult cells. The inappropriate production by cancer cells of certain proteins and other cellular products has been called *ectopic production,* and it has been observed that the pattern of ectopically produced proteins and hormones often resembles more closely that of the embryonic or fetal state than of the adult state of differentiation. This has led to the concept that the expression of these genes in cancer cells results from the derepression of "oncodevelopmental genes," that is, genes that are expressed normally during embryonic development and are usually shut off or only minimally transcribed by differentiated adult cells.

Ectopic Hormone Production by Human Cancers

The first examples of ectopic polypeptide production by human tumors came from the observation of hormonally related syndromes in patients with nonendocrine tumors. In 1928, Brown[43] reported that a patient with small cell carcinoma of the lung had a clinical syndrome manifested by diabetes, hirsutism, hypertension, and adrenal hyperplasia. In other words, the patient had the symptoms of Cushing's disease, which is often caused by excess adrenocorticotropic hormone (ACTH) release. At that time it was not appreciated that tumors could produce ACTH. By 1959, about 40 well-documented reports of Cushing's syndrome in patients with nonendocrine tumors had been reported. It was not until 1961, however, that human nonendocrine gland tumors were observed to produce an

ACTH-like substance, and the term *ectopic ACTH* was coined by Liddle and his colleagues[44] to describe this phenomenon. Although the clinical syndrome associated with ectopic ACTH production is observed in about 2% to 3% of patients with carcinomas of the lung,[45] it has been reported that a high percentage of lung cancers contain detectable amounts of ACTH by radioimmunoassay.[46] The reason for this discrepancy may be that most of the immunoreactive ACTH in the tumor tissue and plasma of cancer patients is present as a precursor form called "big" ACTH, which has only about 4% of the biological activity of ACTH secreted by the normal pituitary gland.[47] It has also been observed that 33% of patients with chronic obstructive pulmonary disease, mostly due to heavy smoking, have elevated plasma immunoreactive ACTH.[48] This, coupled with the observation that the lungs of dogs with atypical hyperplasia (resulting from forced inhalation of cigarette smoke) contained "big" ACTH, whereas lung tissue from dogs with no significant histologic changes did not,[48] suggests that ACTH production may be stimulated in lung tissue undergoing preneoplastic changes.

The ectopic production of parathyroid hormone (PTH) by tumors has also been observed. The first clue to this occurrence was a report in 1936 by Gutman et al.[49] of a patient who had hypercalcemia and hypophosphatemia in association with a nonendocrine tumor not involving bony tissue. Later, it was shown that removal of nonosseous, nonparathyroid tumors in patients with this syndrome corrected the ionic imbalance.[50] Tashjian et al.[51] and Sherwood et al.[52] demonstrated that nonparathyroid tumor extracts from several hypercalcemic patients contained a material that was immunologically similar to PTH. A variety of tumors containing PTH-like material have now been identified; these include certain carcinomas of the lung, kidney, pancreas, and colon as well as the adrenal and parotid glands.

The clue to a possible connection between cancer and the ectopic production of antidiuretic hormone (ADH) came in a report in 1938 by Winkler and Crankshaw,[53] who observed that some patients with lung cancer excreted a very concentrated urine, high in salt content; but the authors did not suggest a possible hormonal explanation for their observation. In 1957, Schwartz et al.[54] described two patients with lung cancer who excreted a hypertonic urine

Table 3-5 Ectopic Hormones Produced by Various Human Cancers

Hormone	Tumors Producing Hormone Ectopically	Associated Clinical Syndrome
ACTH	Carcinomas of lung, colon, pancreas, thyroid, prostate, ovary, cervix; thymoma; pheochromocytoma; carcinoid tumors	Cushing's syndrome
ADH	Carcinomas of lung, duodenum, pancreas, ureter, prostate; thymoma; lymphoma; Ewing's sarcoma	Inappropriate antidiuresis, hyponatremia
Calcitonin	Carcinomas of lung, breast, prostate, bladder, pancreas, liver, esophagus, stomach, colon, larynx, testis; carcinoid tumors; insulinoma; pheochromocytoma; melanoma	No apparent syndrome
Erythropoietin	Hemangioblastoma; uterine myofibroma; pheochromocytoma; carcinoma of liver, ovary	Polycythemia (erthrocytosis)
Gastrin	Carcinoma of pancreas	Zollinger-Ellison syndrome (gastric hypersecretion with intractable peptic ulceration)
Glucagon	Carcinoma of kidney	Hyperglycemia, malabsorption, gastrointestinal stasis
Growth hormone	Carcinomas of lung, stomach, ovary, breast	Hypertrophic pulmonary osteoarthropathy, acromegaly
HCG	Carcinomas of breast, stomach, small intestine, pancreas, biliary tract, liver, colon, rectum, lung, ovary, testis; melanoma	Gynecomastia, precocious puberty
Prolactin	Carcinomas of lung, kidney	Galactorrhea, gynecomastia
PTH	Carcinomas of kidney, lung, liver, adrenal, pancreas, parotid, ovary, testis, spleen, breast	Hypercalcemia
TSH	Carcinomas of lung, breast	Hyperthyroidism

containing high concentrations of sodium, and these investigators attributed their findings to an inappropriate secretion of ADH. Ectopic ADH has been found in several lung tumors (mostly small cell carcinomas) as well as in duodenal, pancreatic, and other carcinomas.

Another example of ectopic hormone production by human tumors is the secretion of the placental hormone human chorionic gonadotropin (hCG). In 1959, Reeves et al.[55] demonstrated by bioassay the presence of an hCG-like substance in a hepatic carcinoma from a boy with precocious puberty. Since that time, hCG has been shown to be produced by a wide variety of human cancer cells both in vivo and in vitro.[56-58]

A summary of the ectopic hormones produced by various human cancers is given in Table 3-5.

Possible Mechanisms of Ectopic Protein Production

The inappropriate expression of polypeptides by tumor cells could result from a rearrangement or mutation in a regulatory gene that leads to the increased transcription of structural genes coding for oncodevelopmental proteins and hormones. The evidence accumulated to date indicates that the amino acid composition of ectopic hormones produced by tumors is the same as that made by the normal hormone-producing cell, indicating that the ectopic product does not simply result from "chaotic" protein synthesis coded for by a scrambled or mutated structural gene.

A second means by which ectopic polypeptide production could occur is by an increased abundance of the mRNA coding for a particular protein. This could occur as a result of an increased "gene dosage" (resulting from gene amplification or gain of chromosomal material, as noted earlier) or of an increased rate of gene transcription, an elevated rate of mRNA processing, or a decreased degradation of mRNA. Evidence is accumulating that mammalian cells contain "leaky" genes, in the sense that there is a very low rate of transcription of many genes in cells that do not make functional amounts of the protein encoded by those genes. For example, normal tissues other than placenta have been found to produce small amounts of hCG,[59] and a small amount of hemoglobin transcription has been found in nonerythroid cells.[60] These results suggest that a low-level transcription of certain oncodevelopmental genes continues in adult differentiated tissues and that ectopic protein production is an expansion of production of these proteins, some of which perhaps provide a selective growth advantage for transformed cells.

Pearse[41] suggested a third means by which ectopic protein production could occur in neoplasms—namely, the clonal expansion of certain cells that produce the ectopic product continuously and that are normally present in only very small numbers in adult tissues. He originated the idea that certain cellular derivatives of embryonic neuroectoderm tissue are present in normal adult tissues. These cells retain the ability to synthesize and secrete certain hormones, and they may proliferate after carcinogenic alteration of normal tissues. The increased proliferation of these clones of cells after malignant transformation, then, could lead to an elevated ectopic production of certain proteins and hormones. Pearse has coined the term "APUD-oma" (amine precursor uptake and decarboxylation) to describe certain features of this class of cells. This definition relates to the ability of these cells to take up and decarboxylate amine precursors involved in the synthesis of certain hormones and neurotransmitters (e.g., epinephrine, norepinephrine, and serotonin). Although this hypothesis could account for ectopic hormone production by tumors arising in certain tissues known to contain these types of cells, it does not explain the general phenomenon of the reexpression of oncodevelopmental products by tumor cells, since many of the latter are not products characteristic of neuroectoderm cells—for example, carcinoembryonic antigen, hCG, alpha-fetoprotein, and placental alkaline phosphatase.

It is interesting that the "leakiness" of gene expression increases with aging. Ono and Cutler[62] found that the amount of globin mRNA present in mouse brain and liver cells increased in aging mice. Thus, the controls regulating gene expression may become less stringent as an organism ages. Since cancer is primarily a disease of aging organisms, it could be that an aging-related loss of stringent gene control contributes to the emergence and clonal expansion of transformed malignant cells.

GENETIC ALTERATIONS IN VARIOUS TYPES OF HUMAN CANCER

Retinoblastoma

Retinoblastoma is a malignant cancer of retinal tissue in the eye and usually occurs at an early age. About 40% of patients have a hereditary form of the disease, and a majority of these children develop tumors in both eyes. This was one of the first tumors in which a clear hereditary predisposition was noted; it led Knudson to propose the hypothesis of two genetic hits, in which one genetic mutation is inherited and a second, somatic mutation occurs after birth (see Chap. 2). The nonhereditary form of the disease results when both genetic events occur after birth.

Genetic linkage studies in the familial form of the disease and evidence for loss of heterozygosity and deletion of a specific chromosomal region led to identification of a region located on the long arm of chromosome 13 (13q14) as the locus involved in retinoblastoma.[63] This led to the isolation of the *RB*1 gene, the first tumor suppressor gene identified. Both deletions and point mutations may occur within this gene, resulting in absence, abnormally sized, or mutated mRNA transcripts.[64] In some of the familial retinoblastomas, inherited mutation of one allele, followed by loss of the normal allele in the tumor cells, has been observed. This type of genetic one-two punch has now been seen in a number of human solid tumors. In the nonhereditary form of the tumor, two distinct somatic events lead to the same result, i.e., development of retinoblastoma, but usually only in one eye.

Germ-line inheritance of an altered *RB* gene also predisposes to other types of childhood cancers such as osteosarcoma, and mutations of this gene have been observed in several other cancers, including breast, lung, and bladder cancers.[65] Yet patients with germ-line mutations of *RB*1 do not have an increased incidence of these latter cancers, indicating that inactivation of *RB*1 is not the crucial or rate-limiting event in these cancers.

One explanation may be that the time of maximum proliferation of the stem cells, i.e., the self-renewal cells, of a given tissue is different for different tissues at various times of life. Stem cells of the retina proliferate most rapidly early in life; then, as the retina develops, the proliferation rate drops off dramatically. The proliferation rate of tissue stem cells for bone begins to increase early in life but peaks in the early teens, the time when most osteosarcomas arise. Breast growth is most pronounced from puberty until early adulthood. Colon stem cells continue to proliferate throughout life. The chance of a second mutational event and of additional mutational events is greater when cells are rapidly proliferating. Thus, a cell type already bearing a mutational change (first hit) is much more likely to undergo a subsequent cancer-initiating mutation (second hit) if the cell population is rapidly replicating its genome for cell division. This could explain why retinoblastomas tend to occur in young children and osteosarcomas in teens or young adults, while breast and colon cancers usually occur in older adults. In the case of the last two cancer types, however, mutations in the *RB* gene cannot be the rate-limiting event. As discussed below, a complex cascade of genetic events seems to be involved in colon and breast cancer. Even in retinoblastomas, other genetic alterations are most likely involved. For example, alterations of chromosome region 6p have also been found in a high percentage of retinoblastoma patients.[65]

Wilms' Tumor

Wilms' tumor is a pediatric kidney cancer or nephroblastoma, arising in primitive stem cells of the developing kidney.[66] Some 5% to 10% of these tumors are bilateral and the remainder are unilateral. In this cancer as in retinoblastoma, there is evidence for an inherited germ-line mutation, but Wilms' tumor does not occur as often in familial clusters, suggesting a more complex array of genetic events than in retinoblastoma.

There is a familial form of the disease that includes Wilms' tumor, aniridia, genitourinary abnormalities, and mental retardation (called the

WAGR syndrome). Individuals with this rare inherited syndrome have a germ-line deletion within band p13 of chromosome 11.[67] A gene called *WT*1 has been mapped to this locus. This *WT*1 gene codes for a zinc finger DNA-binding protein that is primarily expressed in the blastemal cells of the developing kidney—a much more restricted cellular expression than the *RB*1 gene.[65] Moreover, unlike retinoblastoma, the majority of Wilms' tumors express apparently normal *WT*1 mRNA. A few Wilms' tumors have shown DNA rearrangement, small deletions, or a germ-line deletion with subsequent loss of the wild-type allele.[65] In addition, the involvement of a second genetic locus at 11p15 has been shown in Wilms' tumors of patients with another syndrome called Beckwith-Weidemann syndrome. Moreover, some apparently nonhereditary Wilms' tumors have a loss of heterozygosity for gene markers within the 11p15 locus but no involvement of 11p13, while others have deletions in chromosome 11p13 and loss of heterozygosity in markers of 11p15. Still others appear to have no involvement of either 11p13 or 11p15.[65] Clearly, the situation appears to be much more complex for Wilms' tumor than for retinoblastoma and may be more like colon cancer in that multifactorial genetic events are involved in the neoplastic process (see below). Indeed, most human cancers appear to reflect more the case of Wilms' tumors and colon cancer in that a multiplicity of genetic events can lead to the same endpoint, namely, loss of regulation of cell proliferation and differentiation. Many roads lead to Rome.

Li-Fraumeni Syndrome

Li and Fraumeni in 1969 reported data from four families with an extraordinarily high incidence of various kinds of cancer.[68] Over 50 families with this syndrome have now been studied. Members of these families have a high incidence of cancers at an earlier age than would be expected for "sporadic" cancers, i.e., those with no known hereditary factor.[69] For example, breast cancer in family members may occur between ages 15 and 44. A variety of sarcomas (soft tissue sarcomas, osteosarcomas, and chondrosarcomas) may occur in early childhood. Cancers of the brain and adrenal cortex and leukemias have also been reported to occur in high incidence in family members. The inheritance pattern indicates an autosomal dominant trait with high penetrance.[69] Interestingly, the excess risk of developing cancer is greater below age 20 and falls off with increasing age; by age 60, predicted gene carriers do not have a significantly increased risk of developing cancer compared to the general population. It is as if the genetic defect affects developing tissues and once they are past the higher proliferative stage and full tissue differentiation occurs, the risk goes down, similar to what was noted above for the *RB* gene defects.

The primary gene defect in the Li-Fraumeni syndrome has been identified, but others may also be involved.[69] The involved chromosome region is 17p. The involved gene is called p53 because it codes for a 53,000-Da protein. A point mutation in this gene has been identified in both tumor and normal cells from affected family members.[70] The p53 gene acts as a tumor suppressor gene (see Chap. 8) and is expressed ubiquitously in body tissues. Mutations of this gene have been reported in a wide variety of sporadic human cancers, including colon, breast, and lung carcinomas as well as leukemias. A germ-line mutation of p53, however, as seen in Li-Fraumeni syndrome families, appears to predispose individuals to only certain kinds of cancer. For example, although p53 mutations occur in both colon and breast carcinomas, only the latter occurs with increased incidence in Li-Fraumeni syndrome. This suggests that p53 mutations play a primary, crucial role in causation of some cancers and only a secondary, contributing role in others, possibly due to the need to accumulate additional genetic events in some cancers more than others. Also, a different mixture or sequence of genetic events may be involved in hereditary cancers than in their nonhereditary counterparts.

Leukemia

Leukemias arise from stem cells in the bone marrow that proliferate and differentiate into the myeloid (granulocyte) or lymphoid (lymphocyte) lineage cells. They may be rapid (acute) or more indolent (chronic) in onset and progression. Acute leukemias have their peak incidence in children and young adults. These differences suggest different kinds of genetic lesions in acute versus chronic leukemias.

Leukemias as a group have been the most thoroughly studied human cancers for genetic abnormalities, and several abnormalities have been reported[8,71] (Table 3-1). Most of these involve chromosomal translocations that affect expression of genes involved in regulation of cell proliferation and/or differentiation, including transcriptional regulatory factors—of which the helix-loop-helix proteins, LIM-domain-binding proteins, and homeobox (HOX) proteins are examples (see Chap. 5). The kinds of genes affected are different in acute and chronic leukemias. For example, in acute leukemias, chromosomal translocations frequently result in activation of transcription factors that appear to affect differentiation (e.g., *HOX*-11); whereas in chronic leukemia, translocation events affect genes that foster cell proliferation and survival (e.g., *bcl*-2). One concept to explain how alterations of one or several genes can be involved in causation of one clinical type of cancer is the concept that there are "master genes" that can regulate several genes in a cascade or hiearchy of gene control.[71]

This could explain why in a given leukemia, different patterns of genetic abnormalities occur (e.g., in acute T-cell leukemia), even though it appears to be one clinical disease. For example, mutation of a master gene could result in a similar effect as that of individual mutations of several genes in the hierarchical cascade under the master gene's control. Thus, mutation of one or more genes could produce the same end result by hitting a master gene that controls common responder genes (or genes with analogous functions) or by hitting multiple responder genes in the pathways. In some tumor types, by contrast, the same translocations consistently occur. This is the case for Burkitt's lymphoma, with its translocations involving chromosome 8, and for CML, with its t(9; 22) translocation.

A few examples of the genetic events that chromosomal abnormalities in leukemic cells produce are given below.

t(9; 22) Translocations

The reciprocal translocation involving the long arms of chromosomes 9 and 22 is a consistent finding in CML and results in the production of the chimeric protein Bcr-Abl. The Bcr-Abl protein is a 210,000-Da protein present in virtually all cases of CML. This fusion event can be detected by Southern blot analysis of *bcr* rearrangements or by in vitro amplification by polymerase chain reaction (PCR) of a cDNA transcript copied from CML mRNA. In about 5% of CML cases, however, the reciprocal chromosomal translocation involving chromosomes 9 and 22 may not be apparent by karyotypic analysis. In a number of such cases, *bcr-abl* fusion transcripts can be detected by a sensitive two-color fluorescence in situ hybridization technique.[72] This indicates that even in CML patients without an identifiable Ph[1] chromosome, the *bcr-abl* translocation may have occured. Interestingly, in patients with CML, the translocated chromosome 9 appears to be of paternal origin and the translocated chromosome 22 of maternal origin.[73] The translocated regions 9q34 and 22q11 are differentially methylated in normal and transformed cells. Region 9q34 is hypomethylated both in normal and translocated chromosome 9, whereas 22q11 is hypomethylated after translocation. Since hypomethylation of genes is known to be related to activation of their expression (see Chap. 5), this suggests that juxtaposition of two differently methylated gene domains can exert a positive activation effect on the hypermethylated (inactive) domain by the hypomethylated (active) domain.[73] This could also explain the activation of "downstream" genes that may be turned on by the hypomethylation event. If such genes are growth-promoting genes, the cells containing such translocated, reactivated genes would have a proliferative advantage on their normal counterpart cells.

11q23 Translocations

Chromosomal translocations involving chromosome 11, band q 23, have been observed in acute lymphoid and acute myeloid leukemias as well as lymphoma. The gene located at the breakpoint has been termed *MLL*[74] (myeloid-lymphoid leukemia gene). A number of translocations involving 11q23 have been observed: t(4; 11), t(6; 11), t(9; 11), and t(11; 19) are among the most common. By Southern blot analysis, rearrangements of the *MLL* gene have been detected in cells from patients with 11q23 translocations. Northern blot analysis for mRNA indicated that these genes can express multiple mRNA transcripts, some of which appear to be cell lineage–specific.[74] The transcripts of the *MLL* gene found in leukemic cells differ from those seen in normal cells, suggesting that rearrangements of the *MLL* gene (and perhaps other genes in the same chromosomal domain) may have altered function after translocation.

In the case of t(4; 11) translocations, involving 11q23, the resulting translocation results in fusion of the *ALL*-1 gene with a gene termed AF-4 on chromosome 4.[75] The *ALL*-1 gene encodes a protein of 3910 amino acids rich in zinc finger–like domains, suggesting that it is a transcriptional regulator. Fusion of this domain with the AF-4 domain suggests generation of an oncogenic fusion protein similar to Bcr-Abl. The fact that 11q23 breaks are seen in ALL, AML, acute myelomonocytic, and acute monocytic leukemias suggests that the 11q23 region contains genes important for differentiation along lymphoid and myeloid pathways. Depending on where the breakpoint is and which chromosome receives the translocated genes, different types of transformed cell lineages could result.

t(15; 17) Translocations

Although similar chromosomal translocations are seen in a variety of leukemias and lymphomas, the t(15; 17) (q22; q12-21) translocations appear to be exclusively associated with acute promyelocytic leukemia (APL). This translocation is observed in about 90% of APL patients and is diagnostic of the disease.[76] The chromosome 17 breakpoint in APL has been mapped to the retinoic acid receptor α gene (*RAR*α), and the translocation results in a fusion between the *RAR*α gene and a region on chromosome 15 called *PML* (for promyelocyte gene).[77-79]

The *RAR* gene is a member of the steroid/thyroid hormone–receptor superfamily of receptor genes whose gene products regulate gene expression by binding to DNA after interaction with their specific hormone ligands and acting as ligand-dependent transcription factors. In the case of *RAR*, the ligand retinoic acid induces cell differentiation in a wide variety of cell types in vitro and in vivo. The *PML* gene codes for a protein that is a member of a family of transcription factors and recombination-activating gene products (*RAG*-1).[80]

The RAR-PML fusion protein retains weak retinoic acid binding properties, but it appears to block ability of cells to respond to the normal differentiation signal provided by retinoic acid, perhaps by tying up the ligand or by preventing the normal RAR-ligand complex from producing its signal transduction event. This effect is a so-called dominant-negative regulatory event, in which an aberrant gene product may interfere with the function of a normal gene product. A similar effect is seen with mutations of the p53 tumor suppressor gene (see Chap. 8).

One of the exciting new discoveries in cancer therapy resulted when it was observed that treatment with a retinoic acid analogue called all-*trans* retinoic acid could induce remission in APL.[81-83] The all-*trans* retinoic acid can overcome the dominant-negative effect, apparently by activating the ligand binding-domain still present in the RAR-PML fusion product or by somehow reactivating the inhibited normal receptor. This leads to induction of differentiation of APL cells into mature granulocytes, producing clinical remission of the disease. This is the first real clinical demonstration that inducing normal differentiation of differentiation-inhibited cancer cells can reverse the malignant phenotype. This strategy is an attractive one, since it would be quite cell-specific and be relatively nontoxic to patients. Other strategies for inducing normal differentiation in cancer cells are being discovered and some are being tried in the clinic (see Chap. 5).

Chromosomal Abnormalities in Leukemic Patients Exposed to Genotoxic Agents

A number of studies have shown a relationship between exposure to known mutagenic and carcinogenic agents and hematologic malignancies, particularly for acute nonlymphocytic leukemia (ANLL). Mittelman et al.[84] found that 32% of 162 patients with ANLL had occupational exposure to insecticides, solvents, or petroleum products, and 75% of the exposed ANLL patients, as opposed to 32% of ANLL patients with no history of such exposure, had chromosomal abnormalities in their bone marrow cells. Chromosomal abnormalities have also been observed in leukemias and myelodysplastic syndromes that develop after previous treatment with antineoplastic drugs; a number of these drugs are themselves mutagenic and carcinogenic.[85]

The Fourth International Workshop on Chromosomes in Leukemia[86] summarized data from 716 patients with ANLL and reported that chromosomal abnormalities were observed in about 55% of patients with no history of previous anticancer therapy, whereas 75% who had previous exposure to anticancer drugs had abnormalities, usually involving chromosomes 5 and/or 7. Karyotypic abnormalities involving chromosomes 5 and 7 are associated with poor prognosis in ANLL patients. In only 13% of "spontaneous" ANLL cases were these chromosomes involved.

Exposure to benzene is also associated with ANLL, and in these patients chromosomal abnormalities are frequently observed.[87] Chromosomal abnormalities have also been observed in myelodysplastic syndrome (MDS), a preleukemic condition that progresses to ANLL. However, there is no significant difference in the type of chromosomal abnormalities seen between MDS patients who had a history of previous exposure to genotoxic agents and those who did not.[88]

Lymphoma

A number of chromosomal abnormalities have been observed in human lymphomas. The earliest one found and the one first characterized in detail involves a translocation between chromosomes 8 and 14 [t(8; 14)][89] in Burkitt's lymphoma, an aggressive B-cell lymphoma seen primarily in children in parts of Central Africa. Later, two other translocations involving chromosome 8 were observed[90]: t(2; 8) and t(8; 22). About 80% of patients with Burkitt's lymphoma have a t(8; 14)(q24; q32) translocation; 15% have a t(8; 22)(q24; q11) translocation; and 5% have a t(2; 8)(p11; q24) translocation.[91] All of these chromosomes carry genes of the immunoglobulin family: 14q32, 22q11, and 2p11 contain the human immunoglobulin heavy chain, λ light chain, and κ light chain genes, respectively, suggesting that immunoglobulin genes are involved in a key way in Burkitt's lymphoma. Similar translocations have now been found in other B-cell malignancies, including a t(14; 18)(q32; q21) in about 85% of follicular lymphomas, a low-grade B-cell lymphoma common in North America and western Europe, and a t(11; 14)(q13; q32) in CLL, diffuse B-cell lymphoma, and multiple myeloma.[91]

T-cell malignancies are less common than B-cell malignancies and often have chromosomal alterations involving the 14q11 locus, which contains the T-cell-receptor α and δ genes. One or both of these genes are involved in several translocations including t(8; 14)(q24; q11), t(10; 14)(q24; q11), and t(11; 14)(p13; q11).

Thus, in both T- and B-cell malignancies, translocations involving the immunoglobulin superfamily of genes are frequently observed. As noted earlier in this chapter, the t(8; 14) translocation first observed in Burkitt's lymphoma involves translocation of the c-*myc* protooncogene from chromosome 8 to 14 in the locus of the immunoglobin heavy chain gene (IgH).

Chromosome 14q32 translocations are seen in several lymphoid malignancies.[91] For example, about 90% of follicular lymphomas have a t(14; 18)(q32; q21) translocation that involves the J_H region of the IgH locus and a gene on chromosome 18 called *bcl*-2. This leads to deregulation of the *bcl*-2 gene, a gene involved in maintenance of cell proliferation and inhibition of programmed cell death (apoptosis).

A t(11; 14) translocation is found in multiple myeloma and in 10% to 30% of CLL patients. This involves the J_H locus on 14 and a breakpoint cluster on chromosome 11 called *bcl*-1.

In some cases, two or more translocation events may occur, and when they do, a more aggressive disease may result. For example, a t(14; 18) translocation involving the *bcl*-2 gene cluster may result in a low-grade B-cell malignancy; but in the course of the disease, a second translocation event involving the c-*myc* gene on chromosome 8 may occur, leading to progression to more aggressive, high-grade lymphoma or acute leukemia.[91]

Cytogenetic studies have also shown that chromosome 7 band q34-35, which contains the T-cell receptor β gene, is a common site for translocation in T-cell neoplasms. One interesting translocation of this type is t(7; 9)(q34; q34), found in cases of ALL. This involves the TCR-β gene on chromosome 7 and a newly described gene called *TAN*-1 on 9.[92] The sequence of this gene indicates that it encodes a protein that is highly homologous to the *Notch* gene product of the fruit fly *Drosophila*. *Notch* is an integral membrane protein involved in determination of embryonic cell fates in *Drosophila.* In mice and humans, *TAN*-1 is highly expressed and appears to be involved in normal lymphoid differentiation. The *TAN*-1 mRNA found in ALL cells, however, is truncated and appears to contribute to transformation or progression in some T-cell malignancies.

Colorectal Cancer

Colorectal carcinomas have proven to be an excellent model system in which to study the role of hereditary factors, environmental carcinogenesis, oncogene activation, and tumor suppressor gene inactivation in the initiation and progression of human solid tumors.[93,94] Tumors of various stages of dysplasia and malignancy, from benign adenomas to invasive cancers, can be obtained surgically. There is a clear familial predisposition for some of the cases, and a variety of genetic mutational events have been identified in the diseased tissue. In addition, there is a clinically definable progression from benign adenoma to malignant neoplasia, and the cancers that arise are clonal in origin. Thus, the availability of tissue by surgical resection and the monoclonality of the cancers make this an ideal disease to study.

Studies by Vogelstein and his colleagues have led to the model for colorectal carcinogenesis proposed in Figure 3-6. It is important to note that although there is evidence for a "preferred" order of the genetic changes leading from benign adenoma to carcinoma, it is the total accumulation of these changes rather than their specific sequence that is important in the initiation and progression of the neoplastic process.[93,94] For example, more than 90% of the colon carcinomas studied had two or more genetic alterations, and many had four or five alterations, whereas only 7% of early adenomas had more than one of the four component genetic alterations observed in the cancer cells.[93] As early adenomas progressed to intermediate and late adenoma stages, the number of tumors with more than one genetic change went to 25% and 49%, respectively. Some late-stage adenomas had all four of the common genetic mutations, indicating that probably more than these four events are required for progression to malignancy. Another interesting point is that most if not all colorectal cancers appear to have similar (though not necessarily identical) mutational changes irrespective of hereditary factors or environmental exposures.[93]

The high propensity of patients with the autosomally dominant familial adenomatous polyposis syndrome (FAP) to develop colorectal carcinomas has provided an important clue to the predisposing genetic alterations that are involved in the initiation and progression of this type of cancer. FAP is present in about 1 in 10,000 individuals in the United States. Affected individuals may develop hundreds or thousands of adenomatous polyps in their colons and rectums. Only a small percentage of these go on to become malignant, consistent with the multihit hypothesis of cancer discussed earlier. The inherited, affected gene present in FAP has now been cloned and is called *APC* (adenomatous polyposis coli) (reviewed in Ref. 94). The *APC* gene is mutated in the germ line of FAP patients. Most often, these are point mutations or frameshift mutations and, in a few cases, gene deletions. In any case, all the inherited mutations lead to inactivation of the *APC* gene, which is located on chromosomal band 5q21. These data suggest that *APC* is a tumor suppressor gene.

That the *APC* gene is also involved in nonhereditary, sporadic cases of colorectal cancer is supported by the fact that allelic losses of 5q21 are seen in 35% to 60% of patients with no known familial predisposition. The data support the concept that one or both alleles of the *APC* gene are lost or inactivated at an early stage in the development of colorectal cancer, since 5q allelic losses are detected often in small, early adenomas.[95]

Another gene locus that maps to chromosome band 5q21 and is altered in colorectal cancer is called *MCC* (mutated in colorectal cancer). This *MCC* gene appears to be inactivated in about 15% of sporadic colorectal cancers.[96] Loss or inactivation of the *APC* and *MCC* genes may be involved in increasing the rate of proliferation of the affected mucosal cells, thus increasing the chance for subsequent genetic abnormalities to occur.

Another early change that occurs in the ge-

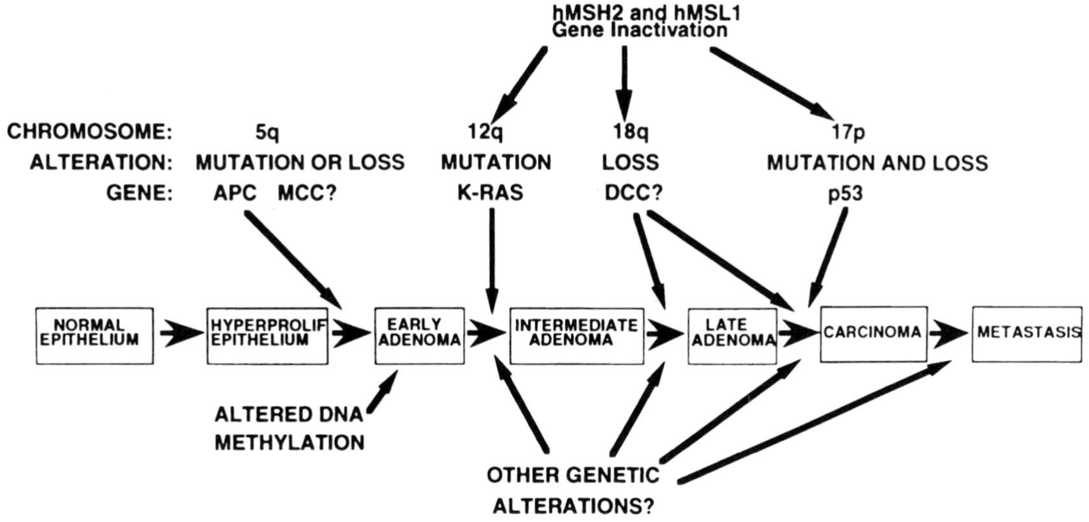

Figure 3-6 A genetic model for colorectal tumorigenesis. Tumongenesis proceeds through a series of genetic alterations involving oncogenes (*ras*) and tumor suppressor genes (particularly those on chromosomes 5q, 17p, and 18q). The three stages of adenomas, in general, represent tumors of increasing size, dysplasia, and villous content. In patients with familial adenomatous polyposis (FAP), a mutation on chromosome 5q is inherited. This alteration may be responsible for the hyperproliferative epithelium present in these patients. In tumors arising in patients without polyposis, the same region may also be lost and/or mutated at a relatively early stage of tumorigenesis. Hypomethylation is present in very small adenomas in patients with or without polyposis, and this alteration may lead to aneuploidy, resulting in the loss of suppressor gene alleles. *ras* gene mutation (usually K-*ras*) appears to occur in one cell of a preexisting small adenoma and through clonal expansion produces a larger and more dysplastic tumor. The chromosomes most frequently deleted include 5q, 17p, and 18q; the putative target of the loss event (i.e., the tumor suppressor gene) on each chromosome is indicated as well as the relative timing of the chromosome loss event. Allelic deletions of chromosome 17p and 18q usually occur at a later stage of tumorigenesis than do deletions of chromosome 5q or *ras* gene mutations. However, the order of these changes is not invariant, and accumulation of these changes, rather than their order with respect to one another, seems most important. Tumors continue to progress once carcinomas have formed, and the accumulated loss of suppressor genes on additional chromosomes correlates with the ability of the carcinomas to metastasize and cause death. (From Fearon and Vogelstein.[93])

nome of early adenomas is loss of methyl groups from DNA. About one-third of DNA sequences from adenomas show loss of methyl groups.[97] Hypomethylation has been correlated with activation of silent genes and with alterations of chromosome condensation (see Chap. 5). The latter may lead to mitotic nondisjunction, resulting in gain or loss of chromosomes. Activation of genes involved in cell proliferation, such as c-*myc* and tyrosine kinase gene products, is also higher in colorectal cancer than in normal colonic mucosa. This would also contribute to the hyperproliferative state resulting from inactivation of tumor suppressor genes.

Activation of the *ras* protooncogenes has been implicated in a number of human cancers, including colorectal carcinomas (see also Chap. 7). Mutations in the K-*ras* gene have been identified in about 50% of colonic adenomas larger than 1 cm in diameter and in 50% of colorectal carcinomas, but in only about 10% of adenomas smaller than 1 cm.[95] Such *ras* mutations, usually in codons 12 or 13, tend to be observed in more dysplastic adenomas and are thought to be related to conversion of cells bearing these mutations into carcinomas.

Allelic losses—sometimes called loss of heterozygosity (LOH) because they are detected by RFLPs—of additional tumor suppressor genes appear to be key to later stages in colorectal tumorigenesis (Fig. 3-6). These losses usually involve only one of the two parental alleles; and the other allele may or may not be inactivated by mutation (see below). These lost genetic loci are thought to contain tumor suppressor genes. Sequentially, the first locus that appears to be lost in colorectal tumor progression is chromosome 18q. This region is deleted in 50% of late adenomas and more than 70% of carcinomas.[95] A candidate tumor suppressor gene located in the 18q band, *DCC* (*d*eleted in *c*olorectal *c*arcinoma),[93] encodes a protein with putative cell adhesion properties. Its expression is reduced or absent in colorectal carcinomas, suggesting its involvement in normal cell-cell or cell–extracellular matrix interactions required to maintain a normal state of differentiation (see Chap. 5).

The most common allelic deletion seen in colorectal cancer involves a large portion of the p region of chromosome 17. Loss of part of 17p is infrequently observed in adenomas at any stage, but it is observed in over 75% of colorectal carcinomas. Furthermore, in several patients in whom tumor progression was observed, 17p allelic losses correlated with progression of individual tumors from adenoma to carcinoma.[93,95] The common segment lost from 17p is now known to contain the p53 tumor suppressor gene.[98] As will be discussed in Chapter 8, loss and/or mutation of p53 alleles has now been observed in a wide variety of human cancers.

Point mutation of one p53 allele and loss of the remaining normal (wild-type) allele is a common finding in colorectal cancer, suggesting that the wild-type p53 gene inhibits later events in colorectal carcinogenesis; when it is lost, this control is abrogated. In some tumors, however, an intermediate stage may occur: i.e., tumor cells may express one mutant and one wild-type p53 gene. Even in this situation, however, there is evidence that the mutant gene product can overwhelm the normal gene product by binding to it or otherwise blocking the function of the normal p53 protein, the so-called dominant negative effect noted above.[93] Nevertheless, the data indicate that colorectal carcinomas with 17p allelic losses are more aggressive than those without them.[99]

Allelic losses in addition to those noted for chromosomes 5q, 18q, and 17p have been observed in colorectal cancer. These include regions from chromosomes 1q, 4p, 6q, 8p, 9q, and 22q.[93,100] One or more of these losses is seen in 25% to 50% of patients. On the average, colorectal cancers contain four or five allelic losses. Patients with more than that number generally have a poorer prognosis. The majority of colorectal carcinomas have 17p and 18q deletions, suggesting that the suppressor genes contained in those bands are crucial to the carcinogenic process. Other regions are lost in a more heterogeneous manner, suggesting that they may represent suppressor genes, any combination of which, if lost, could result in further clonal expansion and uncontrolled proliferation of the affected cells. This heterogenicity in allelic loss patterns could account for some of the observed differences in the biological properties and clinical course of individual patient's tumors.

As noted above, defects in mismatch DNA repair genes have also been detected in colon cancer. The first one discovered was the hMSH2 gene, homologous to the bacterial *mut* S repair gene, found on human chromosome 2p. This genetic defect was found in 60% of cases of hereditary nonpolyposis colon cancer (HNPCC). A second mismatch repair gene defect has also now been found. This gene is located on chromosome 3p21 and appears to be altered in 30% of HNPCC cases.[100a,100b] The gene, called hMLH1, is homologous to the *mut* L mismatch repair gene of *E. coli.* These data strongly suggest that, in addition to activation of oncogenes and inactivation of tumor suppressor genes, defects in DNA repair account for a major predisposing cause of human cancer.

In sum, the data indicate that although there is an overall sequence of genetic alterations in colorectal carcinogenesis, it is the progressive *accumulation* of these alterations that is the most consistent property of colorectal cancers.

Other Gastrointestinal Cancers

Esophageal Cancers

Esophageal adenocarcinomas and squamous cell carcinomas show some of the same genetic alterations observed in colorectal carcinomas, particularly with respect to deletions (observed by loss of heterozygosity) of regions containing putative tumor suppressor genes. In one series of 46 squamous cell carcinomas and 26 adenocarcinomas, LOH occurred in the p53 locus in 55% of cases.[101] Loss of heterozygosity also occurred in 48% of cases at the *RB* gene locus, 66% at *APC*, 63% at *MCC*, and 24% at *DCC*. Of all esophageal tumors for which data were obtainable by RFLP ("informative" samples), 93% had LOH on at least one gene locus and 71% had LOH at more than one locus. Thus, just as for colorectal cancer, accumulation of multiple allelic deletions involving tumor suppressor genes appears to be important for the development and progression of esophageal cancers.

This theme occurs again and again in human cancers as they are studied and, indeed, is a consistent finding in much of human neoplastic disease, contributing to the two-edged sword of oncogene activation and tumor suppressor gene inactivation. Examples of this loss of genes having putative suppressor activity are the 5q, 17p, and 18q deletions noted above for colon and esophageal cancer; the 3p, 13q, and 17p deletions in small cell lung carcinoma; and the 13q, 16q, and 17p deletions in breast cancer.[101]

Gene amplifications have also been observed in esophageal adenocarcinomas, including *HER-2/neu*, c-*met*, and c-*erb*-B,[102] although these were observed in a small percentage of the total samples examined. One difference between colorectal and esophageal cancer is the apparent absence of H-, K-, and N-*ras* mutation in the latter.[103]

Gastric Carcinoma

In a series of 48 human gastric carcinomas, LOH of 17p, containing the p53 gene, was observed in 68% of informative cases.[104] In advanced cancers, LOH was observed on chromosomes 1q (67%), 5q (36%), 7p (33%), 7q (39%), and 17p (73%). Allelic losses were not detected at 1q, 5q, and 7p in poorly differentiated adenocarcinomas, suggesting that these cancers have a somewhat different evolutionary history than well-differentiated gastric carcinomas. Loss of heterozygosity of chromosome 17p, however, was common to all histologic types of gastric carcinoma.

Certain oncogenes are also mutated or overexpressed in gastric carcinomas. There is evidence for a rearrangement and fusion of the *TPR* locus (translocated promoter region) of chromosome 1 to the 5′ region of the *met* protooncogene on chromosome 7.[105] This *TPR-met* rearrangement was first observed in a human osteosarcoma cell line transformed in vitro with *N*-methyl-*N*′-nitro-*N*-nitrosoguanidine (MNNG). Since MNNG is a potent carcinogen that induces gastric tumors in experimental animals, it is significant that this translocation was also found in human gastric carcinomas.

Hepatocellular Carcinoma

A number of genetic alterations have been reported in hepatocellular carcinomas (HCC), perhaps the most interesting of which are losses and mutations of the p53 gene. The data are interesting because of the association of environmental factors such as infection with hepatitis B virus (HBV) and aflatoxin in causing HCC. Thus, the alterations seen in HCC give a clue how viruses and carcinogenic chemicals might specifically target part of the human genome to cause cancer.

Mutations of the p53 gene have been observed in both HBV-antigen-positive and HBV-antigen-negative patients and in both high and low aflatoxin exposure areas (aflatoxin is elaborated by molds that may contaminate peanuts, rice, and other grains). In studies done in China, Taiwan, and the United States, a higher percentage of HBV-antigen-positive patients were found to have p53 gene mutations (37% to 60% versus 25 to 30%).[106,107] Thus, a significant number of HBV-antigen-negative HCC patients also had p53 mutations. Also, in a study done in Japan, only 30% of HCC patients' tumors had p53 gene mutations, and there was no correlation with hepatitis virus infection, sex, age, or underlying liver disease.[108] But p53 mutations were more prevalent in moderately to poorly differentiated HCCs and in association with allelic loss of the other p53 locus. Similarly, although some increase in p53 mutations has been seen in HCCs from patients living in high-aflatoxin-exposure areas compared to low-exposure areas, the difference was only 25% versus 12%, respectively.[109] Taken together, the data suggest that p53 mutations occur late in the disease rather than as an inciting cause and that they are associated with tumor progression.

Loss of heterozygosity on chromosome 16 has also been reported in HCC (52% of informative cases)[110] and occurred more frequently in larger, poorly differentiated, metastatic HCCs. Again, there was no particular correlation with hepatitis B or C virus infection.

Breast Cancer

Sporadic (nonhereditary) breast cancers have shown LOH at one or more of a large number of chromosomal loci, including 1q, 3p, 6q, 16q, 17p, and 18q in 50% or more tumors and at 1p, 7q, 8q, 9q, 11p, 13q, 15q, 17q, and 22q in about 30% of tumors (reviewed in Refs. 111 to 113). These data suggest that a number of tumor suppressor genes are operative in breast tissue.

The p53 gene appears to be affected in both hereditary and sporadic breast cancers, since there is evidence for mutation and/or allelic loss in both types. However, germ-line mutations of p53 appear to be rare in women who develop sporadic breast cancer, even if at an early age, except for those women who have inherited the Li-Fraumeni syndrome trait.[112,114] This suggests that a different multiplicity of factors is at work in breast cancers caused by germ-line mutations of p53 than in those that occur sporadically or by other kinds of heritable mutations.

One such heritable trait for breast cancer susceptibility maps to chromosome 17q21 by RFLP analysis.[115] A specific locus on chromosome 17q has been identified that is associated with a high incidence of familial breast cancer and familial breast/ovarian cancer (reviewed in Ref. 115a). This locus, called BRCA1, is linked to about 45% of affected families with breast cancer only and virtually 100% of breast cancer families that had at least one case of ovarian cancer. These analyses were based on studies of 214 families collected worldwide by the Breast Cancer Linkage Consortium.[115a] An intense hunt for the genes that are affected in the familial breast cancer cases has been under way in several laboratories. This has led to the identification of two such genes, the BRCA1 gene on chromosome 17q[115b] and the BRCA2 gene on chromosome 13q.[115c] The BRCA1 gene is a strong candidate for the breast and ovarian cancer susceptibility gene. It is expressed in numerous tissues, including breast and ovary, and encodes a putative protein of 1863 amino acids, which contains a zinc-finger domain, suggesting it is a DNA-binding protein. BRCA1 mutations, however, do not appear to be critical for the development of the majority of breast and ovarian cancers, which do not have germ-line mutations of this gene. The BRCA2 gene mutations appear to confer a high risk for breast, but not for ovarian cancers. Although it appears that a high number of currently identified high-risk families have mutations in either the BRCA1 or BRCA2 genes, hereditary breast cancer accounts for only about 5% of all cases.[115c] Thus, it is likely that additional gene mutations remain to be discovered.

Accumulation of some of the above genetic alterations correlates with breast cancer progression and metastasis. For example, frequent LOH of chromosomes 3p, 13q, 16q, 17p, and 17q was observed in a study of 219 breast cancers, and those with multiple involvement of these loci had evidence of more aggressive growth patterns based on tumor size, grade, and lymph node metastasis.[116] In another study, LOH of 11p and 17p correlated with a significantly higher incidence of regional lymph node metastasis than tumors without these allelic losses.[117] In addition, patients with tumors that retained both 11p and 17p had a much lower incidence of metastasis than those whose tumors had deletions of both 11p and 17p.

Several oncogenes have also been implicated in breast cancer, including c-*myc*, *HER-2/neu* (c-*erb*-B-2), and *int*-2. These genes have been reported to be amplified in a significant number of breast cancers (reviewed in Ref. 111). Some groups have reported a correlation of *HER-2/neu* amplification and poor prognosis,[118] whereas other studies have not found this to be the case.[119]

Overall, the data for breast cancer support the concept stated for other solid tumors—that is, that multiple but not identical genetic alterations can occur leading to development of a malignancy; that the more genetic alterations there are, the more likely the disease is to have progressed to a higher state of malignancy; and that combinations of tumor suppressor gene deletions and oncogene activation appear to be the most deadly in causing tumor progression and metastasis.

Lung Cancer

Chromosomal deletions and/or LOH of chromosomal regions 3p, 6p, 6q, 9p, 11p, 13q, and

17p have been reported in lung cancer (reviewed in Refs. 120 and 120a). A high frequency of inactivating abnormalities of the tumor suppressor genes *RB* (on 13q) and p53 (on 17p) has been observed. D'Amico et al.[120] have also reported allelic loss of chromosome 5q21 in 40% of informative non–small cell lung carcinomas and 80% of small cell lung carcinomas (SCLC). Interestingly, this is the same region lost in colorectal carcinoma and contains the *APC* and *MCC* putative tumor suppressor genes described above. In another series of SCLC samples, chromosomal losses of 3p, 5q, and 17p were also observed, as well as amplification of c-*myc*1 or N-*myc*.[121]

Again, as commonly seen in other solid tumors, the tumor suppressor gene p53 is frequently altered in lung cancers of all histologic types. In one series, structural abnormalities of the p53 gene were observed in 8 of 14 large cell lung carcinomas, 24 of 58 adenocarcinomas, 25 of 37 squamous cell carcinomas, and 3 of 6 adenosquamous carcinomas.[122] DNA sequencing revealed 45 single base substitutions, 9 deletions or insertions of short nucleotide sequences, and 3 two-base substitutions in 57 tumors examined. Of the mutations observed, 80% occurred in evolutionarily conserved regions, many of which are "hot spots" for mutations in other human tumors (see Chap. 7). Allelic loss of p53 ocurred in 14 of 43 informative cases and, in 11 of the 14, the remaining p53 allele was mutated. Alterations of p53 were observed in a variety of histologic tumor types and clinical stages. Mutations of the *ras* gene were observed in a frequency similar to mutations of the p53 gene (46%).

There is also evidence for Mendelian inheritance of susceptibility to lung cancer. Since not everyone who regularly smokes cigarettes gets lung cancer, it has been suggested for some time that some genetic susceptibility or predisposition may be involved. Indeed, as discussed in Chapter 6, there are some genetic polymorphisms associated with the enzymatic pathways that activate the chemical carcinogens found in cigarette smoke. In a study of 337 families that had members with lung cancer, the results suggested a Mendelian codominant inheritance of an autosomal gene that induces susceptibility to lung cancer at an early age.[123]

Urinary Tract Cancers

Studies of human uroepithelial and renal cell carcinoma cell lines have shown nonrandom chromosomal losses of 10 chromosome arms: 1p, 1q, 3p, 5q, 6q, 9q, 11p, 13q, 17p, and 18q.[124,125] Loss of 3p and 18q both correlated with transformation to high-grade uroepithelial carcinomas.[124]

Chromosomal deletions frequently observed in urinary bladder carcinomas include 9q, 11p, and 17p.[126] The high frequency of allelic loss of 9q is somewhat unusual in that it has not often been seen in other human solid tumors, suggesting that this region may harbor a tumor suppressor gene that is particularly important in bladder epithelial cell proliferation. The data from some studies[126] indicate that loss of chromosome 9q is associated with all grades of bladder cancer and that LOH of 17p is most frequently seen in grade III or IV tumors, indicating a greater correlation of p53 alterations with progressive disease. Decreased expression of the tumor suppressor gene *RB* has also been observed in locally advanced bladder cancer,[127] and decreased *RB* expression correlated with poor survival in a study that found altered *RB* expression in 13 out of 38 patients with invasive tumors versus only one of 10 patients with superficial carcinomas.[128]

Because bladder epithelial cells are shed into the urine to some extent and are easily accessible for biopsy by endoscopy, the study of genetic defects in bladder cancer is facilitated. For example, p53 mutations have been observed in exfoliated cells from patients with bladder cancer.[129] This suggests that such assays could be employed to detect invasive bladder cancer.

Ovarian Cancer

A few differences and a number of similarities to chromosomal alterations in other human solid tumors have been observed in human ovarian carcinomas. For example, in one study of 14 informative cases, 64% had allelic loss at the estrogen receptor locus of chromosome 6q; 75% had LOH at the 17p locus; and 46% had LOH at the H-*ras*-1 locus of chromosome 11p.[130] The LOH of the estrogen receptor region on 6q appears to be associated more with ovarian carcinoma than other cancers, but 6q LOH has also been reported in 60% of informative melanomas.[131]

On the other hand, some of the "universal" tumor suppressor genes are lost or inactivated in ovarian carcinoma. In some ovarian tumors, overexpression of a mutated form of the p53 protein has been observed,[132] a phenomenon observed in a number of other cancers as well (see Chap. 8). In another study, allelic loss of the p53 gene was observed in 16 of 20 ovarian carcinomas and mutations of the p53 gene in 9 of 31 ovarian cancers examined.[133] In 8 of 9 informative cases, p53 mutations were accompanied by loss of the normal allele, and these alterations of the p53 gene were observed in stage I through IV disease, suggesting that p53 alterations occur early in the carcinogenic process for ovarian cancer. Allelic loss of the *RB* gene has also been observed in 30% of ovarian cancers.[134]

Prostate Cancer

Carter et al.[135] found that 54 percent (13/24) of clinically localized prostatic carcinomas and 4 out of 4 metastatic carcinomas had allelic loss of at least one of the following chromosomal loci: 3p, 7q, 9q, 10q, 11p, 13q, 16q, 17p, and 18q. Chromosome loci 10q and 16q had the highest frequency of allelic loss, with 30% of prostate cancers showing these defects, suggesting that 10q and 16q contain tumor suppressor genes important for regulation of prostatic cell proliferation.

Losses of a portion of chromosome 11p in prostate cancer are close to but do not include the Wilms' tumor gene locus *WT*-1.[136] A provocative additional finding in this latter study was that introduction of human chromosome 11 into highly metastatic rat prostate cells by somatic cell hybridization suppressed the metastatic potential of the hybrid cells without suppression of in vivo growth or tumorigenicity of the cells.

The p53 gene is also altered in prostatic cancer. A high level of p53 expression, presumably of a mutated form, has been reported for prostatic carcinomas of high histologic grade with chromosomal aneuploidy and a high cell proliferation rate.[137] Patients with high levels of p53 in their tumors had a significantly poorer survival rate.

CONCLUSIONS

By now the reader is getting the point that multiple genetic lesions are associated with individual human cancers and that many of these defects show up consistently in cancers of very different tissue types. This leads to the conclusion that there are families of tumor suppressor genes, or perhaps what are more appropriately called growth regulatory genes, located on different chromosomes and probably activated in different cell types at different stages of their embryonic development and/or in their tissue renewal stem cells at different stages of their growth and differentiation phases. Thus, in one tissue type, the p53 gene may be important in an early step of stem cell proliferation and differentiation, whereas in other tissues, it may be more important at a later step. This may explain why the p53 gene seems to undergo allelic loss early in the tumorigenesis of one type of cancer (e.g., ovarian carcinoma) but later in another type of cancer (e.g., prostatic cancer). This may also be the case for other tumor suppressor genes, not all of which have yet been identified. Their multiplicity is strongly suggested by the large number of allelic losses that show up consistently in cancer cells.

The other side of the coin is the activation or mutation of cellular oncogenes. More is said about this in Chapter 7; suffice it to say here that oncogene activation is also a frequent phenomenon in human cancer and provides the second edge of the two-edged sword of uncontrolled cell proliferation and loss of ability of cells to differentiate. The potency of these two events occurring simultaneously or sequentially in cells has been shown by experiments in which cotransfection of the *ras* oncogene and a mutated p53 gene into rat cells induced their malignant transformation even if the wild-type p53 gene was still expressed.[138]

One key question is: Which of these events are causes and which are effects of the carcinogenic process? Presumably, those genetic effects that occur early in the carcinogenic process and consistently in a high percentage of cases are associated with the underlying causes of the disease, while those that occur later are associated with progression of a tumor cell into a more invasive, metastatic cancer. Because of the proliferative advantage and higher rate of cell division of cancer cells compared to their normal counterparts and because of the genetic instability of transformed cells (see Chap. 4), additional genetic defects are likely to accumulate in malignant cells as they evolve into more aggressive cancers. Some of the genetic changes that occur may just "be along for the ride," without being involved in a crucial way in the carcinogenic process. If such genes, for example, are located in chromosomal regions that are translocated, deleted, or amplified in cancer cells, their expression could be increased or decreased. This might partly explain the production of certain "ectopic" proteins by cancer cells. These gene products may have nothing to do directly with the cancer process—for example, the production of hCG-β subunit by a variety of nongonodal tumors.[139]

To all of this one must now add the findings coming out of studies of hereditary susceptibility genes. A number of such genes have been found (e.g., the germ-line mutations of p53 in Li-Fraumeni syndrome and the *APC* gene in familial polyposis). Others are yet to be identified. One point that is becoming clearer is that the ability to repair DNA is crucial to protecting the genome from carcinogenic damage. Several genes show up in this category, including p53, which stops cells in the cell cycle until DNA is repaired or targets cells for death if DNA is too damaged to be repaired before the next cell division. There are also the DNA mismatch repair genes hMSH2 and hMLH1. Thus, it is likely that a concatenation of events involving several gene types can lead to the loss of cellular control that produces cancer.

REFERENCES

1. T. Boveri: *Zu Frage der Erstehung Maligner Tumoren.* Jena: Fisher, 1914.
2. P. C. Nowell and D. A. Hungerford: Chromosome studies on normal and leukemic human leukocytes. *JNCI* 25:85, 1960.
3. J. D. Rowley: A new consistent chromosomal abnormality in chronic myelogenous leukemia identified by quinacrine flourescence and Giemsa staining. *Nature* 243:290, 1973.
4. A. A. Sandberg and D. K. Hossfeld: Chromosomes in the pathogenesis of human cancer and leukemia, in J. F. Holland and E. Frei III, eds.: *Cancer Medicine.* Philadelphia: Lea & Febiger, 1973, pp. 151–177.
5. T. Caspersson, L. Zech, and C. Johanssen: Differential binding of alkylating fluorochromes in human chromosomes. *Exp Cell Res* 60:315, 1970.
6. J. P. Chaudhuri, W. Vogel, I. Voiculescu, and U. Wolf: A simplified method of demonstrating Giemsa band pattern in human chromosomes. *Humangenetik* 14:83, 1971.
7. J. D. Rowley: Mapping of human chromosomal regions related to neoplasia: Evidence from chromosomes 1 and 17. *Proc Natl Acad Sci USA* 74:5729, 1977.
8. E. Solomon, J. Borrow, and A. D. Goddard: Chromosome aberrations and cancer. *Science* 254:1153, 1991.
9. J. D. Rowley: Molecular cytogenetics: Rosetta Stone for Understanding Cancer—Twenty-Ninth G.H.A. Clowes Memorial Award Lecture. *Cancer Res* 50:3816, 1990.
10. R. Dalla-Favera, M. Bregni, J. Erikson, D. Patterson, R. C. Gallo, and C. M. Croce: Human c-*myc onc* gene is located on the region of chromosome 8 that is translocated in Burkitt lymphoma cells. *Proc Natl Acad Sci USA* 79:7824, 1982.
11. R. Taub, I. Kirsch, C. Morton, G. Lenoir, D. Swam, S. Aaronson, and P. Leder. Translocation of the c-*myc* gene into the immunoglobulin heavy chain locus in human Burkitt lymphoma and murine plasmacytoma cells. *Proc Natl Acad Sci USA* 79:7837, 1982.
12. F. Mitelman and G. Levan: Clustering of aberrations to specific chromosomes in human neoplasms: II. A survey of 287 neoplasms. *Hereditas* 82:167, 1979.
13. A. Hagemeijer, D. Bootsma, N. K. Spurr, N. Heisterkamp, J. Groffen, and J. R. Stevenson: A cellular oncogene is translocated to the Philadelphia chromosome in chronic myelocytic leukemia. *Nature* 300:765, 1982.
14. N. Heisterkamp, J. R. Stephenson, J. Groffen, P. F. Hansen, A. deKlein, C. R. Bartram, and G. Grosveld: Localization of the c-*abl* oncogene adjacent to a translocation breakpoint in chronic myelocytic leukemia. *Nature* 306:239, 1983.
15. L. C. Chan, K. K. Karhi, S. I. Rayter, N. Heisterkamp, S. Eridani, R. Powles, S. D. Lawler, J. Graffen, J. G. Foulkes, M. F. Greaves, and L. M. Wiedemann: A novel *abl* protein expressed in Philadelphia chromosome positive acute lymphoblastic leukemia. *Nature* 325:635, 1987.

15a. M. O. Diaz, M. M. Le Beau, P. Pitha, and J. D. Rowley: Interferon and c-*ets*-1 genes in the translocation (9; 11) (p 22; q23) in human acute monocytic leukemia. *Science* 231:265, 1986.
16. T. G. Lugo, A.-M. Pendergast, A. J. Muller, and O. N. Witte: Tyrosine kinase activity and transformation potency of *bcr-abl* oncogene products. *Science* 247:1079, 1990.
17. Y. Tsujimoto, J. Yunis, L. Onorato-Showe, J. Erickson, P. C. Nowell, and C. M. Croce: Molecular cloning of the chromosomal breakpoint of B-cell lymphomas and leukemias with the t(11; 14) chromosome translocation: *Science* 224:1403, 1984.
18. Y. Haupt, W. S. Alexander, G. Barri, S. P. Klinken, and J. M. Adams: Novel zinc finger gene implicated as *myc* collaborator by retrovirally accelerated lymphomagenesis in Eμ-*myc* transgenic mice. *Cell* 65:753, 1991.
19. Y. Tsujimoto, J. Gorham, J. Cossman, E. Jaffe, and C. M. Croce: The t(14; 18) chromosome translocations involved in B-cell neoplasms result from mistakes in VDJ joining. *Science* 229:1390, 1985.
20. G. Russo, M. Isobe, R. Gatti, J. Finan, O. Batuman, K. Huebner, P. C. Nowell, and C. M. Croce: Molecular analysis of a t(14; 14) translocation in leukemic T-cells of an ataxia telangiectasia patient. *Proc Natl Acad Sci USA* 86:602, 1989.
21. J. Limpens, D. de Jong, J. H. J. M. van Krieken, C. G. A. Price, B. D. Young, G.-J. B. van Ommen, and P. M. Kluin: *Bcl-2*/J_H rearrangement in benign lymphoid tissues with follicular hyperplasia. *Oncogene* 6:2272, 1991.
22. J. J. Yunis: The chromosomal basis of human neoplasia. *Science* 221:227, 1983.
23. E. R. Fearon, A. P. Feinberg, S. H. Hamilton, and B. Vogelstein: Loss of genes on the short arm of chromosome 11 in bladder cancer. *Nature* 318:377, 1985.
24. J. J. Wasmuth, C. Park, and R. E. Ferrell: Report of the committee on the genetic constitution of chromosome 5. *Cytogenet Cell Genet* 51, 137, 1989.
25. R. Sager, I. K. Gadi, L. Stephens, and C. T. Grabowy: Gene amplification: An example of an accelerated evolution in tumorigenic cells. *Proc Natl Acad Sci USA* 82:7015, 1985.
26. O. S. Frankfurt, J. L. Chin, L. S. Englander, W. R. Greco, J. E. Pontes, and Y. M. Rustum: Relationship between DNA ploidy, glandular differentiation, and tumor spread in human prostate cancer. *Cancer Res* 45:1418, 1985.
27. F. A. Klein, M. W. Herr, W. F. Whitemore, Jr., P. C. Sogani, and M. R. Melamed: Detection and follow-up of carcinoma of urinary bladder by flow cytometry. *Cancer* 50:389, 1982.
28. B. Tribukait, H. Gustafson, and P. L. Espositi: The significance of ploidy and proliferation in the clinical and biological evolution of bladder tumors: A study of 100 untreated cases. *Br J Urol* 54:130, 1982.
29. H. S. Smith, L. A. Liotta, M. C. Hancock, S. R. Wolman, and A. J. Hackett: Invasiveness and ploidy of human mammary carcinomas in short-term culture. *Proc Natl Acad Sci USA* 82:1805, 1985.
30. N. Wake, J. Isaacs, and A. A. Sandberg: Chromosomal changes associated with progression of the Dunning R-3327 rat prostatic adenocarcinoma system. *Cancer Res* 42:4131, 1982.
31. L. S. Cram, M. F. Bartholdi, F. A. Ray, G. I. Travis, and P. M. Kraemer: Spontaneous neoplastic evolution of Chinese hamster cells in culture: Multistep progression of karyotype. *Cancer Res* 43:4828, 1983.
32. I. Henry, C. Bonaiti-Pellié, V. Chehensee, C. Beldjord, C. Schwartz, G. Utermann, and C. Junien: Uniparental paternal disomy in a genetic cancer-predisposing syndrome. *Nature* 351:665, 1991.
33. A. C. Ferguson-Smith, B. M. Cattanach, S. C. Barton, C. V. Beechey, and M. A. Surani: Embryological and molecular investigations of parental imprinting on mouse chromosome 7. *Nature* 351:667, 1991.
33a. S. Rainier, L. A. Johnson, C. J. Dobry, A. J. Ping, P. E. Grundy, and A. P. Feinberg: Relaxation of imprinted genes in human cancer. *Nature* 362:747, 1993.
34. B. Alberts, D. Bray, J. Lewis, M. Raff, K. Roberts, and J. D. Watson: *Molecular Biology of the Cell.* New York: Garland Publishing, 1983, pp 214–215.
35. R. I. Richards and G. R. Sutherland: Dynamic mutations: A new class of mutations causing human disease. *Cell* 70:709, 1992.
36. E. J. Kremer, M. Pritchard, M. Lynch, S. Yu, K. Holman, E. Baker, S. T. Warren, D. Schlessinger, G. R. Sutherland, and R. I. Richards: Mapping of DNA instability at the fragile X to a trinucleotide repeat sequence p(CCG)n. *Science* 252:1711, 1991.
37. Y.-H. Fu, D. P. A. Kuhl, A. Pizzuti, M. Pieretti, J. S. Sutcliffe, S. Richards, A. J. M. H. Verkerk, J. J. A. Holder, R. G. Fenwick, Jr., S. T. Warren, B. A. Oostra, D. L. Nelson, and C. T. Caskey: Variation of the CGG repeat at the fragile X site results in genetic instability: Resolution of the Sherman paradox. *Cell* 67:1047, 1991.
38. The Huntington's Disease Collaborative Research Group: A novel gene containing a trinucleotide repeat that is expanded and unstable on Huntington's disease chromosomes. *Cell* 72:971, 1993.
39. M. Pieretti, F. Zhan, Y.-H. Fu, S. T. Warren, B. A. Ooostra, C. T. Caskey, and D. L. Nelson: Absence of expression of the FMR-1 gene in fragile X syndrome. *Cell* 66:817, 1991.

39a. C. J. Yee, N. Roodi, C. S. Verrier, and F. F. Parl: Microsatellite instability and loss of heteroxygosity in breast cancer. *Cancer Res* 54:1641, 1994.

39b. A. Merlo, M. Mabry, E. Gabrielson, R. Vollmer, S. B. Baylin, and D. Sidransky: Frequent microsatellite instability in primary small cell lung cancer. *Cancer Res* 54:2098, 1994.

39c. V. Shridhar, J. Siegfried, J. Hunt, M. del Mar Alonso, and D. I. Smith: Genetic instability of microsatellite sequences in many non-small cell lung carcinomas. *Cancer Res* 54:2084, 1994.

39d. M. Gonzalez-Zulueta, J. M. Ruppert, K. Tokino, Y. C. Tsai, C. H. Spruck III, N. Miyao, P. W. Nichols, G. G. Hermann, T. Horn, K. Steven, I. C. Summerhayes, D. Sidransky, and P. A. Jones: Microsatellite instability in bladder cancer. *Cancer Res* 53:5620, 1993.

39e. N. M. Mironov, M. A.-M. Aguelon, G. I. Potapova, Y. Omori, O. V. Gorbunov, A. A. Klimenkov, and H. Yamasaki: Alterations of $(CA)_n$ DNA repeats and tumor suppressor genes in human gastric cancer. *Cancer Res* 54:41, 1994.

39f. H. T. Lynch, T. C. Smyrk, P. Watson, S. J. Lanspa, J. F. Lynch, P. M. Lynch, R. J. Cavalieri, and C. R. Boland: Genetics, natural history, tumor spectrum and pathology of hereditary nonpolyposis colorectal cancer: An updated review. *Gastroenterology* 104:1535, 1993.

39g. R. Fishel, M. K. Lescoe, M. R. S. Rao, N. G. Copeland, N. A. Jenkins, J. Garber, M. Kane, and R. Kolodner: The human mutator gene homolog MSH2 and its association with hereditary nonpolyposis colon cancer. *Cell* 75:1027, 1993.

39h. R. Parson, G.-M. Li, M. J. Longley, W-H. Fang, N. Papadopoulos, J. Jen, A. de la Chapelle, K. W. Kinzler, B. Vogelstein, and P. Modrich: Hypermutability and mismatch repair deficiency in RER^+ tumor cells. *Cell* 75:1227, 1993.

40. M. O. Diaz, M. M. LeBeau, P. Pitha, and J. D. Rowley: Interferon and c-*ets*-1 genes in the translocation (9; 11)(p22; q23) in human acute monocytic leukemia. *Science* 231:265, 1986.

41. P. D'Eustachio and F. H. Ruddle: Somatic cell genetics and gene families. *Science* 220:919, 1983.

42. R. White, M. Leppert, D. T. Bishop, D. Barker, J. Berkowitz, C. Brown, P. Callahan, T. Holm, and L. Jerominski: Construction of linkage maps with DNA markers for human chromosomes. *Nature* 313:101, 1985.

42a. D. Cohen, I. Chumakov, and J. Weissenbach: A first-generation physical map of the human genome. *Nature* 366:698, 1993.

43. W. H. Brown: A case of pluriglandular syndrome—Diabetes of bearded women. *Lancet* 2:1022, 1928.

44. G. W. Liddle, J. R. Givens, W. E. Nicholson, and D. P. Island: The extopic ACTH syndrome. *Cancer Res* 25:1057, 1965.

45. L. H. Rees and J. G. Ratcliffe: Ectopic hormone production by non-endocrine tumours. *Clin Endocrinol* 3:263, 1974.

46. R. S. Yalow, C. E. Eastridge, G. Higgins, Jr., and J. Wolf: Plasma and tumor ACTH in carcinoma of the lung. *Cancer* 44:1789, 1979.

47. R. S. Yalow and S. A. Berson: Characteristics of "big ACTH" in human plasma and pituitary extracts. *J Clin Endocrinol Metab* 36:415, 1973.

48. G. Gerwitz and R. S. Yalow: Ectopic ACTH production in carcinoma of the lung. *J Clin Invest* 53:1022, 1974.

49. A. B. Gutman, T. L. Tyson, and E. B. Gutman: Serum calcium, inorganic phosphorus and phosphatase activity. *Arch Intern Med* 57:379, 1936.

50. T. B. Connor, W. C. Thomas, Jr., and J. E. Howard: The etiology of hypercalcemia associated with lung carcinoma. *J Clin Invest* 35:697, 1956.

51. A. H. Tashjian, Jr., L. Levine, and P. L. Munson: Immunochemical identification of parathyroid hormone in non-parathyroid neoplasms associated with hypercalcemia. *J Exp Med* 119:467, 1964.

52. L. M. Sherwood, J. L. H. O'Riordan, G. D. Aurbach, and J. T. Potts, Jr.: Production of parathyroid hormone by non-parathyroid tumors. *J Clin Endocrinol Metab* 27:140, 1967.

53. W. A. Winkler and O. F. Crankshaw: Chloride depletion in conditions other than Addison's disease. *J Clin Invest* 17:1, 1938.

54. W. B. Schwartz, W. Bennett, S. Curelop, and F. C. Barrter: Syndrome of renal sodium loss and hyponatremia probably resulting from inappropriate secretion of antidiuretic hormone. *Am J Med* 23:529, 1957.

55. R. L. Reeves, H. Tesluk, and C. E. Harrison: Precocious puberty associated with hepatoma. *J Clin Endocrinol Metab* 17:1651, 1959.

56. J. L. Vaitukaitis, G. T. Ross, G. D. Braunstein, and P. L. Rayford: Gonadotropins and their subunits: Basic and clinical studies. *Recent Prog Horm Res* 32:289, 1976.

57. M. R. Blackman, B. D. Weintraub, S. W. Rosen, I. A. Kourides, K. Steinwascher, and M. H. Gail: Human placental and pituitary glycoprotein hormones and their subunits as tumor markers: A quantitative assessment. *JNCI* 65:81, 1980.

58. R. W. Ruddon, C. Anderson, K. S. Meade, P. H. Aldenderfer, and P. D. Neuwald: Content of gonadotropins in cultured human malignant cells and effects of sodium butyrate treatment on gonadotropin secretion by HeLa cells. *Cancer Res* 39:3885, 1979.

59. Y. Yoshimoto, A. R. Wolfson, and W. D. Odell: Human chorionic gonadotropin-like substance in nonendocrine tissues of normal subjects. *Science* 197:575, 1977.

60. T. Ono and R. C. Cutler: Age-dependent relaxation of gene repression: Increase of endogenous murine leukemia virus-related and globin-re-

lated RNA in brain and liver of mice. *Proc Natl Acad Sci USA* 75:4431, 1978.
61. A. G. E. Pearse: Common cytochemical and ultrastructural characteristics of cells producing polypeptide hormones (the APUD series) and their relevance to thyroid and ultimobranchial C cells and and calicitonin. *Proc R Soc Lond [Ser B]* 170:71, 1968.
62. T. Ono and R. C. Cutler: Age-dependent relaxation of gene repression: Increase of endogenous murine leukemia virus-related and globin-related RNA in brain and liver of mice. *Proc Natl Acad Sci USA* 75:4431, 1978.
63. W. K. Cavenee, M. Hansen, M. Nordenskjold, E. Kock, I. Maumenee, J. Squire, R. Phillips, and B. Gallie: Genetic origin of mutations predisposing to retinoblastoma. *Science* 228:501, 1985.
64. S. H. Friend, H. R. Bernards, S. Rogelj, R. A. Weinberg, J. M. Rapaport, J. M. Albert, and T. P. Dryja: A human DNA segment with properties of the gene that predisposes to retinoblastoma and osteosarcoma. *Nature* 323:643, 1986.
65. D. A. Haber and D. E. Housman: Rate-limiting steps: The genetics of pediatric cancers. *Cell* 64:5, 1991.
66. D. A. Haber, A. J. Buckler, T. Glaser, K. M. Call, J. Pelletier, R. L. Sohn, E. C. Douglass, and D. E. Housman: An internal deletion within an 11p13 zinc finger gene contributes to the development of Wilms' tumor. *Cell* 61:1257, 1990.
67. V. M. Riccardi, E. Sujansky, A. C. Smith, and U. Francke: Chromosomal imbalance in the aniridia-Wilms' tumor association: 11p interstitial deletion. *Pediatrics* 61:604, 1978.
68. F. P. Li and J. F. Fraumeni, Jr.: Soft-tissue sarcomas, breast cancer, and other neoplasms: A familial syndrome? *Ann Intern Med* 71:747, 1969.
69. J. E. Garber, A. M. Goldstein, A. F. Kantor, M. G. Dreyfus, J. F. Fraumeni, Jr., and F. P. Li: Follow-up study of twenty-four families with Li-Fraumeni syndrome. *Cancer Res* 51:6094, 1991.
70. D. Malkin, F. P. Li, L. C. Strong, J. F. Fraumeni, Jr., C. E. Nelson, D. H. Kim, J. Kassel, M. A. Gryka, F. Z. Bischoff, M. A. Tainsky, and S. H. Friend: Germ line p53 mutations in a familial syndrome of breast cancer, sarcomas, and other neoplasms. *Science* 250:1233, 1990.
71. T. H. Rabbits: Translocations, master genes, and differences between the origins of acute and chronic leukemias. *Cell* 67:641, 1991.
72. D. C. Tkachuk, C. A. Westbrook, M. Andreeff, T. A. Donlon, M. L. Cleary, K. Suryanarayan, M. Homge, A. Redner, J. Gray, and D. Pinkel: Detection of *brc-abl* fusion in chronic myelogeneous leukemia by in situ hybridization. *Science* 250:559, 1990.
73. O. A. Haas, A. Argyriou-Tirita, and T. Lion: Parental origin of chromosomes involved in the translocation t(9; 22). *Nature* 359:414, 1992.
74. N. R. McCabe, R. C. Burnett, H. J. Gill, M. J. Thirman, D. Mbankdollo, M. Kipiniak, E. van Melle, S. Ziemin-van der Poel, J. D. Rowley, and M. O Diaz: Cloning of cDNAs of the *MLL* gene that detect DNA rearrangements and altered RNA transcripts in human leukemic cells with 11q23 translocations. *Proc Natl Acad Sci USA* 89:11794, 1992.
75. Y. Gu, T. Nakamura, H. Alder, R. Prosad, O Canaani, G. Cimino, C. M. Croce, and E. Canaani: The t(4; 11) chromosome translocation of human acute leukemias fuses the ALL-1 gene, related to *Drosophila trithorax,* to the AF-4 gene. *Cell* 71:701, 1992.
76. F. Mitelman: *Catalog of Chromosome Aberrations in Cancer,* 3rd ed. New York: Alan R. Liss, Inc. 1988.
77. A. Kakizuka, W. H. Miller, Jr., K. Umesono, P. R. Warrell, Jr., S. R. Frankel, V. V. V. S. Murty, E. Dmitrovsky, and R. M. Evans: Chromosomal translocation t(15; 17) in human acute promyelocytic leukemia fuses RARα with a novel putative transcription factor, PML. *Cell* 66:663, 1991.
78. H. de Thé, C. Lavau, A. Marchio, C. Chomienne, L. Degos, and A. DeJean: The PML-RARα fusion mRNA generated by the t(15; 17) translocation in acute promyelocytic leukemia encodes a functionally altered RAR. *Cell* 66:675, 1991.
79. A. D. Goddard, J. Borrow, P. S. Freemont, and E. Solomon: Characterization of a zinc finger gene disrupted by the t(15; 17) in acute promyelocytic leukemia. *Science* 254:1371, 1991.
80. D. G. Schatz, M. A. Oettinger, and D. Baltimore: The V(D)J recombination activating gene, RAG-1. *Cell* 59:1035, 1989.
81. M. E. Huang, Y. C. Ye, S. R. Chen, J. R. Chai, J. X. Lu, L. Zhao, L. J. Gu, and Z. Y. Wang: Use of *all-trans* retinoic acid in the treatment of acute promyelocytic leukemia. *Blood* 72:567, 1988.
82. C. Chomienne, P. Ballerini, N. Balitrand, M. E. Huang, I. Krawice, S. Castaigne, P. Fenaux, P. Tiollais, A. Dejean, L. Degos, and H. de Thé: The retinoic acid receptor α gene is rearranged in retinoic acid-sensitive promyelocytic leukemias. *Leukemia* 4:802, 1990.
83. W. H. Miller, Jr., R. P. Warrell, Jr., S. R. Frankel, A. Jakubowsky, J. L. Gabrilove, J. Muindi, and E. Dmitrovsky: Novel retinoic acid receptor-α transcripts in acute promyelocytic leukemia responsive to *all-trans* retinoic acid. *JNCI* 82:1932, 1990.
84. F. Mitelman, P. G. Nilsson, L. Brandt, G. Alimena, K. Gastaldi, and B. Dallapiccola: Chromosome pattern, occupation, and clinical features in patient with acute nonlymphocytic leukemia. *Cancer Genet Cytogenet* 4:197, 1981.
85. R. Knapp, G. Dewald, and R. Pierre: Cytogenetic studies in 174 consecutive patients with

preleukemic and myelodysplastic syndrome. *Mayo Clin Proc* 60:507, 1985.
86. FIWCL Fourth International Workshop on Chromosomes in Leukemia, 1982: Clinical significance of chromosomal abnormalities in acute nonlymphoblastic leukemia. *Cancer Genet Cytogenet* 11:332, 1984.
87. E. P. Cronkite: Chemical leukemogenesis: Benzene as a model, *Semin Hematol* 24:2, 1987.
88. H. Goldberg, E. Lusk, J. Moore, P. C. Nowell, and E. C. Besa: Survey of exposure to genotoxic agents in primary myelodysplastic syndrome: Correlation with chromosome patterns and data on patients without hematological disease. *Cancer Res* 50:6876, 1990.
89. L. Zech, U. Haglund, K. Nilsson, and G. Klein: Characteristic chromosomal abnormalities in biopsies and lymphoid-cell lines from patients with Burkitt and non-Burkitt lymphomas. *Int J Cancer* 17:47, 1976.
90. G. M. Lenoir, G. L. Preud'homaue, A. Bernheim, and R. Berger: Correlation between immunoglobulin light chain expression and variant translocation in Burkitt's lymphoma. *Nature* 298:474, 1982.
91. C. M. Croce: Molecular biology of leukemias and lymphomas, in J. Brugge, T. Curran, E. Harlow, and F. McCormick, eds.: *Origins of Human Cancer.* Cold Spring Harbor, NY: Cold Spring Harbor Laboratory: 1991, pp. 527–542.
92. L. W. Ellisen, J. Bird, D. C. West, A. L. Soreng, T. C. Reynolds, S. D. Smith, and J. Sklar: *Tan-1*, the human homolog of the *Drosophila notch* gene, is broken by chromosomal translocations in T lymphoblastic neoplasms. *Cell* 66:649, 1991.
93. E. R. Fearon and B. Vogelstein: A genetic model for colorectal tumorigenesis. *Cell* 61:759, 1990.
94. E. R. Fearon and P. A. Jones: Progressing toward a molecular description of colorectal cancer development. *FASEB J* 6:2783, 1992.
95. B. Vogelstein, E. R. Fearon, S. R. Hamilton, S. E. Kern, A. C. Preisinger, M. Leppert, Y. Nakamura, R. White, A. M. M. Smits, and J. L. Bos: Genetic alterations during colorectal tumor development. *N Engl J. Med* 319:525, 1988.
96. K. W. Kinzler, M. C. Nilbert, L.-K. Su, B. Vogelstein, T. M. Bryan, D. B. Levy, K. J. Smith, A. C. Preisinger, P. Hedge, D. McKechnie, R. Finniear, A. Markham, J. Groffen, M. S. Bogusi, S. F. Altshul, A. Horii, H. Ando, Y. Miyoshi, Y. Miki, I. Nishisho, and Y. Nakamura: Identification of FAP locus genes from chromosome 5q21. *Science* 253:661, 1991.
97. A. P. Feinberg, C. W. Gehrke, K. C. Kuo, and M. Ehrlich: Reduced genomic 5-methylcytosine content in human colonic neoplasia. *Cancer Res* 48:1159, 1988.
98. S. J. Baker, E. R. Fearon, J. M. Nigro, S. R. Hamilton, A. C. Preisinger, J. M. Jessup, P. vanTuinen, D. H. Ledbetter, D. F. Barker, Y. Nakamura, R. White, and B. Vogelstein: Chromosome 17 deletions and p53 gene mutations in colorectal carcinomas. *Science* 244:217, 1989.
99. S. E. Kern, E. R. Fearon, K. W. F. Tersmette, J. P. Enterline, M. Leppert, Y. Nakamura, R. White, B. Vogelstein, and S. R. Hamilton: Clinical and pathologic associations with allelic loss in colorectal carcinoma. *JAMA* 261:3099, 1989.
100. Y. Fujiwara, M. Emi, H. Ohata, Y. Kato, T. Nakajima, T. Mori, and Y. Nakamura: Evidence for the presence of two tumor suppressor genes on chromosome 8p for colorectal carcinoma. *Cancer Res* 53:1172, 1993.
100a. C. E. Bronner, S. M. Baker, P. T. Morrison, G. Warren, L. G. Smith, M. K. Lescoe, M. Kane, C. Earabino, J. Lipford, A. Lindblom, P. Tannergard, R. J. Bollag, A. R. Godwin, D. C. Ward, N. Nordenskjold, R. Fishel, R. Kolodner, and R. M. Liskay: Mutation in the DNA mismatch repair gene homologue hMLH1 is associated with hereditary non-polyposis colon cancer. *Nature* 368:258, 1994.
100b. N. Papadopoulos, N. C. Nicolaides, Y.-F. Wei, S. M. Ruben, K. C. Carter, C. A. Rosen, W. A. Haseltine, R. D. Fleischmann, C. M. Fraser, M. D. Adams, J. C. Venter, S. R. Hamilton, G. M. Petersen, P. Watson, H. T. Lynch, P. Peltomaki, J.-P. Mecklin, A. de la Chapelle, K. W. Kinzler, and B. Vogelstein: Mutation of a *mut*L homolog in hereditary colon cancer. *Science* 263:1625, 1994.
101. Y. Huang, R. F. Boynton, P. L. Blount, R. J. Silverstein, J. Yin, Y. Tong, T. K. McDaniel, C. Newkirk, J. H. Resau, R. Sridhara, B. J. Reid, and S. J. Meltzer: Loss of heterozygosity involves multiple tumor suppressor genes in human esophageal cancers. *Cancer Res* 52:6525, 1992.
102. J. Houldsworth, C. Cordon-Cardo, M. Ladanyi, D. P. Kelsen, and R. S. K. Chaganti: Gene amplification in gastric and esophageal adenocarcinomas. *Cancer Res* 50:6417, 1990.
103. M. C. Hollstein, L. Peri, A. M. Mandard, J. A. Welsh, R. Montesano, R. A. Metcalf, M. Bak, and C. C. Harris: Genetic analysis of human esophageal tumors from two high incidence geographic areas: Frequent p53 base substitutions and absence of *ras* mutations. *Cancer Res* 51:4102, 1991.
104. T. Sano, T. Tsujino, K. Yoshida, H. Nakayama, K. Haruma, H. Ito, Y. Nakamura, G. Kajiyama, and E. Tahara. Frequent loss of heterozygosity on chromosomes 1q, 5q, and 17p in human gastric carcinomas. *Cancer Res* 51:2926, 1991.
105. N. R. Soman, P. Correa, B. A. Ruiz, and G. N. Wogan: The *TPR*-MET oncogenic rearrangement is present and expressed in human gastric carcinoma and precursor lesions. *Proc Natl Acad Sci USA* 88:4892, 1991.
106. C. C. Hsia, D. E. Kleiner, Jr., C. A. Axiotis, A. Di Bisceglie, A. M. Y. Nomura, G. N. Stemmermann, and E. Tabor: Mutations of p53 gene in hepatocellular carcinoma: Roles of hepatitis B

virus and aflatoxin contamination in the diet. *JNCI* 84:1638, 1992.

107. J.-C. Sheu, G.-T. Huang, P.-H. Lee, J.-C. Chung, H.-C. Chou, M.-Y. Lai, J.-T. Wang, H.-S. Lee, L.-N. Shih, P.-M. Yang, T.-H. Wang, and D.-S. Chen: Mutation of p53 gene in hepatocellular carcinoma in Taiwan. *Cancer Res* 52:6098, 1992.
108. T. Oda, H. Tsuda, A. Scarpa, M. Sakamoto, and S. Hirohashi: p53 gene mutation spectrum in hepatocellular carcinoma. *Cancer Res* 52:6358, 1992.
109. K. H. Buetow, V. C. Sheffield, M. Zhu, T. Zhou, F.-M. Shen, O. Hino, M. Smith, B. J. McMahon, A. P. Lanier, W. T. London, A. G. Redeker, and S. Govindarajan: Low frequency of p53 mutations observed in a diverse collection of primary hepatocellular carcinomas. *Proc Natl Acad Sci USA* 89:9622, 1992.
110. H. Tsuda, W. Zhang, Y. Shimosato, J. Yokota, M. Terada, T. Sugimura, T. Miyamura, and S. Hirohashi: Allele loss on chromosome 16 associated with progression of human hepatocellular carcinoma. *Proc Natl Acad Sci USA* 87:6791, 1990.
111. R. Callahan and G. Campbell: Mutations in human breast cancer: An overview. *JNCI* 81:1780, 1989.
112. C. Coles, A. Condie, U. Chetty, C. M. Steel, H. J. Evans, and J. Prosser: p53 mutations in breast cancer. *Cancer Res* 52:5291, 1992.
113. C. S. Cropp, R. Lidereau, G. Campbell, M. H. Campene, and R. Callahan: Loss of heterozygosity on chromosomes 17 and 18 in breast carcinoma: Two additional regions identified. *Proc Natl Acad Sci USA* 87:7737, 1990.
114. D. Sidransky, T. Tokino, K. Helzlsouer, B. Zehnbauer, G. Rausch, B. Shelton, L. Prestigiacomo, B. Vogelstein, and N. Davidson: Inherited p53 gene mutations in breast cancer. *Cancer Res* 52:2984, 1992.
115. J. M. Hall, M. K. Lee, B. Newman, J. E. Morrow, L. A. Anderson, B. Huey, and M.-C. King: Linkage of early-onset familial breast cancer to chromosome 17q21. *Science* 250:1684, 1990.

115a. D. E. Goldgar, P. Fields, C. M. Lewis, T. D. Tran, L. A. Cannon-Albright, J. H. Ward, J. Swensen, and M. H. Skolnick: A large kindred with 17q-linked breast and ovarian cancer: Genetic, phenotypic, and geneological analysis. *JNCI* 86:200, 1994.

115b. Y. Miki, J. Swensen, D. Shattuck-Eidens, P. A. Futreal, K. Harshman, S. Tavtigian, Q. Liu, C. Cochran, L. M. Bennett, W. Ding, R. Bell, J. Rosenthal, C. Hussey, T. Tran, M. McClure, C. Frye, T. Hattier, R. Phelps, A. Haugen-Strano, H. Katcher, K. Yakumo, Z. Gholami, D. Shaffer, S. Stone, S. Bayer, C. Wray, R. Bogden, P. Dayananth, J. Ward, P. Tonin, S. Narod, P. K. Bristow, F. H. Norris, L. Helvering, P. Morrison, P. Rosteck, M. Lai, J. C. Barrett, C. Lewis, S. Neuhausen, L. Cannon-Albright, D. Goldgar, R. Wiseman, A. Kamb, and M. H. Skolnick: A strong candidate for the breast and ovarian cancer susceptibility gene BRCA1. *Science* 266:66, 1994.

115c. R. Wooster, S. L. Neuhausen, J. Mangion, Y. Quirk, D. Ford, N. Collins, K. Nguyen, S. Seal, T. Tran, D. Averill, P. Fields, G. Marshall, S. Narod, G. M. Lenoir, H. Lynch, J. Feunteun, P. Devilee, C. J. Cornelisse, F. H. Menko, P. A. Daly, W. Ormiston, R. McManus, C. Pye, C. M. Lewis, L. A. Cannon-Albright, J. Peto, B. A. J. Bonder, M. H. Skolnick, D. F. Easton, D. E. Goldgar, and M. R. Stratton: Localization of a breast cancer susceptibility gene, BRCA2, to chromosome 13q12-13. *Science* 265:2088, 1994.

116. T. Sato, F. Akiyama, G. Sakamoto, F. Kasumi, and Y. Nakamura: Accumulation of genetic alterations and progression of primary breast cancer. *Cancer Res* 51:5794, 1991.
117. K.-I. Takita, T. Sato, M. Miyagi, M. Watatani, F. Akiyama, G. Sakamoto, F. Kasumi, R. Abe, and Y. Nakamura: Correlation of loss of alleles on the short arms of chromosomes 11 and 17 with metastasis of primary breast cancer to lymph nodes. *Cancer Res* 52:3914, 1992.
118. M. C. Paterson, K. D. Dietrich, J. Danyluk, A. H. G. Paterson, A. W. Lees, N. Jamil, J. Hanson, H. Jenkins, B. E. Krause, W. A. McBlain, D. J. Slamon, and R. M. Fourney: Correlation between c-*erb*B-2 amplification and risk of recurrent disease in node-negative breast cancer. *Cancer Res* 51:556, 1991.
119. D. J. Shou, H. Ahuja, and M. J. Cline: Protooncogene abnormalities in human breast cancer: c-*erb*-B-2 amplification does not correlate with recurrence of disease. *Oncogene* 4:105, 1988.
120. D. D'Amico, D. P. Carbone, B. E. Johnson, S. J. Meltzer, and J. D. Minna: Polymorphic sites within the *MCC* and *APC* loci reveal very frequent loss of heterozygosity in human small cell lung cancer. *Cancer Res* 52:1996, 1992.

120a. A. Merlo, E. Gabrielson, M. Mabry, R. Vollmer, S. B. Baylin, and D. Sidransky: Homozygous deletion on chromosome 9p and loss of heterozygosity on 9q, 6p, and 6q in primary human small cell lung cancer. *Cancer Res* 54:2322, 1994.

121. I. Miura, S. L. Graziano, J. Q. Cheng, L. A. Doyle, and J. R. Testa: Chromosome alterations in human small cell lung cancer: Frequent involvement of 5q. *Cancer Res* 52:1322, 1992.
122. Y. Kishimoto, Y. Murakami, M. Shiraishi, K. Hayashi, and T. Sekiya: Aberrations of the p53 tumor suppressor gene in human non-small cell carcinomas of the lung. *Cancer Res* 52:4799, 1992.
123. T. A. Sellers, J. E. Bailey-Wilson, R. C. Elston, A. F. Wilson, G. Z. Elston, W. L. Ooi, H. Roths-

child: Evidence for Mendelian inheritance in the pathogenesis of lung cancer. *JNCI* 82:1272, 1990.

124. S.-Q. Wu, B. E. Storer, E. A. Bookland, A. J. Kingelhutz, K. W. Gilchrist, L. F. Meisner, R. Oyasu, and C. A. Reznikoff: Nonrandom chromosome losses in stepwise neoplastic transformation in vitro of human uroepithelial cells. *Cancer Res* 51:3323, 1991.
125. P. Anglard, E. Trahan, S. Liu, F. Latif, M. J. Merino, M. I. Lerman, B. Zbar, and W. M. Linehan: Molecular and cellular characterization of human renal cell carcinoma cell lines. *Cancer Res* 52:348, 1992.
126. A. F. Olumi, Y. C. Tsai, P. W. Nichols, D. G. Skinner, D. R. Cain, L. I. Bender, and P. A. Jones: Allelic loss of chromosome 17p distinguishes high grade from low grade transitional cell carcinomas of the bladder. *Cancer Res* 50:7081, 1990.
127. C. J. Logothetis, H.-J. Xu, J. Y. Ro, S.-X. Hu, A. Sahin, N. Ordonez, and W. F. Benedict: Altered expression of retinoblastoma protein and known prognostic variables in locally advanced bladder cancer. *JNCI* 84:1256, 1992.
128. C. Cordon-Cardo, D. Wartinger, D. Petrylak, G. Dalbagni, W. R. Fair, Z. Fuks, and V. E. Reuter: Altered expression of the retinoblastoma gene product: Prognostic indicator in bladder cancer. *JNCI* 84:1251, 1992.
129. D. Sidransky, A. Von Eschenbach, Y. C. Tsai, P. Jones, I. Summerhayes, F. Marshall, M. Paul, P. Green, S. R. Hamilton, P. Frost, and B. Vogelstein: Identification of p53 gene mutations in bladder cancers and urine samples. *Science* 252:706, 1991.
130. J. H. Lee, J. J. Kavanagh, D. M. Wildrick, J. T. Wharton, and M. Blick: Frequent loss of heterozygosity on chromosomes 6q, 11, and 17 in human ovarian carcinomas. *Cancer Res* 50:2724, 1990.
131. D. Millikin, E. Meese, B. Vogelstein, C. Witkowski, and J. Trent: Loss of heterozygosity for loci on the long arm of chromosome 6 in human malignant melanoma. *Cancer Res* 51:5449, 1991.
132. J. R. Marks, A. M. Davidoff, B. J. Kerns, P. A. Humphrey, J. C. Pence, R. K. Dodge, D. L. Clarke-Pearson, J. D. Iglehart, R. C. Bast, Jr., and A. Berchuck: Overexpression and mutation of p53 in epithelial ovarian cancer. *Cancer Res* 51:2979, 1991.
133. A Okamoto, Y. Sameshima, S. Yokoyama, Y. Terashima, T. Sugimura, M. Terada, and J. Yokota: Frequent allelic losses and mutations of the p53 gene in human ovarian cancer. *Cancer Res* 51:5171, 1991.
134. S.-B. Li, P. E. Schwartz, W.-H. Lee, and T. L. Yang-Feng: Allele loss at the retinoblastoma locus in human ovarian cancer *JNCI* 83:637, 1991.
135. B. S. Carter, C. M. Ewing, W. S. Ward, B. F. Treiger, T. W. Aalders, J. A. Schalken, J. I. Epstein, and W. B. Isaacs: Allelic loss of chromosomes 16q and 10q in human prostate cancer. *Proc Natl Acad Sci USA* 87:8751, 1990.
136. T. Ichikawa, Y. Ichikawa, J. Dong, A. L. Hawkins, C. A. Griffin, W. B. Isaacs, M. Oshimura, J. C. Barrett, and J. T. Isaacs: Localization of metastasis suppressor gene(s) for prostatic cancer to the short arm of human chromosome 11. *Cancer Res* 52:3486, 1992.
137. T. Visakorpi, O.-P. Kallioniemi, A. Heikkinen, T. Koivula, and J. Isola: Small subgroup of aggressive highly proliferative prostatic carcinomas defined by p53 accumulation. *JNCI* 84:883, 1992.
138. L. F. Parada, H. Land, R. A. Weinberg, D. Wolf, and V. Rotter: Cooperation between gene encoding p53 tumor antigen and *ras* in cellular transformation. *Nature* 312:649, 1984.
139. I. Marcillac, F. Troalen, J.-M. Bidart, P. Ghillani, V. Ribrag, B. Escudier, B. Malassagne, J.-P. Droz, C. Lhommé, P. Rougier, P. Duvillard, M. Prade, P.-M. Lugagne, F. Richard, T. Poynard, C. Bohuon, J. Wands, and D. Bellet: Free human chorionic gonadotropin β subunit in gonadal and nongonadal neoplasms. *Cancer Res* 52:3901, 1992.

4

Phenotypic Characteristics of Cancer Cells

Treatment of cells in culture with oncogenic agents has been the easiest means of studying discrete biochemical events that lead to malignant transformation. Studies of cell transformation in vitro, however, have many pitfalls. These "tissue culture artifacts" include overgrowth of cells not characteristic of the original population of cultured cells (e.g., overgrowth of fibroblasts in cultures that were originally primarily epithelial cells), selection for a small population of variant cells with continued passage in vitro, or appearance of cells with an abnormal chromosomal number or structure (karyotype). Such changes in the characteristics of cultured cell populations can lead to "spontaneous" transformation that mimics some of the changes seen in populations of cultured cells treated with oncogenic agents. Thus, it is often difficult to sort out the critical malignant events from the noncritical ones. Malignant transformation can also be induced in vivo, of course, by treatment of susceptible experimental animals with carcinogenic chemicals or oncogenic viruses or by irradiation. (see Chap. 6). Identification of critical biochemical changes in vivo, however, is even more tenuous than that of changes in vitro because it is difficult to discriminate toxic from malignant events and to determine what role a myriad of factors, such as the nutritional state of the animal, hormone levels, or endogenous infections with microorganisms or parasites, might have on the in vivo carcinogenic events. Moreover, tissues in vivo are a mixture of cell types, and it is difficult to determine in which cells the critical transformation events are occurring and what role the microenvironment of the tissue plays. Most studies that are designed to identify discrete biochemical events occuring in cells during malignant transformation have therefore been done with cultured cells, since clones of relatively homogeneous cell populations can be studied and the cellular environment defined and manipulated. The ultimate criterion that establishes whether or not cells have been transformed, however, is their ability to form a tumor in an appropriate host animal.

Over the past 50 years, much scientific effort has gone into research aimed at identifying the phenotypic characteristics of in vitro–transformed cells that correlate with the growth of a cancer in vivo. This research has tremendously increased our knowledge of the biochemistry of cancer cells. However, many of the biochemical characteristics initially thought to be closely associated with the malignant phenotype of cells in culture have subsequently been found to be dissociable from the ability of those cells to produce tumors in animals. Furthermore, individual cells of malignant tumors growing in animals or in humans exhibit marked biochemical heterogeneity, as reflected in their cell surface composition, enzyme levels, immunogenicity, response to anticancer drugs, and so on. This had made it extremely difficult to identify the essential changes that produce the malignant phenotype. Nevertheless, it is very useful to review the characteristics associated with the transformed phenotype to learn more about how malignant cells grow, to determine the variability between cancer cell types in their growth characteristics, and to design the next generation of experiments to determine the essential properties of the malignant state.

In the early 1940s, Earle and his colleagues reported that cultures of fibroblasts obtained from C3H mice could undergo cytologic changes resembling those of malignant cells after treatment for as few as 6 days with the chemical carcinogen 20-methylcholanthrene at a concentration of 1 μg/mL in the culture medium.[1] When these cells were injected back into C3H mice, they produced sarcomas that eventually killed the animals.[2] One of the problems with these early experiments was that untreated cultures of C3H fibroblasts also produced tumors in about 8% of the animals. It was not appreciated until later than some of the apparently normal cells had "spontaneously" transformed in culture to an altered cell type. These observations lay dormant for 20 years until Berwald and Sachs[3] showed that benzo[*a*]pyrene and a combination of benzo[*a*]pyrene plus polyomavirus could transform cultured hamster embryo cells. About that time it was also shown that polyoma-

Table 4-1 Properties of Transformed Malignant Cells Growing in Cell Culture and/or in Vivo

1. Cytologic changes resembling those of cancer cells in vivo, including increased cytoplasmic basophilia, increased number and size of nuclei, increased nuclear/cytoplasmic ratio, and formation of clusters and cords of cells.
2. Alteration in growth characteristics:
   a. "Immortality" of transformed cells in culture. Transformed malignant cells become "immortal" in that they can be passaged in culture indefinitely.
   b. Decreased density-dependent inhibition of growth or loss of "contact inhibition." Transformed cells frequently grow to a higher density than their normal counterparts, and they may "pile up" in culture rather than stop growing when they make contact.
   c. Decreased serum requirement. Transformed cells usually require lower concentrations of serum or growth factors to replicate in culture than nontransformed cells do.
   d. Loss of anchorage-dependence and acquisition of ability to grow in soft agar. Transformed cells may lose their requirement to grow attached to surfaces and can grow as free colonies in a semisolid medium.
   e. Loss of cell cycle control. Tranformed cells fail to stop in G_1, or at the G_1/S boundary in the cell cycle when they are subject to metabolic restriction of growth.
   f. Resistance to apoptosis (programmed cell death).
3. Changes in cell membrane structure and function—including increased agglutinability by plant lectins, alteration in composition of cell surface glycoproteins, proteoglycans, glycolipids, and mucins; appearance of tumor-associated antigens; and increased uptake of amino acids, hexoses, and nucleosides.
4. Loss of cell–cell and cell–extracellular matrix interactions that foster cell differentiation.
5. Loss of response to differentiation-inducing agents and altered cellular receptors for these agents.
6. Altered signal transduction mechanisms, including constitutive rather than regulated function of growth factor receptors, phosphorylation cascades, dephosphorylation mechanisms.
7. Increased expression of oncogene proteins due to chromosomal translocation, amplification, or mutation.
8. Loss of tumor suppressor gene protein products due to deletion or mutation.
9. Genomic imprinting errors that lead to overproduction of growth promoting substances, e.g., IGF-2.
10. Increased or unregulated production of growth factors, e.g., TGF-α, tumor angiogenesis factors, PDGF, hematopoietic growth factors (e.g., CSFs, interleukins).
11. Genetic instability leading to progressive loss of regulated cell proliferation, increased invasiveness, and increased metastatic potential. "Mutator" genes may be involved in this effect.
12. Alteration in enzyme patterns. Transformed cells have increased levels of enzymes involved in nucleic acid synthesis and produce higher levels of lytic enzymes, e.g., proteases, collagenases, glycosidases.
13. Production of oncodevelopmental gene products. Many transformed malignant cells growing in culture or in vivo produce increased amounts of oncofetal antigens (e.g., carcinoembryonic antigen), placental hormones (e.g., chorionic gonadotropin), or placental-fetal type isoenzymes (e.g., placental alkaline phosphatase).
14. Ability to produce tumors in experimental animals. This is the sine qua non that defines malignant transformation in vitro. If the cells believed to be transformed do not produce tumors in appropriate animal hosts, they cannot be defined as "malignant." However, failure to grow in an animal model does not rule out the fact that they may be tumorigenic in a different type of animal.
15. Ability to avoid the host's antitumor immune response.

virus alone[4] and x-rays[5] could induce cell transformation in vitro. Thus, it became possible to study changes associated with malignant transformation in defined cell culture systems.

The general characteristics of transformed malignant cells are listed in Table 4-1. A number of these changes may be characteristic of certain rapidly proliferating cell populations and not be specific for the malignant state. No single criterion is adequate, as there are many exceptions both ways: some nontransformed cells have one or more of these characteristics and some transformed cells lack one or more of them. Taken together, however, these criteria appear to define a cell that will produce a tumor when injected into an appropriate host animal. A number of these properties relate only to malignant cells growing in cell culture. However, a number of the properties of transformed cells in culture are characteristic of cancer growing in an animals or a patient.

The evidence that these properties are found in transformed cells and that they are related to malignancy is discussed in the following sections.

CYTOLOGIC CHARACTERISTICS OF TRANSFORMED CELLS

Cytologic changes resembling those seen in cancer cells growing in vivo have been observed in cells transformed in culture by oncogenic agents, or "spontaneously." Barker and Sanford[6] followed the progression of cytologic changes during the spontaneous transformation of a series of cell lines derived from mouse, hamster, and rat embryos. They found that after long-term culture, most of these cell lines became malignant based on their ability to produce tumors in compatible host animals. Most cell lines undergoing malignant transformation showed a progression of cytologic changes in the following sequence: (1) increase in cytoplasmic basophilia, (2) increase in number and size of nucleoli, (3) increase in nuclear/cytoplasmic ratio, (4) retraction of cytoplasm, and (5) formation of clusters and cords of cells. In some respects, these changes resemble those induced by transformation with such oncogenic viruses as polyoma. Some of the nontumorigenic cultures had a few of the early changes in some of the cells, but all cultures with at least the first four of these five characteristics were tumorigenic in vivo. Based on these criteria, prediction of the in vivo malignancy of a culture was 93% accurate. The tumors resulting from growth of the transformed cells in animals retained the cytologic features and growth pattern of the in vitro cultures.

Examination of a series of 22 tumorigenic and nontumorigenic rat liver cell lines has also shown that a number of cytologic characteristics correlate with in vivo malignancy.[7] Cytologic eval-

Table 4-2 Correlation of Cytologic Changes with Tumorigenicity of Cultured Rat Liver Cells

Cytologic Characteristic	Correlation Coefficient for Tumorigenicity
Cell cording	0.38
Cell clumping	0.35
Lack of smooth colony edge	0.42
Cell separation	0.40
Cytoplasmic basophilia	0.61
High nuclear/cytoplasmic ratio	0.66
Reduced cytoplasmic spreading	0.45
Variation in cell size and shape	0.53
Variation in nuclear size and shape	0.28
Variation in nucleolar size, shape, and number	0.50
Abnormal mitoses	0.15
Thickened nuclear membrane	0.11
Nuclear lobulation	−0.25
Granular chromatin	−0.21
Clumped chromatin	0.00

Source: Data from Montesano et al.[17]

uation and growth in soft agar were the most reliable means of determining the malignant potential of the cultured cells. Cytodiagnosis proved to be accurate 94% of the time. Of the cytologic characteristics that correlated with tumorigenicity (Table 4-2), increased cytoplasmic basophilia and increased nuclear/cytoplasmic ratio showed the highest correlation. Other criteria useful in assessing malignancy were decreased cytoplasmic spreading, variation in cell size and shape (pleomorphism), and variation in nucleolar size and shape. It is interesting that most of the cytologic characteristics of malignant transformed cells in culture are also seen in cancer cells taken from animal or human tumors growing in vivo (see Chap. 1) and are, in fact, some of the criteria used to diagnose cancer.

GROWTH CHARACTERISTICS OF TRANSFORMED CELLS

Immortality of Transformed Cells in Culture

Most normal diploid mammalian cells have a limited life expectancy in culture. For example, normal human fibroblast lines may live for 50 to 60 population doublings, but then viability begins to decrease rapidly unless they transform spontaneously or are transformed by oncogenic agents. However, malignant cells, once they become established in culture, will generally live for an indefinite number of population doublings provided the right nutrients and growth factors are present. It is not clear what limits the life expectancy of normal diploid cells in culture, but transformed cells that become established in culture frequently undergo karyotypic changes, usually marked by an increase in chromosomes (polyploidy), with continual passage. This suggests that cells with increased amounts of certain growth-promoting genes are generated and/or selected for during continual passage in culture. The more undifferentiated cells from cancers of animals or patients also often have an atypical karyology, suggesting that the same selection process may be going on in vivo with progression over time of malignancy from a lower to a higher grade.

Decreased Density-Dependent Inhibition of Growth

It has been known for a long time that "epithelium will not tolerate a free edge,"[8] that is, with a suitable surface to grow on, a "free edge" of epithelial cells will replicate and move until it makes contact with another edge of epithelial cells. For example, if the skin of an amphibian larva is wounded, a rim of epidermis migrates out from the wound until the free edges of this expanding rim meet in the middle of the wound, whereupon epithelial extension abruptly ceases.[9] Fibroblasts have been shown to have a similar sort of contact reaction, as exemplified by chick heart fibroblasts moving in liquid medium on a glass surface.[10] When two primary explants of these cells were placed 1 mm apart, an outgrowth toward each explant occurred until the cells made contact. At that point, outgrowth and movement of the cells stopped and the cells formed a monolayer on the surface. It was concluded that the cells inhibit one another by mutual contact of their cell surfaces, a phenomenon called *contact inhibition*.[10] A variety of malignant transformed cells do not stop replicating when they come into contact. This was first observed for mouse sarcomas S-37 and S-180 and fibrosarcoma MCIM and M_2C,[8] and later for methylcholanthrene-induced primary mouse tumors as well as for Rous sarcoma virus–infected chick cells.[11] and SV40-infected mouse fibroblasts.[12] In the latter case, it was found that mouse 3T3 fibroblasts, when maintained on a transplanation schedule that prevented them from growing to high density between passages, did not form tumors in syngeneic mice; if they were infected with the oncogenic virus SV40, however, they lost contact inhibition, grew to a high density, and became tumorigenic. There is often a good correlation between the ability of mouse fibroblasts to continue to divide in the presence of extensive cell–cell contact in culture and their capability of producing tumors in animals.[10]

There is some disagreement as to whether or not contact inhibition—which, as originally defined, refers primarily to cell movement, not cell replication[8]—and growth inhibition of densely crowded cells are the same. Cells of the baby hamster kidney line BHK 21, for example, have been shown by time-lapse cinematography to have contact inhibition of movement, but they grow to form multilayers of cells with the same cell density as their transformed counterparts that have lost contact inhibition of movement.[13] Thus, contact inhibition of movement and growth inhibition may be independent variables. Because of the discrepancies between contact inhibition and growth inhibition of cultured cells, the term *density-dependent inhibition of growth* has been suggested[13] and is now generally accepted as more precise terminology for the phenomenon of growth inhibition in dense cultures.

There is some evidence that cell–cell contact is a requirement for cessation of cell replication. For example, proliferation of sparsely seeded human embryonic lung fibroblasts on silica beads to which plasma membrane glycoproteins had been covalently coupled is inhibited.[14] This result mimics the situation observed at high cell density and suggests that cell–cell contact via cell surface glycoproteins is involved in cessation of cell proliferation. Interestingly, transformed cells that have decreased receptors for the glycoproteins involved in cell–cell contact are not susceptible to density-dependent inhibition of proliferation.[15]

To some extent, growth inhibition of dense cultures may also be due to a metabolic effect resulting from the close crowding of cells. Local inhibitory conditions could result from depletion of nutrients, oxygen, or growth factors; local changes in pH; or accumulation of inhibitory factors in the immediate milieu of crowded cells. Transformed cells could presumably continue to replicate under these conditions because they are anchored to each other and to the substratum to a lesser degree and because they require lower concentration of growth-promoting factors (see below).

Decreased Serum Requirement

Another property that distinguishes transformed cells from their nontransformed counterparts is the requirement for serum growth factors for replication in culture. Cells transformed by oncogenic viruses have lower serum growth requirements than normal cells.[16] Thus, 3T3 fibroblasts transformed by SV40,[17] polyoma,[18] murine sarcoma virus,[18] or Rous sarcoma virus[19] are all able to grow in culture medium that lacks certain serum growth factors, whereas uninfected cells are not. The infectious process itself, induced in 3T3 cells by SV40 virus, stimulates several rounds of cell mitosis even before the characteristics of the transformed phenotype become apparent, and this effect is proportional to the multiplicity of infection (the number of infectious viruses employed per cell).[19] It seems that the cells must divide a number of times before the viral genome can be integrated into the host cell genome. This latter process is required for the cells to become permanently transformed. The induction of cell division per se, however, does not permanently alter the growth properties of the cells. Only about 10% to 20% of the colonies that grow in medium free of serum growth factor have the typical morphology of SV40-transformed cells.[19] Many of the remaining cells do not look transformed, do not produce virus-specific T antigen, do not lose contact inhibition, show no evidence of the continued presence of functioning virus information, and revert to a state requiring serum factors for growth. Thus, the induction of cell replication under growth-limiting conditions by infection with oncogenic viruses appears to be only a transient event in the majority of infected cells. The clones in which these properties become permanently fixed are malignant. The ability to grow in growth-factor–free medium is not the same property as loss of density-dependent inhibition of growth, even in "transformed" clones, because some clones that produce SV40 T antigen and that retain "rescuable" virus information (e.g., as evidenced by virus production after fusion with another cell type that is not by itself capable of producing virus) grow in growth-factor–free medium and yet retain density-dependent inhibition of growth.

These data indicate that infection of cells by oncogenic viruses itself signals cell proliferation. In many infected cells that is only transient, but in some cells, the viral genome becomes integrated into the host genome, and these cells have the potential of being permanently transformed. The transformed cells are apparently able to provide their own growth factors and thus to need less of the serum factors that are required for the growth of nontransformed cells. There is evidence that certain virally transformed cells do produce their own growth factors (see below).

Loss of Anchorage Dependence

Most freshly isolated normal animal cells and cells from cultures of normal diploid cells do not grow well when they are suspended in fluid or a semisolid agar gel. If these cells make contact with a suitable surface, however, they attach, spread, and proliferate. This type of growth is called anchorage-dependent growth. Many cell lines derived from tumors and cells transformed by oncogenic agents are able to proliferate in suspension cultures or in a semisolid medium (methylcellulose or agarose) without attachment to a surface. This is called *anchorage-independent growth.* This property of transformed cells has been used to develop clones of malignant cells[20] and to assay for viral transforming activity.[21] This technique has been widely used to compare the growth properties of normal and malignant cells.

Another advantage derived from the ability of malignant cells to grow in soft agar (agarose) is the ability to grow cancer cells derived from human tumors to test their sensitivity to chemotherapeutic agents and to screen for potential new anticancer drugs.[22]

The ability of malignant cells to proliferate autonomously in culture without the addition of serum or hematopoietic growth factors (cytokines) can also have prognostic significance. For example, in a study of 114 patients with acute myelogenous leukemia (AML), a high correlation was found between ability of a given patient's AML cells to proliferate autonomously in culture and disease prognosis.[23] Those patients whose cells had high rates of autonomous proliferation had a poorer response to chemotherapy, a reduced probability of complete remission, and a higher frequency of relapse than those patients whose AML cells had low in vitro proliferative capacity. The ability of AML cells to proliferate autonomously in culture appears to relate to the cells' ability to produce their own cytokines (e.g., GM-CSF, G-CSF, M-CSF, interleukin-1, and interleukin-6; see Chap. 9).[23]

As mentioned at the beginning of this chapter, not all the characteristics of transformed cells in culture are necessarily associated with a cell type that is tumorigenic in vivo. Thus, the question is which, if any, of the phenotypic properties of transformed cells growing in culture are most closely correlated with the ability of these cells to produce tumors in animals. By careful growth selection techniques and studies of revertant cell lines that lose one or more properties of the transformed phenotype over time or under different growth conditions, many of the characteristics of the transformed cell type have been shown to be reversible and dissociable from the ability to produce tumors. The one in vitro growth property of cells that is most closely associated with tumorigenicity is the loss of anchorage-dependent growth. Studies of mouse 3T3 cells transformed by SV40 or murine sarcoma virus and revertants of these cells that have lost one or more properties of the transformed parent cell line have shown that the single cellular property consistently associated with tumorigenicity in athymic "nude" mice (mice that have no thymus gland and that do not readily reject tumor xenografts) is the ability of cells to grow in methylcellulose gel, that is, cells that have lost their anchorage-dependence characteristic.[24] Other cellular characteristics of virus-induced transformation, such as decreased density-dependent inhibition of growth and low serum requirement for growth, are often dissociable from tumorigenicity. Montesano et al.[7] also found that, in addition to cytologic characteristics, the highest correlation between in vitro transformation and in vivo tumorigenicity for a wide variety of spontaneously and chemically transformed rat liver cell lines is the ability of cells to grow in soft agar.

It is not clear what alteration of cellular function causes the loss of anchorage dependence, but it presumably relates to some property of cell surface membranes. In addition, nontransformed cells may require more cell-to-cell "feeding" or more transfer of growth factors between cells to grow than transformed cells need, thus preventing nontransformed cells from growing well under conditions in which cells are more isolated.

The intercellular communication between cocultured nontransformed and methylcholanthrene-transformed 3T3 cells has been examined by microinjection of the fluorescent dye Lucifer Yellow into cells and microscopic observation of the movement of dye between cells through their gap junctions.[25] When the dye was microinjected into a carcinogen-transformed cell, it was transferred to other transformed cells but not to adjacent nontransformed cells, even though they were in physical contact. Moreover, when dye was injected into a nontransformed cell, it was transferred to neighboring nontransformed cells but not to adjacent transformed cells. It has also been shown that tumor promoters inhibit intercellular communication between many types of cultured cells.[26] These data suggest that the normal "cross-talk" between cells that regulates cell proliferation is lost when cells are malignantly transformed. It is interesting, though, that the transformed cells can still communicate among themselves, suggesting that they can pass their own signals for growth and proliferation and thus march to their own, but different, drummer.

Even the relationship between loss of anchorage-dependence and tumorigenicity, however, is not a hard and fast rule. A series of human cell lines derived from urinary bladder cancers were found to have many of the properties of the transformed phenotype when they were grown in culture.[27] Two such lines grew in soft agar but were not tumorigenic in nude mice; one tumorigenic line did not grow in agar. Thus, although the cellular alteration associated with loss of anchorage-dependence appears to be closely related to tumorigenicity for many cell types, this characteristic does not appear to be absolutely essential for growth of malignant cells in vivo.

Loss of Cell Cycle Control and Resistance to Apoptosis

Normal cells respond to a variety of suboptimal growth conditions by entering a quiescent phase in the cell division cycle, the G_0 state (see Chap. 5). There appears to be a decision point in the G_1 phase of the cell cycle, at which time the cell must make a commitment to continue into S phase, the DNA synthesis step, or to stop in G_1 and wait until conditions are more optimal for cell replication to occur. If this waiting period is prolonged, the cells are said to be in a G_0 phase. Once cells make this commitment, they must continue through S, G_2, and M to return to G_1. If the cells are blocked in S, G_2, or M for any length of time, they die. The mechanism for this is not clear, but it is as if the cell committed to DNA synthesis and division is vulnerable to any interruption. There appear to be internally signaled events linked to the release of lysosomal lytic enzymes—for example, proteases and nucleases—that are set in motion if cells are blocked in one of the critical phases of cell division. This may well be the mechanism of cytotoxicity of a variety of anticancer drugs that appear to block cells in S, G_2, or M phase (the so-called cell cycle phase–specific agents). These drugs kill primarily dividing cells and not resting ones. The point in G_1 at which normal cells are blocked during growth restriction is common to a variety of restrictive conditions; it is called the "restriction point."[28] Restriction point control may have evolved to protect cells against growth-limiting conditions that would lead to cell death if cells were allowed to continue willy-nilly through the division cycle, using up nutrients and energy as they went, until their resources were exhausted. It is as if cells have a mechanism for going into hiberation until more favorable conditions occur.

A variety of conditions can shift cells into the quiescent phase. These include limitation of serum, of certain essential amino acids such as isoleucine and glutamine, or glucose, and of certain lipids in the growth medium.[29] High cell density, elevated levels of cAMP, or the presence of certain drugs—such as cycloheximide, streptovitacin, or caffeine—can also accomplish this.[28,29] Excess thymidine at dosages used to block cell division and to "synchronize" cell populations in the same cell cycle phase also appears to block cells at the restriction point.[29]

Regulation of the cell cycle in eukaryotic cells is a complex process, discussed in detail in Chapter 10. The cycle is controlled by a series of protein–protein interactions and phosphorylation/dephosphorylation events.[30] A series of cell-cycle phase-specific regulatory proteins called cyclins are synthesized and then degraded at specific points in the cycle. These cyclins associate with cycle-specific kinases to form a complex called *maturation promoting factor* (MPF), first identified in clams and sea urchins.

A kinase activity called cdc2, or $pp34^{cdc2}$ because it is a 34,000-Da phosphoprotein, is activated by its association with cyclin and by its phosphorylation state. A tyrosine kinase called Wee 1 phosphorylates cdc2; this and a second phosphorylation on a threonine in cdc2 keep it in an inactive state. The cdc2 is then activated by a tyrosine phosphatase called cdc 25 to form the active cyclin/cdc2 kinase or MPF complex. After each regulatory step in the cell cycle, the cycle-specific cyclin is degraded, inactivating the cdc2 activity. The most important mitosis phase (M phase)–specific cyclin in eukaryotic organisms, including humans, is called cyclin B_1, and this cyclin must be degraded for cells to exit M phase.

In budding yeast, in which many of the regulatory events of the cell cycle have been worked out, a decision point for $G_1 \rightarrow$ S-phase transition is called *Start.* In addition, the events of the cell cycle of most organisms, from yeast to human, are ordered into specific regulatory steps in which the initiation of later events is dependent on the completion of early events. These control mechanisms regulating dependency in the cell cycle are called *checkpoints.*[31] Loss of checkpoint control may result in cell death, unequal chromosome distribution, and increased susceptibility to DNA-damaging agents. For example, if normal cells accrue DNA damage due to ultraviolet or X-irradiation, they arrest in G_1, so that the damaged DNA can be repaired prior to DNA replication. Another checkpoint in the G_2 phase allows repair of chromosome breaks before chromosomes are segregated at mitosis. There are two well-documented checkpoints, one at the G_1/S boundary and another at the G_2/M boundary. However, studies of cdc mutants in yeast suggest at least six cell-cycle checkpoints at which cell-cycle events stop until an earlier event is allowed to be completed[31,32].

Cancer cells contain mutations that allow them to escape and disobey the signals that control cell proliferation. Thus, one might predict that these mutations are located in genes that are important for controlling cell division. Indeed, mutations in cyclin genes, inappropriate expression of cyclins,[23] and inappropriate expression of the *mos* oncogene, a meiotic cytostatic factor,[34] have been observed in transformed cells.

As noted in Chapter 3, cancer cells accumulate a number of genetic changes in their progress from an early transformed state to an invasive, metastatic cancer. One of the results of this genetic "drift" toward a more progressive state of malignancy is karyotypic alterations that are seen as translocations, aneuploidy, and gene amplification. The increased genetic instability of cancer cells may result from alterations in cell-cycle arrest, which may, in turn, result from defects in cell-cycle checkpoints.[35] This genetic instability is exemplified by the ability of tumor cells to undergo gene amplification at a much higher rate than normal cells.[36] For example, gene amplification during the proliferation of normal human fibroblasts in culture is virtually undetectable (relative frequency of $<10^{-9}$), whereas amplification occurs at a high frequency in transformed cells (10^{-3} to 10^{-5}). Interestingly, loss of the tumor suppressor gene p53 correlates with the ability of cells to undergo gene amplification,[36] suggesting that p53 acts as one of the checkpoint controls in cell division. While normal cells that suffer DNA damage in G_1 do not enter S phase until they have repaired their DNA, cells with mutant or deleted p53 genes enter S phase and replicate the damaged DNA leading to genomic rearrangements, one result of which is gene amplification. The p53 gene product prevents cells with DNA damage from dividing and has been called "the guardian of the genome" (see Chap. 8).

One of the ways in which p53 acts is to induce programmed cell death *(apoptosis)* in cells in which excessive DNA damage occurs. Many cancer cell types have deleted or mutated p53 genes that appear during tumor initiation or progression, depending on the cell type, and escape apoptosis, even though they may contain significant DNA rearrangements or mutations. Thus, cancer cells have evolved mechanisms to bypass cell-cycle checkpoints that prevent this from happening in normal cells. Loss of p53 is one of the mechanisms involved in this, but not the only one (see Chap. 10).

CHANGES IN CELL MEMBRANE STRUCTURE AND FUNCTION

The cell surface membrane (plasma membrane) plays an important role in the "social" behavior of cells, that is, communication with other cells, cell movements and migration, adherence to

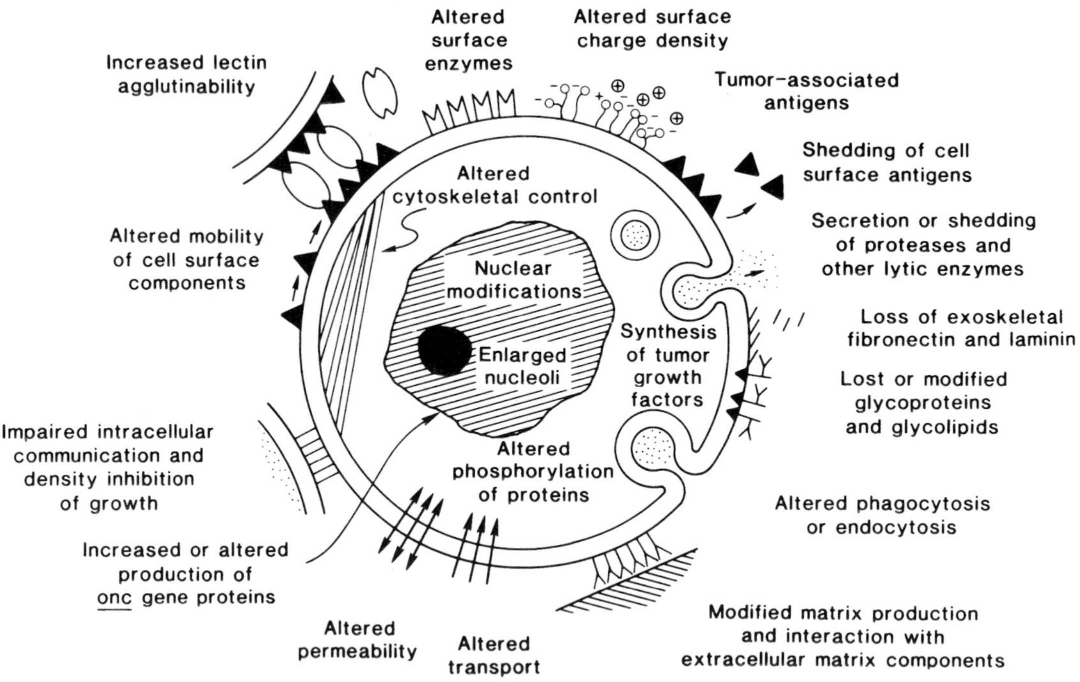

Figure 4-1 Some cellular alterations observed after neoplastic transformation. (Modified from Nicholson.[37])

other cells or structures, access to nutrients in the microenvironment, and recognition by the body's immune system. Alterations of the plasma membrane in malignant cells may be inferred from a variety of properties that characterize their growth and behavior: for example, the loss of density-dependent inhibition of growth, decreased adhesiveness, loss of anchorage-dependence, and invasiveness through normal tissue barriers. In addition, a number of changes in the biochemical characteristics of malignant cells' surfaces have been observed. These include appearance of new surface antigens, proteoglycans, glycolipids, and mucins and altered cell–cell and cell–extracellular matrix communication. Some of the cell membrane alterations observed for transformed cells are depicted in Figure 4-1.

Agglutinability of Transformed Cells

One of the characteristics that defines the plasma membrane of transformed cells is their increased agglutination by such lectins as wheat germ agglutinin (WGA), concanavalin A (Con A), and phytohemagglutinin (PHA). Lectins, from the Latin term *lectus*, meaning "selected," are usually glycoproteins extracted from plants, but they have also been found in bacteria, fungi, fish, snails, and mammalian, including human cells. They bind specifically to certain carbohydrates.[38] Thus, they have been used as probes to determine cell agglutinability as well as the glycoprotein and glycolipid composition and the configuration of the cell surface membranes of normal and transformed cells.

Aub et al.[39] were the first to show that an extract from wheat germ agglutinates several transformed cell lines but not their normal counterparts. Since that time, a wide variety of transformed cell types have been tested for lectin agglutinability and, in general, have been found to have more of this property than normal cells. The evidence for a positive correlation between increased lectin agglutinability and transformation can be summarized as follows[37,40]: (1) A wide variety of malignant cells obtained from tumors in vivo or transformed by oncogenic viruses in culture have increased agglutinability. (2) Revertant lines of transformed cells selected for density-dependent inhibition of growth lose agglutinability. (3) Revertant transformed cells selected for loss of agglutinability regain at least partial growth control. (4) Temperature-sensitive mutants that lose some of their transformed properties at nonpermissive temperatures simultaneously lose their agglutinability. There are exceptions, however, to the generalization that transformed cells agglutinate with lectins more readily than nontransformed cells. A number of nontransformed adult rat and monkey cells are highly agglutinable.[41] Phenotypic revertant cells of polyomavirus-transformed BHK cells that have regained near-normal density-dependent inhibition of growth and that have low tumorigenicity have Con A agglutination properties similar to nontransformed cells, although they still express virus-specific nuclear (T) antigen.[42] Moreover, infection of cells with some nononcogenic viruses also enhances agglutination.[37] Changes in lectin-induced agglutinability also occur in normal cells during the cell cycle. Nontransformed 3T3 cells in the mitotic phase (M) are agglutinated by wheat germ agglutinin to the same extent as polyoma- or SV40 virus-transformed 3T3 cells.[40] The agglutinability of nontransformed 3T3 cells returns to normal levels as the cells enter G_1 phase. Thus, what appears to be a more permanent alteration in some transformed cell lines is a transient event in nontransformed cells.

There are a number of possible explanations for the increased lectin agglutinability of transformed cells. One explanation is that transformed cells simply have more cell-surface binding sites for lectins. This appears not to be the case. In general, nontransformed cells and their transformed counterparts (of the same cell type) have similar numbers of lectin surface binding sites.[43-45] What does appear to be different between the two cell types is the mobility of surface binding sites for the lectins. The ability of the lectin binding sites on transformed cells to move laterally through the cell membrane increases, leading to more concentrated areas of receptors.[46] This produces a series of local "patches" of lectin-receptor complexes. The enhanced mobility of surface lectin receptors has been generally correlated with the increased agglutinability of transformed cells.[37] This enhanced mobility may be due to increased plasma membrane fluidity, which allows more lateral movement of glycoprotein components in the lipid bilayer of the membrane. The intrinsic fluidity of the plasma membranes of normal and transformed cells has been measured by electron-spin resonance and fluorescent-polarization techniques; however, no consistent differences between normal and transformed cells have been observed.[37] Analysis of the lipid composition of cell membranes from the two cell types suggests some differences—for example, a decreased cholesterol/phospholipid ratio in transformed cells—but, again, these findings vary with the cells beings analyzed, and their relationship to agglutinability is not well defined.

The configuration of cell surface structures probably also plays a role in cell agglutination. Most cells have cell surface protrusions that include extensions of a segment of membrane with some enclosed cytoplasm (pseudopodia) and smaller outward-extending folds of membrane (microvilli). These structures could be involved in interactions between cells by increasing the surface area for contact or presenting a more concentrated localization of attachment sites. It is know that a number of transformed cell types have increased numbers of microvilli on their cell surface.[37,47] and it has been proposed that agglutination of transformed cells is due to the abundance of their cell surface microvilli, the formation of which is controlled by intracellular levels of cAMP.[47]

There is evidence that the membrane-associated cytoskeletal system of microfilaments and microtubules is altered in transformed cells. For example, it has been observed that nontransformed 3T3 cells develop an extensive microfilament-microtubule system at confluent cell densities but that transformed 3T3 cells fail to do this after contact in culture.[48] Microfilaments contain large amounts of actin, and some of them form sheaths that run parallel to and immediately under the plasma membrane, with bundles extending into the cell cytoplasm (see Fig. 4-1). These microfilament sheaths appear to act as "anchors" for cell surface receptors, and they have been observed in areas where the cell surface is relatively immobile. It has also been found that Con A–induced redistribution and "patching" of surface lectin receptors are not detectable on confluent 3T3 cells in regions immediately over membrane-associated microfilaments but do occur in regions of the cell surface not associated with microfilament sheaths.[37]

Transformed 3T3 cells, however have only the highly mobile type of lectin receptors.[46] Transformation appears to cause the partial disorganization of the microfilament-microtubule cytoskeletal system, resulting in "the uncoupling of surface receptors from cytoplasmic transmembrane control."[37] Because intracellular cAMP levels appears to be linked to the organization of the cytoskeletal system and also appear to be altered during transformation, these two events may reflect alteration of the same cellular process during malignant transformation.

Of course, lectin binding and the resultant cellular agglutination could be altered during transformation by modification of the lectin receptors themselves, so that the type of receptor that is produced, or the binding affinity of the receptors, is changed. Thus, although the *total* lectin binding of normal and of transformed cells does not seem to differ, the receptors to which the lectins are bound could be different. This, in turn, could affect the availability of other binding sites on lectin molecules to attach to adjacent cells. There is evidence for alterations of surface glycoproteins and glycolipids (see below), that is, molecules involved in the attachment of lectins to cell surfaces. In addition, modified production or loss of external cell surface matrix glycoproteins such as fibronectin that occurs in transformed cells could also affect lectin binding and lectin-induced cell agglutination.

As this evidence indicates, agglutination of cells by lectins reflects a complex series of events, and no one event is clearly associated with the cell surface changes that affect agglutination. Transformed cells *tend* to be more agglutinable than their nontransformed counterparts, but that is not a hard and fast rule.

Lectin binding by cancer cells has also been used to distinguish different malignant characteristics, including metastatic potential. For example, the most consistent change in cell-surface oligosaccharide expression distinguishing high-metastatic-potential from low-metastatic-potential murine lymphoma cells was the ability of the high-metastatic-potential cells to bind soybean agglutinin lectin.[49] Since this lectin binds *N*-acetyl-D-galactosamine residues, it suggests that oligosaccharides bearing this residue are more prevalent or more available for binding on cells of high metastatic potential.

In addition to the plant lectins used to detect various carbohydrate-binding molecules on various cell types, animal and human cells themselves have lectins on their cell surfaces. The plant lectins, which have been the best studied, were generally thought to be bivalent in the sense that they all contained at least two carbohydrate binding sites.[50] More recently, however, some plant and animal lectins have been found that are monovalent. Thus, the term *lectin* has more generally been defined as "a carbohydrate-binding protein other than an enzyme or an antibody."[50] This definition is perhaps too broad because it allows inclusion of carbohydrate-binding proteins with a wide array of biological activities.

Animal cells produce a variety of lectins, some membrane-bound and some soluble, that have been implicated in cell-cell and cell-matrix recognition phenomena (reviewed in Refs. 51 and 52). The first animal lectin discovered was a membrane-bound galactose (Gal) and *N*-acetylgalactosamine (GalNAc)-specific lectin isolated from rabbit hepatocytes, which has become known as the asialo glycoprotein receptor. This receptor is involved in the clearance of desialylated serum glycoproteins that have exposed terminal Gal residues. A similar galactose-specific lectin has been isolated from rat peritoneal macrophages. This lectin appears to mediate binding of the lectin-bearing cell to cells bearing galactose-terminated glycoproteins. Another role for Gal and GalNAc-binding lectins involves the ability of activated macrophages to recognize and kill tumor cells.

Membrane-associated lectins with other carbohydrate-binding specificities have also been isolated (reviewed in Ref. 50). These include a *N*-acetylglucosamine (GlcNAc)-specific lectin from chicken liver, two mannose-6-phosphate–specific lectins from a variety of vertebrate cell types, an L-fucose-specific lectin from rat liver Kupffer cells, and mannose-specific and GlcNAc-specific lectins from macrophages. A high degree of amino acid–sequence homology has been found in the carbohydrate-binding domain of many of the membrane-bound lectins, suggesting an evolutionarily conserved function.

A number of functions have been ascribed to the membrane-bound lectins of animal cells, including (1) phagocytosis by macrophages of microorganisms and parasites bearing mannose or GlcNAc residues, targeted by the lectin with the analogous binding specificity on macrophages; (2) a role in cellular differentiation and organ formation exemplified by the developmentally regulated appearance and cellular distribution of β-galactoside–specific lectins in many developing tissues; (3) migration of lymphocytes based on lectin-carbohydrate interactions between lymphocytes and the cells of lymphoid organs; and (4) determination of metastatic potential of tumor cells. The latter is strongly suggested by the evidence that lectins on human and mouse metastatic cell surfaces are involved in tumor cell emboli formation and adhesion of tumor cell aggregates to capillary endothelial cells. This suggests a therapeutic approach to prevent metastasis by injecting sufficient amounts of the competing sugar to block these tumor cell–tumor cell and tumor cell–endothelial cell interactions. Due to the exquisite specificity of lectins for their binding sugar, such an approach could provide some target cell specificity. This would be particularly true if cancer cells had somewhat different cell surface carboyhydrate moieties compared to their normal counterpart cells, which in many instances is the case (see below).

Alterations in Cell Surface Glycolipids, Glycoproteins, Proteoglycans, and Mucins

Alterations in tumor-associated carbohydrate-containing cellular components were first implicated more than 50 years ago by studies demonstrating that rabbit antisera obtained by immunization of a tumor tissue homogenate and adsorbed on normal tissues had a preferential complement fixation reaction with lipid extracts of the original tumor tissue used as the immunogen.[53,54] In studies 30 years later, Rapport et al.[55] generated rabbit antibodies to a human epidermoid carcinoma that reacted specifically to one glycolipid component of the tumor. This component turned out to be a lipid called lactosylceramide. In another study from the same laboratory, it was found that rabbit antisera directed to a rat lymphosarcoma reacted specifically with a glycosphingolipid fraction of the tumor.[56] Since both lactosylceramide and glycophingolipids are also found in normal tissue, the relationship of these findings to cancer was not clear at the time.

Aberrant glycosylation was first suggested as the basis for the tumor-associated determinants of glycolipids by the finding of a remarkable accumulation of fucose-containing glycolipids found in human adenocarcinomas, some of which were identified as lactofucopentaose-III-ceramide, lactofucopentaose-II-ceramide (Lewis A blood group glycolipid), and lactodifucohexaose and lactodifucooctaose ceramide (Lewis B glycolipid) (reviewed in Ref. 57). These identifications were confirmed once the technique of monoclonal antibodies (MAbs) was used to definitively identify antigens (see Chap. 13). As discussed below, a number of the MAbs with preferential reactivity for tumor cells over normal cells have been shown to react with Lewis blood group antigens such as Le^x, Le^a, and Le^b or their analogues.[57]

The first definitive evidence that altered glycolipids could be generated on cells during neoplastic transformation came from studies of polyoma-transformed hamster embryonic fibroblasts that contained a lactoneotetraosylceramide not present on the untransformed parent cells.[58] In another study, an accumulation of a gangliotriaosylceramide called Gg3 was observed in a Kristen sarcoma virus–induced sarcoma in mice. A rabbit antibody to Gg3 specifically stained the sarcoma cells in vivo but not other mouse tissues.[57]

The biochemical characterization of the aberrant glycosylation of glycoproteins was difficult to demonstrate in earlier studies. However, the presence of high-molecular-weight glycopeptides with altered glycosylation patterns was detected on transformed cells in early studies before they were clearly chemically identified.[59,60] Later, the chemical basis for some of the changes in tumor cell glycoproteins was attributed to the fact that the N-linked oligosaccharides of tumor cells contain more multiantennary structures than the oligosaccharides derived from normal cells.[61]

Tumor-associated carbohydrate antigens can be classified into three groups[57]: (1) epitopes expressed on both glycolipids and glycoproteins, (2) epitopes expressed only on glycolipids, and (3) epitopes expressed only on glycoproteins.

To the first group belongs the lacto-series structure that is found in the most common human cancers, e.g., lung, breast, colorectal, liver, and pancreatic cancers. The common backbone structure for these epitopes is Galβ1→3GlcNAcβ1→Gal (type 1 blood group, see below) or Galβ1→4 GlcNacβ1→3 Gal (type 2 blood group). The second group of epitopes, expressed exclusively on glycolipids, are mostly on the ganglio- or globo-series structures. This series of epitopes is expressed abundantly only on certain types of human cancers such as melanoma, neuroblastoma, small cell lung carcinoma, and Burkitt's lymphoma. The third group of epitopes, seen only on glycoproteins, comprises the multiantennary branches of N-linked carbohydrates and the alterations of O-linked carbohydrate chains seen in some mucins (see below).

Tumor-associated carbohydrate antigens can also be classified by the cell types expressing them as those (1) expressed on only certain types of normal cells (often only in certain developmental stages) and greatly accumulated in tumor cells as a result of blocked synthesis of more complex structures or enhanced synthesis of precursor (incompletely processed) structures; (2) expressed only on tumor cells (e.g., altered blood group antigens or mucins); and (3) expressed commonly on normal cells but present in much higher concentration on tumor cells, (e.g., the GM_3 ganglioside in melanoma and Le^x in gastrointestinal cancer).[57]

A variety of chemical changes in tumor cells have been identified that can explain altered glycosylation patterns. These result from three kinds of altered processes: (1) incomplete synthesis and/or processing of normally existing carbohydrate chains and accumulation of the resulting precursor form, (2) "neosynthesis" resulting from activation of glycosyltransferases that are absent or have low activity in normal cells, and (3) organizational rearrangement of tumor-cell–membrane glycolipids.[62] Similar changes have been noted in the carbohydrate components of glycolipids and of membrane-associated and secreted glycoproteins.[63] Because both the altered carbohydrates of glycolipids and glycoproteins can be variably sialylated, depending on the cell type and the stage of differentiation of the transformed cell, this probably accounts for the conflicting observations of the sialylation patterns reported in the earlier manuscripts.

Interest in the carbohydrate components of cell surface glycolipids, glycoproteins, and proteoglycans has been heightened by the fact that most of the monoclonal antibodies developed to tumor-cell–associated antigens recognize these carbohydrate moieties. Moreover, many of these have turned out to be blood-group–specific antigens or modifications of blood-group–specific antigens, some of which are antigens that are seen at certain stages of embryonic development and thus fit the definition of oncodevelopmental antigens.[62,64]

Developmentally regulated antigens were first identified as carbohydrates by the use of sera from patients with autoimmune hemolytic disorders that were known to contain antibodies to two distinct antigens—designated I and i—of human erythrocytes.[64] Type i antigen was known to be present in human fetal erythrocytes and to decrease during the first year of life, whereas type I antigen was observed to be present in adult erythrocytes. These antigens were subsequently shown to be carbohydrate structures, contained in both glycoproteins and glycolipids, and composed of a core structure of repeating units of *N*-acetyllactosamine with a galactose linked by a β1 → 4 linkage to *N*-acetylglucosamine (Galβ1 → 4GlcNac), typical of the so-called Type 2 blood group chain. In the case of the i antigen, this core structure can be repeated two or more times in an unbranched chain (structure 1 in Table 4-3), whereas I antigen contains branched structures formed by a β1 → 6 linkage of *N*-acetyllactosamine units to internal galactose residues of the repeating sequence (structure 2 in Table 4-3). Presumably, the change from type i to type I antigen results from the developmental stage–specific activation of a branching glycosyltransferase that transfers GlcNAc by means of 1 → 6 linkage to Gal.

A number of these Type 2 blood group substances (Galβ1 → 4GlcNAc) have been observed on animal and human cancer cells. Type 1 blood group substances containing a Galβ1 → 3GlcNAc backbone have also been observed on embryonal cell types and on human cancer cells (Table 4-3).

One of the first clear demonstrations of the

Table 4-3 Differentiation Antigens First Detected by Monoclonal Antibodies and Identified Subsequently as Carbohydrate Structures Based on Type 1 or Type 2 Blood Group Chains

Structure Number	Structure	Designation of Antigen	Cell Association
1	Galβ1→4GlcNAcβ1→3(Galβ1→4GlcNAcβ1→3)⩾2	i	Fetal erythrocytes, human; primary endoderm, mouse
2	Galβ1→4GlcNAcβ1↘6 Galβ1→4GlcNAcβ1↗3 Galβ1→4GlcNAcβ1	I	Adult erythrocytes, human; preimplantation embryos, mouse
3	Galβ1→4GlcNAcβ1↘ 6Gal/GalNAc	I(Ma)	Adult erythrocytes, human; gastric mucosa, nonsecretors; gastric adenocarcinomas, secretors
		M18, M39	Breast epithelium, human
4	Galβ1→4GlcNAc ↑ 1,3 Fucα	SSEA-1	Eight-cell-stage mouse embryo
		VEP8 and VEP9, Myl, VIM-D5, etc.	Myeloid cells, human
		$D_1$56-22 (X hapten, Le^x)	Colorectal cancer, human
5	Galβ1→4GlcNAc ↑ 1,2 Fucα	TRA-1-85 (blood group H)	Human cell lines
6	Galβ1→4GlcNAc ↑ 1,2 ↑ 1,3 Fucα Fucα	C14	Colonic adenocarcinoma, human
		F3	Lung adenocarcinoma, human
		AH6	Gastric cancer, human
		(Y hapten, Le^y)	Embryonal carcinoma cells, human and mouse
7	GalNAcα1→3Galβ1→4/3GlcNAc ↑ 1,2 Fucα	TL5 (blood group A)	EGF receptor of A431 cells
8	Galα1→3Galβ1→4GlcNAc ↑ 1,2 Fucα	E_1 series (blood group B)	Pancreatic cancer
9	Galβ1→3GlcNacβ1→3Galβ1→4Glc/GlcNAc	FC10.2	Embryonal carinoma cells, human; gastric mucosa, nonsecretors; gastric adenocarcinoma, secretors
10	Galβ1→3GlcNac ↑ 1,4 Fucα	CO-514 (blood group Le^a)	Adenocarcinomas, human
11	Galβ1→3GlcNAc ↑ 1,2 ↑ 1,4 Fucα Fucα	NS-10 CO-43 (blood group Le^b)	Adenocarcinomas, human
12	GalNAcα1→3Galβ1→3/4GlcNAc ↑ 1,2 ↑ 1,3/4 Fucα Fucα	G49	EGF receptor of A431 cells
		MH2	Colonic adenocarcinoma, human
		(blood group.ALe^b/Le^y)	
13	Galβ1→3GlcNAc ↑ 2,3 ↑ 1,4 NeuAcα Fucα	19.9 (sialyl Le^a)	Colon cancer, human; normal pancreatic ducts, normal gastric mucins, nonsecretors; gastric cancer mucins, secretors, human; seminal fluid proteins, human

Source: From Feizi.[64]

presence of carbohydrate blood-group–related antigens in specific stages of development and in transformed cells was the observation that a monoclonal antibody, called SSEA-1, developed to F9 murine embryonal carcinoma cells, reacted with the eight-cell stage of the developing mouse embryo but less with cells of earlier or later stages of development.[65] It was later shown that this antibody also reacted with human fetal colon and adenocarcinoma cells.[66] It has now been established that anti-SSEA recognizes a fucosylated form of a Type 2 blood group substance chain—that is, Galβ1 → 4 (Fucα1-3)GlcNAc (structure 4 in Table 4-3)—indicating that, during mouse development, backbone structures similar to the I blood group antigen of human erythrocytes are expressed and are modified by an α1 → 3 fucosylation reaction.[67] This structure occurs only in certain tissues in the adult mouse and only weakly on normal as opposed to carcinomatous human colon tissue.[66] When a second fucose linked α1 → 2 is added to this backbone (structure 6) in Table 4-3), reactivity with anti-SSEA-1 is masked, indicating that expression of cell surface antigen can be markedly altered by the addition or deletion of monosaccharides, and suggesting that this might be a mechanism for the appearance or disappearance of various stage-specific cell surface determinants during development and carcinogenesis.

A wide variety of monoclonal antibodies directed against differentiation-stage–specific epitopes or against tumor-cell–associated antigens recognize carbohydrates that are expressed on Type 1 or Type 2 backbone sequences or their further glycosylated analogues and that are carried on both cell surface glycoproteins and glycolipids. A number of these have turned out to be structures identical to the blood group antigens, A, B, H, Lewis types Le^a, Le^b, and Le^x or their fucosylated or sialosylated analogues. It should be noted that a carbohydrate antigen may be a normal component of one cell type but a tumor-associated antigen in another. For example, structure 3 in Table 4-3 is a normal component of adult human erythrocytes but a tumor-associated antigen in gastric carcinomas of patients who secrete A, B, or H blood group substances from their gastric mucosae. This is presumably because in A, B, or H individuals, who secrete glycoproteins or glycolipids in their gastrointestinal mucins, antigens are normally masked by blood group A, B, or H monosaccharide modifications. When gastric cancers arise in these individuals, their blood group chains are incompletely synthesized and the antigenicity of the blood group substance antigen becomes unmasked.[64]

Incomplete synthesis or processing of cell surface glycoproteins or glycolipids is typical of a number of cancer cell types. For example, in certain transformed mouse cell lines, the early precursors lactosylceramide or GM3 accumulate in the ganglioside biosynthetic pathway and the more complex gangliosides beyond GM3 decrease.[68] In addition to precursor accumulation, "neosynthesis" of fucolipids and their sialosylated derivatives has also been observed in a wide variety of human cancers, particularly epithelial cancers of gastrointestinal tract, lung, and breast (Table 4-4).[62] These "novel" fucolipids have been detected by monoclonal antibodies generated to various tumor cell types, but none of them is specific for a given tumor type. Some have also been detected on embryonic cells or normal stem cells or even in small amounts on cells of the normal tissue of origin of the tumor or on normal cells of other tissues. It should be noted that analogous carbohydrate antigens on normal cells may be cryptic and immunologically undetectable, and their appearance on cancer cells may be due to the activation of fucosyl- or sialosyl-transferases normally unexpressed in nontransformed cells.

Although many of the tumor-associated determinants of the lacto series (Table 4-4) were originally discovered in glycolipids, primarily because glycolipids can be more easily purified to homogeneity,[62] determinants such as Le^x, sialosyl Le^x, and sialosyl Le^a have been found on glycoproteins of various types of human cancer. On the other hand, the saccharide antigens of the globo and ganglio series have been found mainly in glycolipids. Some of these antigenic determinants, such as sialosyl Le^a and sialosyl Le^x, are also found in the plasma of cancer patients and thus may be useful as tumor markers for cancer diagnosis. Increased expression of sialosyl Le^x in colorectal carcinoma tissue has been correlated with the depth of tumor invasion, metastasis, and decreased patient survival.[68a]

Table 4-4 Novel Fucolipids and Fucogangliosides as Human-Tumor-Associated Markers Defined by Specific Monoclonal Antibodies

			Association	Structure
A.	Lacto-series chain: Sialosyl Le[a]	Type 1 blood group chain	Gastrointestinal/ pancreas cancer	Galβ1→3GlcNAcβ1→3Galβ1→R 3 (↑ NeuAcα2) on Galβ1; 4 (↑ Fucα1) on GlcNAcβ1
B.	Lacto-series chain: Difucosyl Y_2 $III V^3 Fuc_2 nLc6$	Type 2 blood group chain	Gastrointestinal/ lung/breast cancer	Galβ1→4GlcNAcβ1→3Galβ1→4GlcNAcβ1→3Galβ1→4Glcβ1→1Cer 3 (↑ Fucα1) on first GlcNAcβ1; 3 (↑ Fucα1) on second GlcNAcβ1
	Sialosyldifucosyl Y_2 $III^3 V^3 Fuc VI^3 NeuAc$		Gastrointestinal/lung breast cancer	Galβ1→4GlcNAcβ1→3Galβ1→4GlcNAcβ1→3Galβ1→4Glcβ1→1Cer 3 (↑ NeuAcα2) on first Galβ1; 3 (↑ Fucα1) on first GlcNAcβ1; 3 (↑ Fucα1) on second GlcNAcβ1
	Sialosyl Le[x]		Gastrointestinal/ lung/breast cancer	Galβ1→4GlcNAcβ1→3Galβ1→R 3 (↑ NeuAcα2) on Galβ1; 3 (↑ Fucα1) on GlcNAcβ1
C.	Globo series		Breast cancer	Galβ1→3GalNAcβ1→3Galα1→4Galβ1→4Glcβ1→1Cer 2 (↑ Fucα1) on first Galβ1
D.	Ganglio series		Small-cell lung carcinoma	Galβ1→3GalNAcβ1→4Galβ1→4Glcβ1→1Cer 2 (↑ Fucα1) on first Galβ1; 3 (↑ NeuAcα2) on second Galβ1

Source: From Hakomori.[62]

Role of Glycosyl Transferases and Oligosaccharide Processing Enzymes

The substitution of additional carbohydrate moieties on blood-group–related structures is not the only aberrant modification of glycoproteins or glycolipids observed in cancer cells. Increased branching of asparagine-linked oligosaccharides and incomplete processing of these oligosaccharides have also been noted in certain cell-surface as well as secretory glycoproteins.[63,69] The general scheme for the biosynthesis and processing of asparagine-linked oligosaccharides is shown in Figure 4-2.[70] The increased activity of specific *N*-acetylglucosaminyl transferases in tumor cells appears to be responsible for the appearance of tri- and tetraantennary structures, whereas the analogous glycoprotein in normal cells is often a biantennary structure (Fig. 4-2). Several distinct *N*-acetylglucosaminyl transferases have been identified,[71] and their substrate-specific sites for addition of GlcNAc moieties are shown in Figure 4-3. In polyomavirus-transformed baby hamster kidney (BHK) cells, there is an increase in GlcNAcβ1→6Manα1→groups and a decrease in the more typical GlcNAcβ1→4Manα1→groups on membrane glycoproteins as well as increase in the activity of GlcNAc transferase V, the enzymatic activity that catalyzes this addition.[63] Moreover, the asparagine-linked oligosacchardies of the placental glycoprotein hormone chorionic gonadotropin (hCG) found in the urine of women with choriocarcinoma have an unusual biantennary structure as well as triantennary sugar chains.[69] Since hCG found in the urine of normal pregnant women does not contain triantennary sugar chains, it appears that increased activity of GlcNac transferases that produce these additional branched structures is also a feature of human cancer cells. Similarly, the extra fucosylations that appear on membrane glycoproteins and glycolipids have been associated with the induction of an unusual α-fucosyltransferase in chemical carcinogen–induced precancerous rat liver and in the resulting hepatomas.[72] These investigations strongly suggest that the regulation of glycosyltransferase genes is important in malignant transformation.

Several studies support this concept. For example, hamster sarcoma virus transformed fibroblasts produced an increase in the average number of *N*-acetyllactosamine type N-linked glycans on glycoprotein substrates due both to an increase in the average number of branches per chain and in the fraction of N-linked glycans containing GlcNAcβ1→3Galβ1→4 polylactosaminyl chains.[73] suggesting an increase in GlcNAc transferase V activity. Easton et al.[74] have shown that induction of N-*ras* gene expression in NIH 3T3 cells produces a 2- to 2.5-fold increase in the branching GlcNAc transferase III and V and a five- to sevenfold increase in the elongating transferases β4-galactosyl transferase and β3-N-GlcNac transferase, leading both to increased branching and elongation of polylactosaminyl chains. In addition, the level of the major chain terminating enzyme α3-galactosyl transferase was somewhat decreased and the levels of α3- and α6-sialyltransferases were elevated. Increases in GlcNAc transferase V activity have also been observed in Rous sarcoma virus transformed BHK cells[75] and in GlcNAc III activity in preneoplastic hepatic nodules during induction and promotion of carcinogenesis rat liver.[76]

A number of other differences in glycosyl transferases have been noted between normal and cancerous tissues. Human ovarian carcinomas have severalfold higher levels of α1,3-fucosyl- and α1,4-fucosyltransferases compared to normal ovarian tissue.[77] Sialyltransferase activity levels have been shown to be higher in metastatic human colon carcinoma cell lines, suggesting that increased sialylation of adhesion molecules may favor implantation of tumor cells into distant tissue sites.[78] Four to 18-fold increases in the enzyme activity that introduces an additional branch into O-linked glycans has been observed in human AML and chronic myelogenous leukemia (CML) cells.[79] Human melanoma cells that express high levels of the gangliosides GM_3 and GD_3 have high levels of the sialyltransferases involved in their synthesis.[80]

All of these data strongly support the idea that glycosylation patterns change during transformation of normal cells into malignant ones. Now, what advantage would a tumor cell have, one might ask, by producing these altered cell-surface glycoproteins and glycolipids? Because cell–cell interactions, adhesion to extracellular matrices, regulation of cell proliferation, and recognition by the host's immune system are all

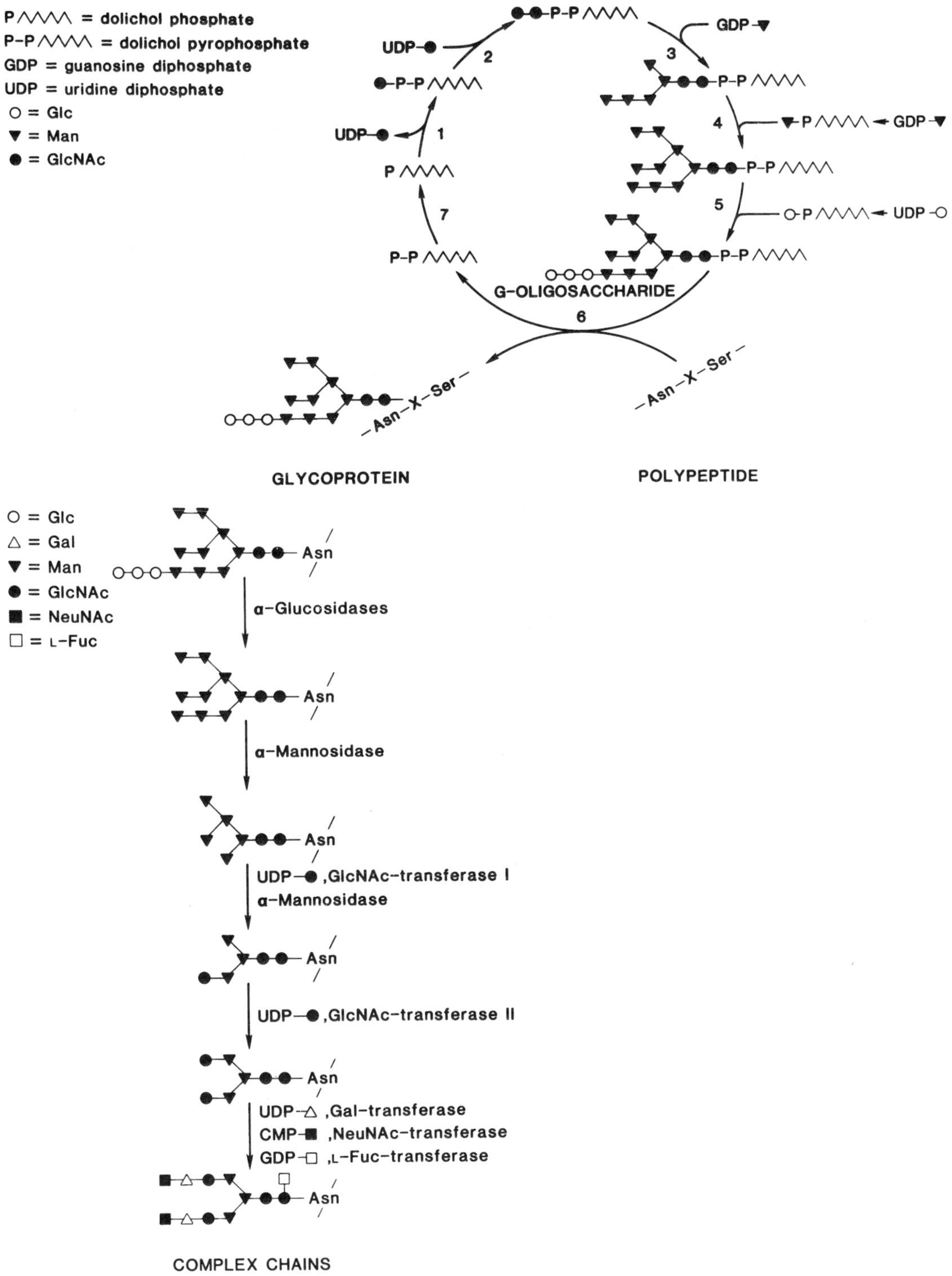

Figure 4-2 Schematic diagram of biosynthesis and processing of asparagine-linked oligosaccharide chains of glycoproteins. (Modified from Sharon and Lis.[70])

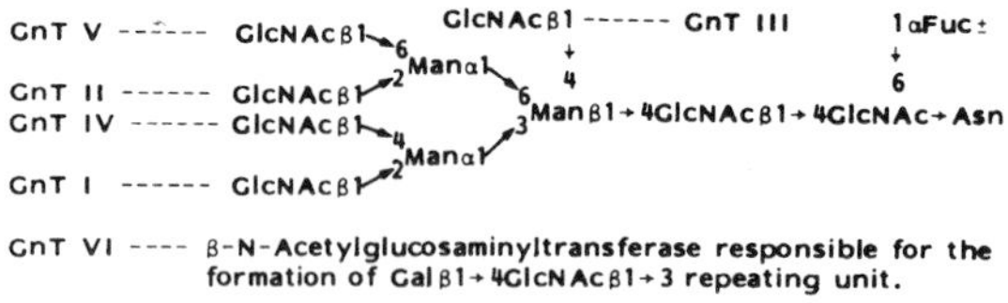

Figure 4-3 A structure showing the possible β-*N*-acetylglucosamine residues that can be attached to the trimannosyl cores of the complex-type asparagine-linked sugar chains. The different β-*N*-acetylglucosaminyltransferases (abbreviated as GnT) responsible for the addition of each β-*N*-acetylglucosamine residue are indicated by the Roman numerals I through VI. (From Yamashita et al.[63])

profoundly affected by the composition of the cell surface, it can quickly be grasped how the entire social behavior of a cell could be altered by such changes. Gangliosides have been shown to be present in the cell attachment matrix for cultured cells, for example, but certain tumor-derived polysialosyl gangliosides inhibit fibronectin-mediated cell attachment; moreover, mutant cell lines deficient in polysialosyl ganglioside production lack fibronectin-dependent cell attachment to a substratum (for review, see Ref. 62). In addition, many of the monoclonal antibodies raised against the epidermal growth factor (EGF) receptor are directed against blood-group–related carbohydrate structures, suggesting that the receptor for EGF is glycosylated with this type of carbohydrate chain. Also, antireceptor antibodies known to bind carbohydrate chains modulate proliferation of cells rich in EGF receptors (for review, see Ref. 64). This raises the intriguing possibility that structures containing blood-group carbohydrates are associated in some key fashion with growth-factor receptors. The well-known mitogenic effect of carbohydrate-binding lectins for lymphocytes is another example of the importance of cell-surface carbohydrates in the recognition of signals for cell proliferation.

There is increasing experimental evidence that gangliosides may modulate the immune response as well. Lymphocyte function and natural killer cell activity can be suppressed by certain gangliosides in vitro, and it has been postulated that circulating levels of polysialosylated gangliosides contribute to the immunosuppressed condition of animals and patients with a high tumor burden.[62] Presumably, the origin of such immunosuppressive carbohydrate-containing molecules would be from the shedding of cell surface glycoproteins and glycolipids by tumor cells.

Additional evidence for the importance of glycosylation patterns of cell-surface glycoproteins and glycolipids in the malignant phenotype comes from the use of glycosylation inhibitors and oligosaccharide-processing inhibitors. For example, tunicamycin, an inhibitor of addition of N-linked glycans to nascent polypeptide chains; castanospermine, an inhibitor of glucosidase (see Fig. 4-2); and KI-8110; an inhibitor of sialyltransferase activity reduce the number of lung metastases in murine experimental tumor models.[81-83] In addition, swainsonine was shown to reduce the rate of growth of human melanoma xenografts in athymic nude mice,[84] and castanospermine was observed to inhibit the growth of N-*fms* oncogene-transformed rat cells in vivo.[85] These results support the hypothesis that the synthesis of highly branched complex-type oligosaccharides is associated with the malignant phenotype and may provide tumor cells with a growth advantage.

Mucins

Mucins are a type of highly glycosylated glycoprotein produced by a variety of secretory epithelial cells. They are 50% to 80% carbohydrate by weight and function to lubricate and protect ductal epithelial cells. They contain O-linked glycans (serine- and threonine-linked) of various lengths and structures, depending on the tissue type in which they are produced. They are made in a wide variety of tissues, including the gastrointestinal tract, lung, breast, pancreas, and ovary; tumors arising in these organs may have altered glycosylation patterns that distinguish them from the normal mucins and renders them immunogenic.

A number of the genes for the core proteins have been cloned and bear the names MUC1, MUC2, and so on (Table 4-5).[86] The core protein of the MUC1 gene has a transmembrane domain and a large extracellular domain made up of tandem repeats of 20 amino acids; variation in the number of these tandem repeats (30 to 100) is

Table 4-5 Mucin Gene Characteristics

Name	Organism	Tissue Expression	Chromosome	Tandem Repeat Sequence
MUC1	Human	Ductal epithelial cells of breast, pancreas, gall bladder, lung, kidney, stomach, prostate, others	1q21	GSTAPPAHGVTSAPDTRPAP
Muc1	Mouse	Same as above	Unknown	DSTSSPVHSGTSSPATSAPE
MUC2	Human	Goblet cells of small intestine, colon, pancreas, bronchus, cervix, gall bladder, others	11p15	PTTTPITTTTTVTPTPTPTGTQT
MUC3	Human	Small intestine, colon	7q22	HSTPSFTSSITTTETTS
MUC4	Human	Tracheobronchial tissues, colon, pancreatic tumors	3q29	TSSASTGHATPLPVTD A V SDT S LH S L L L I P
MUC5B	Human	Submaxillary gland, bronchus	Unknown	Unknown
MUC5C	Human	Goblet cells and submucosal glands of tracheobronchial tissues, colon	11p15	interspersed repeats of TTSTTSAP, GSTPSPVP, and TASTTSGP
MUC6	Human	Stomach, gall bladder, terminal ileum, colon	11p15	SPFSSTGPMTATSFQTTTTYPTPSHPQTTLPTHVPPFSTSL VTPSTGTVITPTHAQMATSASIHSTPTGTI PPPTTLKATGSTHTAPPMTPTTSGTSQAHSSFSTAKTSTS LHSHTSSTHHPEVTPTSTTTITPNPTSTGTSTPVAHTTSA TSSRLPTPFTTHSPPTGS
MUC7	Human	Sublingual gland, submandibular		TTAAPPTPSATTPAPPSSSAPPE
PSM	Pig	Submaxillary gland	Unknown	GAGPGTTASSVGVTETARPSVAGSGTTGTVSGASGSTG SSSGSPGATGASIGQPETSRISVAGSSGAPAVSSGASQAA GTS
BSM	Cow	Submaxillary gland	Unknown	None
FIM-A.1	Frog	Mucous glands of frog skin	Unknown	VPTTPETTT
FIM-B.1	Frog	Mucous glands of frog skin	Unknown	GESTPAPSETT

responsible for the polymorphism observed in the core proteins. Some of the human mucin genes have homologous genes from mouse, cow, pig, and frog (Table 4-5). In each case, though the tandem repeats may vary in sequence and length, the repeats contain high levels of threonine, serine, and proline.

The O-linked sugars are formed through a GalNAC linked to a threonine or serine and extended with polylactosamine side chains that terminate in sialic acid or fucose (reviewed in Ref. 86). Several studies have shown that novel carbohydrate epitopes appear on various carcinomas, and these have been associated with O-GalNAc(Tn) or sialylated O-GalNAc (sialylated Tn) and O-GalNAc-Gal (T) or sialylated T (all linked to serine or threonine) (Fig. 4-4). Comparison of tumor-derived mucins with those of their normal counterpart cells showed that the carbohydrate chains are shorter and more sialylated on the tumor mucins. Even though it appears that the core proteins are the same in tumor cells and normal cells from the same tissue, the antigenic epitopes differ because the carbohydrates are different and different amounts of the core protein are exposed. Moreover, total expression of the mucins is increased in many cancers and upregulated in some normal tissues

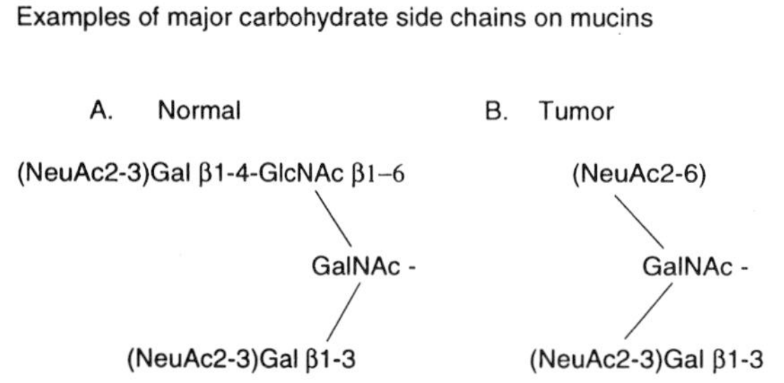

Figure 4-4 Examples of major carbohydrate side chains on mucins. (A) Basic tetrasaccharide and monosialyated products. Also more extended and branched polylactosamine side chains are found and disialylated and fucosylated derivatives. (B) Basic disaccharide (T) and mono and disialylated products.

under different physiologic states (e.g., lactating mammary gland).[86] Although the MUC1-encoded mucin has been the most extensively studied, cancer related alterations in other mucins have been observed. Moreover, it appears that some cells, both normal and cancerous, can express more than one mucin. For example, increased expression of MUC1 has been observed in most adenocarcinomas of the breast, lung, stomach, pancreas, prostate, and ovary. With the exception of prostate cancer, focal aberrant expression of MUC2 and MUC3 was frequently observed in these tumors.[87] In general, however, mucin genes appear to be independently regulated and their expression is organ-and cell-type–specific.[87]

There is evidence for host immune recognition of breast cancer mucin in that cytotoxic T lymphocytes isolated from breast cancer patients recognize a mucin epitope expressed on the breast cancer cells.[88] The immune-recognized epitope involves the core protein that appears to be selectively exposed on breast, ovarian, and other carcinomas. It has also been demonstrated that patients can produce antibodies to cancer mucins (reviewed in Ref. 86), and this is the basis for the proposal that glycopeptides based on the aberrantly processed mucins of cancer cells may have some utility as tumor vaccines (see Chap. 13). Clinical trials to test this possibility are under way.[89]

Some mucin antigens are shed from tumor cells and can be detected in the sera of patients with pancreatic, ovarian, breast, and colon cancer. These include CA19-9, CA125, CA15-3, SPan-1, and DuPan-2, which are currently being used as tumor markers.[90]

Proteoglycans

Another important type of glycosylated protein comprises the proteoglycans. These are high-molecular-weight glycoproteins that have a protein core, to which are covalently attached large numbers of side chains of sulfated glycosaminoglycans as well as N-linked and/or O-linked oligosaccharides. They are categorized based on their glycosaminoglycans into several types, including heparin, heparin sulfate, chondroitin sulfate, dermatan sulfate, and keratan sulfate (Fig. 4-5).[91] The glycosaminoglycans have different repetitive disaccharide units bound to the core protein through a common glycosaminoglycan linkage region: GlcNAcβ1→3Galβ1→3Galβ1→4Xylβ1-O-Ser. The structure of the sulfated glycopeptides from the carbohydrate-protein linkage region of some of the proteoglycans has been determined (reviewed in Ref. 92).

The core proteins of the proteoglycans appear to be similar among the various types; thus, at the protein level, they are not as diverse as the cell surface glycoproteins. Yet the proteoglycans have more complexity of their carbohydrates, resembling in structure a "bottlebrush" with a central core protein and numerous carbohydrate side chains (Fig. 4-6),[93] whereas glycoproteins as a rule resemble more a tall tree with a long trunk (if they are in the denatured form lacking three-dimensional conformation) and branches extending from the trunk.

Proteoglycans interact via their multiple binding domains with many other structural macromolecules, giving them the capacity "to function as a multipurpose 'glue' in cellular interactions".[93] They bind together extracellular matrix (ECM) components such as hyaluronic acid, collagen, laminin, and fibronectin; mediate binding of cells to the ECM; act as a reservoir for growth factors; and "present" growth factors to growth factor receptors on cells (Fig. 4-6). The proteoglycans also act as cell adhesion factors by promoting organization or actin filaments in the cell's cytoskeleton (Fig. 4-7).

The proteoglycans also have growth regulatory effects, part of which relates to their ability to bind growth factors such as FGF and TGF-β

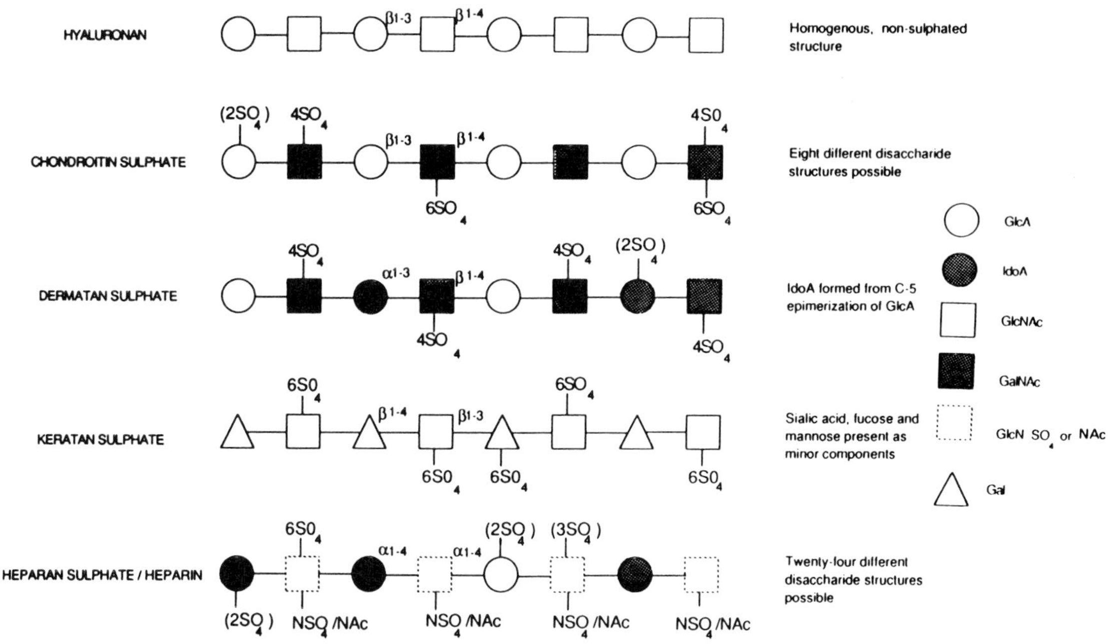

Figure 4-5 Repeating disaccharide structure of glycosaminoglycans. Hyaluronan is synthesized without a covalent link to any protein. Dermatan sulfate is formed from chondroitin sulfate by intracellular epimerization of glucuronate (GlcA) to iduronato (IdoA). 2-sulfated hexuronate residues are more common in dermatan sulfate than in chondroitin sulfate. Heparin is a more extensively epimerized and sulfated form of heparan sulfate. Heparan sulfate frequently contains some chain segments with little or no epimerization or sulfation. Heparin is synthesized only on a mast cell granule proteoglycan, whereas heparan sulfate is found on many cell surfaces and on some matrix proteoglycans. (From Hardingham and Fosang.[91])

(see Chap. 9). The TGF-β released by cells can, in turn, stimulate the synthesis of various proteoglycans, which regulate cell proliferation and differentiation.

In addition to having a role in cell attachment and spreading, growth control, cell–cell recognition, and cellular differentiation, proteoglycans have been shown to undergo both quantitative and qualitative changes during malignant transformation, and alterations have been reported in breast, colon, and liver carcinomas, in glioma cells, and in transformed murine mammary cells and 3T3 fibroblasts (reviewed in Ref. 94). Alterations of proteoglycans (PGs) in adenovirus EIA gene–transfected, immortalized rat fetal intestinal epithelial cells, compared to their normal counterparts, include (1) increased release of PGs from the immortalized cells, (2) higher amounts of membrane-intercalated PGs, (3) release of PGs of smaller hydrodynamic size, (4) increased percentage of heparan sulfate–type PG in the cell-associated fraction, (5) heparan sulfate and chondroitin sulfate of shorter chain length, and (6) synthesis of less sulfated glycosaminoglycan chains.[94]

These results suggest that multiple changes in PG synthesis, cellular localization, and shedding occur early in the oncogenic transformation process. However, a number of these changes may be more related to an increase in cell proliferative capacity than to the malignant phenotype, since immortalized cells are by their very nature programmed to proliferate indefinitely.

Modification of Extracellular Matrix Components

The ECM plays a key role in regulating cellular proliferation and differentiation. In the case of tumors, it is now clear that development of a blood supply and interaction with the mesenchymal stroma (on which tumor cells grow) are involved in their growth, invasive properties, and metastatic potential. This supporting stromal

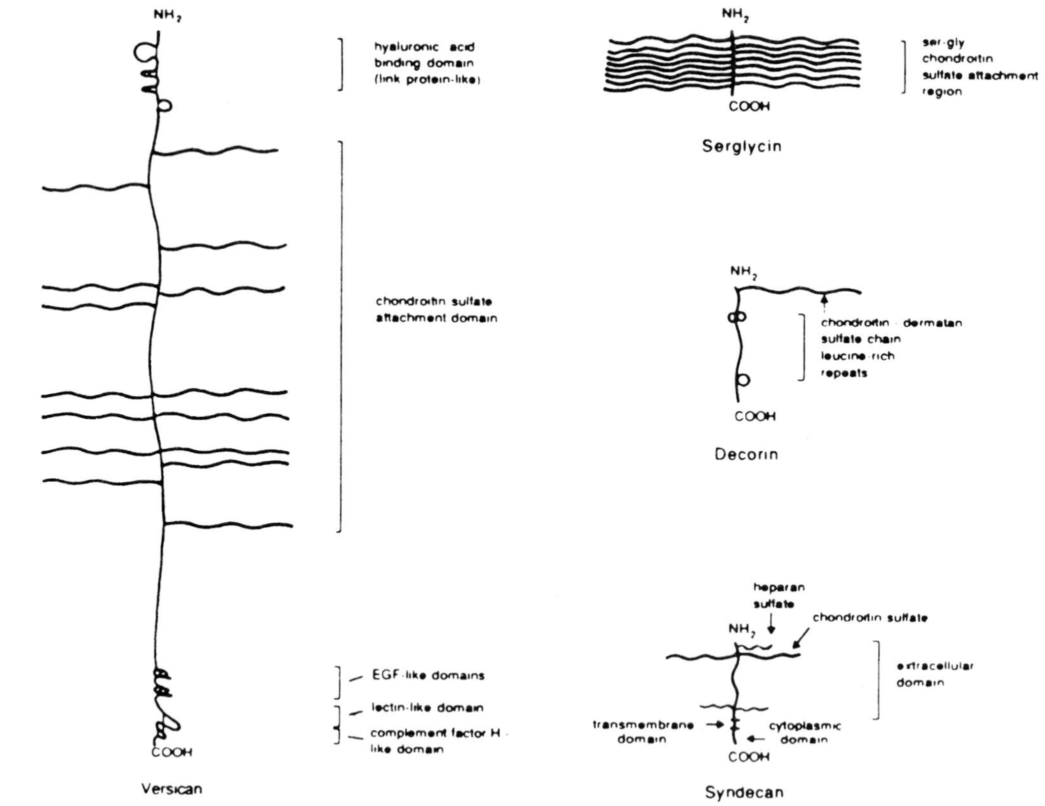

Figure 4-6 Schematic representation of proteoglycan structures. The various structural domains are depicted as folded structures based on the presumed disulfide bonding and/or electron microscopic appearance. *EGF,* epidermal growth factor. (From Ruoslahti.[93])

structure is continuously remodeled by the interaction between the growing tumor and host mesenchymal cells and vasculature.

The ECM—made up of collagen, proteoglycans, and glycoproteins such as fibronectin, laminin, and entactin (see Chap. 5)—forms the milieu in which tumor cells proliferate and provides a partial barrier to their growth. Basement membranes are a specialized type of ECM. These membranes serve as a support structure for cells; act as a "sieving" mechanism for transport of nutrients, cellular metabolic products, and migratory cells (e.g., lymphocytes); and play a regulatory role in cell proliferation and differentiation (reviewed in Ref. 95). The basement membranes (basal laminae) form a sheetlike structure on which rest epithelial cells, endothelial cells, and many types of mesenchymal cells. Basement membranes also prevent the willy-nilly passage of cells across them, but there are mechanisms that permit the passage of inflammatory cells. It is also clear that basement membranes act as regulators of cell attachment, through cellular receptors called integrins. There is also "cross-talk" between epithelial cells and their ECM to create a microenvironment for accurate signal transduction for growth factors and other regulatory molecules. It has been shown, for example, that exogenous reconstituted basement membranes stimulate specific differentiation of a variety of cell types, including mammary cells, hepatocytes, endothelial cells, lung alveolar cells, uterine epithelial cells, Sertoli cells, and Schwann cells (reviewed in Ref. 96).

The basement membrane barrier is only partial because, as will be discussed in Chapter 11, tumor cells can release a variety of proteases, glycosidases, and collagenases that have the abil-

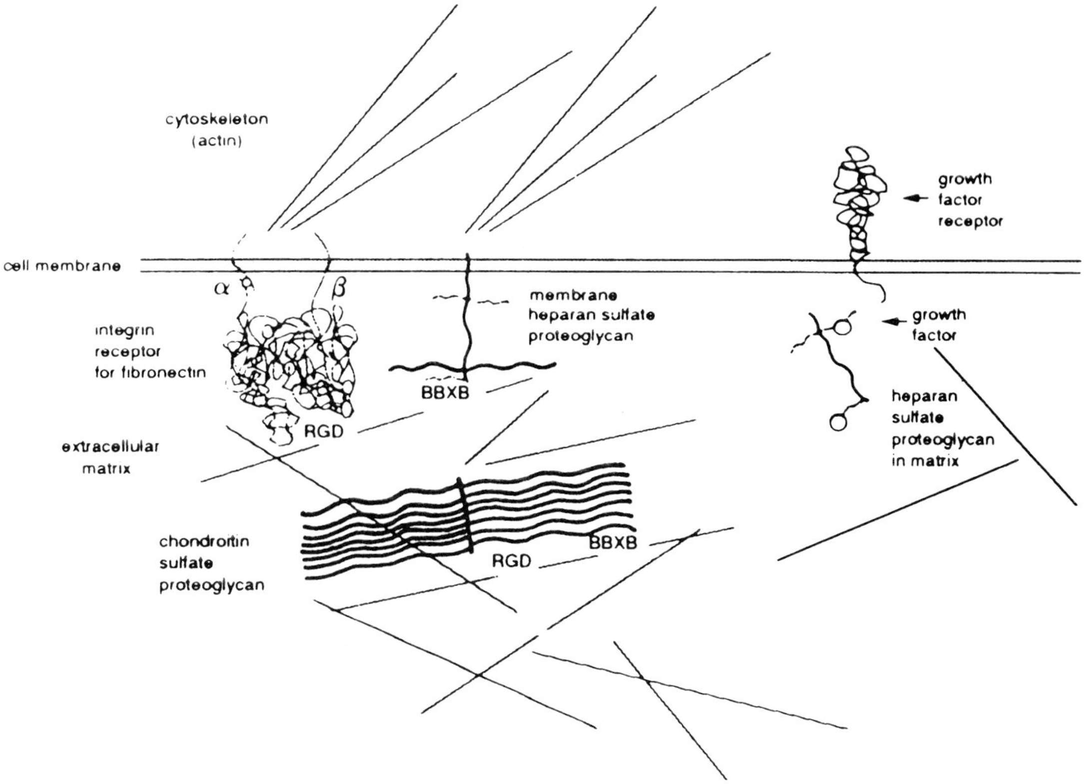

Figure 4-7 Participation of proteoglycans in cell adhesion and growth factor binding. The figure shows an integrin-type receptor binding to the Arg-Gly-Asp (*RGD*) cell attachment site and the membrane-bound heparan sulfate proteoglycan (syndecan) binding to the BBXB (B = basic amino acid, *X* = any amino acid) glycosaminoglycan binding site of fibronectin. Their binding may be inhibited by a chondroitin sulfate proteoglycan that has already bound to the BBXB site of a fibronectin molecule. Also depicted in this figure is the binding of growth factor molecules (*open circles*) to the heparan sulfate chains of a heparan sulfate proteoglycan present in the matrix. The proteoglycan binding is thought to serve to concentrate and localize growth factors, enhancing their availability to growth factor receptors. (From Ruoslahti.[93])

ity to degrade various components of the matrix and thus to invade through tissue barriers and blood vessel and lymph channel walls. In addition, malignant cells themselves have receptors for and/or can produce certain components of the matrix; this capability enables them to bind to vascular endothelium and may be involved in their ability to metastasize (see Chap. 11). Tumor cells may also release polypeptide factors that can modulate the type of proteoglycans produced by host mesenchymal cells. For example, normal fibroblasts have been shown to produce proteoglycans containing an unusual amount of chondroitin sulfate when they are exposed to conditioned growth medium from cultured human colon cancer cells.[97]

Other tumor-cell–modulated changes in the connective tissue surrounding growing tumors have also been observed. For example, invasive rabbit V_2 carcinomas induce production of a high level of hyaluronic acid in the surrounding host animals' connective tissue[98]; this may enhance tumor-cell migration and infiltration. Mammary carcinomas stimulate a 10-fold increase in Type V collagen production by host stromal cells.[99] Cultured mouse epithelial cells transformed by murine sarcoma virus (MSV) or murine leukemia virus (MuLV) produce and shed more fibronectin, and the type of collagen chains deposited in the matrix that these cells lay down is altered.[100]

Although there is morphologic evidence of aberrant basal laminae in epithelial tumors and biochemical evidence of the breakdown or altered production of the ECM components of the basal lamina by tumor cells, the mechanisms for these alterations are not entirely clear. One hypothesis, described previously, is that tumor cells can continually remodel the basal lamina by release of lytic enzymes and/or release of factors that modulate production of ECM components by the host's own cells. Another hypothesis is the malignant cells of epithelial origin (the most common type of human cancer; see Chap. 1) make an abortive attempt to produce their own ECM, and as they proliferate and replace the normal epithelium, a normal basal lamina fails to form under them because of this aberrant synthetic capacity. There is now clear evidence that human tumor cells from a variety of tissue sources can, in fact, produce laminin, one of the key components of the basal lamina of epithelial tissues, and that various tumor cell types differ in the type and amount of laminin produced and secreted.[101,102] Thus, the aberrant matrix of certain tumors may result from the inability of the transformed cells to produce and lay down the required components.

There is, however, good evidence that transformed cells release and shed their cell surface components more rapidly than nontransformed cells. This could come about because of increased cell membrane turnover, perhaps related to the increased rate of cell division and cell loss that many tumor-derived or transformed cell lines and tumor tissues have, or because of the increased amount of degradative enzymes found in many transformed malignant cells. Glycosidases and protease activities are elevated in malignant cells, and this may be one of the mechanisms by which cell surface components are released. Whatever the exact mechanism, the phenomenon is widespread enough among malignant cells to indicate that it is a common (although perhaps not universal) characteristic of the malignant phenotype. From a practical point of view, this effect results in the production of "biologic markers" of cancer that can be taken advantage of clinically. For example, the presence of such cell surface components as carcinoembryonic antigen, alkaline phosphatase isoenzymes, and various glycosyl transferases in the blood of cancer patients can be used to indicate the degree of tumor spread and the tumor burden. These molecules are found on the cell surface or are integral components of the cell membrane, and their elevation in the blood of cancer patients appears to reflect a phenomenon similar to that observed in cell culture studies.

Alteration of Membrane Transport Systems

Transport systems used in the uptake of such needed nutrients as sugars, amino acids, and nucleosides frequently function at higher capacity in transformed cells. Since similar changes occur during stimulation of cell proliferation by addition of growth factors or mitogens to cultures of growth-arrested normal cells, it is not clear that any of these changes are specific for malignant transformation.

Transport of glucose, its nonmetabolizable analogues 2-deoxyglucose and 3-*O*-methylglucose, mannose, galactose, and glucosamine increase with transformation.[103-105] Transport of certain amino acids—such as glutamine, arginine, and glutamic acid—and of amino acid analogues that are not incorporated into protein—such as cycloleucine and α-aminoisobutyric acid—is also increased after transformation of cultured cells.[106,107] In most of these studies, kinetic analyses show that the initial velocity (V_{max}) is increased with no change in the K_m,[104,106,107] indicating that the rate of transport rather than the affinity of the carrier is altered by transformation.

The greater transport capacity of some transformed cells appears to be associated with low cellular levels of cAMP, and it has been postulated that transformation modifies the cell membrane adenylate cyclase, keeping intracellular cAMP levels low.[108] The association of cyclic nucleotide changes; increased transport; elevated DNA, RNA, and protein synthesis; and greater cell proliferation has been termed the pleiotypic response.[109] This is observed when density-dependent inhibition of growth is overcome by addition of fresh serum or certain cell-growth-stimulatory hormones (e.g., insulin) to growth-inhibited cultures of normal cells or by cell transformation with oncogenic viruses. These changes are thus likely to be typical of a "shift up" in the proliferation rate of cells and only secondarily related to the malignant phenotype.

It is now well accepted that tumor-associated antigens are real entities and that they usually represent altered cell-surface components present on and often shed by tumor cells. The development of monoclonal antibody technology and modern radioimaging and radioimmunotherapeutic approaches using monoclonal antibodies tagged with radionuclides has provided a number of new approaches to cancer diagnosis and treatment. In addition, the serologic detection of cancer by the development of sensitive assays for tumor antigens (so-called tumor markers) has provided improved diagnostic tests for certain kinds of cancer. The use of prostate-specific antigen (PSA) and CA-125 for the diagnosis of prostate and ovarian cancer, respectively, are examples of the use of tumor markers in cancer detection (Chap. 1).

ENZYME PATTERNS IN CANCER

The biochemical differences between normal and malignant cells were first discovered in their patterns of enzymatic activity, and much of what we know about the biochemistry of cancer cells originally came from studies on enzyme pathways in malignant cells. In the 1920s, Warburg studied glycolysis in a wide variety of human and animal tumors and found that there was a general trend toward an increased rate of glycolysis in tumor cells, leading to the production of large amounts of lactic acid from glucose.[110] He noted that when normal tissue slices were incubated in nutrient medium containing glucose but without oxygen, there was a high rate of lactic acid production (anaerobic glycolysis); however, if they were incubated with oxygen, lactic acid production virtually stopped. The rate of lactic acid production was higher in tumor tissue slices in the absence of oxygen than in normal tissues, and the presence of oxygen slowed but did not eliminate lactic acid formation. Warburg concluded that cancer cells have an irreversible injury to their respiratory mechanism, which increases the rate of lactic acid production even in the presence of oxygen (aerobic glycolysis). He regarded the persistence of aerobic glycolysis as the crucial biochemical lesion in neoplastic transformation. This idea persisted for many years, and it was not until about 1950, when radioactive tracer techniques became available and were used to study the metabolism of radioactively labeled glucose to carbon dioxide, that it became apparent that tumor tissue could oxidize glucose at a rate comparable to that of normal tissues.[111] It soon became apparent as well that tumor tissues contain functional respiratory enzymes and coenzymes, that the citric acid cycle is functional in these cells, and that they can couple oxidation with ATP formation. Thus, the Warburg hypothesis is untenable in light of what is now known about glycolysis and the pathways of oxidative phosphorylation. However, certain tumor cells do have an increased ability to survive in an anaerobic environment because of their high rate of anaerobic glycolysis. This is probably secondary to loss of growth control and reflects tumor progression and the increased growth fraction (percentage of dividing cells) of many malignant tissues rather than a process in the initiation of malignant transformation.

In the early 1950s, Greenstein formulated the "convergence hypothesis" of cancer, which states that the enzymatic activity of malignant neoplasms tends to converge to a common pattern.[112] Although he recognized some exceptions to this rule, he considered the generalization valid. It is now more fully appreciated that there is tremendous biochemical heterogeneity among malignant neoplasms and that there are many fairly well differentiated cancers that do not have the common enzymatic alterations he suggested. Thus, cancers do not have a universally uniform malignant phenotype as exemplified by their enzyme patterns. But the Greenstein convergence hypothesis was useful in that it stimulated a significant amount of research into the biochemistry of cancer. As is true for the observations made by Warburg, the patterns observed by Greenstein appear to be present primarily in highly undifferentiated, rapidly growing tumors.

Based on work of about 40 years ago, which evolved from studies on the production of hepatic cancer by feeding aminoazo dyes, the Millers advanced the "deletion hypothesis" of cancer.[113] This hypothesis was based on the observation that a carcinogenic aminoazo dye covalently bound liver proteins in animals undergoing carcinogenesis, whereas little or no dye binding occurred with the protein of tumors induced by the dye. They suggested that carcinogenesis resulted from "a permanent alteration or loss of protein essential for the control of growth." Distinct aminoazo dye binding proteins have been identified,[114] but their presence has also been detected in better differentiatied hepatomas.

About 10 year later, Potter suggested that the proteins lost during carcinogenesis may be involved in the feedback control of enzyme systems required for cell division,[115] and he proposed the "feedback deletion hypothesis."[116] In this hypothesis, Potter proposed that "repressors" crucial to the regulation of genes involved in cell proliferation are lost or inactivated by the action of oncogenic agents on the cell, either by interacting with DNA to block repressor gene transcription or by reacting directly with repressor proteins and inactivating them. This perspicacious prediction anticipated the discovery of tumor suppressor proteins such as p53 and RB by about 25 years.

Biochemical studies of cancer were also aided by the so-called minimal-deviation hepatomas developed by Morris.[117] These tumors were originally induced in rats by feeding them the carcinogens fluorenylphthalamic acid, fluorenylacetamide compounds, or trimethylaniline. These hepatocellular carcinomas are transplantable in an inbred host strain of rats and have a variety of growth rates and degrees of differentiation. They range from slowly growing, well-differentiated, karyotypically normal cells to rapidly growing, poorly differentiated polyploid cells. All these tumors are malignant and eventually kill the host. The term "minimal deviation" was coined by Potter[116] to convey the idea that some of these neoplasms differ only slightly from normal hepatic parenchymal cells. The theory was that if the biochemical lesions present in the most minimally deviated neoplasm could be identified, the crucial changes defining the malignant phenotype could be determined. Although this has not been fully realized, studies of these tumors greatly advanced our knowledge of the biochemical characteristics of the malignant phenotype, and they have ruled out many secondary or nonspecific changes that relate more to tissue growth rate than to malignancy. For example, the Warburg and Greenstein hypotheses became untenable as generalizations about cancer after the minimal-deviation hepatomas were thoroughly studied.

In extensive biochemical analyses of the Morris minimal-deviation hepatomas, various enzyme and isoenzyme patterns have been correlated with the degree of malignancy of these tumors.[111,118,119] These data led Weber to formulate the "molecular correlation concept" of cancer, which states that "the biochemical strategy of the genome in neoplasia could be identified by elucidation of the pattern of gene expression as revealed in the activity, concentration, and isozyme aspects of key enzymes and their linking with neoplastic transformation and progression."[118] Weber proposed three general types of biochemical alterations associated with malignancy: (1) transformation-linked alterations that correlate with the events of malignant transformation and are probably altered in the same direction in all malignant cells; (2) progression-linked alterations that correlate with tumor growth rate, invasiveness, and metastatic potential; and (3) coincidental alterations that are secondary events and do not correlate strictly with transformation or progression. Weber maintained that key enzymes, that is, enzymes involved in the regulation of rate and direction of flux of competing synthetic and catabolic pathways, would be the enzymes most likely to be altered in the malignant process. In contrast, "non-key" enzymes, that is enzymes that are not rate-limiting and do not regulate reversible equilibrium reactions, would be of lesser importance. As one would expect, a number of the enzyme activities that Weber and others have found to be altered in malignant cells are those involved in nucleic acid synthesis and catabolism. In general, the key enzymes in the de novo and salvage pathways of purine and pyrimidine biosynthesis are increased and the opposing catabolic enzymes are decreased during malignant transformation and tumor progression. Enzyme changes that occurred in all hepatomas studied, and are by Weber's definition therefore transformation-linked, are listed in Table 4-6. These include increased activities of some of the rate-limiting enzymes in the purine and pyrimidine biosynthetic

Table 4-6 Transformation-Linked Discriminants—Enzymatic Markers of Malignancy

Metabolic Area	Synthetic Pathway (Increased)	Degradative Pathway (Decreased)
Pentose phosphate	Glucose-6-phosphate dehydrogenase	
	Transaldolase	
	PRPP synthetase[a]	
Purine	Glutamine PRPP amidotransferase	Inosine phosphorylase
	Adenylosuccinate synthetase	Xanthine oxidase
	Adenylosuccinase	
	AMP deaminase[a]	
	IMP dehydrogenase[a]	Uricase
Pyrimidine and DNA	Orotate phosphoribosyltransferase	
	Orotidine-5′-phosphate decarboxylase	
	Uridine kinase	Thymidine phosphorylase
	UDP kinase	
	CTP synthetase[a]	

[a]Also linked with progression
Source: From Weber.[119]

pathways and decreased activities of purine and pyrimidine degradative enzymes. Alterations that relate to tumor progression are depicted in Table 4-7; these also include increased activities of enzymes involved in nucleic acid synthesis and decreased activities of the opposing degradative enzymes. Weber maintained that all Morris hepatomas also have decreased glycogen content and decreased conversion of glucose to fatty acids. Progression-linked changes in carbohydrate metabolic pathways include higher concentrations of key glycolytic enzymes (e.g., hexokinase, phosphofructokinase, and pyruvate kinase) and lower concentrations of key gluconeogenic enzymes (e.g., glucose-6-phosphatase, fructose-1, 6-diphosphatase, and pyruvate carboxylase). The concentration of certain enzymes involved in protein synthesis are increased and some involved in amino acid catabolism are decreased, relative to tumor progression. Moreover, enzymes of the pentose phosphate shunt, such as glucose-6-phosphate dehydrogenase, transaldolase, and phosphoribosyl pyrophosphate (PRPP) synthetase, are elevated in all hepatomas. Since these enzymes are involved in supplying PRPP needed for purine and pyrimidine biosynthesis, this is also a predictable alteration.

These data provided a detailed picture of the biochemistry of minimal-deviation hepatomas and indicated which key pathways are more likely to be altered early in neoplastic transformation and which are clearly late or secondary events; however, the molecular correlation idea suffers from a flaw similar to that of the Warburg and Greenstein hypotheses in that the observed changes (although much more detailed and precise) appear to relate primarily to tumor progression and the concomitant loss of differentiated characteristics, increased cellular proliferation rate, and accumulation of karyotypic abnormalities rather than to the crucial molecular events that initiate the malignant process.

The question of why anaplastic, rapidly growing tumors tend to be biochemically alike, whereas better differentiated tumors display a vast array of phenotypic characteristics, has been approached by Knox.[120] He thought that the vast bulk of biochemical components in tumor tissues are "normal" in the sense that they are produced by certain specialized adult normal cells or by normal cells at some stage of their differentiation. In cancer cells, it is the combination and proportions of these normal components that are abnormal. The biochemical diversity of cancer cells, then, would depend on the cell of origin of the neoplasm and its degree of neoplasticity.[120] All too frequently in the biochemical model systems used to study cancer, the fact that a given biochemical component is present or absent or increased or decreased is not considered in relation to the particular cell of origin of a tumor, its differentiation state, or its degree of neoplasticity. Knox and his colleagues associated the degree of biochemical diversity among a series of normal and neoplastic rat tissues by measuring 23 different enzyme activities as well

Table 4-7 Progression-Linked Discriminants—Enzymatic Markers of the Degrees of Malignant Transformation

Nucleic Acid Metabolism	Carbohydrate Metabolism	Protein and Amino Acid Metabolism	Other Metabolic Areas
Pyrimidine, RNA and DNA synthesis, increased Carbamoyl-phosphate synthase (glutamine) Aspartate carbamoyltransferase Dihydro-orotase CTP synthetase Uridine kinase Ribonucleotide reductase Thymidylate synthetase Deoxycytidylate deaminase Thymidine kinase DNA polymerase DNA nucleotidyltransferases tRNA methylase *Pyrimidine catabolism, decreased* Dihydrouracil dehydrogenase *Purine synthesis, increased* IMP dehydrogenase AMP deaminase *Purine catabolism, decreased* Adenylate kinase	*Glucose catabolism, increased* Hexokinase Phosphofructokinase Pyruvate kinase *Glucose synthesis, decreased* Glucose 6-phosphatase Fructose 1,6-diphosphatase Phosphoenolpyruvate carboxykinase Pyruvate carboxylase *Specific phosphorylating enzymes, decreased* Fructokinase Glucokinase *Fructose metabolism, decreased* Thiokinase Aldolase *Isozyme shift, decreased* High-K_m isozymes *Isozyme shift, increased* Low-K_m isozymes *R-5-P utilization, increased* PRPP synthetase	*Protein synthesis, increased* Amino acid incorporation into protein (alanine, aspartate, glycine, serine, isoleucine, valine) Activity of postmicrosomal protein synthesizing system *Enzymes catabolizing amino acids, decreased* Glutamate dehydrogenase Glutamate oxaloacetate transaminase Tryptophan pyrrolase Serotonin deaminase 5-Hydroxytryptophan decarboxylase *Urea cycle, decreased* Ornithine carbamoyltransferase	*Membrane cAMP metabolism* cAMP phosphodiesterase increased Adenylate cyclase decreased *Polyamine synthesis, increased* Ornithine decarboxylase *Polyamine synthesis, decreased* S-Adenosylmethionine synthetase *Ketone-body utilization, increased* Succinyl CoA: acetoacetate CoA transferase *Lipid metabolism, decreased* α-Glycerophosphate dehydrogenase HMG-CoA synthase

Source: From Weber.[119]

as such characteristics as protein content, DNA content, and number of nuclei per gram of tissue. By scoring each of these parameters and calculating the correlation coefficient for their expression between one tissue type and another, they estimated the biochemical similarity or dissimiliarity of two different tissues. By these criteria, it was observed that fetal tissues tend to resemble each other, whereas adult tissues usually do not. Moreover, undifferentiated tumors, regardless of their cell of origin, tend to be biochemically alike, but more differentiated tumors tend to be more like their tissue of origin than like other tumors. For example, a well-differentiated Morris hepatoma is more like adult normal liver than an undifferentiated hepatoma. Similarly, a well-differentiatied rat mammary tumor is more like normal mammary gland than undifferentiated mammary tumors. However, undifferentiated mammary and liver tumors have a similar phenotype.

Taken together, the data on enzyme patterns of cancer cells indicate that undifferentiated, highly malignant cells tend to resemble one another and fetal tissues more than their adult normal counterpart cells, whereas well-differentiated tumors tend to resemble their cell of origin more than other tumors. Of course, between these two extremes, several levels of neoplastic gradation occur, leading to the vast biochemical heterogeneity of tumors. This heterogeneity also exists for tumors of the same tissue type arising in different patients or even in the same patient at different stages of the disease.

Other important classes of enzymes that have important cell-regulatory functions are protein kinases and phosphatases. These enzymes have key functions in signal transduction pathways involving cellular responses to growth factors, hormones, and other external stimuli as well as in cell-cycle regulation and in other key metabolic pathways. A number of oncogene products are protein kinases or interact with protein kinases. The functional state of a number of cellular proteins is determined by their phosphorylation state, and the ability of cells to regulate the phosphorylation state of its proteins is altered in cancer cells. These issues are discussed in detail in Chapters 7 and 9.

PROTEASES IN CANCER

It has long been known that certain types of malignant tumors produce and release proteolytic enzymes. In 1925, Fischer[121] observed that explants of virally induced tumors in chickens lysed plasma clots, whereas explants of normal connective tissues did not. A number of years later, increased fibrinolytic activity was found in human sarcomas.[122] The relationship between fibrinolytic activity and transformation was studied in detail by Reich and his colleagues. They established that fibroblast cultures derived from chicken, hamster, mouse, and rat embryos produced an enzymatic activity capable of hydrolyzing fibrin after the cells were transformed by oncogenic DNA or RNA viruses.[123,124] The fibrinolytic activity was not detected in untransformed embryo or 3T3 cell cultures (but later was found in other untransformed 3T3 lines during their log growth phase).[125] The activity was also low in cultures of cells infected with a variety of nononcogenic viruses. It was subsequently shown that the fibrinolytic activity was produced by the interaction of the serum proteolytic proenzyme plasminogen, present in the cell culture medium, and a factor called plasminogen activator elaborated by the transformed cells.[126,127] Plasminogen activator (PA) is itself a protease of the serine-active site type that activates plasminogen to the active fibrinolytic enzyme plasmin by cleavage of an arginine-valine bond in the carboxy terminal portion of the proenzyme molecule. Plasmin is a normal plasma protein that has a role in the body in dissolving fibrin clots (Fig. 4-8). It is usually activated locally to lyse a clot in the bloodstream, and its action is opposed by a number of circulating antiproteases that prevent widespread proteolysis. The most important plasmin inhibitor is antiplasmin itself, but other inhibitors include α_1-antitrypsin, α_2-macroglobulin, and C_1-inactivator.[128] Production of plasminogen activator by Rous sarcoma virus-transformed cells is associated with expression of the viral genes required for transformation because elevation of PA activity in cells infected with a temperature-sensitive mutant of Rous sarcoma virus occurred only at the "permissive" temperature, that is, at the temperature at which the transformed phenotype was expressed.[129]

Extracts from several human tumors have higher levels of PA than extracts of adjacent control tissues, as well as higher levels of a number of other proteases such as cathepsins and increased activity of other lytic enzymes. The release of these proteases and lytic enzymes plays a role in the metastic potential of cancer cells, as discussed in Chapter 11.

ONCODEVELOPMENTAL GENE PRODUCTS

A number of cancer cell characteristics, such as invasiveness and escape from host immune responses, are also seen in embryonic tissue. For example, the developing trophoblast invades the uterine wall during the implantation step of embryonic development. The developing embryo carries antigenic determinants from both parents, and yet it is not rejected as foreign tissue by the mother. During organogenesis, embryonic cells dissociate themselves from surrounding cells and migrate to new locations, a process not unlike metastasis. All of these, however, are controlled processes with a definitive endpoint, whereas in cancer similar events occur, but in an uncontrolled manner. One of the keys to understanding neoplasia may be to discover what regulates the expression of these embryonic processes.

During the process of carcinogenesis, a number of developmental genes are reexpressed, a phenomenon known as oncodevelopmental gene expression. Which, if any, of these gene products are involved in invasion, metastatic spread, and avoidance of host immunologic defense mechanisms is not yet clear, but it seems reasonable to predict that at least some of them are so involved. The ectopic expression of oncodevelopmental genes was discussed in Chapter 3. A few more examples are given here.

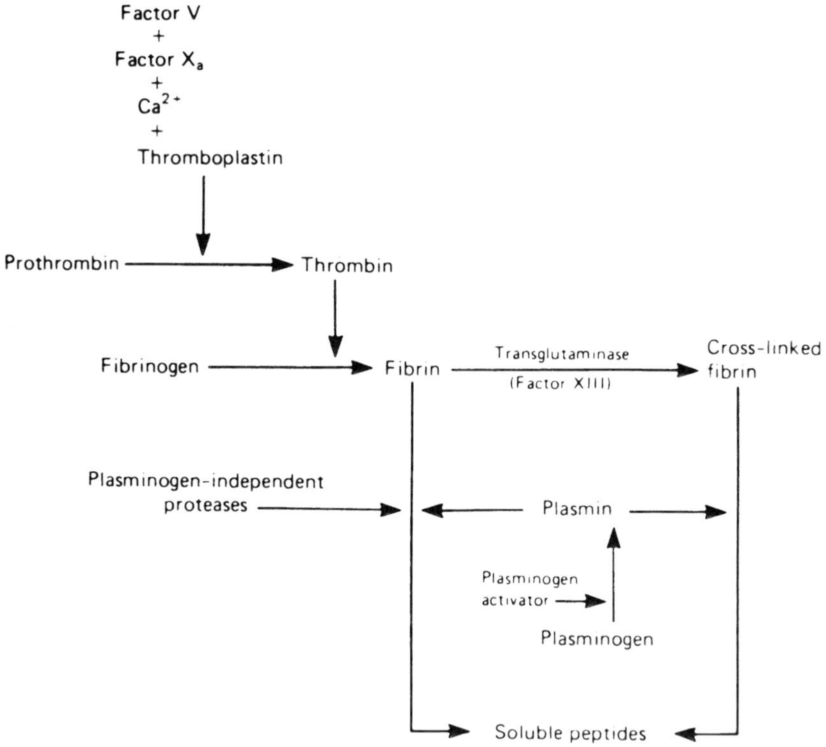

Figure 4-8 Factors and enzymes involved in the blood clotting pathway and fibrinolysis.

A number of the tumor-associated antigens appear on tumor cells as a result of the apparent reexpression (or increased expression) of embryonic genes. These include alpha-fetoprotein (AFP) and carcinoembryonic antigen (CEA). Identified in human fetal cord blood as an alpha-globulin not detectable in adult blood by electrophoresis,[130] AFP is a serum protein produced by the fetal liver and yolk sac during prenatal development.[131] It is similar in size, structure, and amino acid composition to serum albumin. In the human fetus, AFP levels peak at 15 weeks of gestation and then fall rapidly until the normal adult level is reached. Abnormally high levels in the amniotic fluid or maternal serum during gestation are indicative of congenital defects.[131] In 1963, Abelev et al.[132] observed that a fetal alpha-globulin was also present in adult mice bearing hepatocellular carcinomas, and soon after it was observed that human patients with hepatocellular carcinomas also had elevated levels of AFP.[133] Blood levels of AFP also rise during the induction of liver tumors in rodents treated with chemical carcinogens (for review, see Ref. 131). AFP is currently used as a tumor marker in patients with embryonal testicular cancers and hepatocellular carcinomas.

Carcinoembryonic antigen was first described in 1965 by Gold and Freedman.[134] It was at first thought to be unique to the fetal digestive tract and to digestive tract neoplasms. It is now appreciated, however, that CEA, or a protein that is immunologically very similar, is produced in a variety of tissues, particularly those with mucin-secreting epithelium. A high-molecular-weight glycoprotein (about 180,000), and about 60% carbohydrate by weight,[135] CEA is produced and shed into the blood by a variety of carcinomas, in addition to those arising in the digestive tract, and has been useful as a tumor marker, particularly for colon cancer metastatic to liver.

A number of fetal and placental isoenzymes have also been found in malignant tissues, including fetal-type aldolase, glycogen phosphorylase, phosphotransferase, pyruvate kinase, and alkaline phosphatase.[136] As these fetal isoenzymes appear, the adult forms are frequently lost. Similarly, such placental hormones as chorionic gonadotropin and placental lactogen reappear in certain human cancers.

These data indicate that the reexpression of a more embryonic phenotype is a common occurrence during the development of cancer.

PRODUCTION OF TUMORS IN EXPERIMENTAL ANIMALS

The ability of transformed cells to produce tumors after injection into experimental animals is usually taken as the ultimate proof that the cells are malignantly transformed. Although some false negatives can occur (i.e., cells that are actually malignant not "taking" in the animal model for physiologic or immunologic reasons), it is less likely that false positives occur. Thus, the appearance of a tumor after the injection of cells into appropriate host animals is good evidence that they are tumorigenic. One of the problems with the animal test systems, however, is that tumors thus produced may not readily metastasize, rendering one of the most important properties of malignant neoplasia unevaluable. Development of systems to study metastases is discussed in Chapter 11.

Athymic (Nude) Mice

In the case of animal cells, evaluation of their tumorigenicity in vivo is made easier by the availability of syngeneic strains of animals. Such a system is, of course, not available to similarly determine the tumorigenicity of human cells. A unique system to test the neoplastic potential of human cells has been found, however, in the development of the hairless or nude mouse. This strain of mice, which arose during the development of inbred mouse strains, is characterized by lack of hair and lack of a thymus gland.[137,138]

Since the thymus gland is necessary for the production of the T lymphocytes involved in cell-mediated immunity, this strain of mice cannot defend itself against certain types of infection, nor can it easily reject grafts of foreign tissues. Thus, tumor grafts from heterologous species can grow in these animals and immune rejection will not occur. The nude mouse has thus been used as a convenient animal host in which to study the growth of tumors of animal and human origin. In general, the degree of differentiation and phenotypic characteristics of human tumors growing in nude mice reflects that of the original tumor taken from the patient.[139]

SCID Mice

Another immune-deficient mouse model for the transplantation and growth of human tumors is the SCID (severe combined immunodeficient) mouse. These mice have no functional T or B lymphocytes.[140] The immunologic defect is in the stem-cell progenitor pool of T and B cells. Macrophages and natural killer (NK) cell functions are intact. Because SCID mice will not reject xenografts of human lymphocytes or human tumors, they can be repopulated with human T or B cells and serve as hosts for the growth of human tumors. An example of how this animal system can be used is the injection of human colon adenocarcinoma cells by portal vein to establish a model for hepatic metastasis of colon cancer. Such a system has been used to test the effectiveness of lymphokine-activated killer (LAK) cells and cytokines as immunotherapeutic agents[141] (see Chap. 13).

Transgenic Animals

A real advance in understanding how genes are expressed during development, how the mutation or deletion of genes affect growth and development, and how oncogenes participate in the oncogenic process has been the development of transgenic animals. This experimental system comes about by the introduction of exogenous genes or partial gene sequences into early embryonic cells of a developing organism.

The first use of such transgenic technology involved the microinjection of SV40 virus DNA into the blastocyst cavity of mice[142] (see Fig. 4-9[143] for definition of blastocyst). Although SV40 DNA was expressed in a variety of somatic tissues in the fully developed mouse, integration of the viral DNA into the germ-line cells was not demonstrated.

Germ-line transmission of foreign DNA was subsequently demonstrated in 1976 when it was shown that mouse embryos exposed to infectious

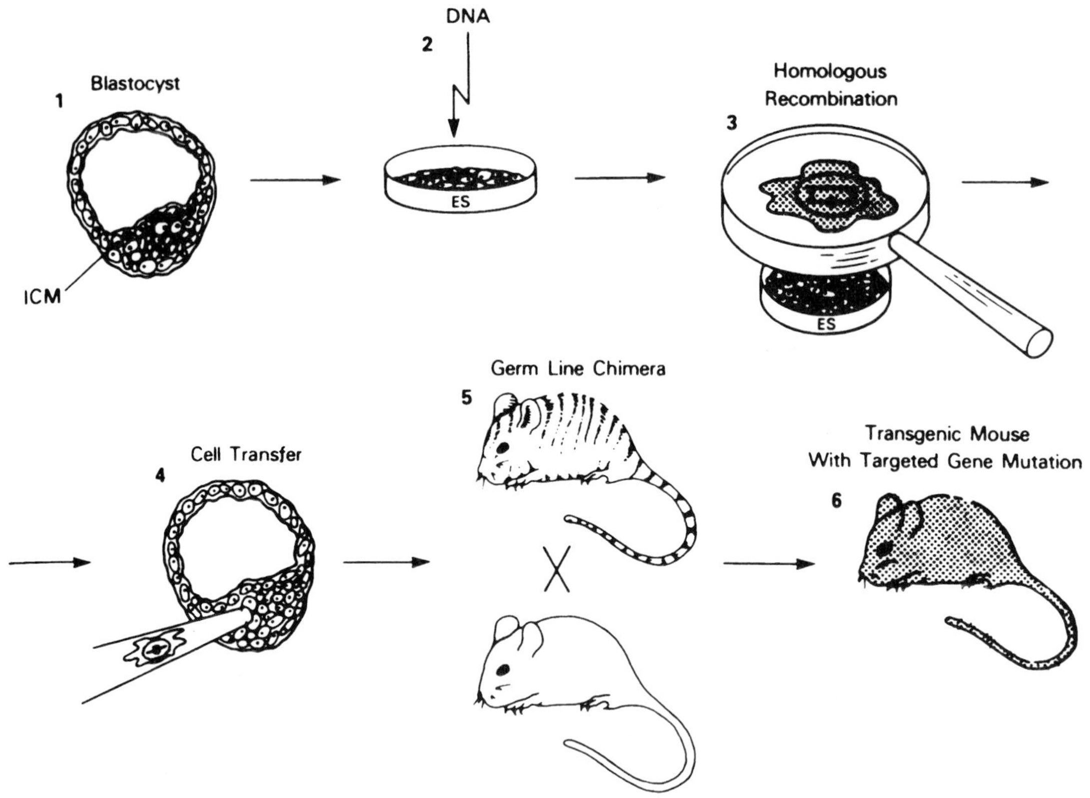

Figure 4-9 Homologous gene targeting. The inner cell mass (ICM) of a mouse blastocyst gives rise to embryonic stem (ES) cell lines, which are targeted by DNA constructs bearing homology to an endogenous mouse gene. ES cells are selected in which the incoming DNA recombined with homologous chromosomal sequences to create a mutation in the target gene. These cells are transferred into a recipient blastocyst, and germ line chimeras are generated. These chimeras are the founders of mouse lines that carry the selected mutation. (From Westphal.[143])

Moloney murine leukemia retrovirus (M-MuLV) incorporated MuLV DNA into germ-line cells and passed it on to subsequent offspring, thus creating the first transgenic mouse strain (reviewed in Ref. 142). One real power of the technique is that DNA becomes fully integrated into the genome, such that the function of such DNA can be studied in the offspring of various matings with animals of a different genotype.

Infection of mouse embryos with retroviruses was the first method for developing transgenic animals, but a more commonly used technique now is the direct microinjection of recombinant DNA into the pronucleus of a fertilized egg. Another technique is the introduction of DNA by transduction or transfection into embryonic stem cells. Each of these techniques is briefly described here (reviewed in Ref. 142).

Microinjection of DNA into the Pronucleus

This technique involves microinjection of cloned DNA directly into a fertilized ovum pronucleus, as noted above. It is proposed that injected DNA molecules, arranged in multiple head-to-tail arrays, associate by homologous recombination with similar sequences in the host's genome before integration, and in most cases multiple DNA copies insert at a single chromosomal site. It is thought that chromosomal breaks, perhaps caused by DNA repair enzymes called into play by the free ends of the injected DNA (which the cell's repair mechanisms may see as a break that needs repairing), open up integration sites. In support of this idea are the observations that rearrangements, deletions, duplications, or translocations occur at the integration sites.

One problem with this approach is that the injected DNA does not always integrate into the host cell's genome. The principal advantage is that it is an efficient way to get germ-line expression in a reasonably predictable manner. A disadvantage is that this method cannot be used to introduce genes at late developmental stages.

Retroviral Infection

In this method, embryonic cells are infected with a retrovirus carrying the gene of interest. Only a single proviral copy is integrated without gene rearrangement occurring. A short duplication of host DNA sequences occurs at the integration site. The primary advantage of this approach is that retrovirus infection can be used to insert genes into embryos at various developmental phases. Also, it is easier to isolate the host DNA sequences flanking the integration site by this method than by microinjection of the pronucleus. Disadvantages include a size limitation of the transduced DNA and a lack of reproducibility of expression of the introduced gene.

Embryonic Stem Cells

Embryonic stem (ES) cells are obtained from blastocysts grown as cultured explants in vitro. These cells can then be injected into host blastocysts, thus providing daughter cells that populate various host tissues, including germ-line cells producing a chimeric animal. Genes can be inserted into ES cells by DNA transfection or retroviral infection (transduction). This approach can be used to generate animals with a selected phenotype and also to generate animals lacking a specific gene by a procedure called "gene knockout" (see below).

Transgenic animals have been used to study a number of questions in biology—for example, to understand when genes are activated during development and what the regulatory promoter or enhancer sequences are that turn genes on during development. Another interesting use is the insertion of genes with regulatory elements modulated by some external signal such as a hormone. For example, hormone-inducible promoters have been attached to genes placed into transgenic animals to allow their expression after injection of the hormone. Moreover, tissue-specific promoters can be attached to genes to foster their expression only in given target tissues in transgenic animals.

It should be noted that transgenic technology is not limited to mice. Exogenous genes have also been microinjected into rabbit, sheep, and pig embryos with some limited success.

Transgenic technology has been very useful for studying the role of oncogene expression in tumorigenesis in experimental animals. The following questions, for example, can be addressed: What tissues are susceptible to the transforming activity of a specific oncogene? Is there some necessary cooperativity between expression of two or more oncogenes in causing oncogenesis in a given tissue? Is there a required order to the expression of oncogenes to cause cancer in a given tissue? Which oncogenes may be involved in various steps in oncogenesis (e.g., initiation, promotion-progression, metastasis)? Are some genes of infectious viruses involved in causation of cancer or other human diseases? What is the effect of oncogene expression on growth and development of the animal? What type of cancer or other conditions result from the knockout of various putative tumor suppressor genes? Examples of the carcinogenic effects of oncogenic expression in transgenic mice are shown in Table 4-8.[144] A transgenic mouse model has been used to show the synergistic effects between oncogenes and growth factors in tumorigenesis. For example, coexpression of the oncogene c-*myc* and transforming growth factor α (TGFα) in the livers of transgenic mice caused a great acceleration in the development of hepatocellular carcinomas in these animals.[145]

Gene Knockout

The alteration or loss of a gene function can also be studied in transgenic animals by homologous DNA combination, as noted above. In the process of homologous gene targeting, the exogenous DNA sequence differs in a defined way from the same gene in the host by, for example, having a base mutation, deletion, etc. The exogenous and host DNA sequences "find one another" and recombine at the regions of base sequence homology, resulting in the integration of the exogenous DNA such that the host's DNA gene sequence is lost by recombination events

Table 4-8 Transgenic Oncogenes as Cancer-Inducing Agents

Oncogene			Induced Cancers			
Oncoprotein	Characteristics	Regulatory Region	Proliferative Preneoplastic Stage?	Tumor Type(s)	Incidence	Comments
c-*myc*	Nuclear protein	Ig	Yes	B and pre-B-cell lymphomas	High	Tumors clonal, by Ig rearrangements
		MMTV LTR[a]	Yes	Mammary adenocarcinomas in females	High	Tumors arise in one of the ten mammary glands after second or third pregnancy
		Whey acidic protein (WAP)	Yes	(as above)	Moderate	(as above)
c-*fos*	Nuclear protein, transcription factor	Human metallothionein promoter, FBJ LTR 3′ region	Yes	Osteosarcomas	Low	Widespread hyperplasia of osteoblasts; infrequent tumors
SV40 large T antigen	Nuclear protein (binds Rb tumor suppressor and p53 oncoprotein)	SV40 early promoter	Yes	Choroid plexus tumors	High	Includes transcriptional activation of latent transgenome
		Elastase	Yes	Pancreatic acinar cell tumors	High	Discernible stages of hyperplasia; focal tumor nodules
		Insulin	Yes	Pancreatic islet (β) cell tumors	High	Typically 1% to 2% of the islets progress to tumors
		αA-crystallin	Yes	Lens cell tumors of the eye	High	Transformation of a cell type not associated with natural cancers
		Atrial natriuretic factor	Yes	Heart tumors	Moderate	Asymmetrical hyperplasia and tumors (in right atrium only)
H-*ras*	21-kDa GTP-binding protein associated with cell membrane	MMTV LTR	Yes	Mammary adenocarcinomas (males and females)	High	Transformation of a mammary epithelial progenitor (in males and females)

		Elastase	Yes	None; lethal neonatal hyperplasia	High	ras involved in exocrine development?
*int*1	(Secreted protein?)	MMTV LTR	Yes	Mammary and salivary-adenocarcinomas (males and females)	High	Typically, a tumor in 1 of the 10 mammary glands
c-*myc* + H-*ras*	(see above)	MMTV LTR	Yes	Mammary adenocarcinoma	High	Accelerated tumor formation, still focal
neu	Mutant form of (a) rat c-*neu*, an EGF-R like membrane tyrosine kinase	(a) MMTV LTR	?	Mammary adenocarcinomas	Very high	Multifocal tumors in every mammary gland (one step; or rapid progression?)
		(b) MMTV LTR	YES	(as above)	High	Sporadic tumors, typically in a few separate glands (minor differences in construct from "one-step" version)
Py-mT	Associated with c-*src* and cytoplasmic membrane	Py early region	?	Endothelial cell tumors	Moderate	Endothelial cells highly susceptible to transformation by middle T
		HSV-TK promoter	?	Endothelial cell tumors and chrondrosarcomas	Variable	(as above)
tat	Nuclear protein (transactivator)	HIV LTR	Yes	Skin sarcomas and hepatomas	Moderate	Similar to Kaposi's sarcoma in AIDS patients
tax	Nuclear protein (transactivator)	HTLV-I LTR	Yes	Neurofibrosarcomas	Moderate	Implicates HTLV-1 in human neurofibromatoses
BPV	Cell surface protein (E5) and nuclear protein (E6)	BPV	Yes	Skin fibrosarcomas	Moderate	Specific cytogenic changes in progression to neoplasia

[a]MMTV LTR, mouse mammary tumor virus long terminal repeat; FBJ-LTR, FBJ murine osteosarcomavirus long terminal repeat; Py-mT, polyomavirus middle T protein; HSV, herpes simplex virus; HIV, human immunodeficiency virus; HTLV-1, human T-cell leukemia virus type 1; BPV, bovine papillomavirus type 1.

Source: Reprinted with permission from Hanahan, Transgenic mice as probes into complex systems. *Science* 246:1265. Copyright 1989 by the American Association for the Advancement of Science.

in some of the cells, thus allowing expression of the altered gene (or loss of function of that gene in the case of a deletion that prevents expression) in selected clones. Such techniques have been used to study the effects of knocking out the hypoxanthine-guanine phosphoribosyl transferase (HPRT) gene in transgenic mice, i.e., by homologous recombination between the native chromosomal DNA and exogenous DNA constructs to generate HPRT-deficient mice.[146,147] The technique of homologous gene targeting is depicted in Figure 4-9. Exogenous DNA targeting an endogenous gene sequence is introduced into ES cells, and cells with the altered gene are selected by an appropriate selection process. The selected cells are then placed into recipient blastocysts and some percentage of chimeric mice containing the altered gene in their germ-line cells are obtained (if all goes well). Using the chimeras with the altered germ-line genes, mutant strains of mice, in which one or both alleles of the target gene are altered, can be obtained.

PHENOTYPE OF NORMAL–MALIGNANT CELL HYBRIDS

Expression of the malignant phenotype results from the transcription of certain sets of genes. Experimental techniques are now available to answer several important questions about this. For example: (1) Can these genes be mapped to a specific chromosome location? (2) Is the expression of these genes dominant or recessive? (3) Can the expression of these genes be suppressed by the presence of normal genes? The original approach to answering these questions was the use of cell fusion experiments in which normal and malignant cells from the same or different species are fused in the presence of inactivated Sendai virus or polyethylene glycol. This procedure allows the formation of heterokaryons (nuclei from both cell types in one cytoplasm), which carry the chromosomes of both cells. Both intraspecies and interspecies heterokaryons can be formed. Ultimately, hybrid cells, in which both sets of chromosomes are present in a single nucleus, may be formed. When malignant cells are fused with normal cells, there is usually a loss of chromosomes of the nonmalignant cell type with continued passage in culture. This results in the formation of "segregant" cells, which have a variable content of chromosomes. Hybrid cells resulting from human–human cell fusion usually have a more stable karyotype, but even for these, there may be some chromosomal segregation. Using this experimental approach, Harris and his colleagues demonstrated that cell hybrids between highly tumorigenic and nontumorigenic mouse cells are nearly always nontumorigenic in animals until tumorigenic segregants are produced by the elimination of some chromosomes.[148-150] These investigators concluded that tumorigenicity behaves as a recessive trait and that a specific chromosome or a combination of chromosomes is involved in the suppression of malignancy. When this chromosome or these chromosomes are lost, the malignant trait is reexpressed. Similar results were obtained by Stanbridge,[151] who fused human cervical carcinoma cells (HeLa) and normal human diploid fibroblasts and found that this suppressed malignancy, as measured by tumor formation in athymic mice. Interestingly, the nontumorigenic human cell hybrids between HeLa and normal cells retained many of the properties of the transformed phenotype—that is, lectin agglutinability, anchorage independence, lowered serum requirement for growth, and loss of density-dependent inhibition of growth—suggesting that tumorigenicity and expression of a significant part of the so-called transformed phenotype are genetically dissociable.[152]

Other investigators, using interspecies cell hybrids between malignant mouse cells and normal human cells[153] or between normal mouse cells and SV40-transformed human cells,[154] found that tumorigenicity was not suppressed. In the latter case, cells derived from tumors in nude mice after injection of the hybrid cells all had human chromosome 7, which is the chromosome carrying the SV40 genome integration site. The tumorigenic cells also expressed SV40 T antigen, suggesting that tumorigenicity, at least in some cell types, is a "positively" controlled dominant trait and depends on expression of the oncogenic virus genome. Contrasting results were obtained in a study of the tumorigenicity of hybrid cells formed by the fusion of a highly malignant human lung cancer cell line and a nontumorigenic mouse cell line.[155] In this case, all the hybrid clones were nontumorigenic. The discrepancy in these experiments probably results from the fact that interspecies cell hybrids are karytotypically unstable, and depending on which chromosomes are lost, variable tumorigenicity can result. Gene dosage may also play a role: for example, if a polyploid cell is fused with a diploid cell, several copies of particular genes may be involved in the expression or suppression of malignancy. Moreover, in interspecies hybrids, the set of genetic complementations may differ from the set that would occur in intraspecies hybrids. The fact that intraspecies hybrids between malignant and normal cells usually express the nonmalignant phenotype, however, suggests that normal negative feedback controls can be imposed on a malignant cell genetically. This has now been borne out by the discovery of tumor suppressor genes (Chap. 8).

The cell hybridization approach has been used to examine another characteristic of the malignant phenotype—namely, the tendency to undergo gene amplification. As noted above under the discussion of loss of cell-cycle control by malignant cells, the ability to amplify genes is associated with tumor progression. In a study by Tlsty et al.,[156] amplification frequency was measured in somatic cell hybrids between normal human fibroblasts and human fibrosarcoma or cervical carcinoma (HeLa) cells. The ability to amplify an endogenous cellular gene [i.e., the multifunctional CAD enzyme responsible for resistance to the drug *N*-(phosphonoacetyl)-L-aspartate behaved as a recessive trait; however, the control of gene amplification segregated as an independent trait from that of immortality in culture and tumorigenicity in animals.

CLONAL DERIVATION OF CANCER

There is significant evidence that many human tumors arise from a single cell of origin and thus are clones of the original transformed cell. This evidence derives from studies showing that (1) in many primary cancers, all the cells have the same abnormal chromosome arrangement—for example, the Philadelphia chromosome in chronic myelocytic leukemia cells; (2) all the cells of a given multiple myeloma produce the same type of immunoglobulin, whereas populations of normal immunoglobulin-producing cells in the body produce many different types; and (3) tumors arising in a woman whose cells are heterozygous for the X chromosome–linked isoenzymes of glucose-6-phosphate dehydrogenase (G-6-PDH) express only one of the two possible isoenzyme types.[157,158] The explanation for the latter finding comes from the observation that only one X chromosome is expressed in the adult female, and its expression is randomly determined. Hence, some females who are heterozygous for the G-6-PDH gene will be mosaics in the sense that some cells will express the A-type and some the B-type isoenzyme. Because malignant tumors arising in women with this characteristic usually have only one of the two isoenzyme types, the implication is that the tumor arose from a single clone of cells. Otherwise, the tumor would be expected to be a mosaic as well. When this criterion is applied a number of human tumors have been identified as of probably single-clone origin.[157] It has also been found that, when multiple tumors of the same type arise in the same patient, all the cells of a given tumor contain only one G-6-PDH isoenzyme, but the cells of other tumors may contain the other isoenzyme.[159] This indicates that different primary tumors can arise from different clones of cells.

Additional methods have more recently been used to determine the clonality of human cancers, and these include X-linked restriction fragment length polymorphisms (RFLPs) to determine which X chromosome is inactivated, immunoglobulin light chain gene analysis, DNA fingerprinting, viral gene integration analysis, cytogenetic analysis for chromosome translocation, and loss of genetic heterozygosity.[160] Using techniques of p53 tumor suppressor gene mutation, loss of heterozygosity of specific chromosomal loci, and X-chromosome inactivation, for example, the clonal origin of human ovarian carcinoma has been verified.[161,162]

Clonal analysis of experimental animal tumors to follow the expression of multiple transformation–related phenotypes has proven useful in determining clonal selection in vivo and in defining which phenotypes cosegregate with the ability of transformed cells to produce tumors in animals. For example, for chemically transformed mouse embryo cells, the phenotypes of morphologic transformation, loss of density-dependent growth inhibition, anchorage independence, and foci formation in culture segregated independently.[163] Anchorage-independence correctly predicted tumorigenicity 81% of the time, and coexpression of anchorage-independence with growth of transformed foci on a confluent "lawn" of wild-type untransformed cells predicted tumorigenicity 91% of the time.

The hypothesis of single-cell clonal origin of cancer does not explain how the phenotypic characteristics of a cancer can change as it grows and metastasizes in a patient. The explanation for this is probably that neoplastic cells are more genetically unstable than their normal counterparts and go through several "evolutionary" changes as the tumor progresses. Nowell[158] has proposed a model to explain this (Fig. 4-10). In this model, tumor initiation is postulated as occurring from the interaction of a carcinogenic agent and a susceptible normal cell. The induced change provides some selective growth advantage over adjacent normal cells. Other cells in the same individual may also undergo "initiation" of transformation after exposure to carcinogens, but presumably they never proliferate (i.e., undergo tumor promotion), or they are destroyed before they progress to a clinically detectable tumor. The proliferation of neoplastic cells, after some lag time, leads eventually to genetically variant cells because of the genetic instability of neoplastic cells. This accounts for the aneuploidy of most advanced cancers. Many of these genetic variants are eliminated because of a metabolic disadvantage or an immunologic reaction of the host, but a few tumor cells with an additional selective advantage proliferate and become predominant. As the tumor progresses, there is sequential selection of sublines of cells with increasingly abnormal karyotype, state of differentiation, invasiveness, and metastatic po-

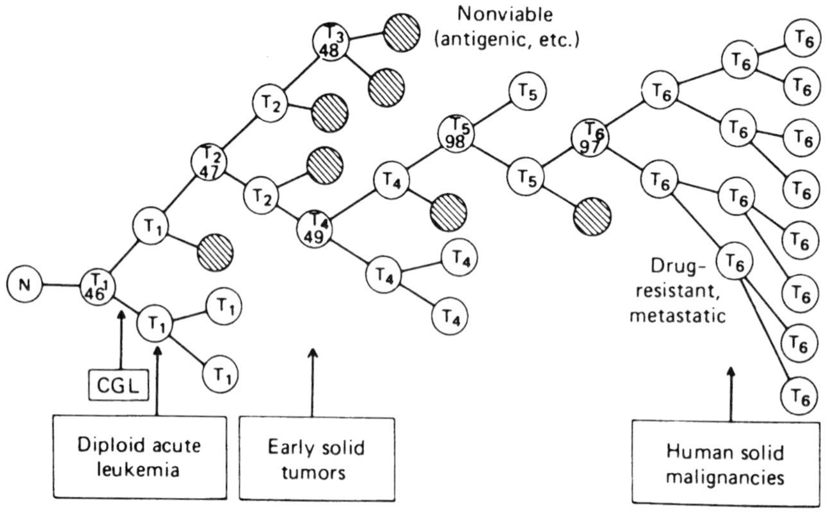

Figure 4-10 Model of clonal evolution in neoplasia. Carcinogen-induced change in progenitor normal cell *(N)* produces a diploid tumor cell (T_1, 46 chromosomes) with growth advantage, permitting clonal expansion to begin. The genetic instability of T_1 cells produces variants (illustrated by changes in chromosome number, T_2 to T_6). Most variants die, because of metabolic or immunologic disadvantage *(hatched circles)*; occasionally one has an additional selective advantage (e.g., T_2, 47 chromosomes), and its progeny become the predominant subpopulation until an even more favorable variant appears (e.g., T_4). The stepwise sequence in each tumor differs (being partially determined by environmental pressures on selection) and results in a different, aneuploid karyotype in each fully developed malignancy (T_6). Biologic characteristics of tumor progression (e.g., morphologic and metabolic loss of differentiation, invasion and metastasis, resistance to therapy) parallel the stages of genetic evolution. Human tumors with minimal chromosome change (diploid acute leukemia, chronic granulocytic leukemia) are considered to be early in clonal evolution; human solid cancers, typically highly aneuploid, are viewed as late in the developmental process. (From Nowell.[158])

tential. Although many advanced cancers share similar genotypic and phenotypic characteristics (e.g., increased ability to synthesize DNA and to proliferate continuously), tumors of the same histologic type from different patients may have many differences because their microenvironments affect the rate of emergence and type of variant sublines. The fact that most advanced human cancers, even those of the same histologic and tissue type, are aneuploid and have unique biochemical and immunologic alterations is discouraging, from a therapeutic point of view, but it helps explain the vast biochemical heterogeneity of human tumors and the difficulty in developing anticancer agents that will specifically kill tumors of a given cell type.

REFERENCES

1. W. R. Earle: Production of malignancy in vitro: IV. The mouse fibroblast cultures and changes seen in living cells. *JNCI* 4:165, 1943.
2. W. R. Earle, A. Nettleship, E. L. Schilling, T. H. Stark, N. R. Straus, M. F. Brown, and E. Shelton: Production of malignancy in vitro: V. Results of injections of cultures into mice. *JNCI* 4:213, 1943.
3. Y. Berwald and L. Sachs: In vitro cell transformation with chemical carcinogens. *Nature* 200:1182, 1963.
4. M. Vogt and R. Dulbecco: Virus-cell interaction with a tumor-producing virus. *Proc Natl Acad Sci USA* 46:365, 1960.
5. C. Borek and L. Sachs: In vitro cell transformation by X-irradiation. *Nature* 210:276, 1966.
6. B. E. Barker and K. K. Sanford: Cytologic manifestations of neoplastic transformation in vitro. *JNCI* 44:39, 1970.
7. R. Montesano, C. Drevon, T. Kuroki, L. Saint Vincent, S. Handleman, K. K. Sanford, D. DeFeo, and I. B. Weinstein: Test for malignant transformation of rat liver cells in culture: Cytology, growth in soft agar, and production of plasminogen activator. *JNCI* 59:1651, 1977.
8. M. Abercrombie and E. J. Ambrose: The surface properties of cancer cells: A review. *Cancer Res* 22:525, 1962.
9. J. W. Lash: Studies on wound closure in urodeles. *J Exp Zool* 128:13, 1955.
10. M. Abercrombie and J. E. M. Heaysman: Social behavior of cells in tissue culture: II. Monolayering of fibroblasts. *Exp Cell Res* 6:293, 1954.
11. H. M. Temin and H. Rubin: Characteristics of an assay for Rous sarcoma virus and Rous sarcoma cells in tissue culture. *Virology* 6:669, 1958.
12. S. A. Aaronson and G. J. Todaro: Basis for the acquisition of malignant potential by mouse cells cultivated in vitro. *Science* 162:1024, 1968.
13. M. G. P. Stoker and H. Rubin: Density-dependent inhibition of cell growth in culture. *Nature* 215:171, 1967.
14. R. J. Wieser and F. Oesch: Contact inhibition of growth of human diploid fibroblasts by immobilized plasma membrane glycoproteins. *J Cell Biol* 103:361, 1986.
15. T. Nakamura, Y. Nakayama, H. Teramoto, K. Nakwa, and A. Ichihara: Loss of reciprocal modulations of growth and liver function of hepatoma cells in culture by contact with cells or cell membranes. *Proc Natl Acad Sci USA* 81:6398, 1984.
16. R. Dulbecco: Topoinhibition and serum requirement of transformed and untransformed cells. *Nature* 227:802, 1970.
17. R. W. Holley and J. A. Kiernan: "Contact inhibition" of cell division in 3T3 cells. *Proc Natl Acad Sci USA* 60:300, 1968.
18. J. Jainchill and G. J. Todaro: Stimulation of cell growth in vitro by serum with and without growth factor. *Exp Cell Res* 59:137, 1970.
19. H. S. Smith, C. D. Scher, and G. J. Todaro: Induction of cell division in medium lacking serum growth factor by SV40. *Virology* 44:359, 1971.
20. F. K. Sanders and B. O. Burford: Ascites tumours from BHK 21 cells transformed in vitro by polyoma virus. *Nature* 201:786, 1964.
21. J. Macpherson and L. Montagner: Agar suspension culture for the selective assay of cells transformed by polyoma virus. *Virology* 23:291, 1964.
22. M. C. Alley, C. M. Pacula-Cox, M. L. Hursey, L. R. Rubinstein, and M. R. Boyd: Morphometric and colorimetric analyses of human tumor cell line growth and drug sensitivity in soft agar culture. *Cancer Res* 51:1247, 1991.
23. B. Löwenberg, W. L. J. van Putten, I. P. Touw, R. Delwel, and V. Santini: Autonomous proliferation of leukemic cells in vitro as a determinant of prognosis in adult acute myeloid leukemia. N Engl J Med 328:614, 1993.
24. S.-S. Shin, V. H. Freedman, R. Risser, and R. Pollack: Tumorigenicity of virus-transformed cells in nude mice is correlated specifically with anchorage independent growth in vitro. *Proc Natl Acad Sci USA* 72:4435, 1975.
25. T. Enomoto and H. Yamasaki: Lack of intercellular communication between chemically transformed and surrounding nontransformed BALB/C 3T3 cells. *Cancer Res* 44:5200, 1984.
26. L. P. Yotti, C. C. Chang, and J. E. Trosko: Elimination of metabolic cooperation in Chinese hamster cells by a tumor promoter. *Science* 206:1089, 1979.
27. C. J. Marshall, L. M. Franks, and A. W. Carbo-

nell: Markers of neoplastic transformation in epithelial cell lines derived from human carcinomas. *JNCI* 58:1743, 1977.
28. A. B. Pardee: A restriction point for control of normal animal cell proliferation. *Proc Natl Acad Sci USA* 71:1286, 1974.
29. A. B. Pardee, R. Dubrow, J. L. Hamlin, and R. E. Kletzien: Animal cell cycle. *Annu Rev Biochem* 47:715, 1978.
30. A. W. Murray: The cell cycle as a cdc2 cycle. *Nature* 342:14, 1989.
31. L. H. Hartwell and T. A. Weinert: Checkpoints: Controls that ensure the order of cell cycle events. *Science* 246:629, 1989.
32. A. W. Murray: Creative blocks: Cell cycle checkpoints and feedback controls. *Nature* 359:599, 1992.
33. T. Hunter and J. Pines: Cyclins and cancer. *Cell* 66:1071, 1991.
34. N. Sagata, N. Watanabe, G. F. Vande Woude, and Y. Ikawa: The c-*mos* protooncogene product is a cytostatic factor responsible for meiotic arrest in vertebrates eggs. *Nature* 342:512, 1989.
35. L. Hartwell: Defects in a cell cycle checkpoint may be responsible for the genomic instability of cancer cells. *Cell* 71:543, 1992.
36. L. R. Livingstone, A. White, J. Sprouse, E. Livanos, T. Jacks, and T. D. Tlsty: Altered cell cycle arrest and gene amplification potential accompany loss of wild type p53. *Cell* 70:923, 1992.
37. G. L. Nicolson: Trans-membrane control of the receptors on normal and tumor cells: II. Surface changes associated with transformation and malignancy. *Biochim Biophys Acta* 458:1, 1976.
38. I. J. Goldstein and C. E. Hayes: The lectins: Carbohydrate-binding proteins of plants and animals. *Adv Carbohydr Chem Biochem* 35:127, 1978.
39. J. C. Aub, B. H. Sanford, and M. N. Cote: Studies on reactivity of tumor and normal cells to a wheat germ agglutinin. *Proc Natl Acad Sci USA* 54:396, 1965.
40. M. M. Burger: Surface changes in transformed cells detected by lectins. *Fed Proc* 32:91 1973.
41. A. Sivak and S. R. Wolman: Classification of cell types: Agglutination and chromosomal properties. *In Vitro* 8:1, 1972.
42. Z. Rabinowitz and L. Sachs: The formation of variants with a reversion of properties of transformed cells: I. Variants from polyoma-transformed cells grown in vivo. *Virology* 38:336, 1969.
43. G. L. Nicolson and M. Lacorbiere: Cell contact-dependent increase in membrane D-galactopyranosyl-like residues on normal, but not virus- or spontaneously transformed murine fibroblasts. *Proc Natl Acad Sci USA* 70:1672, 1973.
44. H. Ben-Bassat, N. Goldblum, N. Manny, and L. Sachs: Mobility of concanavalin A receptors on the surface membrane of lymphocytes from normal persons and patients with chronic lymphocytic leukemia. *Int J Cancer* 14:367, 1974.
45. P. G. Phillips, P. Furmanski, and M. Lubin: Cell surface interactions with concanavalin A: Location of bound radiolabeled lectin. *Exp Cell Res* 86:301, 1974.
46. G. Nicolson: Temperature-dependent mobility of concanavalin A sites on tumour cell surfaces. *Nature New Biol* 243:218, 1973.
47. M. C. Willingham, K. M. Yamada, S. S. Yamada, J. Pouysségur, and I. Pastan: Microfilament bundles and cell shape are related to adhesiveness to substratum and are dissociable from growth control in cultured fibroblasts. *Cell* 10:375, 1977.
48. N. S. McNutt, L. A. Culp, and P. H. Black: Contact-inhibited revertant cell lines isolated from SV40-transformed cells: IV. Microfilament distribution and cell shape in untransformed, transformed, and revertant Balb/c 3T3 cells. *J Cell Biol* 56:412, 1973.
49. M. R. Reese and D. A. Chow: Tumor progression in vivo: Increased soybean agglutinin lectin binding, *N*-acetylgalactosamine-specific lectin expression, and liver metastasis potential. *Cancer Res* 52:5235, 1992.
50. S. H. Barondes: Bifunctional properties of lectins: Lectins redefined. *Trends Biochem Sci* 13:480, 1988.
51 N. Sharon and H. Lis: Lectins as cell recognition molecules. *Science* 246:227, 1989.
52. Y. C. Lee: Biochemistry or carbohydrate-protein interaction. *FASEB J* 6:3193, 1992.
53. E. Witebsky: *Z Immunitätsforsch* 62:35, 1929.
54. L. Hirszfeld, W. Halber, and J. J. Laskowski: *Z Immunitätsforsch* 64:81, 1929.
55. M. M. Rapport, L. Graf, V. P. Skipski, and N. F. Alonzo: Immunochemical studies of organ and tumor lipids. *Cancer* 12:438, 1959.
56. M. M. Rapport, L. Graf, and H. Schneider: Immunochemical studies of organ and tumor lipids: XIII. Isolation of cytolipin K, a glycosphingolipid hapten present in human kidney. *Arch Biochem Biphys* 105:431, 1964.
57. S.-I. Hakomori: Biochemical basis of tumor-associated carbohydrate antigens: Current trends, future perspectives, and clinical applications. *Immun Allerg Clin North Am* 10:781, 1990.
58. C. G. Gahmberg and S. Hakomori: Surface carbohydrates of hamster fibroblasts: I. Chemical characterization of surface-labeled glycosphingolipids and a specific ceramide tetrasaccharide for transformants. *J Biol Chem* 250:2438, 1975.
59. H. C. Wu, E. Meezan, P. H. Black, and P. W. Robbins: Comparative studies on the carbohydrate-containing membrane components of normal and virus-transformed mouse fibroblasts: I. Glucosamine-labeling patterns in 3T3, spontaneously transformed 3T3, and SV-40-transformed 3T3 cells. *Biochemistry* 8:2509, 1969.
60. L. Warren, J. P. Fuhrer, and C. A. Buck: Surface

glycoproteins of normal and transformed cells: A difference determined by sialic acid and a growth-dependent sialyl transferase. *Proc Natl Acad Sci USA* 69:1838, 1972.

61. S. Ogata, T. Muramatsu, and A. Kobata: New structural characteristic of the large glycopeptides from transformed cells. *Nature* 259:580, 1976.
62. S.-I. Hakomori: Aberrant glycosylation in cancer cell membranes as focused on glycolipids: Overview and perspectives. *Cancer Res* 45:2405, 1985.
63. K. Yamashita, Y. Tachibana, T. Ohkura, and A. Kobata: Enzymatic basis for the structural changes of asparagine-linked sugar chains of membrane glycoproteins of baby hamster kidney cells induced by polyoma transformation. *J Biol Chem* 260:3963, 1985.
64. T. Feizi: Demonstration by monoclonal antibodies that carbohydrate structures of glycoproteins and glycolipids are onco-developmental antigens. *Nature* 314:53, 1985.
65. D. Solter and B. B. Knowles: Monoclonal antibody defining a stage-specific mouse embryonic antigen (SSEA-1). *Proc Natl Acad Sci USA* 75:5565, 1978.
66. Z. R. Shi, L. J. McIntyre, B. B. Knowles, D. Solter, and Y. S. Kim: Expression of a carbohydrate differentiation antigen, stage-specific embryonic antigen 1, in human colonic adenocarcinoma. *Cancer Res* 44:1142, 1984.
67. H. C. Gooi, T. Feizi, A. Kapadia, B. B. Knowles, D. Solter, and M. Evans: Stage-specific embryonic antigen involves α1→3 fucosylated type 2 blood group chains. *Nature* 292:156, 1981.
68. R. O. Brady and P. H. Fishman: Biosynthesis of glycolipids in virus-transformed cells. *Biochim Biophys Acta* 355:121, 1974.

68a. S. Nakamori, M. Kameyama, S. Imaoka, H. Furukawa, O. Ishikawa, Y. Saski, T. Kabuto, T. Iwanaga, Y. Matsushita, and T. Irimura: Increased expression of sialyl Lewis^x antigen correlates with poor survival in patients with colorectal carcinoma: Clinicopathological and immunohistochemical study. *Cancer Res* 53:3632, 1993.

69. T. Mizuochi, R. Nishimura, C. Derappe, T. Taniguchi, T. Hamamoto, M. Mochizuki, and A. Kobata: Structures of the asparagine-linked sugar chains of human chorionic gonadotropin produced in choriocarcinoma: Appearance of triantennary sugar chains and unique biantennary sugar chains. *J Biol Chem* 258:14126, 1983.
70. N. Sharon and H. Lis: Glycoproteins: Research booming on long-ignored, ubiquitous compounds. *Chem Eng News* 59:21, 1981.
71. P. A. Gleeson and H. Schacter: Control of glycoprotein synthesis. *J Biol Chem* 258:6162, 1983.
72. E. H. Holmes and S.-I. Hakomori: Enzymatic basis for changes in fucoganglioside during chemical carcinogenesis: Induction of specific α-fucosyltransferase and status of an α-galactosyltransferase in precancerous rat liver and hepatoma. *J Biol Chem* 258:3706, 1983.
73. S. C. Hubbard: Differential effects of oncogenic transformation on N-linked oligosaccharide processing at individual glycosylation sites of viral glycoproteins. *J Biol Chem* 262:16403, 1987.
74. E. W. Easton, J. G. M. Bolscher, D. H. van den Eijnden: Enzymatic amplification involving glycosyltransferases forms the basis for the increased size of asparagine-linked glycans at the surface of NIH 3T3 cells expressing the N-*ras* protooncogene. *J Biol Chem* 266:21674, 1991.
75. M. M. Palcic, J. Ripka, K. J. Kaur, M. Shoreibah, O. Hindsgaul, and M. Pierce: Regulation of *N*-acetylglucosaminyltransferase V activity. *J Biol Chem* 265:6759, 1990.
76. S. Narasimhan, H. Schachter, and S. Rajalakshmi: Expression of *N*-acetylglucosaminyltransferase III in hepatic nodules during rat liver carcinogenesis promoted by orotic acid. *J Biol Chem* 263:1273, 1988.
77. E. V. Chandrasekaran, R. K. Jain, and K. L. Matta: Ovarian cancer, α1, 3-*L*-fucosyltransferase. *J Biol Chem* 267:23806, 1992.
78. B. E. Harvey, C. A. Toth, H. E. Wagner, G. D. Steele, Jr., and P. Thomas: Sialyltransferase activity and hepatic tumor growth in a nude mouse model of colorectal cancer metastases. *Cancer Res* 52:1775, 1992.
79. I. Brockhausen, W. Kuhns, H. Schachter, K. L. Matta, D. R. Sutherland, and M. A. Baker. Biosynthesis of *O*-glycans in leukocytes from normal donors and from patients with leukemia: Increase in *O*-glycan core 2 UDP-GlcNAc:Galβ3GalNAcα-R (GlcNAc to GalNAc) β(1-6)-*N*-acetylglucosaminyltransferase in leukemic cells. *Cancer Res* 51:1257, 1991.
80. S. Ruan and K. O. Lloyd: Glycosylation pathways in the biosynthesis of gangliosides in melanoma and neuroblastoma cells: Relative glycosyltransferase levels determine ganglioside patterns. *Cancer Res* 52:5725, 1992.
81. J. G. M. Bolscher, D. C. C. Schaller, L. A. Smets, H. van Rooy, J. G. Collard, E. A. Bruyneel, and M. M. K. Mareel: Effect of cancer-related and drug-induced alterations in surface carbohydrates on the invasive capacity of mouse and rat cells. *Cancer Res* 46:4080, 1986.
82. I. Kijima-Suda, Y. Miyamoto, S. Toyoshima, M. Itah, and T. Osawa: Inhibition of experimental pulmonary metastasis of mouse colon adenocarcinoma 26 sublines by a sialic-nucleoside conjugate having sialyltransferase inhibiting activity. *Cancer Res* 46:858, 1986.
83. H. E. Wagner, P. Thomas, B. C. Wolf, A. Rapazo, and G. Steele: Inhibition of sialic acid incorporation prevents hepatic metastases. *Arch Surg* 125:351, 1990.

84. J. W. Dennis, K. Koch, S. Yousefi, and I. Van der Elst: Growth inhibition of human melanoma tumor xenografts in athymic nude mice by swainsonine. *Cancer Res* 50:1867, 1990.
85. G. K. Ostrander, N. K. Scribner, and L. R. Rohrschneider: Inhibition of v-*fms*-induced tumor growth in nude mice by castanospermine. *Cancer Res* 48:1091, 1988.
86. S. J. Gendler, A. P. Spicer, E.-N. Lalani, T. Duhig, N. Peat, J. Burchell, L. Pemberton, M. Boshell, and J. Taylor-Papadimithiou: Structure and biology of a carcinoma-associated mucin, MUC1. *Am Rev Respir Dis* 144:S42, 1991.
87. S. B. Ho, G. A. Nichans, C. Lyftogt, P. S. Yan, D. L. Cherwitz, E. T. Gum, R. Dahiya, and Y. S. Kim: Heterogeneity of mucin gene expression in normal and neoplastic tissues. *Cancer Res* 53:641, 1993.
88. K. R. Jerome, D. L. Barnd, K. M. Bendt, C. M. Boyer, J. Taylor-Papadimitriou, I. F. C. McKenzie, R. C. Bast, Jr., and O. J. Finn: Cytotoxic T-lymphocytes derived from patients with breast adenocarcinoma recognize an epitope present on the protein core of a mucin molecule preferentially expressed by malignant cells. *Cancer Res* 51:2908, 1991.
89. G. D. MacLean, M. B. Bowen-Yacyshyn, J. Samuel, A. Meikle, G. Stuart, J. Nation, S. Poppema, M. Jerry, R. Koganty, T. Wong, and B. M. Longenecker: Active immunization of human ovarian cancer patients against a common carcinoma (Thomsen-Friedenreich) determinant using a synthetic carbohydrate antigen. *J Immunother* 11:292, 1992.
90. S. B. Ho and Y. S. Kim: Carbohydrate antigens on cancer-associated mucin-like molecules. *Semin Cancer Biol* 2:389, 1992.
91. T. E. Hardingham and A. J. Sosang: Proteoglycans: Many forms and many functions. *FASEB J* 6:861, 1992.
92. K. Sugahara, I. Yamashina, P. De Waard, H. Van Halbeek, and J. F. G. Vliegenthart: Structural studies on sulfated glycopeptides from the carbohydrate-protein linkage region of chondroitin 4-sulfate proteoglycans of swarm rat chondrosarcoma. *J Biol Chem* 263:10168, 1988.
93. E. Ruoslahti: Proteoglycans in cell regulation. *J Biol Chem* 264:13369, 1989.
94. P. Levy, S. Emami, G. Cherqui, E. Chastre, C. Gespach, and J. Picard: Altered expression of proteoglycans in E1A-immortalized rat fetal intestinal epithelial cells in culture. *Cancer Res* 50:6716, 1990.
95. P. D. Yurchenco and J. C. Schittny: Molecular architecture of basement membranes. *FASEB J* 4:1577, 1990.
96. C. H. Streuli and M. J. Bissell: Expression of extracellular matrix components is regulated by substratum. *J Cell Biol* 110:1405, 1990.
97. R. V. Iozzo: Neoplastic modulation of extracellular matrix: Colon carcinoma cells release polypeptides that alter proteoglycan metabolism in colon fibroblasts. *J Biol Chem* 260:7464, 1985.
98. B. P. Toole, C. Biswas, and J. Gross: Hyaluronate and invasiveness of the rabbit V2 carcinoma. *Proc Natl Acad Sci USA* 76:6299, 1979.
99. S. H. Barsky, C. N. Rao, G. R. Grotendorst, and L. A. Liotta: Increased content of type V collagen of human breast carcinoma. *Am J Pathol* 108:276, 1982.
100. J. Keski-Oja, C. G. Gahmberg, and K. Alitalo: Pericellular matrix and cell surface glycoproteins of virus-transformed mouse epithelial cells. *Cancer Res* 42:1147, 1982.
101. B. P. Peters, R. J. Hartle, R. F. Krzesicki, T. G. Kroll, F. Perini, J. E. Balun, I. J. Goldstein, and R. W. Ruddon: The biosynthesis, processing, and secretion of laminin by human choriocarcinoma cells. *J Biol Chem* 260:14732, 1985.
102. G. P. Frenette, T. E. Carey, J. Varani, D. R. Schwartz, S. E. G. Fligiel, R. W. Ruddon, and B. P. Peters: Biosynthesis and secretion of laminin and laminin-associated glycoproteins by nonmalignant and malignant human keratinocytes: Comparison of cell lines from primary and secondary tumors in the same patient. *Cancer Res* 48:5193, 1988.
103. M. Hatanaka: Transport of sugars in tumor cell membranes. *Biochim Biophys Acta* 355:77, 1974.
104. S. Venuta and H. Rubin: Sugar transport in normal and Rous sarcoma virus-transformed chick embryo fibroblasts. *Proc Natl Acad Sci USA* 70:653, 1973.
105. H. M. Kalckar, D. Ullrey, S. Kijomoto, and S. Hakomori: Carbohydrate catabolism and the enhancement of uptake of galactose in hamster cells transformed by polyoma virus. *Proc Natl Acad Sci USA* 70:839, 1973.
106. D. O. Foster and A. B. Pardee: Transport of amino acids by confluent and nonconfluent 3T3 polyoma virus-transformed 3T3 cells growing on glass cover slips. *J Biol Chem* 244:2675 (1969).
107. K. J. Isselbacher: Increased uptake of amino acids and 2-deoxy-D-glucose by virus transformed cells in culture. *Proc Natl Acad Sci USA* 69:585, 1972.
108. A. Pardee, L. Jiminez de Asua, and E. Rozengurt: Functional membrane changes and cell growth: Significance and mechanism, in B. Clarkson and R. Baserga, eds.: *Control of Proliferation in Animal Cells.* New York: Cold Spring Harbor Laboratory, 1974, pp. 547–562.
109. R. Kram, P. Momont, and G. M. Tomkins: Pleiotypic control by adenosine 3′:5′-cyclic monophosphate: A model for growth control in animal cells. *Proc Natl Acad Sci USA* 70:1432, 1973.
110. O. Warburg: *The Metabolism of Tumors.* London: Arnold Constable, 1930.
111. S. Weinhouse: Glycolysis, respiration, and anom-

alous gene expression in experimental hepatomas: G. H. A. Clowes Memorial Lecture. *Cancer Res* 32:2007, 1972.
112. J. P. Greenstein: *Biochemistry of Cancer.* New York: Academic Press, 1954.
113. E. C. Miller and J. A. Miller: The presence and significance of bound aminoazo dyes in the livers of rats fed *p*-dimethyllaminoazobenzene. *Cancer Res* 7:468, 1947.
114. S. Sorof, B. P. Sani, V. M. Kish, and H. P. Meloche: Isolation and properties of the principal liver protein conjugate of a hepatic carcinogen. *Biochemistry* 13:2612, 1974.
115. V. R. Potter: The biochemical approach to the cancer problem. *Fed Proc* 17:691, 1958.
116. V. R. Potter: Biochemical perspectives in cancer research. *Cancer Res* 24:1085, 1964.
117. H. P. Morris: Studies in the development, biochemistry, and biology of experimental hepatomas. *Adv Cancer Res* 9:227, 1965.
118. G. Weber: Enzymology of cancer cells: Part 1. N Engl J Med 296:486, 1977.
119. G. Weber: Enzymology of cancer cells: Part 2. N Engl J Med 296:541, 1977.
120. W. E. Knox: *Enzyme Patterns in Fetal, Adult, and Neoplastic Rat Tissues,* 2nd ed. Basel: S. Karger, 1976.
121. A. Fischer: Betrag zur Biologie der Gewebezellen: Eine vergleichend-biologische Studie der normalen und maligen Gewebezellen in vitro. *Wilhelm Roux Arch Entwicklungsmech Organ* 104:210, 1925.
122. E. E. Cliffton and C. E. Grossi: Fibrinolytic activity of human tumors as measured by the fibrin plate method. *Cancer* 8:1146, 1955.
123. J. C. Unkeless, A. Tobia, L. Ossowski, J. P. Quigley, D. B. Rifkin, and E. Reich: An enzymatic function associated with transformation of fibroblasts by oncogenic viruses: I. Chick embryo fibroblast cultures transformed by avian RNA tumor viruses. *J Exp med* 137:85, 1973.
124. L. Ossowski, J. C. Unkeless, A. Tobia, J. P. Quigley, D. B. Rifkin, and E. Reich: An enzymatic function associated with transformation of fibroblasts by oncogenic viruses: II. Mammalian fibroblast cultures transformed by DNA and RNA tumor viruses. *J Exp Med* 137:112, 1973.
125. I.-N. Chou, S. P. O'Donnell, P. H. Black, and R. O. Roblin: Cell density-dependent secretion of plasminogen activity by 3T3 cells. *J Cell Physiol* 91:31, 1977.
126. J. Unkeless, K. Danø, G. M. Kellerman, and E. Reich: Fibrinolysis associated with oncogenic transformation: Partial purification and characterization of the cell factor, a plasminogen activator. *J Biol Chem* 249:4295, 1974.
127. J. P. Quigley, L. Ossowski, and E. Reich: Plasminogen, the serum proenzyme activated by factors from cells transformed by oncogenic viruses. *J Biol Chem* 249:4306, 1974.
128. D. Collen: Identification and some properties of a new fast-reacting plasmin inhibitor in human plasma. *Eur J Biochem* 69:209, 1976.
129. G. S. Martin, S. Venuta, M. Weber, and H. Rubin: Temperature-dependent alterations in sugar transport in cells infected by a temperature-sensitive mutant of Rous sarcoma virus. *Proc Natl Acad Sci USA* 68:2739, 1971.
130. C. G. Bergstrand and B. Czar: Demonstration of a new protein fraction in serum from the human fetus. *Scand J Clin Lab Invest* 8:174, 1956.
131. S. Sell and F. F. Becker: Alpha-fetoprotein. *JNCI* 60:19, 1978.
132. G. I. Abelev, S. D. Perova, N. I. Khramkova, Z. A. Postnikova, and I. S. Irlin: Production of embryonal α-globulin by transplantable mouse hepatomas. *Transplantation* 1:174, 1963.
133. Y. Tatarinov: Detection of embryospecific α-globulin in the blood sera of patients with primary liver tumors. *Vopr Med Khim* 10:90, 1964.
134. P. Gold and S. O. Freedman: Specific carcinoembryonic antigens of the human digestive system. *J Exp Med* 121:439, 1965.
135. H. S. Slayter and J. E. Coligan: Electron microscopy and physical characterization of the carcinoembryonic antigen. *Biochemistry* 14:2323, 1975.
136. W. H. Fishman and S. Sell, eds: *Onco-Developmental Gene Expression.* New York: Academic Press, 1976.
137. S. P. Flanagan: "Nude:" A new hairless gene with pleiotropic effects in the mouse. *Genet Res* 8:295, 1966.
138. E. M. Pantelourns: Absence of thymus in a mouse mutant. *Nature* 217:370, 1968.
139. J. Fogh, J. M. Fogh, and T. Orfeo: One hundred and twenty-seven cultured human tumor cell lines producing tumors in nude mice. *JNCI* 59:221, 1977.
140. M. J. Bosma and A. M. Carroll: The SCID mouse mutant: Definition, characterization, and potential uses. *Annu Rev Immunol* 9:323, 1991.
141. H. Takahashi, T. Nakada, and I. Puisieux: Inhibition of human colon cancer growth by antibody-directed human LAK cells in SCID mice. *Science* 259:1460, 1993.
142. R. Jaenisch: Transgenic animals. *Science* 240:1468, 1988.
143. H. Westphal: Transgenic mammals and biotechnology. *FASEB J* 3:117, 1989.
144. D. Hanahan: Transgenic mice as probes into complex systems. *Science* 246:1265, 1989.
145. H. Murakami, N. D. Sanderson, P. Nagy, P. A. Marino, G. Merlino, and S. S. Thorgeirsson: Transgenic mouse model for synergistic effects of nuclear oncogenes and growth factors in tumorigenesis: Interaction of c-*myc* and transforming growth factor α in hepatic oncogenesis. *Cancer Res* 53:1719, 1993.
146. K. R. Thomas and M. R. Capecchi: Site-directed

mutagenesis by gene targeting in mouse embryo-derived stem cells. *Cell* 51:503, 1987.
147. T. Doetschman, R. G. Gregg, N. Maeda, M. L. Hooper, D. W. Melton, S. Thompson, and O. Smithies: Targeted correction of a mutant HPRT gene in mouse embryonic stem cells. *Nature* 330:576, 1987.
148. H. Harris, O. J. Miller, G. Klein, P. Worst, and T. Tachibana: Suppression of malignancy by cell fusion. *Nature* 223:363, 1969.
149. G. Klein, S. Friberg, F. Wiener, and H. Harris: Studies on hybrid cells derived from the fusion of the TA3/H_a ascites carcinoma with normal fibroblasts: I. Malignancy, karyotype and formation of isoantigenic variants. *JNCI* 50:1259, 1973.
150. F. Wiener, G. Klein, and H. Harris: The analysis of malignancy by cell fusion. *J Cell Sci* 15:177, 1974.
151. E. J. Stanbridge: Suppression of malignancy in human cells. *Nature* 260:17, 1976.
152. E. J. Stanbridge and J. Wilkinson: Analysis of malignancy in human cells. Malignant and transformed phenotypes are under separate genetic control. *Proc Natl Acad Sci USA* 75:1466, 1978.
153. R. Kucherlapati and S.-I. Shin: Genetic control of tumorigenicity in interspecific mammalian cell hybrids. *Cell* 16:639, 1979.
154. C. M. Croce, D. Aden, and H. Koprowski: Somatic cell hybrids between mouse peritoneal macrophages and simian virus-40-transformed human cells: II. Presence of human chromosome 7 carrying simian virus 40 genome in cells of tumors induced by hybrid cells. *Proc Natl Acad Sci USA* 72:1397, 1975.
155. D. M. Carney, C. J. Edgell, A. F. Gazdar, and J. D. Minna: Suppression of malignancy in human lung cancer (A549/8) × mouse fibroblast (3T3-4E) somatic cell hybrids. *JNCI* 62:411, 1979.
156. T. D. Tlsty, A. White, and J. Sanchez: Suppression of gene amplification in human cell hybrids. *Science* 255:1425, 1992.
157. P. J. Fialkow: The origin and development of human tumors studied with cell markers. *N Engl J Med* 291:26, 1974.
158. P. C. Nowell: The clonal evolution of tumor cell populations. *Science* 194:23, 1976.
159. S. B. Baylin, S. H. Hsu, D. S. Gann, R. C. Smallridge, and S. A. Well, Jr.: Inherited medullary thyroid carcinoma: A final monoclonal mutation in one of multiple clones of susceptible cells. *Science* 199:429, 1978.
160. J. S. Wainscoat and M. F. Fey: Assessment of clonality in human tumors: A review. *Cancer Res* 50:1355, 1990.
161. C.-H. Mok, S.-W. Tsao, R. C. Knapp, P. M. Fishbaugh, and C. C. Lau: Unifocal origin of advanced human epithelial ovarian cancers. *Cancer Res* 52:5119, 1992.
162. I. J. Jacobs, M. F. Kohler, R. W. Wiseman, J. R. Marks, R. Whitaker, B. A. J. Kerns, P. Humphrey, A. Berchuck, B. A. J. Pander, and R. C. Bast, Jr.: Clonal origin of epithelial ovarian carcinoma: Analysis by loss of heterozygosity, p53 mutation, and X-chromosome inactivation. *JNCI* 84:1793, 1992.
163. G. J. Smith, W. N. Bell, and J. W. Grisham: Clonal analysis of the expression of multiple transformation phenotypes and tumorigenicity by morphologically transformed 10T1/2 cells. *Cancer Res* 53:500, 1993.

5

Alterations of Cellular Differentiation in Cancer

A cancer develops from cells that are capable of dividing. All tissues in the body contain some cells that can divide and renew themselves. A subset of the cell population in any tissue can differentiate into the functional cells of that tissue. The normal process of cellular differentiation ultimately leads to an adult, fully differentiated, "dead-end" cell that cannot, under ordinary circumstances, divide again. These fully differentiated cells are the workhorse cells in most tissues of the body; they are the neurons in the brain controlling ideation and behavior, the liver cells that manufacture enzymes to metabolize substrates needed for growth, produce plasma proteins, and clear the blood of potentially toxic substances, the pancreatic cells that manufacture insulin and the enzymes necessary for digestion, the kidney cells that filter, secrete, or reabsorb substances and fluid in the formation of urine, the polymorphonuclear white blood cells that phagocytize and destroy bacteria, and so on. Under circumstances that are not clearly understood, cells that have the potential to divide can be changed by interaction with carcinogenic agents into a cell type that is capable of continued proliferation and thereby is prevented from achieving the normal state of complete differentiation. The carcinogen-altered cell is said to have undergone malignant transformation. Somehow the genes controlling cell proliferation are locked in the "on" position when they should be in the "off" position, and the genes controlling differentiation are either not expressed or are expressed only imperfectly. What we must know to understand carcinogenesis and to develop ways of preventing or curing cancer, then, is contained in the mechanisms of normal cellular differentiation. Only by understanding these mechanisms can the manner in which cells are altered during malignant transformation be ascertained.

Differentiation is the sum of all the processes by which cells in a developing organism achieve their specific set of structural and functional characteristics. By the acquisition of these special traits, progeny cells are distinguishable from their parent cells and from each other. Somatic cells that share a set or a subset of structural and functional characteristics become organized into tissues in higher organisms. Tissues are arranged as organs, and organs make up the organism. Indeed, cellular differentiation is the sine qua non of multicellular life.

Differentiation requires a progressive restriction of genomic expression in the pathway from the totipotential fertilized ovum to the unipotent cells of specialized organs (Fig. 5-1). The totipotentiality of cells starts to change very early

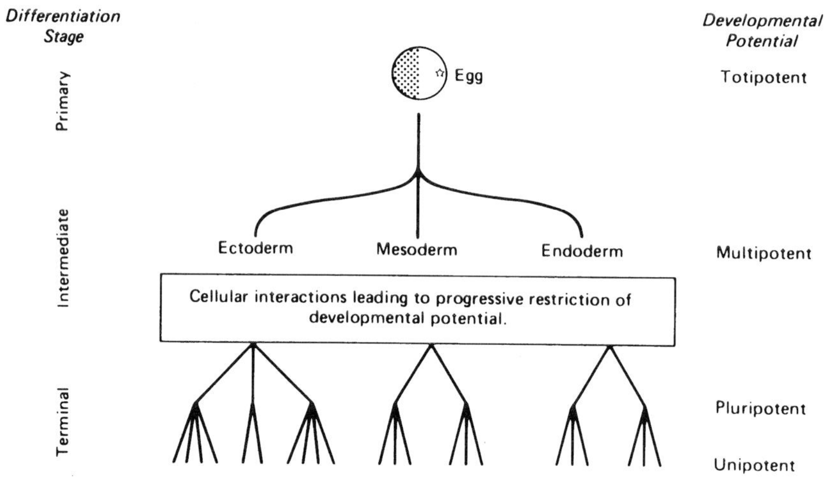

Figure 5-1 Schematic model of the various stages of differentiation in multicellular organisms. (Adapted from Rutter et al.[1])

after fertilization as the developing embryo proceeds through the blastula stage, where it is already evident that certain cells are predestined to form certain tissues. This appears to occur by means of a combination of internal reprogramming (which perhaps occurs as early as the first cell division) and external stimuli by neighboring cells, proximity to the maternal circulation, and gradients of growth factors, oxygen, nutrients, ions, and so on that result from the position in which a cell finds itself as the early embryonic cells continue to divide. The whole microenvironment of the cell determines its developmental destiny. The process of differentiation appears to be fairly permanent in that, as tissues develop, some cells retain the capacity to divide, whereas others divide and then differentiate into cells with a more restricted phenotype.[1] These latter cells are then said to be pluripotent rather than totipotent, that is, they are now committed to develop into one of the cell types peculiar to their tissue of origin. For example, a pluripotent stem cell of the bone marrow may differentiate into an erythrocyte, a polymorphonuclear leukocyte, or a megakaryocyte, but it will not become an intestinal mucosa cell or a liver cell. Embryologists have traditionally defined the commitment of a cell to one general pathway of differentiation rather than another as *determination*. They reserve the term *differentiation* for the final events in which a terminally differentiated cell arises from a pluripotent one. However, this is probably an artificial distinction biochemically, because the total process most likely represents a continuum of biochemical and molecular events leading from a totipotential cell to a terminally differentiated one. The final characterization of differentiation requires the identification of the particular biochemical events that lead to the uniquely specialized adult cell.

MECHANISMS REGULATING THE PROCESS OF DIFFERENTIATION

In order to understand the alterations in cellular differentiation that occur in cancer cells, one must examine the processes regulating normal differentiation. By definition, the process of differentiation requires a heritable alteration in the pattern of gene readout in one of the two progeny cells arising from the same parent cell. Because all the cells in the body are derived from a single cell, the fertilized ovum, this process must entail the expression of characteristics in one progeny cell that are not expressed in the other progeny cell from the same parent, and this process must continue to occur throughout embryonic development to generate the wonderful diversity of cell types present in the adult organism. How this occurs is still something of a mystery, but there are several possible explanations for which there is some experimental evidence. These mechanisms can be classified under four general headings: (1) asymmetric cell divisions, (2) selective changes in gene structure and transcription, (3) selective translation of messenger RNAs, and (4) differential response of cells to "cues" from their microenvironment. These mechanisms are not mutually exclusive; in fact, they most likely act in concert.

In a sense, there is a "European" and an "American" theory of differentiation. The European theory holds that what is most important is who your ancestors were—that is, a heritable genetic change accounts for the divergence of cells in their capacity for differentiation. The American theory asserts that the most important factor is who your neighbors are—that is, the position of cells in the developing embryo and cell–cell interactions determine their fate. In fact, both cell ancestry and cell position are important for phenotypic diversification. There is evidence to support the notion that direct changes in gene structure (both in the number and the arrangement of genes, as well as in the ability of genes to be "open" for transcription into messenger RNA) occur during development. And experimental evidence shows that the microenvironment in which cells grow is crucial to how they differentiate.

It is essential at the outset to distinguish between the regulatory processes that govern the expression of developmentally programmed macromolecules, which are essentially permanent, and those processes that regulate the synthesis of macromolecules in response to physiologic changes, which are usually transient and reversible. An example of the former is the "switching" from fetal to adult hemoglobin gene expression during development of blood-forming (hematopoietic) tissues. An example of the latter is the induction of milk proteins in the lactating breast by hormonal changes in the gravid female.

Asymmetric Cell Division

After fertilization of the ovum, the zygote undergoes some internal rearrangements of both nuclear and cytoplasmic elements. The subsequent cell divisions that occur do not partition all cellular elements equally between daughter cells. This unequal distribution of cytoplasmic elements, of itself, could lead to different cellular phenotypes early in development. This is clearly seen in the patterns of embryonic development in some lower organisms. For example, the sea urchin ovum has an "animal–vegetal" polarity even before fertilization. After fertilization, cortical cytoplasmic elements of the vegetal pole are moved toward the equator of the egg.[2] The first cleavage furrow divides the egg into equal halves along the animal–vegetal axis. The top of the egg (the animal pole) later becomes the site of cilia formation, while the opposite half (the vegetal pole) becomes the site of gastrular invagination and primitive gut formation. At the fourth cleavage, the vegetal pole descendants give rise to four very small cells, the micromeres, which are destined to become progenitors of the larval skeleton.

In the nematode *Caenorhabditis elegans*, cleavage of the fertilized egg leads to a series of asymmetric cell divisions, giving rise to blastomeres that differ in size and developmental potential.[3] By the fourth cleavage (16-cell stage), the five major somatic lineages and the germ line are clearly established. The mechanisms for these asymmetric divisions are not clearly understood, but they most likely involve cytoskeletal elements and appear to be mediated by microfilaments.[3] The fact that these asymmetric divisions lead to different cell lineages with heritable differences in gene expression suggests that cytoplasmic factors can regulate nuclear function and gene structure. In the fruit fly *Drosophila*, for example, there is evidence that cytoplasmic elements called "germ plasm" serve to determine which cells become germ-line cells during embryogenesis.[4]

The molecular basis for the reprogramming of nuclear events by the cytoplasm, so that a heritable change in differentiation potential occurs, is unclear. In the case of sea urchins and amphibians, embryogenesis involves the use of maternal mRNAs during the early cleavage states, with gradual replacement by mRNAs transcribed by embryo cells. An obvious hypothesis, then, is that, as a result of asymmetric cleavage, different early embryonic cell types receive a particular subset of maternal mRNAs that determine the cells' developmental course. This possibility appears to be excluded, however, by the finding (from RNA hybridization experiments) that the sequence content of maternal mRNA is the same in all cells of the 16-cell embryo.[2] Some differences do exist, though, in the amount of new embryonic gene transcription in the micromeres compared to the rest of the 16-cell embryo in sea urchins. These results suggest that the distinction between micromere and other cell lineages is in the time of onset of embryo cell gene transcription, probably because of the different cytoplasmic environment of the various cell types. The micromeres bud off from the vegetal pole with a very small amount of cytoplasm, and this may account for the earlier expression of cellular rather than maternal mRNAs in these cells, perhaps as a result of the presence of less total mRNA per cell or the loss of gene repressors, or the presence of a different set of transcriptional regulators (see below).

That some sort of repression of gene expression (as opposed to a permanent change in the genome) occurs in developing amphibian embryos is also implied by the fact that transplantation of nuclei from later-stage frog embryos or adult cells into enucleated zygotes can result in normal embryogenesis (at least to the swimming tadpole stage).[5-7] Interestingly, similar transplantation experiments with mouse blastomere nuclei from embryos beyond the four-cell stage do not support further embryogenesis.[8] This difference is most likely due to the dependence of embryonic development of amphibians on maternal mRNAs for a considerable length of time and independence from transcriptional activity of the embryonic cells' genome, whereas in mammals the embryonic genome becomes active as early as the two-cell stage and interference with transcription results in immediate arrest of development. Thus, amphibian gene expression in the embryo is delayed compared with that of mammals, and reversing the early differentiation events that have already started in the mammalian cells is therefore more difficult.

Interactions of Regulatory Events in Early Development

During the early events in the morphogenesis of multicellular organisms, clusters of cells develop a polarity; that is, they find a particular position in the embryo. This leads to a series of local events that turn on a specific set of genes. *Pattern formation* is the term used to describe this process of early biological organization, localized in time and space, that occurs during development. Several regulatory factors and processes are involved in these events (reviewed in Refs. 9–11).

The most detailed understanding of early development and pattern formation has been obtained for the fruit fly, *Drosophila melanogaster.* Although less is known about early developmental regulation in vertebrates, what is known indicates a significant amount of commonality between the fruit fly and higher animals. In vertebrates embryos, early development occurs in localized cellular aggregates, e.g., as seen in the blastula stages (Fig. 5-2); whereas in *Drosophila* early development occurs in a syncytium.[11] Nevertheless, a number of regulatory events are very similar. This positioning (axis formation) is thought to be caused by gradients of mRNA and growth factors in the oocyte (Fig. 5-2), leading ultimately to asymmetric disposition of transcriptional regulatory factors that turn on specific genes. These early events define an anterior-posterior and a dorsal-ventral axis in the embryo.

In the case of the fruit fly, axis formation is driven by gene products called *bicoid* and *nanos,* which are primary determinants of anterior and posterior development, respectively (Fig. 5-3). A segmentation pattern then develops and the diversification of these segments is brought about by the products of the so-called homeotic selector genes. These genes were first identified by mutations that caused abnormal development of specific parts of the fly. The first to be identified were genes such as *Bithorax* (which when mutated produced a double thorax) and *Antennapedia* (abnormal antenna development). The products of the homeotic selector genes are transcription factors that can activate or repress gene transcription. The protein products produced by these genes have a 61–amino acid run of homology called the homeodomain or homeobox, which includes a helix-turn-helix motif typical of DNA-binding proteins. The genes coding for these transcription factors have come to be known as homeobox or *Hox* genes and are now known to play a key developmental role in higher animals, including humans (see below).

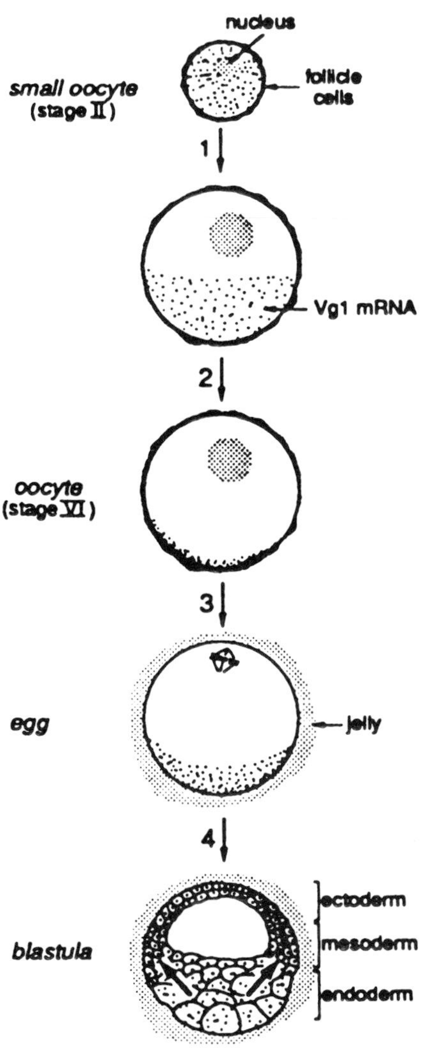

Figure 5-2 Axis formation in *Xenopus* eggs. The primary axis of the egg, the animal-vegetal axis, is formed early in oogenesis. Vgl mRNA, which encodes a TGFβ-like peptide, is synthesized early in oogenesis and is uniformly distributed in small (StII) oocytes. Vgl mRNA is localized in two steps: (1) translocation that is dependent on intact microtubules and (2) anchoring of the mRNA to a subcortical region of the vegetal cytoplasm, dependent on intact microfilaments. After maturation (3), Vgl mRNA is released and distributed to presumptive endoderm during cleavage divisions (4). At the early blastula stage, vegetal endoderm induces mesoderm in overlying cells (arrows). The cells that will give rise to ectoderm, mesoderm, and endoderm are roughly positioned in three layers, top to bottom. (Reprinted with permission from Melton, Pattern formation during animal development. *Science* 252:234. Copyright 1991 by the American Association for the Advancement of Science.)

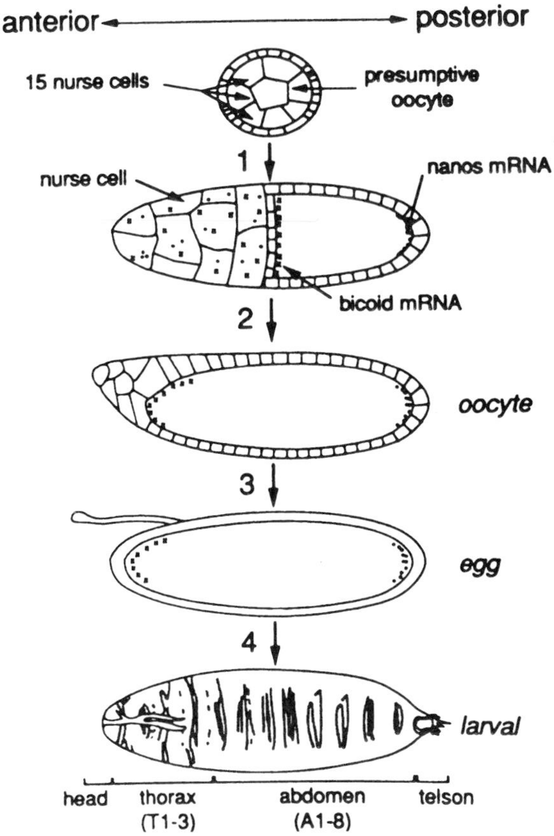

Figure 5-3 Anterior-posterior axis formation in *Drosophila* eggs. An egg chamber is composed of 15 nurse cells and one oocyte, all surrounded by follicle cells. The oocyte grows (1) as nurse cells synthesize various maternal components and deposit them in the oocyte through cytoplasmic bridges. Among the various maternal components provided by transport across these bridges are factors involved in anterior and posterior specification of the embryo, the *bicoid* and *nanos* gene products, respectively. These and other maternal components initiate a cascade of regulatory events that affect the expression of the gap and pair rule genes, which leads in turn to a distinctive larval pattern. Gene products involved in specifying the termini and the dorsal-ventral axis are also active during the periods shown, but are not depicted. (Reprinted with permission from Melton, Pattern formation during animal development. *Science* 252:234. Copyright 1991 by the American Association for the Advancement of Science.)

The homeobox genes are expressed in a restricted manner that correlates with their position to the embryo. Interestingly, these genes occur as clusters and they are expressed in the same direction, such that those genes at the more 3′ end of the cluster function to define the more anterior segments of the fly and those at the 5′ end of the cluster specifically effect posterior segment development.[11]

A key question is: What regulates the pattern of *Hox* gene expression in various positions in the embryo? It has been postulated for some time that gradients of some signaling factor or "morphogen" must be involved. A candidate for such a morphogen is retinoic acid (RA). The evidence for this comes from experiments in which RA produced a polarizing effect when placed at the anterior margin of a developing limb bud in the chick embryo (reviewed in Ref. 9). There, RA mimicked the effect of a polarizing region in that it increased the length of the apical ectodermal ridge. Thus, RA acts like the sought-after morphogen, yet it could merely be acting as an external pharmacologic agent simulating an endogenous morphogen. A true endogenous morphogen would be a molecule that forms a gradient whose concentration directly specifies pattern formation—or, in other words, a specific position-directed regulator of gene expression. Because RA does not strictly meet those requirements, its role as an endogenous morphogen is not clear.[9] However, it is clear that exogenous RA can stimulate the correct temporal sequence of *Hox* gene expression when placed in a developing limb bud, suggesting that indeed RA can regulate *Hox* gene expression. This is supported by evidence that RA can induce *Hox* gene expression in embryonal carcinoma (EC) cells in culture. All four *Hox* gene clusters can be sequentially activated in EC cells in a concentration range of 10^{-5} to 10^{-8} *M* and genes at the 3′ end of the cluster are activated at lower concentrations than those at the 5′ end.[9,9a] As discussed later in this chapter and in Chapter 15, RA and its chemical analogues can also stimulate differentiation of some types of cancer cells and in some situations act as cancer chemopreventive agents.

Getting to Know All the Players

The process of early development is a complicated one, and there are some similarities and some differences among various multicellular organisms. The biochemical signals and genes involved, however, show a lot of evolutionary conservation. Figure 5-4 is a schematic representation of how these biochemical signals interact to induce early development and start the organism well on its way to organogenesis and specific pathways of cellular differentiation. A number of these signals are discussed in more detail below, but an overview is given here.

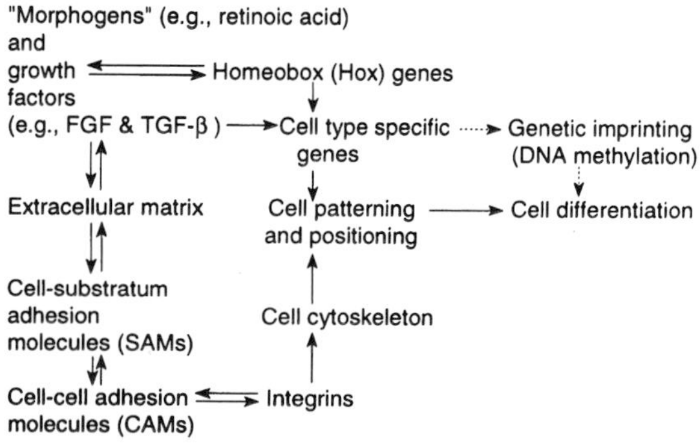

Figure 5-4 Interactions of homeobox genes, morphogens, growth factors, the extracellular matrix, and adhesion factors in development.

In addition to the postulated morphogen RA, various polypeptide growth factors have been shown to play a role in early morphogenesis. For example, in early *Xenopus* development, there is a series of inductive events that involve growth factors whose actions lead to differentiation of mesoderm at the interface between the animal and vegetal poles of the embryo.[11] This induction is most efficiently achieved by a combination of members of the fibroblast growth factor (FGF) and transforming growth factor β (TGF-β) families of growth factors. Apparently, FGF is more active in inducing ventral mesoderm and TGF-β plus FGF in inducing dorsal mesoderm.

Activin B is the member of the TGF-β family that appears to play the most important role in inducing dorsal–ventral orientation in the animal pole and in anterior–posterior axis formation in *Xenopus*. Activin B mRNA is present in the blastula-stage embryo at the time of dorsal–ventral axis determination. Activin A mRNA is not detected until somewhat later, in the gastrulation phase; however, activin A has some of the expected properties of a morphogen in that it has a clear dose-response relationship to the induction of actin genes that are turned on in this phase of development.

A key question for any postulated morphogen is: How does its expression in the tiny space of a developing embryo induce localized effects such as the determination of anterior–posterior and dorsal–ventral orientation? Two possibilities (and maybe more) exist. First, cells in a given area, e.g., the animal pole, may be homogenous and exposed to a gradient of morphogen or growth factor. Second, cells of a given pole may not be homogenous but may differ in some way that can discriminate among them in how they respond to a morphogen; that is, they might have different concentrations of cell surface receptors for the morphogen or different levels of signal transduction events in response to the morphogen. There is some evidence for both possibilities.[10]

In any case, there is good evidence from *Xenopus*, *Drosophila*, and developing chick limb buds for a role for members of the FGF and TGF-β families of polypeptide growth factors in early development.[9-11] The growth factors appear to act in early development by regulating expression of *Hox* genes. For example, growth factors regulate expression of a *Hox* gene called *Xhox3* in *Xenopus* that is required for anterior–posterior patterning. Similar observations have been made in *Drosophila*. Since *Hox* genes themselves code for transcriptional regulators that can turn genes on or off, some of which may code for growth factor–like substances, one can visualize a cascade of events in which a local concentration of growth factor turns on a *Hox* gene, which, in turn, activates another growth factor that turns on another *Hox* gene in a responding cell, suggesting a way that pattern formation could be transmitted from one cell region to another.

The activation of *Hox* genes, however, does not clearly explain how, for example, within a given mesodermal area, different mesodermal cell types arise because *Hox* genes are expressed, albeit perhaps at different times and levels, throughout the mesodermal layer. Thus, additional genes must be expressed in a carefully regulated way to lead to further "subspecialization" or differentiation events. One well-studied example of this is the expression of genes involved in the muscle differentiation pathway, e.g., the myogenic genes *MyoD* and myogenin (see below, under "MyoD").

Other important parameters of morphogenesis include the ability of like cells to cluster together and "talk to each other" and the ability of cells to produce and interact with a specific tissue-type extracellular matrix (ECM). Thus, the ability to regulate cell–cell and cell–ECM (cell–substratum) interactions is also key to normal development and cellular differentiation (Fig. 5-4). Two families of adhesion molecules are involved: cell–cell adhesion molecules, or CAMs, and cell–substratum adhesion molecules, or SAMs.[11-13] The CAMs produce cell–cell contact between like-minded cells, fostering their interactions and cell sorting into homogeneous populations. The CAMs, or cadherins, as they are also called, are large transmembrane proteins that interact through cytoplasmic connections called catenins that link cadherins to the cell cytoskeleton, thus providing an internal signaling process for CAMs that are in contact with the extracellular environment. These interactions are capable of modulating formation of actin cables in the cytoplasm and thus of affecting cell migration and cell surface polarity.[11]

The SAMs are part of the ECM and interact with cells via cell surface receptors called integrins, of which there are several types with different ligand specificities (see below). The integrins are transmembrane αβ heterodimers that interact with SAMs such as fibronectin, laminin, proteoglycans, and collagens in the ECM.

In addition to cell–cell adhesion interactions, there are likely to be cell–cell repulsion events involved in cell migration and cell clustering that provide a tracking mechanism, enabling cells to move and find the right cell clusters to set up organogenesis. Cell surface glycoproteins that appear to play a cell repulsion role have been identified and may represent a whole family of such molecules (reviewed in Ref. 11).

Another function of the ECM is to sequester growth factors such as TGF-β and basic FGF. This could provide a mechanism both for generating localized concentration gradients of growth factors and for presenting these in the appropriate configuration for cell recognition. Growth factor–ECM interactions have been demonstrated in a number of tissues and observed during development and differentiation of a number of tissue types.[11]

Thus, a number of key interactions among growth factors, *Hox* genes, CAMs, SAMs, the ECM, and specific genes involved in cell lineage–specific pathways occur during early development and early differentiation. Next, we will address some of the biochemical mechanisms by which these signals may act to produce organogenesis and terminal cellular differentiation.

Selective Changes in Chromatin Structure-Function

Components of Chromatin

To understand how gene transcription is controlled in eukaryotic cells and how these controls can be upset during oncogenesis, it is necessary to consider how the genetic material is "packaged." Chromatin is basically made up of double-helical DNA and a variety of associated proteins. Usually some RNA is also associated with isolated chromatin, but it does not appear to be part of the basic core structure. Chromatin proteins are divided into two classes: histones and nonhistones. There are five major types of histones in eukaryotic cells: the very lysine-rich histone class H1 ($M_r \cong 22{,}000$), the lysine-rich histones H2a ($M_r = 14{,}000$) and H2b ($M_r = 13{,}700$), and the arginine-rich histones H3 ($M_r = 15{,}000$) and H4 ($M_r = 11{,}300$). These proteins have basic isoelectric points and are highly conserved in the evolutionary scale. There is little variation in amino acid sequence, particularly for the arginine-rich histones, between organisms widely separated on the phylogenetic tree. The estimated mutation rate of histone H4 is 0.06 per 100 amino acid residues per 100 million years, which makes this the most highly conserved protein known.[14] H3 is also highly conserved, but H2a and H2b have undergone more evolutionary changes. From an evolutionary point of view, H1 is the most divergent histone and also has several detectable subfractions in various organisms and even in different tissues of the same organism. In addition, H1 and the other histones can be modified posttranslationally by the addition of methyl, acetyl, or phosphate groups (see below) that appear to modify their function. There may be a difference of 15% to 20% in the amino acid sequence of histone H1 among species.[15] The amino acid substitutions frequently involve interchanges of lysine, alanine, proline, and serine, and some of these substitutions have potentially important functional results. For example, an alanine is substituted for a serine in some species at position 37, a site at which cAMP-dependent phosphorylation of serine is known to occur in hormone-stimulated rat liver cells.[16]

The "core" histones involved in the fundamental structural unit of chromatin called the nucleosome (see below) are the histones H2A, H2B, H3, and H4. All four of these core histones have multiple domain structures consisting of randomly coiled, highly basic amino-terminal regions and globular carboxy-terminal domains.[17] The flexible, randomly coiled amino-terminal "tails" constitute about 20% of the mass of the core histones. H2A and H3 have, in addition, short carboxy-terminal tails.

The histone amino-terminal tails are enriched in the basic amino acids lysine and arginine, which are highly positively charged at physiologic pH and thus able to bind tightly to the negatively charged phosphates on DNA. The binding of histone tails to DNA is modulated by acetylation of lysine residues in the tails, reducing the charge-charge interaction and creating potentially more open regions of DNA. Other important posttranslational modifications of histones, such as phosphorylation and ubiquitination, also occur on the tail regions. These posttranslational modifications of histones modulate chromatin function (see below). For example, the amino-terminal tail of histone H4 contains four lysine residues, at positions 5, 8, 12, and 16, that are sites for in vivo acetylation, which affects chromatin transcriptional activity.[17]

The lysine-rich histone H1 and its variants H1° and H5 (the H1 variant specific to nucleated erythrocytes) are called "linker" histones because they interact with the linker DNA between nucleosomes, sealing two turns of DNA around the nucleosome core.[18]

The chromatin-associated nonhistone proteins (NHP) are a tremendously diverse group of polypeptides. Several hundred polypeptides have been seen on two-dimensional polyacrylamide gels after electrophoresis of NHP extracted from nuclei. They have a ratio of acidic to basic amino acid residues of 1.2 to 1.6, isoelectric points from less than 4 to 9, and a range in molecular weight from about 10,000 to more than 200,000. There is some evidence for a tissue-specific distribution of NHP, but one of the key questions is how many of them are actually involved as structural components of chromatin. This class of proteins contains a number of enzymes, including DNA and RNA polymerases, nucleases, DNA ligase, phosphoprotein kinases, proteases, histone acetyltransferase and methylase, terminal deoxynucleotidyl transferase, and topoisomerases I and II. In addition, the DNA-binding proteins that modulate gene transcription, the so-called transcriptional regulators or transcription factors, belong to this group of proteins (see below).

One abundant subset of proteins in the NHP class are the high-mobility group (HMG), so called because their electrophoretic mobility at low pH is greater than that of most of the other NHP. The HMG proteins were first isolated by Goodwin et al.[19] from calf thymus nuclei. Four distinct proteins in this group—HMG-1, -2, -14, and -17—occur in all animal tissues studied. They are among the more abundant NHP, being present at about 10^6 molecules per nucleus,[20] and they appear to play an important role in the structure and function of chromatin.

The histones are generally stable proteins and are synthesized primarily during the S phase of the cell cycle in conjunction with DNA synthesis. The NHP fraction, however, contains a number of proteins with relatively short intracellular half-lives. Alterations in the pattern of synthesis and turnover of NHP have been observed when cells are induced by hormones or drugs to produce new proteins or are stimulated to progress from a resting to a proliferative state. Regulation of expression of HMG proteins has been correlated with events in cellular differentiation. For example, during muscle development (myogenesis) the levels of HMG-14 and -17 mRNA and protein are downregulated, and reinduction of HMG-14 expression in differentiating myoblasts prevents normal myogenesis.[21]

Chemical Modifications of Chromatin-Associated Proteins

Important postsynthetic chemical modifications occur for both histone and nonhistone chromatin proteins; they affect the binding of these proteins to DNA and, by implication, play a role in the control of DNA replication and transcription. A greater variety of postsynthetic modifications occurs for the histones. Two such modifications—acetylation of ϵ-amino groups of histone lysyl residues and phosphorylation of histone seryl, threonyl, lysyl, and histidyl residues—tend to neutralize positive charges at specific sites in histone molecules. A number of methylations also occur in histones to yield mono-, di-, and trimethyl lysine derivatives, methyl guanidino arginine, and *N*-methyl histidine. These latter changes tend to increase positive charges at specific sites in histones. Not only do these histone modifications occur at specific sites in the molecules, but they also occur at specific times during the cell cycle, further suggesting that they have some important function in the control of gene expression. For example, in synchronized cultures of Chinese hamster ovary cells, the methyllysine content of histones H2a and H3 begins to rise in the early S phase; peaks after termination of DNA and histone synthesis, coincident with the beginning of the mitotic phase; and begins to fall by mid-M phase.[22] The methyllysine content of histone H2b, on the other hand, peaks in the early S phase, coincident with initiation of DNA synthesis, and rapidly falls to its original unmethylated state by late S phase. Acetylation of histones also appears to play a role in the function of chromatin; for instance, it has been shown that acetylation of arginine-rich histones occurs before the large increase in RNA synthesis observed in phytohemagglutinin-stimulated human lymphocytes[23] and posthepatectomized rat liver.[24] In both instances, the acetylation occurs before the increased cell division resulting from the mitogenic stimulation.

As noted above, the positively charged core histone tails bind to the negatively charged phosphodiester backbone of DNA. This interaction is thought to prevent access of transcription factors to DNA promoter/enhancer regions and limit RNA transcription. In support of this idea, it has been shown that the H3 and H4 core histones prevent binding of the transcription factor TFIIIA (see below) to a specific DNA sequence (the gene for 5S RNA) and that acetylation of these two histones allows TFIIIA binding.[25] Moreover, acetylation of histone H4 has been correlated with induction of loop formation in the transcriptionally active regions of the so-called lampbrush chromosomes of oocytes from the amphibian *Triturus cristatus*,[26] a classic model for chromatin activation. In early development of *Xenopus*, it has been shown that an excess of histones represses gene activity that is later overcome by TATA-binding proteins[26a] (see below).

The phosphorylation of histones, particularly H1, has also been associated with various phases of the cell cycle. An increase in H1 phosphorylation occurs at specific sites during S phase and additional sites are phosphorylated in M phase.[27] The first phosphorylations may play a role in DNA replication and the latter in chromosome condensation. At least 50% of H1 molecules are phosphorylated in rapidly dividing cells, and a rapid dephosphorylation of H1 occurs as the cells move into early G_1 phase,[28] indicating specific cell-cycle control of these events. Phosphorylation of H1 also occurs following hormone stimulation of certain tissues, but in this case only about 1% of the total H1 is phosphorylated. In glucagon-stimulated rat liver, a cascade of events occurs, leading to the elevation of intracellular cAMP, the activation of cAMP-dependent histone phosphokinase, and the phosphorylation of lysine 37 in H1.[29] Because the phosphorylation of H1 that occurs during S phase is not cAMP-dependent, there are clearly two different kinds of events, one of which presumably affects DNA synthesis while the other affects transcription of DNA into RNA. Animal tissues also contain specific phosphatases that can remove phosphate groups and thereby regulate the phosphorylation state of histones. Thus, specific phosphorylation/dephosphorylation of histones appears to play a key role in gene activation events.

Phosphorylation is the most important postsynthetic modification of the nonhistone proteins. Phosphorylated NHP are highly heterogeneous, but they have a tissue-specific distribution and a subfraction appears to bind specifically to the DNA of the tissue of origin.[30] Increased phosphorylation of NHP occurs at times of gene activation—for example, after stimulation of lymphocytes with phytohemagglutinin, following hormone treatment of various tissues, and in the actively transcribing chromosome "puffs" of insect chromosomes. The rates of NHP phosphorylation also vary during the cell cycle, being most rapid during periods of high RNA synthesis (G_1 and early S) and decreasing when RNA synthesis is suppressed (G_2 and M).[31] One of the ways in which NHP phosphorylation could be involved in chromatin activation is by destabilizing the charge–charge interactions between positively charged histones and negatively charged phosphates of DNA.[30] This could open up gene sites for active transcription. Similar to the histones, specific phosphorylation/dephosphorylation enzymes regulate the phosphorylation state of NHP. Both cAMP-dependent and cAMP-independent phosphoprotein kinases have been shown to be involved in the phosphorylation of NHP.[30,32] Among the NHP whose functions are regulated by their phosphorylation state are those involved in cell cycle regulation. A kinase called cdc-2 or cdk is activated during the cell cycle by interaction with cycle-phase specific proteins called cyclins. This is discussed in detail in Chapter 10.

The HMG chromosomal proteins are also subject to a number of postsynthetic modifications, including acetylation, methylation, glycosylation, phosphorylation, and ADP-ribosylation. The latter reaction (for review, see Ref. 33) involves the addition of adenosine diphosphoribose groups to carboxyl groups of the proteins, a reaction catalyzed by a chromosomal enzyme, ADP-ribose synthetase. Several ADP-ribosyl groups can be added—up to 50 on some proteins. Histone H1 is also subject to this modification, which is of interest because the nature of the association of H1 histone and HMG proteins with DNA may determine gene activity (see below). Such posttranslational modifications most likely affect the binding of these proteins to DNA. HMG proteins 1, 2, 14, and 17 have been shown to contain ADP-ribosyl groups in intact cells.[34] HMG proteins 14 and 17 are also phosphorylated in intact cells, and the extent of phosphorylation varies with different physiologic conditions[33]; for example, the phosphorylation of HMG 14 is higher in metaphase chromosomes than in interphase chromatin and in the G_2 phase of the cell cycle than in the G_1 phase. Phosphorylation of HMG 17 is increased in the early S phase of the cell cycle and in the log phase of growth over that in stationary-phase cell cultures. Phosphorylated HMG-14 and -17 have also been shown to induce correct spacing in chromatin nucleosomes assembled in vitro, supporting the concept that they are important in determining chromatin structure in intact cells.[35]

HMG-1 can inhibit gene transcription by interaction with the TATA-binding protein[35a] and switch a transcriptional activator to a repressor.[35b]

It should be noted here, although it is discussed in more detail in Chapter 6, that various chemical carcinogens and oncogenic viruses can alter modification of chromatin proteins by directly reacting with them or by affecting the modifying enzymes. Thus, one way oncogenic agents can bring about the alterations of gene readout observed in malignantly transformed cells is by changing the function of chromatin-associated proteins.

Packaging of Chromatin

Chromatin is a complicated structure, and the way it is "packaged" in cells affects its function. The basic organization of chromatin was derived from the results of three kinds of studies: direct visualization by electron microscopy, isolation of chromatin units after nuclease digestion, and neutron and x-ray diffraction analyses. In 1974, Olins and Olins[36] reported the electronmicroscopic visualization of chromatin isolated from rat thymus, rat liver, and chicken erythrocytes. They observed linear arrays of spherical chromatin particles, which they called ν bodies. These are spherical particles about 70 Å in diameter, separated by connecting strands of about 15 Å. This structural arrangement has now been observed for several eukaryotic cells and has come to be known as the "beads on a string" structure of chromatin (Fig. 5-5).[37] The repeating particles of chromatin, the ν bodies of nucleosomes as they are now called, can be obtained by mild nuclease digestion that clips the connecting DNA links between the "beads" to produce subunit monomers, dimers, trimers, or higher oligomers (Fig. 5-6). Endogenous endonuclease,[38] staphylococcal nuclease,[39] pancreatic DNase I,[39] and spleen acid DNase II,[40] all preferentially cut between nucleosomes to produce the typical repeating subunit pattern. The nucleosome core may also be cleaved internally by continued digestion with these nucleases to produce pieces of DNA separated by integral multiples of 10, suggesting that the DNA as it is coiled in the nucleosome has a kind of periodicity, exposing every 10th base pair to nuclease attack. The size of the DNA in nucleosome

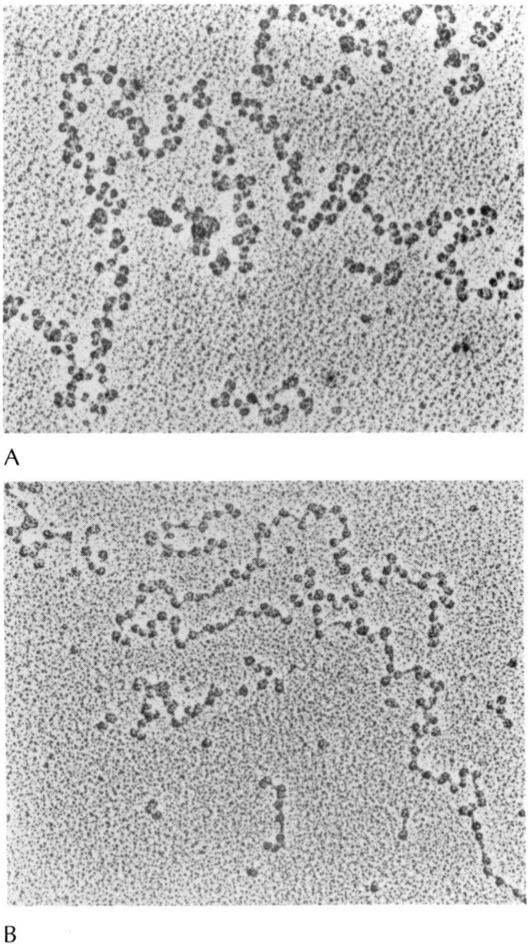

Figure 5-5 Electron micrographs of chromatin samples from rat liver showing the "beads on a string" structure and the effect of histone H1 on chromatin packaging. (A) Chromatin containing all the histones; (B) chromatin from which histone H1 has been selectively removed. (From Thoma and Koller.[37])

monomers varies from about 200 to 240 base pairs, depending on the organism and the tissue from which the nucleosomes were isolated. However, after further nuclease digestion of nucleosomes, a nucleosome core that contains 146 ($\pm$2) base pairs is produced. This number is invariant for all species and tissues studied so far. Thus, the variation in size of DNA in the nucleosome oligomers is due entirely to the variation in length of the spacer DNA between the cores.[41] The DNA in the nucleosome is wrapped around a core of histones, representing an octamer of two each of histones H2a, H2b, H3,

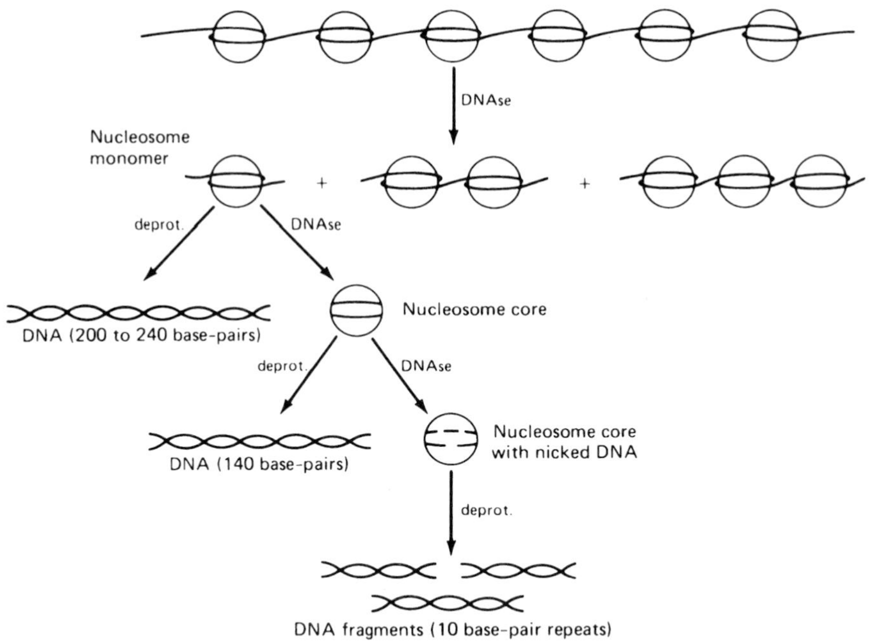

Figure 5-6 Schematic representation of digestion of chromatin by DNase. Chromatin is digested by mild DNase treatment to yield nucleosomes of monomer, dimer, trimer, or higher oligomer size. The DNA content of a nucleosome monomer varies from about 200 to 240 base pairs, depending on the cell type. This can be seen by removing protein from the nucleosome ("Deprot"). Further digestion of nucleosomes by DNase yields the nucleosome core structure which has a DNA content of 146 base pairs. Still further digestion with DNase yields a core structure with DNA nicked at 10-(or multiples of 10) base-pair intervals.

and H4.[42] Histone H1 is not part of the core but binds to the outside of the nucleosomes,[43] and removal of H1 exposes the DNA spacer region between cores to nuclease attack; this suggests that histone H1 is involved in the tight packing of nucleosomes that occurs in native chromatin.[37] This is supported by the fact that if histone H1 is removed from chromatin by selective extraction, the condensed structure of chromatin is converted to a looser, more open configuration (Fig. 5-5). The HMG proteins appear to provide some selectivity to sites of nuclease attack and to be involved in regulation of gene transcription (see below).

The orders of chromatin packaging are illustrated schematically in Figure 5-7. Watson-Crick double-stranded DNA is coiled around histone octamers of H2a, H2b, H3, and H4 like a cord wrapped around two stacked tunafish cans. Data from neutron and x-ray scattering behavior indicate that the nucleosomes are actually cylindrical rather than spherical and have a diameter of 110 Å and a height of about 55 Å.[41] The DNA makes two turns around each core, corresponding to about 70 base pairs per turn. A higher order of nucleosome packing (order II) is assumed by additional coiling of the nucleosome structure in which the H1 histones appear to play a major role by forming bridges between superhelical turns of DNA in adjacent nucleosomes. The HMG proteins are bound to chromatin in a way that provides selectivity of digestion by DNase 1, and this effect appears to require a ratio of only 1 HMG protein molecule for every 10 or 20 nucleosome monomers.[44]

The supercoiled nucleosome is further packed into a solenoidlike structure (order III), with a diameter of about 300 Å and a pitch of 110 Å per turn of six nucleosomes.[45] More condensed packing can be achieved by coiling the solenoidal form into yet another order of helix or by winding a number of solenoids around each other. This can be accomplished by kinks in the DNA

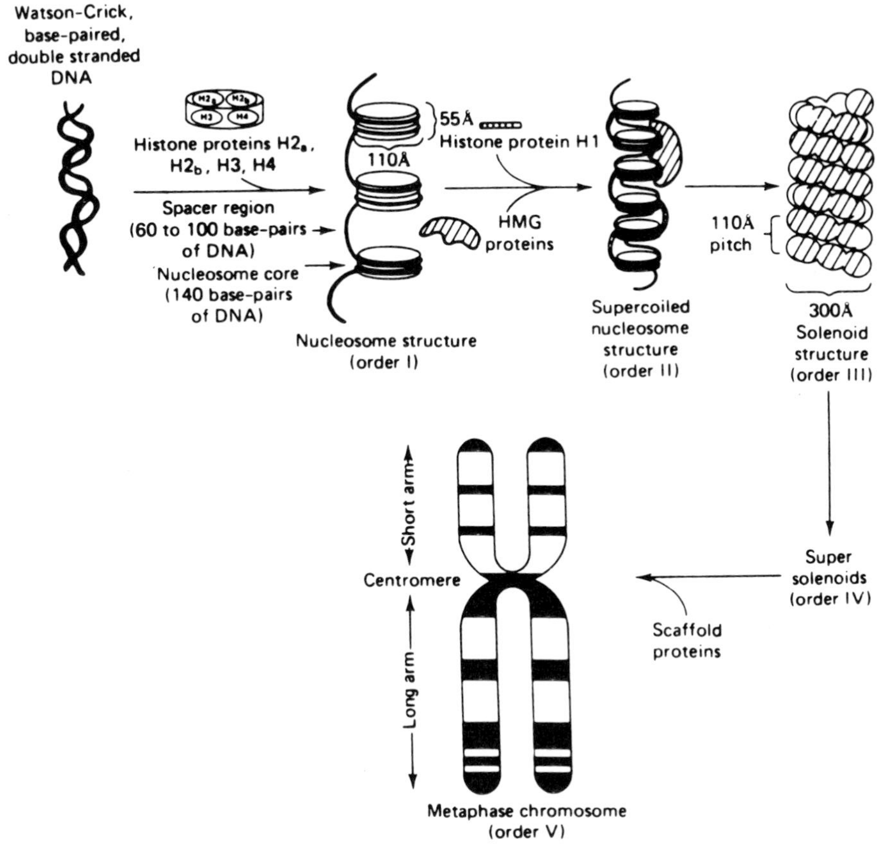

Figure 5-7 Orders of chromatin packaging. DNA and histones $H2_a$, $H2_b$, H3, and H4 interact to form the basic structure of chromatin, the nucleosome (order I). Through additional coiling and bridging of nucleosomes by H1, a supercoiled nucleosome is achieved (order II). This can be further wound into a solenoidlike structure (order III). Two or three further orders of packaging probably occur before the state achieved by the metaphase chromosome (order V) during mitosis can occur.

double helix, in addition to the supercoiling that occurs in achieving the lower orders of packaging.[45] It is likely that two or three higher orders of packaging are achieved before the fully condensed metaphase chromosome is attained. Proteins other than histones appear to be involved in this highest order of packaging because even after the histones (and most of the nonhistones) are removed by extraction from metaphase chromosomes, chromosomal DNA is still highly organized and relatively compact.[46] The structure of histone-depleted chromosomes is due to the presence of a subset of nonhistone proteins that form a central DNA-containing "scaffold" resembling intact chromosomes in size and shape. Even when the DNA is removed from these structures by nuclease digestion, the scaffold remains intact. The scaffolding of the metaphase chromosome contains two prominent high-molecular-weight nonhistone proteins, designated Sc_1 and Sc_2, with molecular weights of 170,000 and 135,000, respectively.[47]

Structure and Function of Interphase Chromosomes

When cells are preparing for cell division and enter the mitotic (M) phase of the cycle, chromosomes condense and become more identifiable as discrete entities. During metaphase, they line up along an axis in preparation for cell division, and it is the metaphase chromosomes that are used in cytogenetic studies to identify chromosomal defects (Chap. 3). However, most of a cell's life is spent in interphase. It is here that the cell spends its time making proteins specific to its differentiated function. Even in interphase, chromatin, though more diffuse and less visible, has a distinct structure that determines how it functions, which genes are transcribed, etc. Orderly, cell-specific transcription involves highly folded chromosomal domains consisting of large sequences of DNA, up to hundreds of kilobases, and this three-dimensional structure contributes to the gene readout that distinguishes the phenotype of different cells.[48] As noted in Chapter 3, some characteristics of cancer cells are discernible in the higher order of organization of chromosomes.

The term *chromatin* usually refers to the lower orders of packaging as observed at the level of nucleosomes or nucleosomes wrapped into solenoids (Fig. 5-7). Each full turn around a solenoid contains about 1200 bases of DNA, less than the length of many transcriptional units.[48] Some single genes may range from 30,000 bases up to 1 million bases. Thus, a transcriptional unit must be able to "read through" several nucleosomes. Superimposed on this solenoid structure are areas of highly condensed or packed solenoids known as heterochromatin. These areas of chromatin are generally transcriptionally inactive. More open regions of solenoidal chromatin are called euchromatin, and these are generally more active in gene transcription.

Noncoding DNA (i.e., "silent" DNA not transcribed in mRNA) makes up more than 90% of the mammalian genome. This DNA includes satellite DNA; long, interspersed repeated elements; and smaller nontranscribed DNA sequences.[48] This apparently inactive DNA most likely plays an important structural role in determining which genes of chromatin are silent and which are transcribed in a given tissue. In support of this is the evidence that regions of noncoding DNA are nonrandomly dispersed and organized in chromosomes of different mammalian species.[48]

Actively transcribed genes are more accessible to the transcription machinery (e.g., transcription factors, RNA polymerase) than highly condensed regions of chromatin. One index of this accessibility or "openness" is the sensitivity to nucleases (see below). The domains of chromatin that are open for gene transcription vary from cell type to cell type and even within the same cell type during the differentiation process. For example, during red blood cell differentiation (erythropoiesis), there is a progressive change of chromatin to a more heterochromatic, less transcriptionally active state as the cells become more and more terminally differentiated. The opposite occurs in neuronal cells. As early neuroblasts differentiate into neurons, chromatin becomes more euchromatic, consistent with the high transcriptional activity of brain cells. A number of events that damage DNA during oncogenesis may also depend on the packaging of chromatin. For example, more open regions of chromatin may be more subject to insertion of viral genes, damage by chemical carcinogens, chromosomal breakage, and translocation.

One problem is trying to understand how chromatin packaged in nucleosomes can be transcribed. In other words, how does the transcriptional machinery get access to DNA wrapped around a core of histones and linked together by H1 histones? One thing is agreed upon, and that is that H1 histone is present at lower concentrations in active chromatin than it is in inactive chromatin, allowing some loosening of the tightly packed array of nucleosomes. But how transcription can proceed from there is unclear. There are several theories.[49-51] These include (1) the displacement of core histones H2A and H2B, allowing a partial release of DNA from the core such that it can more easily complex with RNA polymerase; (2) a "rolling stone" effect, such that RNA polymerase induces a progressive, transient displacement of histone H2A-H2B dimers as it passes through the nucleosome; and (3) the transfer of histone octamers to sites behind the transcribing polymerase once it has passed through a region of DNA. At any rate, some type of opening of the tight nucleosome packing is probably required for efficient gene transcription. An illustration of how this might occur is shown in Figure 5-8. There is evidence that acetylation of histones, particularly H4, and ubiquitination (the addition of ubiquitin to a lysine side chain) of H2A and H2B occur in transcriptionally active regions of chromatin, suggesting that these modifications may open up sites for initiation of transcription (i.e., binding of transcription factors).

No mechanism as yet clearly explains how transcription occurs on chromatin arranged in

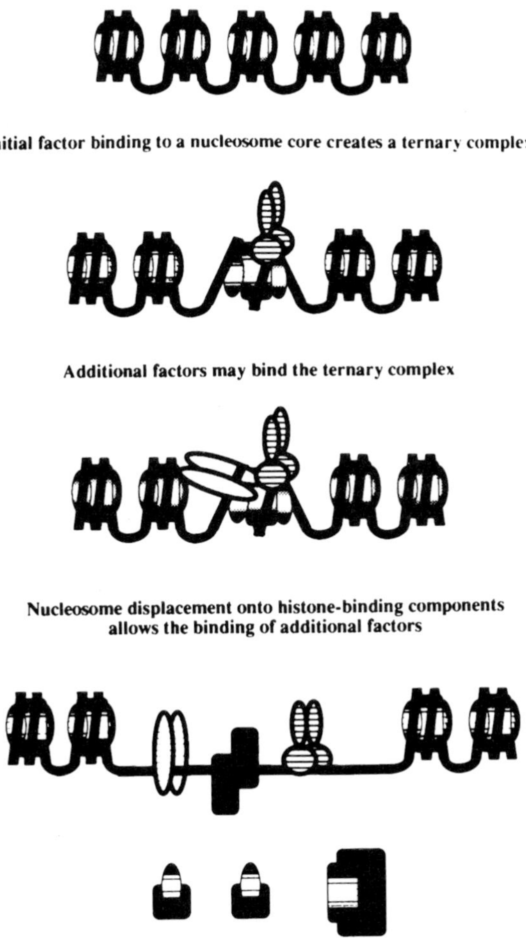

Initial factor binding to a nucleosome core creates a ternary complex
Additional factors may bind the ternary complex
Nucleosome displacement onto histone-binding components allows the binding of additional factors

Figure 5-8 Factor-nucleosome ternary complex formation followed by nucleosome displacement. (From Adams and Workman.[50])

nucleosome arrays. There is even in vitro evidence that no histone octomer dissociation is required for T7 RNA polymerase to transcribe DNA present in nucleosomes.[51]

One might also ask how DNA bundled into nucleosomes replicates itself during S-phase in the cell cycle. The data, taken together, indicate that DNA replication in higher eukaryotes occurs just as it does in simple eukaryotes and bacteria; that is, replication occurs in a bidirectional manner from a specific origin of replication and proceeds via a replication fork, producing one continuous daughter strand and one discontinuous strand (Okasaki fragments) (Fig. 5-9).[52] Newly synthesized DNA is rapidly assembled into nucleosomes. Newly synthesized and assembled histone octamers are randomly distributed to both arms as the replication fork passes through each nucleosome[52] (Fig. 5-9).

Nuclear Matrix

The protein framework of interphase chromatin—the more diffuse chromatin present in nondividing, metabolizing cells—also has an organizational network of nonhistone proteins called the "nuclear matrix."[53] Morphologically, this structure has three elements: a peripheral nuclear lamina, an internal protein network, and a residual nucleolar structure. The peripheral lamina contains three major NHPs of 60,000 to 70,000 molecular weight. The two major scaffold proteins Sc_1 and Sc_2 are also present in the internal network of the nuclear matrix.[54]

Biochemically, the nuclear matrix is prepared from interphase nuclei by a series of extractions, including high salt, that yields a fast-sedimenting complex of DNA and nonhistone proteins. Nuclease treatment of this complex removes most of the chromatin and leaves a residual fraction of DNA that is resistant to further nuclease digestion. This residual DNA appears to be attached to the matrix proteins. The points of attachment of the DNA to the nuclear matrix occur at interspaced sequences and involve binding sites of 50 to 100 kilobases of DNA. Between these attachment sites the chromatin exists in supercoiled loops. It has been suggested that these attachment points are the sites at which DNA replication is initiated.[55] Moreover, genes that are actively expressed are also bound to the nuclear matrix.[56] The transcribable portions of DNA appear to be held in a transcription-dependent form by the nature of their association with the matrix, perhaps by bringing into close proximity promoter sequences and their upstream regulatory elements at the attachment points.[54]

The DNA sequence regions involved in binding to the nuclear matrix have been termed *matrix association regions* (MARs).[57] Enhanced transcription of genes transfected into cells has been observed when MAR sequences are transfected in as part of the gene construct, and linkage of MAR sequences with transcription promoters or enhancer sequences and replication origins has also been found (reviewed in Ref. 57). These sorts of data strongly suggest that

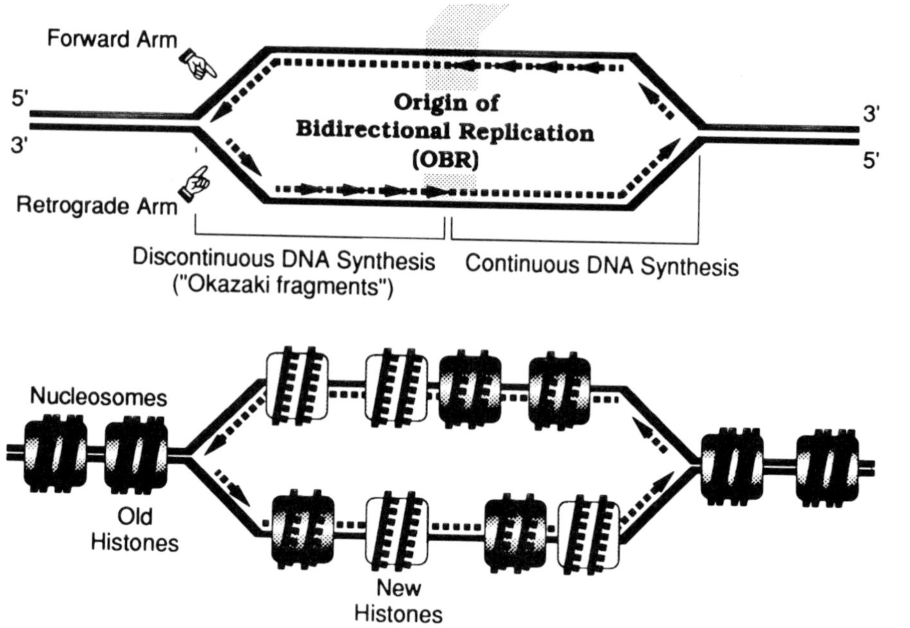

Figure 5-9 Replication fork model with random segregation of prefork histone octamers. (From DePamphilis.[52])

MAR sequences act as *cis* regulatory elements for gene transcription and origins of DNA replication. The MAR sequences appear to be looped out from attachment sites on the matrix and to assume a superhelical state that may be involved in their ability to act as transcription start sites. Such MAR sequences have been used to "fish out" nuclear matrix proteins that bind to MAR and may provide specific MAR attachment sites. One such 120,000-molecular-weight protein, called SP120, has been identified.[57]

It has also been shown that the protein composition of the nuclear matrix depends on the differentiation state.[58] The matrix protein lamin B is expressed throughout differentiation in a number of mammalian cell types. Lamins A and C are absent in undifferentiated embryonal carcinoma cells. Yet all of these are seen in all mammalian nuclear matrices, suggesting that they have a common, fundamental role in nuclear matrix function.[58]

Of interest is the fact that some differences have been observed between normal cells and cancer cells in the composition of their nuclear matrix proteins. For example, it has been reported that prostatic adenocarcinoma cells have a different array of nuclear matrix proteins than normal prostatic cells.[59] In another study, it has been shown that nuclear matrices derived from normal human breast tissue and from breast cancer tissue share some common nuclear matrix proteins but also have some specific differences in their protein composition.[60] Colon cancer tissue has also been shown to contain an array of nuclear matrix proteins different from normal colonic tissue.[60a] Nuclear matrix proteins, most likely derived from cell death and release of nuclear contents, have also been detected in the sera of patients with breast, colon, uterine, lung, ovarian, and prostate cancer at higher levels than in normal sera.[61]

Nuclease Sensitivity

The fact that only about 10% to 20% of the total DNA contained in the chromatin of eukaryotic cells is transcribed[62] suggests that there must be something structurally different about the way active sequences are packaged. Much of the evidence that actively transcribed genes are packaged in a way different from the bulk of nontranscribed DNA has been obtained by the use of endonucleases that can cleave between base pairs within a DNA sequence. Weintraub and Groudine[63] have shown that the globin gene in chick erythrocyte nuclei is preferentially sensitive to digestion with pancreatic DNase (DNase I) but not to digestion with micrococcal nuclease derived from staphylococci. The resistance of globin genes to micrococcal nuclease suggests that these genes are contained in nucleosome-like particles, whereas the sensitivity to DNase I indicates that the chromatin regions containing these genes are conformationally different from most nucleosomes.

Active transcriptional units from some species can actually be observed in the electron microscope—for example, the Balbiani rings on the large chromosome IV in the salivary glands of *Chironomus tentans* and the ribosomal RNA genes in *Oncopeltus fusciatus* embryos.[62] Inactive chromatin isolated from these organisms has a uniform beaded appearance, whereas the areas of active transcription have "streamers" of growing, nascent RNA chains attached, and the nucleosome beads, although present in the area of intense transcription, were decreased in frequency along the chromatin fiber. In addition, a smooth, unbeaded segment is visible in the electron microscope about 500 base pairs 5′ to the region of the growing RNA chains. This area may correspond to the "hypersensitive" DNase sites (see below). A similar change in the frequency of nucleosome packing is also seen in early development of *Oncopeltus*. At 32 hours after fertilization (early blastula), only tightly beaded chromatin is seen, but 6 hours later non-beaded stretches of chromatin appear in the region of ribosomal RNA synthesis before actual RNA chains can be detected biochemically. This is further evidence that the changes in chromatin structure associated with gene activation precede actual transcription.

A change in DNase I hypersensitive sites has been observed in the globin genes during development of chick embryos.[64] Red blood cell precursor cells isolated from 20 to 23-day blastoderms lacked DNase I hypersensitive sites; 5-day-old and 14-day-old chicks had them, but in different places. These experiments support the concept that, as differentiation occurs, a sequence of structural changes occurs in chromatin as the activation of new gene readout is called for by the developmental program. It is notable that a similar sequence of events occurs during development of human hematopoietic tissues,[65] in which the globin genes, both fetal and adult, gain DNase I hypersensitive sites in erythrocytes but not in leukocytes. Interestingly, the sensitive sites occur at the 5′ side of both fetal (γ) and adult (δ and β) genes in early erythrocytes, and both fetal and adult globin genes are expressed. As the embryo develops, however, the 5′ hypersensitive sites disappear from the fetal genes, and the transcription of the fetal genes is turned off, while the transcription of adult globin genes continues and is even augmented. What causes this augmentation is not clear, since the adult genes are already in the active conformation as detected by DNase I sensitivity. It may be that the amount of binding of transcriptional activators or the DNA methylation state controls this second event (see below).

Another interesting point about the packaging of active genes in chromatin is that the conformation detected by DNase I susceptibility is maintained in tissues "committed" to express that gene, whether or not the cells are actually transcribing the gene at the time. For example, the globin gene retains its sensitivity to DNase I in erythroid cells of 18-day-old chick embryos, even though globin mRNA synthesis has stopped.[63] This indicates that a record of gene activity is maintained and passed from parent to progeny cells, a likely mechanism by which commitment to a line of differentiation is maintained and inherited.

Sensitivity of active genes to DNase I appears to be a general phenomenon, but, as noted earlier, it only reflects the *potential* for a gene to be transcribed. That a wide variety of transcriptionally active gene regions can be selectively digested by DNase I has been demonstrated by finding these regions in the ovalbumin gene of hen oviduct,[66] the heat-shock proteins[67] and the histone genes[68] of *Drosophila*, SV40 viral genes,[69] and the herpes simplex virus thymidine kinase gene.[70] In general, the hypersensitive regions have been shown to occur in the 5′ DNA flanking the active genes, but they may also occur in the 3′ flanking regions.[65]

Recent studies also indicate that regions of DNA corresponding to the DNase-I-hypersensitive sites 5′ to active genes are also sensitive to DNases such as S1 nuclease, which preferentially attacks single-stranded DNA.[71] Although it seems unlikely that DNA actually exists as a true single strand in these regions, the sensitivity to

S1 nuclease suggests that the hypersensitive regions could represent loops or "hairpins" or other DNA conformations that could contain promoter sequences serving as specific recognition signals for transcription activator proteins. Along this line of thought, Sweet et al.[70] have shown that the thymidine kinase promoter region of herpes simplex virus is contained entirely within a short hypersensitive region of the nucleosome in virus-infected mouse cells. Moreover, it has been shown that the origins of viral gene transcription in SV40- and polyomavirus-infected cells and the promoter region of *Xenopus* 5S RNA genes are preferentially sensitive to DNase I. Because these regions bind specific transcription factors, it seems likely that the binding of such proteins renders the region more sensitive to DNase attack.

Putting all this information together, one can generate a schematic drawing of where the DNase-hypersensitive sites are in chromatin (Fig. 5-10). If the bulk of DNA in chromatin—that is, the inactive chromatin—has a DNase I sensitivity arbitrarily set as 1 sensitivity unit, regions distal to the 5′ and 3′ ends of active genes will have an intermediate sensitivity of 3 units, and the RNA polymerase–binding regions will have a sensitivity of 30 units, whereas the 5′ and 3′ hypersensitive regions flanking the active, RNA polymerase–binding sequences have a sensitivity of 300 units. If the 5′ or 3′ flanking regions are perturbed by the insertion of viral promoter sequences, such as long terminal repeat (LTR) sequences, a change in DNase hypersensitive sites can be observed. Furthermore, if chicken embryo fibroblasts are infected with Rous sarcoma virus or incubated with a high salt–containing medium for a short time, globin gene expression, which is normally silent in these cells, can be turned on and, what is even more surprising, this increased expression is propagated for up to 20 cell generations. This inappropriate gene expression is accompanied by the appearance of DNase I–hypersensitive sites near the globin genes.[72] This suggests that insertion of transcription-promoting DNA sequences or alteration of protein interactions with chromatin can change cellular gene expression in some semipermanent way. It is interesting to note that globin genes are also inappropriately activated in leukocytes in certain human leukemias.[73]

This raises some issues about how DNA-protein interactions that appear to dictate the readout of genes in a tissue- and differentiation pathway–specific manner can be generated and passed on from cell to cell in a heritable way. Groudine and Weintraub[72] have proposed a model for how this might occur (Fig. 5-11). The basis of the model is that certain tissue-specific nonhistone DNA-binding proteins induce a looped out, single-strandlike conformation of DNA, producing DNase hypersensitivity. At the

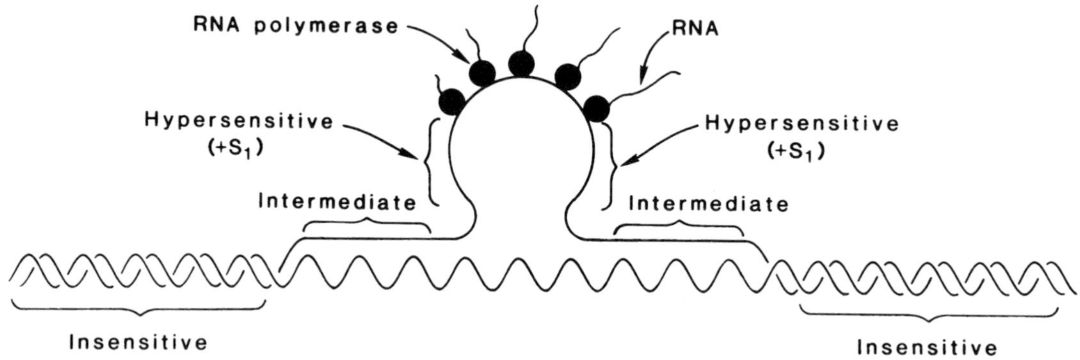

Figure 5-10 Schematic diagram of the degree of DNase I sensitivity of a strand of DNA containing a transcribable gene sequence. "Bulk" DNA tightly wound in nucleosomes is depicted as coils of DNase-I-insensitive DNA. An area of intermediate sensitivity may precede and follow the sequences being transcribed into RNA through RNA polymerase. A "loop out" area of high sensitivity and less densely packed nucleosomes contains the active gene. Hypersensitive areas, also sensitive to S_1 nuclease, may occur at both the 5′ and 3′ ends of the active gene. Relative DNase I sensitivities are 1:3:30:100 for the insensitive, intermediate, high, and hypersensitive regions, respectively.

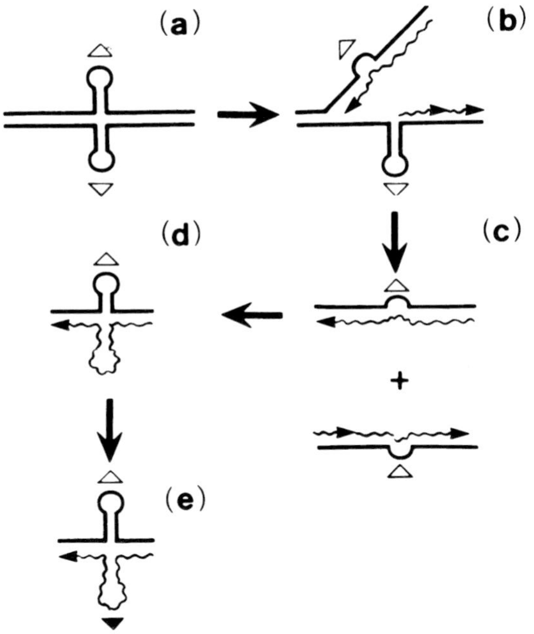

Figure 5-11 A model for propagation of hypersensitive sites, based on the single-stranded character of these sites and a postulated class of single-strand-specific DNA-binding proteins. Triangles: single-strand-specific DNA-binding proteins with the capacity to remain bound to the DNA as the replication fork copies the bases. Wavy lines: newly made DNA strands. (From Groudine and Weintraub.[72])

time of DNA replication, the protein remains bound to the parental DNA strand, presumably to the sugar-phosphate backbone, as the replication fork passes through the replicating DNA. (This would be analogous to the tight binding gene 32 protein of bacteriophage T4.) After the replication fork passes through this region, the binding of the single-strand–specific protein would stabilize the original structure in the parental strand and induce formation of an analogous hairpin loop in the daughter strand. The single strand region in the daughter strand is now in a conformation to accept the single-strand–specific binding protein, and the hypersensitive conformation is maintained for future generations of cells.

This hypothesis has gained additional credence from the work of Felsenfeld and his colleagues, who have shown that there are tissue- and differentiation-state–specific nuclear protein factors that bind to the DNase-hypersensitive 5′-flanking region of the chicken adult β-globin gene.[74] The factors that specifically bind to this region of the β-globin gene are not found in nuclear preparations from erythrocytes of 5-day chick embryos, brain, or oviduct—that is, tissues that do not express the adult β-globin gene. Within the DNA sequence that these specific factors bind are base sequences that are highly sensitive to S1 nuclease and that correspond to binding sites for transcription factors (e.g., SP1 transcription factor); these sequences are also similar to the promoter sequences of SV40 virus. In addition, the DNA region bound by the specific factors contains the consensus sequence common to almost all species' adult β-globin genes.

Methylation Patterns of DNA and Genomic Imprinting

The presence of the methylated base 5-methylcytosine was first reported in 1948 (for review, see Ref. 75). The presence of this minor base was first detected in calf thymus DNA, but since then it has been found in all vertebrate and plant species examined and in a wide variety of other organisms. In mammalian cells, between 2% and 7% (depending on the species and tissue) of all cytosine residues present in the genome are methylated on position 5 in the pyrimidine ring. Methylation of newly formed DNA occurs shortly after replication in proliferating cells and is an enzymatic process catalyzed by DNA methyl transferases utilizing S-adenosylmethionine as the active methyl donor. More than 90% of the 5-methylcytosine residues occur on CpG dinucleotide sequences in DNA (i.e., where C is 5′ in position to G), and they are present in a tissue- and species-specific pattern. However, the pattern of methylation can change during differentiation from one cell type into another or during carcinogenesis, as discussed below.

The nonrandom and tissue-specific distribution of 5-methylcytosine residues (m^5C) in tissues suggested a role for methylation in gene function, and this has been supported by evidence accumulating over the past few years. The basis of this evidence comes from five different kinds of studies: (1) the relationship of the amount of m^5C residues in a gene to its transcriptional activity, (2) changes in DNA methylation patterns during differentiation of specific cell lineages, (3) correlation of transcription of transfected genes into heterologous cell types with the methylation state of those genes, (4) the effects of the drug 5-azacytidine, which blocks DNA methylation, on gene activity; and (5) the close relationship between DNA methylation and genomic imprinting during gametogenesis.

Detection of m^5C residues in DNA is usually based on the sensitivity of DNA to clipping by specific endonucleases that recognize CpG sites depending on whether they are methylated or not. For example, a typical experimental approach uses the fact that the restriction endonuclease Hpa II recognizes a CCGG sequence only when the middle CpG dinucleotide is unmethylated, whereas the endonuclease Msp I recognizes the CCGG sequence and cuts DNA at the middle CpG regardless of the methylation state of the internal cytosine.[75] Using this specific "cutting" technique, total genomic DNA is digested with Hpa II or with Msp I. The DNA fragments are then separated according to size by agarose gel electrophoresis and transferred to nitrocellulose sheets by blotting the gel on the sheets. The fragments containing the gene of interest are detected on the nitrocellulose sheets by hybridization with a radioactively labeled mRNA or cDNA probe. This is the well-known blotting technique developed by E. M. Southern. When the blotting patterns for the Msp I and Hpa II digestions of genomic DNA are compared, one can discern whether or not a gene sequence is cut differentially by the two enzymes. If the patterns are different in different cell types or in cells at different stages of differentiation, it can be deduced in which cell type the gene is methylated (i.e., it is not cut by Hpa II but is by Msp I). Additional evidence for the presence of m^5C can be obtained by sequence analysis of the DNA fragments using the Maxam–Gilbert DNA sequencing technique (see Ref. 75).

The relative amounts of m^5C in active versus inactive genes have been examined using the differential sensitivity to restriction endonucleases as a probe of DNA methylation. McGhee and Ginder[76] first reported that specific CCGG sequences in the region of the chicken β-globin gene are less methylated in erythrocytes and reticulocytes, which produce β-globin, than in oviduct cells, which do not. Mandel and Chambon[77] found a correlation between the expression of the ovalbumin gene in tissues of the chicken and its undermethylation. Similar conclusions have now been reached for a wide variety of genes, including human and rabbit globin genes, the alpha-fetoprotein, immunoglobulin, and metallothionein genes of the mouse, and various virus genes.

Changes in DNA methylation patterns have also been observed during differentiation of various cell lineages. The fact that the pattern of DNA methylation is not identical in all tissues of a given organism strongly suggests that changes in gene methylation occur during tissue differentiation. Razin et al.[78] have found that mouse teratocarcinoma cells induced to differentiate in culture by exposure to retinoic acid undergo a high degree of demethylation of their DNA. Up to 30% of the teratocarcinoma cells' DNA methyl groups are lost after prolonged exposure to the differentiation-inducing agent, and this demethylation can be observed at specific sites in representative genes such as dihydrofolate reductase, β-globin, and histocompatibility genes. This phenomenon appears to mimic developmental processes in vivo because the extent of DNA methylation in mouse embryo yolk sac and placenta is significantly lower than that found in mouse sperm. Adult tissue DNA, however, has a high methylation content similar to sperm DNA. This suggests that, as genes are turned on in the embryo after fertilization, a demethylation process occurs, and as tissues subsequently differentiate, methylation of the genome increases again, most likely as specific genes are shut off in those tissues by the process of genomic restriction. Thus, although large changes in DNA methylation patterns appear to occur during early embryogenesis, the methylation patterns of adult somatic cells seem to be relatively stable.

Mechanisms that have been invoked to explain the role of methylation in gene activity include (1) inhibition of binding of transcription factors by methylated DNA sites, (2) decreased binding of transcription factors by DNA-binding proteins that specifically bind to methylated DNA sequences, and (3) methylation-induced change in chromatin structure that renders it less open for gene transcription.[79,80] There is some evidence for each of these. Levine et al.[79] found that inhibition of promoter activity correlated with the density of methyl CpG sites at the preinitiation domain of the promoter (TATA) box but was not affected by methyl CpG sequences distant from this domain. Their evidence also suggested that a methyl CpG binding protein was involved in this inhibition. Interestingly, some DNA templates were able to establish functional preinitiation complexes even in their methylated state.

A number of proteins that bind to methylated DNA are known (reviewed in Ref. 80). One of these, called MeCP1, binds to DNA containing at least 12 symmetrically methylated CpGs. A second one, MeCP2, can bind to a single methyl CpG pair. Also, MeCP1 has been implicated as an inhibitor of gene transcription and appears to act as a gene "repressor." While the biological role of MeCP2 is less clear, it is found in high concentration in highly methylated heterochromatin.

Methylation of DNA is also capable of inducing structural changes in chromatin. For example, as noted above, transcriptionally active chromatin is DNase I–hypersensitive, whereas the same chromatin when methylated becomes transcriptionally inactive and DNase I–resistant.[81] Moreover, methylated DNA has been shown to be resistant to the restriction enzyme Mspl when it is complexed to H1 histone, while "naked" methylated DNA was digested by Mspl and unmethylated DNA was digested by Mspl whether or not it was complexed to histone H1.[82] These data suggest that histone H1 binding to methylated DNA sequences renders them cryptic, so that these regions would be inaccessible to binding proteins such as transcription factors.

A crucial determinant of the effectiveness of methylation to regulate gene expression is the density of methyl CpGs near the promoter sequence. Weak promoters can be fully repressed by a low level of methylation, but activation of transcription can be restored by interaction with an enhancer sequence.[80] Even strong promoters or promoter/enhancer interactions can be repressed by heavy methylation at a promoter site. The degree of repression appears to be related to the level of MeCP1 binding to the methylated sequence .[80] It should be noted that some CpG-rich regions of the genome, called CpG islands, are typically not methylated, do not bind MeCP1, and are active in transcription.[80] CpG islands can, however, become methylated and transcriptionally inactive; this occurs, for example, during inactivation of one of the X chromosomes in females.

Additional evidence for the role of DNA methylation in gene expression came from experiments in which the pyrimidine antimetabolite 5-azacytidine (5-azaCR) was shown to block DNA methylation and to activate previously silent genes. Early studies of this came from Taylor and Jones,[83] who showed that 5-azaCR could induce differentiation of 10T ½ mouse fibroblasts into muscle and other differentiated cell types, suggesting that demethylation of genes leads to activation of a differentiation program. Later evidence indicates, however, that hypermethylation of certain promoter regions occurs as cells are carried in culture and is not their de novo state in actual tissues. For example, the *MyoD* gene, which triggers muscle differentiation, is normally unmethylated in mouse tissue, indicating that methylation is not the primary regulatory mechanism for this gene in vivo. The *MyoD* gene is, however, methylated in 10 T ½ fibroblasts and becomes demethylated during myogenesis in cultures of these cells.[84]

One thing is certain, however, and that is that maintenance of methylation patterns in the genome is somehow crucial for normal development. Evidence for this comes from the study of transgenic mice in which a homozygous deletion of the methyltransferase that methylates DNA has been introduced.[85] Homozygous mutant mouse embryos resulting from these transgenic manipulations are stunted in developmental growth and die at midgestation. Thus, it appears that methylation-mediated regulation of gene expression is crucial to normal development.

One way that this may occur is through genomic imprinting. This is the process by which the expression of one of the two parental genes is shut off in the embryo. Mammals inherit two complete sets of chromosomes, one from each parent, and thus two copies of every autosomal gene. Both copies of parental genes may be expressed, but sometimes only one of the two parental genes is expressed. (The other allele is said to be "imprinted"). Most of the evidence obtained so far indicates that the mechanism for genomic imprinting involves DNA methylation.[86,86a] Methylation of DNA appears to have developed over evolutionary time as a way for an organism to protect itself from foreign DNA (i.e., "if it ain't us, methylate it"). This prevents foreign DNA from being expressed. Only a few genes have been clearly shown to be imprinted in mammals. Four genes have been shown to be imprinted in the mouse: insulinlike growth factor-2 gene (IGF-2); IGF-2 receptor gene; H19, a gene coding for an embryonic DNA of unknown function; and *Snrpn*, a gene that encodes a ribonucleoprotein that catalyzes RNA splicing (reviewed in Ref. 86 and 86a). In the mouse, the IGF-2 gene and *Snrpn* are exclusively paternal in expression and the IGF-2 receptor gene and H19 are maternal in expression. Since the repressed locus does not express any mRNA, the gene must be switched off. The IGF-2 receptor gene contains a region that is methylated in a developmentally regulated way, as is another transgene, TG-A, artificially introduce into transgenic mice. A fly in the ointment is that the maternal locus of the IGF-2 receptor gene is methylated and yet it is the allele that is expressed.[86] Thus, methylation is probably not the only factor involved in the "turn off" and "turn on" of genes during genomic imprinting.

As noted above, methylation patterns change during development, even though methylation patterns in the gamete may direct how later parent-specific methylation occurs in the embryo. The methylation pattern is heritable and becomes specific for given tissue types in differentiated adult cells. The question of how this can occur then arises. One postulated mechanism is that there are "maintenance methylases" that ensure the heritability of the methylation profile.[75] In this model, gamete DNA is highly methylated, representing a highly repressed genome. During early development, demethylation of multiple genes needed for cell proliferation, cell migration, invasion into the uterine wall, and many "housekeeping" functions (e.g., substrate transport, protein and carbohydrate metabolism, nucleic acid synthesis) are demethylated correlating with their increased transcription. As tissues differentiate, some of these genes become remethylated and turned off while new genes become operational, producing a different specific methylation in each adult somatic tissue. It is of interest that a number of the phenotypic characteristics of early developing embryos, which are turned off as the embryonic tissues differentiate, are similar to those of malignant cancer cells. These include, for example, invasiveness, metastasis (i.e., migration through tissues), and rapid cell proliferation. In cancer cells, the genes controlling these functions are somehow turned back on inappropriately.

Another potential mechanism for maintenance of methylation patterns as parent cells divide into daughter cells is the regeneration of a discrete chromatin structure that allows or disallows methylation at various genetic loci. Such a structure could be determined by the complement of sequence-specific DNA binding proteins and/or the coiling of DNA. Such a model has been proposed by Selker,[87] in which a specific subset of proteins capable of holding chromatin in a given form can associate with DNA and reassociate with it after DNA replication, thereby recapitulating a form that is available (or not available) for methylation.

A peculiarity of DNA methylation is that it creates sites of high mutability.[80,87] Deamination of 5-methylcytidines occurs spontaneously in DNA at a fairly high frequency, producing thymidine, and when DNA replicates, this introduces a CpG → TpG base transition mutation if not repaired. The importance of this is suggested by the fact that CpG → TpG base transitions are thought to produce point mutations that are involved in about one-third of all human genetic diseases.[80] For nature to tolerate this potential for mutation, it must have preserved DNA methylation for a very important reason. The reason probably is that the importance of assuring the fidelity of parental inheritance and of preventing foreign DNA from replicating in an organism's cell must outweigh the danger of mutation. Although this is a teleological argument, it has some attraction.

When methylation patterns go awry, some bad things can happen to cells and organisms—for example, neoplasia.[87a] Abnormal patterns of DNA methylation occur frequently in human tumors. Hypomethylation of three marker genes (alpha and gamma globin and growth hormone) has been observed in human colon cancer tissue compared to normal colon tissue taken from the same patients.[88] Hypomethylation of c-Ha-*ras* and c-Ki-*ras* cellular oncogenes was also seen in primary human colon carcinomas compared to normal adjacent tissue.[89] Benign colonic polyps were also shown to contain hypomethylated genes, suggesting that abnormal DNA methylation may signal an early premalignant event in carcinogenesis.[90]

Hypermethylation of DNA can also occur in human cancers. Hypermethylation of loci on chromosome 17p has been shown to occur in colon carcinomas,[91] renal tumors,[92] and brain tumors[93] and of loci on chromosome 3p in lung carcinomas.[91] These hypermethylations occur in CpG islands in chromosomal regions associated with tumor suppressor genes and regions that lose heterozygosity of allelic expression during tumor progression. This suggests a mechanism by which loss of expression of tumor suppressor genes could occur; i.e., when these CpG-rich gene loci that are normally unmethylated become methylated, they may lose the ability to be transcribed, become subject to chromosomal deletion or rearrangements, and lead to genetic instability in cancer cells, a hallmark of tumor progression. The observation that these hypermethylation events correlate with tumor progression in certain tumors (e.g., colon[91] and renal[92] carcinoma) supports this hypothesis. It is of interest that an important tumor suppressor gene, p53, that is mutated or lost in a wide variety of human cancers (see Chap. 8) has a high prevalence of C-to-T mutations at cytosines methylated in normal tissues.[94] One mechanism for the hypermethylation of DNA sequences in tumor cells may be the tumor-progression–related high expression of DNA methyltransferase observed in human colon cancer.[95]

Another potential role for abnormalities of DNA methylation in human cancer is the relaxation of genomic imprinting and the subsequent biallelic expression of maternal and paternal genes when only one allele should be expressed. This could lead to overexpression of a gene that promotes tumor-cell proliferation. Such an event has been observed for the insulinlike growth factor 2 (IGF-2) gene in Wilms' tumor.[96,97] Normally, the IGF-2 gene is maternally imprinted, i.e., only the paternal gene is expressed, in human tissues. However, in Wilms' tumor tissue, both the maternal and paternal alleles are expressed, potentially exposing cells to a double dose of a growth factor that may be involved in loss of growth control in tumors. Thus, relaxation of genomic imprinting may be another mechanism involved in development of cancer.

Because the methylation state of DNA is related to gene transcriptional activity, one would predict a correlation between methylation and DNase I sensitivity of genes—and, indeed, such a correlation exists. It is of interest that the dinucleotide sequence CpG exists only in vertebrate DNA at one fourth the incidence of that predicted by random distribution. Moreover, CpG sequences (CpG islands) are enriched in certain localized areas of DNA, frequently found in the regulatory regions of genes and, as noted before, frequently found in the methylated state as m^5CpG. CpG clusters in actively transcribed hypoxanthine phosphoribosyl transferase *(hprt)* and glucose 6-phosphate dehydrogenase *(g6pd)* genes in human cells are undermethylated and hypersensitive to DNase I and S1 nuclease.[98] The same is true of murine immunoglobulin-variable region light-chain genes (V_κ) after fusion to the κ locus, and this occurs only in sequences several kilobases upstream from the transcriptional initiation site.[99] Promoter region undermethylation and DNase I hypersensitivity are also characteristic of the constitutively expressed dihydrofolate reductase *(dhfr)* genes in human erythroleukemia and HeLa cells.[100]

As noted earlier, the selective activation of genes in differentiating tissues is a multistep event involving both modifications of chromatin that fix or commit a gene to be expressed and secondary signals such as differentiation factors or transcriptional activating signals that actually turn on the transcription process. Thus, the methylation state of a gene per se does not dictate whether or not a given gene will actually be transcribed. In fact, there appear to be a number of genes for which methylation is not a regulatory event at all. Although it is clear, as noted earlier, that in vitro methylation of DNA followed by transfection into cells inhibits the transcription of several genes, some in vivo evidence raises doubts that demethylation is always a necessary precondition for gene expression. In some cases methylation of a gene appears to have no effect on transcription. For example, in *Xenopus laevis* sperm, ribosomal DNA (rDNA) is highly methylated and yet sperm rDNA is transcribed after injection in the nuclei of *Xenopus* oocytes.[101] Furthermore, methylated genes have not been clearly established in invertebrates, yet these animals exhibit cellular differentiation into organ systems. A number of exceptions also exist for vertebrate cells. For example, complete DNA methylation does not prevent polyoma- or SV40 virus early gene expression in monkey, rat, or mouse cells[102]; developmental activation of the H-2K histocompatibility gene in the mouse occurs even when the gene is methylated[103]; and albumin gene expression is shut off without being methylated in rat hepatoma cells fused to albumin nonexpressing cells.[104] Although it is possible that very selective changes in the methylation state at a few critical sites were not detected in studies of this type, it seems clear that the methylation state of a gene does not always correlate with its expression. It appears that activation of certain genes can begin with the gene still methylated; moreover, demethylation may be a signal for continued expression after activation has occurred and may bring the gene under closer control by secondary, more sophisticated hormonal signals. Where less sophisticated gene regulation is required, this mechanism may be eschewed entirely. This may be why methylation of genes was a relatively late occurrence in animal evolution.

Helical Coiling of DNA

Another level at which the function of DNA as a template for RNA synthesis can be regulated is by the degree of coiling of DNA. Helical DNA can exist in a number of different conformations. This can best be visualized in terms of a stretch of circular DNA, such as might exist in a bacterial plasmid (Fig. 5-12).[105] Such a structure, typically, is "underwound" in that there is an open loop area where the helical turns are not complete. This underwound circular DNA can form what are called "negative supercoils." In bacteria, which have circular chromosomes, the degree of supercoiling depends upon the balance between two opposing enzyme activities: DNA topoisomerase I, which relaxes DNA, and DNA gyrase (also called topoisomerase II), which maintains DNA in a negatively supertwisted state. For a number of prokaryotic genes, supercoiled DNA has been shown to be more readily transcribed than relaxed DNA.[106] However, many genes in bacteria are not activated by supercoiling; some even seem to be inhibited by supercoiling. The mechanism by which the supercoiled conformation favors transcription is not clear but may be related to the fact that this form of DNA arises from an already partially unwound form, thus allowing an entry site for RNA polymerase. It is interesting that in bacteria certain gene promoter regions can be activated either by specific activator proteins, such as the cyclic-AMP–binding protein CAP, or by negative supercoiling. Possibly, CAP activates promoters by locally unwinding DNA, while negative supercoiling accomplishes the same thing by being able to generate segments of DNA in an unwound form.

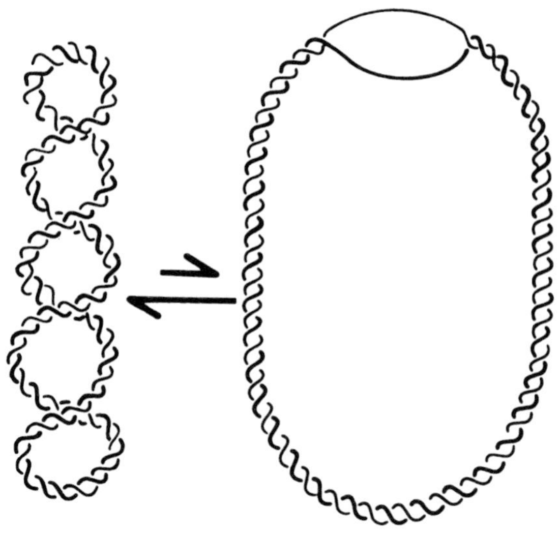

Figure 5-12 Underwound circular duplex DNA *(right)* forms negative supercoils *(left)*. DNA in the form of a closed circle with about one helical turn for every 10 base pairs—the same patch as linear duplex DNA in the B configuration—is said to be "relaxed." By convention, the supercoils formed by underwound DNA are referred to as "negative" and those by overwound DNA as "positive." (From Fisher,[105])

A true DNA-gyrase activity has not yet been found in eukaryotic cells, but DNA topoisomerase-I- and a topoisomerase-II-like activity have been demonstrated in eukaryotic cells.[107] Topoisomerase I has been found in areas of *Drosophila* chromosomes adjacent to binding sites of RNA polymerase,[108] and topoisomerase II has been found in the nuclear matrix of the same organism.[109] Eukaryotic type II topoisomerases differ from the prokaryotic DNA topoisomerase II (DNA gyrase) in that the latter introduces negative supercoils into circular DNA, whereas the former do not. Nevertheless, topoisomerase II in the eukaryotic yeast *Saccharomyces cerevisiae* is required for normal cell mitosis in a manner suggesting that topoisomerase II is required to allow the segregation of intertwined sister chromatids at the time of mitosis.[110] Furthermore, a number of anticancer drugs, such as the DNA-intercalating agents doxorubicin and daunorubicin as well as the nonintercalating antitumor agents of the epipodophyllotoxin class, cause the stabilization of a DNA-topoisomerase II complex in a DNA-nicked form that inhibits DNA replication and cell division.[111] In the case of eukaryotic chromatin, there is some evidence suggesting that the 5′ end of active or activatable genes is supercoiled in regions where the DNA is also hypersensitive to DNase I and S1 nuclease. What is also intriguing is that these regions contain alternating purine–pyrimidine clusters (i.e., CpG) that are potential sites for methylation and Z-DNA formation. The facts that these sequences are found in regulatory regions of various genes—for example, the long terminal repeat (LTR) regions that contain enhancer sequences (see below) for transcription of viral genes—and that the Z-DNA conformation is stabilized by negative supercoiling bring the elements of the story together. We now know a lot about the components, but as yet we do not know how all the pieces fit together or what regulates the cascade of events leading to gene activation.

Taken together, these findings suggest that conformational changes in DNA can have a great effect on the overall transcribability of genes. Because supercoiling of DNA can have profound effects on transcription, conformation transitions of the primary DNA helix structure must be added to the list of genetic regulatory mechanisms. Z-DNA-binding proteins have been identified in eukaryotic cells, and they appear to bind within an area containing regulatory sequences.[112] These proteins may stabilize the Z conformation and, in so doing, play a role in gene regulation.

The ability to adopt the left-handed Z-DNA conformation is not just a property of lower animals' cells. Intervening sequences of three human fetal globin genes contain tracts of alternating purine–pyrimidine sequences, 40 to 60 base pairs in length, that can adopt the Z-DNA conformation when negative supercoiling is induced in the DNA.[113]

Chemical probes can be used to detect the location of unpaired DNA bases that would be expected to occur in underwound DNA or DNA sensitive to the single-strand-specific nuclease S1. One of these probes is bromoacetaldehyde, which reacts specifically with unpaired adenine and cytosine bases in DNA. When erythrocytes from chicken embryos or adult chickens are reacted with this chemical, reaction occurs with chromatin in the S1 nuclease-sensitive region 5′ to the active chicken adult β-globin gene only in cells expressing this gene.[114] It is noteworthy that chloroacetaldehyde, a chemical analogue of bromoacetaldehyde, is a known mutagen and carcinogen. It is likely that bromoacetaldehyde is also carcinogenic. Thus, one way in which chemical carcinogens may alter gene readout is by reacting with unpaired bases of DNA in these crucial supercoiled, S1 nuclease—sensitive areas in 5′ or 3′ regulatory regions of certain genes. These regions appear to be hot spots for genetic recombination events, the kind of events that have been shown to occur in human and animal cancer cells. It is also possible that chemical carcinogens, by reacting covalently with DNA bases (see Chap. 6), could induce local unwinding and negative supercoiling of DNA and thereby activate genes that are not expressed in an adult normal cell.

Telomeres and Telomerase Activity

The biochemical mechanism of DNA replication prevents DNA polymerase from fully replicating the linear ends of DNA at the 5′ end, creating an "end-replication problem" in the production of daughter cells. Because the DNA duplex is antiparallel, each daughter DNA molecule would be shorter on the 5′ end than its parent DNA after replication (reviewed in Ref. 114a). Prokaryotes and viruses have evolved mechanisms to overcome this by "repriming" the ends of chromosomes during replication. Eukaryotic cells, on the other hand, have developed special structures called *telomeres* at the ends of their chromosomes to prevent shortening of crucial DNA sequences.

Telomeres are in a sense "expendable" repeated DNA sequences, packaged with specialized proteins, that occur at the ends of chromosomes. These DNA sequences are highly conserved throughout evolution and are substrates for replication by an enzyme called telomerase, which completes the replication of chromosomal ends. In normal cells, however, this occurs at the expense of losing some of the repeated DNA sequence with successive cell divisions. Telomeres also protect chromosomal ends against faulty recombination events, and since they attach to the nuclear matrix, telomeres may play a general role in regulating DNA replication.

In humans, 5 to 15 kilobases of TTAGGG repeats are found at the ends of all chromosomes. These repeats become shorter as cells mature and undergo senescence, apparently because telomerase activity is diminished.[114b] Telomeres, for instance, are significantly longer in human germ-line cells than in somatic tissues. In nontransformed human cell culture systems, telomeres became shortened by about 65 base pairs per cell generation and underwent senescence, but in cells transformed with SV40 or adenovirus 5, telomeres became stabilized once a telomere length of about 1.5 kilobases was reached and the cells became immortalized.[114b] Telomerase activity was detectable in the immortalized cells but not in the senescing cell populations.

Interestingly, telomerase activity has been detected in human ovarian carcinoma cells but not in normal cells from the same patients.[114c] The cancer cells had short but stable telomeres. These results suggest that progression of the malignant state depends on continued telomerase activity and that telomerase could be a target for new therapeutic approaches to cancer treatment.

Split Genes and RNA Processing

Another important aspect of the transcriptional capability of chromatin is how genes are put together and how the transcriptional products of genes are processed. The isolation and characterization of the restriction endonucleases,[115] enzymes that cleave DNA at specific sequences, have revolutionized our understanding of the organization of genes and led to the ability to clip out specific genes, insert them into bacteria, and produce millions of copies of a specific gene as well as the product of that gene. This has brought about the revolution in molecular biology known as "genetic engineering." One of the startling discoveries that has come from these studies is the finding that many genes of eukaryotic cells are split into a number of pieces. This means that there is yet another order of control interposed in the flow of information from DNA to RNA to protein and another control point that could malfunction in the generation of neoplastic cells. The sequences of DNA converted into functional mRNA are called "exons"; the sequences of DNA intervening between these are called "introns."[116] It has been found that RNA polymerase transcribes a whole gene complex with its introns and exons. The RNA transcript is then "processed" to remove the intron segments and "spliced" to put the pieces of mRNA that are to be translated into protein together (Fig. 5-13).

The mechanisms of the splicing and processing reactions are not entirely understood, but a number of the key components have been recognized. It is interesting that the nucleic acid–

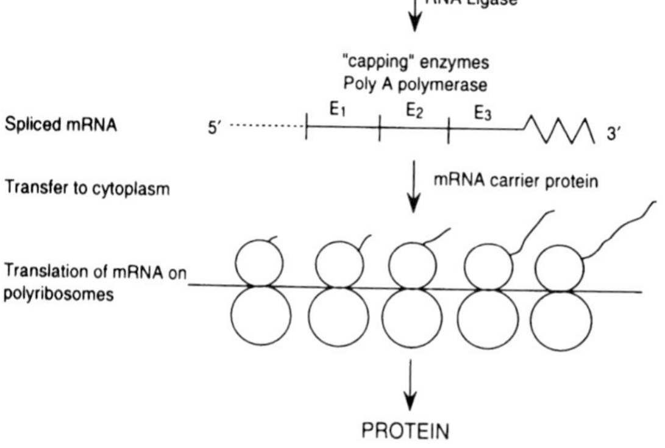

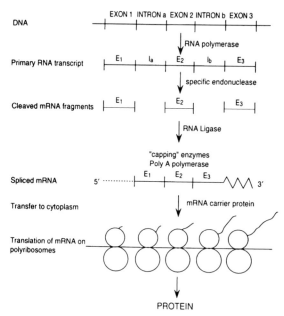

Figure 5-13 Schematic representation of transcription of "split" genes, processing of mRNA, and translation of mRNA into protein.

base sequences at the points of the exon–intron junctions have a specific arrangement (Fig. 5-14). This specific arrangement, known as "Chambon's rule" because it was first clearly demonstrated for the ovalbumin gene in Chambon's laboratory, is that introns begin with a GT and end with an AG base sequence.[117] This same sequence arrangement has now been found in more than 90 exon–intron junctions that have been sequenced in eukaryotic genes. Additional "consensus" sequences have been found around the junctions in the case of many genes. The generality of the GT–AG rule and the presence of other consensus sequences strongly suggest that the splicing enzymes (i.e., the specific endonucleases and RNA ligases—see Fig. 5-13) involved in the processing of precursor messenger RNA molecules into mature mRNA are phylogenetically very old enzymes that have evolved from a single ancestral enzyme system. For example, when the precursor mRNA of chicken ovalbumin is introduced into mouse cells in culture, the pre-mRNA is cut and spliced correctly by the mouse RNA–splicing enzymes.

The primary transcript is further modified by the addition of a methylated guanine "cap" at the 5′ end and of a polyadenylate "tail" at the 3′ end. These modifications appear to play a role in correctly initiating translation of the mRNA into protein and in protecting the mRNA from nuclease digestion. The ligated, capped mature mRNA is transferred by means of a protein carrier[118] through nuclear pores to the cytoplasm where it binds to polyribosomes and is translated into protein.

The recognition signals for the cutting and splicing points in primary RNA transcripts involve the binding of "small nuclear RNA" (snRNA) present in ribonucleoprotein complexes called small nuclear ribonucleoprotein particles (U-snRNPs) so named because they contain a family of uracil-rich snRNAs found in eukaryotic cells. It is now clear that the snRNPs are key factors for RNA sequence recognition during mRNA splicing.[119] Changes in these factors could regulate mRNA splicing patterns as cells differentiate or as cells are exposed to different physiologic stimuli. A general mechanism for mRNA splicing in eukaryotic cells is shown in Figure 5-15.[120] The splicing reaction involves the formation of a "spliceosome" containing snRNPs: U1, U2, U2, U4, U5, and U6. A U2

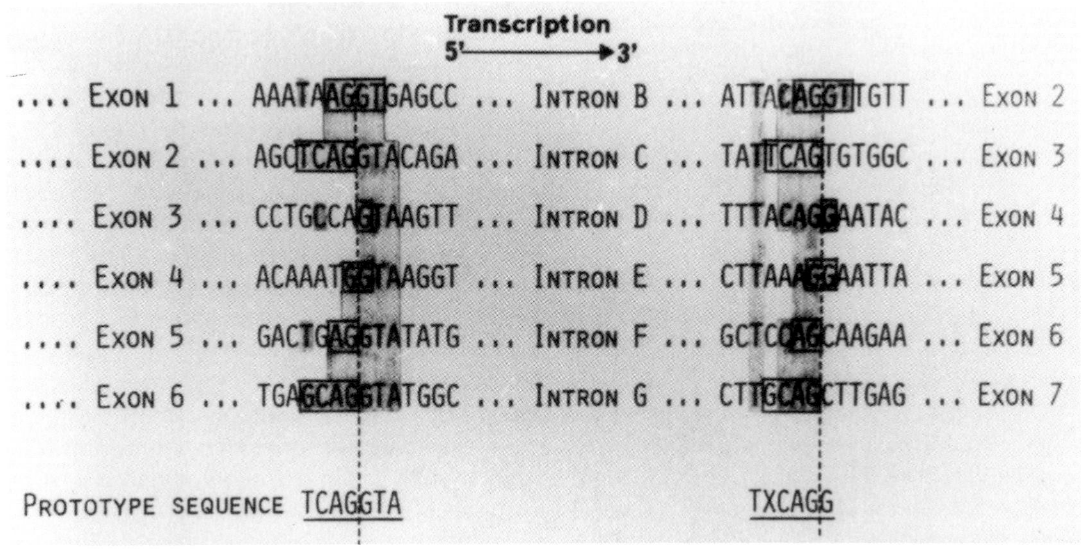

Figure 5-14 The base sequences at the junctions between the exons and introns of the ovalbumin gene. The ovalbumin gene is split into seven exons, and they have been aligned to show their common features. The vertical broken line shows where excision–ligation events could occur. Note that all introns begin with a GT and end with an AG base sequence. The shadowing stresses the similarities of the sequences around the exon–intron junctions. The bottom line shows a prototype sequence for the exon–intron junctions. (From Breathnach et al.[117])

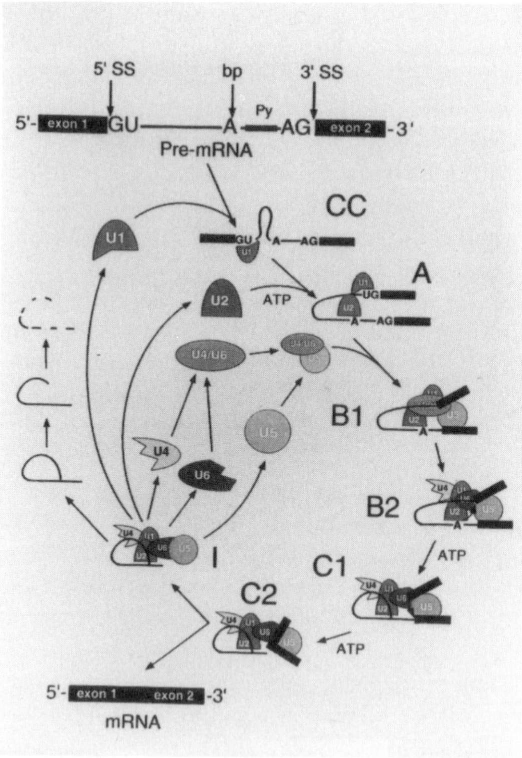

Figure 5-15 Schematic representation of the spliceosome cycle in terms of the role of snRNP particles in Pre-mRNA splicing. Pre-mRNA (*top line*), containing two exons separated by an intron, enters splicing complexes with snRNPs and exits as mRNA (*bottom line*) and excised lariat intron (*left border*). Other non-snRNP factors are required for spliceosome formation but have been omitted for simplicity. CC, A, B1, B2, C1, C2, and I represent complexes within the splicing pathway that have been distinguished biochemically, genetically, or both. *Abbreviations:* 5′ SS, 5′ splice site; 3′ SS, 3′ splice site; bs, branch site; Py, polypyrimidine tract. The individual snRNPs indicated are U1, U2, U4, U5, and U6. (From Sharp.[120])

snRNRNA complex is an intermediate in the formation of the spliceosome. The U1 and U2 snRNAs recognize consensus sequences at the 5′ splice site as an early step in the splicing reaction. U5 snRNA recognizes exon sequences flanking the splice sites. Mutations in the highly conserved AG-containing sequences at the 3′ splice site and deletions in the polypyrimidine sequence that binds U2 snRNP prevent normal splicing[120] and may explain how mutations in these regions could produce an abnormal mRNA.

It is now known that there are three different types of introns: group I, group II, and nuclear pre-mRNA introns.[121] Group I and group II introns have been found in *Tetrahymena,* yeast, *Neurospora, Physarum,* and various other fungi, bacteria, and plants. However, similar elements may exist in higher organisms. Group I and II introns have distinguishing structural features. For example, group II introns have a highly conserved secondary structure with a core sequence and six looped out helixes that define different sequence domains. Nuclear pre-mRNA introns have the conserved 5′ and 3′ sequence motifs described above but do not have the conserved structural sequence motifs of group I and II introns. The group I and II introns are mobile—i.e., they can move around in the genome—and they contain open reading frames that encode proteins. Unlike most other mobile DNA transpositions that involve nonhomologous donor and recipient sequences, group I intron mobility is site-specific, resulting in intron insertion at specific alleles and sometimes involving repeated rounds of insertion.[121] This intron mobility requires a site-specific double-stranded DNA endonuclease encoded by the intron itself. Other proteins encoded by group I introns include RNA splicing enzymes. Some group II–encoded sequences have homology with the enzyme reverse transcriptase, which would allow for production of reverse-transcribed cDNA from processed RNA and could be a mechanism for preserving intron sequences as well as for providing mobility and reinsertion substrates.

Groups I and II introns can be viewed as "ribozymes" in that they produce RNA products that encode products that catalyze their own splicing and play a role in their mobility. In one sense, they can be viewed as "infectious introns" in that they can move around within a genome and possibly even between species.[121a] Group I introns are said to be "homing" introns because they are site-specific and characteristically unidirectional, i.e., they produce nonreciprocal insertions into allelic-related genes that involve intron-encoded site-specific endonucleases.[121,121a] Group II intron "homing" is less well characterized and most likely involves the intron-encoded reverse transcriptase–like activity. The cDNA copies of excised introns' mRNA may be the actual "vectors" for transposition to other sites in the genome and in that sense resemble the transposable elements (transposons) described originally in corn by McClintock (see below). However, both group I intron endonucleases and group II intron reverse transcriptases also function in RNA splicing.[121a]

Self-splicing group I and II introns are thought to be vestiges of primordial evolution. For example, group I introns of a *Cyanobacteria* tRNA gene are found in exactly the same location in the analogous tRNA gene in plant chloroplasts, suggesting that this intron was present prior to the endosymbiotic incorporation of *Cyanobacteria* into plant progenitor cells, which was thought to have occurred about 1 billion years ago.[121a] This intron, however, is absent in the analogous tRNA gene in mitochondria, suggesting that it was sporadically distributed after that time.

Group II introns have only been found in cellular organelles such as chloroplasts and mitochondria and are considered to be the evolutionary precursors of the nuclear pre-mRNA introns that code for snRNAs involved in spliceosome formation and RNA splicing in higher organisms.[121a] In any case, the presence of introns in genes has provided a mechanism for "exon shuffling," alternative mRNA splicing, and gene regulation that may well have conferred selective advantages to the host organisms over evolutionary time.

A question that arises from the discovery of the split-gene arrangement of eukaryotic genes is whether or not multiple protein products could result from one primary transcript depending on the stop-start points of the cutting and splicing mechanisms. The answer is clearly yes. Some examples are the generation in the rat of three different mRNAs coding for different types of the cellular matrix protein fibronectin[122]; two of these mRNAs arise from the utilization of different 3′ splice sites contained within a single exon. Similarly, rat muscle myosin light chains 1 and 3 are produced from a single gene by differential RNA splicing of a single primary gene transcript.[123] In addition, a single mouse α-amylase gene can transcribe two different α-amylase mRNAs in different tissues of the mouse.[124] In all these cases, the splice site sequences follow Chambon's rule of GT at the 5′ end and AG at the 3′ end; however, the sequences around these splice sites vary. This implies tissue-specific and, in some instances, differentiation-specific recognition of these splice sites. Candidates for providing such specific recognition are (1) different splice enzymes activated at different times during differentiation or in different tissues, (2) tissue-specific or differentiation-specific snRNAs that bind to these sites, and (3) protein signals that differentially mark the splice sites. The first alternative seems unlikely in view of the commonality of splicing enzymes noted earlier. The other two candidates, or a combination of them, seem likely possibilities.

The importance of carefully maintaining the correct splicing steps in mRNA processing is indicated by what happens when processing goes awry. For example, in a type of β^+-thalassemia, characterized by anemia with reduced β-globin synthesis, decreased production of normally functioning β-globin mRNA is caused by abnormal splicing of the β-globin gene primary transcript.[125] The β-globin gene from such patients contains a single base substitution 22 base pairs upstream from the 3′ junction between intron 1 and exon 2 of the gene, creating an alternative splice site within intron 1 and resulting in the retention of 19 extra bases from the 3′ end of intron 1 in the mRNA. This abnormally spliced mRNA appears to be less stable and poorly transported to the cytoplasm.

One can visualize how such a delicate and complicated series of events as the correct transcription and splicing of mRNA sequences could be interrupted or upset by agents that interacted with DNA or chromatin proteins. It is interesting that there are a number of guanines around the key exon–intron junction sites. Guanine bases in DNA are one of the primary targets of alkylating-type agents, which frequently are chemical carcinogens (see Chap. 6). Irradiation damage at these key DNA sequences could also disrupt the correct transcription or splicing of mRNA. Oncogenic viruses are known to insert their DNA into the host's DNA at various points. If these alterations occurred in a key functional gene sequence, such as one involved in coding for an enzyme involved in a crucial metabolic step, the result would probably be fatal to the cell; but if such alterations or insertions were at a key control point, such as one involved in switching genes on or off or in the correct processing of exon–intron sequences in mRNA, the cell might survive, although in a way that would allow it to transcribe, process, and translate mRNA abnormally. It has been shown, for example, that there are "intron mutants" in yeast for the mitochondrial gene of cytochrome b, an enzyme in the respiratory chain.[126] These mutations result in an abnormality of RNA splicing that produces an altered but in many cases still functional protein.

Genetic Recombination

Genetic recombination is the process by which new combinations of nucleic acid sequences are generated by shifting around genetic elements in the genome. While most organisms judiciously guard their genetic material, there are times in the life cycle of an organism—during early development in multicellular organisms, for example—when it is advantageous to move genetic information around to achieve a new set of genetic alterations. One obvious example is the ability of microorganisms to mutate and adapt to new environments rapidly—e.g., to gain resistance to antibiotics or to an immunologic challenge from the invaded host. Indeed, a lot of what is known about genetic recombination has been gleaned from bacteria and parasitic organisms like trypanosomes.[127,128]

Since gene rearrangements are potentially dangerous to the viability of an organism, they are well regulated in higher organisms. This is in marked contrast to an organism like the pathogenic yeast *Candida albicans*, which can switch among at least seven phenotypes with a frequency of 10^{-4} per cell division.[127] Willy-nilly, inopportune gene rearrangements can lead, in humans, to inherited disease or cancer. "Incidental" or unprogrammed rearrangements can arise from errors in DNA replication, repair, or recombination, from the movement of mobile elements such as transposons (see below), or from the insertion of viral DNA. Most of these events are deleterious. Programmed gene rearrangements, on the other hand, are part of the normal developmental process and are developmentally regulated.

Genetic recombination is classified into three categories: (1) general recombination between homologous DNA regions, (2) transpositional recombination carried out by transposable elements (transposons) that jump from one chromosomal location to another, and (3) conservative site-specific recombination that produces rearrangements occurring in specific gene sequences within chromosomes (e.g., the gene rearrangements that occur in V(D)J antigen receptor genes of lymphocytes; see below).[128]

The first inklings of genetic recombination came from studies of Bateson and his colleagues in 1905 when they found some non-Mendelian inheritance traits in the sweet pea *(Lathyrus odoratus)* (reviewed in Ref. 129). What they found was that certain combinations of traits occurred more frequently than expected and others less frequently. We now know that this was due to genetic recombination.

General recombination was discovered in *Drosophila* in 1911 by Morgan, who coined the term "crossing over."[130] This term was used to describe the exchange that gave rise to new combinations of linked genetic traits in fruit flies. It involves exchanges of genetic information at equivalent positions along the length of two chromosomes with significant sequence homology. A number of models of how this occurs have been proposed, but the mechanism is not clear.[129]

Transpositional recombination was discovered by McClintock in maize in the 1940s, but her findings were ignored for many years. She was finally recognized for her contribution with the awarding of the Nobel Prize in Physiology and Medicine in 1983.[131] At the time of McClintock's experiments, maize plants provided the best material for locating known genetic traits along a chromosome and also for determining the breakpoints in chromosomes that had undergone various types of gene rearrangements. The crucial experiment began with the growing of about 450 plants, each of which had started its development with a zygote having received a damaged chromosome, with a broken end, from each parent. Such ruptured chromosomes could be produced by x-ray exposure of germinating plants. The seedlings developed from such matings produced totally unexpected phenotypic variants. Variegated seedlings with startlingly different production of chlorophyll were evident, and this altered expression was confined to clearly defined sectors in a given leaf. Mc-

Clintock concluded that the modified expression of the genes regulating chlorophyll production must be related to an event in the precursor cell that gave rise to that sector of the leaf. For this differential gene expression to occur in a localized area of the leaf, all of whose cells had arisen from the same parent cell, some cell component had apparently been unequally segregated at mitosis. McClintock originally called these "controlling elements," but we now know, based on her pioneering work, that these are transposable genetic elements, or "transposons." The mechanism for the heritable segregation of traits in maize was not simply a result of the passage of unrepaired broken chromosomes, because it was clearly demonstrated that broken ends of ruptured chromosomes find each other and fuse in the telophase nucleus. Rather, the selective appearance of different genetic traits is due to the activation of transposable elements that are carried normally in the silent state in the maize genome. Once these elements are activated, their mobility allows them to enter different gene loci and "take over control of action of the gene wherever one may enter."

Movable genetic elements, or transposons, are now known to be present in many organisms, including bacteria, yeast, and the fruit fly *Drosophila*. In *Drosophila*, these elements are present as repeated gene sequences—as many as 30 copies per cell. These repeated sequences are dispersed among several chromosomes and appear to be nomadic. Similar sequences exist in higher organisms, including mammals. In humans, as much as 3% to 6% of the total genome is made up of 300 base pair repeats, called the *Alu* family of sequences. These are likely candidates for mobile DNA elements in humans.[132] The bacterial and maize elements turn genes on and off as they move around in their respective genomes. Moreover, they promote chromosomal rearrangements, thus giving rise to mutations.

Developmental processes involving the breakage and rejoining of DNA at defined sites are now widely known in nature.[133] This leads to the programmed elimination of DNA in some daughter cells and the realignment of rejoined sequences, producing the readout of entirely new patterns of genes. The eliminated DNA may comprise a significant portion of the genome present in the germ line—as much as 10% to 20% in the ciliated protozoa *Tetrahymena*, for example,[134] and is removed from somatic cell nuclei at specific stages in development. In *Tetrahymena* there appear to be more than 5000 rearrangement sites in the genome.[133]

Site-specific recombination, resulting in rearrangements that occur in highly sequence-specific loci in chromosomes, was first described by Campbell[135] for the integration of bacteriophage λ chromosome into its bacterial host's chromosome. In site-specific recombination, specific DNA sequences are bridged by protein–protein interactions between DNA-binding proteins linked to the recombinatorial DNA strands.[129] There are some key enzymes involved in site-directed recombination called "recombinases." These promote breakage and rejoining of DNA via covalent DNA-protein recognition sites surrounding the sites of cleavage and strand exchange.[128] The recombinase family of enzymes includes the integrases and resolvase-invertases, whose activities have been studied in detail in bacterial systems and yeast. Similar enzyme activities exist in higher organisms (see below).

It should be pointed out that the three classes of recombination events noted above share some characteristics. For example, general recombination occurs between homologous DNA molecules, but sequence homology is also important in site-specific recombination.[128] While the specific class of enzymes called recombinases are involved in site-specific recombination, sequence-specific recognition of DNA by cleavage and rejoining enzymes also occurs in general and transpositional recombination. Moreover, though transposons can jump to multiple sites in a target DNA, there are examples of site-specific sequence recognition.[128]

An important example of site-specific genetic recombination in mammalian organisms is that which occurs during the rearrangement of immunoglobulin genes in B-lymphocyte differentiation.

Early in the differentiation pathway for B-type lymphocytes—that is, the cells that produce antibodies as their specific differentiated function—there is a commitment to express only one subset of immunoglobulin molecules, containing one light- and one heavy-chain gene product out of the large numbers of available genes that code for such proteins. The expression of immunoglobulin (Ig) genes is achieved through specific genetic recombinations that occur during the ontogeny of B cells. This specific gene expression results from the splicing of one variable (V) region gene, out of a large pool of V genes, with a constant (C) region gene.

Each Ig molecule contains a heavy (H) and a light (L) (kappa or lambda type) chain, both of which have variable and constant regions that are so designated based on the amount of variation in the amino acid sequence. Both heavy- and light-chain genes undergo splicing to produce a specific type of Ig molecule (Fig. 5-16).

There are several steps in the production of mature B lymphocytes that produce specific types of immunoglobulins. One of the first steps is the chromosomal rearrangement of one V_H, D_H, and J_H gene with a C_μ gene and the transcription of μ mRNA from this spliced gene. A cell at this stage, called a pre-B cell, does not secrete Ig. The next stage of differentiation occurs when one set of V_L and J_L genes is rearranged with its appropriate constant region kappa or lambda light-chain gene to produce an Ig light chain, which combines with the heavy chain to produce a complete Ig molecule of the IgM type. This is expressed on the cell surface but is not secreted, and the cell is now an immature B lymphocyte.

The next stages in B-cell differentiation involve "isotype" switching, in which a V_H–D_H–J_H set becomes associated with different constant region genes. The intermediate stages in B-cell differentiation are also signaled by the appearance of a variety of cell surface proteins, or "markers," involved in regulating B-cell migration, proliferation, and cell recognition. Immature B lymphocytes express membrane-bound IgM and later coexpress IgD molecules that share the same V_H–D_H–J_H and light chains as IgM but have the heavy-chain IgD determinant. Some B cells undergo a further switch (from IgM and IgD) to IgG, IgE, or IgA production by splicing of the V–D–J gene set with the appropriate C_γ, C_ϵ, or C_α gene, a process that involves deleting intervening genes up to the 5′ end of the one that is expressed. The fully mature, terminally differentiated, B lymphocyte, or plasma cell, produces and secretes one subset of antibody molecules.

The changes in gene readout of B cells described in the preceding paragraphs take place in a specific time frame during development. For example, in mice the first heavy chain produced by B lymphocyte precursor cells is the C_μ chain, which is detected in the cytoplasm of fetal liver cells obtained from 11- to 12-day embryos (for review, see Ref. 137). Membrane-associated IgM is found on lymphocytes obtained from livers of 15- to 17-day mouse embryos, and 3 to 5 days after birth IgD molecules appear on the surface of these cells. By 7 weeks of age, more than 90% of the precursor B cells express both IgM and IgD. Further maturation occurs upon antigen stimulation. When B cells are exposed to an antigen and the appropriate growth factors (lymphokines), membrane IgM and IgD are lost and a secretory form of IgM is synthesized or IgM production is replaced by IgG, IgA, or IgE synthesis and secretion by the mechanisms indicated earlier. In the adult organism, multiple clones of functional B lymphocytes continue to be produced, each synthesizing its own specific type of antibody; even during adult life, continual exposure to various antigenic substances in the environment can call forth the expansion of a clone of B cells ready to produce antibody against that antigen. Thus, the tremendous antibody diversity available to the adult organisms is developed through a series of differentiation events resulting from a programmed rearrangement of the genes coding for these proteins. This is a clear example of how cells can change their differentiation phenotype by the mechanism of genetic recombination.

The V(D)J recombination is a complex gene rearrangement event requiring a number of regulatory activities. These include (1) recognition of conserved DNA sequences (called RS sequences) that flank germ-line V, D, or J segments; (2) introduction of site-specific double-strand breaks between the RS sequences and the piece of DNA to be rearranged; (3) in some cases, deletion or addition of nucleotides at coding junctions; (4) polymerization; and (5) ligation.[138]

Two recombinaselike genes, RAG-1 and RAG-2, are involved in V(D)J rearrangements. The RAG gene products are crucial for V(D)J recombination, as shown by the fact that cell lines lacking the RAG genes cannot carry out this

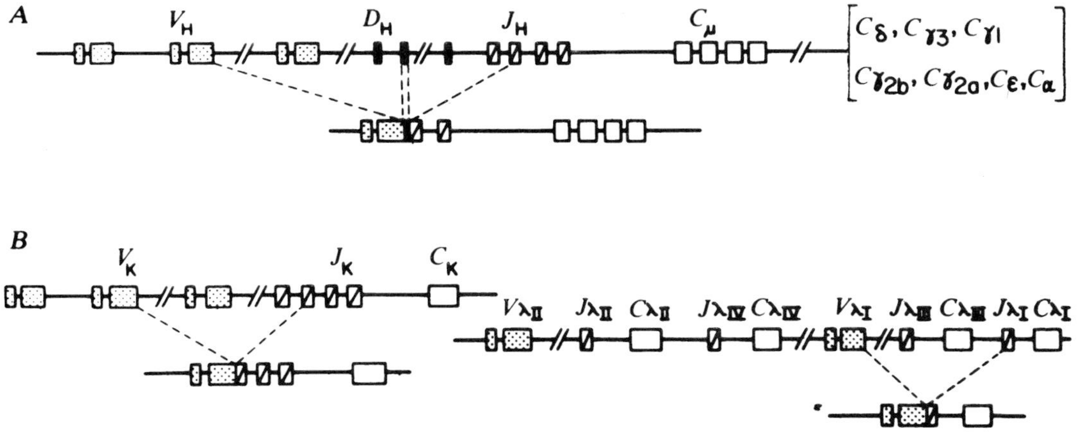

Figure 5-16 A schematic diagram of immunoglobulin gene rearrangement in the mouse; (A) at the mouse heavy (H) chain loci; (B) at the mouse light (κ and λ) chain loci. The three loci indicated are on three different chromosomes. There are probably a few hundred V_κ elements in the κ locus, about the same number of V_H elements in the H locus with perhaps one tenth that number of D_H gene segments. None of these numbers is known precisely, however. The mouse λ locus contains only two V genes and four J-C regions, of which one ($J_{\lambda 4}$-$C_{\lambda 4}$) may be inactive. Immunoglobulin gene expression requires precise fusion of V, D, and J segments in the H locus and precise fusion of V and J in L loci, as shown. The exon shown upstream of each of the V segments encodes the signal sequence for a secreted protein. J-C fusion and removal of other introns occur during RNA processing. The λ-gene rearrangements shown, fusion of $V_{\lambda I}$ with $J_{\lambda I}$, is the one most frequently encountered of this locus. With the exception of Cδ, stable expression of heavy-chain constant-region genes other than the Cμ requires additional rearrangements that move the fused V-D-J segment close to one of the downstream C_H genes. (From Coleclough.[136])

event, whereas cells transfected with the RAG genes can.[138] At least two other genes, both also involved in DNA repair processes, are also involved.[138] Interestingly, cells from immune-deficient SCID mice are impaired in their ability to complete normal V(D)J recombination and also have a defect in double-strand break DNA repair.

Another form of rearrangement involving reverse-transcribed DNA → RNA → DNA events has been observed in human cells. Retrotransposable elements called L1 elements are highly repetitive sequences found in all mammals, including humans, in whom L1 elements make up about 5% of the genome. There appear to be about 3500 six-kilobase L1 elements, of which a subset are actively expressed, reverse transcribed, and transposed to other regions of the genome (reviewed in Ref. 139). The L1 family appears to consist of a small number of "master genes" controlling a larger number of retrotransposably active L1 elements. The L1 elements have some features in common with the repetitive *Alu* sequences noted above and may be involved in the retrotransposition of *Alu* sequences, which themselves lack the encoded reverse transcriptase activity to carry out their own transposable function. If these retrotransposable elements are reinserted into the human genome at an inappropriate place, disease can result. For example, two retrotranspositions of truncated L1 elements into an exon of the factor VIII gene on the X chromosome have been observed in patients with hemophilia A, and L1 insertions have also been found in the dystrophin gene involved in Duchenne's muscular dystrophy and in the *APC* gene, a tumor suppressor gene altered in colorectal cancer (see Ref. 139).

Another rearrangement that can have dire consequences is the activation of the *myc* protooncogene in lymphocytes by a translocation that juxtaposes the *myc* and immunoglobulin genes. This results in the deregulation of the *myc* gene, which thus loses its own regulatory sequences and comes under the influence of the Ig gene regulatory sequences. This may result in the expression of the *myc* gene, normally not expressed in adult lymphocytes, and the production of a type of malignant lymphoma. Whether this translocation is the cause or the result of the carcinogenic process is not yet clear, but it appears to be involved in this process (see Chap. 7).

Gene Amplification

Increasing the number of gene copies ("gene amplification") is a mechanism by which cells meet the demand for increased amounts of certain gene products (e.g., enzymes, structural proteins, ribosomal RNA). Cells also appear to use this mechanism during the process of differentiation to produce high levels of cellular components called for at different developmental stages. That developing organisms use gene amplification as a differentiation mechanism has been known for several years from studies in lower animals. During the maturation of oocytes in amphibians, for example, there is a large amplification of rRNA genes, which disappear into the cytoplasm after the oocyte matures and are no longer active.[140] In *Drosophila* oogenesis, gene amplification occurs on two different chromosomes in response to the need for large amounts of chorion proteins.[141] By the time the ovarian follicle cells degenerate and oogenesis is completed, the number of chorion protein genes has been amplified 16- to 60-fold.

It is clear that gene amplification is not restricted to lower animals. Selective gene amplification in mammalian cells was first documented in 1978 as a mechanism for acquisition of resistance to the anticancer drug methotrexate (MTX).[142] The target enzyme for MTX is dihydrofolate reductase (DHFR), a key enzyme in nucleic acid biosynthesis. This enzyme is inactivated by MTX by forming a very stable drug–enzyme complex. Cells can circumvent this inhibition by producing increased amounts of enzyme; this is done by amplifying the *DHFR* gene. Drug-resistant cell variants can be obtained with as many as 100 to 1000 *DHFR* genes. Many other examples of gene amplification in mammalian cells, including human cells, have been reported. These include the genes for metallothionein, hypoxanthine-guanine phosphoribosyl transferase, hydroxymethylglutaryl coenzyme A reductase, adenosine deaminase, glutamine synthetase, ornithine decarboxylase, and uridylate synthetase (reviewed in Ref. 143). Most of the reports of gene amplification relate to the development of drug resistance in somatic cells undergoing a strong selective pressure provided by a cytotoxic drug. In fact, development of drug resistance and multidrug cross-resistance to anticancer drugs by this mechanism appears to be a widespread and common phenomenon.[144] However, it is clear that gene amplification can occur spontaneously in the absence of selection pressure, and in cultured mammalian cells at least, a twofold increase in gene copy number occurs in 1 out of every 1000 cell replications in the absence of drug.[143] This kind of variation in gene copy number is remarkable in view of the dogma of stringent genomic replication during cell mitosis. However, such amplifications appear to be relatively unstable in the absence of a sustained selection pressure. The extra genes may be extrachromosomal and appear in the cell as minichromosomes known as double minutes (DMs) or incorporated into chromosomes, in which case they often show up as homogeneously staining regions (HSRs). The DMs are usually unstable and may disappear within as few as 20 cell population doublings in the absence of selection pressure. Another point is that DMs, not being part of the chromosomal apparatus and not having a centromere, may be unequally proportioned between daughter cells at mitosis, and thus could provide a way to set up a different differentiation pathway in one of two daughter cells. Whether or not such mechanisms actually play a role in the development of higher organisms is not yet evident.

Amplification of certain genes also appears to be related to the carcinogenic process in certain types of cancer cells. It is known that agents that damage DNA, such as ultraviolet light and the chemical carcinogen *N*-acetoxy-*N*-acetoaminofluorene, enhance amplification of the *DHFR* gene,[144] and carcinogenic agents can also induce amplification of simian virus 40 (SV40) DNA sequences in cultured Chinese hamster ovary (CHO) cells.[145] Moreover, exposure of cells to the tumor-promoting phorbol ester TPA, either at the time of initial exposure to methotrexate or during subsequent cloning of resistant sublines, can enhance *DHFR* gene amplification 100-fold.[146] During tumor progression, tumor cells gain an increased ability to amplify genes as they lose cell cycle control and tumor suppressor gene activity (Chap. 8).

As we see in Chapter 7, cells derived from cancer tissue often have amplified oncogene sequences, including the *myc* and HER-2/*neu* gene sequences. It is not yet established, though, whether amplification of such genes is causally related to the onset of cancer or to tumor progression. Certain experiments suggest that gene amplification may be directly involved in the unregulated growth potential of cancer cells. For example, Levan and Levan[147] have shown that cells taken directly from a mouse tumor (SEWA) contain multiple DMs, suggesting extensive gene amplification. When these cells are grown in culture, they progressively lose DMs with continuous passage. If, however, the cultured SEWA tumor cells are reinjected into a mouse, the cells of the growing tumor again contain multiple DMs. This suggests that growth constraints (immune mechanisms, etc.) indigenous to the mouse induce a stress on the tumor cells such that they produce multiple copies of certain genes to ensure their continued unregulated growth, whereas such constraints are not present in cell culture, and no continued selection pressure exists to maintain the amplified genes.

Gene amplification in somatic cells can occur by means of at least two mechanisms: (1) unequal crossing over of sister chromatids and (2) multiple replication of individual gene sequences during the S (DNA synthetic) phase preceding a cell division. The former mechanism has been shown to occur in bacteria and results in a gain of genes by one sister chromatid and a loss by its pair, followed by multiple mitoses of the cells containing the extra gene copies. The other mechanism, so-called disproportionate replication,[143] is the most likely mechanism for gene amplification in higher organisms, although both mechanisms may occur. A schematic model for this mechanism is shown in Figure 5-17.[148] An example of this is the amplification of the chorion genes in *Drosophila* that results in the generation of multiple gene copies of variable lengths, starting from multiple initiations of replication at specific sites in DNA, producing replication "bubbles" visible by electron microscopy.[149]

Selective Gene Transcription

A sine qua non of differentiation is the selective readout of sets of genes in one cell type that are not transcribed in another cell type. Several mechanisms, discussed below, could come into play to ensure that this happens. What is not clear, however, is *how* these mechanisms for selective restructuring of genes in different cell lineages are selectively triggered. They very likely relate both to differential acquisition of regulatory molecules in the early embryo (discussed above under "Asymmetric Cell Division) and to the various "cues" provided by the microenvironment in which cells find themselves.

Biological Activity of Chromatin

Given the tremendous complexity of the genetic material as it is packaged in eukaryotic cells, it is difficult to see how this material could function during replication of DNA and transcription of DNA into RNA. It seems clear that some "unpackaging" must occur for this to happen. The highest order of packing, namely, the mitotic chromosome, exists only during the mitotic phase of the cell cycle. Cells in G_1 phase contain a much more diffuse chromatin structure. It has been known for some time that the parts of chromatin active in RNA synthesis are packed more loosely (euchromatin) than the more densely packed parts that are not actively transcribing RNA (heterochromatin). For gene transcription to occur, sufficient unpacking must take place to allow association of RNA polymerase with the gene to be transcribed. Polymerase binding must be directed and highly specific, otherwise no selective transcription would occur. This implies that the transcribing chromatin must maintain some order of structure that allows selective "open" sites for RNA polymerase to attach and to initiate RNA synthesis. The proteins associated with DNA accomplish this, since it is unlikely that "naked" DNA could provide such selectivity.

As noted in section above on chromatin structure-function, several interrelated factors are involved in regulating selective transcription of chromatin DNA. Among these are (1) packaging of chromatin as detected by DNase I and S1 nu-

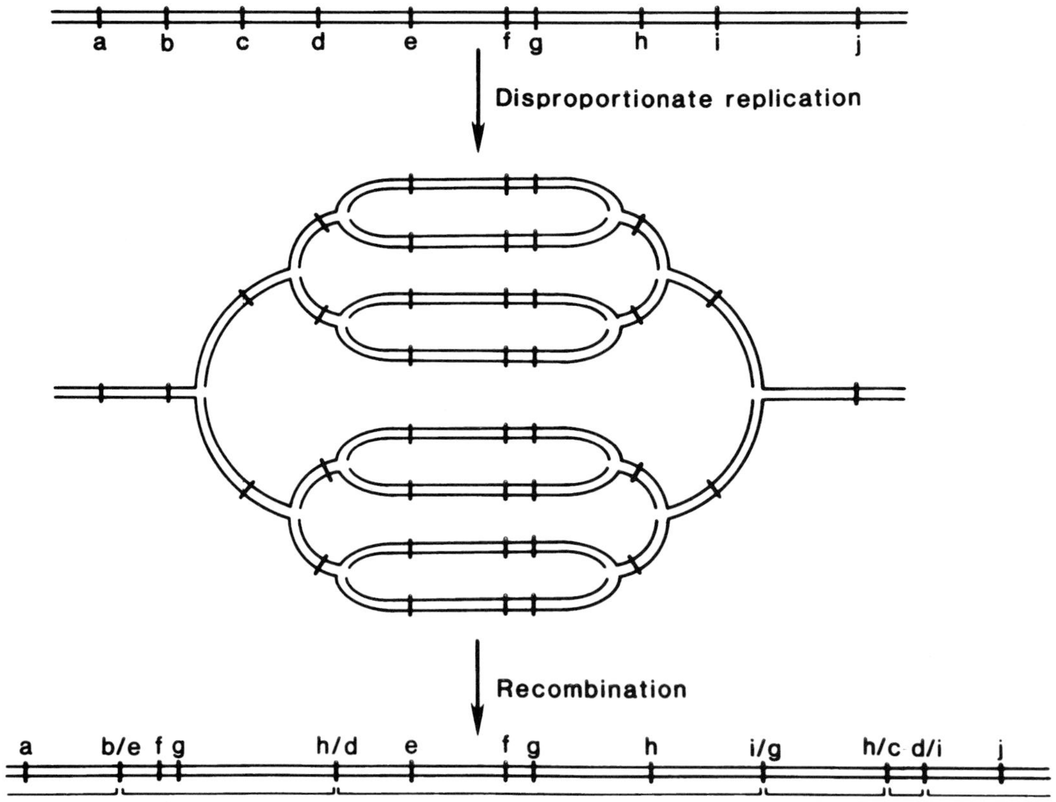

Figure 5-17 Schematic model of gene amplification. Gene amplification proceeds through two steps. Disproportionate replication of a restricted chromosomal region may occur many times in one cell cycle. Three rounds of replication are depicted here. The resulting structure resolves into an irregular linear array by means of homologous recombination between repeated DNA sequences. (Adapted from Roberts et al.[148])

clease digestion, (2) association of H1 histone with nucleosomes, (3) the binding of HMG proteins and other nonhistone proteins to specific sites on chromatin, (4) posttranslational modifications of histones and HMG proteins that affect their binding to DNA, (5) methylation patterns of DNA around transcribable gene sequences, (6) association of DNA with the nuclear matrix, and (7) type of helical coiling of DNA. To summarize briefly, active genes often have specific sites at their 5′ and 3′ boundaries that are very sensitive to DNase I digestion (hypersensitive sites). These areas appear to define active RNA polymerase binding sites, to be "looped out" from the normal coiled nucleosomal structure, and to have some of the characteristics of single-stranded DNA chains (e.g., sensitivity to S1 nuclease attack). Nucleosomes along active genes are associated with HMG proteins 14 and 17 and lack histone H1. The histone octamers of nucleosomes in areas of active genes are enriched in acetylated forms. Also, the degree of coiling of DNA appears to be altered along sequences localized in active chromatin domains, which appear to be demarcated in some instances by an actual change in the direction of DNA winding—that is, from the usual right-handed helix to regions of left-handed winding (so-called Z-DNA). As noted previously, active chromatin domains also appear to be associated with the nuclear matrix, and in higher eukaryotes, DNA sequences found in regions of actively transcribed chromatin often, though not always, contain fewer methylated cytosine bases. How these complex yet apparently interrelated alterations in chromatin are orchestrated and coordinated is not clear, but there must be some regulating signals that dictate the correct temporal sequence of events in the "cascade" of events leading to the turning on of some genes and the turning off of others in the process of differentiation. It is also important to consider which of the mechanisms or factors associated with gene activation are "cause" and which are "effect." For example, some changes in chromatin may precede gene transcription, whereas others may be induced by the process of transcription itself. In the case of differentiation of avian erythrocytes, the appearance of DNase I hypersensitive sites, undermethylation, and association of HMG 14 and 17 with active chromatin precede globin gene expression.[150] Furthermore, during B-lymphocyte differentiation in the mouse, it appears that undermethylation, generation of DNase I hypersensitive sites, loss of H1 histone, and enrichment of HMG protein binding precede the rearrangement and active transcription of immunoglobulin genes.[151] Interestingly, some of the cell lines used in these latter studies were malignant plasmacytoma cells arrested at different stages of differentiation. This raises the intriguing possibility that active regions in chromatin get locked into an on or off configuration by some mechanism—for example, a mutation or an insertion of oncogene sequences (see Chaps. 6 and 7)—that prevents the normal sequential changes in chromatin leading to the selective transcription events needed for further differentiation.

Although cellular differentiation requires the progressive restriction (or the selective amplification) of nuclear genes, there is considerable evidence that the entire genome is retained in all of an organism's cells. If, for example, nuclei from differentiated tissues of the frog, or even from frog renal carcinoma cells, are transplanted into enucleated, fertilized frog eggs, some of the eggs will develop into normal tadpoles. Thus, all the genetic information for the development of an entire organism is present in the nucleus of a differentiated cell.[152,153] Nevertheless, there is evidence that, as the multipotent cells of the early embryo differentiate, a selective transcription of the genome occurs. This is reflected in the fact that specific cells of a given differentiated phenotype produce specific enzymes, hormones, and so on peculiar to that cell type. Studies on DNA–RNA hybridization show that, with the possible exception of mammalian brain, all cells lose transcriptional diversity as they differentiate; that is, the patterns and types of mRNAs (the RNA type translated into protein) transcribed from nuclear DNA become restricted compared with those of the early embryo cells and begin to diverge from cell type to cell type.[154,155] Three "prevalence classes" of mRNAs are transcribed in all eukaryotic cells: a "complex class" of mRNAs, with one or a few gene copies per cell; "moderately prevalent" mRNAs, present at levels up to a few hundred copies per cell; and "superprevalent" mRNAs, represented by more than 10^4 copies per cell.[155] The complex class of mRNAs contains sufficient nucleotide sequence diversity to code for 10^4 to 10^5 different proteins. The moderately prevalent class could code for about 10^3 proteins, and the superprevalent mRNAs contain enough information for 10 to 20 proteins. Most animal cells contain over 10^5 mRNA molecules. About 40% to 45% of the total RNA mass is the complex type, another 40% to 45% is moderately prevalent, and 10% to 20% is superprevalent. A few highly specialized cells, however, contain only one or a few mRNA species. For example, the oviduct of the laying hen contains about 10^5 ovalbumin mRNAs per cell,[156] and mouse and chicken reticulocytes have about 10^5 globin mRNA copies per cell.[157]

Although there are clear differences in the mRNA content of various differentiated cell types of adult multicellular organisms, a high percentage of mRNAs are common to all cell types of a given organism. These represent what are sometimes called "housekeeping" genes—genes that code for enzymes needed in energy generation, transport of amino acids and sugars, nucleic acid synthesis, and other processes. Studies on DNA–RNA hybridization indicate that only about 10% to 20% of the mRNA present in one differentiated cell type is distinct from that transcribed by another cell type of the same organism. For example, mouse liver and mouse kidney show only a 10% to 20% difference in their mRNAs[158]; chicken liver and chicken oviduct have a 15% difference.[159] Only a small percentage of the total cellular DNA, however, is actually ever transcribed into RNA (varying from about 5% of the total DNA for mouse and human liver, kidney, and spleen to about 20% for brain),[160] and only a small percentage of the total

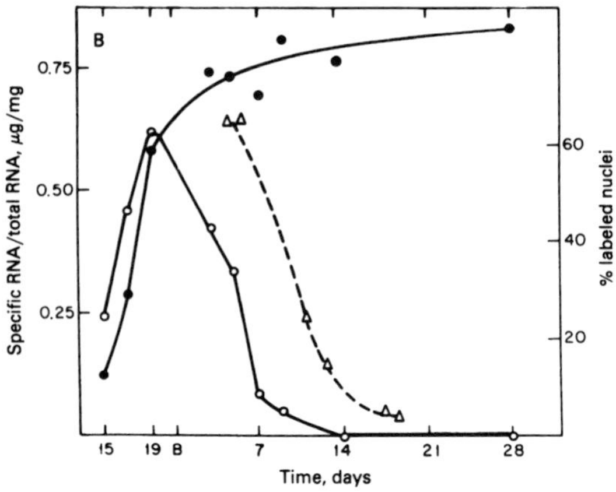

Figure 5-18 Ontogeny of AFP and albumin mRNA levels. Concentrations of alpha-fetoprotein (AFP) (○) and albumin (●) mRNA were determined by dot-blot hybridization at different stages before and after birth (B on the horizontal axis). The percentage of labeled nuclei (Δ) after a 12-hour pulse of [^{3}H]thymidine was determined in mice at different stages after birth. (Adapted from Tilghman and Belayew.[162])

RNA transcribed from nuclear DNA ever reaches the cytoplasm to be translated into protein. Thus, two very stringent filters limit gene expression and provide key checkpoints for cellular development and function. The amount of nuclear RNA (nRNA) that becomes cytoplasmic mRNA is 10% to 20% for the sea urchin embryo, 11% for rat liver, and 18% for mouse brain.[155] Surprisingly, in many organisms the mRNA sequences of one cell type that are specifically absent from the mRNA of another cell type are present in the nRNA of both cell types. Thus, post transcriptional events (e.g., the processing of nRNA to mRNA, the transport of mRNA to the cytoplasm, and the binding of mRNA to polyribosomes) must play an important role in the control of gene expression in eukaryotes. Both transcriptional and posttranscriptional events would be potential sites for alteration during oncogenesis, leading to the emergence of a phenotypically altered, malignant cell.

Good examples of selective gene transcription in mammalian cells are the switching in the expression of fetal to adult hemoglobin in hematopoietic tissues and of alpha-fetoprotein to albumin in developing liver cells. In both cases, the fetal genes and the adult genes appear to be expressed together in fetal cells, but during development, the transcription of fetal genes is switched off and that of the adult genes continues to increase (Fig. 5-18).[161,162] The change in expression of these genes is related to the way in which these genes are arrayed in chromatin.[161] It should be noted that as the liver differentiates as indicated by the switch to the more adult gene product, albumin, the rate of cell proliferation is decreasing, as indicated by decreasing incorporation of [^{3}H] thymidine into DNA. This is another hallmark of differentiating cells, i.e., decreased proliferative capacity.

Cis-acting Regulatory Elements: Promoters and Enhancers

The presence of promoter sequences in DNA for RNA transcription has been known for a long time and was discovered first in bacteria. It is now known that similar regulatory regions are also present in the DNA of eukaryotic cells. The regulation of gene transcription in eukaryotes, however, is more complicated than in prokaryotes and includes higher-level orders of chromatin packaging, methylation, binding of nuclear proteins, and the other mechanisms detailed in the preceding sections of this chapter. There are also regulatory elements on DNA that affect the function of promoters in directing RNA transcription. They act in a *cis* manner in doing so—that is, they directly affect the function of gene sequences on the same DNA strand, even though they may be several hundred base pairs upstream or downstream from the initiation site for RNA transcription. Other mechanisms of gene regulation are said to be *trans*, the term for a factor, a regulatory protein for example, coded for by a distant gene, that modulates transcription of genes not associated directly with DNA sequences in the same strand.

There are two broad categories of *cis*-acting regulatory elements: (1) sequences near the transcription-initiation site, termed promoters by analogy with their prokaryotic counterparts that define the start site and "loading efficiency" for RNA polymerase II, the enzyme that transcribes messenger RNA; and (2) "enhancers" (or activator genes) that are more remotely located from the gene and that increase transcription efficiency by mechanisms not yet perfectly defined but which appear to affect DNA topology in a way that facilitates access of RNA polymerase to initiation sites. A number of things are remarkable about these enhancer sequences. They can exert influence over genes that may be as far as 10,000 base pairs away; they can be effective whether they are in front of (at the 5′ end) or behind (at the 3′ end) the transcribable gene; they function regardless of which way the sequence is oriented (i.e., which direction the sequence reads). Because enhancer elements can act over such large distances, they are candidates for a more regional or global type of gene regulation than individual promoter sequences—the kind of programmatic regulation, in other words, that might be involved in cellular differentiation or neoplastic transformation.

Promoters are organized as a group of "control modules" clustered around the initiation site for RNA polymerase II.[163] Early work on the structural organization of promoters for the herpes simplex virus (HSV) thymidine kinase *(tk)* gene and simian virus 40 (SV40) early transcription genes provided the background for how we think about promoters.[164] They are made up of discrete sequences of 7 to 20 base pairs (bp) of DNA and have recognition sites for transcriptional regulatory proteins (transcription factors; see below). One of the functions of promoters is to position the start site for RNA transcription. For many, but not all, gene promoters, a consensus TATA base–containing sequence, called the TATA box, is this positioning element. Other promoters, typically 30 to 110 bp upstream (toward the 5′ end) from the transcription start site, regulate the frequency of transcription initiation.[163]

Enhancer sequences were first discovered in 1981 by two laboratories studying the regulation of SV40 virus gene transcription. Benoist and Chambon[165] and Gruss et al.[166] described the presence of a *cis*-acting sequence located more than 100 nucleotides upstream (i.e., before the 5′ end) of the cap site of the so-called early SV40 genes, that is, those genes transcribed shortly after infection. Deletion of this sequence reduced early gene expression 100-fold and abolished viral viability. This prototype enhancer is a 72-bp tandem repeat located upstream from three 21-bp, GC-rich repeats and the TATA box.

Subsequently, it was found that SV40 enhancer as well as similar elements from other animal viruses could function not only when linked to their natural genes but also when linked to other genes (so-called enhancer swap experiments). Many other viruses have now been shown to contain enhancer elements in their genomes; these include polyomavirus, BK virus, adenovirus, Moloney sarcoma virus (MSV), bovine papillomavirus, and Rous sarcoma virus (RSV) (for review, see Ref. 167). In the case of some DNA viruses (e.g., polyomavirus, BK virus, adenovirus, and bovine papillomavirus), the enhancer sequences, as in SV40, are present as short tandem repeats of 50 to 100 nucleotides. The enhancer sequences of the RNA retroviruses (e.g., MSV and RSV) have been identified within the so-called long terminal repeat (LTR) regions, the portions of the retroviral genomes known to contain transcriptional regulatory sequences. These sequences can augment transcriptional activity of heterologous genes, and when viral LTRs are integrated into a eukaryotic cell's genome, they can activate cellular genes that come under their influence.[168] Such a process can lead to activation of cellular protooncogenes (see Chap. 7) and may lead to neoplastic cellular transformation.

In addition to viruses, cellular genes have been found to contain enhancer elements. Originally, they were found to be associated with insulin, chymotrypsin, and immunoglobulin (Ig) genes. Enhancers, like promoters, may contain several modules, sometimes called enhansons,[163] and also can bind to positive or negative transcription factors.

Enhancers and promoters have to "talk to each other" if a cell is to coordinate its developmental program as well as its response to environmental cues. Both enhancers and promoters regulate transcription, bind transcription factors, and have similar modular organization. Three different (but not mutually exclusive) mechanisms have been proposed for enhancer–promoter cross-talk: (1) Enhancer and promoter elements may be brought together by binding of distant DNA sequences to bring enhancers and promoters into contact. This binding would be facilitated by transcription factors. (2) Bridging between enhancer and promoter sequences could be brought about by protein–protein interactions between transcription factors binding the two DNA domains. (3) A DNA tracking mechanism could accomplish the interaction by allowing transcription factors or the transcriptional complex to "slide" along DNA, thus conveying the transcriptional signal from one domain to another. Evidence for the latter mechanism has been obtained for regulation of bacteriophage T4 late gene transcription.[169] The DNA-tracking mechanism would predict that the "sliding" of a transcriptional activator complex is related to DNA structural organization rather than a specific DNA sequence. Otherwise, it is difficult to see how passage of such a complex over long sequences of differing base composition could be accomplished. Alternatively, tracking could be accomplished by the DNA-binding or -looping mechanism listed in item 1 above.

Evidence for the protein–protein bridging model comes from experiments of Cullen et al.,[170] who studied the estrogen-induced interaction between enhancer and promoter regions of the prolactin (PRL) gene. Estrogen induces the transcription of the PRL gene by binding to the estrogen receptor (ER), which binds to the estrogen response element (ERE). The ERE is at the 3′ end of a distal enhancer region that is between −1550 and −1578 bp away from the transcription start site. Thus, for the ER complex to function in PRL gene transcriptional regulation, it somehow has to bridge this distance. What Cullen et al. found, by using a unique chromatin ligation assay, was that the distal enhancer and proximal promoter regions of the rat PRL gene are juxtaposed and that estrogen enhanced bridging between these domains two- to threefold over non-estrogen-treated chromatin. It had been previously found that, although the chromatin surrounding the ERE and the promoter became nuclease-sensitive, the intervening regions between the ERE and the promoter remained nuclease insensitive. These data favor the chromatin-binding, protein–protein bridging models of enhancer–promoter interaction.

Another point to be kept in mind is that the modular organization of enhancers and promoters helps explain a way in which cells can modulate their response to external signals and also why there are "strong" and "weak" promoters and enhancers. For example, if a promoter/enhancer contains two different types of modules, recognized by different transcription factors, promoter/enhancer activity might be seen only if both factors were present at the same time in the responding cell, or activity might only be "half maximal" if only one of the two factors were present. Similarly, an enhancer/promoter with three or more modules might be a "stronger" transcriptional regulator than those with only one or two such modules.

Some promoters/enhancers bind transcription factors that are constitutively expressed, i.e., they are continually made in cells and not only when induced by exogenous stimuli like the estrogen-induced ERE response described above. It is not clear why similar regulatory mechanisms would be needed for constitutively expressed genes, which may include general "housekeeping" genes, but it may be that low or baseline levels of expression of some genes must go on continually, so that they are "primed" and ready to go when the cell is stimulated or stressed by exposure to an environmental factor. Examples of such genes include the metallothionein gene, which is involved in protection against heavy metal toxicity and whose promoter contains binding sites for the transcription factor SP-1 (see below), and the immunoglobin genes whose enhancers have binding sites for constitutively expressed transcription factors.[163]

The elegant and complex gene regulation imposed by enhancers and promoters apparently evolved over eons of time. Based on studies of simple prokaryotic and viral genes, it seems likely that single promoter or enhancer regions arose in scattered regions of a genome, perhaps randomly at first, and then gradually by genetic recombination and duplication events found their way into proximity with structural genes on which they imposed their control. A selective advantage to the organism may thus have been gained by imposing a regulated rather than a random response to the environment once the promoter/enhancer elements took on the ability to bind specific proteins.

Transcription Factors

As noted above, the regulation and initiation of transcription in eukaryotic cells is an intricate and complex process. It involves interaction of DNA-binding transcription factors (TFs) with short consensus DNA sequence motifs in enhancer and promoter regions. These interactions produce, by cooperative binding reactions, the formation of a transcription complex (see below) and changes in chromatin structure that foster the binding of RNA polymerase II and initiation of transcription.

Based on experiments initially performed primarily in yeast and later in other eukaryotes, the following "rules" about how eukaryotic TFs work can be formulated[171-173a]:

1. Transcriptional factors are multidomain proteins that have DNA-binding, activation, dimerization (most but not all TFs), nuclear-localization, and ligand-binding domains.
2. The DNA-binding and -activation domains are often interchangeable among TFs.
3. The DNA-binding domains have specific structures and recognize specific DNA sequences, whereas activation domains have less precisely defined structures characterized by an acidic amino acid–rich (i.e., rich in aspartic and glutamic acids), glutamine-rich, or proline-rich sequences.
4. Transcription factors that work in yeast cells will, in general, work in plant, insect, and mammalian cells for genes with the required promoter/enhancer sequences.
5. Transcription factors undergo cooperativity in binding with other components of the transcription complex. This cooperative binding assures specificity and reversibility through multiple, low-affinity interactions.
6. Transcription factors with more potent activation domains (i.e., more activating sites) can act at greater distances on the DNA.
7. Transcription factors have DNA-binding domains with typical structural characteristics such as helix-turn-helix, zinc-finger, and basic region domains (see below). A number of TFs also have a characteristic helix-loop-helix structure (e..g, MyoD). Helix-turn-helix domains form structures that make contact with DNA. Helix-loop-helix (HLH) domains are involved in protein–protein interactions.
8. Many TFs form dimers through leucine-rich "zipper" domains (e.g., Jun–Fos).
9. Transcription factors have nuclear localization motifs that enable them to move into the nucleus after synthesis.
10. The ligand binding domains enable a number of TFs to be activated by hormones, growth factors, developmental morphogens (e.g., retinoic acid), and other exogenous stimulatory agents.

STRUCTURAL MOTIFS OF REGULATORY DNA-BINDING PROTEINS

The DNA-binding proteins that have a regulatory function, i.e., transcriptional activators and repressors, display a number of common structural motifs. These include the helix-turn-helix, zinc-finger, leucine-zipper, and accidic domains mentioned above (Fig. 5-19).[173] The helix-turn-helix motif was the first studied and best characterized structural motif and is found in prokaryotic activator and repressor proteins as well as those of higher organisms. As implied by the name, this class of proteins contains two α helices separated by a β turn. This geometry is common to all types of TFs with this motif even though the primary amino acid sequence can vary. One of the α helices, the DNA-recognition helix, contacts bases in the major groove of DNA (Fig. 5-20).[174] The other helix lies across the ma-

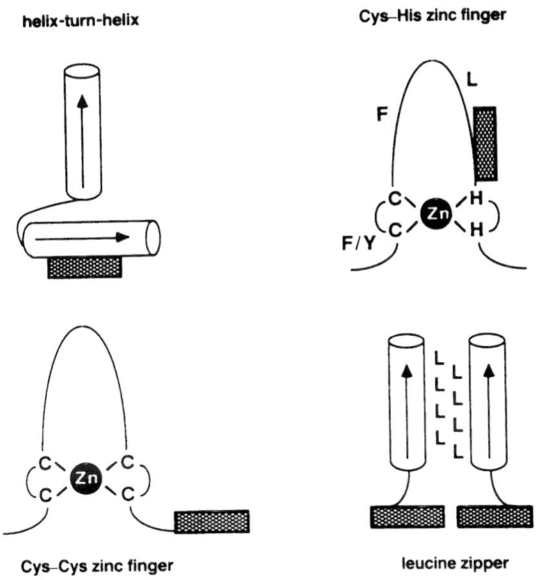

Figure 5-19 Schematic representation of the four structural motifs described in the text. α-Helices are represented by cylinders with arrows indicating directionality; conserved amino acid residues are shown in one-letter code, and zinc ions shown as black circles. The shaded boxes indicate the regions of the proteins proposed to be involved in specific contacts to DNA. (From Struhl.[173])

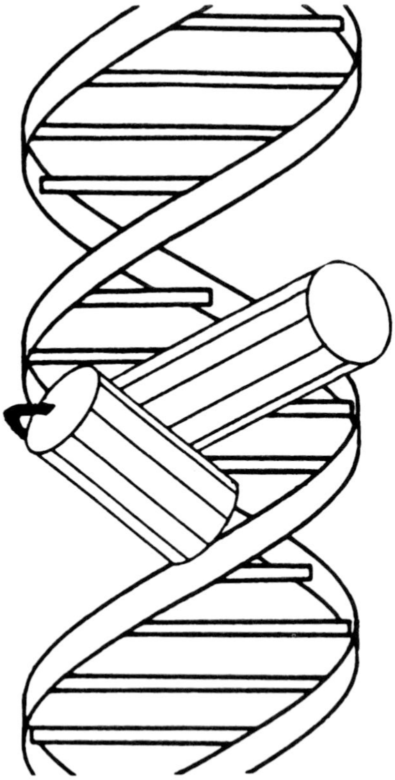

Figure 5-20 The helix-turn-helix with the recognition helix in the major groove of DNA. (Reprinted with permission from Schleif, DNA binding by proteins. *Science* 241:1182. Copyright 1988 by the American Association for the Advancement of Science.)

jor groove. In eukaryotic cells, the helix-turn-helix motif was first described for the homeobox encoded proteins of *Drosophila.* The homeobox proteins contain conserved polypeptide sequences called homeodomains. Homeodomains have now been identified in eukaryotic organisms from yeast to human.

The zinc-finger motif was first proposed for the transcription factor TFIIIA, a "general" TF required for transcription of the 5S RNA genes by RNA polymerase III.[173] TFIIIA has 7 to 11 zinc atoms per molecule and contains 9 repeating units of about 30 amino acids each. Two kinds of zinc fingers have been proposed (Fig. 5-19), one in which each "finger" contains two cysteines and two histidines arranged in a tetrahedral coordination complex that binds zinc, and a second type in which the coordination complex is made up of four cysteines. Similar zinc-finger domains have been found by DNA sequence analysis of organisms from yeast to human, as seen, for example, in the GAL4 transcriptional activator of yeast and the steroid hormone receptor of mammals. Both the zinc-finger region and zinc itself appear to be necessary for DNA binding, and contact with DNA is thought to occur via the outstretched fingers. "Finger swapping" experiments have shown that for four cysteine-containing zinc fingers, the finger motif is essential for DNA binding, but that direct contact with DNA involves additional amino acid sequences.[173]

The leucine-zipper motif is contained in DNA binding proteins such as the yeast GCN4 transcriptional activator and the *jun, fos,* and *myc* oncogene proteins. This structural motif is produced by runs of 4 or 5 leucine residues spaced exactly 7 residues apart, forming interdigitating α helices and leading to protein dimer formation that is important for the transcriptional activity of some TFs, e.g., Jun-Fos dimers (see below). Dimerization appears to involve a "coiled coil" structure consisting of two parallel α helices with hydrophobic contact sites that bring into close proximity a region of each subunit rich in basic amino acids. This interaction provides a bimolecular contact with DNA (Fig. 5-19). Leucine-zipper motifs have also been found in some proteins that contain zinc-finger or helix-turn-helix domains, suggesting that more than one mechanism for DNA contact may be utilized by some TFs.

It has become increasingly clear that transcriptional regulatory mechanisms have been amazingly conserved over evolution. For example, some yeast TFs can function in a variety of other eukaryotic organisms, including vertebrates and vice versa; yeast and mammalian TATA-box TFs are functionally interchangeable, etc. Interestingly, although eukaryotic cells have some related TFs that recognize similar DNA sequences, each organism's TFs may regulate a different set of functions. For example, in yeast, the GCN4 TF activates amino acid biosynthesis and oxygen utilization, while its homologue in mammalian cells, c-Jun, activates a different series of events involved in so-called immediate-early events in response to external signals such as growth factors. Thus, during evolution, the structural genes responding to TFs have diverged, even though the transcriptional regulatory sequences upstream or downstream from them have been remarkably conserved.

Acidic domain motifs, as noted above, are contained in the transcriptional activation regions of the TFs. Again, these domains have several features in common among different organisms of the phylogenetic tree. The negative charge of these domains must be important, because if it is removed, activating function is lost. Yet additional characteristics are also clearly important. These appear to be repeating α-helical structures, amphipathic in character, that favor interactions with other proteins in the transcription machinery.[173] As will be discussed below, the acidic regions of TFs appear to be involved in the interactions that bring an enhancer sequence into proximity with the RNA transcription start sites via binding to TATA-box binding proteins. Other types of activation domains contain glutamine- or proline-rich sequences, but the mechanisms of these different types of TFs is not clearly defined.[173a]

Chapter 7 discusses how this transcriptional machinery can go awry in cancer cells. Suffice it to say here that oncogenic conversion of normal cells into cancer cells involves changes in transcription factors. Such changes are exemplified by conversions of c-*jun* to v-*jun*, c-*fos* to v-*fos*, c-*myb* to v-*myb*, and c-*erb*A to v-*erb*A.

REPRESSORS

Just a word about repressors. Precise regulation of gene expression requires both positive and negative control. To wit, during evolution, two kinds of gene regulation mechanisms have evolved. The negative factors are sometimes called transcriptional downregulators or repressors. Some of these have DNA-binding domains but lack functional activation domains, and they compete with transcriptional activators for binding to regulatory elements in DNA.[175] Some can form heterodimers with activators and then block the ability of TFs to bind to DNA or to activate transcription. Others may bind to TFs already attached to DNA and prevent the interaction of the acidic domain of TFs with the TATA-box binding proteins. Another type of repressor can sequester activators into inactive multimeric complexes.

A somewhat surprising finding is that transcriptional activators and repressors can be encoded by the same gene.[175] One way this can be accomplished is by alternate mRNA splicing such that an activator becomes a repressor. Examples of this include the alternate splicing of the *erb Aa* gene mRNA, which modifies the carboxyl-terminal domain and prevents binding to its ligand, and the alternate splicing of the *fos* B gene mRNA to create a defective activation domain (reviewed in Ref. 175).

Another perhaps surprising finding is that activators and repressors can be coexpressed in cells, and during development of an organism, the balance of expression of activators and repressors can control gene expression in a temporal and tissue-specific developmental pattern. Another key point to keep in mind is that the function of activators and repressors is often regulated by posttranscriptional modifications such as phosphorylation. For example, the ability of the retinoblastoma (RB) gene product to repress gene expression is lost when it becomes phosphorylated.

A point to be kept in mind is that some TFs can activate one gene or set of genes and at the same time repress another gene or set of genes. Such activity has, for example, been shown for some homeobox proteins in *Drosophila* and for steroid hormone receptors in mammals (reviewed in Ref. 176).

Another important concept to be kept in mind is that there is "cross-talk" among TFs. In some cases, they may work together to activate a gene or genes. In other cases, one TF may turn on the gene for a second TF. And in still other instances, one TF may inhibit the action of a second. Some examples will make these points.

The virus-inducible enhancer of the human interferon β (IFN-β) gene has overlapping regulatory elements recognized by the TFs NF-κB, IRF-1, and ATF-c-Jun (reviewed in Ref. 173a). None of these function on their own, and two or more of the TF-activated enhancers are required to turn on the IFN-β gene. This apparently results from protein–protein interactions between the TFs to form a stereospecific complex that brings the enhancer and promoter regions together.

Interleukin-1 (IL-1) is a cytokine that mediates a variety of cell proliferative responses (see Chap. 9). It does so by activating c-*myc* gene expression, which—as we will see—is a TF involved in a variety of cell proliferative mechanisms. The action of IL-1 in turning on c-*myc* is accomplished by activating the TF NF-κB, which activates c-*myc* via NF-κB response elements in the c-*myc* gene.[176a]

Yin-Yang-1 (YY1) is a zinc-finger DNA-binding protein that, depending on the context, can function as an activator, a repressor, or an initiator of transcription (reviewed in Ref. 176b). The repressor actions of YY1 include repression of the adeno-associated virus P5 promoter, the c-*fos* promoter, the human papillomavirus 18 promoter, the LTR of Moloney murine leukemia virus, and the N-*ras* promoter. In contrast, YY1 activates the c-*myc* promoter and promoters of ribosomal protein genes L30 and L32. On the other hand, c-Myc protein binds to YY1 and inhibits both its repressor and activator functions.[176b]

Thus, the interaction of TFs is both complex and context-dependent. Depending on the cell type, the environmental signals, and the state of differentiation, the amount and type of TFs expressed may vary. The requirement for them to form dimers and protein–protein complexes to be active allows them to act in specific ways in different cell types, depending on who their partners are. One way to look at it is that there are multiple competing TFs for the same DNA binding sites, and whoever gets there "the fustest with the mostest" wins.

GENERAL TRANSCRIPTION FACTORS

The regulatory factors involved in initiation of transcription by RNA polymerases are usually divided into two classes: general transcription factors (GTFs) and promoter (or enhancer)-specific transcription factors (STFs).[177-180]

In eukaryotic organisms, three different RNA polymerases are involved in gene transcription. RNA polymerase I (RNA pol I) transcribes ribosomal RNA; RNA pol II transcribes protein coding mRNAs and many small nuclear (sn) RNAs; RNA pol III carries out synthesis of tRNA and 5S RNA. The transcriptional machinery for each of these types of genes has features in common. All three RNA polymerases require formation of a transcription complex containing a TATA-box–like binding protein (TBP) and a series of TBP–associated factors (TAFs) that are somewhat different for each of the polymerase complexes.

The factors for RNA polymerase II have been studied in the most detail, and thus we know the most about them. Many but not all (e.g., many housekeeping genes lack a TATA element) protein encoding genes that use RNA pol II contain a TATA containing DNA sequence near the transcription site. And it is now known that class I (using RNA pol I) and class III (using RNA pol III) also contain TATA-box–like binding proteins (TBPs) in their transcription complexes. In the case of RNA pol I, a TBF-TAF complex called SL1 makes up a key part of the transcription complex. In the case of RNA pol III, transcription requires at least two TAFs as well as the TBP.[181] These various TBP-TAF complexes are specific for each type of gene (class I, II, or III) ensuring that each type of gene forms a complex only with its appropriate RNA polymerase.[178,181]

As noted above, most attention has been focused on mRNA-transcribing class II genes transcribed by RNA pol II. The general transcription factors required for accurate transcription of class II genes include the TATA-binding proteins TFIIA, TFIIB, TFIID, TFIIE, and TFIIF. These are thought to assemble in an ordered fashion on a promoter to form a preinitiation complex with RNA pol II.[180] The first step is binding of TFIID to the TATA box, a step that is facilitated by TFIIA. The binding of TFIIB to this DNA-protein complex fosters recruitment of RNA pol II at the promoter site. This step requires the RAP30 subunit of TFIIF. Also, TFIIE appears to play a role in formation of an active complex. In the case of class II genes, the TATA-box binding protein TBP is a subunit of TFIID, called TFIIDτ.[182]

In order for efficient transcription to occur, promoter (enhancer)-specific transcription factors must come into play. Since the regulatory DNA sequences to which the STFs bind are often a long distance away from the transcription start site, it is not clear how they make contact with the transcription complex. One way this could happen is shown in Figure 5-21.[183] In this model, the activation region of an STF (promoter-specific activator), bound to its promoter or enhancer, binds to a putative target site on an

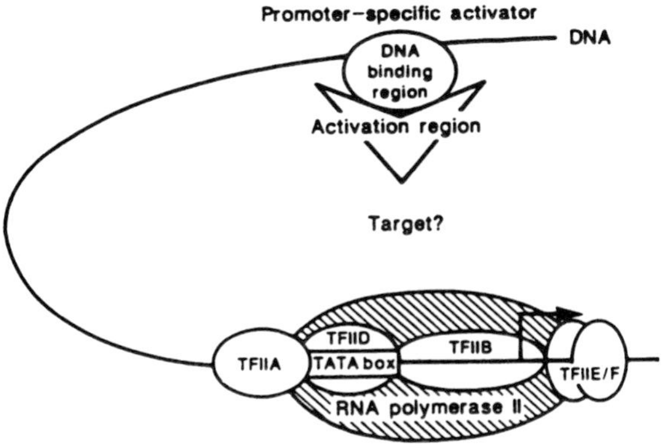

Figure 5-21 A possible mechanism for transcription stimulation by promoter-specific activators. The activator recruits one or more of the general transcription factors to facilitate assembly of a preinitiation complex and then enhances a step following assembly of the general factors into a preinitiation complex. The promoter-specific activator is shown bound to its site on the DNA loop. The site of transcription initiation is indicated by the arrow. (Adapted from Lillie and Green.[183])

already formed transcription complex to stimulate initiation of transcription. In the absence of an activator, nonproductive or inefficient preinitiation complexes could form, producing a low baseline level of transcription. In the presence of activator, TFIID and TFIIB would assemble in highly productive manner and transcription would be increased. An alternative model is one in which the activator interacts with TFIIB-TFIID and helps them assemble into a preinitiation complex. In any case, it is clear that interaction between the activation domain–containing TF and its DNA-binding region with TFIIB is crucial for transcriptional activation.[184]

PROMOTER (ENHANCER)-SPECIFIC TRANSCRIPTION FACTORS

Eukaryotic promoters and enhancers contain a unique array of DNA sequences that bind STFs. The STFs include SP1, AP1, ATF, CTF, USF, Oct1, Gal4, and many more. These are proteins that have DNA binding domains and activation domains as noted above. Many of them also contain specific ligand-binding domains that enable them to recognize and bind to external signals such as hormones and growth factors. In addition, there are families of STFs that induce the expression of tissue-specific genes and play a key role in cellular differentiation. These include the MyoD factors involved in muscle differentiation, Pit-1 involved in gene expression in pituitary cells, and HNF-1 involved in liver cell differentiation. Some promoter-specific transcription factors as well as their DNA-binding sites, size, and the promoters/enhancers that they activate are shown in Table 5-1.[185] Some of these factors are discussed below.

AP-1/Fos/Jun. The AP-1 promoter-specific factor was identified as a trans-acting factor that binds to the SV40 virus enhancer element. The SV40 enhancer was the first such element described and is often used as a prototype because it contains a number of prototypical STF-binding sequences, (e.g., SP-1 and AP-2, -3, -4, and 5).[185] Subsequently, it was found that AP-1 binding sites could bind a variety of transcriptional activators of the Fos/Jun cellular oncogene family as well as the tumor promoter phorbol ester (TPA) binding factor and the glucocorticoid receptor. Thus, AP-1 DNA binding sites are also known as TPA responsive-element (TRE) and the glucocorticoid receptor element (GRE). This family of transcriptional activators form dimers via leucine zippers as noted above and are encoded by a family of genes including *fos*, *fra*-1, *fra*-2, *fos* B, c-*jun*, *junB*, and *junD*.[186] For these factors to be active, they must dimerize. Fos/Jun heterodimers are the most active; Jun-Jun homodimers are weakly active; and Fos-Fos homodimers are difficult to form and are not active. The activity of Fos and Jun appears to be regulated by their phosphorylation state.[187] The AP-1 binding proteins are induced by mitogenic, differentiation-inducing, and neuronal-specific stimuli.[186]

ATF/CREB. The ATF/CRE enhancer sequence (TGACGTCA) was identified as the activating transcription factor (ATF) binding site or the cyclic AMP response element (CRE) that bound the cAMP response element binding protein (CREB). The ATF/CREB subfamily of transcription factors include CREB, CRE-BP1, ATF-3, and ATF-4.[186] This family of STFs has been implicated in cAMP-, calcium-, and virus-induced alterations in gene transcription.

While originally it was thought that Fos/Jun

Table 5-1 Characteristics of Some Specific Transcription Factors

Factor	Binding Site	Size	Promoter/Enhancer	Comments
AP1	T(T/G)AGTCA	47 kDa	SV40/Py enhancers BLE of hMTIIA collagenase stromolysin α_1-anti-trypsin transthyretin MHC-H2^d AdE3	Binding site is the TPA-responsive element (TRE). AP1 sites are bound by Jun/Fos heterodimer.
CREB ATF	TGACGTCA	43 kDa	Somatostatin (CREB) E1A/E2A/E3/E4 Ad early genes (ATF), c-*fos*, hsp70, tyrosine hydroxylase, α-gonadotropin, VIP, fibronectin, HTLV-II LTR, HTLV-I LTR, BLV LTR	ATF/CREB family includes CREB, CRE-BP1, ATF-3, and ATF-4; binding site is the cAMP responsive element (CRE).
MLTF/USF	GGCCACGTGACC	46 kDa	MLP of adenovirus α-fibrinogen mouse Mt1	
CTF/NFI	TGGCT(N_3)AGCCAA	52–66 kDa	α-, β-globin, hsp70, HSV *tk, ras,* Ad2/5 origin AdE3 c-*myc*, albumin	A family of factors, required for transcriptional stimulation and stimulation of adenovirus DNA replication in vitro. Gene has been cloned and recognizes multiple mRNA species. A half-site is sufficient for binding.
SPI	GGGCGG	105 kDa	SV40 early promoter, hMTIIA human ADA, type II procollagen, E1B, HSVIE-3, DHFR, HIV LTR, AdITR	Human Spl cDNA cloned
Octamer binding proteins:	ATTTGCAT		heavy/light Ig, histone H2B, snRNA genes, SV40 enhancer	
Ubiquitous				
OTFI		90 kDa		
OBP100		100 kDa		
NF111/octB1A				
IgNFA1				
octB3				
B-cell–specific				
OTF11		62–58.5 kDa		
IgNFA2/octB2				
octB1B				

and ATF/CREB protein families were distinct sets of STFs that share the basic-region/leucine-zipper motif but have different DNA binding specificities, it is now known that members of each family can cross-dimerize to form active STFs.[186] The three Jun protein family members bind to DNA as homodimers or as heterodimers among themselves or with members of the Fos or ATF/CREB families of proteins.[187] The four members of the Fos family bind to DNA as heterodimers with members of the Jun family or the ATF/CREB family. More than 50 different complexes among the Fos/Jun and ATF-CREB families of proteins have been identified, many of them in intact cells.[187]

There is some fine tuning of gene regulation that goes on by the interactions of the AP-1 and ATF/CREB families of STFs. For example,

Table 5-1 *(continued)*

Factor	Binding Site	Size	Promoter/Enhancer	Comments
E2F	TTTCGCGC	54 kDa	adenovirus E2A E1A enhancer	Binding activity detected in infected but not uninfected HeLa cells; also detected in undifferentiated F9 cells.
AP3	GGGTGTGGAAAG[a]		SV40 enhancer Py enhancer	Overlaps the core motif of SV40. Induced by TPA.
EBP20	TGTGG(A/T)(A/T)(A/T)G CCAAT		MSV enhancer SV40 enhancer Py enhancer	This protein originally purified by virtue of its ability to bind to the SV40 enhancer, also binds to the CCAAT sequence. It has recently been demonstrated that both EBP20 (enhancer binding activity) and CBP (CCAAT binding activity) reside on one polypeptide encoded by a single gene.
TFIID	TATA box		many genes	
AP2	CCCCAGGC	52 kDa	SV40 promoter/enhancer Py enhancer/origin pre-proenkephalin collagenase, mouse $H2K^b$ Ad MLP, human hsp70 hMT11A	
AP4	CAGCTGTGG		SV40 enhancer Py enhancer	
GT11-1B			pre-proenkephalin	
AP5	CTGTGGAATG		SV40 enhancer	
EF,E GT11-C			F441 Py enhancer	
PEA2	GACCGCA		Py enhancer	
EF,C	$GTTGCN_2GGCAAC$		Py enhancer hepatitis B enhancer	
E2aE-Cβ E4EF2	TGGAATT		E2A (E2aE-Cβ) E4 (E4F2)	
E4TFI	GGAAGTG		E4	

[a]This is the SV40 binding site and not a consensus.

Source: Modified from Jones et al.[185]

while the glucocorticoid receptor (GR) and AP1 (Jun/Fos) are primary regulators of two different signal transduction pathways, triggered by glucocorticoids and by mitogens such as TPA, respectively, they can modulate each other's activity. It has been observed that a "composite" glucocorticoid response element (GRE) could bind either GR or a Jun/Fos heterodimer. In the presence of c-Jun, the GRE was active, but it was inactive when a Jun/Fos complex with high Fos content was added.[188] Moreover, there is evidence that GR and Jun/Fos can reciprocally repress one another's transcriptional activation.[189] Overexpression of c-Jun prevented GR-induced activation of genes with a GRE promoter and, conversely, GR could repress genes carrying an AP-1 binding element. These data suggest that members of these two classes of STFs (i.e., GR and Jun/Fos) can modulate one another's activity in either a positive or negative direction depending on the environment of the cell.

In addition, members of the Jun family of STFs can function in opposing ways. For example, when NIH 3T3 cells become growth-inhibited, the level of c-Jun falls and JunD accumulates.[189a] Where resting cells are stimulated by the addition of serum, JunD is degraded and c-Jun synthesis is increased, followed by resynthesis of JunD later in G1. Overexpression of JunD results in cells accumulating in G_0/G_1, while c-Jun overexpression drives cells into S/G_2 and M phase. Also, JunD partially suppresses cell transformation by an activated *ras* oncogene, but c-Jun cooperates with *ras* to transform cells.

AP-1 (Jun/Fos) activity has also been shown to be required for TPA- or EGF-induced transformation of JB6 mouse epidermal cells, and blocking AP-1 activity by introducing a dominant-negative mutant of c-Jun inhibits tumor promoter-induced transformation.[189b]

SP1. The promoter-specific transcription factor SP1 was first detected as a protein in HeLa cells that could bind to and activate the SV40 early promoter.[190] It was later shown that SP1 binds selectively to GC-rich sequence elements (GC boxes) in a wide variety of viral and cellular promoters. A single polypeptide with a molecular weight of 105 kDa, SP1 contains three zinc fingers near the C-terminal end of the protein, which provide the DNA binding domains of the protein. The activation domains are glutamine-rich sequences in the *N*-terminal half of the protein.[191] Although SP1 is generally considered to be a "proximal promoter factor" in that its functional binding sites are usually found within a few hundred base pairs of the transcription start site, there is some evidence that SP1 can also act at distal promoter sites.[191] There is also evidence that SP1 can bind the retinoblastoma gene product (RB) control element (RCE) within the c-*fos*, c-*myc*, and TGF-β1 genes and that SP1 stimulates RCE-dependent transcription in vivo, suggesting that RB may regulate transcription by an interaction with SP1.[192]

Oct-3. A number of transcription factors that are active in early embryonic development, including that of mammals, have been identified. These include proteins that bind particular DNA octamer base sequence motifs. One of these, Oct-3, first detected in mouse embryonal carcinoma (EC) cells and in mouse embryonic stem (ES) cells, specifically binds the DNA octamer motif ATTTGCAT.[193] One of a family of transcription factors known as the POU family (see below), Oct-3 is expressed during early mouse development, from the early one- and two-cell embryo stage up to the point that the inner cell mass and primitive ectoderm differentiate into primitive endoderm and early ectoderm, mesoderm, and endoderm, at which point Oct-3 expression is downregulated.[193] Expression of Oct-3 in EC and ES cells is also downregulated when they are induced to differentiate by addition of retinoic acid. In contrast, homeobox genes are expressed at low levels in undifferentiated EC and ES cells and activated when differentiation is induced, suggesting a reciprocal relationship between the regulation of expression of these families of genes. The decrease of Oct-3 expression as early embryonic cells lose pluripotency suggests that its expression is important for the proliferative, highly undifferentiated state of totipotent and pluripotent stem cells. Studies showing that other members of the POU family, namely Oct-1 and Oct-2, can stimulate DNA replication of adenovirus DNA in vitro and that other octamer motif binding proteins can stimulate SV40 DNA replication in cells support this notion.[193]

The Superfamily of Hormone Receptors. Receptors for steroid hormones (including glucocorticoids, estrogen, and progesterone), thyroid hormones, retinoic acid, and vitamin D3 belong to a superfamily of ligand-binding proteins that can bind to DNA and activate or repress gene transcription. They most likely arose from a common ancestral gene and share a number of common structural features.[194,195] After binding ligand, these intracellular receptors act as dimeric transcription factors to activate or repress target genes by binding to specific DNA promoter/enhancer sequences called hormone response elements (HREs).

This family of receptors all have DNA-binding, ligand-binding, dimerization, nuclear localization, and activation domains (Fig. 5-22). The DNA-binding domains are highly conserved and consist of two zinc-finger–like structures, which recognize HREs, consisting of an inverted TGTTCT palindromic repeat (the glucocorti-

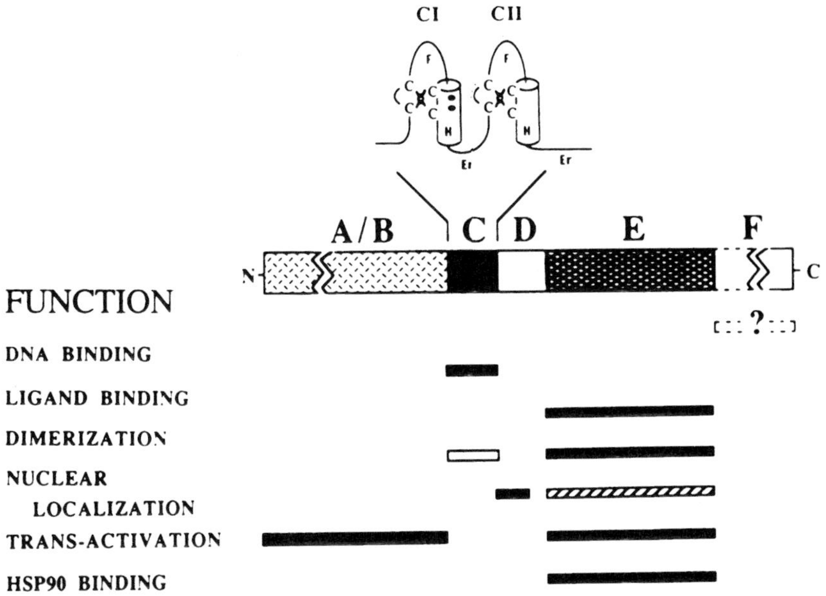

Figure 5-22 Structural and functional organization of steroid nuclear receptors. They are divided into six regions (A–F). The most conserved region (C) comprises two zinc fingers (CI and CII) in which the metal ion (X) is coordinated by two pairs of cysteines. Each of the two fingers (F) is followed by an α-helix domain (H) and an extended region (Er). In a model of the *DNA binding domain-HRE complex,* the first α-helix is taken as the recognition helix located in the DNA major groove. The two amino acids involved in half-HRE sequence discrimination, situated in the α-helix domain (H) of CI, are indicated by two dots. Dimerization involves weak interactions between DNA binding domains C and strong interactions between hormone binding domains E. The nucelar localization signals of the GR are between the DNA binding domain and the hormone binding domain, and within the hormone binding region. The ER lacks the second of these two signals. HSP 90 binding so far was reported only for GR, PR, and ER. The length of the A/B and F domains is variable and the function of the latter is unknown. (From Wahli and Martinez.[194])

coid, mineralocorticoid, androgen, and progesterone receptors) or HREs consisting of an inverted repeat of TGACCT or a closely related sequence (estrogen, thyroid hormone, retinoic acid, and vitamin D3 receptors).[195] The ligand-binding domains also share a number of common features and homology.

Since the DNA-binding and ligand-binding domains are rather well conserved among this family of receptors, one might ask: How is functional diversity achieved? This is an important question, since these hormones have different roles during development and in the adult organism. One explanation is that their ligand-binding domains are sufficiently different as to recognize different hormones, even though their DNA-binding domain may be quite conserved. A second point is that different target cells have a different relative abundance of receptors for a given hormone and/or different, cell-specific posttranslational modifications (e.g., phosphorylation) of receptors. It should also be noted that some hormone receptors such as the thyroid hormone receptor may function as a gene repressor in the absence of ligand and as a gene activator in the presence of ligand.[194] It is also interesting that truncated or mutated forms of members of this superfamily of receptors can act as oncogenic proteins (see Chap. 7).

TISSUE-SPECIFIC TRANSCRIPTION FACTORS

Tissue specific gene expression during early organ development and tissue differentiation is regulated to a great extent at the level of gene transcription. Several transcription factors have been identified that carry out this regulation. These include MyoD involved in muscle differentiation, HNF-1, involved in hepatocyte differentiation, and Pit-1, a pituitary activator of growth hormone and prolactin gene expression during normal ontogeny and necessary for the differentiation of lactotroph, somatotroph, and thyrotroph cells in the anterior pituitary gland. Some of these will be discussed here.

MyoD. The ability to induce muscle cell differentiation in undifferentiated cultured mouse fibroblasts has provided a unique tool to look at lineage-specific events that regulate commitment and terminal differentiation. A family of transcription factors that induce myogenesis has been isolated and cloned, and the way in which the factors work has provided a model for cellular differentiation in multiple organ systems.

The first of these to be identified, MyoD, was initially cloned by subtractive hybridization with cDNAs prepared from mRNA transcripts expressed in myoblasts but not in undifferentiated 10T ½ mouse cells used as the model cell line in which the myogenic pathway can be induced (reviewed in Refs. 196 and 197). A nuclear protein of 318 amino acids, MyoD forms heterodimers of a helix-loop-helix (HLH) motif and binds to muscle-specific enhancers. It is now known that the *MyoD* gene, encoding the MyoD protein, is one of a family of myogenic genes that includes *myogenin, myf*-5, and *mrf*-4. Each of these factors is expressed only in skeletal muscle, but gene knockout experiments indicate that mice lacking MyoD are viable and have normal muscle specific gene activation mechanisms.[197]

Members of the MyoD family are about 80% homologous in a region that includes a basic region followed by a HLH motif in which two amphipathic α helices are separated by an intervening unstructured loop. Such HLH regulatory proteins have been found in mammals, a wide variety of other vertebrates, the fruit fly *Drosophila*, the worm *Caenorhabtidis elegans*, and the sea urchin.[197] Of interest is the fact that myogenic factors from *C. elegans* and sea urchin can activate the myogenic differentiation pathway in mouse 10T ½ cells, indicating the highly conserved nature of these factors. The HLH motif is also typical of the *myc* family of oncogene products as well as their dimer-forming partner Max protein and certain other "E proteins," such as E12, E47, and HEB.

The HLH motif provides an interface for heterodimerization between MyoD and other E proteins, forming a dimer that recognizes a DNA sequence CANNTG (where N can be any base) called the E-box.[196,197] This base sequence motif is in the regulatory region of muscle-specific genes such as muscle creatine kinase and of other cell-type–specific genes regulated by HLH-type proteins. Functional activity of myogenic HLH proteins requires heterodimer formation with E proteins such as E12 and E47.[198] As with other transcriptional factors, MyoD and the HLH family have DNA-binding domains, dimerization domains, and activation domains. In addition, their activity is modulated by their phosphorylation state.

Muscle cell determination and differentiation is dictated by a balance of factors that determine the proliferation potential and the shutdown of proliferation that accompanies differentiation. For example, once myoblasts enter the terminal differentiation pathway, they stop proliferating and fuse with neighboring myoblasts to form multinucleated myotubes. This process is inhibited by growth factors such as FGF, TGF-β, and serum.[197] Moreover, expression of a variety of oncogenes—including *src, ras, fos, jun, fps, erb*A, *myc*,and E1A as well as mitogens such as phorbol esters—inhibit myogenic differentiation.[196] Differentiation can also not be induced in a number of tumor cell types even though MyoD is expressed. Rhabdomyosarcoma cells that are derived from a muscle cell–type malignant tumor differentiate only poorly though they express *myoD.* These data suggest that during the process of oncogenesis, some factors required for normal differentiation are lost or their function inhibited.

Some HLH proteins lack the functional basic regions but can still form dimers with transcriptional factors of the HLH type, thus forming nonfunctional dimers. These proteins can act as negative regulators of E-box–dependent transcription. One of these proteins is the Id (inhibitor of differentiation) protein, which can dimerize with MyoD and other E proteins and render them inactive as inducers of myogenesis and other tissue-specific differentiation processes in which E proteins are involved.[199]

The four members of the MyoD family are first expressed in early embryos when myogenic precursor cells appear in the somites, and they are present in the limb bud during muscle terminal differentiation. Yet each gene is activated at a slightly different time, suggesting that they are expressed in response to somewhat different signals. Alternatively, activation of one of the early myogenic genes could lead to a cascade of timed activation of subsequent genes in the pathway. There is evidence that members of the *myoD* gene family can positively autoregulate one another's expression.[197] The timed expression of these genes produces the right mix of transcription factors that in turn induce the expression of the genes that make a muscle cell a muscle cell.

Factors that inhibit muscle cell differentiation may act in different ways. For example, activated Ras and c-Fos proteins block MyoD transcription; TGF-β inhibits MyoD activity but not its transcription; and c-Jun can block MyoD function by a protein–protein interaction between the leucine-zipper domain of c-Jun and the HLH region of MyoD.[200]

Liver-Specific Transcription Factors. At least six liver-specific TFs are functional in development and terminal differentiation of the liver. These are HNF-1, C/EBP, DBP, HNF-3, HNF-4, and LF-A1 (reviewed in Ref. 201). These act together in development of the hepatocyte phenotype, yet none of these appears to have expression limited to the liver, suggesting that interplay with other environmental or endogenous signals is important.

The HNF-1 TF can bind to the promoters of a variety of liver-specific genes including α-fibrinogen, α-fetoprotein, α_1-antitrypsin, albumin, and transthyretin. HNF-1 binds as a homodimer to an inverted palindrome of the sequence GTTAATNATTAAC. Optimal transcription of liver-specific genes depends on interaction among transcription factors. For example, the albumin gene contains six *cis*-regulatory elements A through F. Basal expression of the albumin gene can be achieved by binding of an ubiquitous TF, known as NF-1, to the C element, but fuller activity is achieved by binding of HNF-1 to the B element and DBP or C/EBP to the D element.[201] Augmented expression is achieved by binding of C/EBP to the A and F elements.

Since the liver expresses over 1000 liver-specific genes, it seems unlikely that the whole panoply of liver gene expression would be controlled by only six transcription factors. Thus, many more likely await discovery. Some additional regulatory diversity may be achieved by molecules similar in structure to HNF-1 or other liver TFs that can form dimers having different DNA sequence specificity. Then, by mixing and matching various dimer motifs, a much wider variety of genes could possibly be regulated.

Pit-1. As noted above, Pit-1 is a pituitary gland–specific TF that regulates the development of hormone-producing cells in the anterior pituitary. Pit-1 is a POU domain TF and Pit-1–activated promoters respond to epidermal growth factor, cAMP, and phorbol esters. The Pit-1 TF is phosphorylated in pituitary cells at two different sites in response to cAMP and phorbol esters, and phosphorylation causes conformational changes in Pit-1 that alter its DNA-binding specificity, with increased binding at some promoter sites and decreased binding of others.[202]

E2A. The E2A gene codes for the E-box transcription factor E12 and E47. These E-box elements are present in the immunoglobin heavy chain enhancer and in genes involved in muscle, pancreas, and B lymphocyte differentiation. As noted above, they are HLH proteins and bind to DNA after forming heterodimers with other E-type proteins. The E2A genes have been found to be mutated in B cell leukemias and are involved in chromosomal translocations that result in chimeric proteins being formed between E2A-encoded proteins and other DNA-binding proteins. One example is a chimera formed by a t(1;19) translocation that brings the N-terminal coding region into proximity with the C-terminal coding region of the transcription factor Pbx1, which is normally not expressed in B cells.[203] This converts a nonactivating DNA-binding protein into a potent transcriptional activator and suggests that this may be involved in the leukemogenic process.

NF-κB. NF-κB was first detected as a protein that could form a complex with a 10-bp site in the κ light chain enhancer called κB (reviewed in Ref. 204). Since it is constitutively expressed in B cells undergoing κ light-chain expression and is crucial for κ gene enhancer function, it was thought to be a tissue-specific transcription factor. Later work, however, revealed that it is ubiquitously expressed and involved in regulation of gene expression in multiple cell types, leading to the idea that it may be a "pleiotropic mediator of inducible and tissue-specific gene control."[204]

Induced by several T-cell mitogens and by antibodies that activate T cells, NF-κB appears to be important for T-cell activation by a variety of mechanisms. It is involved in transcriptional regulation of a variety of genes, including interleukin-2α receptor (IL-2αR), β-interferon, interleukin-6 (IL-6), tumor necrosis factor-α (TNF-α), and lymphotoxin.[204] Several agents induce NF-κB expression, including the *tax* gene product of human T-lymphotropic virus HTLV-1 and the viral *trans*-activators of cytomegalovirus (CMV) and hepatitis B virus. In addition, CMV, SV40, and human immunodeficiency virus (HIV-1) enhancer regions all have NF-κB binding sites, suggesting an important role of NF-κB in viral replication.

POU-Domain–Binding Proteins. Several mammalian transcription factors and developmental TFs in *C. elegans* have a unique structural motif called POU. The POU proteins have a highly conserved N-terminal region of 76 amino acids (the POU-specific domain), a variable linker region 15 to 27 amino acids, and a 60 amino acid homeodomain (the POU homeodomain) that diverges with species of organism (reviewed in Ref. 205). The entire POU domain is required for high-affinity and sequence-specific DNA binding. The POU proteins appear to be able to bind DNA as either a monomer or a dimer and they bind particular octamer (8-bp) sequences. The POU-family proteins include the octamer-binding TFs OCT-3 and Pit-1 noted above as well as other developmental regulators that generate specific cell phenotypes.

Ets1 and Ets2. The Ets proteins are transcription factors that bind to a GGAA purine-rich core DNA sequence seen in promoters or enhancers of various cellular and viral genes. Examples of genes under Ets protein transcriptional regulation include murine sarcoma virus long-terminal repeat (LTR), stromelysin, urokinase-type plasminogen activator, and IL-2 (reviewed in Ref. 206). The *ets* 1 and *ets* 2 genes can act like oncogenes when overexpressed or expressed at an inappropriate time. For example, overexpression of *ets* 1 and *ets* 2 can transform murine fibroblasts, and *ets* 1 is expressed in embryonal neuroectodermal tumors such as neuroblastoma.[206] This theme occurs again and again: a transcription factor that is normally expressed during a certain stage or stages of development and sometimes only in specific embryonic cell types becomes a transforming oncoprotein when activated, mutated, or expressed at the wrong time or in the wrong cell type later in life. This is discussed in detail in Chapter 7.

Expression of *ets* 1 is limited to certain cell types during fetal development of the mouse. Its expression is predominantly seen in lymphoid tissues, brain, and organs like the lung, kidney, and salivary gland when they are undergoing branching morphogenesis.[206] In neonatal development, *ets* 1 is expressed in lymphoid tissues and brain, but in adult mice it is only expressed in lymphoid tissue. It is also expressed in bone during bone formation or remodeling, a role that may be mediated by the Ets 1 transcriptional regulation of metalloproteinases such as stromelysin and of plasminogen activator. Activation of these proteinases most likely is important for the degradation and remodeling of extracellular matrix components during branching morphogenesis and bone deposition.

Expression of *ets* 2, on the other hand, is widespread in all organs of embryonic, neonatal, and adult mice, consistent with a more fundamental role in cell growth control such as regulation of cell mitosis.[206]

HOMEOBOX PROTEINS

Among the many complex things that multicellular organisms must accomplish during development is to induce an anterior–posterior axis. This is accomplished by turning on, in a spatial and temporal pattern, a series of genes called homeotic selector genes or homeobox genes. These genes were originally discovered in the fruit fly *Drosophila,* because when they were mutated, abnormal structural development was seen. It is now known that these genes code for a series of transcription factors and that they are present in organisms as diverse as acorn worms and humans.[207] The homeobox family of genes has been detected in simple hydrozoans (*Sarsia* species), which suggests that they are at least 500 million years old.[208,209]

The first homeotic mutation in *Drosophila* was reported in 1915, but it was not until the advent of DNA technology that the responsible genes were identified (reviewed in Ref. 210 and 210a). These mutations cause dramatic alterations in the architecture of the fruit fly, such as a second pair of wings (a mutation called *bithorax*) or the growth of legs instead of antennae on the head (a mutation called *Antennopedia* or *Antp*). Identification and sequencing of the genes involved in these and other structural defects in *Drosophila* revealed that there was a lot of cross-homology among them. An 180-bp DNA segment, the so-called homeobox region, was highly conserved in all of them. These genes encode the homeobox proteins that have several structural features in common. They share a 60 amino acid motif with a helix-turn-helix structure. A typical structure is that of the *Antp* protein with a flexible amino-terminal arm, three well-defined α-helices, and a fourth, more flexible helix.[210] A recognition helix binds to specific base pairs in the major groove of DNA and a flexible amino-terminal domain contacts DNA bases in the minor groove, while two helices make contact with the DNA backbone.

One interesting feature of the homeobox genes is that they are clustered in a 3′- to 5′ orientation that is exactly in the same order as the anterior–posterior segments of the body whose formation they regulate; that is, the genes located most 3′ in the cluster are expressed in the most anterior segments, and those located toward the 5′ end are expressed more posteriorly (Fig. 5-23). Furthermore, the *Antp*-class homeobox clusters, called HOM-C in *Drosophila,* have remarkably similar organization and expression in the mouse and human genomes, where they are called *Hox* genes. Several loss-of-function and gain-of-function mutations of these genes have been identified in mice and frogs (Table 5-2).

In *Drosophila,* there is a single cluster of homeobox genes; whereas in mice and humans, there are four clusters on four different chromosomes. However, the principles of organization are conserved.[210] Thus, the homeobox genes are a highly conserved class of "master control genes" that regulate the structural orientation of the "body plan" during development.

What genes or functions that the homeobox proteins in turn regulate is not entirely clear, but they have been shown to regulate expression of cell adhesion proteins such as the neural cell ad-

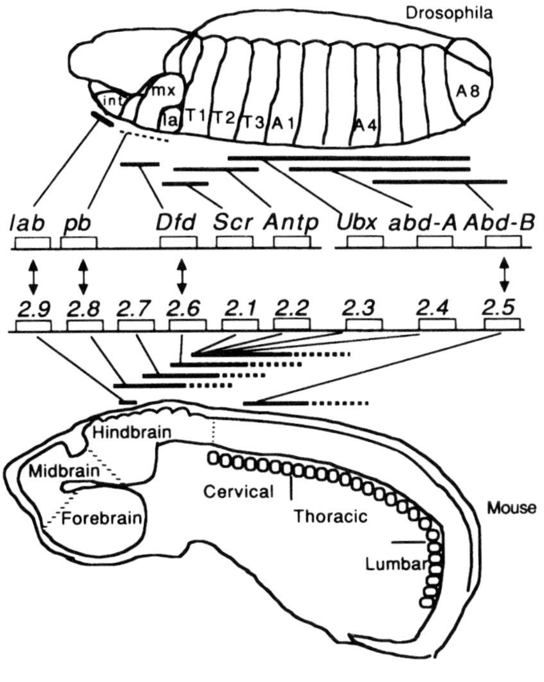

Figure 5-23 Summary of HOM-C and *Hox-2* expression patterns. The upper half of the figure contains a diagram of a 10-hr *Drosophilia* embryo, with the approximate extents of the epidermal expression domains of the *HOM-C* genes *lab* through *Abd-B* indicated by the horizontal bars. The expression domains of these genes also approximately correspond to the indicated limits within the embryonic CNS. The *pb* expression pattern is represented by a thin dotted bar, as *pb* has no detectable function in the embryonic head. int, mx, and la designate intercalary, maxillary, and labial segments, respectively. T1, T2, and T3 indicate thoracic segments 1–3. A1–A8 indicate abdominal segments 1–8. The lower half of the figure contains a schematic diagram of a 12 day mouse embryo, with the approximate extents of *Hox-2* expression domains in the CNS indicated by the horizontal bars. The dotted extensions of the *Hox-2.8* through *Hox-2.5* patterns indicate that these expression domains extend in overlapping fashion to posterior regions of the CNS. *Hox 2.1*, *Hox-2.2*, *Hox-2.3*, and *Hox-2.4* have subtly different boundaries in the posterior regions of the hindbrain: for simplicity, their expression domains are represented together. (From McGinnis and Krumlauf.[209])

Table 5-2 Alterations to *Hox* Expression and the Resulting Phenotypes in Vertebrate Embryos

Gene (Species) (Homologue)	Type/Mechanism of Mutation	Phenotype
Hox-1.6 (mouse)	Loss-of-function: targeted disruption in ES cells	Recessive. Neonatal lethal. Defects concentrated at the level of rhombomeres 4–7 in structures derived from paraxial and head mesoderm, neural crest, placodal, and neuroectoderm; e.g., missing motor nucleus of facial (VII) nerve; alterations to basioccipital and exoccipital bones, the inner ear, and cranial sensory ganglia; in neural crest mostly neurogenic components abnormal.
Hox-1.5 (mouse)	Loss-of-function: targeted disruption in ES cells	Recessive. Neonatal lethal. Defects focused in head and thorax in structures and organs derived from mesoderm, pharyngeal endoderm, neural crest: missing thymus, parathyroids, lesser horn of hyoid bone; altered thyroid, heart, maxilla, mandible, third and fourth branchial arch, circulatory system; in neural crest mostly mesenchymal components abnormal.
Hox-1.1 (mouse)	Gain-of-function: ectopic expression of a *Hox-1.1* transgene from a β-actin promoter	Dominant lethal. Several craniofacial abnormalities, including secondary cleft palate. In axial skeleton, normal vertebrae up to C3, and variations in C1 and C2 consistent with transformations, new proatlas and a vertebral body associated with the atlas.
Hox-1.4 (mouse)	Gain-of-function: ectopic expression of an altered *Hox-1.4* transgene from its own promoter	Dominant lethal. Highly elevated levels of expression in the gut are associated with hyperproliferation of the colon which leads to compaction and death in the adult. Decreases density of innervation by enteric ganglia in gut. Resembles megacolon phenotype associated with Hirschsprung's disease.
XlHbox6 (*Xenopus*) (mouse *Hox-2.5*)	Gain-of-function: injection of *XlHbox6* mRNA in 2-cell *Xenopus* embryos and Einsteck grafts	Dominant alteration to axial properties of mesoderm. Animal cap region exposed to injected mRNA and grafted into blastocoel resulted in secondary axis with tail-like structures, the injected homeodomain protein also respecifies animals caps exposed to growth factors.
XlHbox1 (*Xenopus*) (mouse *Hox-3.3*)	Loss-of-function: injection of antibodies to long form of XlHbox1 protein	Dominant alteration to anterior spinal cord. Hindbrain appears enlarged and extends to more posterior regions relative to pronephros. This is accompanied by XlHbox1 expression in the abnormal region. Dorsal fin defects in neural crest cells that express the protein.
XlHbox1 (*Xenopus*) (mouse *Hox-3.3*)	Gain-of-function: injection of mRNA for long and short XlHbox1 proteins	Dominant alteration to somitic segmentation and myotome markers when long form of protein used. Dominant alteration to anterior spinal cord, giving appearance of expanded hindbrain, similar to antibody injections, but phenotype extends more posterior. Also localized neuronal asymmetry.
Xhox-1A (*Xenopus*) (mouse *Hox-2.6*)	Gain-of-function: injection of mRNA for *Xhox-1A*	Complex dominant phenotypes. One major alteration was the perturbation to paraxial segmentation. Regional variation in the myotome component of somites.

Source: From McGinnis and Krumlauf.[209]

hesion molecule N-CAM, which is important in nervous system development.[211] Cell adhesion molecules are crucial for embryonic development (see below), and regulating their expression may be one way that *Hox* gene proteins control tissue growth and development in a spatiotemporal manner. Moreover, homeodomain proteins have been shown to enhance the DNA-binding activity of growth stimulatory signals such as serum response factor,[212] and this may be another way in which *Hox* gene transcription factors control segmental growth (i.e., by being turned on at a specific time and place when growth in a particular part of the body is called for).

Now, one might ask: What happens when expression of these genes goes awry? Since they are a class of master growth control genes, one might predict dire consequences if they were expressed at the wrong time or in the wrong place. One clue to this is the role that *Hox* genes play in hematopoiesis. The *Hox* 3.3 gene has been shown to be involved in an early step in proliferation of the erythroid colony-forming (CFU-E) precursor cells in red blood cell formation, and this gene is also expressed in erythroleukemia cells.[213] Moreover, a subset of T cell acute lymphocytic leukemias possess a t(10;14) or t(7; 10) chromosomal translocation that involves a Hox gene (*Hox* 11) present on chromosome 10, suggesting that its activation is involved in the leukemogenic process.[214,215]

Posttranscriptional Regulation

After genes are transcribed into mRNA, a whole series of events regulate how an mRNA gets translated into a functional protein. These events include (1) splicing of the high-molecular-weight precursor mRNA (pre-RNA) transcripts into mRNA; (2) capping, polyadenylation, and editing of the mRNA; (3) nuclear-cytoplasmic transport; (4) initiation of translation; (5) alternate translation from overlapping reading frames; (6) turnover of the mRNA; (7) protein folding and processing; (8) posttranslational modifications of the protein; and (9) intracellular translocation of the mature protein, leading to secretion or sequestration into its functional compartment. Some of these events have already been discussed. Others are discussed briefly below.

The mechanisms for splicing of precursor mRNAs were discussed above. Some pre-mRNAs are spliced differently ("alternate mRNA splicing") in different cell types or at different times during development.[216] Splicing of pre-mRNA is under both positive and negative control. Pre-mRNA binding proteins, called splicing factors or SR proteins, that modulate pre-mRNA splicing have been identified.[216,217] These proteins play a role in the alternate splicing of pre-mRNAs and are differentially expressed in a variety of tissues.[217] Moreover, pre-mRNA splice-site selection can be negatively regulated by repressor proteins that bind specifically to pre-mRNA recognition sequences. This has been shown in yeast, *Drosophila,* and mammals.[216]

RNA editing is a posttranscriptional process that produces an mRNA with a nucleotide sequence differing from that of the transcribed DNA.[218,219] It is another mechanism for modulating gene expression. It was first described as a mechanism for mitochondrial gene expression in protozoa, where the insertion or deletion of uridine in an mRNA was observed. Other examples include the conversion of a cytidine to uridine in mammalian apolipoprotein-B mRNA, insertion of two guanosine residues in a paramyxovirus transcript, and conversion of a cytidine to uridine at multiple positions in the mRNA for subunit II of cytochrome C oxidase in wheat mitochondria.[218,219] In these instances, mRNA editing either changes a nontranslated message into a translated one or modifies a translatable message into one that generates a protein with a different amino acid sequence. The mechanism for such mRNA editing is not clear, but it appears to involve an error-prone base-pairing mechanism with a "guide RNA."[220] In some organisms such as trypanosomes, the edited proteins accumulate mutations about two-fold faster than unedited proteins,[220] suggesting that pre-mRNA editing plays a role in the process of evolution.

Nuclear-cytoplasmic transport of mRNA is required to get the message to the polyribosomes where they are translated. This is also a regulated event and requires RNA-binding proteins.[221]

In higher eukaryotic cells, mRNA translation is regulated by structural features of the mRNA as well as by the translation-initiation machinery, including a finely tuned series of initiation factors (reviewed in Ref. 222). Structural features of the mRNA that modulate translation include (1) the m7G cap at the 5′-end, (2) the primary base sequence around the AUG initiation codon, (3) the position of the AUG codon (whether it is first or in a place where a second initiation can occur; see below) (4) secondary structure upstream and downstream from the AUG codon, and (5) length of the leader sequence. The sequence of events, briefly, is as follows. The 40S ribosomal subunit, bearing a methionine transfer RNA (tRNA) for the AUG codon and the appropriate set of initiation factors, attaches to the 5′ end of the mRNA and migrates along the mRNA until it finds the first AUG codon, whereupon a 60S ribosomal subunit joins the 40S subunit and the first peptide bond is formed.

Certain characteristics of the mRNA or of the initiation factors can alter the way in which these translational events occur. For example, in some cases, two different proteins can result from different AUG initiation codons within the same mRNA. This can occur when the first AUG site is, at a particular time or cellular environment, in a less favorable conformation than a second one downstream. Thus, different proteins can be produced from an overlapping reading frame depending on which start site is used.[223] Such alternate production of two proteins by initiation at the first or second AUG codon has been observed for a variety of viral mRNAs and some human oncogenes.[223]

Alterations in levels of initiation factors can also modulate mRNA translation. For example, it has been shown that overexpression of the translation initiation factor eIF-4E in NIH 3T3 or rat-2 fibroblasts cause their tumorigenic transformation,[224] apparently as a result of loss of regulation of initiation of protein synthesis.

The turnover rate of various mRNAs is also an important variable in modulating mRNA translation. Some hormones and external factors that induce gene expression may do so by stabilizing a mRNA with a relatively short cellular half-life. Half-lives of eukaryotic mRNAs vary from a few minutes for highly regulated mRNAs such as cellular oncogenes and rate-limiting enzymes to more than 100 hours for very stable mRNAs, such as certain housekeeping and structural proteins.[225] The average half-time for turnover for mRNA in eukaryotic cells is 10 to 20 hours, whereas the average $t_{1/2}$ for proteins is about 48 to 72 hours, although many turn over faster than that. One notable example is the tumor suppressor protein p53, which in the normal, wild-type form has a half-life of about 1 to 2 hours, whereas its mutated form has a $t_{1/2}$ of about 6 to 8 hours (see Chap. 8). In addition, rates of degradation of mRNA and proteins may change during the cell cycle (e.g., the cell-cycle regulatory cyclins), in response to stress (e.g., heat or environmental shock proteins), availability of nutrients, or during various stages of differentiation (e.g., oocyte mRNAs after early stages of embryogenesis).

ROLE OF THE MICROENVIRONMENT IN CELLULAR DIFFERENTIATION

In a sense, all the mechanisms proposed in this chapter as possible regulators of cell differentiation "beg the question" about what initiates the process because there would still be a requirement for some stimulus or signal to affect one daughter cell differently from its identical twin derived from a parent cell in the early embryo. Thus, either the intensity (or concentration) of the signal varies within the microenvironment bathing the early embryo cells or the position of one of the two daughter cells provides some tropistic stimulation that is translated into a differentiation program that diverges from its twin cell. Part or all of this positional effect could be determined by cell–cell interactions with neighboring cells in the microenvironment. Moreover, the specific differentiation program would have to be maintained in the progeny of the responding cell; that is, it would have to be heritable. This is easier to visualize for gene amplification or recombination events, since a change in the genome would occur, but for a heritable, selective change in gene transcription or translation, a given molecular signal (probably a protein) would have to be present to shut off or stimulate a gene in one daughter cell and be absent in the other daughter cell that will become a different type of cell. Presumably, once this crucial control protein is induced in one cell type, it is synthesized in the progeny of that cell so that it continues to block or stimulate the expression of the same gene in the progeny cells. Thus, either the protein itself or its mRNA would pass from parent to progeny at each mitosis. Selective alteration of gene readout could also occur by the transmission of information by means of RNA that could enter cells and be transcribed by RNA-directed reverse transcriptase to make DNA, which could then be inserted into the genome. Temin suggested that this occurs, based on the evidence from RNA oncogenic virus-infected cells (see Chap. 7).

The total milieu in which a developing cell

finds itself can affect the way it differentiates. The factors that contribute to the microenvironment of cells include (1) positional effects and polarity of a given cell in relation to the rest of the developing cellular mass; (2) interactions of cells with the surrounding extracellular matrix and supportive tissues; (3) direct cell–cell contact; (4) gradients of oxygen, nutrients, and ions in the area of a developing multicellular mass; and (5) presence of specific growth and differentiation factors.

Positional Effects

In embryonic growth and organogenesis, cell development is determined by relative cell position and by substances originating from neighboring cells. For example, in the gastrulation stage of the developing sea urchin, interactions between mesenchymal cells and ectodermal cells at one pole of the embryo play a key role in directing the formation of the primitive gut. Interaction between mesenchymal and epithelial elements is important in the development of many organ systems in higher organisms, including pancreas, salivary glands, lung, kidney, thyroid gland, mammary gland, pituitary, and liver. A requirement for a specific mesenchymal tissue has been noted for several of these tissues. The information transfer between cells could occur by direct cell–cell contact or by hormones or growth factors secreted by one cell type and acting locally. During the development of many organs, cells destined to share structural or functional identity adhere. It has been known for a long time that embryonic cells, dissociated in an appropriate medium, will reaggregate with cells from the same organ type more avidly than with cells of other developing organs. Such organotypic interactions could result from cell-specific recognition of surface molecules, perhaps determined by the distribution of their cell surface glycoproteins. The subsequent formation of desmosomes or junctional complexes between cells could result from the specific localization of such attachment sites. The junctions between cells serve both to provide an orientation to the cell, for example, an apical and basal orientation, and a means of chemical and electrical communication. It seems highly likely that such couplings are crucial in organogenesis. The primitive gut, for example, is composed of a single layer of cells joined in a tubular sheet by cell junctions. The pituitary, thyroid, lungs, pancreas, and other digestive organs are formed by budding in a manner that maintains cell–cell junctions.

Extracellular Matrix

The extracellular matrix (ECM) is a complex intercellular "glue," components of which are produced and secreted by the cells that make up a given tissue. In the case of epithelial tissues, epithelial cells and stromal or mesenchymal cells on which the epithelial cells sit produce various components of the ECM. There is evidence that two-way signals are involved in the interaction between epithelial and mesenchymal cells in the production of the ECM. As will be described below, this interaction is also crucial for regulation of cellular proliferation and differentiation, in which the ECM plays a key role.

The ECM varies to some extent from tissue to tissue but its basic components include collagen, elastin, proteoglycans, and "anchorage" proteins such as laminin, fibronectin, vitronectin, and entactin that serve as attachment sites to the matrix for epithelial cells and fibroblast-type (mesenchymal) cells. Other ECM components that modulate cell-matrix interactions include thrombospondin, tenascin, and osteonectin (also called SPARC).

In epithelial tissues, the ECM is organized into a basal lamina (or basement membrane). The term *basement membrane* has been used for years to describe the densely packed "scaffold" between epithelial cells and mesenchymal cells. This is because, under a light microscope, the histochemically stained, densely packed matrix looked like a membranous structure to early morphologists. We now know that it is not truly a membrane but a complex secretion produced by epithelial and mesenchymal cells. Basal laminae are found at the dermal–epidermal junction of skin; at the base of all lumen-lining epithelia of the digestive, respiratory, reproductive, and urinary tracts; underlying capillary endothelium and venules; around Schwann cells in the nervous system; surrounding fat cells, skeletal muscle, and cardiac muscle cells; and at the base of parenchymal cells of exocrine and endocrine glands. In general, the basal lamina consists of a lamina densa, 20 to 50 nm thick, that runs parallel to the basal side of epithelial cell membranes, separated by a 10-nm-thick, less-dense layer called the lamina lucida. All epithelial tissues "rest" on these ubiquitous membranelike structures that serve both as a support and a barrier for the tissue. During the process of cancer cell invasion and metastasis, this barrier is breached by processes involving the secretion of lytic enzymes by the cancer cells and the ability of cancer cells to form attachments to the basement membrane barrier (see Chap. 11). Some of the structural characteristics of various ECM components are shown in Figure 5-24.[226]

The function of the ECM was originally thought to be primarily if not exclusively one of structural support for tissues. In some tissues the support function is primary, for example, cartilage and ligaments. The type of collagen and amount of elastin determine the tensile strength and elasticity of the tissue. For example, in skin where tensile strength is important, type I collagen, which forms dense sheets of collagen fibrils, predominates, whereas in ligaments, which require more elasticity, significant amounts of elastin are present. The associated proteins, glycoproteins, and proteoglycans also vary from tissue to tissue. Laminin, heparan sulfate-containing proteoglycans, and type IV collagen are usually found together as the supporting structure of epithelial tissues such as the gastrointestinal tract, mammary gland, and endocrine

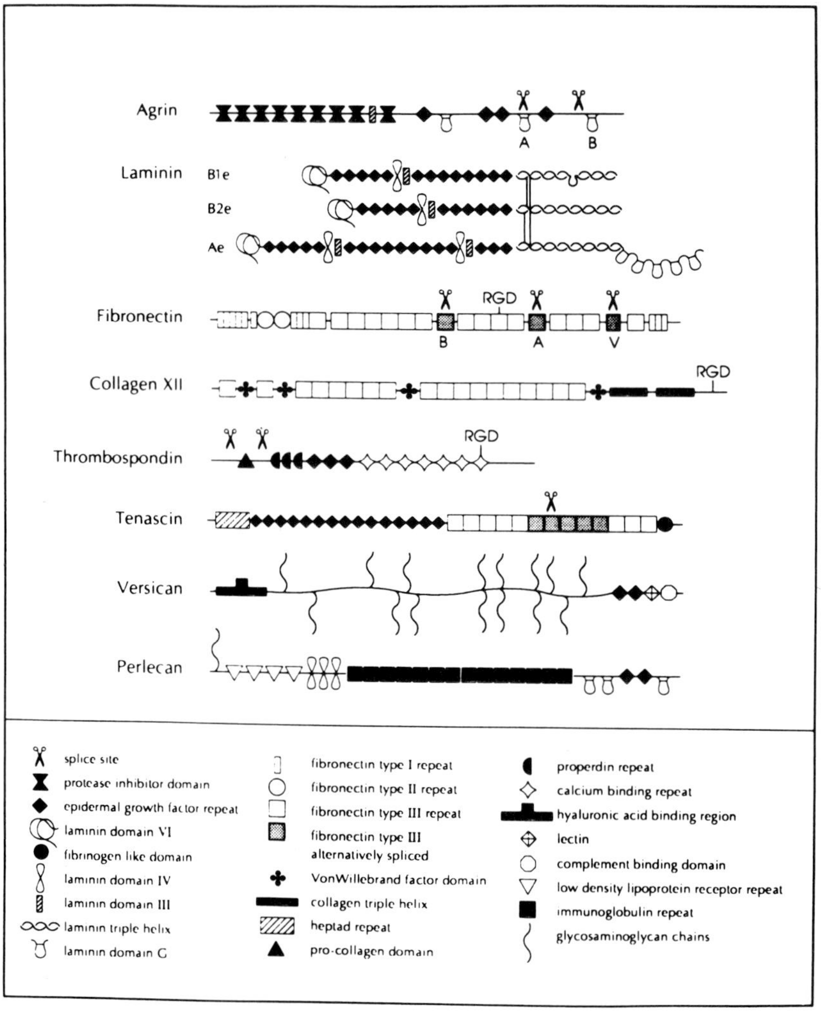

Figure 5-24 ECM molecules. (From Venstrom and Reichardt.[226])

organs. Fibronectin, chondroitin sulfate–containing proteoglycans, and type I collagen are present in tissues like skin. Thus, there are "matrix families" that can be characterized as type I, type II matrix, and so on, based on their collagen types. The proteoglycan component also may differ in various tissues; for example, heparan sulfate from skin is different from that of lung. Therefore, there is some tissue specificity to the ECM, and this appears to provide a signal for tissue differentiation (see below). The type of matrix and its components also change in various disease states.

Laminin

Laminin is a large (MW $\approx$ 900,000) glycoprotein, which appears by the technique of rotary shadowing to be cruciform, providing multiple attachment points for the molecule. Through its cellular receptors and attachment sites to the matrix, it effects binding of cells to the ECM. Laminin was originally isolated from the mouse EHS sarcoma but is now known to be a component of all basement membranes studied.[227] It contains distinct polypeptide chains B1 and B2, both of 220,000 MW, and a 440,000-MW A chain. There is a family of laminin subtypes in different tissues. For example, Ae, Am (merosin), and Ak (K-laminin), Ble, Bls (S-laminin), B1-2 (avian form), B2e, and B2t (truncated) polypeptides have been identified and may form as many as 18 distinct isoforms, all of which contain a similar AB1B2 trimer of cruciform structure.[226] There are demonstrated attachment sites for heparan sulfate localized in the globular terminal domain of the long arm. The short arms appear to be involved in cell binding and binding to type-IV collagen. On the basis of selective protease digestion studies, the globular ends of the short arms appear to be necessary for laminin binding of type-IV collagen, and the intersection region of the three short arms appears to contain a major receptor binding site. The globular ends of mouse laminin are enriched in carbohydrates containing α-D-galactosyl end groups. In addition, the laminin chains contain a number of EGF-like repeats.

Laminin has an important role in many of the biological functions of the ECM, including cellular attachment, migration, growth, and differentiation. A number of the biologically active sites in the molecule have been identified. These include a YIGSR pentapeptide of the B1 chain and an RGD-containing region of the A chain that promote cell attachment, an IKVAV-containing sequence that induces c-*fos* and c-*jun* expression and has a cell proliferation-stimulating effect,[228] and a SIKVAV-containing peptide that stimulates angiogenesis and tumor growth in vivo.[229] Attachment of tumor cells to a laminin-containing ECM also has been reported to foster tumor cell invasion and metastases.[230]

Fibronectin

The glycoprotein fibronectin is one of the anchorage proteins of which laminin and chondronectin are other examples. These glycoproteins play a key role in anchoring cells to the ECM. Fibronectin is a 450,000-MW disulfide-linked dimer that contains five N-linked oligosaccharides and has been localized by immunohistochemical studies in extracellular spaces and basement membranes. It exists in both a cell-associated and a circulating form (present in plasma). These forms are very similar but not identical biochemically; the cell-associated form also exists as a mixture of larger polymers. Fibroblasts, myoblasts, undifferentiated chondrocytes, and endothelial cells have been shown to biosynthesize fibronectin in cell culture, as have certain epithelial cells and macrophages. In general, these results are consistent with the distribution of fibronectin by immunofluorescence in tissue sections; but some cultured lines appear to make fibronectin in culture whereas their counterparts in vivo do not appear to. This discrepancy could be accounted for by overgrowth of contaminating fibroblasts in the cell culture systems.

Fibronectin can bind to a number of macromolecules important for cell adhesion and interaction, such as collagen, heparan sulfate, hyaluronic acid, the enzyme transglutaminase, and cell membranes themselves. There are separate binding domains in the molecule for these activities. Cell-associated fibronectin is also involved in cell spreading, cell movement, and cell proliferation of a wide variety of cultured cell types. Cell proliferation is promoted by the presence of an ECM, and fibronectin, by favoring attachment and spreading of cells on their substratum, has a positive effect on cell proliferation of cultured cells. For normal cells to survive in culture, anchorage to a substratum is important, and the requirements for attachment and for serum or growth factors are interrelated (see below). Fibronectin also affects cellular differentiation in several cell culture systems. Both myogenic and chondrogenic cell lines lose fibronectin during differentiation, but differentiation of early mesenchymal cells into fibroblasts is associated with a high level of fibronectin production. Thus, the amount of fibronectin surrounding a cell type appears to change as that cell type achieves its more differentiated state and acquires the type of ECM that goes with it.

Thrombospondins

Thrombospondins (TSPs) are homotrimeric glycoproteins of about 450,000 MW. Originally identified in human platelets, they have been found to be synthesized and secreted by a wide variety of cell types.[231] Four related thrombospondin genes have been identified (reviewed in Ref. 226). Most often, TSPs have been associated with proliferating cells, and, in vivo, have been localized to areas of cell proliferation and migration in developing embryos (reviewed in Ref. 232). Several carcinoma, melanoma, and leukemic cell lines synthesize high levels of TSP, which has also been shown to support cell attachment and motility and to activate plasminogen. A 140-kDa domain of TSP has been found to mediate chemotaxis and to be related to the metastatic potential of human squamous carcinoma cells.[232]

Tenascins

Tenascins are a family of three related ECM proteins made up of disulfide-linked multimers. They contain EGF-like and fibronectin-like domains.[226] The tissue distribution of tenascin is widespread, and in the embryo, expression correlates with areas of mesenchyme–epithelial interaction, cell migration, and tissue differentiation. In the nervous system, tenascin is located in the pathways of migrating neural crest cells, suggesting that it functions as a substrate for cell migration.

Proteoglycans

Proteoglycans are complex macromolecules composed of a core protein and multiple glycosaminoglycan (GAG) chains. These GAGs are made up of repeating disaccharide units of sugar residues extended into long chains. They are usually bound to protein core molecules, but at least one, hyaluronic acid (repeating unit = glucuronic acid-*N*-acetylglucosamine) is not linked covalently to a protein core and makes up a key part of the matrix in most tissues, although its content varies from tissue to tissue and during development.

A key difference between the glycoproteins and proteoglycans is the amount of carbohydrate bound to the protein. Whereas in glycoproteins 30% to 40% of the molecular weight may be carbohydrate, in proteoglycans the carbohydrate may account for 90% of the molecular weight. The proteoglycans are categorized by the number, length, and composition of the GAG chains (for review, see Ref. 233). Figures 5-25 and 5-26 illustrate a typical proteoglycan monomer structure from cartilage and the composition of various carbohydrate chains linked to a protein core. Note that the GAG chain keratan sulfate and the so-called O-linked oligosaccharides are linked to the core by a covalent bond between serine or threonine residues and *N*-acetylgalactosamine (GalNAc), whereas the GAG chain of chondroitin sulfate forms a linkage between serine and xylose, and the so-called N-linked oligosaccharides are bound by an asparagine–GlcNAc linkage. The monomer proteoglycan molecules are linked noncovalently to hyaluronic acid through a "link protein" of about 50,000 MW to form a complex, cross-linked meshwork.

Proteoglycans are large molecules. For example, a typical cartilage proteoglycan core protein has a molecular weight of around 250,000 and is about 300 nm in length, placing it among the largest of the matrix proteins. The GAG chains extend out from the core protein, giving a "bottle brush" appearance by electron microscopy, because the repeating disaccharide backbones are stiff and contain large amounts of charge-repelling negative groups (e.g., carboxyl, sulfate, and sialic acid).

As is typical of protein–carbohydrate molecules, the proteoglycan core protein is biosyn-

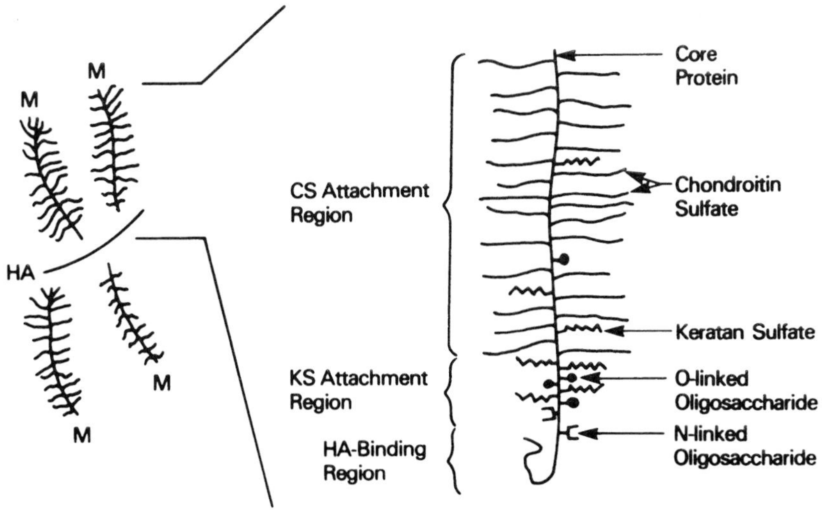

Figure 5-25 Cartilage proteoglycan, monomer structure. Four distinct monomers (M) with extended glycosaminoglycan (GAG) chains are associated with a central strand of hyaluronic acid (HA), as is shown schematically in the line drawing to the left. The chemical structure of a typical cartilage proteoglycan monomer is shown schematically at the right. (From Hascall and Hascall.[233])

thesized in the rough endoplasmic reticulum (RER) of secreting cells. While in the RER, N-linked oligosaccharides are added cotranslationally. The nascent molecules are then translocated to the Golgi apparatus (perhaps by means of a receptor-mediated event involving specific "shuttle" vesicles), where the O-linked oligosaccharides and GAG chains are added. The molecules are then packaged into secretory vesicles and released by exocytosis to the cell surface. The secreted proteoglycans are organized extracellularly into highly specific aggregates with hyaluronic acid and collagen (which has been produced, processed, and secreted by similar mechanisms).

The function of proteoglycans varies from tissue to tissue. Proteoglycans occupy a large hydrodynamic volume and are reversibly compressible. This characteristic is crucial for weight-bearing tissue such as articular cartilages, in which collagen fibrils provide a network of tensile strength, and the interspersed proteoglycans provide a viscous gel that absorbs compressive forces.

The proteoglycan–collagen matrix in other tissues varies depending on the function. In epithelial tissues, the basal lamina contains an array of proteoglycans in the so-called laminae rarae. These proteoglycans also contain the GAG heparan sulfate (repeating disaccharide unit = glucuronic acid sulfate–galactosamine disulfate) linked to core protein. The intactness of the basement membrane and the composition of its proteoglycans are important for the development of epithelial tissues. Heparan sulfate–containing proteoglycans are also found anchored to the plasma membranes of most if not all cell types. These molecules remain attached to plasma membranes by receptors for the proteoglycans and appear to be important for cell–cell adhesion.

The cell surface–associated heparan sulfate proteoglycans (HSPGs), in contrast to the ECM-associated forms, are very diverse. There are two general families of cell surface HSPGs: a *syndecan*like family of four distinct types, which are cell membrane–spanning, and a type linked to the cell surface via a covalent linkage to glycosyl phosphatidylinositol (reviewed in Ref. 234). The different families of HSPGs have different metabolic fates and highly regulated spatiotemporal expression that correlates with their role in development. For example, syndecan is expressed on many embryonic epithelial tissues, is lost from the epithelial cell surfaces during early morphogenesis, is reexpressed in later morpho-

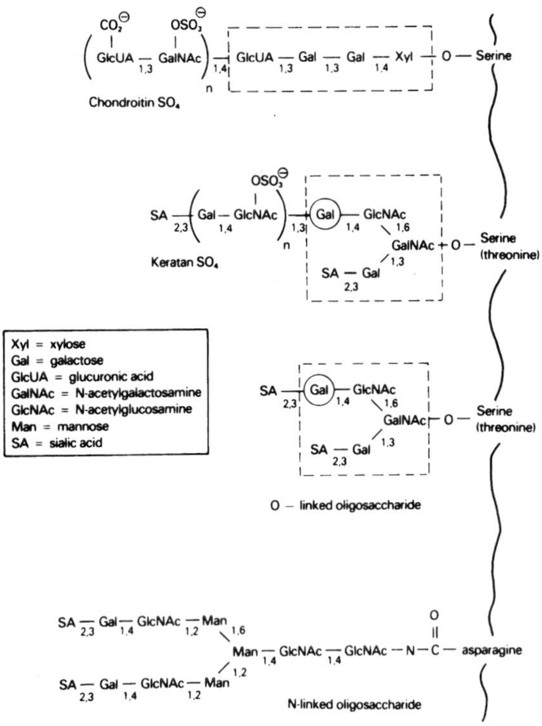

Figure 5-26 Structures of the complex carbohydrates attached to cartilage proteoglycans. The repeating disaccharide structures of chondroitin sulfate and keratan sulfate are indicated along with the specialized attachment regions by which the chains are covalently bound to the core protein (*shown in the boxes*). The O-linked oligosaccharides are related to the linkage region structure for keratan sulfate, as indicated by the circled, similarly located galactose residues. The exact structures of the N-linked oligosaccharides for proteoglycans remain to be determined, but they are related to the structure shown, which is common for glycoproteins. (From Hascall and Hascall.[233])

genesis, and is lost again during terminal differentiation.[234] Fibroglycan, another cell membrane-spanning HSPG, is expressed only on the surface of mesenchymal cells and persists during development into connective tissue.[234]

The ECM-associated and cell surface–associated HSPGs are antigenically distinct,[235] and HSPG synthesis can be induced by growth factors such as TGF-β and basic FGF; but different HSPGs are subject to different types of transcriptional and posttranscriptional regulation.[234,236]

The HSPGs bind an impressive array of growth factors, ECM components, protease inhibitors, and enzymes, which is indicative of their important role in development, terminal differentiation, tissue remodeling, and wound healing.[234] Indeed, the HSPGs are essential cofactors in cell–ECM interactions, control of growth factor activity, and regulation of protease activity.

Role of the Extracellular Matrix in Cell Migration, Cell Proliferation, Differentiation, and Cancer

In the early embryo, cellular migration plays a major role in organogenesis (for review, see Ref. 237). Such is the case, for example, with primordial germ cells, which arise in the embryo some distance from the location of the early gonad and must reach their destination by interstitial migration through other cell layers. Similarly, the development of hematopoietic organs and of the nervous system requires migration of cells through the embryo. The migratory patterns of these early committed cell types can be followed by specific markers that distinguish them. For example, neural crest cells have high levels of the enzyme acetylcholinesterase, which can be used to identify the location of these cells. A cell surface marker peculiar to hematopoietic precursor cells can be used as a marker for that tissue.

The role of cellular adhesion and ECM components has been studied in some detail for the developing nervous system in avian species. Primordial neural crest cells have a neural cell adhesion molecule (N-CAM) on their surface that is lost as the neural crest cells proliferate and initiate migration. In so doing, they also lose their epithelial arrangement and rupture the basement membrane to which they were attached. The mechanism for this is not clear but appears to involve the local release of proteolytic activities such as plasminogen activator. The migrating cells send out filopodia and penetrate the adjoining ECM, which is rich in hyaluronic acid at this stage. The hyaluronic acid-rich ECM seems to provide the substratum for cell migration because it is inhibited by hyaluronidase introduced into the cell-free space through which the neural crest cells migrate. The ECM at this time also contains a large amount of fibronectin and smaller amounts of types I and III collagens. Laminin and type IV collagen are found only in the basement membranes limiting the pathway of crest cell migration. Neural crest cells have been shown to attach and spread on fibronectin, but only poorly on laminin and collagen. Interestingly, both hyaluronic acid and fibronectin decrease significantly in the ECM surrounding the neural cells once the cells reach their destination and aggregate to form ganglia, and, at the same time, the cells again start to produce the adhesion molecule N-CAM. Thus, there must be some complex interaction of the primordial neural crest cells with their microenvironment that allows a carefully controlled switching of gene expression as the cells proliferate, migrate, and take up a new residence in the target tissue.

Neural crest cells migrate in a direction away from the neural primordium; this has been postulated to be a result of one or more mechanisms[237]: *chemotaxis,* a response of the migratory neural crest cells to a chemoattractant released by the target tissue; *haptotaxis,* a mechanism based on the affinity of crest cells for the ECM substrate on which they move (presumably directionality would be provided by a change in the ECM along the route or an increasing affinity of the cell surface for the ECM as the cells shed or gain some membrane component); and *contact guidance,* a mechanism involving the movement of cells through some preexisting "tunnels" or tracks in the ECM. However, another key factor seems to be the population pressure in the expanding neural cell mass that drives proteins into the migratory pattern. Quite remarkably, the migrating cells exhibit a selective, "controlled invasion" of mesenchymal tissues. For example, in an in vitro system, the migrating crest cells will invade developing gut wall mesenchyme but not limb bud mesenchyme.

Several features of this process remind one of the malignant neoplastic process—namely, invasion of surrounding tissues and "metastatic" spread to other areas of the organism. The difference is that in the developing embryo this is a controlled process with a finite endpoint, carefully limited by switching on and off at the appropriate times whatever genetic mechanisms drive the cells' behavior. In malignant cancer, these same processes occur, but in an uncontrolled manner. It will be most instructive to learn in more detail how these controlled genetic switches work, since everything we learn about the biology of cancer cells suggests that they take advantage of normal cellular processes, or at least of processes that are normal for some stage of development. A case in point is the cellular oncogenes (see Chap. 7) that now appear to have been present in eukaryotic evolution all the way back to yeast and that have been "captured" by certain oncogenic viruses and used to transform infected cells.

Although neural crest cells do not invade blood vessels, some normal embryonic cells do. One example is the primordial germ cells, which have been shown in avian embryos to travel from the primordial germinal crescent, via the embryonic bloodstream, to the gonadal primordium. A second example is the migration of embryonic hematopoietic cells from their tissue of origin to the newly developing blood-cell-forming organs, such as the thymus and bone marrow. This process appears to involve both chemotactic factors and cell–cell recognition markers. A chemotactic factor has been isolated from the embryonic avian thymus.[238] In addition, entry of lymphocytes into lymph nodes occurs through specific cell surface recognition markers located on the surface of endothelial cells of the venules lining the nodes. An antibody prepared against this marker (MEL-14) blocks the homing of lymphocytes into lymph nodes in vivo.[239]

The ability to invade blood vessels is also a cancer cell-like property. Again, it is intriguing that this is a "normal" property of certain embryonic cells at a certain point in their differentiation. Moreover, cell recognition signals may also be involved in the metastatic process because metastasis in a number of cases seems to involve more than a random process and actually to be organ site–specific for subclones of cells from some tumors, that is, there appears to be some homing mechanism involving cell recognition signals between tissues and tumor cells (see Chap. 11).

The ECM also exerts a permissive effect on cell proliferation, as evidenced by cell culture studies. For example, when bovine ovarian granulosa or adrenal cortical cells are grown on plastic culture dishes, they will not replicate even in the presence of serum, and their proliferation is totally dependent on the addition of fibroblast growth factor (FGF).[240] When these same cell types are grown on an ECM extracted from corneal endothelial cells, however, they proliferate in low serum and without added FGF. Since the cells plated on plastic dishes do not respond to factors present in serum or plasma but do respond when grown on ECM, it was postulated that close contact of the cells with the ECM restores their sensitivity to growth factors present in serum or plasma.[240] The mechanism of this is not clear but may involve an effect of the matrix on cell shape or morphology such that the cell surface receptors for growth factors are more expressed or more readily available for binding of the factors. This idea is supported by the findings that vascular endothelial cells can assume different shapes, depending on the substratum on which they are grown, and that attached, flattened cells respond to serum factors whereas spheroidal cells do not.[241] Another point to be considered is that growth factors bind avidly to ECM; thus the ECM itself may serve as a reservoir of growth-promoting factors.

In 1953, Grobstein[242] proposed the idea that epithelial cell–mesenchymal interactions are important in cell differentiation. This has since been borne out by numerous studies. Tissue-specific ECMs have been found to enhance cell survival and better maintain differentiated functions in cultures of liver, pancreatic islet cells, and prostatic epithelial cells of the rat[243]; in mouse mammary gland[244,245]; and in human keratinocytes.[245] Other examples of ECM-induced changes in cell differentiation include peritoneal macrophages, pheochromocytorma cells, endothelial cells, bone cells, and myoblasts (reviewed in Ref. 246). An example of how expression of tissue-specific functions by cultured cells requires growth on the appropriate matrix is the expression of liver-specific proteins such as albumin, alpha$_1$-antitrypsin, and cytochrome P-450 by cultured hepatocytes, which is favored by growth on a type IV matrix containing type IV collagen, laminin, and liver derived heparan sulfate-containing proteoglycans. Reid et al.[243] have shown that this increase in expression of liver-specific genes involves formation of gap junctions between the cells and an increased transcription and stabilization of the mRNAs-coding for these proteins. Interestingly, if normal hepatocytes are grown on type I collagen alone, they begin to "fetalize" within a few days and to express the more embryonic phenotype characterized by production of alpha-fetoprotein and carcinoembryonic antigen. Moreover, hepatoma cells respond to growth on normal type IV matrix by expressing a more normal phenotype, but this is only a transient response compared to that of normal hepatocytes. This may be because the tumor cells degrade matrix factors required for normal gene expression.[243] It should be noted that proliferation of normal hepatocytes also requires the addition of epidermal growth factor (EGF), insulin, glucagon, prolactin, growth hormone, and linoleic acid, whereas the replication of hepatoma cells occurs with a much reduced or no requirement for such factors. Clearly, there is a synergism between hormones or growth factors and the ECM in regulating cell proliferation and differentiation. The requirements for growth factors depends on the type of substratum on which the cells grow. Thus, hormones and growth factors do not operate alone in regulating cell function but act in an integrated fashion with ECM components (see below).

Another difference between normal and tumor cells is the requirement for tissue-specific ECM. Whereas normal cells require a tissue-specific ECM for optimal growth and expression of differentiated gene functions, tumor cells often do not, and will grow equally well on ECM from a number of tissues.

Another example is the differentiation of mouse mammary cells in culture that can be induced to produce the differentiation-specific markers β-casein and whey acidic protein.[245] The production of these proteins is dependent on the type of ECM on which the mammary cells are grown. It is intriguing that the ECM-mediated induction of β-casein is a gene transcription event that works in concert with the response to lactogenic hormones.[245] This and similar experiments[246] make two important points clear: (1) the ECM can have a direct effect on gene transcription via triggering some intracellular signaling system, and (2) exogenous factors such as hormones and growth factors can act as cofactors with ECM components in inducing gene transcription.

How this might occur is depicted in Figure 5-27. The ECM components act as "ligands" (lig ECM) by binding to their cell surface receptors (called integrins, see below), triggering a trans-

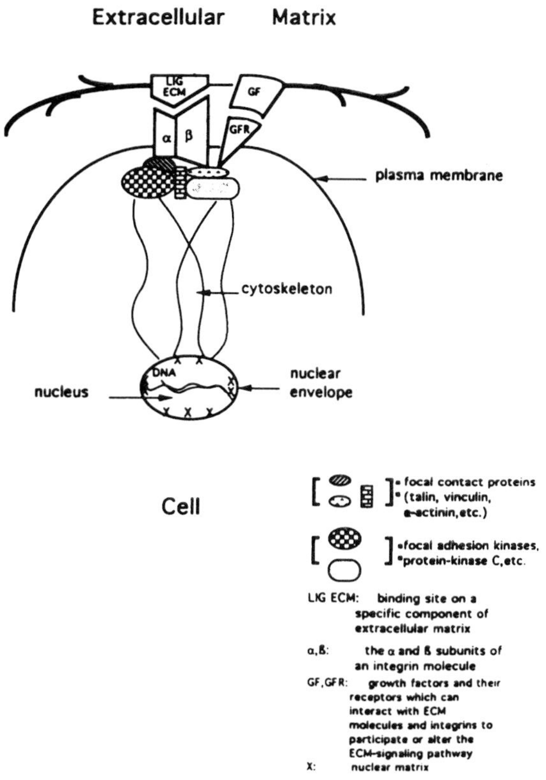

Figure 5-27 General hypothesis for the mechanism by which extracellular matrix regulates gene expression and differentiation. Extracellular molecules bind to their specific receptors (here shown as an integrin), triggering changes in the cytoplasmic domain of the receptor, which in turn causes appropriate assembly of the focal contact proteins and multiple phosphorylation of other intracellular components. These changes can bring about rearrangement of the cytoskeleton, which by connecting with the nucleus can trigger differential interaction of the chromatin with the nuclear matrix or differential activation of transcription factors. (From Lin and Bissell.[245])

membrane signal transduction event that most likely involves protein kinase pathways and alteration of cytoskeletal proteins,[246] and leading to activation of gene transcription in the nucleus. Growth factors and hormones can impinge on the same pathway via their receptors, producing a synergistic effect on the signaling pathway. In addition, ECM components can bind and act as a storage depot as well as a "presentation vehicle" for growth factors. For example, fibronectin, thrombospondin, collagen type IV, and HSPGs have all been shown to bind TGF-β,[247] which also stimulates the production and deposition of ECM components and acts to regulate proliferation of epithelial tissues by this mechanism.

The role of the ECM in metastasis is discussed in some detail in Chapter 11. It should be mentioned here that the integrity and function of the ECM is crucially involved in determining the metastatic potential of malignant cells. As suggested earlier, the ECM is usually the first barrier that cancer cells must pass through to invade lymphatic channels or blood vessels. Cells with high metastatic potential secrete lytic enzymes—for example, proteases, glycosidases, heparanases, and type IV collagenase. Second, the ability of metastatic cells to recognize and bind to the basement membranes of tissues in which these cells are about to set up housekeeping appears to be a key to their ability to select metastatic sites. This may be because metastatic cells have empty or overexpressed receptors for ECM components such as laminin[248] or because these cells synthesize an ECM component or components that provide attachment to capillary endothelial and other cellular surfaces.[249] In any case, there does appear to be some tissue specificity to the metastatic sites selected by certain tumor cell types.

Integrins

The integrins are a family of heterodimeric cell surface proteins, consisting of non-covalently associated α and β subunits, that act as receptors for a wide variety of ECM components and mediate both cell–substratum and cell–cell adhesion.[250-252] These receptor molecules all have a large extracellular ligand binding domain, a membrane spanning region, and a cytoplasmic domain that makes contact with the signal transduction machinery in the cytoskeleton. There are at least 14 distinct α subunits and 8 or more β subunits that can associate in various combinations that determine the ligand binding specificity for various ECM proteins such as laminin, fibronectin, thrombospondin, and collagen. Individual cells may have multiple integrins that bind a particular ECM protein, and some integrins may recognize and bind different regions of the same ECM protein, suggesting some redundancy in the response network. However, there is evidence for sequence divergence of the intracellular domains of the α subunits and for

Table 5-3 Cell–Cell Adhesion Molecules

Receptor	Family	kDa	Distribution	Ligand	Function
L-PAM-1	Integrin α4m/βp	160/130	Lymphocytes	?	Lymphocyte–EC adhesion
L-PAM-2/VLA-4	Integrin α4/β1	160/130	Leukocytes, melanoma cells	VCAM-1 INCAM-110	Lymphocyte–EC adhesion, Tumor–EC adhesion
CD11/CD18 (LEU-CAMs)	Integrin	Heterodimer			Cell–cell adhesion
LFA-1(CD11a)	α_L/β_2	177/95	All leukocytes	ICAM-1	Leukocyte–leukocyte adhesion, leukocyte–EC adhesion
MAC-1(MO-1, CR3, CD11b)	α_m/β_2	165/95	Neutrophils, monocytes, some lymphocytes	iC3b, Fibrinogen Factor X, ICAM-1, LPS	Cell–cell adhesion, phagocytosis, complement binding
gp150/95 (CR4, CD11c)	α_x/β_2	150/95	Granulocytes, monocytes	iC3b	Complement binding
I-CAM-1	IgG superfamily	76–114	Many cells; induced by inflamation	LFA-1	Leukocyte adhesion
CD2 (LFA-2, OKT11)	IgG superfamily	40	All T cells	LFA-3	T cell adhesion
LFA-3	IgG superfamily	50–77	Widespread	CD2	T cell adhesion
EndoCAM, PECAM-1 (CD31)	IgG superfamily family	130	EC, platelets, leukocytes	?	Initiation of EC–EC adhesion, platelet–monocyte–EC adhesion
VCAM-1	IgG superfamily	90	Activated EC	VLA-4 (α_4/β_1)	Adhesion of WBCs to EC activated by IL-1 or TNF
Ca^{2+}-Independent CAMs	IgG superfamily				Developmentally regulated cell–cell adhesion
N-CAM (D2, BSP-2)		180, 140, 120	Brain, muscle, heart, kidney	Homophilic binding	
Ng-CAM (L1, NILE)		200	Neural and glial cells	Homophilic binding (neural-neural) Heterophilic binding (neural-glial)	

alternative splicing of α subunits in a tissue-specific and developmentally regulated manner, suggesting that different integrins that bind the same ECM ligand may in fact transduce different intracellular signals.[246] One thing is clear, though: the integrins are the conduit for communication between the ECM and the intracellular environment.

Alterations in integrin receptor expression have been observed in chemically transformed human cells,[253] in human colonic cancer,[254] and in human breast cancer.[255] Also, modulation of the expression of the integrin receptor for laminin by estrogen and progestins has been seen in human breast cancer cell lines.[256]

Cell–Cell Interactions

The importance of cell–cell interactions in cell differentiation processes has been alluded to for hepatocyte cultures, in which it was observed that factors favoring formation of gap junctions between cells stimulate expression of tissue differentiation-specific markers. Studies using the transfer of the fluorescent dye Lucifer yellow between cells in developing *Xenopus laevis* embryos bear directly on this point.[257] At the 32-cell stage of development it was found that dye transfer via gap junctions occurs more between cells in one position of the embryo than in others. Furthermore, a progressive restriction of

Table 5-3 *(continued)*

Receptor	Family	kDa	Distribution	Ligand	Function
Ca^{2+}-dependent CAMs	Cadherin			Homophilic binding	Developmentally regulated cell–cell adhesion
N-Cadherin (A-CAM, N-Cal-CAM)		135	Brain, muscle, lens		
E-Cadherin (L-CAM, Arc-1, Uvomorulin, Cell CAM 120/80)		120–125	Epithelium		
P-Cadherin		130	Placenta, epithelium, mesothelium		
H-CAM (CD44, Hermes antigen, ECMR III)	Cartilage link protein	90	Lymphocytes	Endothelial cell muscosal addressin (MECA 367 Ag)	Homing RTR-lymphocytes to HEV cell–substratum adhesion RTR
$gp90^{MEL-14}$ (murine) Leu 8 (human)	Lec-CAM	90	Lymphocytes	HEV carbohydrate?	Homing RTR-lymphocyte to HEV of peripheral LNs
$gp100^{MEL-14}$	Lec-CAM	100	Neutrophils	Endothelial cell, carbohydrate?	Initial adhesion of neutrophil to EC adjacent to inflammation
ELAM-1	Lec-CAM	115	Activated EC	?	Adhesion of WBCs to EC activated by LPS, IL-1, or TNF
GMP140 (PADGEM)	Lec-CAM	140	Platelets, neutrophils, monocytes, EC	?	Adhesion of activated platelets to monocytes and neutrophils Adhesion of WBCs to activated EC
CD36 (GPIV)	?	88	EC, platelets, monocytes	Thrombospondin	Platelet–monocyte adhesion

Source: From Albelda and Buck.[252]

dye transfer occurs from the 16-cell through the 32- and 64-cell stages, followed by an increase in the early blastula stage. Injection of an antibody to gap junction protein at the 8-cell stage blocks dye transfer and prevents electrical coupling of cells.[258] Moreover, the injected embryos have substantial morphologic defects in the brain and eyes. This is not a result of cell death, as no differences in cell viability were observed, nor was it a nonspecific effect because injection of preimmune serum or unrelated antibody were without effect.

Cell–cell interactions are mediated by a family of molecules called cell adhesion molecules or CAMs, which act as both receptors (on one cell) and ligands (for another cell). The expression of CAMs is programmed during development to provide positional and migratory information for cells. Some CAMs require calcium for their function (the cadherins) and others do not (e.g., N-CAM). Cadherins also appear to play a key role in maintaining structural integrity and polarity, as well as in formation of gap junctions, in adult epithelial tissues.[252] A number of cell–cell adhesion molecules are shown in Table 5-3.

GROWTH AND DIFFERENTIATION FACTORS

It is clear that other key components in the microenvironment that determine and direct the

Table 5-4 Some Factors Promoting Growth and/or Differentiation of Specific Cell Types

Cell Type	Factor
Pancreas	Mesenchymal factor
Erythroid cells	Erythropoietin
Granulocyte–monocyte precursor cells	Colony-stimulating factors (e.g., GM-CSF)
Megakaryocyte precursor cells	Thrombopoietin
Thymus-dependent lymphocytes	Thymosin
Mammary gland	Estrogen, progesterone, prolactin
Ovary	Follicle-stimulating hormone
	Luteinizing hormone
	Estrogen
	Progesterone
Neural tissue	Nerve growth factor (NGF)
Epidermis (and other epithelial tissues)	Epidermal growth factor (EGF)
Fibroblasts	Fibroblast growth factor (FGF), platelet-derived growth factor (PDGF), insulin
Endothelial cells	Angiogenesis factors
Many cell types	Insulin-like growth factors (IGF)

differentiation of specific "committed" cell types are growth and differentiation factors. These are often polypeptides of relatively low molecular weight (6000 to 15,000 Das), produced and secreted by tissues distant from the target organ, although they may be produced by cells in the target organ itself, or even by the target cells themselves. The terms endocrine, paracrine, and autocrine, respectively, are applied to these three different cases. Steroid hormones (e.g., estrogen, progesterone) and larger glycoprotein hormones (e.g., follicle-stimulating and luteinizing hormones) also act as growth and differentiating factors for certain tissues. Examples of some growth/differentiation factors are listed in Table 5-4. These factors are discussed in more detail in Chapter 8.

IONIC AND PH EFFECTS

It has been known for some time that mitogenic effects of growth-stimulatory agents are accompanied by a transient increase in monovalent cation influx into cells and a transient increase in intracellular pH in a wide variety of cell types, including human cells. In addition, divalent cations, especially calcium, are not only important for cell adhesion and cell attachment to the substratum but also affect cell behavior. Extracellular calcium concentration has been shown to be an important determinant of cell proliferation and differentiation in cell cultures of murine or human epidermal keratinocytes.[257] In the presence of 0.02 to 0.09 m*M* Ca^{2+}, keratinocytes proliferate rapidly in monolayer cultures, but when the extracellular Ca^{2+} concentration is raised to 1.2 m*M*, cell-to-cell contact occurs and proliferation ceases; the cells also express skin differentiation markers, become stratified, and produce keratin. Carcinogen-transformed mouse skin cells, however, are resistant to the induction of differentiation by high Ca^{2+}. Calcium ions are also essential for the fusion of chick myoblasts (undifferentiated muscle cells) into myotubes in muscle differentiation.[258] The precise molecular mechanisms involved in these Ca^{2+}-regulated effects are not known, but they may be modulated by a 17,000 molecular weight Ca^{2+}-binding protein called calmodulin that is present in most cells. A calmodulin–calcium complex is involved in the activation of several enzyme systems, including protein kinase, phosphodiesterase, and adenylate cyclase, which controls cellular levels of 3′,5′-cyclic AMP.

Intracellular pH (pHi) also plays a role in control of cell differentiation in multicellular organisms. "Arousal" from dormancy to activity in sea urchin eggs is regulated in part by a moderate alkalinization of intracellular pH (about 0.4 pH unit). Even larger increases in pHi accompany the transitions between dormancy and active development in the brine shrimp *Artemia salina* (about 1 pH unit).[259] This change in pHi appears to be the fundamental regulator of the transition between dormancy and activation of embryonic development. Similarly, differentiation of the slime mold *Dictyostelium discoideum* is inhibited by agents that block the ATP-dependent proton "pump" (Na^+/H^+-ATPase) and stimulated by agents that increase intracellular pH.[260] The participation of protons and the key role of proton pumps in energy metabolism suggest that pHi may be a primitive and fundamental signal in the regulation of cell proliferation and differentiation.

Thus, one can visualize how local gradients of metal ions and H^+ in the microenvironment could affect the differentiation potential of cells in selected areas of the developing embryo and how this could provide some of the "polarity" effect seen in developing organisms, that is, how a cell in one area could behave differently from one in another.

CYCLIC NUCLEOTIDES

Cycle AMP has an important role as an intracellular signal in cell proliferation and differentiation. Suffice it to say here that cAMP appears to be important in the differentiation of certain cell types. One interesting model illustrating the effect of cAMP on cell differentiation is the slime mold *D. discoideum.* This organism starts its life cycle as individual ameba that feed on bacteria. When this food source is no longer available, the amebas aggregate to form a multicellular organism, which ultimately differentiates into a fruiting body. Coordinated cell movement is essential to both aggregation and formation of the fruiting body. Aggregation is stimulated by a chemotactic phenomenon. A "pacemaker" cell begins to secrete cAMP, which has been identified as the chemotactic factor.[261,262] Surrounding cells are drawn to the signaling cell and, after contact has been made, start to secrete the chemoattractant themselves in a pulsatile manner, attracting more cells to the aggregated core, which in turn begin to produce cAMP in a "relay" effect that results in spiral waves of inward cell movement to the aggregating core of cells. After aggregation is completed, the fruiting body is formed.

These effects have been shown to be mediated by receptors for cAMP, the levels of which are regulated by secreted cAMP. Expression of this receptor, CAR1, is induced during early development and is dependent on low level oscillations of cAMP.[263] After aggregate formation, there is an increase in extracellular cAMP levels and this downregulates CAR1 expression.

Cyclic AMP also appears to be involved in the expression and maintenance of several differentiated functions of cultured cells from higher organisms. For example, dibutyryl cAMP stimulates neurite formation in neuroblastoma cells,[264] collagen synthesis in Chinese hamster ovary cells,[265] and the appearance of a variety of phenotypic changes characteristic of a more differentiated phenotype in transformed cells.[266] In cultured chick myoblasts, there is a burst of cAMP production just prior to cell fusion,[267] one of the first steps in the differentiation of myoblasts into mature muscle cells.

Cellular responses in the slime mold to the cAMP signal include an increase in intracellular cGMP, a transient increase in Ca^{2+} uptake, a decrease in extracellular pH, activation of adenylate cyclase, and an increase in cAMP-dependent protein kinase. It is generally assumed that the intracellular effects of cAMP in eukaryotes are mediated by cAMP-dependent protein kinases. These, in turn, phosphorylate cellular proteins, including nuclear proteins, that regulate gene expression. This signal transduction mechanism will be discussed in detail in Chapter 8.

PROLIFERATION VERSUS DIFFERENTIATION

An organism could not develop, of course, without vigorous cell replication. Nor could it survive without continued cell division. However, there is normally a well-controlled balance between cell division, cell differentiation, and cell death. It is this delicate balance that is disrupted in cancer tissue. Cell differentiation usually produces, ultimately, a cell that no longer has the capacity to divide, but many cells in the process of differentiating continue to divide. Hence, the two processes are not mutually exclusive.

The period following fertilization up to late blastula is a period of intense cell division with very little cell growth between mitoses. In amphibians, for example, the cell cycle prior to the blastula stage is abbreviated; there is no intermitotic G_1 phase and G_2 is short.

During blastulation the G_2 period is prolonged, and a short G_1 appears. Thereafter, the G_1 period between S (DNA replication) phases lengthens until the tail-budding stage is reached, at which time the cell cycle approaches that of adult proliferating cells. The cell cytoplasm appears to exert control over the timing of nuclear DNA synthesis, since nuclei from adult tissues in which DNA synthesis is rare can be induced to synthesize DNA by injection into unfertilized ova.

In mammalian cells, also, growth arrest is coordinated with expression of the differentiated phenotype, for example, in hematopoietic cells and in epithelial cells of the skin and gastrointestinal tract as well as in such cell culture systems as the preadipocyte mouse 3T3 lines that can be induced to differentiate into fat cells.[268] Cells transformed by carcinogenic agents or oncogenic viruses lose this ability to become growth arrested and to become terminally differentiated.

REVERSIBILITY OF DIFFERENTIATION

It is generally agreed that the malignant neoplastic cell is less differentiated than the normal adult cell in the organ from which the cancer originates. However, for a long time there has been a controversy among cancer biologists on the question of whether neoplastic transformation results from dedifferentiation of normal adult type cells or from the abnormal differentiation of committed but not yet mature stem cells in tissues. It can be concluded, however, that malignant neoplastic transformation occurs only in cells that are capable of dividing. Hence, brain tumors are not likely to arise from a mature neuron, nor are leukemic cells likely to arise from a terminally differentiated polymorphonuclear leukocyte. It is more likely that the dividing, committed stem cell of a given tissue would be most affected by oncogenic agents. However, it is difficult in some tissues to define the stem-cell population. There are clearly identifiable stem cells in the bone marrow, skin, and gastrointestinal epithelium, but not in the liver or brain, and yet it is well known from animal studies that partial removal of the liver will lead to the initiation of DNA synthesis and partial regeneration of the hepatic parenchyma. Obviously, some cells in the liver cell population can respond to a mitotic stimulus (hepatectomy) by initiating DNA synthesis and cellular replication. Do these cells undergo dedifferentiation to a more primitive cell type, or does a small, and histologically ill defined population of less mature liver parenchymal cells respond? And which of these mechanisms would be involved in the generation of a hepatocellular carcinoma after exposure to a carcinogenic agent? To answer these questions, it would be helpful to know if dedifferentiation actually occurs under normal circumstances.

The answer is that, in general, commitment to a given pathway of cellular differentiation appears to be irreversible, but the steps of terminal differentiation may be reversible. An example of the conservation of cellular commitment to a differentiation pathway is provided by the experiments of Yaffé and his colleagues, who showed that muscle cells converted into a permanently dividing line of undifferentiated myoblasts by treatment with the carcinogen 3-methylcholanthrene can pass through as many as 600 population doublings and still, after appropriate manipulation, fuse into multinucleated cells and differentiate into striated muscle cells.[269] Another example comes from the work of Hadorn,[270,271] who demonstrated by experiments with fruit flies that the imaginal disks, the undifferentiated primordial structures of the larvae from which the tissues of the adult fly develop, can be grafted onto the abdomen of adult flies, where they continue to grow but do not differentiate. The imaginal disks can be transferred from fly to fly and carried for a number of years in this fashion. However, if parts of the disk are grafted into a larva, they will differentiate during metamorphosis into different parts of the adult fruit fly, indicating that a commitment to a particular pathway of differentiation is maintained through many cell generations.

Reversibility of terminal differentiation, however, has been demonstrated. In certain amphibians, the lens of the eye can be produced by cells of the iris.[272] If a piece of the dorsal part of the iris is grafted into an eye from which the lens has been removed, the iris cells undergo cell division leading to the formation of one pigmented and one nonpigmented cell. The nonpigmented cells divide to produce a lens. Another example from amphibians is the complete regeneration of the retina of an eye from pigmented retinal cells that dedifferentiate, divide to form unpigmented cells, and then reform an entire new retina.[273] A somewhat similar example from mammals is the stimulation of small peripheral blood lymphocytes by antigens or certain plant substances called lectins. Certain lectins (e.g., concanavalin A and phytohemagglutinin) can stimulate mature-looking, nondividing T lymphocytes to divide to form larger, less mature lymphoblasts. The end result, however, of these events is a cell type that still has the differentiated properties of T cells. Thus, the chain of events is similar to those given for amphibians; that is, a mitogenic stimulus leads a nondividing mature cell to dedifferentiate in the sense that it is converted to an actively proliferating cell. After a certain number of mitoses, some of the progeny will differentiate into mature functional cells, leading to replacement of the tissue (e.g., lens) or to a functional response (e.g., T cell functions). Whether or not a similar dedifferentiation can occur in the tissues of animals and humans as a result of exposure to carcinogenic agents is not clear, but the fact that it can occur in certain organisms suggests that this is a possibility. At any rate, it is clear that the end result of carcinogenesis is usually the generation of a less mature cell type. Thus, if similar events are involved in carcinogenesis, they would differ from normal tissue regeneration in that cellular maturation does not occur.

INDUCTION OF DIFFERENTIATION IN CANCER CELLS

A cancer is made up of cells derived from normal tissues of the body. A crucial difference between cancer cells and normal cells is the lack of normal differentiation. As noted above, a key question is how does this come about—is it due to the dedifferentiation of adult cells or to a blockade of differentiation of undifferentiated cells? The answer to this question has important implications for cancer diagnosis and treatment. For example, the earliest carcinogenic changes observed in an adult differentiated cell would be expected to produce a cell with altered, but very similar, characteristics compared to those of the normal adult cell. Thus, the initial differences between normal and carcinogen-altered cells in metabolism, enzyme patterns, cell surface architecture, and other aspects, might be very subtle. Because cancer diagnosis, as well as the design of agents to selectively destroy cancer cells, depends on differences between normal and cancer cells, the strategies developed for diagnosis and treatment would be somewhat different for this situation than if the earliest malignant change occurred in a more primitive, proliferating stem cell type. Although it is clear that cancer cells ultimately become less differentiated than their adult counterparts, it is not clear how this occurs. Pierce and his colleagues have championed the idea that normal tissue stem cells, those with renewal capacity, are the cells of origin of malignant tumors.[274] This concept holds that cancer is an abnormality of tissue renewal resulting in the production of a partially differentiated cell type with unlimited proliferative capacity. However, this abnormality is not necessarily an irreversible or complete blockade of differentiation. A number of examples are known of malignant cancer cells that can further differentiate and, in some cases, even become nonmalignant. If Pierce's hypothesis is correct, it would be logical to search for the mechanism of carcinogenesis among the cells involved in tissue renewal. The only problem with this is that not all tissues have a definable stem-cell compartment. The liver is one such tissue although there is some evidence for a periportal "oval cell" population that may be the liver stem-cell type.[275] Since hepatocellular carcinomas involving liver parenchymal cells do occur, the hypothesis would seem to have to include the idea that a damaged hepatic parenchymal cell may become a proliferative "stem" cell capable of self-renewal and of being a target for carcinogenic agents. For instance, hepatocellular carcinomas are known to occur more frequently in livers that were previously damaged by chemical toxins or hepatitis virus.

The best evidence that the committed tissue stem cell is the target for oncogenic agents comes from studies of oncogenic virus-induced leukemias. It has been shown, for example, that the target cell for mouse Friend virus-and avian erythroblastosis virus-induced erythroleukemias is an erythroid precursor cell.[276,277] In vitro transformation of chicken bone marrow cells with the avian erythroblastosis virus (AEV) gives rise to rapidly growing cells expressing the characteristics of early progenitors of the erythroid lineage—namely, the so-called burst-forming units–erythroid (BFU–E).[278] AEV-transformed BFU–E cells further differentiate to a hemoglobin-synthesizing cell type (colony forming units–erythroid, CFU–E), at which point they become arrested in the differentiation pathway. Normally, the erythroid differentiation pathway proceeds from the committed but pluripotent marrow stem cell CFU–L–M → BFU–E → CFU–E → erythroblast → mature red blood cell. Thus, the transformed cell type can advance partway through the pathway, but, because some key signal is not perceived, it goes no further.

Another concept to explain the generation of cancer has been formulated by Potter in the phrase "oncogeny as blocked ontogeny."[279] Ontogeny is viewed as a specific time-dependent sequence of events that involves switching some genes from "off" to "on" and vice versa. Using the liver as an example, it can be seen that the phasing in of the enzymes characteristic of the adult liver occurs over a time course characteristic for each enzyme. During this programmed sequence of differentiation, the enzymes that characterize fetal liver are gradually phased out; these include the enzymes for DNA replication and cell proliferation. If a similar sequence of differentiation events occurs in damaged, replicating liver cells exposed to a carcinogenic agent, they could be blocked from completing differentiation, leaving some of the differentiation-specific genes off and some of the genes for proliferation on. The result would be a partially differentiated cell that was still capable of continued proliferation and that would not be subject to the same negative feedback controls that regulate adult cell function. Presumably, the blockade of differentiation could occur at one or more steps in the differentiation process. Thus, this hypothesis could explain the large diversity of chemical carcinogen-induced rat hepatomas, ranging from "minimally deviated," highly differentiated hepatomas to anaplastic, highly malignant varieties.[279]

The mammary gland, another organ without a clear stem cell compartment, is a common site of cancer in humans as well as in animals. As noted in Chapter 2, pregnancy, lactation, and hormonal manipulation affect the incidence of mammary cancer in women and in animals. It is interesting that during the estrous cycle in rats, a population of proliferating cells can be detected (by ^{3}H-thymidine labeling of mammary cell DNA) at the terminal end buds of the mammary ducts.[280] Two proliferating cell populations have been identified: one that synthesizes DNA preferentially at proestrus and early estrus and another, including the epithelial cells, that synthesizes DNA primarily during middle and late estrus–metestrus. Because most mammary cancers are of epithelial origin, the hormonally responsive cell types, identified by their ability to synthesize DNA during the estrous cycle, may represent the target cells for carcinogenic agents in the mammary gland.

The stem cell concept in cancer biology also has an important implication for cancer treatment. Because normal stem cells, by definition, have self-renewal capacity and are subject to two kinds of signals that induce opposing processes—that is, proliferation and terminal differentiation—one might be able to tip the balance of a tumor stem cell population away from the proliferation pathway and into the differentiation pathway if these signals and the mechanisms involved in the cellular response to them could be characterized. Another important implication for anticancer therapy is that the renewal stem cells, the so-called clonogenic cells, of tumors are the cells that must be destroyed for a cancer to be cured. Assays have been developed, based on the ability of these clonogenic cells to proliferate in soft agar outside the body, to estimate the sensitivity of various human cancers to chemotherapeutic agents (for review, see Ref. 281). Estimates of the frequency of clonogenic cells in a human tumor range from 0.001% to 1.0% of the total cell population. Within a given tumor, there is also considerable heterogeneity in the phenotypic characteristics, including growth rate, degree of expression of differentiation markers, and response to drugs. Nevertheless, it is the clonogenic stem-cell population at which curative cancer therapy is aimed. The characteristics, and indeed the fate, of most cells within the tumor (i.e., the nonrenewal cells) is of little consequence to long-term tumor growth. Why, then, is cancer so difficult to treat since the clonogenic population appears to be small? There are several potential answers to this question, and the correct ones probably vary from tumor type to tumor type—perhaps even from patient to patient; however, factors such as the ability of drugs to penetrate the tumor, development of drug-resistant clones of cells, and the lack of selectivity of most currently available chemotherapeutic agents for tumor over normal stem cells are clearly involved.

The observation that certain tumors can partially differentiate under appropriate conditions is instructive because it helps to pinpoint at what level the carcinogenic stimulus blocked further differentiation and to suggest mechanisms for overcoming the differentiation blockade. If the latter could be done by practical, nontoxic means, a great improvement in cancer therapy could result. A number of unique and interesting model systems for the study of these events are described next.

Teratocarcinoma

Teratocarcinomas are highly malignant tumors that occur in both animals and humans. The term is derived from a Greek word for monster, and in many of these tumors a variety of differentiated tissues can sometimes be observed, such as muscle, bone, hair, and skin. These tumors usually arise in the gonads and, in humans, are often metastatic by the time of diagnosis.

An animal model system was developed by Stevens and Little,[282] who first noted that testicular teratocarcinomas developed spontaneously in about 1% of the inbred mouse strain 129. The tumors were observed as early as 1 week after birth, and they contained undifferentiated proliferating cells as well as differentiated cells of a variety of tissues. When male fetuses of strain 129 were examined, it was found that the developing teratocarcinomas could be identified as early as 15 days of gestation. All these teratocarcinomas were found among the primordial germ cells of the testicular seminiferous tubules, suggesting that the cell of origin was a primordial germ cell. This was supported by experiments in which the undifferentiated, primordial genital ridges of 11- to 13-day-old male fetuses were grafted into an extrauterine site, such as the liver or kidney, of adult mice. In a high percentage of these animals, teratocarcinomas developed. When genital ridges were grafted into the testes of adult mice of the same strain, tumors developed in about 75% of the grafts.[283] Interestingly, if the genital ridges of 13-day-old or older fetuses were similarly grafted, the incidence of teratocarcinomas decreased,[284] suggesting that once the germ cells have matured to a certain point they are prevented from undergoing further neoplastic progression in the environment of adult tissues.

When teratocarcinoma cells are injected into the peritoneal cavity of strain 129 mice, large cystic bodies called embryoid bodies develop.[285] These contain carcinoma cells and differentiated cells, suggesting that both carcinoma cells and differentiated tissues can arise from a single precursor cell. This idea is supported by an experiment by Kleinsmith and Pierce,[286] who injected single mouse teratocarcinoma cells into adult mice and observed that tumors containing teratocarcinoma cells and differentiated cell types developed. More definitive proof, however, that malignant tumor cells could differentiate into normal tissues is derived from a series of experiments in which teratocarcinoma cells from a strain of mice with certain genetic markers are injected directly into blastocysts (4½-day-old embryos) of pseudopregnant foster mothers.[287,288] These blastocysts develop into complete normal mice with tissues containing the genetic markers of the teratocarcinoma cells as well as those of the blastocyst (Fig. 5-28). The progeny mice, for example, have a coat color that is a mixture of hair colors from both the tumor-derived and the normal parent strains. In addition, a variety of genetic markers, such as hemoglobin and immunoglobulin type, isoenzyme patterns, and skin pigmentation patterns, also show that these mice are genetic mosaics (chimeras) containing genetic information from the teratocarcinoma cells as well as from the normal blastocyst cells. Surprisingly, none of the progeny mice develop teratocarcinomas, indicating that the microenvironment of the blastocyst is apparently able to convert the neoplastic potential of a carcinoma cell to a normal pathway of differentiation. Additional experiments in which single teratocarcinoma cells were injected into

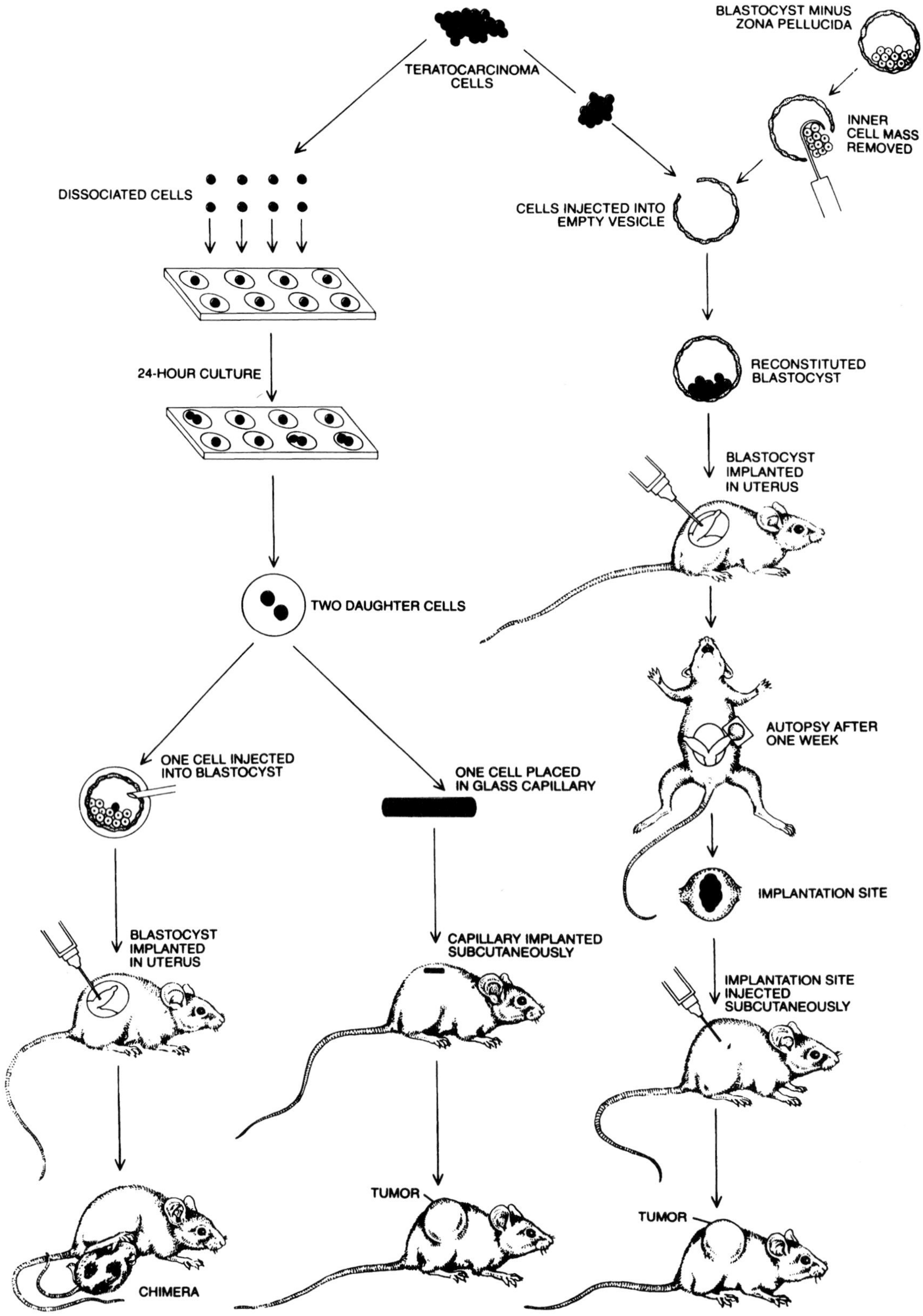

TERATOCARCINOMA CELLS
BLASTOCYST MINUS ZONA PELLUCIDA
INNER CELL MASS REMOVED
DISSOCIATED CELLS
CELLS INJECTED INTO EMPTY VESICLE
24-HOUR CULTURE
RECONSTITUTED BLASTOCYST
BLASTOCYST IMPLANTED IN UTERUS
TWO DAUGHTER CELLS
AUTOPSY AFTER ONE WEEK
ONE CELL INJECTED INTO BLASTOCYST
ONE CELL PLACED IN GLASS CAPILLARY
IMPLANTATION SITE
BLASTOCYST IMPLANTED IN UTERUS
CAPILLARY IMPLANTED SUBCUTANEOUSLY
IMPLANTATION SITE INJECTED SUBCUTANEOUSLY
TUMOR
TUMOR
CHIMERA

blastocysts showed that a single tumor cell contributes progeny cells to all the major tissues of the chimeric adult mouse.[290] Thus, the teratocarcinoma cells obviously retain the totipotency equivalent to that of normal embryonic cells. These results are not due to an alteration of the carcinoma cells during experimental manipulation because if single carcinoma cells are cultured and allowed to proliferate before daughter cells from this single carcinoma cell are injected either into blastocysts or under the skin of an adult mouse, the same experimental results are obtained—namely, those cells injected into the blastocyst develop into normal mice, whereas those injected subcutaneously produce teratocarcinomas. Thus, the embryonic microenvironment of the blastocyst apparently provides signals for normal differentiation that are not present in adult subcutaneous tissue.

The ability of embryonal carcinoma (EC) cells, the stem cells of teratocarcinomas, to differentiate into nonmalignant cell types depends on cell–cell contact between the EC cells and the blastocele surface (inner surface) of the trophoectoderm surrounding the blastocyst.[291] This induction of differentiation does not hold for all types of cancer cells, however. When murine melanoma or lymphoid leukemia cells are injected into a blastocyst prior to implantation into mice, tumors still develop[292a]; this suggests that an embryonic microenvironment or "field" appropriate to a given cell type may be required to induce differentiation of aberrant cells. The analogous normal cell types for melanocytes or lymphoid cells would not have developed yet in the early blastocyst, and thus the inductive field necessary for their differentiation is presumably not present. It is interesting that in this same series of experiments,[291] neuroblastoma cells were partially regulated and induced to differentiate by injection into a blastocyst prior to implantation. Because neurulation follows blastulation by only a few days, the inductive environment for neural tissue development could be beginning to emerge and could have a partial effect on the expression of the malignant phenotype of neuroblastoma cells.

The results of these experiments have two other important implications. First, at least some forms of cancer can develop as a result of altered gene *function*, rather than permanently altered gene *content*. The development of neoplasia in the case of the teratocarcinoma model system is clearly an example of this phenomenon. If a permanent genetic mutation had induced this tumor, it does not seem likely that a single carcinoma cell could give rise to normal adult animals. Second, these findings have important therapeutic implications. If the neoplastic state is reversible for some malignant tumors, perhaps it will be possible to design therapies that will stimulate more normal differentiation in cancer cells and bring about reversion to a nonmalignant state.

In the case of mouse teratocarcinoma cells, induction of differentiation into cell types expressing various differentiation markers has been observed. An embryonal carcinoma cell line (F9) derived from a murine testicular teratocarcinoma has been well studied in this regard. The F9 cells have been shown to differentiate after exposure to retinoic acid, cAMP analogues, hexamethylbisacetamide, or sodium butyrate.[292b,293] In another murine teratocarcinoma cell line, P19, exposure to dimethyl sulfoxide (DMSO) induces the formation of cardiac and skeletal muscle cells, whereas retinoic acid induces development of neuronal and glial tis-

Figure 5-28 Developmental potential of mouse teratocarcinoma cells. Teratocarcinoma cells are dissociated with a protein-digesting enzyme and are cultured singly. Some tumor cells divide to form two daughter cells, one of which is injected into a blastocyst of a different mouse strain. The blastocyst is placed in the uterus of a pseudopregnant "foster mother," where it gives rise to a healthy chimeric mouse (*left*) that has tumor-cell-derived tissues in the coat (*color*) and in internal organs. The other daughter cell, in a glass capillary, is implanted under the skin of an adult mouse and develops into a teratocarcinoma (*middle*), showing that a single tumor cell can either differentiate normally or remain malignant, depending on the microenvironment. To demonstrate that the presence of embryonic cells is required for the reversion of malignancy, teratocarcinoma cells are injected into a blastocyst from which the inner cell mass has been removed (*right*); the reconstituted blastocyst is placed in the uterus of a foster mother, where it implants. When the implantation site is removed and transplanted under the skin of a mouse, it forms a teratocarcinoma. (From Illmensee and Stevens.[289])

sues[294]; this suggests that different chemical signals can induce alternative differentiation programs in the same cell population. After exposure to retinoic acid, various differentiation-specific markers have been observed in F9 cells, including the extracellular matrix proteins laminin and type IV collagen, developmental stage-specific antigens, and the mouse major histocompatibility complex antigen, H-2.[295] An important factor for the rationale of using such agents to treat cancer is the observation that, when animals bearing embryonal carcinomas were treated by intratumor injection of retinoic acid for 4 consecutive days, all the animals survived for up to 16 weeks, whereas all the untreated control tumors grew and killed the host animals within 25 days. The tumors that remained in the treated animals were found histologically to be benign cystic teratomas.[296]

It should be noted here that the ability of various embryonal cell lines to be transplanted into mice creates a number of marvelous opportunities for molecular biologists to study gene expression during development and carcinogenesis. Embryonal carcinoma cells, for example, can be fused to various other cell types, including human cancer cells, and then implanted into blastocysts to find out which sets of genes are necessary to express the malignant rather than the differentiated phenotype. Single "cancer genes" can also be transferred into fertilized ova and implanted into foster mothers to create "transgenic mice"; in this process, one can try to answer questions about the role of individual genes in development of malignancy in whole animals (see Chap. 4). Finally, mutagenized genes can be transferred by the same mechanism to answer questions about how various mutations affect normal development or the development of malignancy. Indeed, the potential of these techniques appears to be limited only by the imagination of the investigator.

Myelocytic Leukemia

Leukemia results from the interruption of normal bone marrow cell differentiation, and, as noted previously, the target cell for leukemogenic agents appears to be a committed stem cell (Fig. 5-29). After it was demonstrated that normal stem cells of the neutrophil and monocyte-macrophage series can be grown in culture in semisolid agar and can be stimulated to differentiate by appropriate growth factors (see Chap. 9), similar studies were initiated with leukemic cells derived from these cell types. It has been demonstrated, for example, that cells from a spontaneously arising myeloid (granulocytic) leukemia of the SL strain of mice grow in semisolid agar as colonies of blast cells exhibiting little evidence of further differentiation.[297] However, if "conditioned" medium, obtained from cultures of normal mouse embryo cells, is added to the cultures of leukemic cells, the number of colonies increases, with a high percentage containing differentiating neutrophils and monocytes. Conditioned medium from cultures of other normal cells also does this. Interestingly, conditioned medium from cultures of leukemic cells stimulates growth of leukemic colonies, but not of normal granulocyte/monocyte-macrophage stem cell colonies, which suggests that the growth factor produced by leukemic cells might be different from that produced by normal cells. It has also been observed that leukemic cells induced to differentiate by growth in conditioned medium have a markedly reduced capacity to produce leukemia upon injection into syngeneic mice compared with leukemic cells cultured in the absence of conditioned medium.[298]

These early studies showed that soluble growth factors could induce differentiation of hematopoietic stem cells. Now, a wide variety of hematopoietic growth and differentiation factors are known, such as the colony-stimulating factors G-, M-, and GM-CSF and the interleukins (see Chap. 9). In addition to these, a number of exogenous chemicals and vitamins or vitamin analogs have been shown to stimulate differentiation of hematopoietic cells.

One of the early models used in these studies was a cultured cell line called HL-60 derived from a human acute promyelocytic leukemia (APL). Several agents have been shown to induce differentiation in this cell line, including a number of anticancer drugs, sodium butyrate, dimethyl sulfoxide, hexamethylene bisacetamide, vitamin D_3, phorbol esters, and retinoic acid analogs.[299-301] Based on studies of Breitman et al.,[302] showing that all-*trans*-retinoic acid and 13-*cis*-retinoic acid could effectively induce differentiation in HL-60 cells, a number of groups

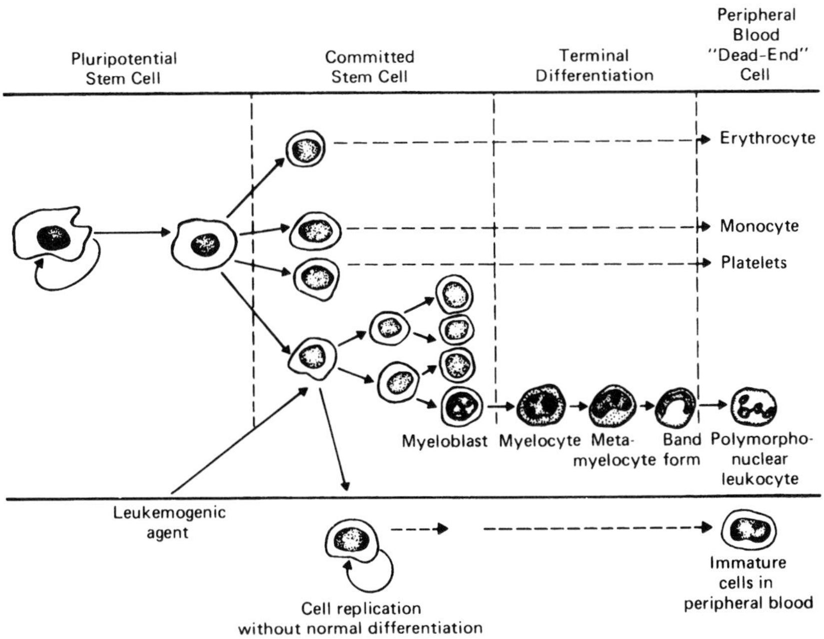

Figure 5-29 Differentiation of bone marrow cells and target cell for leukemogenic agents. In this schematic diagram, the details of myeloid cell differentiation only are shown. The other cell types (e.g., erythroid) go through similar maturation steps. The pluripotent stem cell has a self-renewal capacity. Cells in the committed stem cell compartment continue to proliferate, but as the cells terminally differentiate, they lose the ability to proliferate and become "dead-end" cells. Normally, only mature cells are seen in the peripheral blood. Interaction of leukemogenic agents with a committed stem cell leads to the production of an altered cell that fails to differentiate normally and retains its proliferative capacity. These altered cells continue to proliferate in the bone marrow, crowding out other cell types and spilling out into the peripheral blood.

studied the effects of these agents in fresh human leukemic cells in culture and showed some effectiveness in inducing differentiation. This led to a number of chemical trials in China, France, Japan, and the United States (reviewed in Ref. 303). To sum up the results of these trials, 50% to 94% of patients (depending on the study) have had a complete remission induced. The remission correlated with differentiation of immature leukemic cells into mature granulocytes and was dependent on the presence of an altered retinoic acid receptor (RAR-α) in these cells. The altered receptor results from a t(15;17) chromosomal translocation that produces a chimeric receptor fusion protein called PML/RAR-α that is defective in producing a response to retinoic acid and its analogues. Although it seems paradoxical that cells bearing a defective receptor would respond to a ligand for whose action it is responsible, the explanation for this appears to be that the defective receptor has a "dominant-negative" effect on the normal RA receptor in that it may bind RA but not produce a response. The high concentrations of RA analogues used in cell culture and in therapy may overcome this effect by inducing the expression of more normal RAR-α and/or by increasing the amount of ligand binding to the remaining normal RAR-α and overcoming the blocking effect.[303,304] However, some evidence suggests that the PML/RAR-α fusion protein has a direct effect in blocking differentiation and favoring survival of APL cells,[305] perhaps by acting as a transcription factor for proliferation-inducing genes. In this case, RA analogues might act by binding to this chimeric gene product and inactivating it.

Unfortunately, the clinical remissions in APL patients induced by all-*trans*-RA (the one most commonly used in the clinical trials) are transient and the drug has multiple side effects.[303]

Moreover, patients develop resistance to the drug and do not respond to subsequent doses of all-*trans*-RA. This resistance appears to be due to a marked increase in plasma clearance rates, possibly by a drug-induced increase in the cytochrome P-450 that metabolizes the drug.[303] Nevertheless, this remains the best example to date for the clinical feasibility of differentiation induction as a viable treatment modality. Clinical trials of similar protocols with other differentiation inductive agents are also under way for other human tumors.

Erythroleukemia

The murine virus-induced erythroleukemia provides another unique model system in which to study the relationship between differentiation and malignancy. This disease is characterized by the proliferation of primitive hematopoietic cells and an abnormal number of circulating erythroid cells, and it has a human counterpart, di Guglielmo's disease.

Mouse erythroleukemia is induced by an oncogenic virus initially recovered from the spleen of Swiss mice inoculated with a filtrate from mouse ascites tumor cells.[306] This virus, named Friend virus after the investigator who first reported it, produces an erythroblastosis (an increase in production of undifferentiated cells of the erythroid series) and an infiltration of the spleen and liver by these cells after injection into susceptible mice. As mentioned before, the target cell for Friend virus appears to be an early erythroid precursor cell. Unlike myeloid leukemia, the cultured erythroleukemia cells do not respond to the normal differentiation factor—in this case, erythropoietin. They do, however, further differentiate when exposed to a variety of chemical substances. The first agent observed to induce differentiation was DMSO.[307] This discovery came about serendipitously when, during the course of experiments to determine the effect of superinfecting erythroleukemia cells with Friend virus, DMSO was added to the culture medium. Dimethyl sulfoxide had been shown to enhance infectivity of cells by a number of viruses, and it was therefore added in the hope that it would do the same here. It was noted that, after the erythroleukemia cells were cultured for about 4 days in the presence of DMSO, the harvested cell pellets turned red. This was found to be due to the induction of hemoglobin synthesis in these cells. It is now known that a whole sequence of events, typical of further erythroid differentiation, is initiated by exposure of murine erythroleukemia (MEL) cells to DMSO, including accumulation of globin mRNA, synthesis of α- and β-globin, increase in enzymes of the heme synthetic pathway, appearance of characteristic erythrocyte membrane proteins and morphologic characteristics of a more mature erythroid cell, and a limited increase in capacity for further cell division.[308] Moreover, the malignancy of DMSO-treated MEL cells is less than that of untreated MEL cells after injection into syngeneic mice.[307] A number of other agents that also induce MEL cell differentiation have been discovered (Table 5-5).

Addition of the inducing agents listed in Table 5-5 to cultures of MEL cells leads to a program of differentiation with many but not all of the characteristics of normal erythroid differentiation, including accumulation of globin mRNA, synthesis of hemoglobin, and the appearance of membrane changes characteristic of more mature erythroid cells. However, treatment of these malignant cells with inducing agents does not lead to complete maturation. When MEL cells are treated in vitro with differentiation-inducing agents, they do not proceed to the nonnucleated stage characteristic of normal erythropoiesis, and the pattern of gene expression is not identical to that of differentiating normal erythroid cells.[308] The mechanisms by which these agents induce differentiation are not well defined, and the different classes of agents appear to act by means of somewhat different mechanisms. One of the potent inducers of MEL cell differentiation, hexamethylenebisacetamide (HMBA), appears to act by blocking cell cycle progression through G_1 via a maintenance of underphosphorylated retinoblastoma protein and changes in cyclin levels.[309]

Some agents that induce differentiation in MEL cells may do so by affecting the plasma membrane of these cells.[308] The polar-planar class of agents affect the phase-transition temperature of membrane phospholipids, produce an increase in plant lectin agglutinability, and

Table 5-5 Agents Active as Inducers of Murine Erythroleukemia Cell Differentiation

Generally Strong Inducers	Generally Weak Inducers
Polar-planar compounds	*Polar-planar compounds*
Dimethyl sulfoxide	2-Pyrrolidinone
1-Methyl-2-piperidone	Propionamide
N,N-dimethylacetamide	Pyridine-*N*-oxide
N-methylpyrrolidinone	Piperidone
N-methylacetamide	Pyridazine
N,N-dimethylformamide	Dimethylurea
N-methylformamide	*Antibiotics and antitumor agents*
Acetamide	Vincristine
Triethylene glycol	5-Fluorouracil
Polymethylene bisacetamides (n = 2–8)	Mithramycin
Hexamethylene bispropionamide	Cycloheximide
Tetramethylurea	X-irradiation
Antibiotics and antitumor agents	UV-irradiation
Bleomycin	Adriamycin
N-dimethylrifampicin	Cytosine arabinoside
Actinomycin D	Mitomycin C
Actinomycin C	Hydroxyurea
Purine and purine derivatives	*Diamines*
Hypoxanthine	Cadaverine
1-Methylhypoxanthine	*Fatty acids*
2,6-Diaminopurine	Acetate
6-Mercaptopurine	Propionate
6-Thioguanine	Butyrate
6-Amino-2-mercaptopurine	Isobutyrate
2-Acetylamino-6-mercaptopurine	*Other*
Fatty acids	Hemin
Butyrylcholine	Methy isobutyl xanthine
Other	
Ouabain	

Source: From Marks and Rifkind.[308]

produce changes in cell volume that may be related to membrane transport systems. Ouabain, a drug that inhibits plasma membrane Na^+-K^+-Mg^{2+}-activated ATPase, also induces differentiation of MEL cells. Local anesthetics, such as cocaine and tetracaine, that counteract the effects of polar–planar agents on the phase transition temperature of membrane phospholipids inhibit DMSO-induced MEL differentiation. Taken together, these data suggest that differentiation-promoting agents produce membrane effects that are involved in the stimulation of cell differentiation.

The preceding results with the ATPase inhibitor ouabain suggest that alterations in ion flux across plasma membranes may relate to the induction of differentiation in MEL cells. This has proved to be the case. An increase in cellular Ca^{2+} has been shown to be required for DMSO-induced terminal differentiation of MEL cells.[310] (1) DMSO causes a rapid increase in Ca^{2+} exchange in MEL cells; (2) commitment of MEL cells is blocked by the calcium chelating agent EGTA, and this effect is reversed if excess Ca^{2+} is added back to the culture medium; (3) MEL differentiation is blocked by amiloride, a drug that blocks an Na^+/Ca^{2+} antiport ion transporter system, and this effect is reversed by the calcium ionophore A23187, an agent that carries Ca^{2+} across cell membranes; and (4) A23187 itself accelerates DMSO-stimulated MEL cell differentiation.

There is, as noted above, evidence that not all of the inducing agents act by the same mechanism. First, agents of widely differing chemical classes with widely varying effects on cells can induce differentiation. Second, variant MEL cell lines that are resistant to a particular class of agents will still differentiate when another type of agent is added to the culture medium. Nevertheless, the MEL cell model has been useful in defining agents and mechanisms for the induction of differentiation of cancer cells, and has led to the identification of agents that are being used in clinical trials for the treatment of cancer.

CONCLUSIONS

The data reviewed in this chapter lead to several conclusions:

1. The target cell in tissues for carcinogenic agents is often (if not always) a committed but not yet fully differentiated stem cell or a cell that has the capacity to divide when further stimulated by mitogenic agents.
2. Cancer cells are cells that have not achieved a fully differentiated state. They are blocked at some earlier stage of maturation. Although they may exhibit some of the characteristics of normal differentiated cells, they maintain the capacity to divide that most normal differentiated cells lose as they mature.
3. The microenvironment in which malignant cells grow is involved in the maintenance of the malignant state. For example, teratocarcinoma cells, when placed in a normal embryonic environment, differentiate into normal tissues but, when inoculated into adult liver, kidney, or skin, produce malignant tumors.
4. The blocked differentiation of cancer cells is not totally irreversible. Some types of cancer cells can be induced to differentiate to a more mature, less malignant state by exposure to various growth factors or drugs.
5. Some of the agents that induce differentiation and reduce malignancy of cultured tumor cells in vitro only transiently stop malignant cell growth in vivo. This may be due to pharmacologic problems (e.g., poor absorption of the agent into the bloodstream, low penetration into tissues, rapid metabolism, or rapid excretion of the agent). In addition, the continual presence of the inducer might be required until the cells are past a critical stage of differentiation. There is also the possibility that the adult stroma in which the cancer cells arise does not usually favor differentiation of these cells. At any rate, the possibility that the malignant state of certain types of cancer cells is potentially reversible holds out the hope that therapies aimed at stimulating differentiation are possible.

REFERENCES

1. W. J. Rutter, R. L. Pictet, and P. W. Morris: Toward molecular mechanisms of developmental processes. *Annu Rev Biochem* 42:601, 1973.
2. E. H. Davidson, B. R. Hough-Evans, and R. J. Britten: Molecular biology of the sea urchin embryo. *Science* 217:17, 1982.
3. S. Strome and W. B. Wood: Generation of asymmetry and segregation of germ-line granules in early *C. elegans* embryos. *Cell* 35:15, 1983.
4. K. Illmensee and A. P. Mahowald: The autonomous function of germ plasm in a somatic region of the *Drosophila* egg. *Exp Cell Res* 97:127, 1976.
5. R. Briggs and T. J. King: Transplantation of living nuclei from blastula cells into enucleated frogs' eggs. *Proc Natl Acad Sci USA* 38:456, 1952.
6. J. B. Gurdon, R. A. Laskey, and O. R. Reeves: The developmental capacity of nuclei transplanted from keratinized skin cells of adult frogs. *J Embryol Exp Morphol* 34:93, 1975.
7. M. A. D. Barardino and N. J. Hoffner: Gene reactivation in erythrocytes: Nuclear transplantation in oocytes and eggs of *Rana. Science* 219:862, 1983.
8. J. McGrath and D. Solter: Inability of mouse blastomere nuclei transferred to enucleated zygotes to support development in vitro. *Science* 226:1317, 1984.
9. C. J. Tabin: Retinoids, homeoboxes, and growth factors: Toward molecular models for limb development. *Cell* 66:199, 1991.

9a. A. Simeone, D. Acampora, L. Arcioni, P. N. Andrews, E. Boncinelli, and F. Mavilio: Sequential activation of *Hox* 2 homeobox genes by retinoic acid in human embryonal carcinoma cells. *Nature* 346:763, 1990.

10. D. A. Melton: Pattern formation during animal development. *Science* 252:234, 1991.
11. L. Reid: From gradients to axes, from morphogenesis to differentiation. *Cell* 63:875, 1990.
12. G. M. Edelman: Morphoregulatory molecules. *Biochemistry* 27:3533, 1988.
13. M. Takeichi: Cadherins: a molecular family important in selective cell–cell adhesion. *Annu Rev Biochem* 59:237, 1990.
14. R. J. DeLange and E. L. Smith: Histone function and evolution as viewed by sequence studies. *Ciba Found Symp* 28:59, 1975.

15. S. C. R. Elgin and H. Weintraub: Chromosomal proteins and chromatin structure. *Annu Rev Biochem* 44:725, 1975.
16. T. A. Langan, S. C. Rall, and R. D. Cole: Variation in primary structure at a phosphorylation site in lysine-rich histones. *J Biol Chem* 246:1942, 1971.
17. L. Hong, G. P. Schroth, H. R. Matthews, P. Yau, and E. M. Bradbury: Studies of the DNA binding properties of histone H4 amino terminus. *J Biol Chem* 268:305, 1993.
18. D. Krylov, S. Leuba, K. van Holde, and J. Zlatanova: Histones H1 and H5 interact preferentially with crossovers of double-helical DNA. *Proc Natl Acad Sci USA* 90:5052, 1993.
19. G. H. Goodwin, C. Sanders, and E. W. Johns: A new group of chromatin-associated proteins with a high content of acidic and basic amino acids. *Eur J Biochem* 38:14, 1973.
20. J. M. Walker, G. H. Goodwin, and E. W. Johns: The similarity between the primary structures of two non-histone chromosomal proteins. *Eur J Biochem* 62:461, 1976.
21. J. M. Pash, P. J. Alfonso, and M. Bustin: Aberrant expression of high mobility group chromosomal protein 14 affects cellular differentiation. *J Biol Chem* 268:13632, 1993.
22. P. Byvoet, G. R. Shepherd, J. M. Hardin, and B. J. Noland: The distribution and turnover of labeled methyl groups in histone fractions of cultured mammalian cells. *Arch Biochem Biophys* 148:558, 1972.
23. B. G. T. Pogo, V. G. Allfrey, and A. E. Mirsky: RNA synthesis and histone acetylation during the course of gene activation in lymphocytes. *Proc Natl Acad Sci USA* 55:805, 1966.
24. B. G. T. Pogo, A. O. Pogo, V. G. Allfrey, and A. E. Mirsky: Changing patterns of histone acetylation and RNA synthesis in regeneration of the liver. *Proc Natl Acad Sci USA* 59:1337, 1968.
25. D. Y. Lee, J. J. Hayes, D. Pruss, and A. P. Wolffe: A positive role of histone acetylation in transcription factor access to nucleosomal DNA. *Cell* 72:73, 1993.
26. J. Sommerville, J. Baird, and B. M. Turner: Histone H4 acetylation and transcription in amphibian chromatin. *J Cell Biol* 120:277, 1993.
26a. M.-N. Prioleau, J. Huet, A. Sentenac, and M. Méchali: Competition between chromatin and transcription complex assembly regulates gene expression during early development. *Cell* 77:439, 1994.
27. S. C. R. Elgin and H. Weintraub: Chromosomal proteins and chromatin structure. *Annu Rev Biochem* 44:725, 1975.
28. R. Balhorn, R. Chalkley, and D. Granner: Lysine-rich histone phosphorylation: A positive correlation with cell replication. *Biochemistry* 11:1094, 1972.
29. T. A. Langan: Phosphorylation of liver histone following the administration of glucagon and insulin. *Proc Natl Acad Sci USA* 64:1274, 1969.
30. L. J. Kleinsmith: Phosphorylation of non-histone proteins in the regulation of chromosome structure and function. *J Cell Physiol* 85:459, 1975.
31. V. G. Allfrey, A. Inoue, J. Karn, E. M. Johnson, and G. Vidali: Phosphorylation of DNA-binding nuclear acidic proteins and gene activation in the HeLa cell cycle. *Cold Spring Harbor Symp Quant Biol* 38:785, 1973.
32. L. E. Rikans and R. W. Ruddon: Partial purification and properties of a chromatin-associated phosphoprotein kinase from rat liver nuclei. *Biochem Biophys Acta* 422:73, 1976.
33. G. M. Walton and G. N. Gill: Identity of the in vivo phosphorylation site in high mobility group 14 protein in HeLa cells with the site phosphorylated by casein II kinase in vitro. *J Biol Chem* 258:4440, 1983.
34. S.-I. Tanuma and G. S. Johnson: ADP-ribosylation of nonhistone high mobility group proteins in intact cells. *J Biol Chem* 258:4067, 1983.
35. D. J. Tremethick and H. R. Drew: High mobility group proteins 14 and 17 can space nucleosomes in vitro. *J Biol Chem* 268:11389, 1993.
35a. H. Ge and R. G. Roeder: The high mobility group protein HMG1 can reversibly inhibit class II gene transcription by interaction with the TATA-binding protein. *J Biol Chem* 269:17136, 1994.
35b. N. Lehming, D. Thanos, J. M. Brickman, J. Ma, T. Maniatis, and M. Ptashne: An HMG-like protein that can switch a transcriptional activator to a repressor. *Nature* 371:175, 1994.
36. A. L. Olins and D. E. Olins: Spheroid chromatin units (ν bodies). *Science*183:330, 1974.
37. F. Thoma and Th. Koller: Influence of histone H1 on chromatin structure. *Cell* 12:101, 1977.
38. D. Hewish and L. Burgoyne: Chromatin substructure: The digestion of chromatin DNA at regularly spaced sites by a nuclear deoxyribonuclease. *Biochem Biophys Res Commun* 52:504, 1973.
39. M. Noll: Subunit structure of chromatin. *Nature* 251:2491, 1974.
40. D. Oosterhof, J. Hozier, and R. Rill: Nuclease action on chromatin: Evidence for discrete, repeated nucleoprotein units along chromatin fibrils. *Proc Natl Acad Sci USA* 72:633, 1975.
41. G. Felsenfield: Chromatin. *Nature* 271:115, 1978.
42. R. K. Kornberg: Chromatin structure: A repeating unit of histones and DNA. *Science* 184:868, 1974.
43. J. P. Baldwin, P. G. Boseley, E. M. Bradbury, and K. Ibel: The subunit structure of the eukaryotic chromosome. *Nature* 253:245, 1975.
44. S. Weisbrod, M. Groudine, and H. Weintraub: Interaction of HMG 14 and 17 with actively transcribed genes. *Cell* 19:289, 1980.

45. J. T. Finch and A. Klug: Solenoidal model for superstructure in chromatin. *Proc Natl Acad Sci USA* 73:1897, 1976.
46. K. W. Adolph, S. M. Chevy, J. R. Paulson, and U. K. Laemmli: Isolation of a protein scaffold from mitotic HeLa cell chromosome. *Proc Natl Acad Sci USA* 74:4937, 1977.
47. C. D. Lewis and U. K. Laemmli: Higher order metaphase chromosome structure: Evidence for metalloprotein interactions. *Cell* 29:171, 1982.
48. L. Manuelidis: A view of interphase chromosomes. *Science* 250:1533, 1990.
49. K. E. van Holde, D. E. Lohr, and C. Robert: What happens to nucleosomes during transcription? *J Biol Chem* 267:2837, 1992.
50. C. C. Adams and J. L. Workman: Nucleosome displacement in transcription *Cell* 72:305, 1993.
51. T. E. O'Neill, J. G. Smith, and E. M. Bradbury: Histone octamer dissociation is not required for transcript elongation through arrays of nucleosome cores by phage T7 RNA polymerase in vitro. *Proc Natl Acad Sci USA* 90:6203, 1993.
52. M. L. DePamphilis: Origins of DNA replication in metazoan chromosomes. *J Biol Chem* 268:1, 1993.
53. R. Berezney and D. Coffey: Identification of a nuclear protein matrix. *Biochem Biophys Res Commun* 60:1410, 1974.
54. J. Mirkovitch, M.-E. Mirault, and U. K. Laemmli: Organization of the higher-order chromatin loop: Specific DNA attachment sites on nuclear scaffold. *Cell* 39:223, 1984.
55. D. M. Pardoll, B. Vogelstein, and D. S. Coffey: A fixed site of DNA replication in eucaryotic cells. *Cell* 19:527, 1980.
56. E. M. Ciejek, M.-J. Tsai, and B. W. O'Malley: Actively transcribed genes are associated with the nuclear matrix. *Nature* 306:607, 1983.
57. K. Tsutsui, K. Tsutsui, S. Okada, S. Watarai, S. Seki, T. Yasuda, and T. Shohmori: Identification and characterization of a nuclear scaffold protein that binds the matrix attachment region DNA. *J Biol Chem* 268:12886, 1993.
58. N. Stuurman, A. M. L. Meijne, A. J. van der Pol, L. deJong, R. van Driel, and J. van Renswoude: The nuclear matrix from cells of different origin. *J Biol Chem* 265:5460, 1990.
59. A. W. Partin, R. H. Getzenberg, M. J. CarMichael, D. Vindivich, J. Yoo, J. I. Epstein, and D. S. Coffey: Nuclear matrix protein patterns in human benign prostatic hyperplasia and prostate cancer. *Cancer Res* 53:744, 1993.
60. P. S. Khanuja, J. E. Lehr, H. D. Soule, S. K. Gehani, A. C. Noto, S. Choudhury, R. Chen, and K. J. Pienta: Nuclear matrix proteins in normal and breast cancer cells. *Cancer Res* 53:3394, 1993.
60a. S. K. Keesee, M. D. Meneghini, R. P. Szaro, and Y.-J. Wu: Nuclear matrix proteins in human colon cancer. *Proc Natl Acad Sci USA* 91:1913, 1994.
61. T. E. Miller, L. A. Beausang, L. F. Winchell, and G. P. Lidgard: Detection of nuclear matrix proteins in serum from cancer patients. *Cancer Res* 52:422, 1992.
62. S. Weisbrod: Active chromatin. *Nature* 297:289, 1982.
63. H. Weintraub and M. Groudine: Chromosomal subunits in active genes have an altered conformation. *Science* 193:848, 1976.
64. M. Groudine and H. Weintraub: Activation of globin genes during chicken development. *Cell* 24:393, 1981.
65. M. Groudine, T. Kohwi-Shigematsu, R. Gelinas, G. Stamatoyannopoulus, and T. Papayannopoulou: Human fetal to adult hemoglobin switching: Changes in chromatin structure of the β-globin gene locus. *Proc Natl Acad Sci USA* 80:7551, 1983.
66. A. Garel and R. Axel: Selective digestion of transcriptionally active ovalbumin genes from oviduct nuclei. *Proc Natl Acad Sci USA* 73:3966, 1976.
67. M. A. Keene, V. Corces, K. Lowenhaupt, and S. C. R. Elgin: DNase I hypersensitive sites in *Drosophila* chromatin occur at the 5″ ends of regions of transcription. *Proc Natl Acad Sci USA* 78:143, 1981.
68. B. Samal, A. Worcel, C. Louis, and P. Schedl: Chromatin structure of the histone genes of D. melanogaster. *Cell* 23:401, 1981.
69. W. A. Scott and D. J. Wigmore: Sites in simian virus 40 chromatin which are preferentially cleaved by endonucleases. *Cell* 15:1511, 1978.
70. R. W. Sweet, M. V. Chao, and R. Axel: The structure of the thymidine kinase gene promoter: Nuclease hypersensitivity correlates with expression. *Cell* 31:347, 1982.
71. A. Larsen and H. Weintraub: An altered DNA conformation detected by S1 nuclease occurs at specific regions in active chick globin chromatin. *Cell* 29:609, 1982.
72. M. Groudine and H. Weintraub: Propagation of globin DNAse I-hypersensitive sites in absence of factors required for induction: A possible mechanism for determination. *Cell* 30:131, 1982.
73. A. M. Gianni, R. D. Favera, E. Polli, I. Merisio, B. Giglioni, P. Comi, and S. Ottolenghi: Globin RNA sequences in human leukaemic peripheral blood. *Nature* 274:610, 1978.
74. B. M. Emerson, C. D. Lewis, and G. Felsenfeld: Interaction of specific nuclear factors with nuclease-hypersensitive region of the chicken adult β-globin gene: Nature of the binding domain. *Cell* 41:21, 1985.
75. A. Razin and A. D. Riggs: DNA methylation and gene function. *Science* 210:604, 1980.
76. J. D. McGhee and G. D. Ginder: Specific DNA methylation sites in the vicinity of the chicken β-globin genes. *Nature* 280:419, 1979.
77. J. L. Mandel and P. Chambon: DNA methyl-

ation: Organ specific variations in methylation pattern within and around ovalbumin and other chicken genes. *Nucleic Acid Res* 7:2081, 1979.

78. A. Razin, C. Webb, M. Szyf, J. Yisraeli, A. Rosenthal, T. Navek-Many, N. Sciaky-Gallili, and H. Cedar: Variations in DNA methylation during mouse cell differentiation in vivo and in vitro *Proc Natl Acad Sci USA* 81:2275, 1984.
79. A. Levine, G. L. Cantoni, and A. Razin: Methylation in the preinitiation domain suppresses gene transcription by an indirect mechanism. *Proc Natl Acad Sci USA* 89:10119, 1992.
80. A. Bird: The essentials of DNA methylation. *Cell* 70:5, 1992.
81. I. Keshet, J. Lieman-Hurwitz, and H. Cedar: DNA methylation affects the formation of active chromatin. *Cell* 44:535, 1986.
82. M. Higurashi and R. D. Cole: The combination of DNA methylation and H1 histone binding inhibits the action of a restriction nuclease on plasmid DNA. *J Biol Chem* 266:8619, 1991.
83. S. M. Taylor and P. A. Jones: Multiple new phenotypes induced in 10T1/2 and 3T3 cells treated with 5-azacytidine. *Cell* 17:771, 1979.
84. P. A. Jones, M. J. Wolkowicz, W. M. Ridout III, F. A. Gonzales, C. M. Marziasz, G. A. Coetzee, and S. J. Tapscott: De novo methylation of the MyoD1 CpG island during the establishment of immortal cell lines. *Proc Natl Acad Sci USA* 87:6117, 1990.
85. E. Li, T. H. Bestor, and R. Jaenisch: Targeted mutation of the DNA methyltransferase gene results in embryonic lethality. *Cell* 69:915, 1992.
86. D. P. Barlow: Methylation and imprinting: From host defense to gene regulation? *Science* 260:309, 1993.

86a. A. Razin and H. Cedar: DNA methylation and genomic imprinting. *Cell* 77:473, 1994.

87. E. U. Selker: DNA methylation and chromatin structure: A view from below. *Trends Biochem Sci* 15:103, 1990.

87a. S. Rainier and A. P. Feinberg: Genomic imprinting, DNA methylation, and cancer. *JNCI* 86:753, 1994.

88. A. P. Feinberg and B. Vogelstein: Hypomethylation distinguishes genes of some human cancers from their normal counterparts. *Nature* 301:89, 1983.
89. A. P. Feinberg and B. Vogelstein: Hypomethylation of *ras* oncogenes in primary human cancers. *Biochem Biophys Res Commun* 111:47, 1983.
90. S. E. Goelz, B. Vogelstein, S. R. Hamilton, and A. P. Feinberg: Hypomethylation of DNA from benign and malignant human colon neoplasms. *Science* 228:187, 1985.
91. M. Makos, B. D. Nelkin, M. I. Lerman, F. Latif, B. Zbar, and S. B. Baylin: Distinct hypermethylation patterns occur at altered chromosome loci in human lung and colon cancer. *Proc Natl Acad Sci USA* 89:1929, 1992.
92. M. Makos, B. D. Nelkin, R. E. Reiter, J. R. Gnarra, J. Brooks, W. Isaacs, M. Linehan, and S. B. Baylin: Regional DNA hypermethylation at D17S5 precedes 17p structural changes in the progresion of renal tumors. *Cancer Res* 53:2719, 1993.
93. M. Makos, B. D. Nelkin, V. R. Chazin, W. K. Cavenee, G. M. Brodeur, and S. B. Baylin: DNA hypermethylation is associated with 17p allelic loss in neural tumors. *Cancer Res* 53:2715, 1993.
94. W. N. Rideout, III, G. A. Coetzee, A. F. Olumi, and P. A. Jones: 5-methylcytosine as an endogenous mutagen in the human LDL receptor and p53 genes. *Science* 249: 1288, 1990.
95. W. S. El-Deiry, B. D. Nelkin, P. Celano, R.-W. C. Yen, J. P. Falco, S. R. Hamilton, and S. B. Baylin: High expression of the DNA methyltransferase gene characterized human neoplastic cells and progression stages of colon cancer. *Proc Natl Acad Sci USA* 88:3470, 1991.
96. S. Rainier, L. A. Johnson, C. J. Dobry, A. J. Ping, P. E. Grundy, and A. P. Feinberg: Relaxation of imprinted genes in human cancer. *Nature* 362:747, 1993.
97. O. Ogawa, M. R. Eccles, J. Szeto, L. A. McNoe, K. Yun, M. A. Maw, P. J. Smith, and A. E. Reeve: Relaxation of insulin-like growth factor II gene imprinting implicated in Wilms' tumour. *Nature* 362:749, 1993.
98. S. F. Wolf and B. R. Migeon: Clusters of CpG dinucleotides implicated by nuclease hypersensitivity as control elements of housekeeping genes. *Nature* 314:467, 1985.
99. E. L. Mather and R. P. Perry: Methylation status and DNase I sensitivity of immunoglobulin genes: Changes associated with rearrangement. *Proc Natl Acad Sci USA* 80:4689, 1983.
100. T. Shimada and A. W. Nienhuis: Only the promoter region of the constitutively expressed normal and amplified human dihydrofolate reductase gene is DNase I hypersensitive and undermethylated. *J Biol Chem* 260:2468, 1985.
101. A. P. Bird: DNA methylation-How important in gene control? *Nature* 307:503, 1984.
102. M. Graessmann, A. Graessmann, H. Wagner, E. Werner, and D. Simon: Complete DNA methylation does not prevent polyoma and simian virus 40 virus early gene expression. *Proc Natl Acad Sci USA* 80:6470, 1983.
103. K. Tanaka, E. Appella, and G. Jay: Developmental activation of the H-2K gene is correlated with an increase in DNA methylation. *Cell* 35:457, 1983.
104. M.-O. Ott, L. Sperling, and M. C. Weiss: Albumin extinction without methylation of its gene. *Proc Natl Acad Sci USA* 81:1738, 1984.
105. L. M. Fisher: DNA supercoiling and gene expression. *Nature* 307:686, 1984.
106. G. R. Smith: DNA supercoiling: Another level for regulating gene expression. *Cell* 24:599, 1981.

107. G. C. Glikin, I. Ruberti, and A. Worcel: Chromatin assembly in *Xenopus* oocytes: In vitro studies. *Cell* 37:33, 1984.
108. G. Fleischmann, G. Pflugfelder, E. K. Steiner, K. Javaherian, G. C. Howard, J. C. Wang, and S. C. R. Elgin: *Drosophila* DNA topoisomerase I is associated with transcriptionally active regions of the genome. *Proc Natl Acad Sci USA* 81:6958, 1984.
109. M. Berrios, N. Osheroff, and P. A. Fisher: In situ localization of DNA topoisomerase II, a major polypeptide component of the *Drosophila* nuclear matrix fraction. *Proc Natl Acad Sci USA* 82:4142, 1985.
110. C. Holm, T. Goto, J. C. Wang, and D. Botstein: DNA topoisomerase II is required at the time of mitosis in yeast. *Cell* 41:553, 1985.
111. C. L. Chen, L. Yang, T. C. Rowe, B. D. Halligan, K. M. Tewey, and L. F. Liu: Nonintercalative antitumor drugs interfere with the breakage-reunion reaction of mammalian topoisomerase II. *J Biol Chem* 259:13560, 1984.
112. A. G. Herbert, J. R. Spitzner, K. Lowenhaupt, and A. Rich: Z-DNA binding protein from chicken blood nuclei. *Proc Natl Acad Sci USA* 90:3339, 1993.
113. M. W. Kilpatrick, J. Klysik, C. K. Singleton, D. A. Zarling, T. M. Jovin, L. H. Hanau, B. F. Erlanger, and R. D. Wells: Intervening sequences in human fetal globin genes adopt left-handed Z helices. *J Biol Chem* 259:7268, 1984.
114. T. Kohwi-Shigematsu, R. Gelinas, and H. Weintraub: Detection of an altered DNA conformation at specific sites in chromatin and supercoiled DNA. *Proc Natl Acad Sci USA* 80:4389, 1983.
114a. M. Z. Levy, R. C. Allsopp, A. B. Futcher, C. W. Greider, and C. B. Harley: Telomere end-replication problem and cell aging. *J Mol Biol* 225:951, 1992.
114b. C. M. Counter, A. A. Avillion, C. E. LeFeuvre, N. G. Stewart, C. W. Freider, C. B. Harley, and S. Bacchetti: Telomere shortening associated with chromosome instability is arrested in immortal cells which express telomerase activity. *EMBO J* 11:1921, 1992.
114c. C. M. Counter, H. W. Hirte, S. Bacchetti, and C. B. Harley: Telomerase activity in human ovarian carcinoma. *Proc Natl Acad Sci USA* 91:2900, 1994.
115. D. Nathans: Restriction endonucleases, simian virus 40, and the new genetics. *Science* 206:903, 1979.
116. W. Gilbert: Why genes in pieces? *Nature* 271:501, 1978.
117. R. Breathnach, C. Benoist, K. O'Hare, F. Gannon, and P. Chambon: Ovalbumin gene: Evidence for a leader sequence in mRNA and DNA sequences at the exon-intron boundaries. *Proc Natl Acad Sci USA* 75:4853, 1978.
118. H. Schwartz and J. E. Darnell: The association of protein with the polyadenylic acid of HeLa cell messenger RNA: Evidence for a "transport" role of a 75,000 molecular weight polypeptide. *J. Mol Biol* 104:833, 1975.
119. E. J. Sontheimer and J. A. Steitz: The U5 and U6 small nuclear RNAs as active site components of the spliceosome. *Science* 262:1989, 1993.
120. P. A. Sharp: Split genes and RNA splicing: Nobel lecture. *Cell* 77:805, 1994.
121. P. S. Perlman and R. A. Butow: Mobile introns and intron-encoded proteins. *Science* 246:1106, 1989.
121a. A. M. Lambowitz and M. Belfort: Introns as mobile genetic elements. *Annu Rev Biochem* 62:587, 1993.
122. J. W. Tamkun, J. E. Schwarzbauer, and R. O. Hynes: A single rat fibronectin gene generates three different mRNAs by alternative splicing of a complex exon. *Proc Natl Acad Sci USA* 81:5140, 1984.
123. M. Periasamy, E. E. Strehler, L. I. Garfinkel, R. M. Gubits, N. Riuz-Opazo, and B. Nadal-Ginard: Fast skeletal muscle myosin light chains 1 and 3 are produced from a single gene by a combined process of differentiatial RNA transcription and splicing. *J Biol Chem* 259:13595, 1984.
124. R. A. Young, O. Hagenbüchle, and U. Schibler: A single mouse α-amylase gene specifies tow different tissue specific mRNAs. *Cell* 23:451, 1981.
125. Y. Fukumaki, P. K. Ghosh, E. J. Benz, Jr., V. B. Reddy, P. Lebowitz, B. G. Forget, and S. M. Weissman: Abnormally spliced messenger RNA in erythroid cells from patients with β+-thalassemia and monkey cells expressing a cloned β+-thalassemic gene. *Cell* 28:585, 1982.
126. B. Dujon: Mutants in a mosaic gene reveal functions for introns. *Nature* 282:777, 1979.
127. P. Borst and D. R. Greaves: Programmed gene rearrangements altering gene expression. *Science* 235:658, 1987.
128. P. D. Sadowski: Site-specific genetic recombination: Hops, flips, and flops. *FASEB J* 7:760, 1993.
129. R. D. Camerini-Otero and P. Hsieh: Parallel DNA triplexes, homologous recombination, and other homology-dependent DNA interactions. *Cell* 73:217, 1993.
130. T. H. Morgan: An attempt to analyze the constitution of the chromosomes on the basis of sex-limited inheritance in *Drosophila. J Exp Zool* 11:365, 1911.
131. B. McClintock: The significance of responses of the genome to challenge. *Science* 226:792, 1984.
132. C. W. Schmid and W. R. Jelinek: The Alu family of dispersed repetitive sequences. *Science* 216:1065, 1982.
133. Y. Meng-Chao, J. Choi, S. Yokoyama, C. F. Austerberry, and C.-H. Yao: DNA elimination in *Tetrahymena:* A developmental process involving extensive breakage and rejoining of DNA at defined sites. *Cell* 36:433, 1984.
134. M.-C. Yao and M. A. Gorovsky: Comparison of

the sequence of macro- and micronuclear DNA of *Tetrahymena puriformis*. *Chromosoma* 48:1, 1974.
135. A. M. Campbell: Episomes. *Adv Genet* 11:101, 1962.
136. C. Coleclough: Chance, necessity, and antibody gene dynamics. *Nature* 303:23, 1983.
137. E. L. Mather, K. J. Nelson, J. Haimovich, and R. P. Perry: Mode of regulation of immunoglobulin μ- and δ-chain expression varies during B-lymphocyte maturation. *Cell* 36:329, 1984.
138. G. E. Taccioli, G. Rathbun, E. Oltz, T. Stamato, P. A. Jeggo, and F. W. Alt: Impairment of V(D)J recombination in double-strand break repair mutants: *Science* 260:207, 1993.
139. B. A. Dombroski, A. F. Scott, and H. H. Kazazian, Jr.: Two additional potential retrotransposons isolated from a human L1 subfamily that contains an active retrotransposable element. *Proc Natl Acad Sci USA* 90:6513, 1993.
140. P. R. Gross: Biochemistry of differentiation. *Annu Rev Biochem* 37:631, 1968.
141. D. V. de Cicco and A. C. Spradling: Localization of a *cis*-acting element responsible for the developmentally regulated amplification of *Drosophila* chorion genes. *Cell* 38:45, 1984.
142. F. W. Alt, R. E. Kellems, J. R. Bertino, and R. T. Schimke: Selective multiplication of dihydrofolate reductase genes in methotrexate-resistant variants of cultured murine cells. *J Biol Chem* 253:1357, 1978.
143. R. T. Schimke: Gene amplification in cultured animal cells. *J Biol Chem* 263:5989, 1988.
144. T. D. Tlsty, P. C. Brown, and R. T. Schimke: UV radiation facilitates methotrexate resistance and amplification of the dihydrofolate reductase gene in cultured 3T6 mouse cells. *Mol Cell Biol* 4:1050, 1984.
145. S. Lavi and S. Etkin: Carcinogen-mediated induction of SV40 DNA synthesis in SV 40 transformed Chinese hamster embryo cells. *Carcinogenesis* 2:417, 1981.
146. A. Varshavsky: Phorbol ester dramatically increases incidence of methotrexate-resistant colony-forming mouse cells: Possible mechanisms and relevance to tumor promotion. *Cell* 25:561, 1981.
147. G. Levan and A. Levan: Transitions of double minutes into homogeneously staining regions and C-bandless chromosomes in the SEWA tumor, in R. T. Schimke, ed.: *Gene Amplification.* New York: Cold Spring Harbor Laboratory, 1982, p. 91.
148. J. M. Roberts, L. B. Buck, and R. Axel: A structure for amplified DNA. *Cell* 33:53, 1983.
149. Y. N. Osheim and O. L. Miller, Jr.: Novel amplification and transcription activity of chorion genes in *Drosophila melanogaster* follicle cells. *Cell* 33:543, 1983.
150. H. Weintraub, H. Beug, M. Groudine, and T. Graf: Temperature sensitive changes in the structure of globin chromatin in lines of red cell precursors transformed by tsAEV virus. *Cell* 28:931, 1982.
151. S. M. Rose and W. T. Garrard: Differentiation-dependent chromatin alterations precede and accompany transcription of immunoglobulin light chain genes. *J Biol Chem* 259:8534, 1984.
152. J. B. Gurdon: The developmental capacity of nuclei taken from intestinal epithelium cells of feeding tadpoles. *J Embryol Exp Morphol* 10:622, 1962.
153. T. J. King and M. A. Di Berardino: Transplantation of nuclei from the frog renal adenocarcinoma: I. Development of tumor nuclear-transplant embryos. *Ann NY Acad Sci* 126:115, 1965.
154. A. I. Caplan and C. P. Ordahl: Irreversible gene repression model for control of development. *Science* 201:120, 1978.
155. E. H. Davidson and R. J. Britten: Regulation of gene expression: Possible role of repetitive sequences. *Science* 204:1052, 1979.
156. R. D. Palmiter: Rate of ovalbumin messenger ribonucleic acid synthesis in the oviduct of estrogen-primed chicks. *J Biol Chem* 248:8260, 1973.
157. J. A. Hunt: Rate of synthesis and half-life of globin messenger ribonucleic acid. *Biochem J* 138:499, 1974.
158. N. D. Hastie and J. O. Bishop: The expression of three abundance classes of messenger RNA in mouse tissues. *Cell* 9:761, 1976.
159. R. Axel, P. Feigelson, and G. Schutz: Analysis of the complexity and diversity of mRNA from chicken liver and oviduct. *Cell* 7:247, 1976.
160. B. J. McCarthy, J. T. Nishiura, D. Doenecke, D. S. Nasser, and C. B. Johnson: Transcription and chromatin structure. New York: *Cold Spring Harbor Symp Quant Biol* 38:763, 1974.
161. M. Groudine, T. Kohwi-Shigematsu, R. Gelinas, G. Stamatoyannopoulus, and T. Papayannopoulow: Human fetal to adult hemoglobin switching: Changes in chromatin structure of the β-globin gene locus. *Proc Natl Acad Sci USA* 80:7551, 1983.
162. S. M. Tilghman and A. Belayew: Transcriptional control of the murine albumin/α-fetoprotein locus during development. *Proc Natl Acad Sci USA* 79:5254, 1982.
163. W. S. Dynan: Modularity in promoters and enhancers. *Cell* 58:1, 1989.
164. S. McKnight and R. Tijan: Transcriptional selectivity of viral genes in mammalian cells. *Cell* 46:795, 1986.
165. C. Benoist and P. Chambon: In vivo sequence requirements of the SV40 early promoter region. *Nature* 290:304, 1981.
166. P. Gruss, R. Dhar, and G. Khoury: Simian virus 40 tandem repeated sequences as an element of the early promoter. *Proc Natl Acad Sci USA* 78:943, 1981.
167. H. R. Schöler and P. Guss: Specific interaction between enhancer-containing molecules and cellular components. *Cell* 36:403, 1984.

168. G. Khoury and P. Gruss: Enhancer elements. *Cell* 33:313, 1983.
169. D. R. Herendee, G. A. Kassavetis, and E. P. Geiduschek: A transcriptional enhancer whose function imposes a requirement that proteins track along DNA. *Science* 256:1298, 1992.
170. K. E. Cullen, M. P. Kladde, and M. A. Seyfred: Interaction between transcription regulatory regions of prolactin chromatin. *Science* 261:203, 1993.
171. M. Ptashne: How eukaryotic transcriptional activators work. *Nature* 335:683, 1988.
172. A. D. Frankel and P. S. Kim: Modular structure of transcription factors: Implications for gene regulation. *Cell* 65:717, 1991.
173. K. Struhl: Helix-turn-helix, zinc-finger, and leucine-zipper motifs for eukaryotic transcriptional regulatory proteins. *Trends Biochem Sci* 14:137, 1989.
173a. R. Tjian and T. Maniatis: Transcriptional activation: A complex puzzle with few easy pieces. *Cell* 77:5, 1994.
174. R. Schleif: DNA binding by proteins. *Science* 241:1182, 1988.
175. N. S. Foulkes and P. Sassone-Corsi: More is better: Activators and repressors from the same gene. *Cell* 68:411, 1992.
176. M. Levine and J. L. Manley: Transcriptional repression of eukaryotic promoters. *Cell* 59:405, 1989.
176a. D. J. Kessler, M. P. Duyao, D. B. Spicer, and G. E. Sonenshein: NF-κB-like factors mediate interleukin 1 induction of c-myc gene transcription in fibroblasts. *J Exp Med* 176:787, 1992.
176b. A. Schrivastava, S. Saleque, G. V. Kalpana, S. Artandi, S. P. Goff, and K. Calame: Inhibition of transcriptional regulator Yin-Yang-1 by association with c-*myc*. *Science* 262:1889, 1993.
177. L. Weis and D. Reinberg: Transcription by RNA polymerase II: Initiator-directed formation of transcription-competent complexes. *FASEB J* 6:3300, 1992.
178. P. A. Sharp: TATA-binding protein is a classless factor. *Cell* 68:819, 1992.
179. G. Gill and R. Tjian: A highly conserved domain of TFIID displays species specificity in vivo. *Cell* 65:333, 1991.
180. A. Barberis, C. W. Müller, S. C. Harrison, and M. Ptashne: Delineation of two functional regions of transcription factor TFIIB. *Proc Natl Acad Sci USA* 90:5628, 1993.
181. A. K. P. Taggart, T. S. Fisher, and B. F. Pugh: The TATA-binding protein and associated factors are components of pol III transcription factor TFIIIB. *Cell* 71:1015, 1992.
182. K. Hisatake, R. G. Roeder, and M. Horikoshi: Functional dissection of TFIIB domains required for TFIIB-TFIID-promoter complex formation and basal transcription activity. *Nature* 363:744, 1993.
183. J. W. Lillie and M. R. Green: Activator's target in sight. *Nature* 341:279, 1989.
184. S. G. E. Roberts, I. Ha, E. Maldonado, D. Reinberg, and M. R. Green: Interaction between an acidic activator and transcription factor TFIIB is required for transcriptional activation. *Nature* 363:741, 1993.
185. N. C. Jones, P. W. J. Rigby, and E. B. Ziff: Trans-acting protein factors and the regulation of eukaryotic transcription: Lessons from studies on DNA tumor viruses. *Genes and Development.* Vol. 2. New York: Cold Spring Harbor Laboratory, 1988, p. 267.
186. T. Hai and T. Curran: Cross-family dimerization of transcription factors Fos/Jun and ATF/CREB alters DNA binding specificity. *Proc Natl Acad Sci USA* 88:3720, 1991.
187. C. Abate, S. J. Baker, S. P. Lees-Miller, C. W. Anderson, D. R. Marshak, and T. Curran: Dimerization and DNA binding alter phosphorylation of Fos and Jun. *Proc Natl Acad Sci USA* 90:6766, 1993.
188. M. I. Diamond, J. N. Miner, S. K. Yoshinaga, and K. R. Yamamoto: Transcription factor interactions: Selectors of positive or negative regulation from a single DNA element. *Science* 249:1266, 1990.
189. R. Schüle, P. Rangarajan, S. Kliewer, L. J. Ransone, J. Bolado, N. Yang, I. M. Verma, and R. M. Evans: Functional antagonism between oncoprotein c-Jun and the glucocorticoid receptor. *Cell* 62:1217, 1990.
189a. C. M. Pfarr, F. Mechta, G. Spyrou, D. Lallemand, S. Carillo, and M. Yaniv: Mouse JunD negatively regulates fibroblast growth and antagonizes transformation by ras. *Cell* 76:747, 1994.
189b. Z. Dong, M. J. Birrer, R. G. Watts, L. M. Matrisian, and N. H. Colburn: Blocking of tumor promoter-induced AP-1 activity inhibits induced transformation in JB6 mouse epidermal cells. *Proc Natl Acad Sci USA* 91:609, 1994.
190. M. R. Briggs, J. T. Kadonaga, S. P. Bell, and R. Tjian: Purification and biochemical characterization of the promoter-specific transcription factor, SP1. *Science* 234:47, 1986.
191. A. J. Courey, D. A. Holtzman, S. P. Jackson, and R. Tjian: Synergistic activation by the glutamine-rich domains of human transcription factor Sp1. *Cell* 59:827, 1989.
192. A. J. Udvadia, K. T. Rogers, P. D. R. Higgins, Y. Murata, K. H. Martin, P. A. Humphrey, and J. M. Horowitz: Sp-1 binds promoter elements regulated by the RB protein and Sp-1-mediated transcription is stimulated by RB coexpression. *Proc Natl Acad Sci USA* 90:3265, 1993.
193. M. H. Rosner, M. A. Vigano, P. W. J. Rigby, H. Arnheither, and L. M. Staudt: Oct-3 and the beginning of mammalian development. *Science* 253:144, 1991.
194. W. Wahli and E. Martinez: Superfamily of steroid nuclear receptors: Positive and negative regulators of gene expression. *FASEB J* 5:2243, 1991.
195. K. Umesono, K. K. Murakami, C. C. Thompson,

and R. M. Evans: Direct repeats as selective response elements for the thyroid hormone, retinoic acid, and vitamin D_3 receptors. *Cell* 65:1255, 1991.

196. H. Weintraub, R. Davis, S. Tapscott, M. Thayer, M. Krause, R. Benezra, T. K. Blackwell, D. Turner, R. Rupp, S. Hollenberg, Y. Zhuang, and A. Lassar: The myoD gene family: Nodal point during specification of the muscle cell lineage. *Science* 251:761, 1991.
197. D. G. Edmondson and E. N. Olson: Helix-loop-helix proteins as regulators of muscle-specific transcription. *J Biol Chem* 268:755, 1993.
198. A. B. Lassar, R. L. Davis, W. E. Wright, T. Kadesch, C. Murre, A. Voronova, D. Baltimore, and H. Weintraub: Functional activity of myogenic HLH proteins requires heterooligomerization with E12/E47-like proteins in vivo. *Cell* 66:305, 1991.
199. R. Benezra, R. L. Davis, D. Lockshon, D. L. Turner, and H. Weintraub: The protein Id: A negative regulator of helix-loop-helix DNA binding proteins. *Cell* 61:49, 1990.
200. E. Bengal, L. Ransone, R. Scharfmann, V. J. Dwarki, S. J. Tapscott, H. Weintraub, and I. M. Verma: Functional antagonism between c-Jun and MyoD proteins: A direct physical association. *Cell* 68:507, 1992.
201. D. B. Mendel and G. R. Crabtree: HNF-1, a member of a novel class of dimerizing homeodomain proteins. *J Biol Chem* 266:677, 1991.
202. M. S. Kapiloff, Y. Farkash, M. Wegner, and M. G. Rosenfeld: Variable effects of phosphorylation of Pit-1 dictated by the DNA response elements. *Science* 253:786, 1991.
203. M. A. Van Duk, P. M. Voorhoeve, and C. Murre: Pbx1 is converted into a transcriptional activator upon aquiring the N-terminal region of E2A in pre-B-cell acute lymphoblastoid leukemia. *Proc Natl Acad Sci USA* 90:6061, 1993.
204. M. J. Lenardo and D. Baltimore: NF-κB: A pleiotropic mediator of inducible and tissue-specific gene control. *Cell* 58:227, 1989.
205. M. N. Treacy, X. He, and M. G. Rosenfeld: I-POU: a POU-domain protein that inhibits neuron-specific gene activation. *Nature* 350:577, 1991.
206. I. Kola, S. Brookes, A. R. Green, R. Garber, M. Tymms, T. S. Papas, and A. Seth: The Ets1 transcription factor is widely expressed during murine embryo development and is associated with mesodermal cells involved in morphogenetic processes such as organ formation. *Proc Natl Acad Sci USA* 90:7588, 1993.
207. J. W. Pendleton, B. K. Nagai, M. T. Murtha, and F. H. Ruddle: Expansion of the *Hox* gene family and the evolution of chordates. *Proc Natl Acad Sci USA* 90:6300, 1993.
208. M. T. Murtha, J. F. Leckman, and F. H. Ruddle: Detection of homeobox genes in development and evolution. *Proc Natl Acad Sci USA* 88:10711, 1991.
209. W. McGinnis and R. Krumlauf: Homeobox genes and axial patterning. *Cell* 68:283, 1992.
210. W. J. Gehring: The homeobox in perspective. *Trends Biochem Sci* 200:277, 1992.
210a. R. Krumlauf: *Hox* genes in vertebrate development. *Cell* 78:191, 1994.
211. F. S. Jones, B. D. Holst, O. Minowa, E. M. De Robertis, and G. M. Edelman: Binding and transcriptional activation of the promoter for the neural cell adhesion molecule by HoxC6 (Hox-3.3). *Proc Natl Acad Sci USA* 90:6557, 1993.
212. D. A. Grueneberg, S. Natesan, C. Alexandre, and M. Z. Gilman: Human and *Drosophila* homeodomain proteins that enhance the DNA-binding activity of serum response factor. *Science* 257:1089, 1992.
213. K. Takeshita, J. A. Bollekens, N. Hijiya, M. Ratajczak, F. H. Ruddle, and A. M. Gewirtz: A homeobox gene of the antennapedia class is required for human adult erythropoiesis. *Proc Natl Acad Sci USA* 90:3535, 1993.
214. M. Hatano, C. W. M. Roberts, M. Minden, W. M. Crist, and S. J. Korsmeyer: Deregulation of a homeobox gene, HOX11, by the t(10;14) in T cell leukemia. *Science* 253:79, 1991.
215. T. N. Dear, I. Sanchez-Garcia, and T. H. Rabbitts: The HOX11 gene encodes a DNA-binding nuclear transcription factor belonging to a distinct family of homeobox genes. *Proc Natl Acad Sci USA* 90:4431, 1993.
216. T. Maniatis: Mechanisms of alternative pre-mRNA splicing. *Science* 251:33, 1991.
217. A. M. Zahler, K. M. Neugebauer, W. S. Lane, and M. B. Roth: Distinct functions of SR proteins in alternative Pre-mRNA splicing. *Science* 260:219, 1993.
218. J. M. Gualberto, L. Lamattina, G. Bonnard, J.-H. Weil, and J.-M. Grienenberger: RNA editing in wheat mitochondria results in the conservation of protein sequences. *Nature* 341:660, 1989.
219. P. S. Covello and M. W. Gray: RNA editing in plant mitochondria. *Nature* 341:662, 1989.
220. L. F. Landweber and W. Gilbert: RNA editing as a source of genetic variation. *Nature* 363:179, 1993.
221. R. D. Klausner and J. B. Harford: Cis-trans models for posttranscriptional gene regulation. *Science* 246:870, 1989.
222. M. Kozak: Structural features in eukaryotic mRNAs that modulate the initiation of translation. *J Biol Chem* 266:19867, 1991.
223. M. Kozak: Bifunctional messenger RNAs in eukaryotes. *Cell* 47:481, 1986.
224. A. Lazaris-Karatzas, K. S. Montine, and N. Sonenberg: Malignant transformation by a eukaryotic initiation factor subunit that binds to mRNA 5′ cap. *Nature* 345:544, 1990.
225. J. L. Hargrove and F. H. Schmidt: The role of mRNA and protein stability in gene expression. *FASEB J* 3:2360, 1989.
226. K. A. Venstrom and L. F. Reichardt: Extracellular matrix 2: Role of extracellular matrix mol-

ecules and their receptors in the nervous system. *FASEB J* 7:996, 1993.

227. R. Timpl, J. Engel, and G. R. Martin: Laminin-A multifunctional protein of basement membranes. *Trends Biochem Sci,* June 1983, p. 207.
228. S. Kubota, K. Tashiro, and Y. Yamada: Signaling site of laminin with mitogenic activity. *J Biol Chem* 267:4285, 1992.
229. M. C. Kibbey, D. S. Grant, and H. K. Kleinman: Role of the SIKVAV site of laminin in promotion of angiogenesis and tumor growth: An in vivo matrigel model. *JNCI* 84:1633, 1992.
230. R. Fridman, M. C. Kibbey, L. S. Royce, M. Zain, T. M. Sweeney, D. L. Jicha, J. R. Yannelli, G. R. Martin, and H. K. Kleinman: Enhanced tumor growth of both primary and established human and murine tumor cells in athymic mice after coinjection with matrigel *JNCI* 83:769, 1991.
231. J. Lawler and R. O. Hynes: The structure of human thrombospondin, and adhesive glycoprotein with multiple calcium-binding sites and homologies with several different proteins. *J Cell Biol* 103:1635, 1986.
232. R. Yabkowitz, P. J. Mansfield, V. M. Dixit, and S. J. Suchard: Motility of human carcinoma cells in response to thrombospondin: Relationship to metastatic potential and thrombospondin structural domains. *Cancer Res* 53:378, 1993.
233. V. C. Hascall and G. K. Hascall: Proteoglycans, in E. D. Hay, ed.: *Cell Biology of Extracellular Matrix.* New York: Plenum Press, 1981, p. 39.
234. G. David: Integral membrane heparan sulfate proteoglycans. *FASEB J* 7:1023, 1993.
235. J. L. Stow and M. G. Farquhar: Distinctive populations of basement membrane and cell membrane heparan sulfate proteoglycans are produced by cultured cell lines. *J Cell Biol* 105:529, 1987.
236. V.-M. Kähäri, H. Larjava, and J. Uitto: Differential regulation of extracellular matrix proteoglycan (PG) gene expression. *J Biol Chem* 266:10608, 1991.
237. N. M. LeDouarin: Cell migrations in embryos. *Cell* 38:353, 1984.
238. S. Ben Slimane, F. Houlier, G. Tucker, and J. P. Thiery: In vitro migration of avian hemopoietic cells to the thymus: Preliminary characterization of a chemotactic mechanism. *Cell Diff* 13:1, 1983.
239. W. M. Gallatin, I. L. Weissman, and E. C. Butcher: A cell surface molecule involved in organ specific homing of lymphocytes. *Nature* 304:30, 1983.
240. D. Gospodarowicz, D. Delgado, and I. Vlodavsky: Permissive effect of the extracellular matrix on cell proliferation in vitro. *Proc Natl Acad Sci USA* 77:4094, 1980.
241. J. Folkman and A. Moscona: Role of cell shape in growth control. *Nature* 273:345, 1978.
242. C. Grobstein: Inductive epithelio-mesenchymal interactions in cultured organ rudiments of the mouse. *Science* 118:128, 1953.
243. L. M. Reid and D. M. Jefferson: Cell culture studies using extracts of extracellular matrix to study growth and differentiation in mammalian cells, in J. P. Mather, ed.: *Mammalian Cell Culture.* New York: Plenum Press, 1984, p. 239.
244. C. H. Streuli, N. Bailey, and M. J. Bissell: Control of mammary epithelial differentiation: Basement membrane induces tissue-specific gene expression in the absence of cell-cell interaction and morphological polarity. *J Cell Biol* 115:1383, 1991.
245. C. Q. Lin and M. J. Bissell: Multi-faceted regulation of cell differentiation by extracellular matrix. *FASEB J* 7:737, 1993.
246. R. L. Juliano and S. Haskill: Signal transduction from the extracellular matrix. *J Cell Biol* 120:577, 1993.
247. R. Bützow, D. Fukushima, D. R. Twardzik, and E. Ruoslahti: A 60-kD protein mediates the binding of transforming growth factor-β to cell surface and extracellular matrix proteoglycans. *J Cell Biol* 122:721, 1993.
248. E. G. Hayman, E. Engvall, and E. Ruoslahti: Concomitant loss of cell surface fibronectin and laminin from transformed rat kidney cells. *J Cell Biol* 88:352, 1981.
249. D. R. Pitelka, S. T. Hamamoto, and B. N. Taggart: Basal lamina and tissue recognition in malignant mammary tumors. *Cancer Res* 40:1600, 1980.
250. E. Ruoslahti and M. D. Pierschbacher: New perspectives in cell adhesion: RGD and integrins. *Science* 238:491, 1987.
251. R. D. Hynes: Integrins: A family of cell surface receptors. *Cell* 48:549, 1987.
252. S. M. Albelda and C. A. Buck: Integrins and other cell adhesion molecules. *FASEB J* 4:2868, 1990.
253. S. Dedhar and R. Saulnier: Alterations in integrin receptor expression on chemically transformed human cells: Specific enhancement of laminin and collagen receptor complexes. *J Cell Biol* 110:481, 1990.
254. V. Cioce, V. Castronovo, B. M. Shmookler, S. Garbisa, W. F. Grigioni, L. A. Liotta, and M. E. Sobel: Increased expression of the laminin receptor in human colon cancer. *JNCI* 83:29, 1991.
255. P. Clezardin, L. Frappart, M. Clerget, C. Pechoux, and P. D. Delmas: Expression of thrombospondin (TSPI) and its receptors (CD36 and CD51) in normal, hyperplastic, and neoplastic human breast. *Cancer Res* 53:1421, 1993.
256. V. Castronovo, G. Taraboletti, L. A. Liotta, and M. E. Sobel: Modulation of laminin receptor expression by estrogen and progestins in human breast cancer cell lines. *JNCI* 81:781, 1989.
257. J. R. Stanley and S. H. Yuspa: Specific epidermal protein markers are modulated during calcium-

induced terminal differentiation. *J Cell Biol* 96:1809, 1983.
258. K. A. Knudsen and A. F. Horwitz: Tandem events in myoblast fusion. *Dev Biol* 58:328, 1977.
259. W. B. Busa and J. H. Crowe: Intracellular pH regulated transitions between dormancy and development of brine shrimp *(Artemia salina)* embryos. *Science* 221:366, 1983.
260. J. D. Gross, J. Bradbury, R. R. Kay, and M. J. Peacey: Intracellular pH and the control of cell differentiation in *Dictyostelium discoideum. Nature* 303:244, 1983.
261. M. Darmon, P. Brachet, and L. H. Pereira da Silva: Chemotactic signals induce cell differentiation in *Dictyostelium discoideum. Proc Natl Acad Sci USA* 72:3163, 1975.
262. B. M. Shaffer: Secretion of cyclin AMP induced by cyclin AMP in the cellular slime mold *Dictyostelium discoideum. Nature* 255:549, 1975.
263. J. M. Louis, C. L. Saxe III, and A. R. Kimmel: Two transmembrane signaling mechanisms control expression off the cAMP receptor gene CAR1 during *Dictyostelium* development. *Proc Natl Acad Sci USA* 90:5969, 1993.
264. K. N. Prasad and A. W. Hsie: Morphological differentiation of mouse neuroblastoma cells induced in vitro by dibutyryl adenosine 3′,5′ cyclic monophosphate. *Nature New Biol* 233:141, 1971.
265. A. W. Hsie, C. Jones, and T. T. Puck: Further changes in differentiation state accompanying the conversion of Chinese hamster cells to fibroblastic form by dibutyryl adenosine cyclic 3′:5′-monophosphate and hormones. *Proc Natl Acad Sci USA* 68:1648, 1971.
266. I. Pastan and M. Willingham: Cellular transformation and the "morphologic phenotype" of transformed cells. *Nature* 274:645, 1978.
267. R. J. Zalin and W. Montague: Changes in adenylate cyclase, cyclic AMP, and protein kinase levels in chick myoblasts, and their relationship to differentiation. *Cell* 2:103, 1974.
268. R. E. Scott, D. L. Florine, J. J. Wille, Jr., and K. Yun: Coupling of growth arrest and differentiation at a distinct state in the G_1 phase of the cell cycle: G_D. *Proc Natl Acad Sci USA* 79:845, 1982.
269. D. Yaffé: Cellular aspects of muscle differentiation in vitro, in A. A. Moscona and A. Monroy, eds.: *Current Topics in Developmental Biology.*Vol 4. New York: Academic Press, 1969, p. 37.
270. E. Hadorn: Regulation and differentiation with Field-districts in imaginal discs of *Drosophila. J Embryol Exp Morphol* 1:213, 1953.
271. E. Hadorn: Problems of determination and transdetermination. *Brookhaven Symp Biol* 18:148, 1965.
272. T. Yamada: Cellular and subcellular events in Wolffian lens regeneration, in by A. A. Moscona and A. Monroy, eds.: *Current Topics in Developmental Biology*. Vol. 2. New York: Academic Press, 1967, p. 247.
273. L. S. Stone: The role of retinal pigment cells in regenerating neural retinae of adult salamander eyes. *J Exp Zool* 113:9, 1950.
274. G. B. Pierce: Differentiation of normal and malignant cells. *Fed Proc* 29:1248, 1970.
275. S. Sell: Is there a liver stem cell? *Cancer Res* 50:3811, 1990.
276. T. A. Kost, M. J. Koury, W. D. Hankins, and S. B. Krantz: Target cells for Friend virus-induced erythroid bursts in vitro. *Cell* 18:145, 1979.
277. T. Graf, N. Ade, and H. Beug: Temperature-sensitive mutant of avian erythroblastosis virus suggests a block of differentiation as a mechanism of leukemogenesis. *Nature* 275:496, 1978.
278. J. Samarut and L. Gazzolo: Target cells infected by avian erythroblastosis virus differentiate and become transformed. *Cell* 28:921, 1982.
279. V. R. Potter: Recent trends in cancer biochemistry: The importance of studies on fetal tissue. *Can Cancer Conf* 8:9, 1969.
280. R. Dulbecco, M. Henahan, and B. Armstrong: Cell types and morphogenesis in the mammary gland. *Proc Natl Acad Sci USA* 79:7346, 1982.
281. R. N. Buick and M. N. Pollak: Perspectives on clonogenic tumor cells, stem cells, and oncogenes. *Cancer Res* 44:4909, 1984.
282. L. C. Stevens, Jr., and C. C. Little: Spontaneous testicular teratomas in an inbred strain of mice. *Proc Natl Acad Sci USA* 40:1080, 1954.
283. L. C. Stevens: Experimental production of testicular teratomas in mice. *Proc Natl Acad Sci USA* 52:654, 1964.
284. L. C. Stevens: Development of resistance to teratocarcinogenesis by primodial germ cells in mice. *JNCI* 37:859, 1966.
285. G. B. Pierce and F. J. Dixon, Jr.: Testicular teratomas: I. The demonstration of teratogenesis by metamorphosis of multipotential cells. *Cancer* 12: 573, 1959.
286. L. J. Kleinsmith and G. B. Pierce: Multipotentiality of single embryonal carcinoma cells. *Cancer Res* 24:1544, 1964.
287. R. L. Brinster: The effect of cells transferred into the mouse blastocyst on subsequent development. *J Exp Med* 140:1049, 1974.
288. B. Mintz and K. Illmensee: Normal genetically mosaic mice produced from malignant teratocarcinoma cells. *Proc Natl Acad Sci USA* 72:3585, 1975.
289. K. Illmensee and L. C. Stevens: Teratomas and chimeras. *Sci Am* 240:121, 1979.
290. K. Illmensee and B. Mintz: Totipotency and normal differentiation of single teratocarcinoma cells cloned by injection into blastocysts. *Proc Natl Acad Sci USA* 73:549, 1976.
291. G. B. Pierce, D. Aguilar, G. Hood, and R. S. Wells: Trophoectoderm in control of murine embryonal carcinoma. *Cancer Res* 44:3987, 1984.

292a. G. B. Pierce, C. G. Pantazis, J. E. Caldwell, and R. S. Wells: Specificity of the control of tumor formation by the blastocyst. *Cancer Res* 42:1082, 1982.
292b. S. Strickland, K. K. Smith, and K. R. Marotti: Hormonal induction of differentiation in teratocarcinoma stem cells: Generation of parietal endoderm by retinoic acid and dibutyryl cAMP. *Cell* 21:347, 1980.
293. P. A. McCue, M. L. Gubler, M. I. Sherman, and B. N. Cohen: Sodium butyrate induces histone hyperacetylation and differentiation of murine embryonal carcinoma cells. *J Cell Biol* 98:602, 1984.
294. M. W. McBurney, E. M. V. Jones-Villeneuve, M. K. S. Edwards, and P. J. Anderson: Control of muscle and neuronal differentiation in a cultured embryonal carcinoma cell line. *Nature* 299:165, 1982.
295. B. B. Knowles, S. Pan, D. Solter, A. Linnenbach, C. Croce, and K. Heubner: Expression of H-2, laminin and SV40 T and TASA on differentiation of transformed murine teratocarcinoma cells. *Nature* 288:615, 1980.
296. W. C. Speers and M. Altmann: Chemically induced differentiation of murine embryonal carcinoma in vivo: Transplantation of differentiated tumors. *Cancer Res* 44:2129, 1984.
297. Y. Ichikawa: Differentiation of a cell line of myeloid leukemia. *J Cell Physiol* 74:223, 1969.
298. Y. Ichikawa: Further studies on the differentiation of a cell line of myeloid leukemia. *J Cell Physiol* 76:175, 1970.
299. J. A. Sokoloski, G. P. Beardsley, and A. C. Sartorelli: Mechanism of the induction of the differentiation of HL-60 leukemia cells by antifolates. *Cancer Commun* 1:199, 1989.
300. T. R. Breitman and R. He: Combinations of retinoic acid with either sodium butyrate, dimethyl sulfoxide, or hexamethylene bisacetamide synergistically induce differentiation of the human myeloid leukemia cell line HL60. *Cancer Res* 50:6268, 1990.
301. H. Aihara, Y. Asaoka, K. Yoshida, and Y. Nishizuka: Sustained activation of protein kinase C is essential to HL-60 cell differentiation to macrophage. *Proc Natl Acad Sci USA* 88:11062, 1991.
302. T. R. Breitman, S. E. Selonick, and S. J. Collins: Induction of differentiation of the human promyelocytic leukemia cell line (HL-60) by retinoic acid. *Proc Natl Acad Sci USA* 77:2936, 1980.
303. R. P. Warrell, Jr., H. De Thé, Z.-Y. Wang, and L. Degos: Acute promyelocytic leukemia. *N Engl J Med* 329:177, 1993.
304. C. L. Sawyers, C. T. Denny, and O. N. Witte: Leukemia and the disruption of normal hematopoiesis. *Cell* 64:337, 1991.
305. F. Grignani, P. F. Ferrucci, U. Testa, G. Talamo, M. Fagioll, M. Alcalay, A. Mencarelli, F. Grignani, C. Peschle, I. Nicoletti, and P. G. Pelicci: The acute promyelocytic leukemia-specific PML-RARα fusion protein inhibits differentiation and promotes survival of myeloid precursor cells. *Cell* 74:423, 1993.
306. C. Friend and J. R. Haddad: Tumor formation with transplants of spleen or liver from mice with virus-induced leukemia. *JNCI* 25:1279, 1960.
307. C. Friend, W. Scher, J. G. Holland, and T. Sato: Hemoglobin synthesis in murine virus-induced leukemic cells in vitro: Stimulation of erythroid differentiation of demethyl sulfoxide. *Proc Natl Acad Sci USA* 68:378, 1971.
308. P. A. Marks and R. A. Rifkind: Erythroleukemic differentiation. *Annu Rev Biochem* 47:419, 1978.
309. H. Kiyokawa, V. M. Richon, G. Venta-Perez, R. A. Rifkind, and P. A. Marks: Hexamethylenebisacetamide-induced erythroleukemia cell differentiation involves modulation of events required for cell cycle progression through G_1. *Proc Natl Acad Sci USA* 90:6746, 1993.
310. R. L. Smith, I. G. Macara, R. Levenson, D. Houseman, and L. Cantley: Evidence that a Na^+/Ca^{2+} antiport system regulates murine erythroleukemia cell differentiation. *J Biol Chem* 257:773, 1982.

6

Causes of Cancer

So far in this book we have discussed the phenomenology of cancer: its general pathology, its apparent relationships to environmental factors, and its phenotypic characteristics at the cellular and subcellular level. Perhaps the most important question in cancer biology is what *causes* the cellular alterations that produce a cancer. The answer to this question has been elusive. If the actual cause of these alterations were known, the elimination of factors that produce cancer and the development of better treatment modalities would be likely to follow.

We observed in earlier chapters that a cancerous growth has a number of predictable properties. The incidence rates of various cancers are strongly related to environmental factors and life-style, and cancers have certain growth characteristics, among which are the ability to grow in an uncontrolled manner, to invade surrounding tissues, and to metastasize. Also, when viewed microscopically, cancer cells appear to be less well differentiated than their normal counterparts and to have certain distinguishing features, such as large nuclei and nucleoli. Most cancers arise from a single clone of cells, the precursor of which has been altered by insult with a carcinogen. There is frequently a long latent period, in some cases 20 years or more, between the initiating insult and the appearance of a clinically detectable tumor. During this time, cellular proliferation must occur, but it may originally be limited by host defenses or lack of access to the host's blood supply. During the process of tumor progression, however, escape from the host's defense mechanisms and vascularization of the growing tumor ultimately occur.

The genetic instability of cancer cells leads to the emergence of a more aggressively growing tumor frequently characterized by the appearance of poorly differentiated cells with certain properties of a more embryonic phenotype. During tumor progression, considerable biochemical heterogeneity becomes manifest in the growing tumor and its metastases, even though all the neoplastic cells may have arisen originally from a single deranged cell. Any theory that seeks to explain the initiation of cancer and its progression must take these observations into consideration.

In this chapter, we examine what is known about various chemical, physical, and viral carcinogenic agents and discuss the putative mechanisms by which they cause cancer.

CHEMICAL CARCINOGENESIS

Historical Perspectives

Carcinogenic chemicals and irradiation (ionizing and ultraviolet) are known to affect DNA and to be mutagenic under certain conditions. Thus, one of the long-standing theories of carcinogenesis is that cancer is caused by a genetic mutation.

The evidence that chemicals can induce cancer in humans has been accumulating for more than two centuries (for review, see Ref. 1). The first observation associating chemicals with cancer was made by John Hill, who in 1761 noted that nasal cancer occurred in people who used snuff excessively. In 1775, Percival Pott reported a high incidence of scrotal skin cancer among men who had spent their childhood as chimney sweeps. One hundred years later, von Volkman, in Germany, and Bell, in Scotland, observed skin cancer in workers whose skin was in continuous contact with tar and paraffin oils, which we now know contain polycyclic aromatic hydrocarbons. In 1895, Rehn reported the development of urinary bladder cancer in aniline dye workers in Germany. Similar observations were later made in a number of countries and established a relationship between heavy exposure to 2-naphthylamine, benzidine, or 4-aminobiphenyl and bladder cancer. Thus, the first observations of chemically induced cancer were made in humans. These observations led to attempts to induce cancer in animals with chemicals. One of the first successful attempts was made in 1915, when Yamagiwa and Ichikawa induced skin carcinomas by the repeated application of coal tar to the ears of rabbits. This and similar observations by other investigators led to a search for the active carcinogen in coal tar and to the conclusion that the carcinogenic agents in tars are the polycyclic aromatic hydrocarbons. Direct evidence for that came in the 1930s from the work of Kennaway and Hieger, who demonstrated that synthetic 1,2,5,6-dibenzanthracene is a carcinogen, and from the identification of the carcinogen 3,4-benzpyrene in coal tar by Cook,

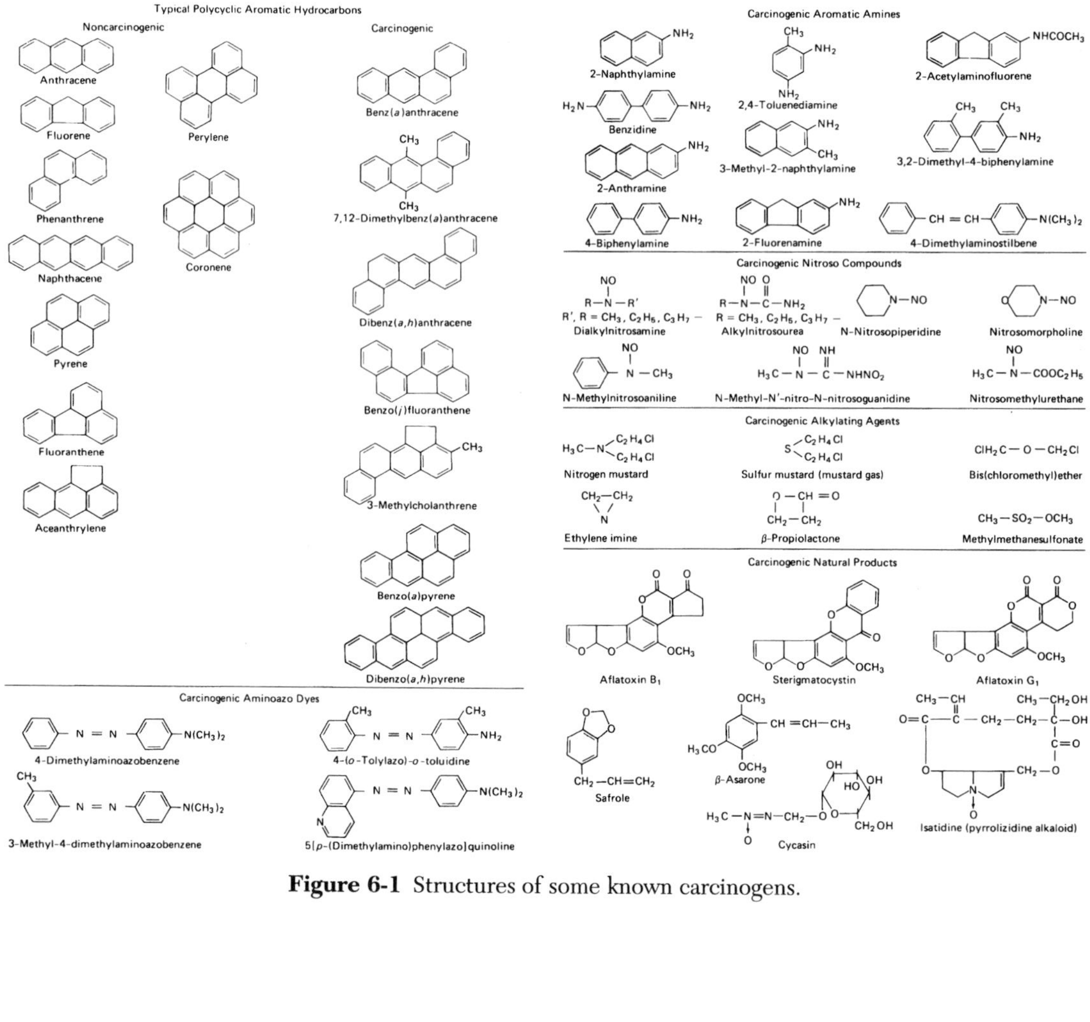

Figure 6-1 Structures of some known carcinogens.

Hewitt, and Hieger. Induction of tumors by other chemical and hormonal carcinogens was described in the 1930s, including the induction of liver tumors in rats and mice with 2′,3-dimethyl-4-aminoazobenzene by Yoshida, of urinary bladder cancer in dogs with 2-naphthylamine by Hueper, Wiley, and Wolfe, and of mammary cancer in male mice with estrone by Lacassagne. The list of known carcinogenic chemicals expanded in the 1940s with the discovery of the carcinogenicity of 2-acetylaminofluorene, halogenated hydrocarbons, urethane, beryllium salts, and certain anticancer alkylating agents. Since the 1940s, various nitrosamines, intercalating agents, nickel and chromium compounds, asbestos, vinyl chloride, diethylstilbestrol, and certain naturally occurring substances, such as aflatoxins, have been added to the list of known carcinogens. The structures of some known carcinogens are shown in Figure 6-1.

Metabolic Activation of Chemical Carcinogens

As studies on the reactions of carcinogens with cellular macromolecules progressed, it became apparent that most of these interactions resulted from covalent bond formation between an electrophilic form of the carcinogen and the nucleophilic sites in proteins (e.g., sulfur, oxygen, and nitrogen atoms in cysteine, tyrosine, and histidine, respectively) and nucleic acids (e.g., purine or pyrimidine ring nitrogens and oxygens). Frequently, the parent compound itself did not interact in vitro with macromolecules until it had been incubated with liver homogenates or liver microsomal fractions. These studies led to the realization that metabolic activation of certain carcinogenic agents is necessary to produce the "ultimate carcinogen" that actually reacts with crucial molecules in target cells. With the exception of the very chemically reactive alkylating agents, which are activated in aqueous solution at physiologic pH (e.g., *N*-methyl-*N*-nitrosourea) and the agents that intercalate into the DNA double helix by forming tight noncovalent bonds (e.g., daunorubicin), most of the known chemical carcinogens undergo some metabolic conversions that appear to be required for their carcinogenic action. Some examples of these metabolic conversions are given next.

Donors of Simple Alkyl Groups

Included in this group are the dialkylnitrosamines, dialkylhydrazines, aryldialkyltriazenes, alkylnitrosamides, and alkylnitrosimides. The alkylnitrosamides and alkylnitrosimides do not require enzymatic activation because they can react directly with water or cellular nucleophilic groups. The alkylnitrosamines, alkylhydrazines, and alkyltriazenes, however, undergo an enzyme-mediated activation step to form the reactive electrophile (Fig. 6-2). These agents are metabolically dealkylated by the mixed-function oxidase system in the microsomal fraction (endoplasmic reticulum) of cells, primarily liver cells. The monoalkyl derivatives then undergo a nonenzymatic, spontaneous conversion to monoalkyldiazonium ions that donate an alkyl group to cellular nucleophilic groups in DNA, RNA, and protein.[2]

2-Acetylaminofluorene

The metabolic interconversions of this compound were studied in detail by the Millers and their colleagues.[1,3] In 1960, it was shown that AAF is converted to a more potent carcinogen, *N*-hydroxy-AAF, after the parent compound was fed to rats. Although both AAF and *N*-hydroxy-AAF are carcinogenic in vivo, neither compound

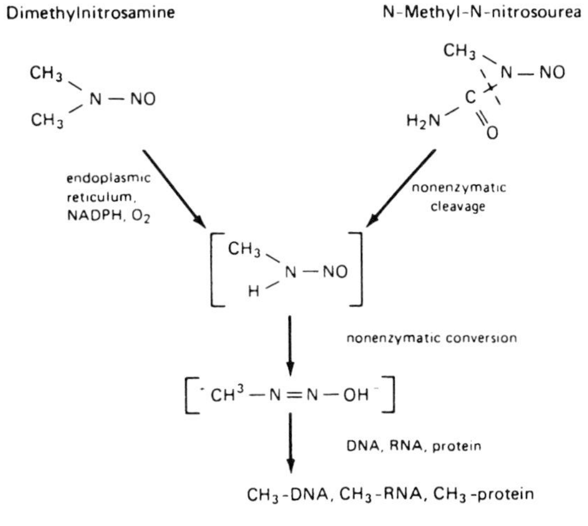

Figure 6-2 The enzymatic and nonenzymatic activations of dimethylnitrosamine and *N*-methyl-*N*-nitroso reactive nucleophiles.

reacted in vitro with nucleic acids or proteins, suggesting that the ultimate carcinogen was another, as yet unidentified metabolite. Subsequent studies showed that N-hydroxy-AAF is converted in rat liver to a sulfate, *N*-sulfonoxy-AAF, by means of a cytosol sulfotransferase activity (Fig. 6-3). This compound reacts with nucleic acids and proteins and appears to be the ultimate carcinogen in vivo. It is also highly mutagenic, as determined by assays of DNA-transforming activity (see below).

Other enzymatic conversions of AAF occur in rat liver, for example, *N*-hydroxy-AAF is converted to *N*-acetoxy-AAF, *N*-acetoxy-2-aminofluorene and the *O*-glucuronide (conjugate with glucuronic acid). These enzymatic reactions may also be involved in the conversion of AAF to carcinogenic metabolites, especially in nonhepatic tissues, which often have low sulfotransferase activity for *N*-hydroxy-AAF. The acetyltransferase-mediated activity converts *N*-hydroxy-AAF to *N*-acetoxy-2-aminofluorene, which is also a strong electrophile and may be the ultimate carcinogen in nonhepatic tissues.

Other Aromatic Amines

Electrophilic forms of the aromatic amines result from their metabolic activation, and the positively charged nitrenium ion that is formed from naphthylamine and aminobiphenyl compounds has been implicated as the ultimate urinary bladder carcinogen in dogs and humans (Fig. 6-4). Hydroxylamine derivatives of these compounds are formed in the liver and then converted to a glucuronide. The glucuronide conjugate is excreted in the urine, where the acid pH can convert it back to hydroxylamine and subsequently to a protonated hydroxylamine, which rearranges to form a nitrenium ion by a loss of water. The electrophilic nitrenium ion can then react with nucleophilic targets in the urinary bladder epithelium.

Polycyclic Aromatic Hydrocarbons

In 1950, Boyland[4] suggested that the carcinogenicity of polycyclic aromatic hydrocarbons (PAHs) was mediated through metabolically

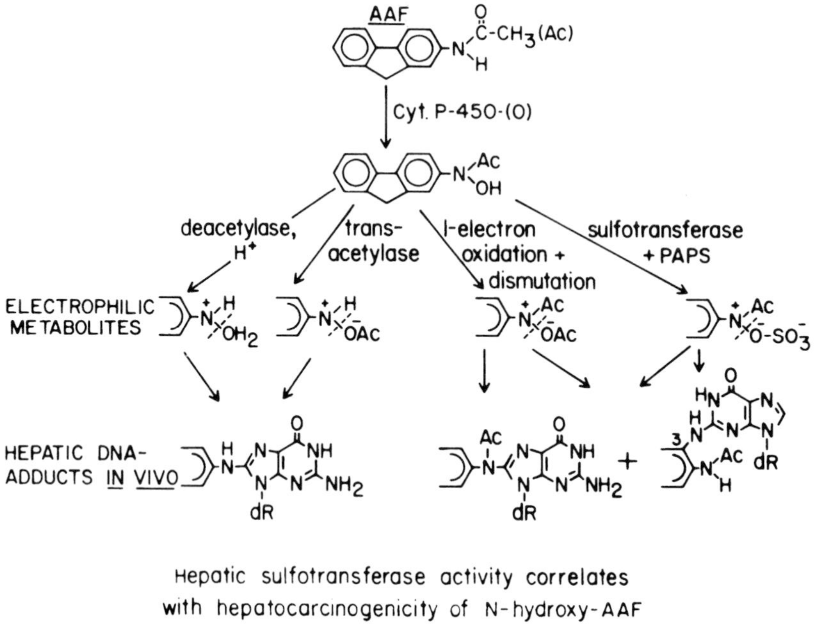

Figure 6-3 The major metabolic pathway in rat liver for activation of AAF to the strong electrophile *N*-sulfonoxy-AAF. 3′-Phosphoadenosine-5′-phosphosulfate (PAPS) is the active sulfate donor for the sulfotransferase activity. (Data from Miller.[3])

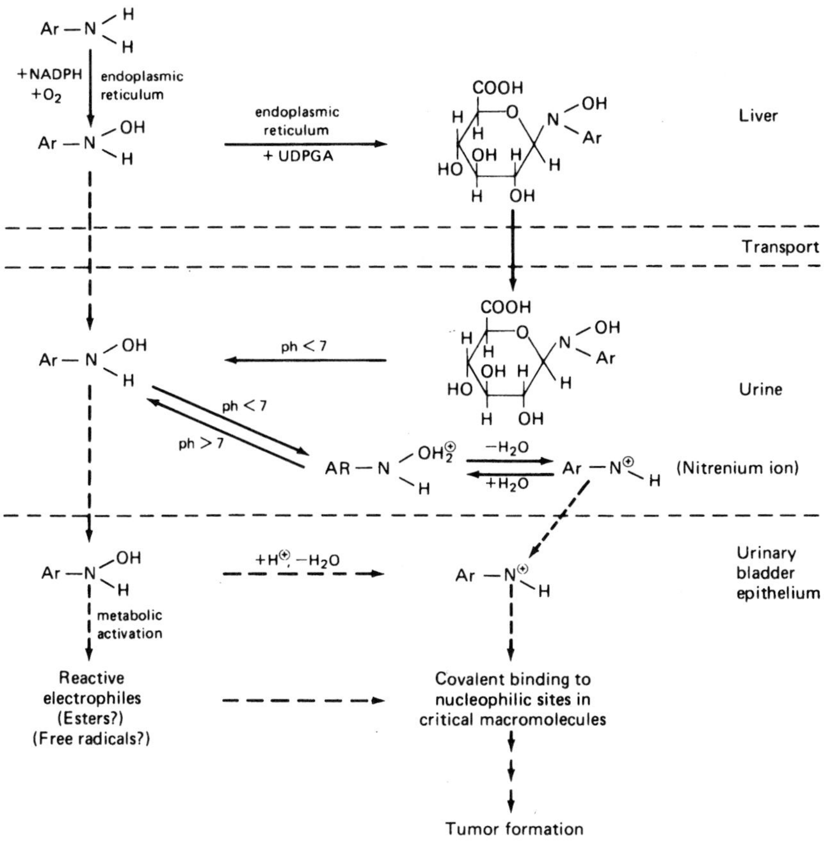

Figure 6-4 Formation and transport of possible proximate and ultimate carcinogenic metabolites of arylamines involved in the induction of urinary bladder cancer; Ar, aryl substituent; UDPGA, uridine diphosphoglucuronic acid. (From Miller.[1])

formed epoxides. It was originally thought that the key expoxide formation involved the K-region of the hydrocarbon ring structure.[5] However, subsequent studies demonstrated that K-region epoxides had little carcinogenicity in vivo. An extensive amount of work has gone on since the 1950s to characterize the metabolism and carcinogenic potential of the PAHs (for review, see Ref. 6). It is now generally accepted that conversion of PAH to dihydrodiol epoxides is a crucial pathway (though not the only one; see below) in the formation of the ultimate carcinogen (Fig. 6-5). Studies from a number of laboratories, for instance, have indicated that 7β, 8α-dihydroxy-9α,10αepoxy-7,8,9,10-tetrahydrobenzo[*a*]pyrene is an ultimate mutagenic and carcinogenic metabolite of benzo[*a*]pyrene.[7–9] The evidence is strong that the analogous metabolites of other PAHs that are similar to benzo[*a*]pyrene are also the ultimate carcinogens of these compounds.

When the various metabolites of benzo[*a*]pyrene (BP) were tested for mutagenicity in the Ames test, using tester strains of *Salmonella typhimurium,* and in Chinese hamster V-79 cells (see below), and tested for carcinogenicity by injection into newborn mice, it became clear that the most potent mutagen/carcinogen was the (+)enantiomer of BP 7,8-diol-9,10-epoxide-1 (see Fig. 6-5).[6] When the (+)- and (−)-enantiomers were reacted with double-stranded DNA, it was determined that the (+)enantiomer was covalently bound preferentially to the 2-amino group of guanine and that the (+) form was bound to a 20-fold greater extent than the (−) form;[10] this result accords with the greater mu-

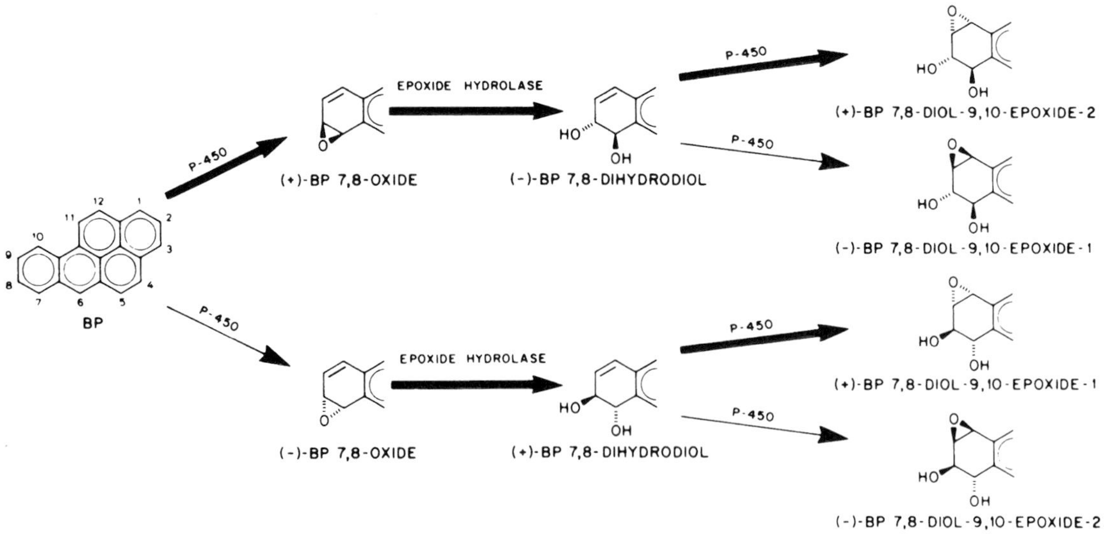

Figure 6-5 Formation and absolute stereochemistry of benzo[a]pyrene metabolites responsible for the carcinogenicity of the parent hydrocarbon. The heavy arrows indicate the major metabolic pathways. (From Conney.[6])

tagenic and carcinogenic effect of this isomer. Surprisingly, both BP 7,8-diol-9,10-epoxide-1 and BP 7,8-diol-9,10-epoxide-2 are weak carcinogens in the mouse skin-painting assay (see below) compared with BP itself, most likely because they are not well absorbed or undergo rapid solvolysis (the epoxides are unstable in aqueous solution) to biologically inactive tetraols. However, when they are injected into newborn mice, they are more potent than the parent compound. The predominant tumors formed were lung adenomas or adenocarcinomas, malignant lymphomas, and liver tumors, the target tissue depending on the dose of carcinogen and duration of treatment.[6] Using synthetically prepared compounds, it was shown that BP 7,8-dihydrodiol and BP 7,8-diol-9,10-epoxide were, respectively, 15-fold and 40-fold more active at equimolar doses than the parent compound BP in producing pulmonary tumors in newborn mice. The diol epoxide was about 2.5-fold more active than the diol, supporting the concept that the diol is a proximate carcinogen and the diol-epoxide the ultimate carcinogen of BP. The other potential metabolites of BP were either much less active or inactive in this system. It is disconcerting to note that liver P-450 systems acting in vivo predominantly form the more active carcinogenic (+)enantiomers of BP 7,8-diol-9,10-epoxides from BP.[6]

Subsequent studies with various benz[*a*]anthracene analogues, methylcholanthrene, and other PAHs have shown that the dihydrodiol epoxides are more carcinogenic than the parent compounds. The structural similarities between the various active carcinogenic intermediates of the PAH led Jerina and Daly[11] to postulate the so-called "bay region theory" of PAH carcinogenesis. The hypothesis is that, for a compound to be highly mutagenic and carcinogenic, diol epoxides must be generated in the bay region formed by the polycyclic structures of the molecules (Fig. 6-6). This hypothesis has now been extensively tested with about a dozen PAH compounds and, in general, has been shown to be true. It should be noted, however, that some carcinogenic hydrocarbons do not possess a bay region and presumably have somewhat different structural requirements to be active as carcinogens.

Additional evidence that formation of the diol epoxides is crucial for tumor induction comes from studies with antagonists of their metabolic formation and with inducers of the inactivation of the diol expoxide compounds. For example, riboflavin 5′-phosphate and flavin adenine nucleotide enhance the conversion of BP 7,8-diol-9,10-epoxide-2 to tetraols and antagonize the mutagenic actions of the ultimate carcinogen.[6] Similarly, a plant phenol, ellagic acid, reacts with and inactivates the diol epoxide, resulting in decreased mutagenicity and decreased carcinogenicity of the ultimate carcinogen.

The effects of environmental agents that modulate the P450 system in vivo on PAH carcinogenesis are not clearly known, but presumably ingestion of drugs, the smoking of cigarettes, exposure to halogenated hydrocarbons, and diet can influence the metabolism, and thus the carcinogenicity of PAH.

An alternate hypothesis is that PAHs are activated to their ultimate carcinogen by a one-electron oxidation pathway (Fig. 6-7).[12] Cavalieri and his colleagues.[12,13] have shown evidence that the one-electron oxidation pathway is important because (1) the predicted DNA adducts result from activation of BP and other potent PAH carcinogens by rat liver microsomes, (2) the activation by this pathway more clearly fits the known carcinogenic potency of various PAHs, and (3) adducts formed by application of BP to mouse skin in a carcinogenic assay result from one-electron oxidation.

It should be noted that although a number of interactions of chemical carcinogens with DNA and other cellular macromolecules have been observed, there is still no formal proof that the major reaction products detected in cells exposed to these agents are the ones actually involved in carcinogenesis. It could be that some minor or as yet undetected reaction is the crucial one. Moreover, because carcinogenesis is a multistage process involving initiation, a lag time, promotion, and tumor progression, multiple actions of a carcinogen—or alternatively, the actions of multiple carcinogens—appear to be necessary to produce a clinically detectable malignant neoplasm.

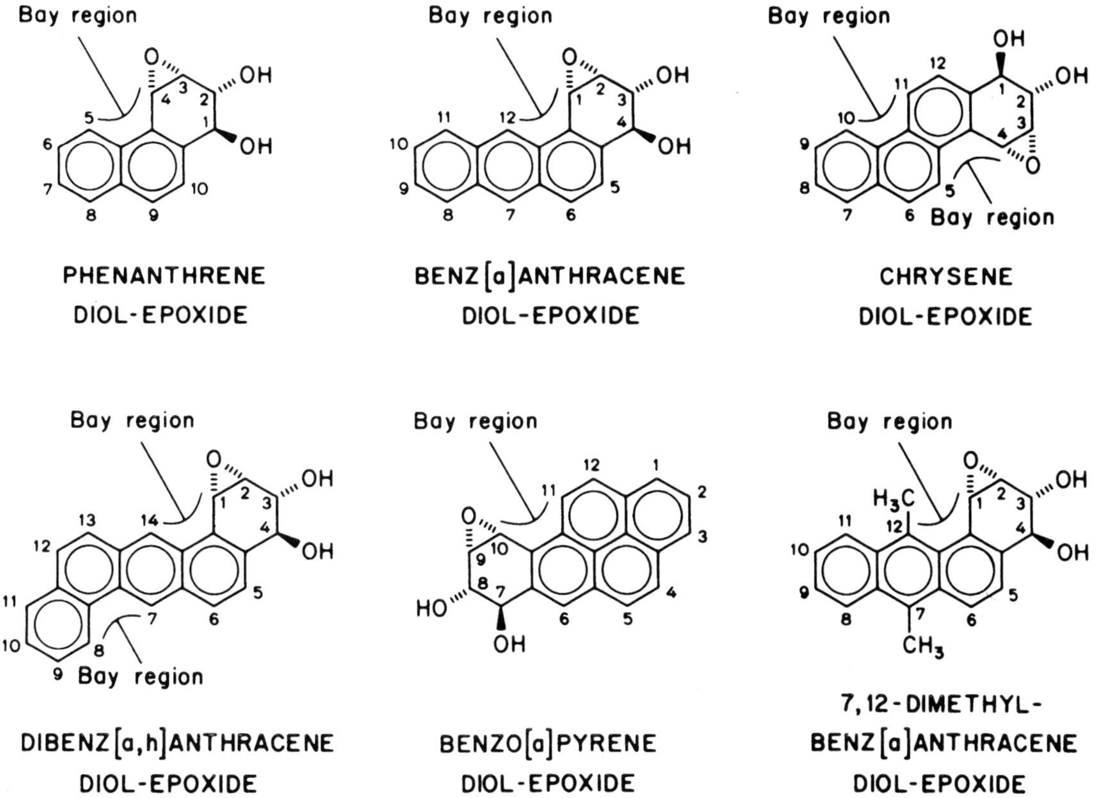

Figure 6-6 Structures of bay-region diol-epoxides of several polycyclic aromatic hydrocarbons. (From Conney.[6])

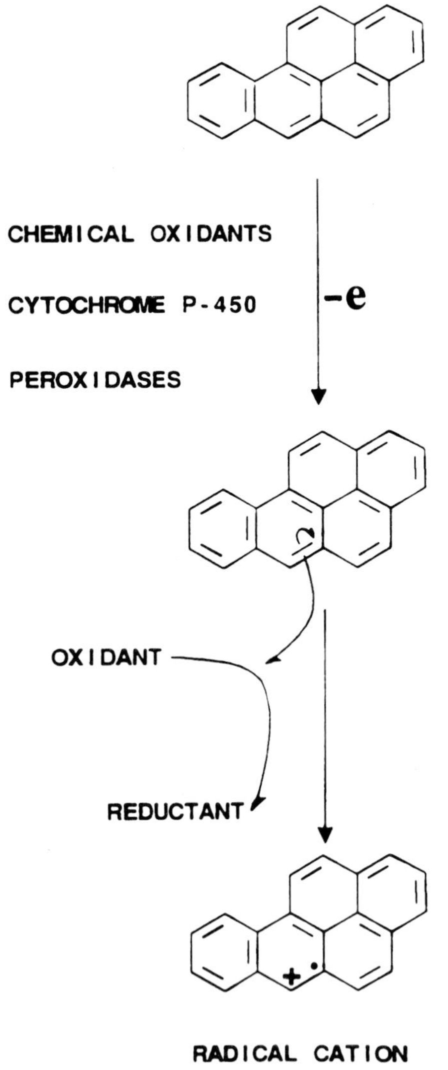

Figure 6-7 One-electron oxidation of BP leading to its radical cation, with charge mainly localized at C-6. (From Cavalieri and Rogan.[12])

The cytochrome P450 family of mixed function oxidases is important in the activation of a number of chemical carcinogens to their ultimate carcinogenic metabolites. These include 2-acetylaminofluorene, other arylamines, aflatoxin B, nitrosamines, and polycyclic aromatic hydrocarbons.[14] One of these, cytochrome P450A1 or CYP1A1 (aryl hydrocarbon hydroxylase) is of particular interest because of its association with cigarette smoking. CYP1A1 is inducible in various extrahepatic tissues by the PAHs contained in cigarette smoke. This has led a number of investigators to examine the relationship between inducibility of CYP1A1 and susceptibility to lung cancer. Early studies of Kellerman et al.[15] showed a correlation between aryl hydrocarbon hydroxylase induction in peripheral blood lymphocytes and the incidence of lung cancer. Subsequent studies by various investigators showed mixed results (reviewed in Ref. 16); however, more recent studies have shown the formation of DNA-BP adducts in pulmonary tissue from cigarette smokers, and a high level of CYP1A1 expression was observed in lung tissue from cigarette smokers compared to nonsmokers.[16] In addition, CYP1A1 was elevated in about half the lung cancers from smokers compared to only 25% of lung cancers from nonsmokers. A genetic polymorphism of CYP1A1 combined with a genetic deficiency in glutathione S-transferase (which detoxifies the electrophilic metabolites of PAHs) is associated with an increased risk of cigarette smoking–induced lung cancer.[16a]

Another important enzyme in the carcinogen activation pathway is the microsomal epoxide hydrolase that catalyzes the stereoselective hydration of alkene and arene oxides to *trans*-dihydrodiols. This enzyme is inducible by various xenobiotics, including some chemical carcinogens and phenobarbital, and its level is increased in hepatic nodules and hepatomas induced by chemical carcinogens.[17]

DNA Adduct Formation

Since most chemical carcinogens react with DNA and are mutagenic, interactions with DNA have been viewed as the most important reactions of these agents with cellular macromolecules. Reaction of chemical carcinogens with DNA is the simplest mechanism that explains the induction of a heritable change in a cell leading to malignant transformation; thus, many investigators view this as the most plausible mechanism for initiation of carcinogenesis. Representative agents from virtually all classes of chemical carcinogens have been shown to affect DNA in some way, and a number of distinct biochemical reaction products have been identified after treatment of cells in vivo or in culture with carcinogenic agents.

The principal reaction products of the nitrosamines and similar alkylating agents with DNA are N-7 and O^6 guanine derivatives. However, the extent of O^6 alkylation of DNA guanine residues correlates better with mutagenic and carcinogenic activity than the quantitatively greater N-7 alkylation of guanine residues (see below). Reactions also occur with other DNA bases and these may also be important in subsequent mutagenic or carcinogenic events. Aflatoxin also forms adducts of guanine at the N-7 position after metabolic activation. The principal reaction product of AAF with cellular DNA is the C-8 position of guanine, just as it is for RNA. Other carcinogenic aromatic amines, such as *N*-methyl-4-aminoazobenzene, also produce C-8 substituted guanine residues as their major nucleic acid reaction product (adduct). Polycyclic aromatic hydrocarbons, after activation, also react with DNA and RNA, forming adducts involving the 2-amino group of guanine, but other reaction products derived from guanine, adenine, and cytosine have also been observed.[12]

The potential biological consequences of DNA base-adduct formation by chemical carcinogens are several. In some cases, it may stabilize an intercalation reaction in which the flat planar rings of a polycyclic hydrocarbon are inserted between the stacked bases of double-helical DNA and distort the helix, leading to a frameshift mutation during DNA replication past the point of the intercalation.[18] Alkylated bases in DNA can mispair with the wrong base during DNA replication—for example, O^6 methylguanine pairs with thymine instead of cytosine during DNA replication, leading to a base transition (i.e., GC→AT) type of mutation during the next round of DNA replication.[19] Many of the base adducts formed by carcinogens involve modifications of N-3 or N-7 positions on purines that induce an instability in the glycosidic bond between the purine base and deoxyribose. This destabilized structure can then undergo cleavage by DNA glycosylase, resulting in loss of the base and creation of an apurinic site in DNA.[20] This "open" apurinic site can then be filled by any base, but most commonly by adenine, during subsequent DNA replication. This substitution can result in a base transition (purine–pyrimidine base change, but in the same orientation, e.g., GC→AT) or a base transversion (inverted purine-pyrimidine orientation, e.g., GC→TA). Finally, interaction with some carcinogens has been shown to favor a conformational transition of DNA from its usual double-helical B form to a Z-DNA form. This could alter the transcribability of certain genes, since B→Z conformational transitions are thought to be involved in regulating chromatin structure (see Chap. 5).

Another interesting point is that interaction of chemical carcinogens with DNA or chromatin does not appear to be a random process. For example, when the ultimate carcinogen of BP—that is, its diol epoxide metabolite, is reacted with cloned chicken β-globin DNA, it preferentially binds in a 300-base-pair sequence immediately 5′ to the RNA cap site.[22] Since this region is thought to contain sequences involved in regulating gene transcription, its alteration by a chemical carcinogen could change the function of genes downstream from the regulatory sequences. Moreover, treatment of the large polytene chromosomes of *Chironomus* with the ultimate carcinogen BP diol epoxide in vitro or administration of the parent unmetabolized compound in vivo to *Chironomus* larvae, demonstrates that the carcinogen binds preferentially to areas most active in gene transcription.[23] DNA in transcribing regions associated with the nuclear matrix also appears to be a preferential target for carcinogen binding. Taken together, these data indicate that the specificity of carcinogen binding is determined to some extent by the base sequence of DNA, its location within the nucleus (e.g., association with nuclear matrix), and the structure of chromatin, with active, "open" sites being favored.

Interaction of Chemical Carcinogens with Oncogenes and Tumor Suppressor Genes

Cellular oncogenes and tumor suppressor genes are two of the critical DNA targets for chemical carcinogens, leading to activation of oncogenes and the inactivation of suppressor genes. This is discussed more in Chapters 7 and 8, but a few examples will be given here.

Carcinogens can activate cellular oncogenes (protooncogenes) by a variety of mechanisms, including base substitution (point) mutations, chromosomal translocations, and gene amplification. One fairly common example is the activation of *ras* protooncogenes by chemical and physical carcinogens in both cultured mammalian cells and animal models (reviewed in Ref. 24). H-*ras* and K-*ras* protooncogene mutations, for example, have been observed in rodent models of skin, liver, lung, and mammary carcinogenesis. The observed mutations in the tumors correlate with expected base adducts formed by the carcinogen: G→A base transitions with alkylating agents (e.g., NMU and MNNG), G→T transversions for BP, A→T transversions for 7,12 dimethylbenzanthracene, G→T transversions and G→A transitions for aflatoxin B1.[24] These mutations appear to reflect similar base substitution mutations in human tumors, which vary by tumor site. For example, G→T transversions of the K-*ras* oncogene are more common in lung tumors and G→A transitions predominate in colonic tumors. This may be instructive as to the type of carcinogens that cause tumorigenesis in these organs. Thus, BP is a suspected carcinogen in cigarette smoke, and it causes G→T transversions like those seen in *ras* in lung tumors, whereas alkylating agents such as nitrosamines are thought to be involved in GI tract carcinogenesis in humans, and these agents mostly cause G→A base transitions, as seen in colon cancer.

Introduction of apurinic sites in DNA by carcinogens is another way that oncogenes can be mutated. Mouse skin tumors are found to have activated c-H-*ras* oncogenes, often caused by point mutations at codons 12 and 13 exon 1 and codons 59 and 61 in exon 2. Mutagenesis by the noncoding apurinic sites can produce G:C → T:A amd A:T → T:A transversions. These mutations occur by translesional DNA replication with more frequent insertion of A opposite the apurinic site. Mutations in the c-H-*ras* gene of mouse skin papillomas induced by several aromatic hydrocarbons have been examined to elucidate the relationship among DNA adducts, apurinic sites, and *ras* oncogene mutations.[24a] Dibenzo[*a,l*]pyrene (DB[*a,l*]P), DB[*a,l*]P-11,12-dihydrodiol, DB[*a,l*]P-8,9-dihydrodiol, 7,12-dimethylbenz[*a*]anthracene, and 1,2,3,4-tetrahydrobenz[*a*]anthracene were consistently found to induce a CAA to CTA mutation in codon 61 of the c-H-*ras* oncogene. Benzo[*a*]pyrene induced a GGC to GTC mutation in codon 13 in 54% of tumors and a CAA to CTA mutation in codon 61 in 15%. The four tumors induced by 6-fluorobenzo[*a*]pyrene exhibited no mutation in exons 1 and 2 of the c-H-*ras* oncogene. The pattern of mutations induced by each hydrocarbon correlated with its profile of DNA adducts. For example, both DB[*a,l*]P and 7,12-dimethylbenz[*a*]anthracene primarily form DNA adducts at the N-7 of adenine that are lost from the DNA by depurination, generating apurinic sites. Thus, these results support the hypothesis that unrepaired apurinic sites generated by loss of hydrocarbon-DNA adducts are responsible for transforming mutations leading to papillomas in mouse skin.

The best-documented example of a tumor suppressor gene being inactivated during carcinogenesis is the p53 gene. Mutations of the p53 gene have been observed in animal tumors and in a wide variety of human cancers. Most of the mutations are point mutations involving "hot spots" in exons 5 through 8. Interestingly, these are the most highly conserved domains of these exons. In human colon tumors, the majority of the mutations are G→A transitions (just as for *ras*); however, other types of base alterations of p53 are seen in other human cancers.

Perhaps the most interesting observation is the finding of a high incidence of p53 point mutations in hepatocellular carcinomas in patients from parts of China and southern Africa where exposure to aflatoxin B1 is endemic.[25,26] Most of these are at a single site, the third base of codon 249, and are G→T transversions. Moreover, hepatocellular carcinomas from areas of low aflatoxin exposure appear only rarely to have this mutation.

Tumor Initiation, Promotion, and Progression

The idea that the development of cancer is a multistage process arose from early studies of virus-induced tumors and from the discovery of the cocarcinogenic effects of croton oil. Rous and his colleagues found that certain virus-induced skin papillomas in rabbits regressed after a period of time and that papillomas could be made to reappear if the skin was stressed by punching holes in it or by applying such irritant substances as turpentine or chloroform. These findings led Rous and his associates to conclude that tumor cells could exist in a latent or dormant state and that the tumor induction process and subsequent growth of the tumor involved different mechanisms, which they called *initiation* and *promotion*.[27] The term *cocarcinogen* was coined by Sall and Shear, who discovered that a basic fraction of creosote oil enhanced the production of mouse skin tumors by BP.[28] In 1941, Berenblum[29] reported that in mice receiving a single skin painting of a carcinogen, such as methylcholanthrene, only a small number of animals developed papillomas, but if the same area of skin was later painted repeatedly with croton oil, which by itself is not carcinogenic, almost all the animals developed skin carcinomas. Taken together, the data of these investigators suggested a multistage mechanism for carcinogenesis.

Studies of the events involved in the initiation and promotion phases of carcinogenesis were greatly aided by the identification of agents that have primarily an initiating activity, such as urethane or a low dose of a "complete" carcinogen (see below), and by the purification of the components of croton oil that have only a promoting activity. Diesters of the diterpene alcohol phorbol were isolated from croton oil and found to be the tumor-promoting substances.[30,31] Of these, 12-*O*-tetradecanoylphorbol-13-acetate (TPA) is the most potent promoter (Fig. 6-8).[32]

A scheme used to study the initiation-promotion phases of mouse skin carcinogenesis is depicted in Figure 6-9. Typically, tumor initiation is brought about by the single application of an initiator, such as urethane, or a subcarcinogenic dose of an agent with both initiating and promoting activity, such as the polycyclic hydrocarbon BP; promotion is carried out by repeated application of a phorbol ester, such as TPA (e.g., three times a week).[31,33] Benign papillomas begin to appear at 12 to 20 weeks and by about 1 year, 40% to 60% of the animals develop squamous cell carcinomas. If the promoting agent is given alone, or before the initiating agent, usually no malignant tumors occur.

The progression stage of carcinogenesis is an extension of the tumor promotion stage and re-

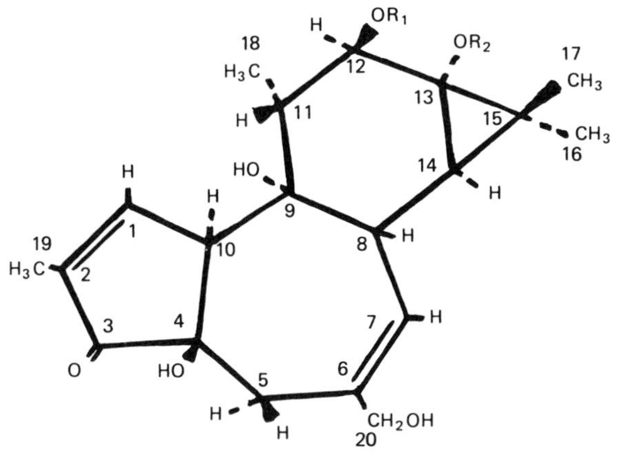

Compound	Substituents (positions) 12 (R_1)	13 (R_2)	Promoting activity on mouse skin
Phorbol	H	H	–
TP	CO–$(CH_2)_{12}$–CH_3	H	–
PT	H	CO–$(CH_2)_{12}$–CH_3	+ +
TPA	CO–$(CH_2)_{12}$–CH_3	$COCH_3$	+ + + +
4α–PDD	CO–$(CH_2)_8$–CH_3	CO–$(CH_2)_8$–CH_3	–

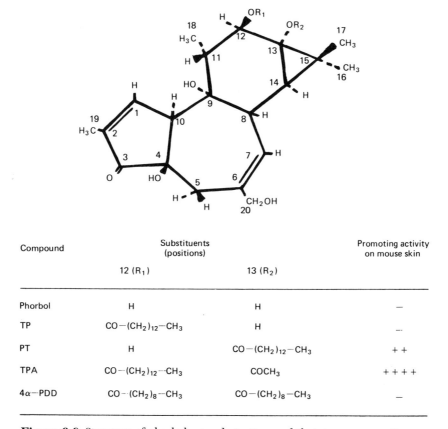

Figure 6-8 Structure of phorbol ester derivatives and their tumor-promoting activity on mouse skin; TP, PT, tetradecanoyl phorbol analogues; TPA, tetradecanoyl phorbol acetate; 4α-PDD, phorbol didecanoate. (From Berenblum.[32])

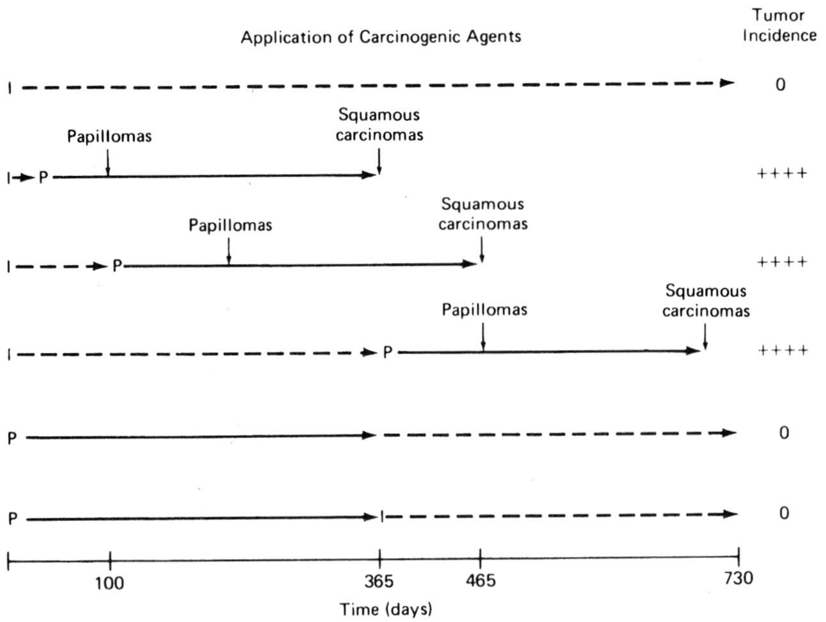

Figure 6-9 Scheme of initiation-promotion phases of induction of carcinogenesis in mouse skin. Initiation is caused by the single application of a subcarcinogenic dose of an agent such as 7,12-dimethylbenz[a]anthracene, benzo[a]pyrene, or urethane. Promotion is carried out by repeated application (e.g., three times a week) of an agent such as the phorbol ester TPA. Papillomas develop within 12 to 20 weeks, squamous carcinomas, about 1 year. Solid lines indicate continual application of agent; dotted lines indicate the duration of time without exposure to agents. Note that promoter may be added up to 1 year after a single application of the initiating agent and tumors still occur; I, initiator; P, promoter.

sults from it in the sense that the cell proliferation caused by promoting agents allows the cellular damage inflicted by initiation to be propagated, and the initiated cells are clonally expanded. The propagation of damaged cells in which genetic alterations have been produced leads to the production of more genetic alterations. This genetic instability is the hallmark of the progression phase of carcinogenesis and leads to the chromosomal translocations and aneuploidy that are frequently seen in cancer cells.[34] Such alterations in the genome of the neoplastic cell during the progression phase lead to the increased growth rate, invasiveness, and metastatic capability of advanced neoplasms.

Evidence for multistage induction of malignant tumors has also been observed for mammary gland, thyroid, lung, and urinary bladder and in cell culture systems (for review, see Ref. 1); thus, it seems to be a general phenomenon. The experimental evidence is consistent with the observed clinical history of tumor development in humans after exposure to known carcinogens—that is, initial exposure to a known chemical or physical carcinogen, a long lag period during which exposure to promoting agents probably occurs, and finally the appearance of a malignant tumor.

Several characteristics of tumor initiation, promotion, and progression provide some insight into the mechanisms involved in these processes. Initiation can occur after a single, brief exposure to a potent initiating agent. The actual initiation events leading to transformation into a dormant tumor cell appear to occur within one mitotic cycle, or about 1 day for the mouse skin system.[32] Furthermore, initiation appears to be irreversible; the promoting agent can be given for up to a year later and a high percentage of tumors will still be obtained. Thus, the initiation phase requires only a small amount of time, is irreversible, and must be heritable because the initiated cell conveys the malignant alteration to its daughter cells. All these properties are consistent with the idea that the initiation event involves a genetic mutation, although other "epigenetic" explanations are possible (see "Validity of Tests for Carcinogenicity," below). The promotion phase, on the other hand, is a slow, gradual process and requires a more prolonged exposure to the promoting agent. Promotion occupies the greater part of the latent period of carcinogenesis, is at least partially reversible, and can be arrested by certain anticarcinogenic agents (see Chap. 15). Tumor promotion is a cell proliferation phase that propagates the initiated damage and leads to the emergence of an altered clone of cells. Most promoting agents are mitogens for the tissue in which promotion occurs. Tumor progression requires continued clonal proliferation of altered cells, during which a loss of growth control and an escape from host defense mechanisms become predominant phenotypic traits. This allows growth to progress to a clinically detectable tumor.

The tumor progression phase is also thought to be irreversible because of the pronounced changes in the genome that occur. Agents that are "pure" progression-causing agents are hard to identify, but the free radical–generating agent benzoylperoxide appears to be a progression-inducing agent during experimental epidermal carcinogenesis.[35]

It should be noted that some potent carcinogens are "complete carcinogens" in that at certain doses they can by themselves induce a cancer. Such agents include polycyclic aromatic hydrocarbons, nitrosamines, certain aromatic amines, and aflatoxin B1. When thees agents are given in sufficient dose to animals during cancer-causing protocols, they can cause DNA damage and produce tissue necrosis, which is itself enough to stimulate several rounds of cell proliferation in response to the tissue damage. In this situation, the promotion-progression phases are often collapsed in time, resulting in the production of aneuploid malignant cells.[34]

Mechanisms of Tumor Initiation

Initiation of malignant transformation of normal cells by a carcinogenic agent involves a permanent, heritable change in the gene expression of the transformed cell. This could come about by either direct genotoxic or mutational events, in which a carcinogenic agent reacts directly with DNA, or by indirect or "epigenetic" events that modulate gene expression without directly reacting with the base sequence of DNA. Most investigators favor the mutational theory of carcinogenesis—that is, that the initiating events involve a direct action on the genome.

The mutational theory depends on three kinds of evidence:

1. Agents that damage DNA are frequently carcinogenic. As discussed previously, chemical carcinogens are usually activated to form electrophilic agents that form specific reaction products with DNA. The extent of formation of some of these reaction products, for example, alkyl-O^6-guanine, has been shown to correlate with mutagenicity and carcinogenicity of certain chemical agents. Ultraviolet and ionizing radiation also interact with DNA at doses that are carcinogenic.
2. Most carcinogenic agents are mutagens. A number of in vitro test systems using mutational events in microorganisms have been developed to rapidly screen the mutagenic potential of various chemical agents. One of the best known of these, the Ames test, is based on certain charcteristics of specially developed strains of the bacterium *Salmonella typhimurium.* The tester strain, a mutant line that requires exogenous histidine for its growth (his^- auxotroph), has a poor excision repair mechanism and an increased permeability to exogenously added chemicals. Using this system, together with a liver microsomal fraction that has the capacity to activate most chemical carcinogens metabolically, Ames and his colleagues have shown that about 90% of all carcinogens tested are also mutagenic.[36] Moreover, few noncarcinogens show significant mutagenicity in this test system.

   Malignant transformation can be induced in a variety of cultured mammalian cells by agents that are mutagenic for the same cells. For example, carcinogenic polycyclic hydrocarbons cause mutations, as measured by induction of resistance to 8-azaguanine, ouabain, or elevated temperature, in Chinese hamster V79 cells if the cells are cocultured with lethally irradiated rodent cells that can metabolize the hydrocarbons to their electrophilic, active metabolite.[37,38] In these studies, mutagenicity was obtained with the carcinogenic hydrocarbons 7,12-dimethylbenz[*a*]anthracene, BP, and 3-methylcholanthrene. There was no mutagenicity with a noncarcinogenic hydrocarbon, and the degree of mutagenicity was related to the degree of carcinogenicity of the chemicals in vivo.

3. Incidence of cancer in patients with DNA repair deficiencies is increased. In individuals with certain recessively inherited disorders, the prevalence of cancer is significantly higher than in the general population.[39] The connecting link between these disorders is the inability to repair certain kinds of physical or chemical damage to DNA. The high incidence of cancer in these diseases constitutes the best available evidence for a casual relationship between mutagenicity and carcinogenicity in humans.

Xeroderma pigmentosum (XP) is characterized by extreme sensitivity of the skin to sunlight and is the most widely studied of the repair-deficient human diseases. Virtually 100% of affected individuals will eventually develop some form of skin cancer. In addition, heterozygotes who carry the XP gene but do not have the disease appear to have a higher incidence of nonmelanoma skin cancer.[40] All individuals with XP are defective in repair of ultraviolet damage to DNA, and most of them have a defect in the excision repair pathway (see "DNA Repair Mechanisms," below). The repair defect ranges from 50% to 90% repair efficiency in cells from different patients, and there is good correlation between the severity of the molecular defect and the extent of the disease. The defect in most patients appears to be at the nicking or incision step of excision repair, although patients in one complementation group have normal excision repair and are defective in postreplication repair.[39] The XP cells are also less efficient at repairing chemically induced damage to their DNA.

Patients with ataxis telangiectasia (AT) have a progressive cerebellar ataxia, abnormalities of the blood vasculature, and immune deficiencies. Their cells are threefold to fourfold more sensitive to X-irradiation and to certain chemicals, including actinomycin D, mitomycin C, and methyl methanesulfonate.[39] The repair defect in these cells is not well understood, but about half the cell lines derived from AT patients are defective in repair replication following irradiation under anoxic conditions. Patients with AT are more prone to develop leukemia and some other cancers.

Fanconi's anemia (FA), another rare, autosomal recessive, inherited disease, is characterized by growth retardation, multiple congenital abnormalities, abnormal skin pigmentation, and hematologic insufficiency. Cells from affected individuals have more spontaneous chromosome aberrations than normal and show more chromosomal damage after treatment with such DNA cross-linking agents as nitrogen mustard or mitomycin C.[39] Thus, FA cells presumably have a defect in the repair of cross-linked bases in DNA; they also appear to have a slight deficiency in the repair of γ-ray- or ultraviolet-induced damage, and patients with FA are at increased risk to develop cancer.

Bloom's syndrome (BS) is a rare autosomal recessive disorder manifested clinically by deficient growth, sun sensitivity, immunodeficiency, and an increased propensity to develop cancer. Cells from BS patients exhibit a marked degree of genetic instability, as indicated by the increased number of chromosomal aberrations and chromatid interchanges. Cultured fibroblasts from BS patients have a higher spontaneous mutation rate when grown in culture than normal cells do, and circulating lymphocytes from BS patients have an eightfold to ninefold higher frequency of resistance to 6-thioguanine than age- and sex-matched controls, suggesting a higher mutation rate in vivo.[41] Lymphoblastoid cell lines from patients with a type of BS characterized by a high rate of sister chromatid exchange (SCE) are more highly tumorigenic, after brief exposure in vitro to 4-nitroquinoline-*N*-oxide and *N*-methyl-*N*′-nitrosoguanidine (MNNG) in the nude mouse assay, than are lymphoblastoid cells from normal individuals.[42]

Although the mutational theory is the simplest explanation of a heritable change in a cell that could produce a cancer-initiating event, not all initiating agents are mutagenic in the test systems used, and some do not react directly with DNA. For example, malignant tumors can be produced by agents that do not interact with DNA or by certain cells placed in abnormal tissue locations. Prolonged administration of the estrogenic compound diethylstilbestrol (DES) has been shown to produce renal adenocarcinomas in male hamsters. The estrogen-dependent induction of these tumors is inhibited by the simultaneous administration of testosterone or progesterone.[43] Malignant neoplasms arise when rat ovarian tissue is transplanted into normal rat spleen, presumably because of the hormonal imbalance thus induced.[44] Tumors can also be induced in rats by the insertion of plastic or metal films, depending only on the physical state (i.e., solid versus porous or fibrous form) of the inserted material. It is unlikely that, in any of these cases, mutational events resulting from interaction with nuclear DNA could have produced the tumors, although it has been shown that an oxidative metabolite of DES can induce SCE in cultured hamster embryo cells.[45] SCE was also observed in human hepatoma cells that metabolize DES but not in cell lines that do not.[46] Presumably, this occurs as the result of the formation of phenoxy radical intermediates from DES metabolites by a peroxidase-mediated reaction, and DNA strand breakage is produced by these oxygen radicals.

In cases in which direct interaction with DNA does not seem to occur, it is likely that regulation of DNA expression is altered by some indirect mechanism. These epigenetic changes that result in malignant transformation could result from (1) interaction of carcinogenic agents with membrane proteins that regulate cell surface receptors for growth factors or that control feedback regulation of cell proliferation in the cell's microenvironment, (2) reaction with RNA molecules involved in translation of proteins, (3) binding to regulatory proteins that control gene transcription, or (4) interaction with proteins involved in cell cycle regulation. These changes could be "heritable," at least for several cell generations, because each daughter cell receives a certain complement of the parent cell's RNA and protein. In any case, the ultimate result is a cell that is genetically unstable, since tumor progression leads to the appearance of cells that are genetically different from the normal cells of the tumor's tissue of origin.

Endogenous Carcinogenesis

An important question arises: What is the source of mutations in the human genome that leads to cancer? One might argue that the answer is obvious. We live in a sea of carcinogens: PAHs from automobile exhaust, industrial pollution, pesticide residues in foods, chlorinated organic compounds in drinking water, etc. Furthermore, epidemiologists argue that almost 30% of human cancers are related to cigarette smoking. Yet, a significant amount of cancers occur in people with no clear evidence of exposure to clearly defined carcinogens. For most cancers of the pancreas, ovary, kidney, and breast, for example, there are no clear geographic or genetic risk factors. Thus, if cancer is initiated through a mutation or a series of mutations, how might these arise?

One possibility is that "spontaneous" mutations arising from an inherent error rate in the fidelity of DNA replication and/or repair could give rise to mutations, some of which by chance could be in key genes that are involved in regulation of cell proliferation and differentiation. The spontaneous or background mutation rate in human somatic and germ-line cells has been estimated to be 1.2 to 1.4×10^{-10} mutations/base pair/cell division[47] (i.e., one mistake in 10 billion base pairs per each cell division). Since there are about 10^{14} cells in the adult human (with a genome of 2×10^{9} base pairs) and they undergo an estimated 10^{16} cell division cycles in a normal life span, 2.8×10^{15} single base (point) mutations could arise in a lifetime.[47] If a single mutation could produce a cancer, this would lead to 2.8×10^{15} spontaneously arising cancer cells in a lifetime, a highly unlikely proposition. Several considerations moderate this wildly excessive number. For example, most base changes are repaired; not all base changes are in coding regions and some are silent (not producing an altered protein); not all mutations produce a cancer cell (they may not occur in key oncogenes or tumor suppressor genes); more than one mutation is necessary to produce a cancer cell (in one study, 25% to 50% of human colonic cancers, for example, contained nine or more mutations).[48]

If at least two mutations are required (e.g., as the Knudson model predicts), the required mutation rate would be the square of 1.4×10^{-10}. In this case, the expected number of spontaneously arising cancer cells in an individuals' lifetime would be about 300,000. Even this seems high. And even if it were true, most of these cells would die or be eliminated by immune mechanisms. If more than two mutations are required to produce a cancer, then the number becomes much smaller, e.g., 5.5 cancers per 100,000 individuals for three mutations. Thus, if this latter assumption is correct, i.e., that three or more mutations are required to produce a cancer, then the spontaneous mutation rate could not by itself explain the number of cancers in the human population. Moreover, the type and distribution of spontaneous mutations differ from those of chemically induced mutations in cells and from those of a number of mutations found in human cancers.[49] Thus, although spontaneous mutations may contribute to the causation of human cancer, they are unlikely by themselves to cause the initiation and progression events that lead to most invasive, metastatic neoplasms.

Several potential mechanisms exist for spontaneous mutations in human cells. These include depurination, deamination, damage to DNA by oxygen radicals, and errors in DNA replication.[50] Depurination is the most common potentially mutagenic event, occurring at a rate of about 10,000 depurination events per cell per day.[50] This results from breakage of the *N*-glycosidic bond connecting a purine base to the deoxyribose-phosphate backbone of DNA and creates a gap in the base sequence. When DNA polymerase encounters such a gap during DNA replication, it may insert the wrong base, usually an adenine, in place of the missing base. Obviously, this could not happen very often or the mutation rate in the human genome would be higher than it is.

Deamination of cytidine to uridine occurs at 1/500 the rate of depurination or about 20 events per cell per day. Since uridine base pairs with adenine during DNA replication, this could lead to a G→A transition. Also, deamination of methylcytosine can occur producing thymidine, which, if not repaired, could produce a G→A transition.

The rate of damage to DNA produced by oxygen radicals, which are continuously generated in cells by normal metabolism, is not clear, but it could be as high as or higher than depurination. Measurements in human urine of 8-hydroxydeoxyguanosine and thymine glycol, which are oxidative breakdown products of DNA, suggest that 10,000 oxygen radical-induced alterations in DNA could occur per cell per day.[51] However, cells have stringent mechanisms to protect themselves against free radicals generated by cellular metabolism. These include superoxide dismutase (SOD), catalase, and glutathione generating systems. Oxygen radical damage, if unrepaired, can produce single and double strand breaks in the DNA backbone.

Errors in DNA replication or repair could result from any of the mechanisms described above. In addition, errors introduced by the DNA replicating machinery itself can occur. DNA polymerases, though usually incredibly accurate, can make some mistakes. Using a ΦX bacteriophage DNA replication system, Kunkel and Loeb[52,53] have determined the error rate of mammalian DNA polymerases. DNA pol α, the major DNA replicating polymerase in eukaryotic cells, has an error rate of 1/30,000 to 1/200,000 bases depending on the method of purification; pol β, a major repair enzyme, has a 1/5000 error rate; and pol δ, a polymerase with proof reading ability, has an error rate of 1/500,000. The most frequent error for all polymerases are single base substitutions and "minus one" base frameshifts. These error rates, determined in vitro, probably overestimate the error rate in vivo, however, because purified enzymes were used and repair enzymes that are part of the intracellular DNA replication complex may have been lost. Nevertheless, errors in DNA replication are another potential source for spontaneous mutations leading to cancer.

The facts that the actual cancer rate in the population cannot be readily explained by the background spontaneous mutation rate and that genetic instability increases with tumor progression has led to the hypothesis that malignant cells have a way to increase their error rate and that they gain some selective advantage therefrom. Thus, the term *mutator phenotype* has been used to describe this phenomenon.[47] By this is meant the ability of tumor cells to direct their own mutation rate or, more precisely, to allow a rapid accumulation of errors that favor their survival. One way this could occur is by alterations in DNA polymerases involved in DNA replication and repair,[47] although evidence for such altered enzymes is not conclusive. It has also been proposed that DNA microsatellite instability such as that associated with the h*MSH2* and h*MLH1* inherited mutations in hereditary nonpolyosis colon cancer (see Chap. 3) is associated with the mutator phenotype.[53a,53b]

One thing is clear, though, and that is that cancer is, in general, a disease of aging. The average age at time of diagnosis of a malignant tumor is about 65. Moreover, the incidence of a number of adult solid cancers increases with the fourth to sixth power of age.[54] This, plus the fact that aneuploidy and other genetic alterations increase during tumor progression—which may occur over many years (up to 15 or 20 years for some cancers)—supports the notion that it is the accumulation of genetic errors over time that is most dangerous for the human genome.

Mechanisms of Tumor Promotion

Tumor-initiating agents most likely act by interacting with DNA to induce mutations, gene rearrangements, or gene amplification events that produce a genotypically altered cell. What happens next is that the initiated cells undergo a clonal expansion under the influence of promoting agents that act as mitogens for the transformed cell type. As will be discussed later, these promoting actions appear to be mediated by cell membrane events, although a direct action of promoters on DNA has also been proposed. It is important to note that multiple clones of cells are likely to be initiated by a DNA-damaging agent in vivo and that, through a rare second event, one or a small number of these clones progresses to malignant cancer.

It may be useful to think of the promotion phase as the stage of cell proliferation and clonal expansion induced by mitogenic stimuli and of the progression phase as the gradual evolution of genotypically and phenotypically altered cells that occurs because of the genetic instability of the progressing cells. This process leads to the development of cell heterogeneity within a tumor, an idea first described by Foulds[55] and later expanded by Nowell[56] (see Chap. 4). During the progression phase, which can take many years in humans, individual tumors develop heterogeneity with respect to their invasive and metastatic characteristics, antigentic specificity, state of cellular differentiation, and responsiveness to hormones, drugs, and immune-modulating agents. Presumably, some powerful selection process goes on to favor the growth of one progressing cell type over another. This may be because a certain cell type develops a growth advantage in the host's tissues over its peers, as proposed by Nowell, or because the host's immunologic defense system can recognize and destroy some cell types better than others, thus providing the selection pressure for expansion of one clone over another, or because of a combination of those factors. Experimental evidence supports such a selection of tumor cells growing in vivo. For example, Trainer and Wheelock[57] have shown that during the growth of L5178Y lymphoma cells in mice a continual selection of cells with a decreasing ability to be killed by cytolytic T lymphocytes (CTL) "armed" against the tumor occurs, until an "emergent phenotype" appears that is highly resistant to the CTL cells.

The isolation and characterization of tumor-promoting agents have provided the tools to study the mechanisms of tumor promotion in vitro and in vivo. The reader is reminded that these agents are primarily defined by their ability to promote skin carcinogenesis in the mouse skin-painting assay, and the mechanisms by which they do this may or may not be relevant to the mechanism of tumor promotion/progression during carcinogenesis in other organs in experimental animals or in humans. Nevertheless, the study of these compounds has been extremely useful in determining the biochemical actions of tumor promoters. Of the promoting agents examined, the phorbol esters have been the most widely studied and thus will be described in some detail here. Still, one must ask: What are the "phorbol esters" in human carcinogenesis? Most likely they are factors to which we are continually exposed in our diet, in cigarette smoke, and in other kinds of environmental agents. This leads to a second question: Do all these agents act through the same receptor or, if not, through the same biochemical steps? The answer is not known, but the list of potential promoters in the human environment is so large it seems unlikely that they all act by means of the same proximal ("receptor") mechanism. More likely, they act through different steps in a cascade leading to the same endpoint—namely, clonal expansion of initiated cells and progressive selection of genetically variant populations of tumor cells.

Tumor-promoting phorbol esters produce a

wide variety of biochemical changes in cells. A number of these changes may be related to the ability of these agents to promote the growth of initiated tumor cells in vivo. Many of the cellular changes induced by phorbol esters are reminiscent of characteristics of the transformed phenotype (see Chap. 4). The effects of phorbol esters on cultured cells include (1) induction of ornithine decarboxylase, 5′-nucleotidase, ATPase, and plasminogen activator activities; (2) stimulation of sugar transport, DNA synthesis, and cell proliferation; and (3) alteration of cell morphology with a loss of cell surface fibronectin and the appearance of a diffuse pattern of actin-containing cytoskeletal elements (for review, see Ref. 58). In addition, phorbol esters stimulate anchorage-independent growth of adenovirus-transformed cells[59] and inhibit the terminal differentiation of chicken myoblasts[60] and chondroblasts,[61] murine lipocytes,[62] erythroleukemia cells,[63] and neuroblastoma cells.[64] Tumor-promoting phorbol esters also transform mouse embryo fibroblasts treated with ultraviolet light,[65] and enhance the transformation of human lymphocytes by Epstein-Barr virus.[66] These cell culture effects are exerted by low concentrations (nanomolar range) of phorbol esters and there is generally a correlation between the potencies of phorbol esters for the cell culture effects and their potencies as promoters in mouse skin carcinogenesis. Phorbol esters share a number of biological properties with epidermal growth factor (EGF) and may act by mechanisms similar to EGF.

An interesting observation suggests that TPA can induce neoplastic transformation of fibroblasts from humans genetically predisposed to cancer.[67] In these experiments, fibroblasts derived from individuals with familial adenomatosis of the colon and rectum were treated with TPA in culture and then injected into athymic mice. Cultures treated with TPA produced tumors in the mice, whereas untreated cultures did not, indicating that the fibroblasts from adenomatosis patients exist in an "initiated" state due to the dominant mutation that produces the disease and that this dominantly inherited trait can be induced to undergo malignant progression by treatment with promoting agents alone. This observation supports the idea that initiation of cancer is a mutagenic event, and it has profound implications for human cancer. For example, if the promoting agents present in our environment could be identified and exposure to them eliminated or significantly diminished, could human cancer be prevented? This approach could conceivably be more effective than eliminating exposure to initiating agents, since exposure to them need be only very short, and is irreversible. Thus, completely preventing exposure to initiating agents over a lifetime is probably not practical, but if the promotion phase takes 15 to 20 years, expanding it to 30 to 40 years would mean that most individuals could have a life expectancy approaching normal before they developed a fatal cancer.

Chronic application of TPA to mouse skin indicates that a number of target cells may be involved in the tumor-promotion phase of skin carcinogenesis.[68] When TPA is applied twice weekly to the skin of mice for several weeks, cell damage, edema, and acute inflammation occur in both the epidermis and dermis during the first week. By 3 weeks, epidermal hyperplasia occurs, accompanied by chronic inflammation in the dermis and hyperplasia of the hair follicles. These features remain until TPA treatment ceases. Although many of these changes regress within 2 weeks after TPA application ceases, an increased number of hair follicles, capillary vessels, and mast cells in the dermis as well as an increased dermal thickness remain; this indicates that the target cells for TPA are not confined to the epidermis and suggests that chronic tissue irritation, as evidenced by the hyperplastic and inflammatory responses, plays a role in tumor promotion. Although the mechanism of this damage is not totally clear, it has been demonstrated that tumor-promoting phorbol esters produce a number of changes in cell membranes and can indirectly damage chromatin by generating oxygen radicals.

Various tumor-promoting phorbol esters, including TPA, stimulate Na^+-outside/H^+-inside exchange across cell membranes, leading to a transient intracellular alkalinization that precedes mitogenic events and appears to be coupled to at least some of the phorbol-ester-induced changes in gene expression in target cells.[68a–70] The Na^+/H^+ exchange induced by TPA also produces cell swelling,[71] and this could lead to further changes in cell functions and integrity. Other membrane effects of tumor-promoting phorbol esters are inhibition of gap-junctional intercellular communication,[72] phosphorylation of cell surface receptors for EGF, IGF-1, insulin, and transferrin, leading to decreased ligand-receptor binding and increased receptor internalization,[73] reorganization of actin and vinculin elements in the cytoskeleton,[74] and structural rearrangement of the nuclear matrix–intermediate filament scaffold.[75]

Tumor-promoting phorbol esters also appear to be able to alter cellular gene expression by indirectly altering DNA structure and chromosomal proteins by generating oxygen radicals (for review, see Ref. 76). It has been shown that TPA induces chromosomal alterations in a variety of human cell types, and these effects are inhibited by the addition of antioxidants. Also, TPA stimulates poly-ADP-ribosylation of chromosomal proteins in human monocytes, an effect frequently stimulated by DNA strand breakage and one that could modify gene expression. Further evidence for the role of oxygen radicals in tumor promotion comes from the observations that O_2^-, H_2O_2, and certain organic hydroperoxides promote carcinogenesis in chemical- or irradiation-initiated cells. In contrast, antioxidants such as butylated hydroxytoluene and butylated hydroxyanisole inhibit transformation of initiated, TPA-treated mouse cells. Some of the changes in gene expression induced by TPA may be due to oxygen radical generation, since induction of ornithine decarboxylase by TPA is blocked by the antioxidant enzymes catalase and superoxide dismutase. These DNA-damaging effects of tumor promoters would be expected to induce chromosomal breaks and gene rearrangements. It has been demonstrated that TPA, during a single-step selection assay for methotrexate resistance in cultured mouse fibroblasts, causes a 100-fold increase in the incidence of MTX-resistant colonies, an effect shown to be due to MTX-gene amplification in these cells.[77] Thus, tumor promoters could also alter gene expression in initiated cells that already have damaged DNA and a propensity for genetic instability.

The DNA-damaging effects of tumor-promoting agents seem to be incompatible with the view that the tumor-promotion phase of carcinogenesis is at least partially reversible. However, most cells have mechanisms to protect themselves against the generation of oxygen radicals, and the ability of agents like TPA to produce oxygen-radical-mediated damage in normal cells may be relatively low. This effect would be expected to be increased in cells whose DNA was already damaged or in cells whose oxygen-radical-scavenging mechanisms are compromised. Moreover, the effects of oxygen-radical-induced damage may be cumulative over time; this could explain the long duration of the tumor promotion/progression phase. It could also explain, at least in part, why cancer is a disease of aging, since aged individuals would have accumulated many more "hits" on their genetic material over time, and there is some evidence that the ability to scavenge free radicals decreases in senescent cells. It is interesting that cells from patients with hereditary diseases, such as ataxia telangiectasia, Fanconi's anemia, and Bloom's syndrome, which are all characterized by increased cancer incidence, are hypersensitive to damage by agents that induce oxygen radical formation. For example, increased oxygen tension causes an excessive amount of chromosomal aberrations in cells from patients with Fanconi's anemia.[76] Moreover, the serum of patients with AT or Bloom's syndrome contains DNA-breaking ("clastogenic") factors; this effect, which can be observed when such serum is added to cultures of normal human cells, is inhibited by addition of superoxide dismutase to the cultures.

In addition to stimulating cell proliferation and altered gene expression, phorbol ester tumor promoters induce the secretion of plasminogen activator and type IV collagenase by human fibroblasts.[78] Because proteases and collagenases released by tumor cells would foster degradation of the growth-limiting basal lamina, release of such enzymes may be another way in which tumor promoters foster tumor expansion and ultimately invasion into underlying tissues. TPA also induces angiogenesis, at least in vitro, as evidenced by its ability to cause cultured endothelial cells to infiltrate into an underlying collagen matrix and form an extensive network of capillary-like structures.[79]

Many of the effects of tumor-promoting phorbol esters are thought to be due to their ability to activate a calcium-dependent protein kinase known as protein kinase C. How this effect of phorbol esters was discovered is an intriguing story in cell biology. Since phorbol esters had been observed to induce several changes in cell membrane function, a number of investigators began to look for cell membrane receptors that would specifically bind these agents. By using [^{3}H]phorbol-12,13-dibutyrate (PDB), several laboratories found evidence for high-affinity, saturable, specific receptors for phorbol esters in membrane preparations or intact cells from avian, rodent, and human tissues.[80–84] Subcellular localization studies have shown specific PDB binding in both cell membrane and cytosolic fractions.[85–87]

In general, the ability of phorbol esters to compete with PDB binding to its specific receptors correlates with their activities as mitogens in cell cultures and as tumor promoters in mouse skin. It also appears that the class of receptors identified by specific PDB binding is responsible for the TPA-like effects of certain nonphorbol promoters; thus, the receptors have been termed "phorboid receptors."[88] Two other types of promoters, teleocidins and aplysiatoxins, are structurally quite different from phorbol esters, yet are potent inhibitors of PDB binding.[88] Structural conformation studies of these latter two classes of promoters indicate that they can assume conformations remarkably similar to a corresponding region in phorbol esters; this suggests that the analogous sites of all three types of compounds represent the crucial area for phorboid receptor binding. This kind of information raises the intriguing possibility that antagonists of the phorboid, tumor-promoting receptor could be designed. There must be other receptors for tumor promotion, however, since agents that do not compete for PDB binding can also act as tumor promoters,[88,89] including "complete" carcinogens such as polycyclic aromatic hydrocarbons, which, upon continuous application, can act as both initiators and promoters of mouse skin carcinogenesis and which act on a different receptor. Notably, glucocorticoid hormones can block the tumor-promoting effects of both TPA and PAHs, suggesting a common link in the tumor-promotion cascade "downstream" from binding at cell surface receptors.

Another action of TPA noted in a variety of cells was its ability to stimulate phosphorylation of a number of cellular substrates, including growth factor receptors (see Chap. 9). This led to the search for a phorbol ester–stimulated kinase. The characteristics of this kinase activity indicated that it was cAMP independent, stimulated by Ca^{2+}, and inhibited by Ca^{2+} chelators. The characteristics of the phorbol-ester-stimulated kinase and that of a phospholipid-dependent, Ca^{2+}-dependent kinase first identified by Nishizuka and his colleagues[90]—called protein kinase C (PKC)—were noted to be remarkably similar. Furthermore, the tissue distribution and intracellular localization of PDB-binding and PKC are similar; their activities both depend on Ca^{2+} and phospholipids; PDB-binding and PKC activities copurify,[85–87,91] and, finally, the dissociation binding constant (K_d) for PDB binding is the same as the activation constant (K_a) for PKC.[90] All of this information led to the conclusion that the phorboid receptor and PKC are one and the same. Other nonphorbol-tumor promoters such as mezerein, teleocidin, and aplysiatoxin also stimulate PKC activity. Because PKC is activated by Ca^{2+} and a variety of hormones and growth-promoting substances, it may serve as a common link in the pleiotypic response of cells to mitogens, including tumor promoters.

The mechanism for TPA activation of PKC has now been worked out (for review, see Refs. 90 and 92). Interaction of TPA with its receptor kinase favors binding of the inactive cytosolic form to the cell membrane, where it is activated. TPA acts as diacylglycerol (DAG) does, and can substitute for it by increasing the affinity of PKC for Ca^{2+} and phosphatidylserine, thereby fostering the translocation of PKC from cytosol to plasma membrane and causing its activation. Part of TPA chemically resembles DAG (Fig. 6-10). TPA thus acts synergistically with Ca^{2+}-mobilizing agents such as those that activate the inositol phospholipid turnover cascade (see Chap. 9). Unlike diacylglycerol, TPA and similar phorbol esters have a long half-life in cellular membranes, and this may explain how they can provide a prolonged signal for cell proliferation, unregulated by the normal feedback mechanism provided by turnover of DAG and the subsequent inactivation of PKC.

At least nine different isozymes of PKC have been cloned (Chap. 9) and intensive research is

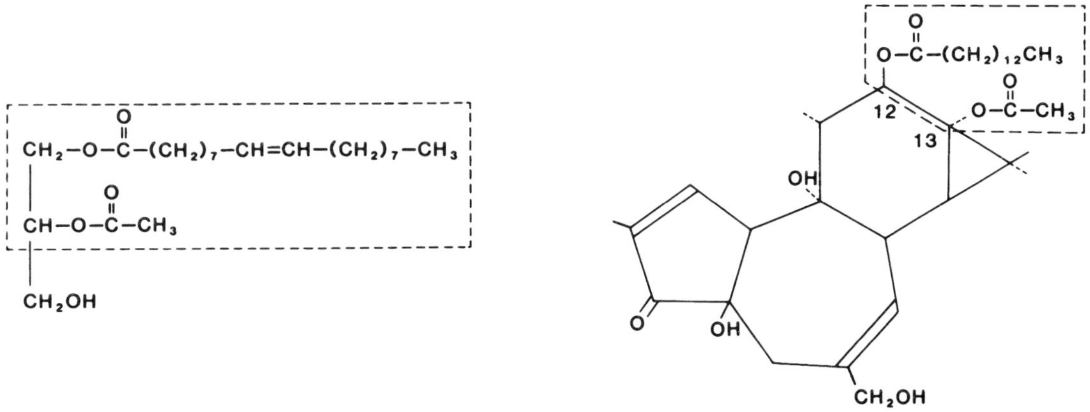

Diacylglycerol

Phorbol ester

Figure 6-10 Structures of a typical diacylglycerol (1-oleoyl-2-acetylglycerol; *left*) and the tumor-promoting phorbol ester (1-O-tetradecanoyl-13-acetate, TPA; *right*). TPA contains a diacylglycerol-like structure in its molecule, as shown by the dotted squares. (From Nishizuka.[90])

being carried out to define the functional role of each PKC isotype.[93,94] There are a number of cellular substrates for TPA-stimulated PKC in addition to the growth factor receptors noted earlier. Among these are a 40-kDa protein and a 20-kDa protein in platelets activated by TPA[90] (the 20-kDa protein has turned out to be myosin light chain), the cytoskeleton protein vinculin,[95] an 80-kDa protein in quiescent mouse 3T3 cells,[96] two 28-kDa "stress" proteins in quiescent rat embryo fibroblasts,[97] a heterogeneous 17- to 20-kDa protein in human leukemia cells and virally transformed murine fibroblasts,[98] and DNA topoisomerase II from *Drosophila.*[99] In cell homogenates in vitro, PKC has broad substrate specificity, phosphorylating serine and threonine residues in many endogenous substrates. Thus, it is not yet clear which of these substrates are involved in a crucial way with the tumor-promoting activity of the phorbol esters, but TPA stimulates a mitogenic cascade of events mediated by activation of the AP-1 (Jun/Fos) transcription factor (Chap. 9). TPA also can activate adenylyl cyclases via PKC.[100]

It is of interest that the tumor promoters TPA and teleocidin, as well as synthetic diacylglycerol, stimulate phosphorylation of the Rous sarcoma gene product pp60^{src} [101] (see Chap. 7), which is itself a tyrosine kinase. A common substrate for TPA-stimulated kinase activity and pp60^{src} is the 40S ribosomal protein S6, the phosphorylation of which has been associated with an increase in cellular protein synthesis and reentry of G_0/G_1-arrested cells into a proliferative state.[102] Phosphorylation of S6 is also stimulated in cells treated with a variety of hormones and growth factors, including insulin, IGF, EGF, PDGF, and prostaglandin $F_2\alpha$. Phosphorylation of S6 stimulated by TPA may, in fact, be indirect, by means of the phosphorylation/activation of a second kinase (e.g., pp60^{src}), because TPA also stimulates tyrosine phosphorylation of some substrates[103] even though PKC itself is a serine/threonine kinase, not a tyrosine kinase. These results suggest that common protein kinases may be activated by diverse stimuli such as growth factors, hormones, and tumor promoters. These kinase activities may be activated either directly, by a specific set of ligands, or indirectly, by phosphorylation by a second kinase, which in turn is activated by a different set of ligands.

Phorbol esters also stimulate the activity of some transcription factors such as c-Jun[104], NF-κB[105], and Oct-T1[105a] and therefore may play a direct role in activating a number of gene transcriptional events.

Mechanisms of Tumor Progression

Tumor progression is characterized by karyotypic instability, aneuploidy and various chromosomal abnormalities, and emergence of the invasive, metastatic phenotype. The progression phase of tumor growth is most likely irreversible because of the chromosomal alterations that accompany it. It is not clear whether there are progression stage–specific agents that are distinct from the initiating and promoting agents described above, but, as noted above, benzoyl peroxide appears to act as a distinct "progressor agent" in experimental skin carcinogenesis.[35] If there are progression-inducing agents, they are presumably DNA-damaging agents, since one of the characteristics is alteration of the genome. On the other hand, proliferation of damaged (initiated) cells by itself is probably sufficient to produce the evolution of cells with a greater and greater number of genomic mistakes due to the continued replication of damaged, mutated DNA.

Activation of additional oncogenes and loss of tumor suppressor gene function frequently accompanies the tumor progression phase. For example, in human colonic cancer, activation of K-*ras* and loss of p53 occur in adenomatous polyps at different times during progression to adenocarcinoma (see Chap. 5). Additional evidence for loss of tumor suppressor gene function as a characteristic of tumor progression comes from gene knockout experiments in transgenic mice. Using the mouse "skin painting" model of tumor initiation (with DMBA) and promotion (with TPA), Kemp et al.[106] found that p53 null mice had much more rapid progression from papillomas to carcinomas than mice with both normal p53 alleles, but in neither heterozygous mice with one normal p53 allele nor in p53 null mice was there an increase in the overall number of papillomas. Interestingly, most of the carcinomas that occurred in all groups of mice had activating mutations of H-*ras*. The authors concluded that absence of p53 does not augment the frequency of initiation or the rate of promotion but greatly enhances tumor progression. These experimental facts fit with the events in human colonic carcinogenesis, in which p53 loss is a later event and correlates with progression to invasive adenocarcinoma.

Experimental Models for the Study of Carcinogenesis

A number of models for the study of carcinogenesis have been developed over the years. Two of the most useful have been the initiation-promotion model of mouse skin carcinogenesis (the "skin painting" model) and the induction of liver cancers in rats. Each of these is described here.

MOUSE EPIDERMAL CELLS

The classic model of carcinogenesis is the single application of an initiating agent such as a polycyclic aromatic hydrocarbon followed by the continuous application of a promoting agent like TPA to the backs of shaved mice. Much of what we know about tumor initiation, promotion, and progression has come from this model system.

Initiation and promotion during mouse skin carcinogenesis produce multiple benign squamous papillomas. A few squamous cell carcinomas eventually arise from the papillomas over many months. However, malignant conversion can be speeded up by exposure of papilloma-bearing mice to mutagens, which activates oncogenes such as H-*ras* and causes loss of tumor suppressor genes such as p53, as noted above.

Yuspa and his colleagues[107] have developed a cell culture model that takes advantage of the fact that newborn mouse keratinocytes can be grown in culture under conditions that allow continuous proliferation (growth in medium with 0.05 m*M* Ca^{2+}) or differentiation (growth in medium with >0.1 m*M* Ca^{2+}). Newborn mouse keratinocytes can be initiated by transfection of v-H-*ras* into the cells. These initiated cells produce papillomas when injected into mice and have a high proliferation rate in culture medium with 0.05 m*M* Ca^{2+}. However, even though they don't proliferate in >0.1 m*M* Ca^{2+}, they fail to differentiate as normal keratinocytes do. When v-H-*ras* transformed keratinocytes were exposed in culture to mutagens such as BP diol epoxide, MNNG, 4-nitroquinoline-*N*-oxide, or *N*-acetoxy AAF, some cell foci appear that now proliferate in >0.1 m*M* Ca^{2+} but still fail to differentiate. These cells produce squamous carcinomas when injected into mice. Thus, this model allows for the study of promotion events (v-H-*ras* initiated cells) as well as progression events (v-H-*ras* initiated, mutagen-treated cells) as distinct events. The production of characteristic biochemical markers for these stages provides a useful adjunct to this model system. For example, keratin 8 is a cell surface marker expressed on v-H-*ras* transformed but not "progressed" keratinocytes; keratin 13 is expressed on cells that have undergone progression and produce squamous carcinomas in vivo.

The mouse skin carcinogenesis model is also a useful one in which to study the role of diet and chemopreventive agents in carcinogenesis (see also Chap. 15). For example, calorie-restricted diets have been shown to reduce the number and size of papillomas during and following promotion with TPA in DMBA-initiated SENCAR mice.[108] Furthermore, the latency period for occurrence of carcinomas was increased and the total number of carcinomas was decreased. Application of apigenin, a plant alkaloid,[109] retinoic acid,[110] and prostratin, a nonpromoting phorbol ester[111] have been shown to inhibit the promotion phase (appearance of papillomas) of mouse skin carcinogenesis.

LIVER CARCINOGENESIS MODEL

Multistage carcinogenesis has also been observed for liver tissue. For example, Peraino et al.[112] observed that a 3-week exposure of rats to AAF in the diet produced only a small number of hepatomas after several months, but if the animals were subsequently treated with phenobarbital for several months after carcinogen feeding was discontinued, a high incidence of hepatomas was noted. Similar results have been obtained by Kitagawa et al.,[113] who fed rats a nonhepatocarcinogenic dose of 2-methyl-*N*,*N*-dimethyl-4-aminoazobenzene for 2 to 6 weeks, and then a dietary administration of phenobarbital for 70 weeks. By 72 weeks, many large hepatocellular carcinomas developed in the phenobarbital-treated animals, whereas only a few small tumor nodules were observed in the rats that were not given phenobarbital. Thus, the action of phenobarbital appears to be analogous to that of TPA in the mouse skin system—that is, it "fixes" the damage to cells induced by an initiating agent and causes a clone of cells arising from a damaged cell to proliferate. However, whereas TPA stimulates DNA synthesis and hyperplasia in skin, phenobarbital produces only a transient and relatively small increase in DNA synthesis in liver. Perhaps that is all that is needed to fix the carcinogenic damage and to allow for the initial proliferation of a damaged clone of cells. Once the damaged clone is present, it could undergo alteration due to its genetic instability and gradually progress to a detectable malignant tumor. This idea is supported by the experiments of Pitot et al.,[114] who treated rats with a single dose of diethylnitrosamine by intubation 24 hours after partial hepatectomy (partial removal of the liver), which stimulates DNA synthesis and cell proliferation in the remaining tissue. If the animals were then treated, starting 8 weeks later, with phenobarbital in the diet for 6 months, many small phenotypically heterogeneous foci characterized by glucose-6-phosphatase-deficient areas, ATPase-deficient areas, and gamma glutamyltranspeptidase-containing areas developed in the liver. Many of these animals also had hepatomas, for which the enzyme-altered foci appear to represent the early stage of neoplastic development. Thus, phenobarbital, in this case, appears to have stimulated the replication of dormant initiated cells, which, in the absence of the promoter, would not have proliferated. If each enzyme-altered focus observed in these experiments were a clone derived from a single cell, about 10^4 to 10^5 cells in the liver were "initiated" by diethylnitrosamine, and a very small number of these subsequently underwent clonal proliferation during phenobarbital feeding.[114] Thus, the conversion of these abnormal foci or early nodules, as they have been called, to a malignant neoplasm is a rare event.

The study of the life history of liver nodules induced in rats by a variety of carcinogenic protocols and their rare conversion to cancer has provided some significant insights into the carcinogenic process in general (for review, see Ref. 115). Tumor-initiating agents interact with DNA, but this interaction alone will not produce a malignant tumor. The initiated cells must undergo several rounds of replication before they become cancerous. This mitogenic signal could be provided by continuous exposure to an initiating agent for some period of time if the agent is itself also a mitogen or if it is cytotoxic and causes significant cell death, leading to liver cell proliferation. Partial hepatectomy or chronic phenobarbital treatment can also provide the needed mitogenic stimulus for clonal expansion of damaged cells. The proliferation of liver cells produces hyperplastic nodules, most of which undergo "redifferentiation" and disappear within a few weeks. At 5 or 6 weeks after initiation with any of a number of initiating agents, the liver may contain up to 1000 hyperplastic nodules. By 7 to 8 weeks postinitiation, the nodules begin to regress, and by 10 weeks they have almost all disappeared. A few, however, remain, perhaps 1% of the original number. These enlarge slowly, have an increased mitotic index, and may proceed to a malignant neoplasm. The "early" nodules and "late" nodules differ considerably in their biological behavior after transplantation into the spleens of syngeneic recipient rats.[115] The cells of early nodules grow slowly and do not progress to invasive cancer. In contrast, late persistent nodules proliferate, retain their hyperplastic nodular appearance, and within 18 months become hepatocellular carcinomas.

The early and late nodules share some morphologic and biochemical characteristics that distinguish them from normal liver. The difference between early and late nodules is apparently that the late nodules lose the capacity to redifferentiate. The nodules form "clumps" or glandlike structures two or three cells thick, compared with the single-cell-thick, organized lobular array of normal liver. The predominant source of their blood supply is shifted from the portal vein to the hepatic artery, consistent with changes seen in hepatocellular carcinomas. In addition to the decreased glucose-6-phosphatase and ATP-ase enzymes noted earlier, the hyperplastic foci also have a decreased concentration of cytochrome P-450, cytochrome b_5, and sulfotransferase. They also show, in addition to increased γ-glutamyltranspeptidase, increases in UDP-glucuronyltransferase, glutathione S-transferase, and epoxide hydrolase.[115] These morphologic and biochemical changes are seen with several carcinogenic protocols, including continuous exposure to the complete carcinogens 2-acetylaminofluorene or 3′-methyl-4-dimethylaminoazobenzene. They are also seen with initiation by diethylnitrosamine, 2-AAF, or dimethylhydrazine, followed by promotion with 2-AAF plus partial hepatectomy or treatment with phenobarbital, carbon tetrachloride, or choline-deficient, low-methionine diets.

Multiple cell types may be involved in hepatocarcinogenesis in the rat. For example, Evarts et al.[116] found expression of biochemical markers that correlate with carcinogenic changes in liver to vary among cell types. In livers of carcinogen-treated (diethylnitrosamine or AAF), partially hepatectomized rats, α-fetoprotein and placental glutathione S-transferase (GST-P) appeared in oval (OV-6) cells at the periphery of the periductal space, whereas γ-glutamyltranspeptidase-positive cells were largely periductal cells, which also expressed transforming growth factor β1. Proliferation of TGF-β1, desmin-positive cells (Ito cells) in proximity to OV-6 cells, which are thought to be the liver stem cell population, suggests an interaction between cell types during liver carcinogenesis.

In contrast to γ-glutamyltranspeptidase and GST-P, whose expression correlates with early steps in liver carcinogenesis, the multidrug resistance protein (P-glycoprotein) becomes elevated in large hyperplastic nodules and in hepatocellular carcinomas.[117] This suggests that P-glycoprotein expression is associated with a more progressive malignant phenotype during liver carcinogenesis, which is of interest in light of the frequent drug resistance of human liver cancers.

Validity of Tests for Carcinogenicity

There is quite a bit of debate among scientists and regulatory agencies about how to assess the carcinogenic hazards of chemicals, both synthetic and natural, in our environment. Much of this debate has spilled over into the media, generating a sort of "carcinogen-of-the-month club." This has generated much confusion among the public. Indeed, as one observer put it: "Cancer news is a health hazard."[118]

For many years, the prevailing view among cancer epidemiologists has been that 60% to 90% of human cancers are attributable to environmental and life-style factors, including cigarette smoking, diet, ultraviolet irradiation, sexual practices, parasitic and viral infections, industrial pollution, and more recently, pesticides.[119,120]

The implication of all this is that most cancers

are preventable. This has led to the prevailing view, adopted by federal regulatory agencies, that as much as possible of all carcinogens should be eliminated from the environment. One outcome of this view is the famous or infamous, depending on your point of view, Delaney Clause to the Food and Drug Act. The Delaney Clause interdicts the use of any food additive in processed foods (interestingly, it pays no attention to pesticide residue on nonprocessed foods) that is found to be carcinogenic at any dose in one or more animal species. Thus, methods to detect carcinogens have become a major issue for the food and pharmaceutical industries. The question, then, became: How can carcinogens be identified before they are found retrospectively to cause human cancer?

The classic approach has been long-term studies in rodents, exposing them to a "maximum tolerated dose" (MTD) for the life span of the animal. An MTD is defined as the highest dose that can be given without causing severe weight loss or other signs of life-threatening toxicity. This kind of testing is very expensive (estimated range up to $1 to $2 million per compound) and time-consuming (3 to 4 years). Not only that, the doses used in these tests are usually orders of magnitude higher than most humans would ever be exposed to. Moreover, these tests fail to take into account the difference in pharmacokinetics, drug metabolism, and excretion mechanisms between mice or rats and humans. This has led to some fascinating snafus. Saccharin is a good example. Sodium saccharin, the artificial sweetener, was shown to increase the incidence of bladder cancer in rats when administered beginning at conception or at birth and continuing through an animal's lifetime. Hence, it was banned as a food additive. The problem is that (1) humans would have to consume 25 kg of sodium saccharin a day to achieve the cancer-causing dose equivalent to that in male rats and (2) humans do not have the same excretion patterns as male rats.[121] It was later found that male rats are considerably more susceptible to the saccharin-caused bladder cancer than female rats, mice, hamsters, monkeys, and, based on epidemiologic studies, humans. The difference among these species is that male rats excrete large amounts of protein in the urine, which the other species do not, and silicate microcrystals form in the male rats' urinary bladders as a result. This leads to a chronic irritation of the uroepithelium, increased cell proliferation, and a hyperplasia eventually leading to carcinoma.[121] These effects are not observed in humans. The realization of these species and dose-related differences eventually resulted in the rerelease of saccharin as a sweetener, albeit with warnings being required on the package.

The example of saccharin and certain other chemical carcinogens raises several important questions about carcinogenicity testing; for example: What is an appropriate dose and time frame to use in animal carcinogenicity assays? What effects might lower dose but lifetime exposure to a chemical have on humans? Is there a "threshold" dose for a given agent, below which human exposure is safe? Are there cheaper, more accurate, faster ways to find out if a chemical is carcinogenic? How does one evaluate the relative risk of exposure to natural chemicals in food and water compared to synthetic chemicals? Is there an acceptable risk-benefit ratio for economically important chemicals? Can society afford the cost of removing every trace of a potentially carcinogenic substance from the environment? Are there dietary or other factors that can be used to supplement diets to prevent or delay cancer even if one is exposed to known or unknown carcinogens? Can highly susceptible individuals be identified so that they can be advised to avoid certain employment or activities? Can exposure levels be determined that could lead to removal of substances from an environment and/or careful follow-up of individuals so exposed?

These are all difficult questions to answer, and there is considerable debate about the answers, but some steps have been taken to answer some of them. As noted in Chapter 2, the Ames test for mutagenesis in bacteria *Salmonella* is one of the widely used short-term tests that has been developed. It is fast, inexpensive, and reasonably accurate. However, the Ames test does not detect certain kinds of chemical carcinogens, e.g., those that do not bind to DNA such as chloroform and dibromochloromethane, and it overpredicts carcinogenicity for others. In one survey of 224 chemicals, the Ames test had a sensitivity of only 54% (percentage of true carcinogens identified) and a specificity of 70% (true nega-

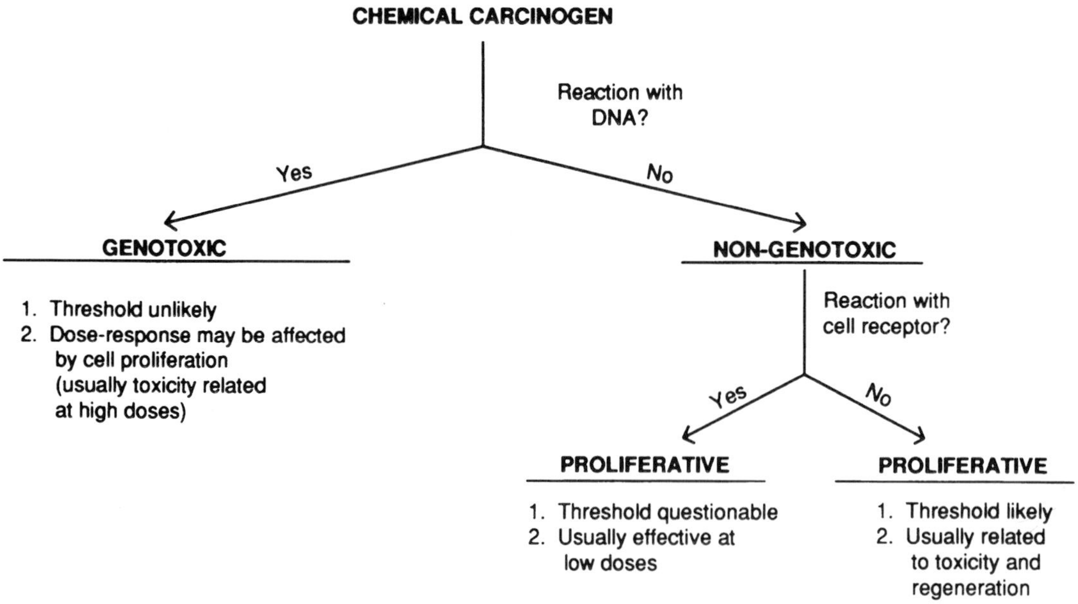

Figure 6-11 Proposed classification scheme for carcinogens. The effect of genotoxic chemicals can be accentuated if cell proliferative effects are also present. Nongenotoxic chemicals act by increasing cell proliferation directly or indirectly, either through interaction with a specific cell receptor or nonspecifically by (*a*) a direct mitogenic stimulus, (*b*) causing toxicity with consequent regeneration, or (*c*) interrupting physiological process. Examples of the latter mechanism include thyroid stimulating hormone stimulation of thyroid cell proliferation after toxic damage to the thyroid and viral stimulation of proliferation after immunosuppression. (From Cohen and Ellwein.[121])

tives, i.e., chemicals correctly identified as noncarcinogens).[122]

Other short-term tests include induction of resistance to antimetabolites in cultured cell lines, and chromosomal breakage in exposed cultured cells (see Chap. 2). Using the data from short-term tests, particularly the Ames test, in combination with determination of a "chemically alerting" structure, (i.e., a chemical containing structural elements known to be carcinogenic such as electrophile producing side chains), Ashly and Tennant[123] showed a high correlation between those compounds that were structurally alerting and those that were mutagenic. When taken together, a reasonable correlation with carcinogenicity was predicted, based on carcinogenicity in animal tests.

One of the problems with the use of the MTD approach to carcinogenicity testing in animals is that, in addition to being a dose that is often an order of magnitude more than a dose to which humans are likely to be exposed, the MTD is a dose that is often sufficient to kill cells and induce a proliferative tissue-repair response in target organs. This increased cell proliferation puts cells at risk for propagating an unrepaired lesion produced by the spontaneous background mutation rate (see above) or by exposure to an environmental agent. This has led Cohen and Ellwein[121] to propose a biological model for carcinogenesis based on whether a chemical is genotoxic or nongenotoxic (Fig. 6-11). The key feature of this model is that an agent can increase the incidence of cancer either by damaging a cell's DNA (genotoxic) or by stimulating cell proliferation (nongenotoxic), increasing the likelihood for a spontaneous genetic error to occur during DNA replication. They provide as evidence for this a number of examples of genotoxic chemicals, e.g., 2-acetylaminofluorene, that cause different effects in different tissues depending on the proliferative response that they produce in a given tissue. They also cite nongenotoxic chemicals that can act at high enough doses to induce cell damage and a mitogenic response that increases the genetic error rate. In the example of 2-AAF, which is both genotoxic and able to stimulate cell proliferation, a threshold for the carcinogenic effect is unlikely. In the case of nongenotoxic chemicals, a threshold effect is likely.

Weinstein and others[124,125] have argued that there is definitive evidence that spontaneous mutation or endogenous DNA damage is carcinogenic and that there is no consistent correlation between the inherent growth fraction (percent proliferating cells) in a tissue and cancer incidence in that tissue. It seems, however, that it must be conceded that cell proliferation has to occur in order for a cancer to develop and that the self-renewal stem cells of a tissue are the most likely targets for carcinogens.

Unfortunately, there is as yet no well-defined way to simulate whole organism carcinogenicity testing other than in whole organisms, where the parameters of absorption, tissue distribution, metabolism (activation and deactivation), pharmacokinetics, and excretion can be evaluated. Furthermore, it can be argued that a full dose range, including doses that stimulate cell proliferation, of potential carcinogens should continue to be tested in animals.[125] The difficulty comes in translating these data to realistic human exposure levels. Thus, computers will still not be able to replace good judgment and common sense.

Ames and his colleagues[126,127] have argued that natural chemicals are as likely as synthetic chemicals to be carcinogenic in various tests and that, based on the actual levels of human exposure (other than industrial workers or farmers, for example, who might have high exposure rates to industrial chemicals or pesticides), natural chemicals are at least as dangerous as synthetic ones. In high-dose tests, 30% to 50% of both natural and synthetic chemicals are carcinogens, mutagens, teratogens, and clastogens (DNA-damaging agents).[127] This has led to the conclusion that natural and synthetic chemicals are equally likely to be positive in animal cancer tests and that at the low doses of most human exposures, "the comparative hazards of synthetic pesticide residues are insignificant."[126] Interestingly, Ames et al.[127] note that attempts to introduce naturally pest-resistant strains of celery and potatoes failed because of the increase of natural pesticides with human toxicity in the resistant strains.

As one can imagine, the conclusions of Ames et al. have elicited cries of protest from environmentalists who may tend to believe that all natural things are good and all synthetic things are bad. The balance is somewhere in the middle. It would be unfortunate, however, for industries to take Ames's conclusions as a license to pollute. There are lots of other environmentally important reasons not to pollute the earth, in addition to the possible carcinogenic effects of pollution.

IRRADIATION CARCINOGENESIS

A number of the points made about chemical carcinogenesis are also true for radiation-induced carcinogenesis. Both x-rays and ultraviolet (UV) radiation, for example, produce damage to DNA. As with chemical carcinogens, this damage induces DNA repair processes, some of which are error-prone and may lead to mutations. The development of malignant transformation in cultured cells after irradiation requires cell proliferation (Fig. 6-12) to "fix" the initial damage into a heritable change and then to allow clonal proliferation and expression of the typical transformed phenotype.[128] Fixation appears to be complete after the first postirradiation mitotic cycle. In the case of mouse C3H/10 T ½ cells, expression of radiation-induced transformation requires an additional 12 rounds of cell division. Thus, as in the case of chemical carcinogenesis, a promotion phase is required for full expression of the initiated malignant alteration. Moreover, when low doses of chemical carcinogens and x-rays are used together, these two types of agents act synergistically to produce malignant transformation.[128]

When cells are exposed to UV light in the 240- to 300-nm range, nucleic acid bases acquire excited energy states, producing photochemical reactions between DNA bases (for review, see Ref. 129). The principal products in DNA at biologically relevant doses of UV light are cyclobutane dimers formed between two adjacent pyrimidine bases in the DNA chain. Both thymine–thymine and thymine–cytosine dimers are formed. That formation of these dimers is linked to mutagenic events is shown by the facts that (1) mutants defective in repair of cyclobutane pyrimidine dimers have increased sensitivity to the lethal and mutagenic effects of UV light and (2) photore-

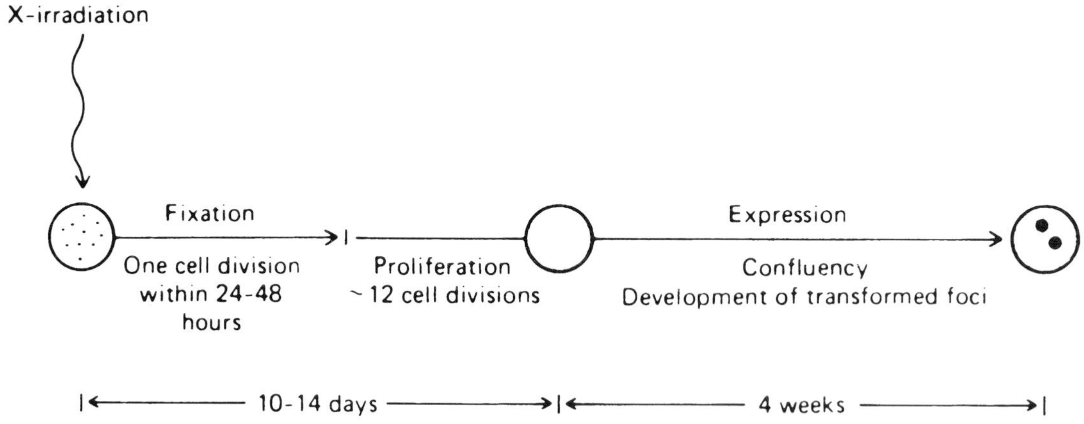

Figure 6-12 Development of malignant transformation in mouse 10 T½ cells following X-irradiation. About 300 viable cells were seeded on a 100-mm Petri dish *(left)* and irradiated. The cells were allowed to form a confluent monolayer *(center)*, at which time normal cell division ceases. This takes 10 to 14 days and requires about 13 rounds of cell division. Dense, piled-up, transformed foci eventually appear overlying the confluent monolayer of normal cells *(right)*. (From Little.[128])

activation repair of these dimers reverses the lethal and mutagenic effects of UV light. Repair of UV-induced damage to DNA takes place through mechanisms similar to those described below, namely, incision by UV-specific endonucleases, excision of a region of single-stranded DNA containing the pyrimidine dimers, replication of the excised region by a repair DNA polymerase, and ligation of the repaired region to the remaining ends of the DNA strand.[129]

Heavy exposure to sunlight induces similar changes in human skin, and the degree of exposure to sunlight is closely related to the incidence of skin cancer (see Chap. 2). Whether continuing exposure to UV rays in sunlight is the promoting agent in skin cancer or additional promoting events are required is not clear, but it seems that UV irradiation is a complete carcinogen, just as some chemicals are—that is, it has both initiating and promoting activities. Patients who can not efficiently repair UV-induced damage, such as those with xeroderma pigmentosum, have a much higher risk of developing malignant skin tumors.

The history of radiation carcinogenesis goes back a long way (reviewed in Ref. 130). The harmful effects of x-rays were observed soon after their discovery in 1895 by W. K. Röntgen. The first observed effects were the acute ones, such as reddening and blistering of the skin within hours or days after exposure. By 1902, it became apparent that cancer was one of the possible delayed effects of x-ray exposure. These cancers—which include leukemia, skin cancers, lymphomas, and brain tumors—were usually seen in radiologists only after long-term exposure before adequate safety measures were adopted. This led to the idea of a safe threshold for radiation exposure. The hypothesis that small doses of radiation might also cause cancer was not adopted until the 1950s, when data from atomic bomb survivors in Japan and certain groups of patients treated with x-rays for noncancerous conditions, such as enlarged thyroids, were analyzed. These and other data (Fig. 6-13) led to the concept that the incidence of radiation-induced cancers might increase as a linear, nonthreshold function of dose. Thus, the debate about a safe threshold or lack thereof pertains to radiation carcinogenesis just as it does to chemical carcinogenesis.

In the case of radiation carcinogenesis, the damage to DNA, and hence its mutagenic and carcinogenic effect, is due to the generation of free radicals as the radiation passes through tissues (Fig. 6-14). The amount of radical formation and hence the amount of DNA damage depend on the energy of the radiation. In general, x-rays and gamma rays have a low rate of linear energy transfer, generate ions sparsely along their tracks, and penetrate deeply into tissue. This contrasts with charged particles, like pro-

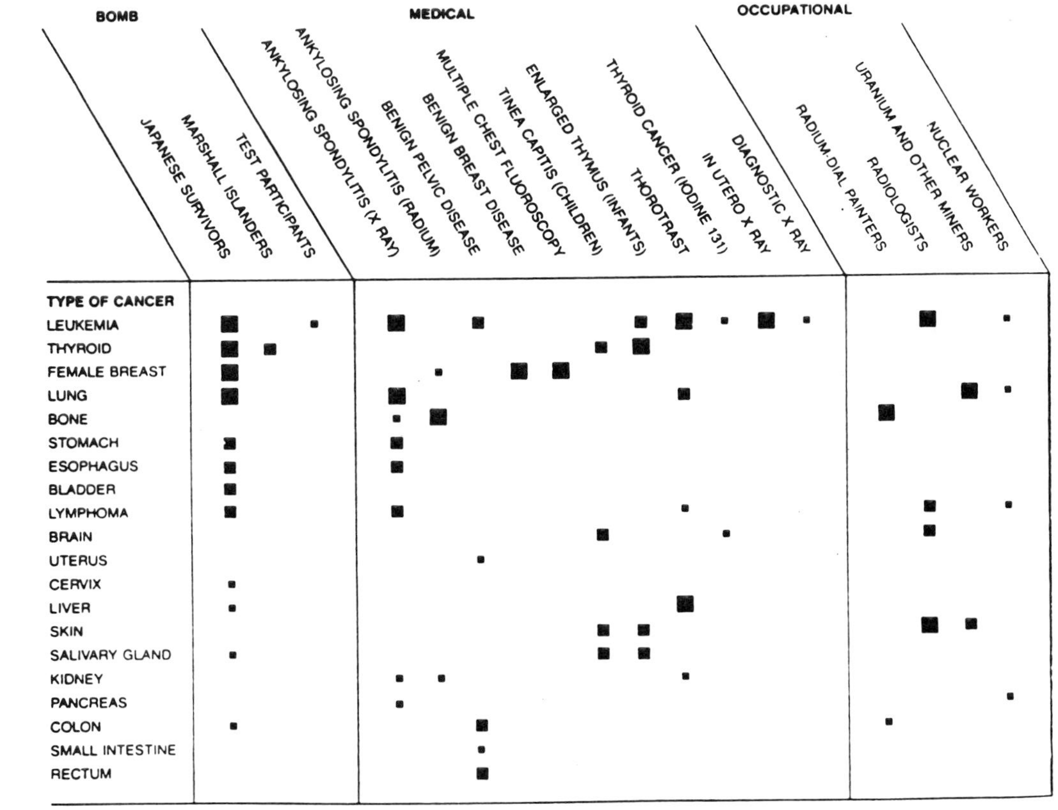

Figure 6-13 Incidence of cancer associated with radiation is charted for people exposed to radiation in three different circumstances: the atomic bombings of Japan, radiation therapy, and occupational exposure. A strong association of a certain type of cancer with a certain type of exposure is indicated by the large squares, a meaningful association by the squares of medium size and a suggestive but unconfirmed association by the small squares. The three groups are the ones in which the exposure to radiation is fairly well documented. (From Upton.[130])

tons and α particles, that have a high linear energy transfer, generate many more radical ions locally, and have low penetration through tissues. The damage to DNA can include single- and double-strand breaks, point mutations due to misrepair deletions, and chromosomal translocations.[130–132] The molecular genetic events that follow radiation damage to cells include (1) induction of early response genes such as c-*jun* and Egr-1, (2) induction of later-response genes such as tumor necrosis factor-α, FGF, and PDGF-α, (3) activation of interleukin-1 and protein kinase c(PKC),[133] and (4) activation of oncogenes such as c-*myc* and K-*ras*.[134] Induction of these genes may be involved in the cellular responses to irradiation and in the longer-range effects that lead to carcinogenesis. At any rate, the production of clinically detectable cancers in

Ionization of water by radiation, hv

$$H_2O \xrightarrow{hv} H_2O^+ + e^-$$

a) Reaction of H_2O^+ to form a hydronium ion plus a hydroxyl radical

$$H_2O^+ + H_2O \longrightarrow H_3O^+ + \boxed{OH^\cdot}$$

b) Solvation of free electron in a cage of water molecules

$$e^- + [H_2O]_n \longrightarrow \boxed{e^-_{ag}}$$

c) Reaction of e^-_{ag} with water to form a hydrogen radical

$$e^-_{ag} + H_2O \longrightarrow OH^- + \boxed{H^\cdot}$$

Figure 6-14 The formation of the three major water radical species (square boxes) by the action of ionizing radiation.

humans after known exposures generally occurs after long latent periods. Estimates of these latent periods are 7 to 10 years for leukemia, 10 to 15 years for bone, 27 years for brain, 20 for thyroid, 22 for breast, 25 for lung, 26 for intestinal, and 24 for skin cancers.

OXYGEN FREE RADICALS, AGING, AND CANCER

The diseases of aging include cardiovascular disease, decline in the function of the immune system, brain dysfunction, and cancer. Cancer increases with about the fifth power of age in lower animals such as rats and also in humans, but occurs over a much shorter time span in rats, who have a sevenfold higher metabolic rate than humans. Cancers that are increasing in mortality rate in people over age 65 include lung, brain, melanoma, multiple myeloma, non-Hodgkin's lymphoma, kidney, esophagus, ovary, prostate, breast, pancreas, and larynx.[135] People living in the United States who are 65 or older have 10 times the risk of those under age 65 for developing cancer, and survival rates also decrease with increased age.

Part of the increase in cancer incidence with aging could be due to an accumulation of damage to DNA over a lifetime of exposure to carcinogenic substances. Another, perhaps more likely possibility is that cellular damage produced by endogenous oxidants accumulates over time and the body's ability to repair this damage decreases with age. There is a fair amount of circumstantial evidence to support this latter hypothesis (reviewed in Refs. 136 to 138). Oxidative damage to DNA, proteins, lipids, and other macromolecules accumulates with age. Oxidation products that are formed during normal metabolic processes in cells include superoxide ($O_2^{\bar{\cdot}}$), hydrogen peroxide (H_2O_2), and hydroxyl radical ($\cdot OH$). These are also produced in cells by radiation, and they are capable of damaging DNA and producing mutagenesis. Of these, the hydroxyl radical appears to be the primary DNA damaging species, but it has a short half-life and high reactivity so it must be generated in close proximity to DNA.[139] This may occur in the cell nucleus by an interaction of H_2O_2 with chromatin-bound metals such as Fe^{2+} by the following reaction:

$$Fe^{2+} + H_2O_2 \rightarrow Fe^{3+} + .OH + OH.^-$$

Singlet oxygen, which is produced by lipid peroxidation or by the respiratory bursts from neutrophils, is also mutagenic and has a much longer half-life than the hydroxyl radical. Lipid peroxidation can also give rise to mutagenic products such as lipid epoxides, hydroperoxides, alkanyl and peroxyl radicals, and α, β unsaturated aldehydes.[136]

Cells have multiple mechanisms to protect themselves from oxidative damage, including superoxide dismutase, catalase, glutathione peroxidase, and glutathione S-transferases. In addition, DNA damaged by oxidants is subject to repair (see below); oxidized proteins are degraded by proteases; and lipid peroxides are destroyed by glutathione peroxidase. Nevertheless, some oxidative damage and misrepair may persist, and the ability to carry out these repair mechanisms decreases with aging. It is estimated that the human genome suffers about 10,000 "oxidative hits" to DNA per cell per day.[136] Mutations accumulate with age in the rat so that an "old" rat (2 years old) has twice as many DNA lesions per cell as a young rat. Furthermore, the frequency of somatic mutations found in human lymphocytes is about ninefold higher in the aged than in neonates.[140] How much of this mutation frequency is due to oxidative damage of DNA is not clear, but a number of altered bases have been observed in cells undergoing oxidative stress. These include hydroxymethyl uracil, thymine glycol, 8-hydroxyguanine, 8-hydroxyadenine, and formamido derivatives of altered purines.[141] Some of these products appear in the urine and may be an index of oxidative damage. They are also produced by exposure of DNA to ionizing radiation and oxygen radical generators. 8-hydroxyguanine appears to be the most frequently altered base to result from oxidative damage to DNA; if this base is left unrepaired in DNA, it produces G→T transversion.[141]

Endogenous oxidants can also damage proteins and lipids. Oxygen free radicals catalyze the oxidative modification of proteins, leading to an age-related increase in carbonyl content of cellular proteins.[137] For example, there is a significant increase in carbonylated proteins in human erythrocytes from older individuals, and the carbonyl content of proteins in cultured human skin fibroblasts increases exponentially with the age of the fibroblast donor.[137] In addition, the protein carbonyl content of fibroblasts from individuals with premature aging (progeria and Werner's syndrome) is higher than that of age-matched controls. There is also an age-related decrease in neutral alkaline protease activity that degrades oxidized proteins. Thus, the end result is an increased retention of damaged proteins with aging. How much oxidative protein damage contributes to diseases of aging and cancer is not clear, but treatment of gerbils with the radical trapping agent *tert*-butyl-α-phenylnitrone inhibits age-related increases in oxidized protein in the brain and blocks age-related memory loss (as measured by a radial-arm maze test).[142] Oxygen radical damage to lipids leads to fluorescent lipid oxidation products that appear to result from cross-links between proteins and lipid peroxidation products, and these also increase with age.[136]

Caloric or protein restriction in the diet slows oxidative damage to proteins and DNA and decreases the rate of formation of neoplasms in rodents. Similar results are seen by dietary supplementation with antioxidants such as tocopherol (vitamin E), ascorbate (vitamin C), and carotenoids such as β-carotene, leading to the hypothesis, supported by epidemiologic data in humans, that dietary intake of such substances could decrease the incidence of human cancer[136] (see also Chap. 15). It should also be noted that one of the major sources of *exogenous* oxidant exposure is the oxides of nitrogen found in cigarette smoke.[136]

DNA REPAIR MECHANISMS

Not all interactions of chemicals and irradiation with DNA produce mutations. In fact, all cells have efficient repair mechanisms that repair such lesions. DNA repair mechanisms include sets of enzymes that survey DNA for specific kinds of damage, remove the altered portion of DNA, and then restore the correct nucleotide sequence. The important role of DNA repair in human cancer has been established by the finding that a number of inherited defects in DNA repair systems predispose individuals to getting cancer. These diseases include xeroderma pigmentosum, ataxia telangiectasia, Fanconi's anemia, Bloom's syndrome, Cokayne's syndrome, and hereditary retinoblastoma.[143]

There are several types of DNA repair systems, a number of which have been preserved from bacteria to humans. These include[143–145] (1) abnormal precursor degradation, e.g., the hydrolysis of the oxidized nucleotide triphosphate 8-hydroxy-dGTP to its nucleotide 8-OH-dGMP, preventing incorporation into DNA; (2) a visible light-activated photoreactivation repair mechanism for removal of UV-induced cyclobutane pyrimidine dimers; (3) strand break repair via an action of DNA ligase, exonuclease, and polymerase activities; (4) base excision repair that recognizes simple base alterations such as cytosine deamination to uracil and requires the action of (a) a purine or pyrimidine glycosylase that breaks the deoxyribose-base bond, (b) an endonuclease to cleave at the abasic site, (c) a phosphodiesterase to clip away the "naked" abasic site, (d) DNA polymerase, and (e) DNA ligase to refill and reclose the site; (5) nucleotide excision repair that recognizes bulky DNA base adducts, pyrimidine dimers, and base cross-links and requires the concerted action of enzymes and recognition factors (see below), and (6) O^6-alkylguanine-DNA alkyltransferase that recognizes and removes small alkyl adducts from DNA.

Excision repair is the most general DNA repair mechanism in higher organisms. Base excision repair removes damage such as deaminated bases, oxidized or ring-opened bases generated by hydroxyl or superoxide radicals, and abnormally methylated bases such as 3-methyladenine.[144] Nucleotide excision repair requires sequential steps of (1) preincision recognition of damage; (2) incision of the damaged DNA strand at or near the damaged site; (3) excision of the damaged site and local removal of nucleotides in both directions from the defect in the affected DNA strand; (4) repair replication to replace the excised region, using the undamaged strand as template; and (5) ligation to join the repaired sequence of nucleotides at its 3′ end to the contiguous DNA strand.[143]

DNA repair is usually very accurate, but if re-

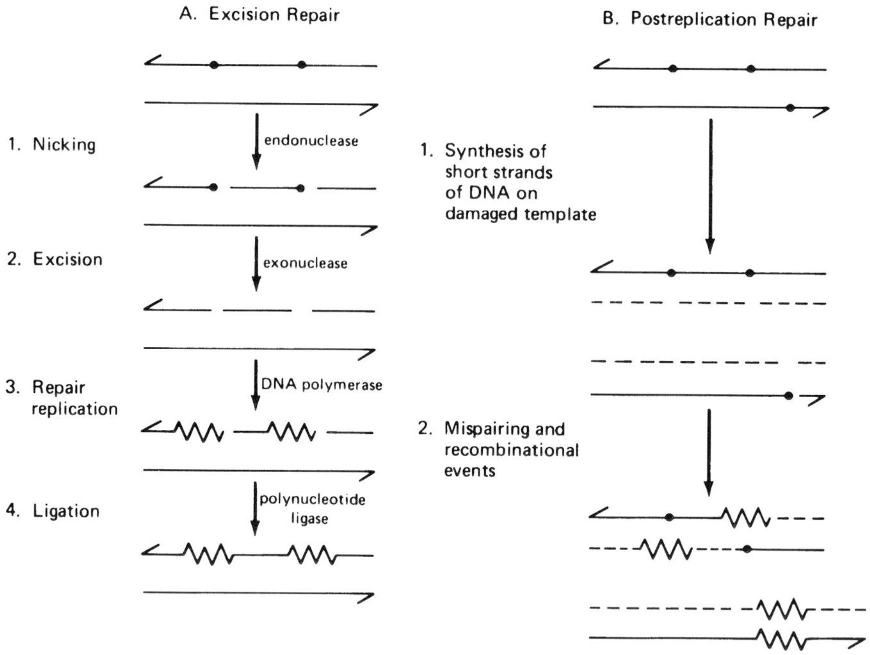

Figure 6-15 Scheme of DNA repair mechanisms. Excision repair is error free, whereas postreplication repair is subject to the possibility of introducing an incorrect base into a daughter strand by mispairing or recombinational events. Not all postreplication repair produces errors, however, as is shown for replication of the lower strand in *B*. Solid circles indicate the damaged bases; jagged lines show newly synthesized base sequences; dashed lines indicate synthesis of daughter strands.

pair cannot occur prior to or during DNA replication, it may be error-prone (Fig. 6-15). This error-prone postreplication repair seems to be brought into play by certain types of agents or when a cell is so overwhelmed by damage that it cannot handle by excision repair before the cell enters S phase during the next round of cell division. In this case, the new DNA is synthesized on templates that still contain damaged bases, leading to mispairing or recombinational events that transfer damaged bases to daughter strands. For example, in mammalian cells, 5% to 30% of UV-induced thymidine dimers are transferred from parental to daughter strands during postreplication repair.[146]

Nucleotide excision repair (NER) of DNA in eukaryotic cells requires several gene products. Some of these gene products appear to be identical or highly homologous in yeast, rodents, and humans.[147,148] A number of defects in the NER system have been found by studying mutations in cells from xeroderma pigmentosum (XP) patients, in whom at least nine different kinds of mutations (i.e., nine different complementation groups) have been found.[143] Some of these XP genes have been cloned and found to be highly homologous to yeast RAD genes required for excision repair in *Saccharomyces cerevisiae.*[147–150] Some of the cloned human genes also correct repair defects in mutant rodent cells and are called ERCC (excision-repair cross-complementing) genes (Table 6-1).

The RAD2 gene in *S. cerevisiae* shares remarkable sequence homology to the XPG gene of the G complementation group of xeroderma, which is the same as the ERCC 5 gene for correcting repair defects in humans. The RAD2 gene product has been shown to have the ability to act as a single-stranded DNA endonuclease, directly implicating the RAD2 gene and its XPG human homologue as an important component in the incision of a damaged DNA strand during excision repair.[148] Two other human DNA repair defect diseases, Cockayne's syndrome (CS) and PIBIDS, which is a photosensitive form of the brittle-hair disease trichothiodystrophy (TTD), also have genetic defects that may correlate to genes in yeast. It is peculiar, however, that pa-

Table 6-1 Eukaryotic Nucleotide-Excision-Repair—Genes and Mutants

Human mutant (and gene)[a]	Rodent mutant[b]	Yeast gene
XP-A (XPA)	?	RAD14
XP-B (XPB)	ERCC3	RAD25 (SSL2)
XP-C (XPC)	?	RAD4[c]
XP-D (XPD)	ERCC2	RAD3
XP-E	?	?
XP-F	ERCC4?	RAD1?
XP-G (XPG)	ERCC5	RAD2
CS-A	?	?
CS-B (CSB)	ERCC6	RAD26
TTD-A	?	?
?	ERCC1	RAD10
?	?	RAD7
?	?	RAD16
? (HHR23A)	?	RAD23
? (HHR23B)	?	RAD23

[a]XPA (etc.) is the term for the gene encoding a protein that corrects XP-A.

[b]Rodent complementation groups are numbered. The human genes correcting the defect of rodent mutants are designated ERCC (for excision repair cross-complementing) genes, the number of referring to the number of the corrected group.

[c]Level and type of homology not conclusive.

Source: Modified after Bootsma and Hoeijmakers.[147]

tients with PIBIDS have all the symptoms of CS as well as some of those of TTD, and patients with CS and PIBIDS do not appear to have a higher-than-expected incidence of cancer. This is one of the curious examples of a remarkable clinical heterogeneity among patients with the same apparent genetic defects, that is, the same mutation in different individuals giving rise to different clinical syndromes. Such heterogeneity has been observed in cystic fibrosis as well. In the case of NER defects, mutations in one gene may give rise to symptoms of XP, combined XP and CS, or PIBIDS.[147] These kind of results indicate that identifying mutations is only a first step in understanding the mechanism of the disease process. A mutant gene may act differently in one cell type than in another cell type, depending on its interaction with cell-specific transcription factors or *cis*-regulatory elements, interaction of its gene product with other gene products expressed at different levels in different cell types, or posttranscriptional and posttranslational mechanisms in different cell types that regulate the expression and/or function of the protein coded for by the mutant gene in question. One candidate for this latter point is the way protein folding is regulated. Proteins must fold into particular conformations to have biological activity, and some mutations observed in human genetic diseases appear to affect protein folding and intracellular translocations processes. Such defects are seen in certain forms of cystic fibrosis, α_1-anti-trypsin deficiency, and certain glycogen storage diseases.

It has become increasingly important to understand the mechanisms of DNA repair and the defects associated with repair processes because gene mutations identified in association with hereditary forms of colon cancer have been found to be mutations in DNA repair enzymes (Chap. 3). This intriguing discovery highlights the importance of DNA repair processes in maintaining the integrity of the human genome and in protecting the organism from genetic alterations that can lead to cancer. It also highlights the importance of studying basic biochemical and genetic functions in lower phylogenetic systems because often that is where the important functions of genes are first discovered. The discovery of the human hereditary nonpolyposis colonic cancer genes is a case in point. Without the knowledge of how DNA mismatch repair genes work in yeast and bacteria and a somewhat serendipitous finding based on a gene bank search, identification of the function of the colon cancer genes as mutated DNA repair genes may have taken many months or years longer.[151]

Two other points about DNA repair systems should be made here. One is that the DNA damage recognition and repair complex contains proteins that function in concert with transcription factors (e.g., XPBC/ERCC3 is part of the basal transcription factor TFIIH) and RNA polymerase II.[147] The RNA pol II inhibitor α-amanitin, for example, has been shown to inhibit the NER process. The XPBC/ERCC3 gene product appears to act as DNA-melting or unwinding factor in the process of transcription-coupled repair.[147] A second point is that actively transcribed genes are repaired more rapidly than inactive genes. For example, lesions in the active dihydrofolate reductase (DHFR) gene induced by UV damage or alkylating agents is repaired much more rapidly and completely than overall genome repair in the same cells.[143] Moreover, actively transcribed protooncogenes or tumor suppressor genes, although perhaps more subject to damage by carcinogens, are also avidly repaired on the transcribing strand. Thus, mutations may accumulate on the nontranscribed strand. Mutations in protooncogenes that are not actively transcribed in a given cell type may not become evident until such time as they become expressed. This might partly explain the long latency time of certain forms of human cancer.

The last repair mechanism mentioned above, namely O^6-alkylguanine-DNA alkyltransferase (AGT), is important for the repair of alkyl adducts in chemical carcinogen-damaged DNA.[145,151a] As both an alkyltransferase and an alkyl acceptor protein, AGT transfers, for example, a methyl group from O^6-methylguanine to an internal cysteine, forming S-methylcysteine in the protein and regenerating unalkylated G in the DNA strand. This transferase can repair O^6-methyl G and O^4-methyl T and is inhibited by O^6-benzylguanine.[152] This inhibitory effect has been taken advantage of to show the importance of O^6-alkylations in carcinogen-induced DNA damage and mutations as well as to augment the cytoxicity of alkylating anticancer drugs such as BCNU. The AGT mechanism is most active in the repair of smaller alkyl adducts—for example, O^6-methyl G is repaired about three times faster than O^6-ethyl G—and it is likely that even larger adducts are repaired primarily by excision repair.[145] The importance of AGT in preventing chemical carcinogenesis in vivo has been shown by the prevention of *N*-methyl-*N*-nitrosourea induced thymic lymphomas in transgenic mice bearing the human AGT gene.[151a]

VIRAL CARCINOGENESIS

Historical Perspectives

It has long been suspected that various forms of cancer, particularly certain lymphomas and leukemias, are caused or at least "cocaused" by transmissible viruses. This theory has had its ups and downs during the first half of this century, and it was not generally accepted that viruses can cause malignant tumors in animals until the 1950s. The known carcinogeneic effects of certain chemicals, irradiation, chronic irritation, and hormones did not fit with the idea of an infectious origin of cancer. In early experiments, the basic assay to determine whether cancer could be induced by a transmissible agent was done by attempting to transmit malignant disease by inoculation of filtered extracts prepared from diseased tissues. If the disease occurred in animals inoculated with such filtrates, it was assumed to be caused by a virus. In 1908, Ellermann and Bang[153] transmitted chicken leukemia by cell-free, filtered extracts and thus were among the first to demonstrate the viral etiology of this disease. In 1911, Rous[154] induced sarcomas in chickens by filtrates obtained by passing tumor extracts through filters that were impermeable to cells and bacteria. These findings remained dormant for two decades until Shope showed, in 1933, that the common cutaneous papillomas of wild rabbits in Kansas and Iowa were caused by a filterable agent.[155] It was later found that when these tumors were transplanted subcutaneously they became invasive squamous cell carcinomas.[156] In 1934, Lucké observed that kidney carcinomas commonly found in frogs in New England lakes could be transmitted by lyophilized cell-free extracts.[157] Two years later, Bittner demonstrated the transmission of mouse mammary carcinoma through the milk of mothers to offspring.[158] This was the first documented example of transmission of a tumor-inducing virus from one generation to another.

Drawing on the experiments of Bittner, Gross postulated that mouse leukemia was also caused by a virus and that occurrence of the disease in successive generations of mice was due to transmission of virus from parents to offspring.[159] The proof of this hypothesis eluded Gross for a number of years until he was prompted, by evidence based on transmission of Coxsackie viruses to newborn mice, to attempt inoculation of mice less than 48 hours old. Using this approach, he successfully transmitted mouse leukemia by injecting filtered extracts prepared from organs of inbred AK or C58 mice, which have a high incidence of "spontaneous" leukemia, or from embryos of these mice, into newborn C3H mice, which have a very low incidence of leukemia. These experiments demonstrated for the first time that mouse leukemia is caused by a virus and that the virus is transmitted in its latent form through the embryos. This led to the isolation of a mouse leukemia virus.[160] The isolated virus was also found to induce leukemias and lymphomas in inbred strains of mice. Electron-microscopic studies[161] showed that the mouse leukemia virus is spheroid, has a diameter of about 100 nm, and contains a dense, centrally located "nucleus" separated from the external envelope by a clear circular zone. The Gross mouse leukemia virus was classified as a "type C" virus, a term now used to describe a wide variety of RNA-containing oncogenic viruses of similar morphology (Fig. 6-16).[162]

The RNA oncoviruses have been classified by morphologic criteria. Intracytoplasmic type A particles were initially observed in early embryos of mice and in certain murine tumors. These A particles are noninfectious, bud into intracellular membranes rather than through the plasma membrane, and thus stay within the cell. They have an active reverse transcriptase and exist as a proviral form in chromosomal DNA. Type B viruses have spikes on their outer envelope, bud from cells, and have been identified primarily in murine species, mouse mammary tumor virus (MMTV) being an example. Type C viruses have been found widely distributed among birds and mammals, can induce leukemias, sarcomas, and other tumors in various species, and have certain gene sequences that are homologous to "transforming" sequences isolated from various human tumors (see Chap. 7). Another subgroup, type D RNA oncoviruses, have been isolated from primate species but their oncogenic potential is not well established. The subtypes of RNA tumor viruses, known as *Retroviridiae,* share a genetically related genome containing a *gag-pol-env* gene sequence coding for virus internal structural proteins, the special type of RNA-directed DNA polymerase called reverse transcriptase, and viral envelope proteins, respectively. Thus, they most likely share a common evolutionary heritage.[163] However, distinct subclasses of retrovirus evolution, based on *pol* gene sequence homologies, have been found: one major pathway gives rise to mammalian type C viruses and a second to A, B, D, and avian type C oncoviruses.[163] A more recent addition to the retrovirus classification is the human T-cell leukemia virus (HTLV), isolated from patients with certain forms of adult-T cell leukemias (discussed later). The *pol* gene of HTLV appears to have evolved from a progenitor common to the types A, B, D, and avian C oncoviruses rather than from the mammalian C type.[163] If true, this

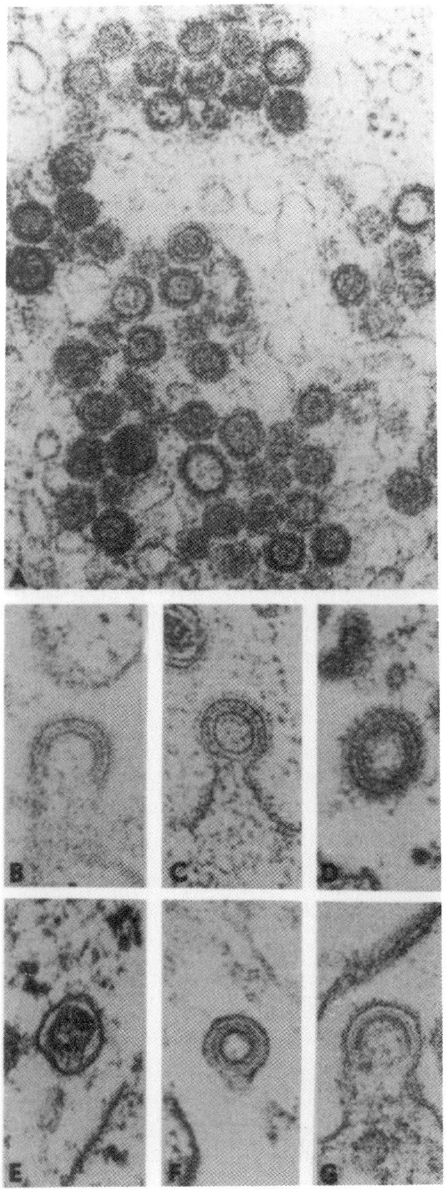

Figure 6-16 Electron micrographs of RNA tumor viruses [magnifications: *A*, ~ × 100,000; *B* to G ~ × 140,000]. (*A*) Group of intracytoplasmic type A particles in a mouse mammary tumor. (*B*) Type B particle budding from a mouse mammary tumor cell. (*C*) Late bud of type B particle from mouse mammary tumor. (*D*) Free, immature, type B particle with spikes on surface, electron lucent center in the nucleoid, and spokes radiating from outer surface of the nucleoid to inner surface of the envelope. (*E*) Type C particle in extracellular space. (*F*) Late bud of type C particle in tissue of human embryo kidney cells infected with a strain of Rauscher leukemia virus. (*G*) Type C particle budding from human embryo lung cell in tissue culture infected with feline leukemia virus. (From Maugh.[162])

would be unusual because most mammalian type C viruses share antigenic determinants among several *gag, pol,* and *env* gene products, suggesting a common progenitor for this subclass of retroviruses.

Unlike most infectious viruses, oncoviruses can be transmitted through the germ line of animal species, and thus these viral genes can be passed from one generation to the next, often in a silent form. The widespread distribution of gene sequences homologous to those of oncoviruses throughout the animal kingdom suggests that these sequences are evolutionarily very old. One of the most important observations of Bittner and Gross was that an oncogenic virus could be transmitted vertically from parent to offspring. In the case of mouse leukemia, it became apparent that a mouse born to AK or C58 parents receives at birth the genetic information for malignant disease. These findings suggest that, later in the life of a carrier animal, the expression of virally coded genes, perhaps triggered by exposure to chemicals, irradiation, hormone imbalance, or chronic irritation, becomes activated, causing leukemia. In the case of the mouse (and perhaps of humans), the activation of latent oncogenic viruses may not occur during the life span of the carrier animal, and the animal may remain disease free, even though it carries and transmits the viral genome to its progeny. How this type of vertical transmission could occur is explained by later findings that the genome of oncogenic viruses is integrated into the host cell's genome prior to cellular transformation (see Chap. 7).

Another oncogenic virus was discovered in the tissue extracts of leukemic AK mice after it was noted that when newborn C3H mice were injected with such extracts, some of the C3H mice developed parotid gland tumors rather than leukemia.[159] Some of these mice also developed cutaneous sarcomas, mammary carcinomas, and other malignancies. Eddy, Stewart, and their colleagues found that virus produced by cultured cells after infection with extracts from leukemic AK mouse organs caused parotid gland tumors and a variety of other neoplasms in mice, hamsters, and rats. They suggested that this multiple-tumorigenic agent be called polyoma virus.[164]

Since the early pioneering work in this field, a number of other oncogenic viruses have been identified and characterized. These include (1) feline leukemia virus, shown by inoculation of cell-free extracts from leukemic cats into newborn kittens; (2) SV40 virus, shown to be latent and harmless in the rhesus monkey but to induce leukemias and sarcomas after inoculation into newborn hamsters; (3) adenoviruses, which cause the common cold in humans and induce sarcomas in newborn hamsters and rats; and (4) such herpesviruses as *Herpes saimiri,* which is indigenous in the New World squirrel monkey and may induce lymphosarcomas and leukemias when inoculated into certain species of monkeys. Table 6-2 lists some of the different types of RNA and DNA oncogenic viruses.

Role of Viruses in the Causation of Human Cancer

To prove a causal relationship between a putative cancer-causing virus and human cancer is not a simple task. Such proof relies on evidence that is to a fair extent circumstantial. This evidence includes (1) epidemiologic data showing a correlation between living in an area of endemic viral infection and a type of cancer; (2) serologic evidence of antibody titers to viral antigens in patients with a given cancer type; (3) evidence for insertion of viral DNA into a cancer-bearing host's cell genome; (4) evidence for a consistent chromosomal translocation, particularly those involving an oncogene, in virally infected patients; (5) data showing that viral infection of cells in culture or transfection of viral genes into cells causes cell transformation and the ability of such cells to produce tumors in nude mice; and (6) development of cancers of the suspected target organ in transgenic mice produced by embryonic gene transfer of viral genes.

Based on this sort of evidence, some human cancers are considered to be caused by viral infection either directly or indirectly. By "directly," I mean that the viral gene(s) can themselves cause cells to become malignant (sometimes also requiring the loss of a tumor suppressor gene). By "indirectly," I mean that viral infection may simply cause the progression of malignant cell growth by producing an im-

Table 6-2 Examples of Oncogenic Viruses

Virus	Species of Isolation
I. Oncogenic RNA viruses	
A. Acute-acting type[a]	
Rous sarcoma	Chickens
Fujinami sarcoma	Chickens
Reticuloendotheliosis	Chickens/turkeys
Avian erythroblastosis	Chickens
Avian myeloblastosis	Chickens
Avian myelocytomatosis	Chickens
Moloney sarcoma	Mice
Abelson leukemia	Mice
FBJ osteosarcoma	Mice
Harvey/Kirsten sarcoma[b]	Rat
Rat sarcoma	Rat
Feline sarcoma	Cat
Woolly monkey sarcoma	Woolly monkey
B. Chronic type[c]	
Avian leukosis	Chickens
Mouse leukemia[d]	Mice
Feline leukemia	Cat
Bovine leukemia	Cow
Gibbon ape leukemia[e]	Gibbon ape
Mouse mammary tumor	Mice
Human T lymphotropic viruses	Human
II. Oncogenic DNA viruses	
A. Papovaviruses	
Papillomaviruses	Rabbit, man, dog, cow, and others
Polyomaviruses	Mouse
SV40	Monkey
JC	Human
BK	Human
B. Adenoviruses	Human,[f] monkey, birds, cow
C. Herpesviruses	
Epstein-Barr virus	Human
Lucké carcinoma	Frog
Marek's disease	Chicken

[a]Acute, transform cells in vitro, rapid disease induction in vivo, carry "transforming *onc* gene" related to cell gene. Most are replication defective but can be isolated free of helper virus.

[b]These hybrid viruses, created experimentally, contain mouse helper virus and rat "src" sequences.

[c]Chronic, no transformation in vitro, long latency period in vivo, no evidence of transforming gene. These all appear to be horizontally transmitted; in some cases, related sequences are found in cell DNA.

[d]The Friend leukemia virus complex contains a defective genome, codes for a smaller envelope glycoprotein not incorporated into virions, does not transform cells in vitro, and perhaps should be placed in a separate category.

[e]Viruses show a distant relationship to mouse DNA, but not that of primates; this indicates "ancient" horizontal transmission.

[f]There are 31 members of the human adenovirus group and at least 12 induce tumors in newborn animals and/or transform cells in vitro.

munodeficiency state (e.g., the occurrence of non-Hodgkin's lymphoma in HIV-infected patients) or by stimulating the proliferation of already transformed cells. Sometimes viral infection acts in concert with other infectious agents or chemical carcinogens. Such is the case for malarial infection of Epstein-Barr virus (EBV) infected patients and for aflatoxin exposure of individuals bearing the hepatitis B viral genome in their liver cells (see below). The types of human cancer thought to be caused by viral infection and the strength of the epidemiologic association are shown in Table 6-3.[165]

Association of Epstein-Barr Virus and Human Cancer

Epstein-Barr virus has been linked to four different types of human cancer: Burkitt's lymphoma (BL), nasopharyngeal carcinoma (NPC), B-cell lymphomas in immunosuppressed individuals such as HIV infected patients, and some cases of Hodgkin's lymphoma.[166] The evidence is strongest for an association with BL and NPC.

Epstein-Barr virus infection does not by itself cause cancer. On average, across the world, about 90% of the population may be infected by the time they reach adulthood. In some endemic areas, the incidence rate approaches 100%. In developing countries, EBV infection often occurs in young childhood. In more affluent societies, EBV infection tends to occur as the "kissing age" of adolescence or young adulthood is reached and manifests itself as infectious mononucleosis. In developing countries, particularly in equatorial Africa, concomitant or subsequent infection with the malarial parasite induces B-cell proliferation and an immunodeficiency state that leads to malignant transformation and progression. There is a consistent chromosomal translocation involving immunoglobulin genes, usually on chromosome 14, and sequences within or adjacent to the c-*myc* gene locus on chromosome 8 (see Chaps. 3 and 7).

The role of EBV in NPC is less well characterized, but the evidence for an association includes high serum antibody titers against EBV antigens and the presence of EBV DNA in NPC cells. Similar evidence suggests an association

Table 6-3 Human Cancer Viruses and Cancers with Which They Are Associated, Strength of Association, and Necessary Preconditions

Virus	Cancer	Strength of Epidemiologic Association	Required Precondition
HBV	Hepatocellular carcinoma	Strong	None
HTLV-I	T-cell lymphoma	Strong	None
EBV	BL	Strong	Chronic malaria
EBV	High-grade lymphoma	Strong	HIV
HPV	Cervical cancer	?Consistent	?None
EBV	NPC	Inconsistent	—
HSV-2	Cervical cancer	Inconsistent	—

Source: From Henderson.[165]

between EBV infection and induction of some B-cell lymphomas and some Hodgkin's disease cases in immunosuppressed individuals, although the exact role of EBV remains to be elucidated.[166]

Hepatitis Virus and Hepatocellular Carcinoma

Epidemiologic evidence strongly points to a link between chronic hepatitis B virus (HBV) infection and hepatocellular carcinoma (HCC). In areas where HBV infection is endemic, such as Taiwan, Senegal, South Africa, Hong Kong, China, and the Philippines, the incidence of HCC is much higher than in countries where HBV infection is less common (reviewed in Ref. 166). Hepatocellular carcinoma usually appears after decades of chronic liver involvement due to HBV-induced liver cell damage and regeneration. Hepatitis B virus DNA can be found in the majority of liver cancers from patients in high-risk areas and a specific piece of HBV DNA called HBx, which encodes a transcription factor, is found in HCC cells and can induce liver tumors in transgenic mice.[167] Some chromosomal modifications are also observed in HCC. These include alterations of the short arm of chromosome 11, deletions in the long arm of chromosome 13, and point mutations of the p53 gene on chromosome 17. These latter mutations are particularly interesting because such mutations are seen in areas where there is concomitant exposure to aflatoxins in foods, suggesting a joint role of HBV infection and chemical carcinogens in causing HCC.

There is also some evidence that chronic infection with hepatitis C virus, which also may cause chronic liver injury and regeneration, may be a causative agent for HCC and may be a factor particularly in HBV-negative cases. There are currently a number of countries in which widespread vaccination against HBV is occurring, and time will tell whether such vaccination lowers the incidence of HCC in these areas.

Papillomaviruses and Cervical Cancer

A large class of papillomaviruses that are pathogenic for humans (HPV) have been identified. More than 60 genotypic subtypes have been isolated.[165] Only two HPV subtypes have been closely associated with cervical cancer, HPV 16 and HPV 18. The evidence for this association is the following (reviewed in Ref. 166): (1) viral DNA is found in about 90% of cervical cancers; (2) in most cases a specific piece of the viral DNA is integrated into the host's genome; (3) the vast majority of all HPV-positive cervical cancers contain cells that express two specific gene transcripts, called E6 and E7; (4) E6 and E7 genes of high-risk HPV (e.g., HPV 16 and 18) but not of low-risk HPV subtypes immortalize human cells in culture; (5) E6- and E7-expressing cells frequently undergo a progression to aneuploidy and gene amplification in culture; (6) the E6 and E7 oncoproteins bind to and inactivate or degrade the p53 and RB tumor suppressor gene proteins, respectively; and (7) uterine cervical dysplasia can be induced in mice by inoculation of a recombinant retrovirus bearing the E6 and E7 genes of HPV 16 into the vaginas of mice.[168]

HTLV-I and Adult T-Cell Leukemia

The retrovirus HTLV-I was first identified in interleukin-2 (IL-2)-stimulated T lymphocytes from two patients, one with a T-cell lymphoma and the other with T-cell leukemia.[169,170] This was the first evidence that a retrovirus could cause malignancy in humans, although it had long been suspected that retroviruses might do so, based on numerous examples in animals.

Since the original isolation of a human T-cell leukemia/lymphoma virus, several similar isolates have been made from patients with T-cell neoplasms in different parts of the world, including Japan, Africa, the Caribbean basin, England, and the Netherlands.[171] The isolates from patients with adult T-cell leukemia-lymphoma (ATLL), named HTLV-I, were found to have several characteristics in common. In 1982, a new subgroup of HTLV, called HTLV-II, was identified.[172] The DNA sequence of the two subgroups is clearly different, but there is significant homology. Moreover, the two types of viral genomes encode a very similar p24 core protein and share a common mechanism of gene activation, indicating that they are members of the same family. The HTLV-I and -II viruses are retroviruses with reverse transcriptase and an RNA genome of about 9 kilobases. DNA sequences homologous to HTLV DNA are not found in the genome of normal human cells; thus, they are exogenous, not endogenous, genomic sequences, in contrast with the c-*onc* genes that are homologous to v-*onc* genes of retroviruses. The HTLV viruses do not appear to carry their own *onc* gene. They appear to be similar to the chronic-acting retroviruses (such as avian leukosis virus) in this respect, and they have an LTR-*gag-pol-env*-LTR gene arrangement typical of other retroviruses. HTLV-I and -II can transform normal human T cells in culture and the transformed cells contain at least one proviral DNA copy, transcribe viral DNA, and make low levels of viral proteins.[172]

In addition to the genes common to all retroviruses, the HTLV-I genome contains a 1.6-kilobase sequence at the 3′ terminal region that encodes at least two *trans*-acting regulator proteins: a 40,000-Da protein product of a gene called *tax* and a 27,000-Da protein product of a gene called *rex*. Transgenic mice bearing the *tax* gene develop multiple mesenchymal tumors at about 3 months of age.[173]

A high percentage of patients with ATLL and certain T-cell lymphomas have antigens to HTLV-I proteins in their serum, and their tumor cells contain one or more copies of the HTLV genome.[171,174,175] But patients with childhood cancers, non-T-cell leukemias and lymphomas, myeloid leukemias, Hodgkin's disease, and solid tumors do not have evidence of HTLV antibodies in their serum, and healthy individuals in nonendemic areas are also antibody-negative.[174] However, almost 50% of relatives of ATLL patients and about 12% of healthy blood donors in endemic areas have been reported to be antibody-positive. These data indicate the T-cell specificity of the neoplastic transformation process induced by HTLV and suggest horizontal spread of the virus among people. Major clusters of HTLV-related T-cell cancers are found in areas of high endemic infection with HTLV, such as southwestern Japan, the Caribbean basin, and certain areas of South America and Africa.[171] As is the case with Burkitt's lymphoma, however, only certain people infected with the virus get ATLL. Thus, other predisposing factors must exist.

Horizontal transmission among individuals appears to require prolonged and intimate contact with an HTLV-positive person. Cell-to-cell transmission can also be demonstrated in cell culture systems when HTLV-producing cells are cocultured with normal T lymphocytes.[171] The HTLV-infected, transformed cells often produce infectious HTLV, but virus-nonproducing transformed cells, containing the integrated viral genome as a provirus, are also observed.

Only about 1 out of 25 to 30 infected individuals will eventually develop ATLL, and HTLV-I DNA is consistently demonstrated in the ATLL cells from these patients. There appears to be a latency period of several decades between primary infection and leukemia development.[166]

The question obviously arises: Where did these viruses come from and are they a recently evolved class of viruses? Although there is no definitive evidence for HTLV infections in humans before the late 1970s, HTLV viruses have probably been around for a long time—perhaps hundreds or thousands of years. Most likely, there was an animal vector originally, and recently the virus may have undergone some evolutionary change that made it more infectious or more cytopathic for humans. Analysis of serum samples from subhuman primates, for example, shows that several Old World monkey species—including Japanese and Chinese macaques, African green monkeys, and baboons—are seropositive for HTLV.[176] In Japan, where HTLV is endemic in humans in certain areas, the distribution of virus in primates is much more widespread, suggesting independent entry of HTLV-like virus into the primate and human populations and arguing against current transmission between primates and humans. Because of the widespread infection of African Old World primates with HTLV-like viruses, it has been proposed that the origin of HTLV was in Africa and that spread to other countries may have occurred by means of explorers who introduced infected primates or had contact with infected primates in Africa.[176]

In conclusion, it is fair to say that infections with oncogenic viruses are clearly associated with certain kinds of human cancer. However, even in those cases in which viral infection appears to be a predisposing factor, viral infection itself is insufficient to cause cancer. In all cases, there are other contributing factors, which include cell-type–specific mitogenic stimulation, suppression of the immune response, and, possibly, genetic factors. It is also clear, though, that a combination of infection with certain oncogenic viruses, chronic mitogenic stimulation of the virus-infected cells, and a concomitant immune deficiency state have a high propensity to induce the cancerous process in human beings.

It should be noted that the human immunodeficiency virus (HIV) that causes AIDS is also a slow infectious retrovirus virus (lentivirus) that is a T-cell lymphotropic virus with some genomic similarities to HTLV. Human immunodeficiency virus infects $CD4^+$ cells and causes disease by its immunosuppressive effects. Patients with AIDS are at high risk to develop Kaposi's sarcoma and non-Hodgkin's lymphoma. These cancers most likely arise because of the immunosuppressive effects of the AIDS virus rather than the direct transforming activity of the virus, although some direct cell-transforming effects have been observed.

REFERENCES

1. E. C. Miller: Some current perspectives on chemical carcinogenesis in humans and experimental animals: Presidential address. *Cancer Res* 38:1479, 1978.
2. P. D. Lawley: Carcinogenesis by alkylating agents, in C. E. Searle, ed.: *Chemical Carcinogens.* Monograph No. 173. Washington, DC: American Chemical Society, 1976, pp. 83–244.
3. E. C. Miller and J. A. Miller: Searches for ultimate chemical carcinogens and their reactions with cellular macromolecules. *Cancer* 47:2327, 1981.
4. E. Boyland: The biological significance of metabolism of polycyclic compounds. *Biochem Soc Symp* 5:40, 1950.
5. A. Pullman and B. Pullman: Electronic structure and carcinogenic activity of aromatic molecules. *Adv Cancer Res* 3:117, 1955.
6. A. H. Conney: Induction of microsomal enzymes by foreign chemicals and carcinogenesis by polycyclic aromatic hydrocarbons: G. H. A. Clowes Memorial Lecture. *Cancer Res* 42:4875, 1982.
7. E. Huberman, L. Sachs, S. K. Yang, and H. V. Gelboin: Identification of mutagenic metabolites of benzo(a)pyrene in mammalian cells. *Proc Natl Acad Sci USA* 73:607, 1976.
8. D. M. Jerina, H. Yagi, O. Hernandez, P. M. Dansette, A. W. Wood, W. Levin, R. L. Chang, P. G. Wislocki, and A. H. Conney: Synthesis and biological activity of potential benzo(a)pyrene metabolites, in R. Frendenthal and P. W. Jones, eds.: *Carcinogenesis—A Comprehensive Survey,* Vol. 1. New York: Raven Press, 1976, pp. 91–113.
9. I. B. Weinstein, A. M. Jeffrey, K. W. Jennette, S. H. Blobstein, R. G. Harvey, C. Harris, H. Antrup, H. Kasai, and K. Nakanishi: Benzo(a)pyrene diol epoxides as intermediates in nucleic acid binding in vitro and in vivo. *Science* 193:592, 1976.
10. T. Meehan and K. Straub: Double-stranded DNA stereoselectively binds benzo(a)pyrene diol epoxides. *Nature* 277:410, 1979.
11. D. M. Jerina and J. W. Daly: Oxidation at carbon, in D. V. Parke and R. C. Smith, eds.: *Drug Metabolism—From Microbe to Man.* London: Taylor and Francis, 1977, pp. 13–32.
12. E. L. Cavalieri and E. G. Rogan: The approach to understanding aromatic hydrocarbon carcinogenesis: The central role of radical cations in metabolic activation. *Pharmacol Ther* 55:183, 1992.
13. E. G. Rogan, P. D. Devanesan, N. V. S. RamaKrishna, S. Higginbotham, N. S. Padma-

vathi, K. Chapman, E. L. Cavalieri, H. Jeong, R. Jankowiak, and G. J. Small: Identification and quantitation of benzo(a)pyrene-DNA adducts formed in mouse skin. *Chem Res Tox* 6:356, 1993.
14. F. P. Guengerich: Roles of cytochrome P-450 enzymes in chemical carcinogenesis and cancer chemotherapy. *Cancer Res* 48:2946, 1988.
15. G. Kellermann, C. R. Shaw, and M. Luyten-Kellerman: Aryl hydrocarbon hydroxylase inducibility and bronchogenic carcinoma. *N Engl J Med* 289:934, 1973.
16. T. L. McLemore, S. Adelberg, M. C. Liu, N. A. McMahan, S. J. Yu, W. C. Hubbard, M. Czerwinski, T. G. Wood, R. Storeng, R. A. Lubet, J. C. Eggleston, M. R. Boyd, and R. N. Hines: Expression of CYP1A1 gene in patients with lung cancer: Evidence for cigarette smoke-induced gene expression in normal lung tissue and for altered gene regulation in primary pulmonary carcinomas. *JNCI* 82:1333, 1990.
16a. K. Nakachi, K. Imai, S.-i. Hayashi, and K. Kawajiri: Polymorphisms of the CYP1A1 and glutathione S-transferase genes associated with susceptibility to lung cancer in relation to cigarette dose in a Japanese population. *Cancer Res* 53:2994, 1993.
17. V. D-H. Ding, R. Cameron, and C. B. Pickett: Regulation of microsomal, xenobiotic epoxide hydrolase messenger RNA in persistent hepatocyte nodules and hepatomas induced by chemical carcinogens. *Cancer Res* 50:256, 1990.
18. M. E. Hogan, N. Dattagupta, and J. P. Whitlock, Jr.: Carcinogen-induced alteration of DNA structure. *J Biol Chem* 256, 1981.
19. J. S. Eadie, M. Conrad, D. Toorchen, and M. D. Topal: Mechanism of mutagenesis by O^6-methylguanine. *Nature* 308:201, 1984.
20. L. A. Loeb: Apurinic sites as mutagenic intermediates. *Cell* 40:483, 1985.
21. S. Needle: New twists to DNA and DNA-carcinogen interactions. *Nature* 292:292, 1981.
22. T. C. Boles and M. E. Hogan: Site-specific carcinogen binding to DNA. *Proc Natl Acad Sci USA* 81:5623, 1984.
23. P. D. Kurth and M. Bustin: Site-specific carcinogen binding to DNA in polytene chromosomes. *Proc Natl Acad Sci USA* 82:7076, 1985.
24. C. C. Harris: Chemical and physical carcinogenesis: Advances and perspectives for the 1990s. *Cancer Res* 51:5023s, 1991.
24a. D. Chakravarti, J. C. Pelling, E. C. Cavalieri, and E. G. Rogan: Relating aromatic hydrocarbon-induced DNA adducts and c-H-*ras* mutations in mouse skin papillomas: The role of apurinic sites. *Biochemistry*, submitted.
25. I. C. Hsu, R. A. Metcalf, T. Sun, J. Welsh, N. J. Wang, and C. C. Harris: p53 gene mutational hotspot in human hepatocellular carcinomas from Qidong, China. *Nature* 350:427, 1991.
26. B. Bressac, M. Kew, J. Wands, and M. Ozturk: Selective G to T mutations of p53 gene in hepatocellular carcinoma from southern Africa. *Nature* 350:429, 1991.
27. W. F. Friedewald and P. Rous: The initiating and promoting elements in tumor production: An analysis of the effects of tar, benzpyrene, and methylcholanthrene on rabbit skin. *J Exp Med* 80:101, 1944.
28. R. D. Sall and M. J. Shear: Studies in carcinogenesis: XII. Effect of the basic fraction of creosote oil on the production of tumors in mice by chemical carcinogens. *JNCI* 1:45, 1940.
29. I. Berenblum: The mechanism of carcinogenesis: A study of the significance of carcinogenic action and related phenomena. *Cancer Res* 1:807, 1941.
30. E. Hecker: Isolation and characterization of the cocarcinogenic principles from croton oil. *Methods Cancer Res* 6:439, 1971.
31. B. L. Van Duuren: Tumor-promoting agents in two-stage carcinogenesis. *Prog Exp Tumor Res* 11:31, 1969.
32. I. Berenblum: Sequential aspects of chemical carcinogenesis: Skin, in F. F. Becker, ed.: *Cancer: A Comprehensive Treatise.* New York: Plenum Press, 1975, pp. 323–344.
33. R. K. Boutwell: Some biological aspects of skin carcinogenesis. *Prog Exp Tumor Res* 4:207, 1964.
34. H. C. Pitot and Y. P. Dragan: Facts and theories concerning the mechamisms of carcinogenesis. *FASEB J* 5:2280, 1991.
35. J. F. O'Connell, A. J. P. Klein-Szanto, D. M. DiGiovanni, J. W. Fries, and T. J. Slaga: Enhanced malignant progression of mouse skin tumors by the free-radical generator benzoyl peroxide. *Cancer Res* 46:2863, 1986.
36. J. McCann, E. Choi, E. Yamasaki, and B. N. Ames: Detection of carcinogens as mutagens in the *Salmonella*/microsome test: Assay of 300 chemicals. *Proc Natl Acad Sci USA* 72:5135, 1975.
37. E. Huberman and L. Sachs: Cell-mediated mutagenesis of mammalian cells with chemical carcinogens. *Int J Cancer* 13:326, 1974.
38. E. Huberman and L. Sachs: Mutability of different genetic loci in mammalian cells by metabolically activated carcinogenic polycyclic hydrocarbons. *Proc Natl Acad Sci USA* 73:188, 1976.
39. R. B. Setlow: Repair deficient human disorders and cancer. *Nature* 271:713, 1978.
40. M. Swift and C. Chase: Cancer in families with xeroderma pigmentosum. *JNCI* 62:1415, 1979.
41. H. J. Vijayalaxmi, J. H. Ray, and J. German: Bloom's syndrome: Evidence for an increased mutation frequency in vivo. *Science* 221:851, 1983.
42. Y. Shiraishi, T. H. Yosida, and A. A. Sandberg: Malignant transformation of Bloom syndrome B-lymphoblastoid cell lines by carcinogens. *Proc Natl Acad Sci USA* 82:5102, 1985.

43. H. Kirkman: *Estrogen-Induced Tumors of the Kidney in the Syrian Hamster.* Bethesda, MD: National Cancer Institute Monograph No. 1, 1959, p. 137.
44. M. S. Biskind and G. S. Biskind: Development of tumors in the rat ovary after transplantation into the spleen. *Proc Soc Exp Biol Med* 55:176, 1944.
45. T. Tsutsui, G. H. Degan, D. Schiffman, A. Wong. H. Maizumi, J. A. McLachlan, and J. C. Barrett: Dependence on exogenous metabolic activation for induction of unscheduled DNA synthesis in Syrian hamster embryo cells by diethylstilbestrol and related compounds. *Cancer Res* 44:184, 1984.
46. S. K. Buenaventura, D. Jacobson-Kram, K. L. Dearfield, and J. R. Williams: Induction of sister chromatid exchange by diethylstilbestrol in metabolically competent hepatoma cells but not in fibroblasts. *Cancer Res* 44:3851, 1984.
47. L. A. Loeb: Mutator phenotype may be required for multistage carcinogenesis. *Cancer Res* 51:3075, 1991.
48. E. R. Fearon and B. Vogelstein: A genetic model for colorectal tumorigenesis. *Cell* 61:759, 1990.
49. B. S. Strauss: The origin of point mutations in human tumor cells. *Cancer Res* 52:249, 1992.
50. L. A. Loeb: Endogenous carcinogenesis: Molecular oncology into the twenty-first century—Presidential Address. *Cancer Res* 49:5489, 1989.
51. C. Richter, J-W. Park, and B. N. Ames: Normal oxidative damage to mitochondrial and nuclear DNA is extensive. *Proc Natl Acad Sci USA* 85:6465, 1988.
52. T. A. Kunkel and L. A. Loeb: Fidelity of mammalian DNA polymerase. *Science* 213:765, 1981.
53. T. A. Kunkel: The mutational specificity of DNA polymerase-β during in vitro DNA synthesis: production of frameshift, base substitution and deletion mutations. *J Biol Chem* 260:5787, 1985.
53a. N. P. Bhattacharyya, A. Skandalis, A. Ganesh, J. Groden, and M. Meuth: Mutator phenotypes in human colorectal carcinoma cell lines. *Proc Natl Acad Sci USA* 91:6319, 1994.
53b. L. Loeb: Microsatellite instability: Marker of a mutator phenotype in cancer. *Cancer Res* 54:5059, 1994.
54. B. N. Ames, R. L. Saul, E. Schwiers, R. Adelman, and R. Cathcart: Oxidative DNA damage as related to cancer and aging: The assay of thymine glycol, thymidine glycol, and hydroxymethyluracil in human and rat urine, in R. S. Sohal, L. S. Birnbaum, and R. G. Cutler, eds.: *Molecular Biology of Aging: Gene Stability and Gene Expression.* New York: Raven Press, 1985.
55. L. Foulds: The experimental study of tumor progression: A review. *Cancer Res* 14:327, 1954.
56. P. C. Nowell: The clonal evolution of tumor cell populations. *Science* 194:23, 1976.
57. D. L. Trainer and E. F. Wheelock: Characterization of L5178 Y cell phenotypes isolated during progression of the tumor-dormant state in DBA/2 mice. *Cancer Res* 44:2897, 1984.
58. D. B. Rifkin, R. M. Crowe, and R. Pollack: Tumor promoters induce changes in the chick embryo fibroblast cytoskeleton. *Cell* 18:361, 1979.
59. P. B. Fisher, J. H. Bozzone, and I. B. Weinstein: Tumor promoters and epidermal growth factor stimulate anchorage-independent growth of adenovirus-transformed rat embryo cell. *Cell* 18:695, 1979.
60. R. Cohen, M. Pacifici, N. Rubernstein, J. Biehl, and H. Holtzer: Effect of tumor promoter on myogenesis. *Nature* 226:538, 1977.
61. M. Lowe, M. Pacifici, and H. Holtzer: Effects of phorbol-12-myristate-13-acetate on the phenotypic program of cultured chrondroblasts and fibroblasts. *Cancer Res* 38:2350, 1978.
62. L. Diamond, T. O'Brien, and A. Rovera: Inhibition of adipose conversion of 3Y3 fibroblasts by tumor promoters. *Nature* 269:247, 1977.
63. E. Fibach, R. Gambari, P. A. Shaw, G. Maniatris, R. C. Reuben, S. Sassa, R. A. Rifkind, and P. A. Marks: Tumor promoter–mediated inhibition of cell differentiation: Suppression of the expression of erythroid functions in murine erythroleukemia cells. *Proc Natl Acad Sci USA* 76:1906, 1979.
64. D. Ishii, E. Fibach, H. Yamasaki, and I. B. Weinstein: Tumor promoters inhibit morphological differentiation in cultured mouse neuroblastoma cells. *Science* 200:556, 1978.
65. S. Mondal and C. Heidelberger: Transformation of C3H/10T ½ CL8 mouse embryo fibroblasts by ultraviolet irradiation and a phorbol ester. *Nature* 260:710, 1976.
66. N. Yamamoto and H. zur Hausen: Tumor promoter TPA enhances transformation of human leukocytes by Epstein-Barr virus. *Nature* 280:244, 1979.
67. L. Kopelovich, N. E. Bias, and L. Helson: Tumor promoter alone induces neoplastic transformation of fibroblasts from human genetically predisposed to cancer. *Nature* 282:619, 1979.
68. C. M. Aldaz, C. J. Conti, I. B. Gimenez, T. J. Slaga, and A. J. P. Klein-Szanto: Cutaneous changes during prolonged application of 12-*O*-tetradecanoylphorbol-13-acetate on mouse skin and residual effects after cessation of treatment. *Cancer Res* 45:2753, 1985.
68a. P. M. Rosoff, L. F. Stein, and L. C. Cantley: Phorbol esters induce differentiation in a pre-B-lymphocyte cell line by enhancing Na^+/H^+ exchange. *J Biol Chem* 259:7056, 1984.
69. J. M. Besterman, W. S. May, Jr., H. LeVine III, E. J. Cragoe, Jr., and P. Cuatrecasas: Amiloride inhibits phorbol ester–stimulated Na^+/H^+ exchange and protein kinase C: An amiloride analog selectively inhibits NA^+/H^+ exchange. *J Biol Chem* 260:1155, 1985.

70. I. Sussman, R. Prettyman, and T. G. O'Brien: Phorbol esters and gene expression: The role of rapid changes in K^+ transport in the induction of ornithine decarboxylase by 12-*O*-tetradecanoyl-13-acetate in BALB/c 3T3 cells and a mutant cell line defective in Na^+K^+Cl-cotransport. *J Cell Biol* 101:2316, 1985.
71. S. Grinstein, S. Cohen, J. D. Goetz, A. Rothstein, and E. W. Gelfand: Characterization of the activation of Na^+/H^+ exchange in lymphocytes by phorbol esters: Change in cytoplasmic pH dependence of the antiport. *Proc Natl Acad Sci USA* 82:1429, 1985.
72. C.-C. Chang, J. E. Trosko, H.-J. Kung, D. Bombick, and F. Matsumura: Potential role of the src gene product in inhibition of gap-junctional communication in NIH/3T3 cells. *Proc Natl Acad Sci USA* 82:5360, 1985.
73. W. S. May, N. Sahyoun, S. Jacobs, M. Wolf, and P. Cuatrecasas: Mechanism of phorbol diester-induced regulation of surface transferrin receptor involves the action of activated protein kinase C and an intact cytoskeleton. *J Biol Chem* 260:9419, 1985.
74. M. Schliwa, T. Nakamura, K. R. Porter, and U. Euteneuer: A tumor promoter induces rapid and coordinated reorganization of actin and vinculin in cultured cells. *J Cell Biol* 99:1045, 1984.
75. E. G. Fey and S. Penman: Tumor promoters induce a specific morphological signature in the nuclear matrix-intermediate filament scaffold of Madin-Darby canine kidney (MDCK) cell colonies. *Proc Natl Acad Sci USA* 81:4409, 1984.
76. P. A. Cerutti: Prooxidant states and tumor promotion. *Science* 227:375, 1985.
77. J. Barsoum and A. Varshavsky: Mitogenic hormones and tumor promoters greatly increase the incidence of colony-forming cells bearing amplified dihydrofolate reductase genes. *Proc Natl Acad Sci USA* 80:5330, 1983.
78. T. Solo, T. Turpeenniemi-Hujanen, and K. Tryggvason: Tumor-promoting phorbol esters and cell proliferation stimulate secretion of basement membrane (Type IV) collagen-degrading metalloproteinase by human fibroblasts. *J Biol Chem* 260:8526, 1985.
79. R. Montesano and L. Orci: Tumor-promoting phorbol esters induce angiogenesis in vitro. *Cell* 42:469, 1985.
80. P. E. Driedger and P. M. Blumberg: Specific binding of phorbol ester tumor promoters. *Proc Natl Acad Sci USA* 77:567, 1980.
81. R. D. Estensen, D. K. DeHoogh, and C. F. Cole: Binding of [^{3}H]12-*O*-tetradecanoyl-13-acetate to intact human peripheral blood lymphocytes. *Cancer Res* 40:1119, 1980.
82. M. Shoyab and G. J. Todaro: Specific high affinity cell membrane receptors for biologically active phorbol and ingenol esters. *Nature* 288:451, 1980.
83. V. Solanki and T. J. Slaga: Specific binding of phorbol ester tumor promoters to intact primary epidermal cells from SENCAR mice. *Proc Natl Acad Sci USA* 78:2549, 1981.
84. A. D. Horowitz, E. Greenebaum, and I. B. Weinstein: Identification of receptors for phorbol ester tumor promoters in intact mammalian cells and of an inhibitor of receptor binding in biological fluids. *Proc Natl Acad Sci USA* 78:2315, 1981.
85. J. J. Sando and M. C. Young: Identification of a high affinity phorbol ester receptor in cytosol of EL4 thymoma cells: Requirement for calcium, magnesium, and phospholipids. *Proc Natl Acad Sci USA* 280:2642, 1983.
86. K. L. Leach, M. L. James, and P. M. Blumberg: Characterization of a specific phorbol ester aporeceptor in mouse brain cytosol. *Proc Natl Acad Sci USA* 80:4208, 1983.
87. C. L. Ashendel, J. M. Staller, and R. K. Boutwell: Protein kinase activity associated with a phorbol ester receptor purified from mouse brain. *Cancer Res* 43:4333, 1983.
88. I. B. Weinstein, S. Gattoni-Celli, P. Kirschmeier, W. Hsiao, A. Horowitz, and A. Jeffrey: Cellular targets and host genes in multistage carcinogenesis. *Fed Proc* 43:2287, 1984.
89. K. L. Leach and P. M. Blumberg: Modulation of protein kinase C activity and [^{3}H]phorbol 12,13-dibutyrate binding by various tumor promoters in mouse brain cytosol. *Cancer Res* 45:1958, 1985.
90. Y. Nishizuka: The role of protein kinase C in cell surface signal transduction and tumor promotion. *Nature* 308:693, 1984.
91. J. E. Niedel, L. J. Kuhn, and G. R. Vandenbark: Phorbol diester receptor copurifies with protein kinase C. *Proc Natl Acad Sci USA* 80:36, 1983.
92. R. M. Bell: Protein kinase C activation by diacylglycerol second messengers. *Cell* 45:631, 1986.
93. Y. Nishizuka: Intracellular signalling by hydrolysis of phospholipids and activation of protein kinase C. *Science* 258:607, 1992.
94. E. Berra, M. T. Diaz-Meco, I. Dominguez, M. M. Municio, L. Sanz, J. Lozano, R. S. Chapkin, and J. Moscat: Protein kinase C isoform is critical for mitogenic signal transduction. *Cell* 74:555, 1993.
95. D. K. Werth, J. E. Niedel, and I. Pastan: Vinculin, a cytoskeletal substrate of protein kinase C. *J Biol Chem* 258:11423, 1983.
96. E. Rozengurt, M. Rodriquez-Pena, and K. A. Smith: Phorbol esters, phospholipase C, and growth factors rapidly stimulate the phosphorylation of a Mr 80,000 protein in intact quiescent 3T3 cells. *Proc Natl Acad Sci USA* 80:7244, 1983.
97. W. J. Welch: Phorbol ester, calcium ionophore, or serum added to quiescent rat embryo fibroblast cells all result in the elevated phosphory-

lation of two 28,000-Dalton mammalian stress proteins. *J Biol Chem* 260:3058, 1985.
98. N. Feuerstein, M. Nishikawa, and H. L. Cooper: Cell-free studies on the phosphorylation of the 17,000–20,000 Dalton protein induced by phorbol ester in human leukemic cells and evidence for a similar event in virally transformed murine fibroblasts. *Cancer Res* 45:3243, 1985.
99. N. Sahyoun, M. Wolf, J. Besterman, T.-S. Hsieh, M. Sander, H. LeVine III, K.-J. Chang, and P. Cuatrecasas: Protein kinase C phosphorylates topoisomerase II: Topoisomerase activation and its possible role in phorbol ester-induced differentiation of HL-60 cells. *Proc Natl Acad Sci USA* 83:1603, 1986.
100. O. Jacobowitz, J. Chen, R. T. Premont, and R. Iyengar: Stimulation of specific types of G_s-stimulated adenylyl cyclases by phorbol ester treatment. *J Biol Chem* 268:3829, 1993.
101. K. L. Gould, J. R. Woodgett, J. A. Cooper, J. E. Buss, D. Shalloway, and T. Hunter: Protein kinase C phosphorylates pp60^{src} at a novel site. *Cell* 42:849, 1985.
102. J. M. Trevillyan, O. Perisic, J. A. Traugh, and C. V. Byus: Insulin- and phorbol ester-stimulated phosphorylation of ribosomal protein S6. *J Biol Chem* 260:3041, 1985.
103. G. Grunberger, Y. Zick, S. I. Taylor, and P. Gorden: Tumor-promoting phorbol ester stimulates tyrosine phosphorylation in U-937 monocytes. *Proc Natl Acad Sci USA* 81:2762, 1984.
104. V. Adler, C. C. Franklin, and A. S. Kraft: Phorbol esters stimulate the phosphorylation of c-Jun but not v-Jun: Regulation by the N-terminal δ domain. *Proc Natl Acad Sci USA* 89:5341, 1992.
105. S. Ghosh and D. Baltimore: Activation in vitro of NF-κB. *Nature* 344:678, 1990.
105a. A. K. Bhargava, Z. Li, and S. M. Weissman: Differential expression of four members of the POU family of proteins in activated and phorbol 12-myristate 13-acetate-treated Jurkat T cells. *Proc Natl Acad Sci USA* 90:10260, 1993.
106. C. J. Kemp, L. A. Donehower, A. Bradley, and A. Balmain: Reduction of p53 gene dosage does not increase initiation or promotion but enhances malignant progression of chemically induced skin tumors. *Cell* 74:813, 1993.
107. D. Morgan, D. Welty, A. Glock, D. Greenhalgh, H. Hennings, and S. H. Yuspa: Development of an in vitro model to study carcinogen-induced neoplastic progression of initiated mouse epidermal cells. *Cancer Res* 52:3145, 1992.
108. D. F. Birt, J. C. Pelling, L. T. White, K. Dimitroff, and T. Barnett: Influence of diet and calorie restriction on the initiation and promotion of skin carcinogenesis in the Sencar mouse model. *Cancer Res* 51:1851, 1991.
109. H. Wei, L. Tye, E. Bresnick, and D. F. Birt: Inhibitory effect of apigenin, a plant flavonoid, on epidermal ornithine decarboxylase and skin tumor promotion in mice. *Cancer Res* 50:499, 1990.
110. A. Leder, A. Kuo, R. D. Cardiff, E. Sinn, and P. Leder: v-Ha-ras transgene abrogates the initiation step in mouse skin tumorigenesis: Effects of phorbol esters and retinoic acid. *Proc Natl Acad Sci USA* 87:9178, 1990.
111. Z. Szallasi and P. M. Blumberg: Prostratin, a nonpromoting phorbol ester, inhibits induction by phorbol 12-myristate 13-acetate of ornithine decarboxylase, edema, and hyperplasia in CD-1 mouse skin. *Cancer Res* 51:5355, 1991.
112. C. Peraino, R. J. M. Fry, E. Staffeldt, and W. E. Kisielski: Effects of varying the exposure to phenobarbital on its enhancement of 2-acetylaminofluorene-induced hepatic tumorigenesis in the rat. *Cancer Res* 33:2701, 1973.
113. T. Kitagawa, H. C. Pitot, E. C. Miller, and J. A. Miller: Promotion by dietary phenobarbital of hepatocarcinogenesis by 2-methyl-*N*,*N*-dimethyl-4-aminoazobenzene in the rat. *Cancer Res* 39:112, 1979.
114. H. C. Pitot, L. Barsness, T. Goldsworthy, and T. Kitagawa: Biochemical characterization of stages of hepatocarcinogenesis after a single dose of diethylnitrosamine. *Nature* 271:456, 1978.
115. E. Farber: Cellular biochemistry of the stepwise development of cancer with chemicals: G. H. A. Clowes Memorial Lecture. *Cancer Res* 44:5463, 1984.
116. R. P. Evarts, H. Nakatsukasa, E. R. Marsden, C.-C. Hsia, H. A. Dunsford, and S. S. Thorgeirsson: Cellular and molecular changes in the early stages of chemical hepatocarcinogenesis in the rat. *Cancer Res* 50:3439, 1990.
117. G. Bradley, R. Sharma, S. Rajalakshmi, and V. Ling: P-glycoprotein expression during tumor progression in the rat liver. *Cancer Res* 52:5154, 1992.
118. S. R. Lichter: Why cancer news in a health hazard. *Wall Street Journal,* November 12, 1993.
119. J. Higginson: Changing concepts in cancer prevention: Limitations and implications for future research in environmental carcinogenesis. *Cancer Res* 48:1381, 1988.
120. F. Perera and P. Boffetta: Prespectives on comparing risks of environmental carcinogens. *JNCI* 80:1282, 1988.
121. S. M. Cohen and L. B. Ellwein: Genetic errors, cell proliferation, and carcinogenesis. *Cancer Res* 51:6493, 1991.
122. A. Hay: Testing times for the tests. *Nature* 350:555, 1991.
123. J. Ashby and R. W. Tennant: Definitive relationships among chemical structure, carcinogenicity and mutagenicity for 301 chemicals tested by the U.S. NTP. *Mutat Res* 257:229, 1991.
124. I. B. Weinstein: Mitogenesis is only one factor in carcinogenesis. *Science* 251:387, 1991.
125. R. L. Melnick: Does chemically induced hepa-

tocyte proliferation predict liver carcinogenesis? *FASEB J* 6:2698, 1992.

126. B. N. Ames, M. Profet, and L. Swirsky Gold: Dietary pesticides (99.99% all natural). *Proc Natl Acad Sci USA* 87:7777, 1990.
127. B. N. Ames, M. Profet, and L. Swirsky Gold: Nature's chemicals and synthetic chemicals: Comparative toxicology. *Proc Natl Acad Sci USA* 87:7782, 1990.
128. J. B. Little: Radiation carcinogenesis in vitro: Implications for mechanisms, in H. H. Hiatt, J. D. Watson, and J. A. Winsten, eds.: *Origins of Human Cancer.* Cold Spring Harbor, NY: Cold Spring Harbor Laboratory, 1977, pp. 923–939.
129. W. A. Haseltine: Ultraviolet light repair and mutagenesis revisited. *Cell* 33:13, 1983.
130. A. C. Upton: The biological effects of low-level ionizing radiation. *Sci Am* 246:41, 1982.
131. A. J. Grosovsky, J. G. De Boer, P. J. De Jong, E. A. Drobetsky, and B. W. Glickman: Base substitutions, frameshifts, and small deletions constitute ionizing radiation-induced point mutations in mammalian cells. *Proc Natl Acad Sci USA* 85:185, 1988.
132. L. H. Lutze, R. A. Winegar, R. Jostes, F. T. Cross, and J. E. Cleaver: Radon-induced deletions in human cells: Role of nonhomologous strand rejoining. *Cancer Res* 52:5126, 1992.
133. R. R. Weichselbaum, D. E. Hallahan, V. Sukhatme, A. Dritschilo, M. L. Sherman, and D. W. Kufe: Review: Biological consequences of gene regulation after ionizing radiation exposure. *JNCI* 83:480, 1991.
134. S. J. Garte, F. J. Burns, T. Ashkenazi-Kimmel, M. Felber, and M. J. Sawey: Amplification of the c-*myc* oncogene during progression of radiation-induced rat skin tumors. *Cancer Res* 50:3073, 1990.
135. *Cancer Statistics Review 1973–1989.* National Cancer Institute, NIH publication No. 92-2789, 1992.
136. B. N. Ames, M. K. Shigenaga, and T. M. Hagen: Oxidants, antioxidants, and the degenerative diseases of aging. *Proc Natl Acad Sci USA* 90:7915, 1993.
137. E. R. Stadtman: Protein oxidation and aging. *Science* 257:1220, 1992.
138. L. D. Youngman, J.-Y. K. Park, and B. N. Ames: Protein oxidation associated with aging is reduced by dietary restriction of protein or calories. *Proc Natl Acad Sci USA* 89:9112, 1992.
139. H. Joenje: Genetic toxicology of oxygen. *Mutat Res* 219:193, 1989.
140. S. A. Grist, M. McCarron, A. Kutlaca, D. R. Turner, and A. A. Morley: In vivo human somatic mutation: Frequency and spectrum with age. *Mutat Res* 266:189, 1992.
141. T. M.Reid and L. A. Loeb: Mutagenic specificity of oxygen radicals produced by human leukemia cells. *Cancer Res* 52:1082, 1992.
142. J. M. Carney, P. E. Starke-Reed, C. N. Oliver, R. W. Lundum, M. S. Cheng, J. F. Wu, and R. A. Floyd: Reversal of age-related increase in brain protein oxidation, decrease in enzyme activity, and loss in temporal and spatial memory by chronic administration of the spin-trapping compound *N*-tert-butyl-alpha-phenylnitrone. *Proc Natl Acad Sci USA* 88:3633, 1991.
143. V. A. Bohr, D. H. Phillips, and P. C. Hanawalt: Heterogeneous DNA damage and repair in the mammalian genome. *Cancer Res* 47:6426, 1987.
144. T. Lindahl: Instability and decay of the primary structure of DNA. *Nature* 362:709, 1993.
145. A. E. Pegg: Mammalian O^6-alkylguanine-DNA alkyltransferase: Regulation and importance in response to alkylating carcinogenic and therapeutic agents. *Cancer Res* 50:6119, 1990.
146. R. B. Setlow, F. E. Ahmed, and E. Grist: Xeroderma pigmentosum: Damage to DNA is involved in carcinogenesis, in H. H. Hiatt, J. D. Watson, and J. A. Winsten, eds.: *Origins of Human Cancer.* Cold Spring Harbor, NY: Cold Spring Harbor Laboratory, 1977, pp. 889–902.
147. D. Bootsma and J. H. J. Hoeijmakers: Engagement with transcription. *Nature* 363:114, 1993.
148. Y. Habraken, P. Sung, L. Prakash, and S. Prakash: Yeast excision repair gene RAD2 encodes a single-stranded DNA endonuclease. *Nature* 366:365, 1993.
149. D. Scherly, T. Nouspikel, J. Corlet, C. Ucla, A. Bairoch, and S. G. Clarkson: Complementation of the DNA repair defect in xeroderma pigmentosum group G cells by a human cDNA related to yeast RAD2. *Nature* 363:182, 1993.
150. E. C. Friedberg: Xeroderma pigmentosum, Cockayne's syndrome, helicases, and DNA repair: What's the relationship? *Cell* 71:887, 1992.
151. J. E. Cleaver: It was a very good year for DNA repair. *Cell* 76:1, 1994.
151a. L. L. Dumenco, E. Allay, K. Norton, and S. L. Gerson: The prevention of thymic lymphomas in transgenic mice by human O^6-alkylguanine-DNA alkyltransferase. *Science* 259:219, 1993.
152. S. M. O'Toole, A. E. Pegg, and J. A. Swenberg: Repair of O^6-methylguanine and O^4-methylthymidine in F344 rat liver following treatment with 1,2-dimethylhydrazine and O^6-benzylguanine. *Cancer Res* 53:3895, 1993.
153. V. Ellermann and O. Bang: Experimentelle Leukämie bei Hühnern. *Centralbl Bacteriol* 46:595, 1908.
154. P. Rous: A sarcoma of the fowl transmissible by an agent separable from the tumor cells. *J Exp Med* 13:397, 1911.
155. R. E. Shope: Infectious papillomatosis of rabbits. *J Exp Med* 58:607, 1933.
156. P. Rous and J. W. Beard: The progression to carcinoma of virus-induced rabbit papillomas (Shope). *J Exp Med* 62:523, 1935.
157. B. A. Lucké: A neoplastic disease of the kidney

of the frog, *Rana pipiens*. *Am J Cancer* 20:352, 1934.
158. J. J. Bittner: Some possible effects of nursing on the mammary gland tumor incidence in mice. *Science* 84:162, 1936.
159. L. Gross: Viral etiology of cancer and leukemia: A look into the past, present, and future: G. H. A. Clowes Memorial Lecture. *Cancer Res* 38:485, 1978.
160. L. Gross: Development and serial cell-free passage of a highly potent strain of mouse leukemia virus. *Proc Soc Exp Biol Med* 94:767, 1957.
161. L. Dmochowski: Viruses and tumors in the light of electron microscope studies: A review. *Cancer Res* 20:997, 1960.
162. T. H. Maugh II: RNA viruses: The age of innocence ends. *Science* 183:1181, 1974.
163. I.-M. Chiu, R. Callahan, S. R. Tronick, J. Schlom, and S. A. Aaronson: Major pol gene progenitors in the evolution of oncoviruses. *Science* 223:364, 1984.
164. B. E. Eddy, S. E. Stewart, and W. Berkeley: Cytopathogenicity in tissue cultures by a tumor virus from mice. *Proc Soc Exp Biol Med* 98:848, 1958.
165. B. E. Henderson: Establishment of an association between a virus and a human cancer. *J Natl Cancer Inst* 81:320, 1989.
166. H. zur Hausen: Viruses in human cancers. *Science* 254:1167, 1991.
167. C.-M. Kim, K. Koike, I. Saito, T. Miyamura, and G. Jay: HBx gene of hepatitis B virus indices liver cancer in transgenic mice. *Nature* 351:317, 1991.
168. T. Sasagawa, M. Inoue, H. Inoue, M. Yutsudo, O. Tanizawa, and A. Hakura: Induction of uterine cervical neoplasias in mice by human papillomavirus type 16 E6/E7 genes. *Cancer Res* 52:4420, 1992.
169. B. J. Poiesz, F. W. Ruscetti, A. F. Gazdar, P. A. Bunn, J. D. Minna, and R. C. Gallo: Detection and isolation of type-C retrovirus particles from fresh and cultured lymphocytes of a patient with cutaneous T-cell lymphoma. *Proc Natl Acad Sci USA* 77:7415, 1980.
170. B. J. Boiesz, F. W. Ruscetti, M. S. Reitz, V. S. Kalyanaraman, and R. C. Gallo: Isolation of a new type-C retrovirus (HTLV) in primary uncultured cells of a patient with Sézary T-cell leukemia. *Nature* 295:268, 1981.
171. R. C. Gallo and F. Wong-Staal: Current thoughts on the viral etiology of certain human cancers. *Cancer Res* 44:2743, 1984.
172. V. S. Kalyanaraman, M. G. Sarngadharan, M. Robert-Guroff, D. Blayney, D. Golde, and R. C. Gallo: A new subtype of human T-cell leukemia virus (HTLV-II) associated with a T-cell variant of hairy cell leukemia. *Science* 215:571, 1982.
173. M. Nerenberg, S. H. Hinrichs, R. K. Reynolds, G. Khoury, and G. Jay: The *tat* gene of human T-lymphotropic virus type I induces mesenchymal tumors in transgenic mice. *Science* 237:1324, 1987.
174. R. C. Gallo, V. S. Kalyanaraman, M. G. Sarngadharan, A. Sliski, E. C. Vonderheid, M. Maeda, Y. Nakao, K. Yamada, Y. Ito, N. Gutensohn, S. Murphy, P. A. Bunn, Jr., D. Catovsky, M. F. Greaves, D. W. Bayney, W. Blattner, W. F. H. Jarrett, H. zur Hausen, M. Seligmann, J. C. Brouet, B. F. Haynes, B. V. Jegasothy, E. Jaffe, J. Cossman, S. Broder, R. I. Fisher, D. W. Golde, and M. Robert-Guroff: Association of the human type C retrovirus with a subset of adult T-cell cancers. *Cancer Res* 43:3892, 1983.
175. F. Wong-Staal, B. Hahn, V. Manzari, S. Colombini, G. Franchini, E. P. Gelmann, and R. C. Gallo: A survey of human leukemias for sequences of a human retrovirus. *Nature* 302:626, 1983.
176. H.-G. Guo, F. Wong-Staal, and R. C. Gallo: Novel viral sequences related to human T-cell leukemia virus in T cells of a seropositive baboon. *Science* 223:1195, 1984.

7

Oncogenes

HISTORICAL PERSPECTIVES

The Provirus, Protovirus, and Oncogene Hypotheses

Much of our understanding of the molecular mechanisms involved in cellular transformation by RNA oncogenic viruses is an outgrowth of the seminal work of Temin and Baltimore and their colleagues (for review, see Refs. 1 and 2). In the early 1960s, Temin demonstrated that mutations in the Rous sarcoma virus (RSV) genome of RSV-infected chicken cells could be induced at a high rate, that mutation of an RSV gene present in an infected cell often changes the morphology of the infected cell, and that the virus genome was stably inherited by subsequent progeny cells. This led to the idea that viral genetic information was contained in a regularly inherited structure of the host cell as a "provirus" and that this provirus was integrated into the host cell's genome. The problem with the provirus hypothesis was that there was no known way for the RNA of the tumor virus to be converted into a form of DNA that could be integrated into the host's DNA. The central dogma of molecular biology at the time was that genetic information was transferred *only* from DNA to RNA to protein. When actinomycin D, a drug that specifically blocks DNA-directed RNA synthesis, was added to RSV-producing cells, virus production was blocked, indicating to Temin that the flow of genetic information for RSV could go from RNA to DNA to RNA to protein. Further experiments showed that new DNA synthesis was, in fact, required for RSV production to occur and that new RSV-specific DNA was present in infected cells. On the basis of these results, Temin proposed the DNA provirus hypothesis in 1964. The basis of the hypothesis is that the RNA of the infecting RSV acts as a template for the synthesis of viral DNA, which is then integrated as a provirus into host cell DNA, where it can subsequently act as a template for the synthesis of progeny RSV RNA. He then set about trying to obtain evidence for an RNA-directed DNA polymerase, which would have to be present if his idea were correct.

The DNA provirus hypothesis was largely ignored for about 6 years until Temin and his colleagues[3] and Baltimore and his coworkers,[4] working independently, demonstrated the presence of a virus-contained RNA-directed DNA polymerase activity, which came to be known as reverse transcriptase. Although the discovery of reverse transcriptase explained how the DNA-provirus mechanism could work, formal proof of the hypothesis was not obtained until it was demonstrated that radioactively labeled RSV RNA hybridized to the DNA of infected chicken cells to a much greater extent than to the DNA of uninfected cells[5] and that DNA obtained from

RSV-infected cells could transfect uninfected cells, leading to the production of complete RSV.[6]

In an extension of the DNA provirus hypothesis, Temin later proposed the protovirus theory, in which he postulated that the genome of oncogenic viruses arose during evolution in part from normal cellular DNA that had perhaps been altered by some exogeneous carcinogen. This theory would help explain the known hybridization of oncogenic viral nucleic acid sequences with normal cellular DNA (see discussion of *src* gene, below).

The normal cellular homologues of viral oncogenes have come to be known as cellular "protooncogenes" rather than protovirus genes because it is now clearly established that their origin is cellular and that they have been present in cells over a vast range of evolution. Such conservation implies a central role for these genes in normal cellular function, and it is likely that their oncogenicity derives from a rare event, such as translocation, amplification, or mutation of a key nucleotide sequence. Although the term *protooncogene* has found wide acceptance, it is somewhat misleading, because it is not reflective of the role of these genes in normal cell differentiation and function.

The highly oncogenic viruses presumably arose from genetic recombination events between viruses of low oncogenicity and an evolutionarily stable set of nucleotide sequences of cellular origin, the combination of which has produced a highly transforming viral genome (Fig. 7-1). Since many of these viruses are replication-defective, they do not form complete virus unless the cells are coinfected with a "helper" virus. Thus, recombination between these replication-competent helper viruses and cellular genes, some of which may be involved in regulating cell proliferation, may have produced the highly oncogenic virus strains.

Certain essential features of these theories—namely, that normal cells contain sequences (cellular protooncogenes) homologous to those of oncogenic viruses and that these sequences can be activated during carcinogenesis (perhaps by a mutation or chromosomal rearrangement)—are now well established. Whether or not oncogenic viruses pick up by transduction "cancer genes" mutated long ago in evolution by some carcinogen, as originally proposed in the protovirus hypothesis, is not clear. What is clear is that cellular homologues of genes carried by oncogenic viruses are present in untransformed cells spanning the evolutionary scale from yeast to humans. How these sequences came to reside in viruses appears to involve recombination events at the DNA rather than the RNA level, followed by transcription and splicing of mRNA coded by these genes, and packaging into the retrovirus. This will be described in more detail next for the *src* gene.

The *src* Gene

The cellular origins of oncogenes (*onc* genes) were first clearly established for the Rous sarcoma virus oncogene v-*src*, which is derived from its cellular progenitor (c-*src*). The identification of this gene sequence is an elegant story in molecular biology, and the methods used to detect it have been used as a precedent in the discovery of other *onc* gene sequences. We will describe these experiments here in some detail because of their prototypical significance.

If one assumes that the transforming oncogenic sequences are not necessary for virus replication, as in the case of RSV, but are "extra" genes that provide some selective advantage for cell proliferation, then cells transformed with oncogenic viruses should contain some nucleic acid sequences not present in nontransforming viruses of the same class. Moreover, this idea predicts that a transforming virus should code for a transformation-specific protein that is not needed for viral replication but has a particular function in transformed host cells.

It has been known for some time that the transforming avian sarcoma viruses (of which RSV is an example) contain more genetic information than transformation-defective strains of these viruses.[7] The transformation-defective variants of avian sarcoma viruses do not induce sarcomas in animals or transform fibroblasts in culture, and they lack 10% to 20% of the genetic information (RNA) contained in the parent transforming viruses; yet these "defective" viruses can infect cells and replicate perfectly well. Hence, their genetic deletion does not appear to affect virus-replicative functions. Taking advantage of these transformation-defective strains,

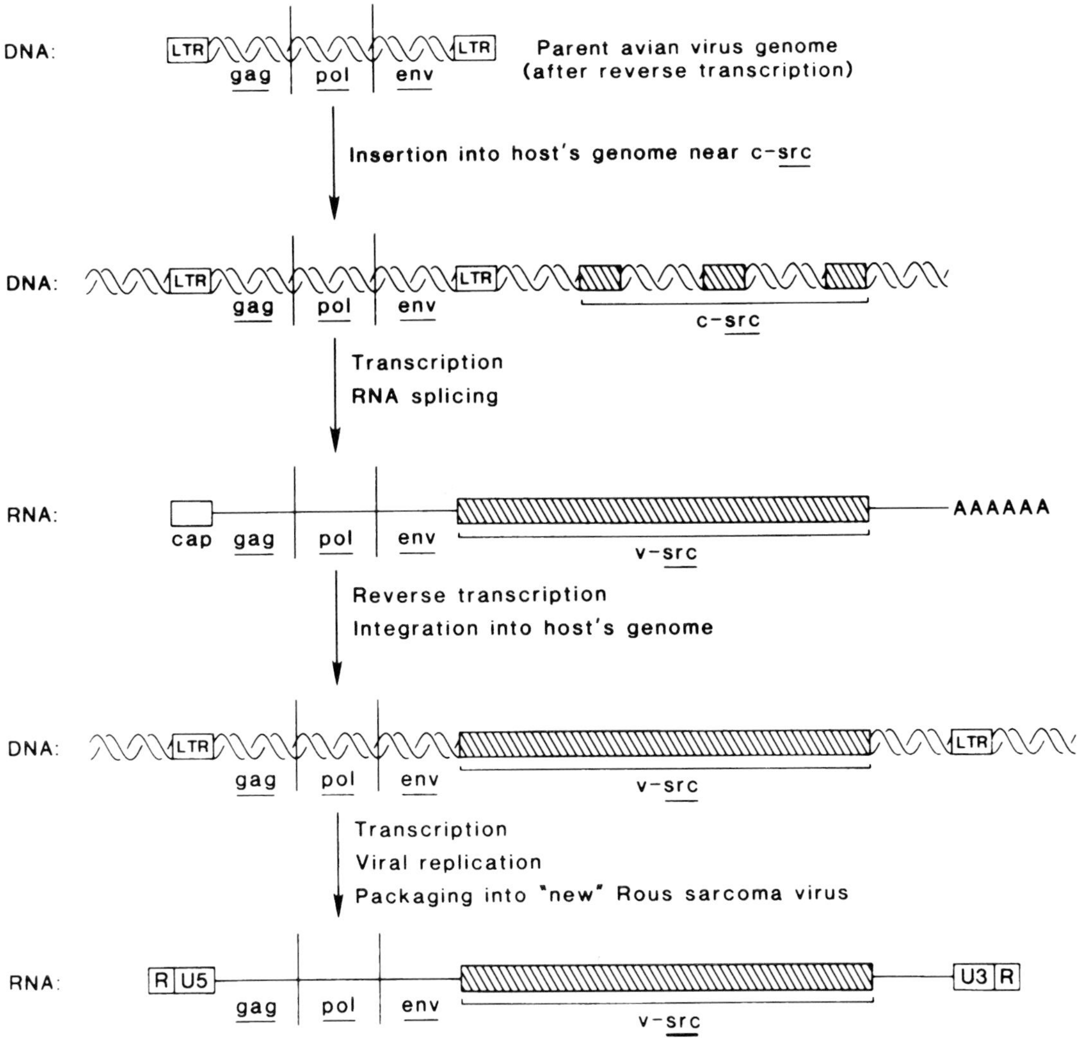

Figure 7-1 Schematic diagram for the generation of oncogenic viruses form cellular protooncogenes. The example shown is for the generation of the RNA tumor virus Rous sarcoma virus (RSV) from an avian virus lacking the *src* gene. The cellular c-*src* gene is a split gene containing introns, whereas v-*src* of RSV is a continuous gene. Thus, generation of v-*src* from c-*src* involves RNA splicing after transcription of the viral–cellular recombinant gene containing c-*src*. The "chimeric" RNA is then packaged into the virus. The parent virus was probably a chronic-acting virus of low oncogenic potential. LTR, long terminal repeat regions; *gag*, genes coding for viral core structure proteins; *pol*, genes coding for reverse transcriptase; *env*, genes coding for viral envelope proteins. Double wavy lines represent DNA; single lines represent RNA; cross-hatched boxes depict exons of c-*src* and single gene of v-*src*; R denotes a short segment of RNA repeated at each end of the molecule; U5 and U3 denote unique noncoding regulatory sequences at the 5′ and 3′ ends of the viral RNA. Replication of the viral genome generates duplicated RU5 and U3R ends, which make up the LTR segments.

Stehelin et al.[8] designed an experiment to isolate the portion of the genome related to the transforming activity of these viruses (Fig. 7-2). They used RNA isolated from a transforming strain of RSV called Prague strain subgroup C as a template to synthesize radioactively labeled DNA pieces complementary to the transforming virus RNA, in vitro, using reverse transcriptase. The complementary DNA pieces synthesized in vitro were then hybridized to the RNA of a transformation-defective strain of the same virus. Since those pieces of DNA that contained the trans-

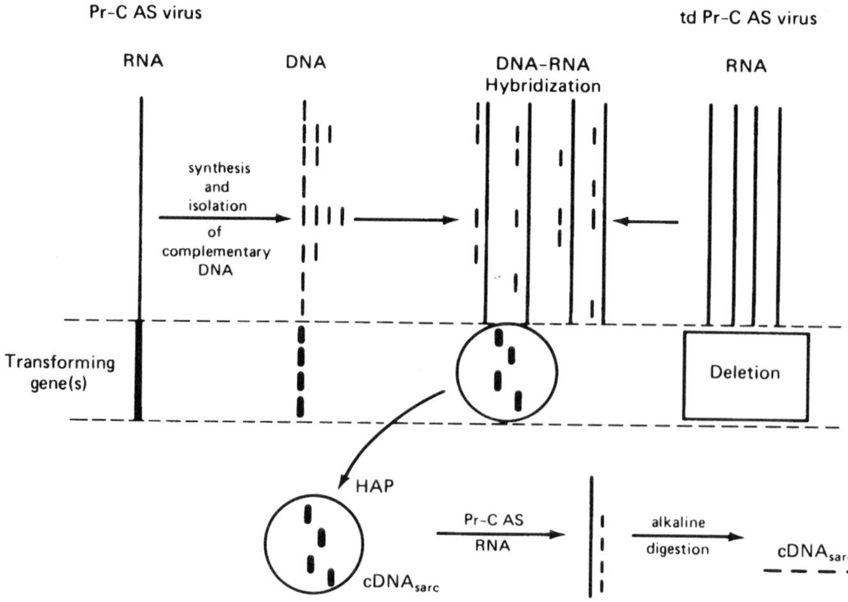

Figure 7-2 Strategy for the preparation of $cDNA_{sarc}$. Radioactive DNA complementary to the genome of the Prague-C strain of avian sarcoma virus (Pr-C AS) was synthesized with detergent-activated virions, released from template RNA by hydrolysis with RNase, purified by chromatography on hydroxylapatite (HAP), and then annealed with a large excess of 70S RNA from the transformation-deficient deletion mutant td Pr-C AS virus. The DNA that failed to anneal ($cDNA_{sarc}$) was purified by chromatography on HAP; it corresponded to the nucleotide sequences deleted from the genome of the parental sarcoma virus in the genesis of the td variant. The $cDNA_{sarc}$ was further purified by annealing with Pr-C AS RNA and rechromatographed on HAP, followed by alkaline digestion of the RNA to yield purified $cDNA_{sarc}$. (Modified from Stehelin et al.[8])

formation-specific sequences would not be represented in the defective strain's RNA, they would not hybridize to it (hybridization requires a significant amount of homology between nucleic acid base pairs). Thus, these transforming DNA pieces did not form double-stranded duplexes with the RNA and were separated from the nontransforming DNA pieces that did form them by chromatography on hydroxyapatite, which binds only the double-stranded nucleic acid hybrids. The transforming pieces were finally purified by hybridization with RNA from the transforming virus and separated from the RNA by hydrolysis in alkali, leaving radioactively labeled, transforming DNA. This DNA, originally designated "cDNA sarc" for complementary DNA-bearing sarcoma-producing gene sequences, was then used as a probe to determine whether these sequences are present only in transformed cells or in normal cells as well.

The answer to this question was surprising. When the cDNA sarc was used as a probe, it was found that sequences homologous to, though not completely identical with, cDNA sarc were present in uninfected avian cells as well as in normal salmon, mouse, calf, and human cells. DNA sequences homologous to v-*src** were also subsequently found in *Drosophila* and other lower organisms. Since the evolutionary separation of birds from teleosts and mammals occurred 400 million years ago, this suggests that a portion of the *src* gene has been conserved in cells of higher organisms for several evolutionary epochs. Moreover, RNA sequences corresponding to cDNA *src* were also found in the cellular RNA of normal and neoplastic avian cells, indicating that the gene is not only present but is also transcribed in normal as well as in transformed cells.[9]

*In current terminology, the specific transforming gene of avian sarcoma viruses has been designated v-*src*; the DNA sequences of normal cells that correspond to v-*src* are called c-*src*. A similar designation has been used for other oncogenes as well—for example, v-*ras* and c-*ras*, v-*myc* and c-*myc*, etc.

The amount of sarcoma-specific RNA and its intracellular location have also been found to be essentially identical in RSV-transformed cells and RSV-transformed-revertant (transformed phenotype lost) cells, supporting the conclusion that neither the presence of the *src* gene not its transcription is specific to transformed cells.[10] Thus, at this point in these studies, it was concluded that if the *src* gene is related to transformation, it must be at the level of translation of *src* gene RNA into protein or the effect of the *src* gene protein on cellular targets.

Detection of the *src* gene product was accomplished by using serum from rabbits bearing avian sarcoma virus-induced tumors to immunoprecipitate proteins from uninfected and ASV-transformed chicken and hamster cells grown in culture.[11] A 60,000-MW protein, designated $p60^{src}$, was precipitated by this serum from transformed chicken and hamster cells, but was not observed in uninfected cells or in cells infected with nontransforming viruses. Also, $p60^{src}$, was synthesized in cell-free systems programmed by the addition of the 3′-one-third of the ASV viral RNA, the region that contains the *src* gene.[12] This $p60^{src}$ protein has turned out to be a phosphoprotein (thus, its phosphorylated form is called $pp60^{src}$) that has protein kinase activity of the cAMP-independent type.[13–15] The site of phosphorylation by this kinase activity is on tyrosine residues rather than serine or threonine residues, the usual phosphorylation sites of previously discovered cellular protein kinases.[16] This raised the possibility that phosphorylation of specific cellular proteins was involved in host cell transformation by ASV. However, as discussed in Chapter 9, it is now known that receptors for a number of normal growth factors also have tyrosine kinase activity.

ONCOGENE FAMILIES

A number of oncogene sequences have now been identified and characterized (Table 7-1). New sequences continue to be found in various eukaryotic organisms, including humans. This might give one the impression that there are innumerable oncogene sequences in nature. However, there is now clear evidence that these genes exist as families or even "superfamilies" of related sequences that have been derived from a much smaller number of ancestral genes. This conclusion is drawn from the repeated appearance of the same *onc* genes in a variety of independent viral isolates. For example, the *myc* gene has been found in the genomes of four isolates of avian myelocytomatosis virus, and *fes*, which was originally discovered in feline sarcoma virus, has also been found in a chicken sarcoma virus and termed *fps*. Both *fes* and *fps* turn out to be members of the *src* gene family, which also includes the *onc* genes, *mos*, *raf*, *erb*-B, *yes*, and *abl*. A common ancestor for all these genes has been postulated based on the relatedness of their conserved sequences. The following family tree has been proposed by Mark and Rapp.[17]

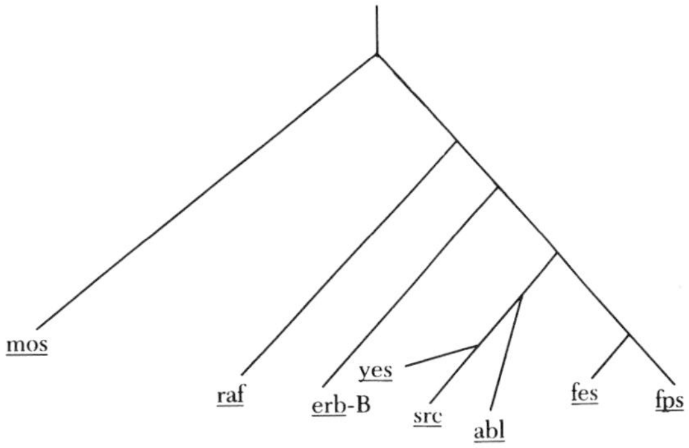

These are very old genes indeed. *fps* and *fes*, for example, appear to have diverged about 200 million years ago.

Similarly, a large family of *ras*-related oncogenes is rampant in nature. This family includes the *ras* gene group: Harvey *ras* (Ha-*ras*), Kirsten *ras* (Ki-*ras*), neuroblastoma-derived *ras* (N-*ras*), and the *rho* gene family. Thus, the *ras*-like genes may be members of a superfamily of genes that diverged from a common ancestral gene hundreds of millions of years ago.[18] Again, the "ancientness" of these genes and their highly conserved sequences suggest that they are very important in the economy of eukaryotic cells. Indeed, as will be discussed later, several functional groupings can be devised based on the products of these genes. For example, oncogene products can be functionally grouped into those that have tyrosine kinase activity (e.g., the *src* family), growth factor–like activity (e.g., *sis*), growth factor receptor domains (e.g., *erb*-B) and

Table 7-1 Some Viral Oncogenes[a]

Oncogene	Animal Retrovirus	Species of Origin
abl	Abelson murine leukemia virus	Mouse
fos	FBJ osteosarcoma virus	Mouse
int-1	Mouse mammary tumor virus	Mouse
int-2	Mouse mammary tumor virus	Mouse
mos	Moloney murine sarcoma virus	Mouse
raf[b]	3611 murine sarcoma virus	Mouse
fes[b]	ST feline sarcoma virus	Cat
fgr	Gardner-Rasheed feline sarcoma virus	Cat
fms	McDonough feline sarcoma virus	Cat
kit	H-Z feline leukemia virus	Cat
fes[b]	Fujinami sarcoma virus	Chicken
*erb*A	Avian erythroblastosis virus	Chicken
*erb*B	Avian erythroblastosis virus	Chicken
ets	E26 virus	Chicken
mil (*mht*)[b]	MH2 virus	Chicken
myb	Avian myeloblastosis virus	Chicken
myc	MC29 myelocytomatosis virus	Chicken
ros	UR II avian sarcoma virus	Chicken
ski	Avian SKV770 virus	Chicken
src	Rous sarcoma virus	Chicken
yes	Y73 sarcoma virus	Chicken
sis	Simian sarcoma virus	Woolly monkey
Ha-*ras*	Harvey murine sarcoma virus	Rat
Ki-*ras*	Kirsten murine sarcoma virus	Rat
neu/*erb*B (HER-2/*neu*)	None	Rat
rel	Reticuloendotheliosis virus	Turkey
hst	None	Human
met	None	Human
N-*ras*	None	Human
N-*myc*	None	Human
L-*myc*	None	Human
trk	None	Human

[a]The names of the viral oncogenes are loosely derived from the names of the viruses in which they were identified or from the types of cancers they cause (*src* from Rous sarcoma virus or *ras* from rat sarcoma, for example). Additional transforming genes, some related to the viral oncogenes and some not, have been identified. In addition, the early region genes EIA and EIB of adenoviruses, the T antigens of SV 40 and polyomaviruses, and E6 and E7 proteins of papillomaviruses are considered oncogenes.

[b]*fes* and *fps* are feline and avian versions of the same oncogene; *raf* and *mil*(*mht*) are murine and avian oncogene counterparts.

those that function as DNA-binding (e.g., *myc* and *myb*), RNA-binding (e.g., *mil*), or guanine-nucleotide-binding (e.g., the *ras* family) proteins.

It seems highly likely that not all oncogene-related sequences have been discovered yet. Because their discovery has most often occurred by their appearance in viral isolates, it could be that a much larger pool of proto-*onc* genes exists but that they are not readily available for transduction by retroviruses. Readily transducible proto-*onc* genes are probably those that are active in many cell types or organisms and at many stages of growth and thus are readily available for retroviral integration, such as genes required for cell proliferation or basic common metabolic functions. Moreover, the close homology between viral *onc* genes and cellular proto-*onc* genes does not necessarily mean that all cellular proto-*onc* genes *cause* malignant transformation of cells, even though the term implies that they do.

It should be noted that two types of sequence conservation have occurred during evolution of the *onc* genes: conservation of nucleic acid sequence, as measured by nucleic acid hybridization, and conservation of the amino acid sequence despite many base changes in nucleotide sequence—that is, conservative base changes that do not alter the sequence of the protein significantly. For example, of 15 *onc* genes originally defined as distinct based on lack of homology in nucleic acid hybridization experiments, several were later found to be related when the amino acid sequence of the gene product was used as a criterion of homology. A case in point is that of the Ha-*ras* and Ki-*ras* oncogenes; these appeared to be only weakly related by DNA–DNA hybridization, and yet the amino acid sequence of the p21 gene product encoded by these genes is 80% homologous.[19]

The finding of various *onc* genes, particularly the *ras* genes, in lower organisms such as *Xenopus laevis, Drosophila melanogaster,* and the yeast *Saccharomyces cerevisiae*[20] provides a powerful tool to look at the function of *onc* genes and how mutations affect their function. *S. cerevisiae* contains two genes (RAS-1 and RAS-2) that are closely related to mammalian *ras* and that code for proteins 90% homologous to the *ras* p21 protein at their amino termini, but the yeast and mammalian proteins are dissimilar at their carboxyl termini. Alteration of both RAS genes is lethal to yeast cell survival, but disruption of only one of them is not.[21,22] Notably, a high percentage of yeast cells remain viable if the yeast RAS genes are replaced by a human *ras* gene.[23] Yeast cells survive and grow well if a chimeric *ras* gene, made up of the first part of the human gene and the second part of the yeast RAS-2 gene, is transfected into the yeast cells. However, when the RAS-2 yeast gene is mutated by a single base substitution at a specific site, the yeast do not sporulate, fail to accumulate carbohydrates, and have poor viability, but do have a high content of cAMP, which reflects the increased GTP-binding activity of the p21 protein and thus its increased ability to stimulate adenylate cyclase.[24] Interestingly, this mutation is equivalent to a mutation in a human *ras* gene found in human bladder cancer (see below). Moreover, when a deletion mutant of the yeast RAS-1 gene is further altered by a point mutation analogous to the one that increases the transforming activity of mammalian *ras* genes, the yeast RAS gene so altered induces transformation in cultured mouse 3T3 cells.[25] The yeast RAS-1 gene is larger than the mammalian *ras* gene, primarily in the region coding for the carboxyl terminus of the protein product, and deletion of this part of the gene enhances transforming activity, perhaps by augmenting the GTP-binding activity, since the amino terminal region of the p21 gene product is where the GTP-binding and GTPase activity reside.[26] These results indicate that mammalian and yeast *ras*-like genes have similar biologic functions. The ability of human *ras* genes to function in yeast allows the biological effects of mutations in human *ras* genes to be tested and determination of which mutations are likely to be important for the altered function of these genes in cancer cells.

CELL-TRANSFORMING ABILITY OF *onc* GENES

Historical Perspectives

Another approach for discovering oncogenes used the procedure of gene transfer (DNA transfection). Several laboratories have reported that DNA segments from a variety of animal and human tumors can cause transformation of cultured NIH/3T3 mouse fibroblasts (for review, see Refs. 27 and 28). Using probes developed to the oncogenes of retroviruses, it has been found that these transforming DNA segments contain sequences homologous to known v-*onc* genes. This startling discovery led to the concept that activation of cellular *onc* genes can occur either by recombination with retroviral genomes, as described earlier, or by some sort of somatic mutational event leading to activation or abnormal expression of cellular proto-*onc* genes. There is now experimental evidence to show that point mutations, gene amplification, and chromosomal translocation events can lead to activation or increased transcription of cellular proto-*onc* genes.

The rationale for attempting to find transforming or "cancer genes" in the cellular DNA of malignantly transformed cultured cells and of tumors goes back to experiments performed in the 1940s by Avery et al.,[29] who were the first to demonstrate that DNA isolated from a virulent strain of pneumococci could transform a nonvirulent strain into a virulent one with the cellular markers characteristic of the latter bacterial type. These findings, coupled with experiments such as those of Hill and Hillova,[6] who showed that DNA from RSV-infected cells could transform cells as well as produce complete RSV, led to the idea that DNA from cells transformed by chemical carcinogens or DNA from cancerous cells themselves might be able to transform normal cells into malignant cells. In the late 1970s, a number of laboratories began experimenting with this idea. The first demonstration that DNA from cells transformed with chemical carcinogens could transform other cells came from the work of Weinberg and his colleagues, who showed that DNA from 3-methylcholanthrene-(3-MC) transformed mouse fibroblasts could morphologically transform a line of "normal" 3T3 mouse fibroblasts known as the NIH/3T3 line,[30] which has become the "gold standard" for testing for transforming DNA (but is not without its problems, as will be discussed below). These experiments were made possible by the development of procedures to transfer intact DNA into whole cells. The procedure originally used was that of calcium phosphate precipitation of the DNA, which in the crystalline form thus produced can enter cells growing in culture and become integrated into chromosomal DNA. In a typical experiment, about 10^6 recipient cells are exposed to 20 μg of donor DNA; in the NIH/3T3 line one gets approximately 0.1 to 1 transformed cell per microgram of DNA. This amounts to 20 transformed cells per cell dish in a typical experiment. These cells are visible because the original transformants multiply to give little colonies or foci of transformed cells that pile up on one another instead of growing as flat monolayers of cells as normal fibroblasts do. In addition to this morphologically detectable transformation, the altered cells, when plucked from the culture dishes, will grow in suspension in soft agar, a typical characteristic of transformed cells (see Chap. 4), and they form tumors when inoculated into mice. When DNA from untransformed NIH/3T3 cells is used in the transfection assay, the recipient cells are not morphologically transformed, nor are they tumorigenic. Thus, treatment of "normal" fibroblasts with the chemical carcinogen 3-MC somehow changes the cells' DNA so that it now carries the genetic information to induce a malignant phenotype in cells into which it is transfected. Subsequently, other laboratories repeated and confirmed these results, thus adding chemical carcinogen-alteration of DNA to retroviral DNA as a means to induce, after integration into a cell's genome, malignant transformation. Other chemically activated transforming DNAs include those extracted from ethylnitrosourea-induced rat neuroblastomas, 7,12-dimethylbenz[a]anthracene-(DMBA) induced mouse bladder carcinomas, benzo[a]pyrene-(BP) induced rabbit bladder carcinoma,[31] and *N*-methyl-*N*′-nitro-*N*-nitrosoguanidine-(MNNG) transformed human cells.[32] As discussed in more detail later, DNA from a wide variety of animal and human cancers has now also been shown to contain segments that will transform NIH/3T3 cells, thus establishing a link between chemically induced transformation in vitro and spontaneously arising cancers in vivo, and showing clearly that alteration of DNA can induce a malignant phenotype in cells.

Identification of the transforming DNA segments contained in chemically transformed cells, or in cancer cells, involved a series of elegant experiments by laboratories that took three different but related fundamental approaches. Shih and Weinberg began with the DNA of a human bladder carcinoma cell line known as EJ, which contains transforming DNA segments.[33] (Closely related human bladder carcinoma lines T24 and J82 have been studied by other groups and shown to contain the same transforming sequences.) They transformed mouse fibroblasts with partially sheared bladder carcinoma DNA of about 30 kilobases in length and then used the transformed mouse cells' DNA to transfect a second cycle of mouse cells. A DNA "library" was constructed from the DNA of these secondarily transformed cells by cloning restriction endonuclease-cleaved DNA fragments in bacteriophage grown in the bacteria *E. coli.*[27] With this technique, each bacteriophage-induced plaque in the *E. coli* culture carries a different fragment of cellular DNA, which in turn can be assayed for transforming activity or hybridized with a radioactively labeled DNA or RNA probe to detect specific nucleotide sequences. The DNA of the second-cycle transformed mouse cells carried only a small amount of human DNA, which was detected by hybridization with a radioactive probe specific for human *Alu* repetitive nucleotide sequences. By screening the library with the *Alu*-specific probe, Shih and Weinberg detected the phage clone carrying human DNA that could transform NIH/3T3 cells.

Cooper and his colleagues took a different experimental approach.[34] They constructed a genomic library from the DNA of a chicken B-cell lymphoma known to carry a transforming sequence. DNA of what was originally several thousand phage clones was screened for transforming activity by analyzing "sublibraries" of 10 clones each. Any positive sublibrary was further divided into 10 sublibraries, which in turn were screened for transformation-positive clones. Eventually, this exponentially narrowing search led to the isolation of a single-phage clone carrying the transforming DNA.

Wigler's group[35] began by linking a bacterial marker gene to each segment of DNA derived from the human T24 bladder carcinoma cell line. They used these DNA fragments to transform mouse fibroblasts in a stepwise fashion. A DNA library was constructed from this serially transferred DNA using a defective bacteriophage vector that would proliferate only if the bacterial marker gene linked to the human bladder DNA was present. This greatly simplified the task of finding the phage clones positive for human DNA, and from among the viable phage clones, several that carried transforming activity were detected and later found to carry a common transforming sequence.

These procedures produced a tremendous enrichment of transforming DNA sequences. For example, whereas it took, on the average, 2 μg of the original total bladder cancer DNA to produce one colony of transformed NIH/3T3 cells, the same amount of isolated, cloned transforming DNA induced about 50,000 transformed foci.[27]

These cloning experiments strongly suggest that a single gene sequence was responsible for the transforming activity in each case. Further evidence for this was obtained by the use of specific endonucleases that cut DNA at specific base sequences. If, for example, a specific endonuclease destroyed transforming activities obtained from a number of clones, whereas another endonuclease did not, this would suggest that the transforming activity from the different clones was the same. Such evidence has been obtained for the transforming activities of DNA isolated from four different chemically transformed mouse fibroblast lines.[36] Specific patterns of restriction-endonuclease sensitivity were subsequently also found for a wide variety of animal and human tumor cells. All of these data point to the conclusion that transforming DNA isolated in each tumor type is carried in a single or small number of genes. Moreover, similar patterns of susceptibility to restriction-endonuclease cleavage have been observed among certain types of human cancers, suggesting that the same or similar *onc* genes may be activated in these cancers. DNA sequence analysis of the isolated transforming DNA sequences has now confirmed this, and it appears that the same transforming genes are activated in neoplasms of the same differentiated cell type, whether the neoplasm was virally or chemically induced or occurred spontaneously.

The next questions were: What were these transforming genes and did they correspond to any known proto-*onc* gene or retroviral *onc* genes? The answers came quickly, and the answer to the second question was a resounding affirmative. Given the probes developed to the c-*onc* and v-*onc* genes, it was a straightforward experiment to test their sequence homology against the cloned transforming genes isolated from various neoplasms and transformed cell lines. Initially, probes developed to the v-*onc* sequences *src*, *myc*, *fes*, *ras*, *erb*, *mos*, *myb*, and *sis* were used to test, by nucleic acid hybridization, sequence homology to the isolated transforming sequences (for review, see Ref. 28). These experiments had some surprising results: the transforming genes of human bladder and lung carcinoma detected by DNA transfection in the NIH/3T3 transformation assay were homologous to the *ras* genes of v-Ha-*ras* and v-Ki-*ras*, respectively.[37–39] Other human carcinomas and human tumor cell lines also possess the Ki-*ras* gene, including carcinomas of the lung, pancreas, colon, gallbladder, and urinary bladder, as well as a rhabdomyosarcoma.[40] In addition, a third *ras*-like gene has been found in the transforming sequences from a human neuroblastoma that is weakly homologous to both v-Ha-*ras* and v-Ki-*ras*.[41] This transforming gene represents a third member of the *ras* gene family, and has been designated N-*ras*. The involvement of different *ras* genes in different types of human cancers suggests that members of the *ras* gene family may be involved in some general way in regulating the phenotypic characteristics of a variety of human malignant neoplasms.

Whether this is a cause or effect of the malignant transformation events is still a matter of some debate (see below). Nevertheless, the activation of cellular *ras* genes in human cancers provides the first direct link between the transforming genes of retroviruses and human cancer.

That human cancer transforming genes are indeed induced by the activation of cellular proto-*onc* genes has been shown by the following types of experiments. Hybridization analysis of restriction-endonuclease-digested cellular DNAs from human bladder and lung carcinomas and from normal human cells with cloned probes of v-Ha-*ras* and v-Ki-*ras* sequences and with cloned probes of the biologically active transforming gene from human bladder cancer has shown that the activated transforming genes of bladder and lung carcinomas are homologous to the *ras* proto-*onc* genes of normal cells (for review, see Ref. 28). Furthermore, when viral transcriptional promoter LTR sequences from murine or feline retroviruses are linked to the *ras* proto-*onc* gene isolated from normal human cells, oncogenic transformation of NIH/3T3 mouse fibroblasts is achieved and an increased expression of the p21 gene product of the proto-*onc ras* gene is observed in the transformed cells,[42] suggesting that elevated expression of a "normal" proto-*onc* gene can induce oncogenic transformation. However, as we shall see later, it is also possible to activate proto-*onc* (c-*onc*) genes by other mechanisms, including somatic mutation and gene amplification.

A wide variety of human tumors have now been examined for expression of cellular c-*onc* genes by DNA-RNA hybridization using v-*onc* gene cDNA probes. Expression (transcription into RNA) of genes homologous to v-*onc* genes in human tumors occurs in a variety of leukemias and lymphomas, carcinomas, various sarcomas, neuroblastoma, teratocarcinoma, and choriocarcinoma.

The DNA transfection experiments suggest that transforming ability is a dominant trait; in other words, if a transforming *onc* gene is transfected into or activated in a normal cell, it captures the cell's genetic machinery and turns it into a cancer cell. This conclusion is most likely wrong for the following reasons. The NIH/3T3 cell is already two-thirds a cancer cell. Indeed, subpopulations of "untransformed" NIH/3T3 cell cultures are tumorigenic and metastatic under the right conditions,[43] although transformation with a *ras* gene markedly increases the malignant potential of these cells. Moreover, as will be further discussed later, transfection with at least two *onc* genes is needed to transform normal diploid fibroblasts in culture, supporting the idea that malignant transformation is a multistep process. Finally, cell hybridization between malignant and normal cells indicates that the hybrid cells formed are more likely than not to be nontumorigenic (see Chap. 4). Thus, expression of the complete malignant phenotype is not likely to be due to insertion or activation of a single "cancer gene" and in most cases also appears to involve the loss of tumor suppressor genes (Chap. 8).

Activation, Oncogenic Potential, and Mechanism of Action of Some Individual Oncogenes

Activation of c-*onc* genes can take place by means of all the mechanisms described in Chapters 3 and 5 for activation of genes during cell transformation or tissue differentiation. These mechanisms include point mutations, gene rearrangement, gene amplification, and increased transcription due to alterations in chromatin packaging. In addition, insertion of retrovirus enhancer regions (LTRs) next to c-*onc* genes or mutation in c-*onc* gene coding sequences can alter their function. Some of these mechanisms have been identified only by test tube or animal experiments, but all could potentially be involved in c-*onc* gene activation during carcinogenesis in humans. The reason for thinking this is that protooncogenes are present in all human cells, just as they are in animal cells, and they apparently have to be activated by some endogenous (e.g., faulty repair of oxidative damage from normal cellular processes) or exogenous agent (e.g., ultraviolet light, chemical carcinogens) to trigger the cancer process.

A number of genetic lesions have been observed in human tumors (Table 7-2,[44] and these most likely are part and parcel of the carcinogenic process, although direct proof that they are cause rather than effect can only be deduced from cell culture– and transgenic animal–type experiments, where an altered gene can be in-

Table 7-2 Protooncogenes and Human Tumors: Some Consistent Incriminations

Protooncogene[a]	Neoplasm(s)	Lesion
abl	Chronic myelogenous leukemia	Translocation
erbB-1	Squamous cell carcinoma; astrocytoma	Amplification
erbB-2 (*neu*)	Adenocarcinoma of breast, ovary and stomach	Amplification
gip	Carcinoma of ovary and adrenal gland	Point mutations
gsp	Adenoma of pituitary gland; carcinoma of thyroid	Point mutations
myc	Burkitt's lymphoma	Translocation
	Carcinoma of lung, breast and cervix	Amplification
L-*myc*	Carcinoma of lung	Amplification
N-*myc*	Neuroblastoma: small cell carcinoma of lung	Amplification
H-*ras*	Carcinoma of colon, lung and pancreas; melanoma	Point mutations
K-*ras*	Acute myelongenous and lymphoblastic leukemia; carcinoma of thyroid; melanoma	Point mutations
N-*ras*	Carcinoma of genitourinary tract and thyroid; melanoma	Point mutations
ret	Carcinoma of thyroid	Rearrangement
ros	Astrocytoma	?
K-*sam*	Carcinoma of stomach	Amplification
sis	Astrocytoma	?
src	Carcinoma of colon	?
trk	Carcinoma of thyroid	Rearrangement

[a]The list is intended to be representative but not exhaustive. It includes loci that suffer demonstrable damage and loci that are abnormally expressed for reasons not yet known.

Source: From Bishop.[44]

troduced and its effect on cell transformation or tumor development directly observed.

Oncogenes and their normal cellular counterparts, the protooncogenes, can be classified by their function into several different categories (Table 7-3).[45] A number of these genes encode growth factors, e.g., *sis* (PDGF B-chain), *int*-2 and *hst* (FGF-like factor). These oncogene growth factors can stimulate tumor cell proliferation by paracrine or autocrine mechanisms but by themselves may not be sufficient to sustain the transformed phenotype.

A second type of oncogene codes for altered growth factor receptors, many of which have associated tyrosine kinase activity. These include the *src* family of oncogenes, *erb* B (EGF receptor), and *fms* (CSF-1 receptor). For some of these receptorlike, tyrosine kinase–associated membrane proteins the actual ligand is not known (e.g., *trk*, *met*, and *ros*). A third receptor class that does not have associated tyrosine kinase activity is the *mas* gene product (angiotensin receptor) and the α_{1B} adrenergic receptor.

A fourth class of oncogene products are membrane-associated, guanine nucleotide–binding proteins such as the Ras family of proteins. These proteins bind GTP, have associated GTPases, and act as signal transducers for cell surface growth factor receptors. The transforming *ras* oncogenes have been mutated in such a way to render them constitutively active by maintaining them in a GTP binding state, most likely because of a defect in the associated GTPase activity.

A fifth category are the cytoplasmic oncoproteins with serine/threonine protein kinase activity. These include the products of the *raf*, *pim*-1, *mos*, and *cot* genes. A prototype of this class is the c-Raf protein, which is activated by a variety of tyrosine kinase–associated receptors. There is evidence that c-Raf acts as an intermediate in the signaling pathway between Ras and the cell nucleus by activating the mitogen-activated protein (MAP) kinase cascade (see below). The oncogenic form of Raf has lost part of its regulatory amino terminal sequence and appears to be constitutively active. c-Crk is also a cytoplasmic protein, and it appears to act by stabilizing tyrosine kinases associated with the Src family of oncoproteins.

A seventh, large class of oncogenes are those that code for nuclear transcription factors such

Table 7-3 Functions of Cell-Derived Oncogene Products[a]

Class 1—Growth Factors	
sis	PDGF B-chain growth factor
int-2	FGF-related growth factor
hst (KS3)	FGF-related growth factor
FGF-5	FGF-related growth factor
int-1	Growth factor?
Class 2—Receptor and Nonreceptor Protein-Tyrosine Kinases	
src	Membrane-associated nonreceptor protein-tyrosine kinase
yes	Membrane-associated nonreceptor protein-tyrosine kinase
fgr	Membrane-associated nonreceptor protein-tyrosine kinase
lck	Membrane-associated nonreceptor protein-tyrosine kinase
fps/fes	Nonreceptor protein-tyrosine kinase
abl/bcr-abl	Nonreceptor protein-tyrosine kinase
ros	Membrane associated receptor-like protein-tyrosine kinase
erbB	Truncated EGF receptor protein-tyrosine kinase
neu	Receptorlike protein-tyrosine kinase
fms	Mutant CSF-1 receptor protein-tyrosine kinase
met	Soluble truncated receptor-like protein-tyrosine kinase
trk	Soluble truncated receptorlike protein-tyrosine kinase
kit (W locus)	Truncated stem-cell receptor protein-tyrosine kinase
sea	Membrane-associated truncated receptor-like protein-tyrosine kinase
ret	Truncated receptorlike protein-tyrosine kinase
Class 3—Receptors Lacking Protein Kinase Activity	
mas	Angiotensin receptor
α1β	Adrenergic receptor
Class 4—Membrane-Associated G Proteins	
H-*ras*	Membrane-associated GTP-binding/GTPase
K-*ras*	Membrane-associated GTP-binding/GTPase
N-*ras*	Membrane-associated GTP-binding/GTPase
gsp	Mutant activated form of G_s α
gip	Mutant activated form of G_i α

as *myc*, *myb*, *fos*, *jun*, *erb*A, and *rel*. For a number of these, the oncogenic alteration that makes them transforming oncoproteins is a mutation that leads to loss of negative regulatory elements (e.g., for *jun*, *fos*, and *myb*), and in other cases (e.g., *erbA* and *rel*), the activating mutations cause the loss of their active domains, producing a mutant protein that prevents the activity of the normal gene product—a so-called dominant-negative mutation. It is interesting that mutations of the tumor suppressor gene p53, in sort of a "reverse twist," produce a dominant-negative effect by producing a protein that in this case prevents the action of a tumor suppressor function (see Chap. 8).

Characteristics of some individual *onc* gene products are given below.

ras

The most frequently detected alterations in oncogenes in both animal tumor model systems and in human cancers are mutations in the *ras* family of oncogenes.[46–48] Members of the *ras* gene family are Ha-, Ki-, and N-*ras*. Activated, tumorigenic *ras* genes differ from their normal cellular counterparts by having acquired a single point mutation at codon 12, 13, 61, or 117[46] or by alternate mRNA splicing.[49]

One of the most intensively studied animal systems for *ras* oncogene mutation is the mouse skin model. A number of mouse skin carcinomas initiated by application of carcinogens such as N-methyl-*N*′-nitro-*N*-nitrosoguanidine (MNNG), methylnitrosourea (MNU), 3-methylcholanthrene (MCA), and 7,12-dimethylbenz[*a*]anthracene (DMBA) contain mutated Ha-*ras* genes. Interestingly, these agents cause somewhat different mutations, and the mutations seen in skin papillomas are not the same as those seen in carcinomas,[46] suggesting that distinct mutagenic events may be involved in the initiation phases and in the promotion-progression phases of skin tumorgenesis. For example, in

Table 7-3 *(continued)*

Class 5—Cytoplasmic Protein-Serine Kinases	
raf/mil	Cytoplasmic protein-serine kinase
pim-1	Cytoplasmic protein-serine kinase
mos	Cytoplasmic protein-serine kinase (cytostatic factor)
cot	Cytoplasmic protein-serine kinase?
Class 6—Cytoplasmic Regulators	
crk	SH-2/3 protein that binds to (and regulates?) phosphotyrosine-containing proteins
Class 7–Nuclear Transcription Factors	
myc	Sequence-specific DNA-binding protein
N-*myc*	Sequence-specific DNA-binding protein?
L-*myc*	Sequence-specific DNA-binding protein?
myb	Sequence-specific DNA-binding protein
lyl-1	Sequence-specific DNA-binding protein?
p53	Mutant form may sequester wild-type p53 growth suppressor
fos	Combines with c-jun product to form AP-1 transcription factor
jun	Sequence-specific DNA-binding protein; part of AP-1
erbA	Dominant negative mutant thyroxine (T_3) receptor
rel	Dominant negative mutant NF-κB-related protein
vav	Transcription factor?
ets	Sequence-specific DNA-binding protein
ski	Transcription factor?
evi-1	Transcription factor?
gli-1	Transcription factor?
maf	Transcription factor?
pbx	Chimeric E2A–homeobox transcription factor
*Hex*2.4	Transcription factor?
Unclassified	
dbl	Cytoplasmic truncated cytoskeletal protein?
bcl-2	Plasma membrane signal transducer?

[a]The table is somewhat selective and obviously incomplete. These oncogenes were originally detected as retroviral oncogenes or tumor oncogenes. Others were identified at the boundaries of chromosomal translocations and at sites of retroviral insertions in tumors, or were found as amplified genes in tumors and shown to have transforming activity.

Source: From Hunter.[45]

one study the alkylating agents MNNG and MNU caused G→A transitions at codon 12, whereas MCA caused G→T transversions at codon 13 and A→T transversions at codon 61 in papillomas.[46] Only G→T mutations were seen in MCA-induced carcinomas, suggesting that the cells bearing that mutation were the ones destined to progress to carcinoma. In the case of DMBA-induced skin tumors, a codon 61 A→T transversion can be seen well before the appearance of papillomas, indicating that this mutation is an early event in initiation.[50] Mutations of *ras* gene family members have also been observed in chemically induced liver,[51] pancreatic,[52] and mammary carcinomas[53] and in ultraviolet-induced skin cancers[54] in rodents.

It is of interest that the cell-transforming ability of mutant *ras* alleles prevails in some cultured cells, although the normal alleles continue to be expressed,[55] suggesting that the mutant alleles are dominant. But as noted above, this does not seem to be the case in the real world of normal diploid cells. An explanation for this apparent conundrum is provided by a study by Finney and Bishop,[56] who found that rat fibroblasts into which a mutant *ras* gene had been introduced by homologous recombination continued to express the mutant and remaining normal alleles equally yet were not neoplastically transformed. However, spontaneously transformed cells did arise at low frequency, and these cells had an amplified mutant allele, indicating that overexpression of the mutant allele could "swamp out" the normal allele.

Structural studies by x-ray crystallography suggest the reason why the mutated Ras protein has altered function. The normal function of the p21 protein coded for by c-*ras* is to interact with tyrosine kinase receptors and other proteins (see below) to activate a signal transduction pathway. To do this, p21 has to bind GTP (the "on" signal) and then hydrolyze it (the "off" signal). Ras p21 mutated at codons 12 or 61 has a structure that disfavors its ability to bind to the GTPase activating protein (GAP), thus keeping p21 in the GTP-bound, activated mode.[57,58]

There is a lot of evidence to indicate that the action of a single oncogene is insufficient to cause neoplastic transformation of cells, particularly normal diploid cells. Instead, the action of at least two oncogenes appears to be required. In the case of *ras*, cooperativity with the *myc* oncogene, for example, has been shown to be required to induce neoplastic transformation of primary embryo fibroblasts or prostate tissue organ cultures.[59] In addition, v-Ha-*ras* and v-*fos* coinfected keratinocytes produced squamous cell carcinomas in nude mice, whereas v-*fos* alone produced only skin hyperplasia, and v-Ha-*ras* alone produced only papillomas.[60]

Mutations in members of the *ras* gene family are found in a wide variety of human cancers, including colon, pancreas, lung, breast, skin, thyroid, bladder, liver, and kidney carcinomas, as well as in seminomas, melanomas, and some forms of leukemia (reviewed in Ref. 61). Most of these mutations are in codons 12, 13, or 61. While *ras* gene mutations can be found in a variety of human tumor types, the incidence of such mutations varies greatly. For example, the highest incidences are found in carcinomas of the pancreas (90%), colon (50%), lung (30%) and thyroid (50%) and in myeloid leukemia (30%), while a much lower incidence is found in urinary bladder (6%)[62] and ovarian (15%)[63] cancers. Some diagnostic advantage may be taken of the fact that some cancers have a high incidence of *ras* gene mutations. For example, 8 of 9 cases of colorectal cancer shown, by biopsy of the tumor, to have Ki-*ras* mutations at codons 12 or 13 also had detectable mutations in sloughed cells obtained in the feces.[64] Similarly, Ki-*ras* gene mutations were found in codon 12 in the pancreatic juice of 7 of 7 patients with pancreatic carcinomas,[65] suggesting a way to differentiate between chronic pancreatitis and pancreatic carcinoma for this difficult-to-diagnose cancer.

While the oncogenic potential of the *ras* gene family of oncoproteins was described in the early 1980s, it has taken a decade to define how the Ras proteins function normally in cells. Results from several laboratories working on different aspects of the problem, and on organisms as diverse as yeast, the worm *Caenorhabditis elegans*, the fruit fly *Drosophila*, and mammalian cells have made rapid strides in working out a pathway for Ras function. The power of genetic manipulations in lower organisms and the fact that the Ras pathway has been conserved over eons of time have made this possible. The essential features of the pathway are shown in Figure 7-3 (reviewed in Refs. 66 to 69). It is now known that the Ras system is an ubiquitous signal transduction pathway that is part of the signal transduction system for multiple tyrosine kinase receptors like those for the growth factors EGF, PDGF, and FGF (see Chap. 9). When a ligand such as EGF binds its receptor, a receptor dimerization occurs, triggering an autophosphorylation event and a interaction with some "adaptor" proteins called Grb 2 and Sos. Grb 2 is the mammalian cell homologue of GDP-releasing factor called Sem 5 that was identified in *C. elegans* and shown to be involved in Ras activation in that system. Grb 2 complexes with the tyrosine kinase receptor and the Sos protein ("son of sevenless," originally identified in *Drosophila* as part of the Ras system) via the Src protein homology domains called SH2 and SH3, respectively. GRB-2 SH3 domains bind to proline-rich sequences present in Sos, allowing recruitment of Sos to the plasma membrane where Ras is bound. Exactly how the complex of receptor, Grb 2, and Sos activates Ras is not clear, but the activation results in a release of GDP and binding of GTP. This activates Ras, which in turn complexes with Raf, inducing a conformational change in Raf that activates Raf kinase activity. Activated Raf then phosphorylates a second kinase called Mek (formerly called mitogen-activated protein kinase kinase or MAPKK). Mek phosphorylates and thereby activates a third kinase, MAP kinase (MAPK), which has among its substrates various transcription factors such as Jun/Fos (AP-1) that regulate gene expression.

This "kinase cascade" is a crucial signaling pathway regulating growth and differentiation in

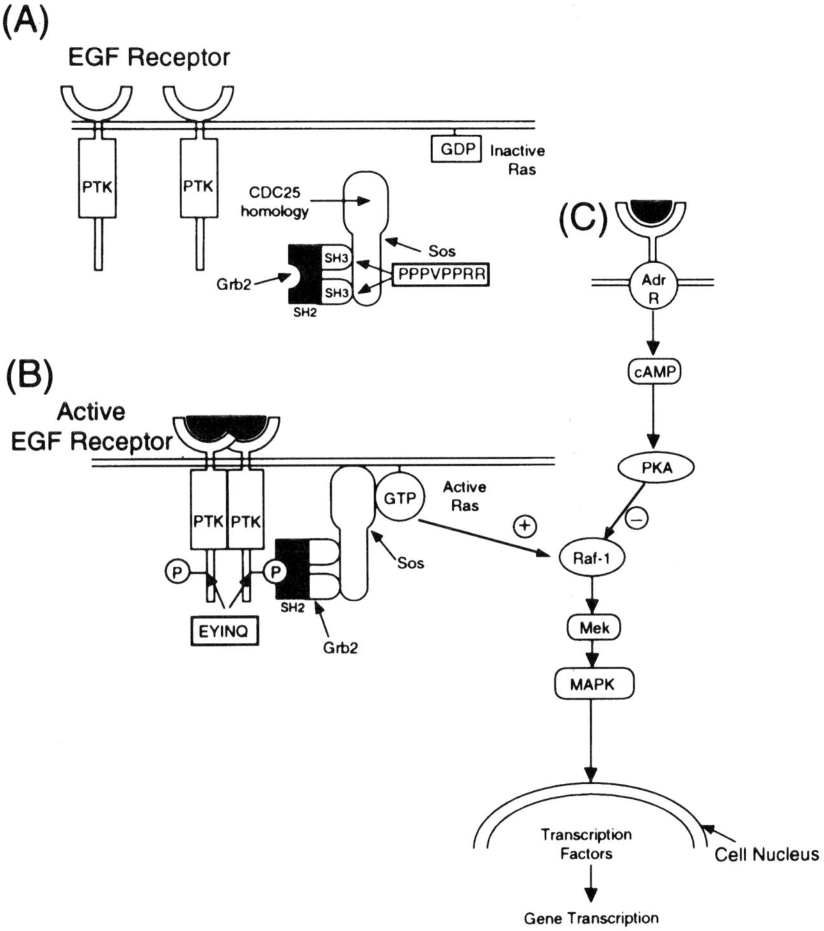

Figure 7-3 Mechanism of Ras activation by the EGF receptor. (A) The SH3 domains of Grb2 are bound to a proline-rich region in the carboxy-terminal tail of Sos. (B) In response to EGF-receptor activation and autophosphorylation, the Grb2-Sos complex binds to tyr1068 at the carboxy-terminal tail of the EGF receptor thus relocating Sos to the plasma membrane. The interaction between the quanine nucleotide releasing activity of Sos (cdc25 homology) and membrane-bound Ras leads to the exchange of GDP for GTP. Formation of GTP-bound Ras leads to activation of a kinase cascade including Raf, Mek-MAPK, and other pleiotropic responses essential for cell growth and differentiation. Consensus sequences are shown for the phosphorylation sites on the EGF receptor and for the SH3 binding sites on Grb2 for Sos. (C) In some cell types (e.g., fibroblasts), cAMP has a negative growth effect stimulated via certain receptors (e.g., the adrenergic receptor, Adr R) and its action on protein kinase A(PKA). This may be via phosphorylation of Raf-1. (Modified from Schlessinger.[66])

many cell types. One might ask, then: How is cell specificity obtained and how does this system go awry in cancer? Specificity may be obtained by the different array or concentration of ligand receptors on a cell's surface and by the cell-specific adaptor proteins that form complexes with Ras and modulate its activity. For example, in some cell types, cAMP-activated protein kinases appear to have a negative modulatory role on Raf. As noted above, the reason that this system is on "fast-forward" in cancer cells appears to be that the mutated Ras protein observed in so many types of cancer tissue does not bind GAP well and hence loses its ability to be downregulated by GTP hydrolysis, which is the "off" mechanism in normal cells.

myc

The *myc* gene was discovered by looking for the cell-transforming sequence of the avian retrovirus MC29[70] and later identified in vertebrate genomes. A number of studies have shown that *myc* plays a key role in cell proliferation and differentiation events. Deregulation of c-*myc* can occur by either gene rearrangement or amplification, and both have been observed in human cancers.

The other members of the *myc* gene family are N-*myc*, cloned in 1983 and found to be amplified in human neuroblastoma; L-*myc*, identified in 1985 and highly expressed in small-cell lung cancer; B-*myc*, encoding a truncated version of a Myc protein and primarily expressed in brain; s-*myc*, a putative suppressor of neoplastic transformation (reviewed in Ref. 71).

Expression of c-Myc protein is higher in proliferating cells and falls as terminal differentiation proceeds. In model systems for differentiation such as murine erythroleukemia cells, 3T3-L1 pre-adipocytes, or F9 teratocarcinoma cells, continued expression of c-*myc* from a transfected gene blocks differentiation (reviewed in Ref. 71), suggesting that downregulation of c-Myc is necessary for cessation of cell division during terminal differentiation.

Rearrangement of *myc* is one of the classic examples of *onc* gene activation by chromosomal translocation (Chap. 3). One of the clear examples of this is the B cell lymphoma known as Burkitt's lymphoma (BL), a primarily pediatric disease of high incidence in equatorial Africa. A number of types of translocation events have been found in various cell lines derived from patients with this disease, including translocations between chromosomes 8 and 14 (the most common), 8 and 2, and 8 and 22. This led Klein to postulate that the intrachromosomal breakages and rearrangements that in normal lymphoid cells are involved in V- and C-region joining of immunoglobulin genes (see Chap. 5) would, in transformed lymphoid cells, bring an *onc* gene and the Ig gene segments together in such a way that the *onc* gene would be derepressed by coming under the influence of the normal Ig gene promoters.[72] This turned out to be a particularly perspicacious idea, for it was soon found that the *myc* oncogene was located on chromosome 8 in humans and was in a region translocated to chromosomes 2, 14, or 22 in Burkitt's lymphoma cells.[73,74] Similar translocations between chromosome 15 (the chromosome bearing the *myc* gene in the mouse) and chromosome 12 (location of Ig genes in the mouse) have been observed in murine plasmacytoma cells,[75–77] the mouse cell type equivalent to Ig-producing B-cell malignancies of humans.

Normally, c-*myc* expression is carefully regulated. Expression is low in resting normal lymphoid cells and is turned on in proliferating lymphoid cells—for example, cells stimulated by lectins or antigens—suggesting that the c-*myc* gene product plays a critical role in DNA synthesis and mitosis. In BL cells, however, c-*myc* expression is not regulated and becomes "constitutive" (i.e., not turned on or off in response to normal metabolic stimuli), perhaps because of the loss or alteration of parts of exon 1, which appear to be involved in regulation of c-*myc* expression. In summary, there may be a variety of mechanisms, involving different types of gene rearrangements of c-*myc*, that lead to c-*myc* gene derepression and unregulated expression so that the cell has a continued transcription of c-*myc* and cannot return to the resting, nonproliferating state.

Members of the *myc* gene family are also deregulated by gene amplification. In fact, the first reported *onc* gene amplification was for the c-*myc* gene in a human promyelocytic leukemia cell line, HL-60, and in primary leukemic cells taken from the patient from whom the cell line has been derived.[78] The gene was amplified about 20 times in these cells. The c-*myc* gene was later found to be amplified in human cell lines derived from colon carcinoma[79] and small-cell lung carcinoma (SCLC).[80] In the latter case, the highest degree of c-*myc* amplification (20- to 76-fold) was found in the SCLC cell types with the least differentiated and most highly malignant phenotype.[80] Amplification, rearrangements, and deregulated, enhanced expression of c-*myc* have also been observed in chemically induced rodent tumors.[81]

One or more of three members of the *myc* gene family—c-*myc*, N-*myc*, and L-*myc*—have been shown to be amplified in various human cancers and in cell lines derived from them. In one study of SCLC, all three *myc* gene family members were found to be amplified, and this amplification was found to occur more commonly in tumors from patients after treatment with chemotherapy.[82] Myc protein levels are also elevated in colon carcinomas.[83] Amplification of c-*myc* also occurs in breast carcinomas, and elevated c-*myc* appears to be an independent prognostic marker of overall survival.[84] It is of interest that treatment with hydroxyurea of human cancer cell lines that have amplified *myc* genes—present as extrachromosomal double minute elements (DMs)—decreases the *myc* copy number and the tumorigenicity of these cells in nude mice,[85] suggesting a way to eliminate amplified *myc* genes in vivo and perhaps improve patient survival.

The c-Myc protein product of the c-*myc* gene is a DNA-binding, nuclear phosphoprotein that has all the characteristics of a transcription factor. It has a transcriptional activation domain, a DNA binding domain, a nuclear localization signal, a site for phosphorylation by a nuclear protein kinase, a helix-loop-helix (HLH) motif, and a leucine-zipper motif typical of transcription factors that have to form dimers to be active (reviewed in Ref. 71).

For a long time, the dimerization partner of c-Myc was a missing link in trying to understand its action as a transcription factor. For a while it was thought that c-Myc formed homodimers with itself, but that did not seem to occur inside cells. This problem was solved when Blackwood and Eisenman[86] cloned a human gene coding for a protein they called Max, which dimerized with c-Myc. A mouse homologue of Max called Myn was cloned a short time later.[87] Max can dimerize with c-Myc, N-Myc, or L-Myc but not other basic HLH-leucine-zipper proteins.[86] Max can bind DNA as a homodimer but does not seem to be able to activate transcription, suggesting that it lacks a transcriptional activation domain. This ability to bind DNA and not activate genes most likely explains the ability of Max-Max homodimers to antagonize the ability of Myc-Max heterodimers to stimulate gene transcription. Myc-Max heterodimers bind to the consensus sequence CACGTG, to which Max homodimers also bind.[88] Myc alone does not form homodimers efficiently and does not bind to DNA except at high concentrations in vitro. Thus, Myc acts as a transcriptional activator that requires dimerization with Max, and Max homodimers act as a repressor of Myc-Max action. It has also been found that Max overexpression in cells represses transcription of reporter genes bearing the CACGTG regulatory sequence, and Max-induced repression is relieved by overexpression of c-Myc.[88,89] Moreover, the oncogenic transforming ability of c-Myc requires dimerization with Max.[90] It should be noted that c-*myc*-induced cellular transformation, as opposed to a number of other oncogenes, results from alternately high expression of a normal coding sequence rather than activation by a point mutation, as occurs with *ras* for example. Nevertheless, the full oncogenic potential of c-*myc* involves cooperation with other oncogenes such as *ras*.[91]

Myc's action as a transcription factor includes induction of ornithine decarboxylase, cyclin A, and cyclin E, all of which are involved in cell proliferation. Somewhat paradoxically, increased c-Myc production in some cell types (e.g., B lymphocytes) is associated with programmed cell death (apoptosis) (see Chap. 10). As noted in Chapter 5, the transcription factor NF-κB regulates c-*myc* expression in a number of cell types.

src

As noted above, *src* was the first transforming oncogene discovered. It exemplifies yet again another way in which a cellular protooncogene is altered to become a transforming oncogene. In this case, data comparing the normal cellular gene product of c-*src* to its viral, transforming counterpart from v-*src* showed that the normal protein $pp60^{c\text{-}src}$ differed from $pp60^{v\text{-}src}$ in that different isolates of the latter contained a number of scattered single amino acid differences between residues 1-514 (but only one common one: Thr 338→Ile) as well as truncations and alterations at the carboxyl terminus (reviewed in Ref. 92). This implied that the C-terminus of $pp60^{src}$ plays an important role in regulating its transforming ability. Since $pp60^{src}$ is a tyrosine kinase and appears to regulate cell function by its ability to phosphorylate key cellular substrates, this led to the idea that the C-terminus somehow regulates the kinase activity of the Src protein. A clue that this was correct emerged when it was observed that Src tyrosine kinase activity is enhanced by phosphatases.[93] It was subsequently found that phosphorylation of tyrosine-527 in the C-terminus of Src is inhibitory, and dephosphorylation of this residue stimulates

Src kinase activity (reviewed in Ref. 94). In most transforming Src mutants, Tyr-527 is either missing or underphosphorylated compared to wild-type Src. In addition, it was found that phosphorylation of Try-416 in the catalytic domain of the kinase activity was required for full activity. Mutations that activate Src have mapped to the kinase domain itself, to SH2 and SH3 domains, and to the C-terminus,[94] all of which appear to produce constitutive activation of Src kinase activity. A model to explain the conformational changes that the activating dephosphorylation/phosphorylation events may play in Src activity has been proposed (Fig. 7-4). A tyrosine kinase called Csk (C-terminal Src kinase) has been identified that can phosphorylate the C-termini of Src and all its family members, and Csk, when overexpressed, inhibits the cell-transforming ability of high levels of Src (reviewed in Ref. 94).

There are at least nine members of the *src* gene family: *yes*, *fgr*, *lyn*, *lck*, *fyn*, *hck*, *yrk*, *crk* and *src* itself.[94] Moreover, alternate translational initiation codons and tissue-specific mRNA splicing results in more than 14 different Src-type proteins being expressed selectively in various cell types.[94] They all have tyrosine kinase activity, an N-terminal myristorylation signal presumably required for their association with cell membranes, SH2 and SH3 domains, a kinase domain, and a C-terminal regulatory "tail." The structural relationships of all these proteins strongly suggests that they are all subject to the same regulatory mechanisms, yet they appear to have very cell-specific functions (see below).

c-Src is present at low levels in most cell types, but high levels are found in neural tissue, platelets, lymphocytes, monocytes, and chromaffin cells (reviewed in Ref. 95). High levels are also found in human neuroblastoma, small-cell lung, colon, and breast carcinomas, and rhabdomyosarcoma.[95,96]

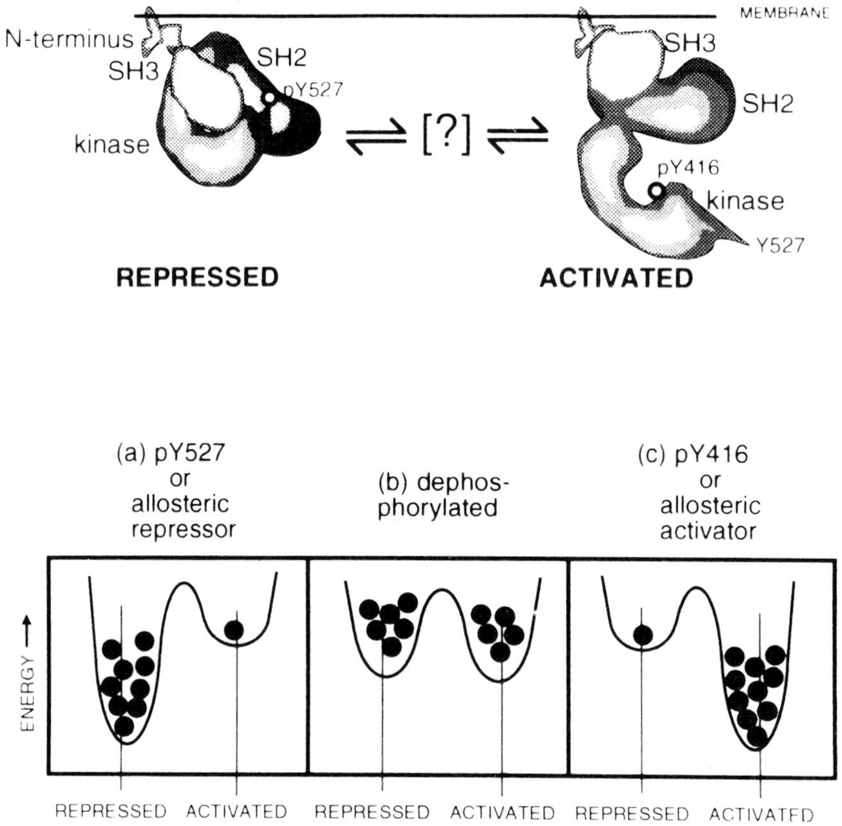

Figure 7-4 Activation of Src by changes in conformation stabilized by changes in phosphorylation. Phosphorylation of Tyr-527 represses kinase activity and phosphorylation of Tyr-416 activates it. (From Cooper and Howell.[94])

Mention should be made here of the Src homology domains SH2 and SH3 because, even though these domains are found in several proteins that interact with tyrosine kinase receptors and other proteins in signal transduction pathways, they are part of the sequence of the *src* gene-encoded protein.

The SH2 domain is a conserved motif of about 100 amino acids found in a diverse array of proteins involved in signal transduction (Fig. 7-5).[97] A number of these proteins also contain SH3 domains, which are sequences about 50 amino acids long. These domains are found in a variety of *onc* gene tyrosine kinases, phospholipase C (PLC-γ), intermediates in the guanine nucleotide exchange pathway involving Ras (see above), GAP, phosphatidylinositol (PI)3′-kinase, and an ever-increasing list of other proteins that are involved in various phosphorylation/dephosphorylation cascades.

A typical scenario is as follows: an external ligand such as a growth factor or hormone binds to its receptor, inducing dimerization and autophosphorylation of the receptor. These steps create binding sites for SH2 domains which recognize phosphotyrosine (or phosphoserine/phosphothreonine) adaptor proteins that are the linkers to the next step in a signal transduction pathway. In the case of the Ras pathway, as noted above, this linker is the Sem5/GRB2 protein. Specificity is provided by the fact that high-affinity binding of an SH2 domain requires recognition of a phosphotyrosine within a specific amino acid sequence.[97] For example, PI3′-kinase, Ras-GAP, and PLC-γ each bind to different autophosphorylated sites on the PDGF-β receptor (reviewed in Ref. 97). Thus, the binding of SH2-containing adaptor proteins to a given receptor depends on the amino acid sequence (and presumably the peptide conformation) at the autophosphorylation site.

The SH3 domains of adaptor proteins recognize, in a context/conformation-dependent way, guanine nucleotide exchange factors such as Sos in the Ras pathway and GTPase-activating proteins. Both SH2 and SH3 domains appear to be involved in the regulation of *onc* gene product tyrosine kinase activities. For example, deletion of the SH3 domain from the c-Abl protooncogene protein activates its kinase activity and ren-

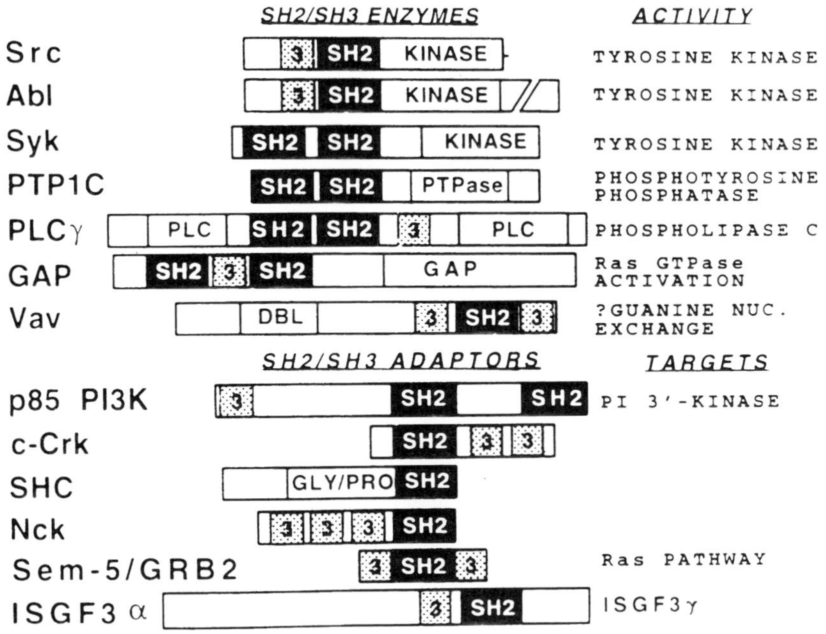

Figure 7-5 Structures of selected SH2-containing proteins. These polypeptides are divided into those with intrinsic enzymatic activity and those without known catalytic domains, which may act as adaptors to couple tyrosine kinases to downstream targets. No catalytic activity has been shown for Vav, but it contains a region of homology to Dbl (indicated as DBL), which has guanine nucleotide exchange activity. ISGF3α is three closely related proteins of 84, 91, and 113 kDa. 3, SH3 domain; PTPase, phosphototyrosine phosphatase domain; GLY/PRO, glycine-proline-rich region. (From Pawson and Gish.[97])

ders it transforming; moreover, alterations of the amino acid sequence of the SH2 domain of c-Abl also render it transforming.[98] These results indicate that both SH2 and SH3 domains play a role in regulating the activity of oncogene protein tyrosine kinases.

jun and *fos*

The role of c-*jun* and c-*fos* as transcription factors was covered in Chapter 5. Here, we focus on the transforming ability of the Jun and Fos proteins when they are expressed in an inappropriate way or at inappropriate times in the life of a cell.

The oncogene v-*jun* was originally discovered as the transforming gene of avian sarcoma virus 17 that can induce fibrosarcomas in chickens via its p65 *gag-jun* fusion protein.[99] Both v-Jun and c-Jun form dimers with c-Fos and act as transcription factors that bind to AP-1 regulatory sequences. Overexpression of c-Jun in rat embryo fibroblasts cotransfected with c-Ha-*ras* gives rise to immortalized cell lines that grow in soft agar and produce tumors in nude mice.[100] However, v-Jun has even greater transforming ability, and v-Jun has significantly greater transcriptional activity than c-Jun. Sequence analysis of c-Jun and v-Jun has revealed that v-Jun has three amino acid substitutions in the C-terminus and deletion of a 27 amino acid run, called the δ region, from the N-terminus.[101] The δ region contains a negative regulatory domain, the lack of which renders v-Jun more active as a transcription factor, apparently because it lacks a site for binding of a cellular factor that modulates c-Jun activity. Since the leucine-zipper motif of Fos is required for transformation by overexpression of c-Jun or expression of v-Jun, it can be concluded that transcriptional activation of AP-1-regulated genes is involved in the cellular transformation events.

Other members of the *jun* gene family include *jun*B and *jun*D, which share significant sequence homology and which can all bind to AP-1 sites in the presence of c-Fos. However, members of the *jun* family are expressed at different levels in different cell types and their production responds differently to extracellular signals. Moreover, they differ in their transforming ability. For example, JunB is less potent than c-Jun in transforming rat embryo cells cotransfected with *ras*, and cotransfection of *jun*B with c-*jun* into *ras*-activated cells decreases transformation compared to c-*jun/ras* transfected cells.[102]

The *fos* oncogene was detected in two independent isolates of mouse osteosarcoma viruses: FBJ-MUSV, isolated from a spontaneous osteosarcoma, and FBR-MuSV, isolated from a radiation-induced osteosarcoma.[103] Both viruses can induce chondrosarcomas or osteosarcomas when inoculated into newborn mice and are able to induce transformation in mouse fibroblast cell lines. The FBJ-encoded v-Fos protein is similar to its cellular homologue c-Fos except for a frameshift mutation at its C-terminus, whereas the FBR v-Fos has N-terminal and C-terminal truncations, internal deletions, and several single amino acid changes (reviewed in Ref. 104). Sequences in the C-terminal half of c-Fos have a regulatory role in its activity, and alteration of this domain in v-Fos correlates with its transforming activity.[104] An amino acid substitution at residue 138 (Glu→Val) activates the cell immortalizing activity of v-Fos.

ets

The v-*ets* oncogene was first identified in the E26 acutely transforming retrovirus of the chicken. The E26 virus is unique in that it produces a transforming fusion protein containing v-Ets and v-Myb (see below). It is now clear that a large family of *ets*-related oncogenes is present in vertebrate and lower organisms (Table 7-4).[105] Ets proteins have a conserved DNA-binding domain, but the DNA-binding motif differs from other DNA-binding proteins in that the typical zinc-finger, leucine-zipper, and helix-turn-helix motifs appear to be absent. The Ets-family proteins are transcription factors, and some of them (e.g., Ets-1 and Ets-2) cooperate with the Jun/Fos (AP-1) transcription factor, while others (e.g., Elk-1 and SAP-1) act by forming complexes with the serum response factor (SRF) (reviewed in Ref. 105). An Ets binding site (EBS) has been identified in a number of regulatory elements of genes, and all of these contain a consensus GGAA or GGAT sequence. EBS sequences have been found in regulatory elements for interleukin-2, SV40, HTLV-1, stromelysin-1, c-*fos*, T-cell receptor α genes, and a number of

Table 7-4 The *ets* Gene Family

Protein	Source	Molecular Mass (kDa)	Amino-Acid Homology to ETS Domain of Ets-1 (%)	Human Chromosomal Location	Expression/Features
Ets-1	Human Mouse Chicken	39–52 63 54/68	100	11q23	Elevated expression in thymus and endothelial cells; phosphorylated; alternatively spliced; positively autoregulates transcription
Ets-2	Human Chicken	58/62	90	21q22	Expression induced following macrophage differentiation and T-cell activation; alternatively spliced; phosphorylated
Erg	Human	41/52	70	21q22	Alternatively spliced; 98% homologous to Fli-1
Fli-1	Human Mouse	 51	68	11q23	Activated by proviral insertion of Friend MuLV: 98% homologous to Erg
Elk-1	Human	60	76	Xp11.2	ETS domain located in the amino terminus of the protein; forms temary complex with SRF; shows three regions of homology with SAP-1
SAP-1 a/b	Human	58/52	75	ND	SRF accessory protein 1, which, like Elk-1, forms a temary complex with SRF over the *c-fos* SRE; contains three regions of homology to Elk-1, including the ETS domain, which is located in the amino terminus of the protein; the two isoforms, SAP-1a and SAP-1b, differ in their carboxyl termini
Spi-1/PU1	Human Mouse	30	38	11p11.22	Activated in Friend erythroleukaemia by proviral insertion of SFFV; normal expression of the PU-1 transcription factor is restricted to B cells and macrophages
E74A/B	*Drosophila*	110/120	50	*Drosophilia* chromosome 3L74EF	E74A is induced by ecdysone and regulates the expression of E74B, which is also *ets*-related
Elf-1	Human	68	50	ND	The ETS domain is the human homologue of the E74A protein of *Drosophilia;* binds to the NF-AT and NFIL-2B sites in the interleukin-2 promoter and the human immunodeficiency virus 2 LTR
GABP-α	Rat	51	82	ND	High-level expression in rat thymus; complexes with GABP-β, which contains ankyrin repeats, and is related to the Notch protein
D-Elg	*Drosophila*	15	64	*Drosophila* chromosome 3R97D	Contains only a DNA-binding domain; maternally expressed message and also expressed throughout embryogenesis
PEA3	Mouse	68	63	ND	Expressed in mouse brain and epididymis and in fibroblast and epithelial cell lines; down-regulated in embryonic cell lines in response to retinoic-acid-induced differentiation
TCF1-α	Human	55	ND	ND	Very limited homology to ETS domain exists within the HMG box of this factor; expression is restricted to the thymus and is induced following T-cell activation; regulates activity of the TCRα enhancer

Abbreviations: ND, not determined; MuLV, murine leukemia virus; SFFV, spleen focus forming virus; SRF, serum response factor; NF-AT, nuclear factor of activated T cells; NFIL-2B, nuclear factor of interleukin 2B; LTR, long terminal repeat; GABP-β, GA-binding protein; PEA3, polyomavirus enhancer activator 3; HMG, high mobility group; TCR, T-cell receptor.

Source: From Macleod et al.[105]

other genes. The v-Ets protein differs from c-Ets in that v-Ets has a different C-terminus that appears to alter its DNA-binding affinity and make it a transforming protein. Since Ets-1 and Ets-2 bind to the promoters of the stromelysin-1, collagenase, and urokinase plasminogen activator genes, it has been postulated that Ets proteins play an important role in modulating degradation of the extracellular matrix and that this may play a role in cancer metastasis.[105]

As for other oncogenes, the transforming ability of v-*ets* is complemented by other oncogenes. For example, v-*ets* and v-*myb* coexpression results in an increased transformation of erythroid cells compared to either gene individually.[106] The human *ets*-1 and *ets*-2 genes are translocated in some forms of acute leukemia; both t(11;4) and t(21;8) translocations have been observed, suggesting that these genes may be deregulated by these gene rearrangements.

bcr-abl

The v-*abl* oncogene was identified as the transforming gene of the Abelson murine leukemia virus and was shown to have tyrosine kinase activity.[107] In comparison to its cellular homologue c-*abl*, the v-*abl* kinase activity is a deregulated chimeric protein.

An intriguing discovery led to our understanding of the role of the c-*abl* tyrosine kinase in human cancer when it was observed that the c-*abl* gene was translocated from its normal position on chromosome 9 into a sequence called *bcr* (breakpoint cluster region) on chromosome 22, producing the Philadelphia chromosome seen in chronic myelogenous leukemia (CML) and some forms of acute lymphocyte leukemia (ALL) in humans (reviewed in Refs. 108 and 109). The hybrid *bcr/abl* gene of the Philadelphia chromosome produces a hybrid 210-kDa phosphoprotein in CML cells and a 185-kDa phosphoprotein in ALL cells. Both of these proteins have the same c-Abl component, in which exon 1 of the c-*abl* gene has been lost, but differ in the Bcr component because the 185 kDa form results from a breakpoint further downstream, resulting in the loss of certain exons of *bcr*. Both Bcr/Abl forms have deregulated tyrosine kinase activity and both forms cause hematopoietic malignancies when placed as transgenes into transgenic mice, although the 185-kDa form tends to produce lymphoid malignancies[110] and the 210-kDa form favors production of myeloid leukemias[109] in the recipient mice.

Since both the p210 and p185 Bcr/Abl proteins as well as the p160 transforming v-Abl protein have lost the exon 1 component of the c-*abl* gene product, it was assumed that this was all that was essential for their transforming activity. But this turned out not to be the case. Additional key changes in the transforming proteins are the substitution of the Bcr sequence in place of the exon 1 component and mutation or deletion of parts of the SH3 domain of the Abl protein. This alteration has been shown to upregulate the Abl tyrosine kinase activity.[108] Overexpression of c-Abl itself can also lead to excess tyrosine kinase activity in cells, but this does not seem to be a factor in human cancer. As a matter of fact, overexpression of c-*abl* in NIH 3T3 cells inhibits growth by causing cell cycle arrest, suggesting that c-*abl* has itself a tumor suppressor function like RB and p53.[110a]

The deletion or mutation of regulatory components of oncogene products are by now, no doubt, becoming familiar to the reader, and this is indeed another example. There is a cellular inhibitor that interacts with normal c-Abl to regulate its activity. The ability of this normal regulator to bind to c-Abl is inhibited by the substitution of the Bcr component for the exon 1 component of c-Abl. Some of the details of this loss of a regulatory mechanism have now been worked out.[108,111] Amino acid sequences within the first exon of Bcr activate the transforming potential of Bcr/Abl. Bcr/Abl forms complexes with our old friend GRB2, which binds to Bcr/Abl via its SH2 domain by interacting with a sequence containing a phosphorylated tyrosine at position 177 in the Bcr first exon. This Bcr/Abl-GRB2 interaction activates the Ras signal transduction pathway described above. If tyrosine-177 of Bcr is mutated to phenylalanine, the binding of Bcr/Abl to GRB2 is blocked, the Ras pathway is not activated, and the ability of Bcr/Abl to transform primary bone marrow cultures is abrogated.[111] Thus, it appears that the normal regulatory mechanism for the c-Abl tyrosine kinase, and presumably its protein substrates, is substituted by one that activates the Ras system to an inappropriate degree, leading to stimulation of the kinase cascade favoring cell proliferation.

myb

As noted above, the E26 chicken acute transforming virus contains a fusion protein of two transforming oncogenes, v-*ets* and v-*myb*. v-*myb* is also found in avian myeloblastosis virus (AMV). Both of these retroviruses block monocyte-macrophage differentiation of infected cells. These viruses encode a truncated version of c-Myb, the normal protooncogene, that is a highly conserved nuclear phosphoprotein involved in the differentiation of hematopoietic cells (reviewed in Ref. 112). c-Myb is expressed in immature hematopoietic cell lineages and downregulated during terminal differentiation of blood-forming cells. Constitutive expression of c-Myb blocks this differentiation event. Both c-Myb and v-Myb act as transcription factors, but v-Myb has lost a phosphorylation site for the nuclear kinase CK-II, which site when phosphorylated in c-Myb inhibits its binding to DNA.[112] Expression of c-Myb is important for the response of T-lymphocytes to PHA and antigenic stimulation, and it appears to be deregulated by overexpression in human T-cell leukemia since down-regulation of its expression by antisense oligonucleotides blocks DNA synthesis in T-leukemia cells taken from patients.[113]

bcl-2

The *bcl*-2 protooncogene is activated by a common chromosomal translocation observed in non-Hodgkin's B-cell lymphomas, namely the t(14;18)(q32;q21) translocation. This event juxtaposes the *bcl*-2 gene (so called because it was identified in B-cell lymphomas) from chromosomal locus 18q21 with the immunoglobin heavy chain (IGH) locus at 14q32, resulting in increased expression of the *bcl*-2 gene.[114] Thus, activation of this gene is similar to that of *myc* gene activation in Burkitt's lymphoma in that the abnormally high levels of expression result from the placement of the gene under the influence of the IGH enhancer. Among oncogenes, *bcl*-2 has a unique action in that it enhances lymphoid cell survival by inhibiting programmed cell death (apoptosis) (see below) rather than stimulating cell proliferation. Thus, its primary action appears to allow B cells to accumulate by prolonging their survival.

The *bcl*-2 gene undergoes translocation in approximately 85% of follicular lymphomas, 20% of diffuse large-cell lymphomas, and 10% of B-cell chronic lymphocytic leukemias.[114] Experiments with transgenic mice suggested that B cells are the primary target for abnormal expression of this gene since the most pronounced effect in the mice receiving the *bcl*-2 transgene was clonal expansion of B cells, with a lesser expansion of the T-cell pool.[115] However, a survey of other tissue indicates that Bcl-2 is expressed in a variety of cell lineages, such as gastrointestinal epithelium, skin, and developing nervous system tissue (primarily in the stem cell or proliferating compartments of these tissues), suggesting a broader role for Bcl-2 in sustaining the progenitor cells of various cell lineages (reviewed in Ref. 116).

The *bcl*-2 gene encodes two proteins, one of 26 kDa called Bcl-2α and one of 22 kDa called Bcl-2β, which result from alternative splicing of the mRNA.[114] Other *bcl*-2-related genes have also been identified, *bcl*-X_S and bcl-X_L, which can render cells resistant to apoptosis and prevent *bcl*-2 overexpression from preventing apoptosis, respectively.[117] Thus, there are both positive and negative regulators of Bcl-2 function. Another point that should be made is that not all mechanisms involved in induction of apoptosis can be prevented by Bcl-2. For example, Bcl-2 blocks apoptosis in hematopoietic cell lines deprived of IL-3, IL-4, or GM-CSF as growth factors but not in cell lines dependent on IL-2 or IL-6,[118] and Bcl-2 can prevent apoptosis in embryonic neurons deprived of NGF but not those dependent on another growth factor.[119] These data suggest that there are multiple mechanisms for inducing and/or preventing apoptosis. One way to explain this is that cells may have a variety of mechanisms to regulate *bcl*-2 gene expression or Bcl-2 protein function. The *bcl*-X_L gene, noted above, and a similar gene product called Bax[116] appear to be such regulators of Bcl-2 function. While the mechanism by which Bcl-2 prevents apoptosis is not clear, the observation that its overexpression prevented cell death in irradiated, *bcl*-2-transfected lymphoid cell lines and suppressed lipid peroxidation in these cells suggests that Bcl-2 stimulates an antioxidant pathway at sites of oxygen free radical generation.[120]

NF-κB/*rel*

NF-κB was originally identified as a nuclear protein that is bound to the κB site in the immunoglobin enhancer in B-lymphoid cells.[121] It was later found that NF-κB is involved in the regulation of a large number of genes in different cell types (reviewed in Ref. 122). NF-κB was the first oncogene transcription factor whose functional regulation was found to depend on its cellular localization rather than its level of transcription. NF-κB is held in an inactive form in the cell cytosol in a complex with an inhibitor protein IκB. After treatment of cells with a variety of agents that induce NF-κB activation—including phorbol ester, lipopolysaccharide, tumor necrosis factor α, double-stranded DNA, and IL-1—the IκB protein is phosphorylated by a cytosolic kinase. This phosphorylation dissociates IκB from NF-κB, releasing the latter, which is then translocated to the nucleus via a now exposed translocation signal. Once in the nucleus, NF-κB can bind to its regulatory DNA sequences and induce the transcription of several genes including cytokines, cytokine receptors, MHC antigens, serum amyloid A protein, and viral gene expression of HIV-1, cytomegalovirus, and SV40.[122]

NF-κB consists of two heterodimeric proteins of 50 kDa (p50) and 65 kDa (p65). The p50 subunit contains the DNA binding site, and p65 is required for binding to IκB. Interestingly, it was later found that NF-κB has high sequence homology with the protooncogene c-*rel* and its viral counterpart v-*rel* as well as with the gene *dorsal* in *Drosophila*.[122,123] Part of this homology includes DNA-binding and dimerization domains, and the dimerization of p50/p65 is apparently required for gene activation.[124]

The oncogene v-*rel* is carried by the reticuloendotheliosis virus strain T, which causes acute leukemia in turkeys. The transforming and immortalizing properties of v-*rel* are related to some small deletions and 14 amino acid substitutions that distinguish it from c-*rel*. The protooncogene c-*rel* has been cloned from turkey, mouse, and human cells, and it appears to require cytosol-to-nucleus translocation to be active, just as NF-κB does.

It is now known that the NF-κB/Rel proteins are a family of transcription factors, whose members share homologous DNA-binding and dimerization domains. In lymphoid tissues of transgenic mice, the NF-κB p50/Rel B protein heterodimer constitutively activates a reporter gene, whereas the NF-κB p50/p65 heterodimer activates the reporter gene in mouse embryo fibroblasts, suggesting that different members of the NF-κB/Rel family of transcriptional activators are involved in gene activation in a tissue-specific manner.[124]

*erb*A

Yet another functional type of oncogene is represented by the *erb*A gene, originally identified as the v-*erb*A gene in avian erthryoblastosis virus (AEV), which induces erythroleukemias and fibrosarcomas in chickens.[125] AEV carries two oncogenes, v-*erb*A and v-*erb*B, that cooperate in the transforming action of AEV. The v-*erb*B gene product turned out to be a truncated version of the EGF receptor with tyrosine kinase activity (see below). In contrast, the v-*erb* A gene product was found to be an altered version of the thyroid hormone receptor family of DNA binding proteins.[126] Other members of this family include the receptors for steroid hormones, retinoic acid, and vitamin D3, all of which have ligand-binding, DNA-binding, and transactivating domains. Though there is some sequence homology in the DNA-binding domains and some overlap in their consensus sequence binding sites, all these receptors bind to specific response elements, called T3 response element (T3RE), glucocorticoid response element (GRE), retinoic acid response element (RARE), or vitamin D3 response element. These hormone receptors can either stimulate or inhibit gene expression, depending on the cell type and the response element involved. Thus, they are "content-dependent" in their action.

The v-*erb*A oncogene is a highly mutated version of its cellular homologue c-*erb*A, having truncations at both the *N*- and *C*-termini and several scattered point mutations. This results in loss of binding of thyroid hormone T3, but it retains its DNA-binding capacity.

Two mechanisms of transforming action have been proposed for v-*erb*A. One is that it acts as a dominant-negative repressor of ligand hormone receptors such as T3 receptor or retinoic acid receptor, which can act as growth slowing, differentiating agents in certain cell types. Evidence for this comes from the observation that, when T3 receptor and v-ErbA are coexpressed in the same cells, v-ErbA functions as an antagonist of T3 action.[126] A second proposed action for the transforming ability of v-*erb*A is based on data showing that the v-ErbA oncoprotein blocks a retinoic acid receptor–induced function in slowing cell proliferation.[127,128] This appears to occur via repression of AP-1 transcriptional activation. Recall that AP-1 is the Fos/Jun transcriptional activation complex that is turned on by a number of mitogenic signals via the Ras-Raf-Mek-MAP kinase pathway. Thus, RA, by binding with its receptor, shuts this mechanism off by interaction with the AP-1 complex, inhibiting cell proliferation, and v-ErbA reverses this by blocking the ability of RAR to carry this out.[127]

sis

The *sis* oncogene encodes an oncoprotein that mimics a growth factor. We now consider this and other oncoproteins that mimic growth factors or growth factor receptors.

In July 1983, two startling papers appeared, one in *Science*[129] and the other in *Nature*,[130] proposing a direct link between growth-regulating factors and oncogene products. These reports indicated for the first time how an oncogene product could directly stimulate cell proliferation.

As described in Chapter 9, platelet-derived growth factor (PDGF) is made up of dimers of two distinct chains, PDGF-A and PDGF-B, with subunits of about 14,000 to 17,000 Da; it is derived from platelets and is a potent mitogen for connective tissue and glial cells in culture. The amino acid sequence of PDGF was reported in May 1983 by Antoniades and Hunkapiller.[131] Russell Doolittle, who was establishing a computer bank of known protein sequences, plugged in the data on PDGF soon after reading the report of its sequence. What fell out was the sequence of the simian sarcoma virus (*sis*) oncogene, which had previously been sequenced by Aaronson and his colleagues.[132] The homology was strong: 87% of 70 amino acids in the sequence of PDGF-B were homologous to the *sis* oncogene product, and what discrepancies there were could be explained by expected species variation, since PDGF was isolated from outdated human platelet preparations, and simian sarcoma virus (SSV) was isolated from a fibrosarcoma of a woolly monkey. The cloned transforming gene of SSV, v-*sis*, is known to produce a 28,000-Da gene product in transformed cells.[131] Similarly, Waterfield et al.,[130] who were also working on the sequence of PDGF, found a region of 104 contiguous amino acids virtually identical to the predicted sequence of $p28^{sis}$, the 28,000-Da protein isolated from cells transformed by the cloned transforming gene v-*sis* of SSV. This led to the speculation that continued production of this growth factor by v-*sis*-transformed cells could account for the malignant phenotype induced by this gene.

Since the cellular homologue (c-*sis*) of the viral *sis* oncogene is present in the human genome as a single gene, in all likelihood SSV or one of its ancestors picked up the normal cellular gene coding for PDGF. Because PDGF primarily stimulates the proliferation of cells of connective tissue origin—such as fibroblasts, smooth muscle cells, and glial cells—it was logical to look for expression of *sis* in cell lines derived from cancers of connective tissue cells. It was found that a PDGF-like product is often produced by tumor cells of connective tissue origin, whereas cancers derived from epithelial cells usually do not make it. Furthermore, cell lysates and conditioned medium of SSV-transformed cells growing in culture contain a PDGF-like mitogenic factor[133,134] that can be partially neutralized by anti-PDGF antibodies.[133] The PDGF-like material produced by SSV-transformed cells binds to cells with PDGF receptors in a manner competitive with PDGF, and the ability of SSV-transformed cells to grow in nude mice correlates with the production of $p28^{sis}$ by the cells growing in culture.[134] Anti-PDGF antibodies also inhibit the growth of high-PDGF-producing SSV-transformed cell lines.[134] Interestingly, SSV-induced tumors appear to be restricted to the cell types that have PDGF receptors (e.g., gliomas or fibrosarcomas).[135]

The v-*sis* gene actually encodes a 271 amino acid protein whose *N*-terminal 51 amino acids are derived from the viral envelope protein and the remainder is derived from c-*sis*, the cellular homologue of v-*sis*. It is the c-*sis* gene that encodes a polypeptide precursor of the B chain of PDGF. The production of the v-*sis* gene product is now known to be more complicated than originally thought (reviewed in Ref. 136). In SSV-transformed cells, the v-*sis* gene product is synthesized as a 36-kDa glycoprotein with one N-linked oligosaccharide chain. It then forms a 72-kDa dimer that is proteolytically processed sequentially into p68, p58, and p44 forms, the last of which is secreted but most of which remains bound to the cell surface. Part of the p44 form is cleaved into a 27-kDa form, which most likely accounts for the earlier observation of a product of about 28 kDa being secreted by SSV infected cells. The high affinity of the secreted v-*sis* gene product for the cell surface suggests a way in which autocrine stimulation of v-*sis*–transformed cells could occur, i.e., by release of the v-*sis* protein and immediate binding to cell surface PDGF receptors. However, other data suggest that the "autocrine loop" is not extracellular but intracellular, in that a p27 form is also generated intracellularly, and it has been postulated that this form binds nascent PDGF receptors, creating a signal-transducing signal without ever having to exit the cell.[136]

*erb*B

A second link between oncogenes and growth factors came from studies of the structure of the EGF receptor. Downward et al.[137] reported that the amino acid sequence of six peptides derived from EGF receptor isolated by immunoaffinity purification from cultured human epidermoid carcinoma A431 cells and from human placenta was identical, at 74 of 83 residues sequenced, to the transforming protein of the v-*erb*B oncogene of avian erythroblastosis virus (AEV). However, the ErbB sequence was missing a large segment of the amino terminal end of the EGF receptor. It is now known that the v-erbB oncogene encodes a truncated EGF receptor, containing only the transmembrane hydrophobic region and the cytoplasmic tyrosine kinase domain but not the cell surface domain associated with ligand binding. These results suggested that the v-*erb*B oncogene could transform cells through an uncontrolled receptor function in which, even in the absence of ligand binding, a constitutive expression of receptor function could occur. It has been difficult, however, to demonstrate a chronic, constitutive activation of EGF receptors in AEV-transformed cells, partly because the tyrosine kinase activity of the *erb*B gene product appears to be quite low.[135] It is possible that a very specific subset of substrates are phosphorylated by ErbB kinase activity and that they are difficult to detect; it is also possible that the transforming activity of v-*erb*B is pronounced only in the presence of other activated oncogenes.

*erb*B-2 (HER-2/*neu*)

The *neu* oncogene was initially identified in rat neuroblastomas and a human homologue called c-*erb*B-2 (also called Her-2) was later found to be amplified in some human adenocarcinomas (reviewed in Ref. 138). ErbB-2 is similar to the ErbB oncoprotein discussed above in that it is an altered, truncated version of the EGF receptor and has intrinsic tyrosine kinase activity that can carry out autophosphorylation as well as other phosphorylation steps. A difference is that ErbB is a 170-kDa protein while ErbB-2 is a 185-kDa protein. A single amino acid change ($Val^{664} \rightarrow Glu^{664}$) in the transforming rat *neu* gene (compared to its normal c-*neu* counterpart) significantly increases its autophosphorylating protein kinase activity and turns it into a potent transforming gene.[139] The transforming potential of the *neu* oncogene can also be activated by overexpression due to gene amplification or deletion of part of the N-terminal extracellular domain.[138–140] It is noteworthy that overexpression of c-*neu* and the EGF receptor in rodent fibroblast lines act synergistically to induce transformation, whereas overexpression of either gene alone does not do it or does it weakly, suggesting that overexpression of two normal cellular kinases can activate the neoplastic transformation process.[138]

Amplification of the Her-2/*neu* gene has been implicated as a factor in the progression of human cancer, particularly breast cancer, and in transgenic mice, expression of *neu* driven by a mouse mammary tumor virus (MMTV) promoter induces mammary tumors.[141,142] While increased tyrosine kinase activity is the presumed mechanism for the transformation of cells by the *neu* oncogene, it is interesting that human breast epithelial cells transformed by transfection with *neu* exhibited an increase in specific protein tyrosine phosphatases.[140] This was likely due to a cellular compensatory response to an increased protein tyrosine phosphate "load." Indeed, there is evidence to indicate that increased tyrosine phosphatase activity (particularly for two called LAR and PTIB) can counteract the transforming potential of tyrosine kinase oncogenes and may act as tumor suppressor genes. In support of this idea are the findings that the colorectal tumor suppressor gene product DCC has some structural homology to LAR[143] and that the *LAR* gene maps to a region on chromosome 1p32–33 that is thought to contain a breast cancer tumor suppressor gene.[144]

For human breast cancer patients, a correlation of relapse and poor survival with amplification of the HER-2/*neu* oncogene has been observed.[145] There is some controversy, however, whether this holds for node-negative patients and even how significant a prognostic marker it is for node-positive patients.[146] More recent data, obtained using a sensitive immunostaining technique for expression of the HER-2/Neu protein (rather than the Southern blot analysis for amplified DNA used in the earlier studies), indicated that overexpression is an independent indicator of increased risk of developing recurrent disease in women with node-negative breast cancer.[147]

Overexpression of HER-2/Neu, determined by immunohistochemical staining of tissues, has also been shown to be associated with poor survival in advanced ovarian cancer[148] and gastric cancer.[149] Overexpression of HER-2/*neu* mRNA, as determined by Northern blot analysis, has also been shown to be a marker for intrinsic drug resistance in non-small cell lung carcinoma cell lines.[150]

A number of other oncogenes that have growth factor or growth factor receptor actions have been discovered. It is likely that many more will be. Indeed, whenever there is a mutation, translocation, amplification, or other means of overexpression, it is possible, perhaps even highly likely, that control of cell proliferation will be deregulated, leading to a hyperplastic proliferation of cells with the concomitant increased chance of malignancy due to the genetic instability that may follow rapid cell division. A summary of characteristics of some growth factor (GF) or GF receptor–like oncogenes is listed below.

fms

The v-*fms* oncogene is contained in feline sarcoma viruses. It and its c-*fms* protoncogene counterpart represent different forms of the hematopoietic colony stimulating factor M-CSF receptor. c-*fms* codes for the normal receptor and the v-*fms* protein product has scattered point mutations and deletions and substitutions in the C-terminus, which activate its transforming, tyrosine kinase activity.[151]

kit

The c-*kit* protooncogene is the normal cellular homologue of v-*kit* found in the H-Z4 feline sarcoma virus. It codes for a transmembrane tyrosine kinase and is the receptor for stem cell factor (SCF). c-*kit* plays a key role in hematopoiesis.[152] Its role in human cancer remains to be defined.

trk

This oncogene was first discovered in a human colon carcinoma biopsy. It is also a transmembrane tyrosine kinase. It becomes activated by chromosomal rearrangement, resulting in replacement of its extracellular domain by unrelated sequences or by other recombination events.[153] The product of the protooncogene c-*trk* is now known to be a crucial component of the nerve growth factor (NGF) receptor. High levels of c-*trk* expression and normal gene copy numbers of N-*myc* are associated with a favorable prognosis for patients with neuroblastoma.[154]

met

The *met* protooncogene encodes $p190^{met}$, a membrane-associated tyrosine kinase that is the receptor for hepatocyte growth factor (HGF), also known as scatter factor. When bound to its receptor, HGF stimulates cell motility, extracellular matrix invasion, and, in some cells, a cell proliferative response.[155] In response to HGF binding to p190^{met}, autophosphorylation, activation of the phosphoinositol pathway, and of Src-like kinases ensue. Although the function of *met* in human cancer is unclear, it may be postulated that inappropriate activation or overexpression of *met* is related to generation of an invasive and metastatic phenotype.

CELLULAR *onc* GENE EXPRESSION DURING NORMAL EMBRYONIC DEVELOPMENT

As previously noted, the ubiquitousness of c-*onc* genes in vertebrate organisms and their conservation through eons of evolution suggest an important role in growth and development of the normal organism. These genes were almost certainly conserved not because they can produce uncontrolled proliferation but because they play some key role in development. A variety of studies support this thesis.

The expression of eight cellular *onc* genes was examined during embryonic and fetal development of the mouse using four avian (v-*myc*, v-*erb*, v-*myb*, and v-*src*), two murine (v-*mos* and v-Ha-*ras*), one feline (v-*fes*), and one primate (v-*sis*) viral cDNA probes to detect homologous sequences in cellular mRNA from various stages of development.[156] Five homologous c-*onc* genes detected by these probes were expressed during embryonic development: c-*sis* expression peaked on about day 8 of prenatal development and continued to be expressed at lower levels throughout gestation; c-*myc*, c-*erb*A, and c-*src* expression peaked in the latter half of fetal development; and c-Ha-*ras* was expressed throughout embryonic development of the mouse. Although sequences homologous to v-*myb*, v-*mos*, and v-*fes* are contained in the mouse genome, transcription of these genes was not detected during development. In some cases, expression of certain c-*onc* genes is linked to the development of specific tissues. For example, c-*fos* expression has been detected in placenta, c-*abl* in developing male germ cells and lymphoid tissues, and c-Ha-*ras* in a variety of developing tissues including bone, brain, GI tract, kidney, lung, skin, spleen, testis, and thymus,[157] suggesting a protean role for this gene.

There is a spatial and temporal pattern to the expression of c-*myc* and c-*sis* in developing human placenta. Expression of both genes peaks in first-trimester placental tissue and declines thereafter, in parallel with the release of platelet-derived growth factor (PDGF) and expression of PDGF receptors on cytotrophoblast cells.[158] Both genes are expressed most abundantly in the cytotrophoblast, suggesting that human placenta has autocrine regulation, with the ability to both produce PDGF and respond to it by increasing expression of c-*myc*, the expression of which correlates with cell proliferation in a number of tissues.

Expression of other c-*onc* genes has also been linked to differentiation of specific cell types. c-*fos* and c-*fms* gene transcription is turned on in differentiating human monocytes.[159–161] In contrast, c-*myc* gene expression is decreased when cultured human promyelocytic leukemia cells (HL-60) are induced to differentiate in culture; when the stimulus for differentiation is removed, c-*myc* mRNA is elevated again,[162] suggesting that c-*myc* expression correlates with the proliferative phase and c-*fms* and c-*fos* expression to a later differentiating phase of monocyte development. Going along with this idea is the finding that in regenerating liver, induced to undergo rapid cell proliferation in response to partial hepatectomy, there is a rapid onset of c-*myc* transcription that increases 10- to 15-fold above the normal resting level within 1 to 3 hours after partial hepatectomy and rapidly declines after 4 hours posthepatectomy.[163] In the same experimental protocol, c-Ha-*ras* transcription has been shown to increase 12 hours after partial hepatectomy, peak at 36 hours, and return to control levels by 72 hours.[164]

Continued expression of *onc* genes, however, can block terminal differentiation of normal cells. Evidence for this is the fact that when the v-*src* gene under viral promoter control is introduced into cultures of mouse bone marrow cells, a dramatic increase occurs in the self-renewing stem cell (CFU-S) compartment, with a decrease in the appearance of mature granulocytes.[165] Similarly, normal mouse skin keratinocytes, when infected with Kirsten or Harvey sarcoma virus, do not progress through a complete maturation program when the v-*ras* gene is expressed.[166] When these cells are induced to differentiate by addition of calcium ions, they progress only to an early reversible stage of differentiation; if subsequently treated with the tumor-promoting phorbol ester TPA, such cultures revert back to a less mature cell type.

DNA TUMOR VIRUSES

The oncogenic DNA viruses consist of three main groups: papovaviruses, adenoviruses, and herpesviruses. Examples of papovaviruses are the papillomaviruses of rabbits (Shope) and other species including humans (human wart virus), SV40 virus of monkeys, and polyomavirus of the mouse. Adenoviruses have been isolated from various animal species, and a number of them have been shown to be tumorigenic in newborn animals. Oncogenic viruses of the herpesvirus class include Epstein-Barr virus, suspected of causing Burkitt's lymphoma and nasopharyngeal carcinoma in humans, the virus that causes Lucké frog renal carcinoma, and a leukemogenic virus in chickens (Marek's disease).

SV40 and Polyoma

The papovaviruses are small icosahedrons containing 3 to 5 $\times$ 10^6 Da of DNA, enough to code for three to six proteins. Of this group, SV40 and polyomavirus have been studied the most. The SV40 virus was discovered in 1960 in rhesus monkey kidney cell cultures used to produce the early polio vaccine.[167] The virus was inadvertently inoculated into thousands of people before its presence became known. Later it was shown that SV40 could produce tumors after injection into weanling hamsters[168] and that it could also transform human cells in culture.[169] However, no human disease, including cancer, has been shown to be caused by SV40 virus, even in this inadvertent human experiment. Since SV40 and the other papovaviruses contain DNA, the flow of genetic information goes directly from DNA to RNA to protein without requiring reverse transcriptase, as do the RNA viruses.

A number of things are known about the molecular biology of SV40 gene expression.[170] Similar molecular events occur during infection and cellular transformation with other papovaviruses. The virus enters the cell by the action of its coat proteins. Viral DNA then enters the nucleus of the cell, and it is transcribed in two "waves" to produce "early" and "late" mRNAs. Transcription of the early SV40 genes is required for synthesis of viral proteins that are involved in the replication of SV40 DNA. The early-region genes also contain the information needed for cell transformation and code for the intranuclear T antigen. Late mRNA is transcribed after viral DNA replication and codes for the viral structural proteins. The SV40 viral DNA is covalently integrated into transformed host cell DNA, and the integrated sequences can be portions as well as full copies of the SV40 genome. The DNA cleavage takes place before integration occurs at various nucleotide sequences in both the viral and host cell DNA, depending on the type of transformed host cell. Mature SV40 can be rescued by a variety of methods from many transformed cells that do not produce virus under usual culture conditions. Transformed cells produce early viral mRNA and T antigen but do not replicate viral DNA and do not produce late viral mRNA or late structural viral proteins. Thus, T antigen is required for initiation of viral DNA synthesis, for the accompanying induction of host cell DNA synthesis, and for both the establishment and maintenance of the transformed state. Because T antigen appears to be the transforming protein of the transforming papovaviruses, a considerable amount of research has gone into characterizing this protein in both SV40 and polyomavirus.

Originally, the T antigen of SV40 was thought to be one protein of about 100,000 MW. This would account for most of the coding capacity of the early gene region. Later it was shown that in vitro cell-free translation of early viral mRNA isolated from infected cells also produced a protein of 17,000 MW that was immunoprecipitable with antiserum to T antigen.[171] These forms are called large and small T antigen. The gene coding for both forms of T antigen is called the A gene. The two mRNAs that code for large and small T have the same 5′ and 3′ ends, and thus appear to have arisen from differential splicing of the A gene transcription product.[172] For transformation to occur after SV40 infection, expression of the large T antigen appears to be a crucial event.

The SV40 large T antigen in virus-infected

cells regulates not only SV40 gene transcription but also the transcription of cellular genes such as thymidine kinase, ribosomal RNA genes, and a whole subset of other cellular genes, the transcripts of which are elevated in SV40-transformed mouse cells.[173] Activation of cellular genes may be a general feature of oncogenesis induced by DNA viruses, as opposed to RNA oncogenic viruses, the latter of which carry their own activated transforming genes. However, under certain circumstances, SV40 viral sequences can transform cells. For example, when various segments of SV40 early region DNA are linked to a retroviral vector from Moloney murine leukemia virus (MoLV), containing only the LTR region and other regulatory sequences required for MoLV viral propagation, it has been shown that vectors carrying SV40 large T antigen as the only SV40 sequence can induce morphologic transformation of primary or established mouse and rat lines with high efficiency.[174] These latter authors thus argue that expression of large T antigen by itself is capable of transforming cells, and this conclusion is supported by experiments in transgenic mice. Brinster et al.[175] have microinjected fertilized mouse eggs with plasmids containing SV40 early-region genes and a fusion gene coding for metallothionein, known to be expressed in transgenic animals. SV40 T antigen mRNA was detected at high levels only in tissues showing histopathologic changes, including thymus, kidney, and brain, but the highest levels were seen in brain tumors (of the choroid plexus) that developed in these animals. Later experiments showed that large T antigen expression is sufficient to induce the choroid plexus tumors and does not require the metallothionein fusion gene.[176] In fact, when the SV40 enhancer region is present to direct T antigen expression, tumors specifically occur in the choroid plexus, but when the SV40 enhancer region is deleted and substituted by a metallothionein–human-growth-hormone fusion gene, an entirely different pattern of pathology ensues: transgenic mice bearing this hybrid gene develop peripheral neuropathies, hepatocellular carcinomas, and pancreatic islet cell adenomas.[177]

The DNA from SV40 T antigen fusion gene–induced tumors, compared with the DNA of unaffected tissues from the same animals, shows structural rearrangements, changes in DNA methylation patterns, and, frequently, SV40 gene amplification,[175] indicating that the enhancer or promoter sequence attached to the T antigen gene has a key role in directing the tissue specificity of T antigen expression and its tumorigenic potential, probably by directing how the T antigen gene is inserted into DNA—that is, by allowing it to be placed in an active, transcribable conformation. Similar results have been obtained by Hanahan,[178] who showed that transfer of recombinant genes made up of regulatory sequences of the insulin gene, fused with sequences of SV40 large T antigen, into fertilized mouse eggs produced tissue-specific expression of large T in β cells of the transgenic mice pancreases, inducing β-cell tumors in these animals. These data strongly suggest that tissue-specific expression of viral-transforming gene sequences, directed by enhancer or promoter elements that have a tissue-specific expression, can produce very specific target-cell oncogenesis.

The transforming T antigen of polyomavirus bears some similarity to that of SV40, but its crucial elements are coded for by a separate gene, the *hr-t* gene, which maps in a position analogous to that deleted in certain early gene deletion mutants of SV40.[172,179] A third type of T antigen has been isolated from polyoma-transformed cells.[180] This has been called middle T antigen and has a molecular weight of 55,000. It contains peptides not found in either small or large T antigen and is altered by *hr-t* gene deletions. Cells infected with mutant polyomaviruses that synthesize large but not middle T antigen have a normal phenotype and do not induce tumors in vivo, indicating that middle T antigen is more important for the transformation of polyoma-infected cells.

Transformation of normal rat cells with genetic recombinant plasmids, derived from polyomavirus, that allow selective expression of large T, middle T, or small T antigens indicates that middle T alone is sufficient to transform established lines (already immortalized), but not primary rat embryo fibroblasts, and that large T lacks intrinsic oncogenic activity but can decrease serum dependence of growth for both normal and transformed cells.[181] Polyoma large T antigen appears to increase the efficiency of transformation by polyoma, but fully transformed colonies can be obtained in the absence of active large T antigen.[182] Furthermore, a recombinant DNA clone consisting of a replication-defective murine leukemia virus vector, the polyoma early region promoter, and the middle-T gene can transform NIH/3T3 cells,[183] indicating that polyoma middle T antigen can act like a dominant transforming gene similar to RNA retrovirus v-*onc* genes in already immortalized cells. The transforming ability of middle T appears to relate to its phosphorylation state, since introduction of a mutation that inserts a phenylalanine for a tyrosine at residue 315 in middle T decreases phosphorylation as well as transforming activity.[184]

Table 7-5 Known T-Antigen/Cellular Protein Interactions

Cellular Protein	T Antigen
c-Src protein	Polyoma middle
c-Yes protein	Polyoma middle
c-Fyn protein	Polyoma middle
61-kDa protein (c-Src-like?)	Polyoma middle
RB protein	SV40 large Polyoma large
120-kDa protein (RB-like?)	SV40 large
p53 protein	SV40 large
85-kDa phosphatidyl inositol kinase	Polyoma middle
36-kDa protein (catalytic subunit type 2A protein phosphatase)	Polyoma middle Polyoma small SV40 small
63-kDa protein (regulatory subunit type 2A protein phosphatase)	Polyoma middle Polyoma small SV40 small
27- to 29-kDa protein(s)	Polyoma middle
DNA polymerase alpha	SV40 large
AP2 transcription factor	SV40 large
25- to 28-kDa transcription factor (octamer binding proteinlike?)	SV40 large
43-kDa transcription factor	SV40 large
73-kDa heat shock protein	Polyoma middle SV40 large
Tubulin	SV40 small SV40 large

Source: From Schreier and Gruber.[185]

Papovaviruses similar to SV40 and polyomavirus have been isolated from human patients.[170] These viruses (JC and BK) can also induce tumors in newborn hamsters and transform animal cells in culture. Their transforming ability also seems to depend on expression of T antigen.

The transforming large T antigen of SV40 and middle T antigen interact with a number of cellular proteins (Table 7-5).[185] These interactions are involved in the transforming activity of these viral antigens (reviewed in Ref. 185). For example, the binding of polyoma middle T antigen (TAg) to the c-Src protein increases its kinase activity about 20-fold, and middle TAg mutants lacking the ability to bind c-Src are transformation-deficient. Middle TAg also interacts with phosphoinositol kinase, another important component of the signal transduction system (see Chap. 9), and this association correlates with the ability of middle TAg to mediate transformation. SV40 large TAg binds to under- or nonphosphorylated RB protein, indicating a way that SV40 large TAg could prevent this tumor suppressor protein from blocking entry of cells into the cell division cycle.[186] SV40 large T also binds to the tumor suppressor protein p53, an action that also appears to mediate cell transformation events. Large TAg of SV40 also interacts with transcription factors such as AP-2, an action that appears to be involved in the turn-on of a gene for nucleic acid synthesis.

Papillomaviruses

Papillomaviruses also belong to the papovavirus family, but they are somewhat larger than SV40 or polyomaviruses and have a somewhat larger genome (5×10^6 Da).[187] They induce benign epithelial tumors in various animal species, including humans, and are sometimes known as "wart viruses." The skin and mucosal tumors induced in animals usually regress, but at least three papillomaviruses have oncogenic potential: the Shope papillomavirus, the bovine fibropapillomavirus, and the bovine alimentary tract papillomavirus. In humans, papillomaviruses are associated with skin warts, anal and genital warts (condylomata acuminata), and oral and laryngeal papillomas. Certain subtypes of human papillomaviruses are strongly associated with cervical carcinoma.

There are 67 distinct human papillomaviruses (HPVs). Of these, a subgroup of about 20 are associated with anogenital tract lesions. Some of them cause condyloma acuminata but are considered low-risk (e.g., HPV-6 and -11) because they rarely cause malignancy. Others (e.g., HPV-16, -18, -31, and -33) are considered of high risk because they are associated with high-grade squamous intraepithelial lesions and invasive carcinomas of the uterine cervix. Of the high-risk HPVs, types 16 and 18 have been most intensively studied. The HPVs express transforming oncoproteins called E6 and E7, and those of the most tumorigenic types (HPV-16 and -18) have potent cell-transforming actions. In primary cervical carcinomas and cervical cancer cell lines, the viral genomes of high-risk HPV types are frequently found integrated into the host cells' genome, allowing active transcription of the E6 and E7 mRNAs (reviewed in Ref. 188). However, although expression of E6 and E7 from high-risk HPVs can immortalize primary epithelial cells in culture, a fully transformed phenotype is observed only after numerous cell passages. Moreover, only a relatively low percentage of women infected with high-risk HPVs develop invasive cervical cancer, although a high percentage of cervical cancers are positive for HPV.[189]

These data indicate that other factors in addition to HPV infection are important in the causation of cervical cancer. While there are some additional associated epidemiologic risk factors,[189] what appears to happen at the level of the cell is a series of progressive events involving genetic instability of cells transformed by high-risk HPVs. One way this could happen is by association of the oncoproteins E6 and E7 with the tumor suppressor genes p53 and RB. E6 proteins translated from high-risk HPV-16 and -18 E6 genes bind to p53 and cause its degradation by an ubiquitin-mediated process.[190] Since normal p53 is involved in protecting cells from genetic damage from a variety of DNA-damaging agents (such as irradiation or chemicals) by causing cell-cycle arrest and allowing time for DNA repair (see Chap. 8), it seems logical that destruction or inactivation of p53 could account for the neoplastic progression seen with chronic HPV infection. It has, in fact, been shown directly that the HPV-16 E6 gene transfected into human cervical epithelial cells disrupts p53-mediated cellular response to DNA damage induced by actinomycin D.[188]

While both E6 and E7 have cell-transforming properties, expression of both is required for efficient immortalization of cells.[190] Thus, a one-two punch appears to be needed. The second punch is provided by E7's ability to bind to and disrupt the action of another tumor suppressor gene, *RB*, which is involved in cell-cycle regulation. In this regard E7 shares a property with SV40 T antigen (see above) and adenovirus oncoprotein E1A (see below). There are regions of amino acid sequence similarity between these three proteins that are involved in binding to RB.

It has also been shown that expression of HPV-16 E6 and E7 oncogenes in transgenic mice causes a high incidence of preneoplastic skin lesions and subsequent development of skin carcinomas.[191] Moreover, infection of nonmetastatic mouse tumor cell lines with a retrovirus bearing inserted E6 and E7 genes from HPV-16, but not HPV-6, converted these cells into metastatic ones.[192] This is consistent with the finding that HPV-16 DNA is frequently found in sites of cervical carcinoma metastasis.

Adenoviruses

The oncogenicity of adenoviruses was first observed by Trentin et al.[193] in 1962. They showed that adenovirus type 12 could produce tumors on inoculation into newborn hamsters. Of the 31 adenovirus serotypes isolated from humans, three (types 12, 18, and 31) are highly oncogenic in newborn rodents; five (types 3, 7, 14, 16, and 21) are less oncogenic, producing fewer tumors after a longer latent period; and other types (e.g., 1, 2, 5, and 6) do not induce tumors by direct inoculation into animals but can transform cultured rodent cells that produce tumors upon injection into animals.[194] As in the case of the papovaviruses, at least part of the adenovirus genome becomes incorporated into the host genome during transformation, and expression of a virus-induced nuclear T antigen is required for transformation. Another similarity is that the transcription of integrated viral DNA preferentially involves "early" sequences and correlates with production of the mRNA molecules detected in infected cells before the onset of viral DNA synthesis. Thus, the process of virus-induced transformation, involving transcription of early DNA into early mRNA, which, in turn, is translated into a T antigen involved in the initiation and maintenance of the transformed state, is common to the oncogenic papovaviruses and adenoviruses. The adenovirus E1A early-region gene can induce immortalization of cells in culture and act in concert with a transforming c-*ras* gene or the polyoma middle-T gene to transform cultured primary diploid cells.[195] The adenovirus E1 gene region encodes the E1A and E1B proteins responsible for the oncogenic properties of these viruses, although the E4 region of adenovirus 9 is involved in the production of mammary fibroadenomas, as shown in mice infected with a recombinant virus containing the E4 gene region of that virus.[196] Expression of the E1A region alone can immortalize primary cultures of rodent cells, but coexpression of E1B is required for complete transformation. Activated Ha-*ras* or polyoma middle T antigen can substitute for E1B to complement E1A in transformation assays, and polyoma large TAg, members of the *myc* family, or mutated p53 can replace E1A to complement E1B in similar assays (reviewed in Ref. 197).

Cellular targets for the E1A and E1B proteins have been identified. E1A binds to and inactivates RB, and E1B complexes with and disrupts the action of p53. Whyte et al.[197] have shown that the regions of the E1A gene product that bind RB are precisely the ones required for E1A-mediated cell transformation, strongly suggesting that inactivation of RB by E1A accounts in a crucial way for the cell transforming activity of E1A. However, E1A is a multifunctional protein: it acts as a transcriptional activator for a number of genes, stimulates DNA synthesis, and induces the production of an epithelial cell growth factor. Hence, its biochemical effects on cells are multifactorial, and a number of these actions could be involved in the loss of growth control seen in E1A-expressing cells.

Hepatitis B Virus (HBV)

Human HBV infects live cells and causes acute and chronic hepatitis. Chronic HBV infection is a high risk factor for developing hepatocellular carcinoma. The small DNA genome of this virus encodes four genes. The product of a gene called HBVx codes for a protein, pX, that is a transcriptional activator of viral and cellular genes, including N-*myc* and NF-κB.[198,199]

The pX protein itself does not appear to be able to bind DNA directly but acts via complex formation with the transcription factors CREB and ATF-2.[199] This action as a component of a transcriptional activation event may account, in part, for the transforming ability of HBV.

Herpesviruses

The other class of DNA viruses with oncogenic potential are the herpesviruses.[200] These viruses are larger than the papovaviruses and adenoviruses and have a genome that contains information for at least 50 proteins. Hence, discerning which of these gene products is the transforming protein(s) has been difficult. Herpesviruses infect humans and nearly all animal species investigated so far. Humans are subject to infection with five viruses of this class: herpes simplex virus 1 (HSV-1), herpes simplex virus 2 (HSV-2), herpes zoster virus (HZV), cytomegalovirus (CMV), and Epstein-Barr virus (EBV). Human herpesviruses have been under suspicion for some time as causative agents for certain cancers: EBV has been implicated as the responsible agent in Burkitt's lymphoma and nasopharyngeal carcinoma, and HSV-1 and HSV-2 have been suspected as contributing to the cause of cancer of the uterine cervix and possibly of other urogenital and oropharyngeal tumors.

Epstein-Barr virus can immortalize B lymphoid cells in culture, and, in so doing, expresses a variety of EBV-determined nuclear antigens (EBNA 1–6). The EBNA-2 protein is involved in the immortalization of B lymphocytes and is localized in the cell nucleus, where it functions as a transcription factor to enhance the expression of several viral and host genes. Expression of the EBNA-2 protein blocks the antiproliferative effect of α-interferon on B cells (reviewed in Ref. 201). Interferons may be acting as tumor suppressor genes for B cells, an action which is overcome by EBNA-2.

REFERENCES

1. H. M. Temin: The DNA provirus hypothesis: The establishment and implications of RNA-directed DNA synthesis. *Science* 192:1075, 1976.
2. D. Baltimore: Viruses, polymerases, and cancer. *Science* 192:632, 1976.
3. H. M. Temin and S. Mizutani: RNA-dependent DNA polymerase in virions of Rous sarcoma virus. *Nature* 226:1211, 1970.
4. D. Baltimore: Viral RNA-dependent DNA polymerase. *Nature* 226:1209, 1970.

5. P. E. Neiman: Rous sarcoma virus nucleotide sequences in cellular DNA: Measurement by RNA-DNA hybridization. *Science* 178:750, 1972.
6. M. Hill and J. Hillova: Virus recovery in chicken cells tested with Rous sarcoma cell DNA. *Nature New Biol* 237:35, 1972.
7. P. H. Duesberg and P. K. Vogt: RNA species obtained from clonal lines of avian sarcoma and from avian leukosis virus. *Virology* 54:207, 1973.
8. D. Stehelin, R. V. Guntaka, H. E. Varmus, and J. M. Bishop: Purification of DNA complementary to nucleotide sequences required for neoplastic transformation of fibroblasts by avian sarcoma viruses. *J Mol Biol* 101:349, 1976.
9. D. H. Spector, K. Smith, T. Padgett, P. McCombe, D. Roulland-Dussoix, C. Moscovici, H. E. Varmus, and J. M. Bishop: Uninfected avian cells contain RNA related to the transforming gene of avian sarcoma viruses. *Cell* 13:371, 1978.
10. R. A. Krzyzek, A. F. Lau, A. J. Faras, and D. H. Spector: Post-transcriptional control of avian oncornavirus transforming gene sequences in mammalian cells. *Nature* 269:175, 1977.
11. J. S. Brugge and R. L. Erikson: Identification of transformation-specific antigen induced by an avian sarcoma virus. *Nature* 269:346, 1977.
12. A. F. Purchio, E. Erikson, J. S. Brugge, and R. L. Erikson: Identification of a polypeptide encoded by the avian sarcoma virus *src* gene. *Proc Natl Acad Sci USA* 75:1567, 1978.
13. M. S. Collett and R. L. Erikson: Protein kinase activity associated with the avian sarcoma virus *src* gene product. *Proc Natl Acad Sci USA* 75:2021, 1978.
14. A. D. Levinson, H. Oppermann, L. Levintow, H. E. Varmus, and J. M. Bishop: Evidence that the transforming gene of avian sarcoma virus encodes a protein kinase associated with a phosphoprotein. *Cell* 15561, 1978.
15. R. L. Erikson, M. S. Collett, E. Erikson, and A. F. Purchio: Evidence that the avian sarcoma virus transforming gene product is a cyclic AMP-independent protein kinase. *Proc Natl Acad Sci USA* 76:6260, 1979.
16. T. Hunter and B. M. Sefton: Transforming gene product of Rous sarcoma virus phosphorylates tyrosine. *Proc Natl Acad Sci USA* 77:1311, 1980.
17. G. E. Mark and U. R. Rapp: Primary structure of v-raf: Relatedness to the *src* family of oncogenes. *Science* 224:285, 1984.
18. P. Madaule and R. Axel: A novel *ras*-related gene family. *Cell* 41:31, 1985.
19. R. W. Ellis, D. De Feo, T. Y. Shih, M. A. Gonda, H. A. Young, N. Tsuchida, D. R. Lowry, and E. M. Scolnick: The p21 *src* genes of Harvey and Kirsten sarcoma viruses originate from divergent members of a family of normal vertebrate genes. *Nature* 292:506, 1981.
20. D. De Feo-Jones, E. M. Scolnick, R. Koller, and R. Dhar: *ras*-Related gene sequences identified and isolated from *Saccharomyces cerevisiae. Nature* 306:707, 1983.
21. K. Tachell, D. T. Chaleff, D. De Feo-Jones, and E. M. Scolnick: Requirement of either of a pair of *ras*-related genes of *Saccharomyces cerevisiae* for spore viability. *Nature* 309:523, 1984.
22. T. Kataoka, S. Powers, C. McGill, O. Fasano, J. Strathern, M. Broach, and M. Wigler: Genetic analysis of yeast RAS 1 and RAS 2 genes. *Cell* 37:437, 1984.
23. T. Kataoka, S. Powers, S. Cameron, O. Fasano, M. Goldfarb, J. Broach, and M. Wigler: Functional homology of mammalian and yeast RAS genes. *Cell* 40:19, 1985.
24. T. Toda, I. Uno, T. Ishikawa, S. Powers, T. Kataoka, D. Broek, S. Cameron, J. Broach, K. Matsumoto, and M. Wigler: In yeast, RAS proteins are controlling elements of adenylate cyclase. *Cell* 40:27, 1985.
25. D. De Feo-Jones, K. Tatchell, L. C. Robinson, I. S. Sigal, W. C. Vass, D. R. Lowy, and E. M. Scolnick: Mammalian and yeast *ras* gene products: Biological function in their heterologous systems. *Science* 228:179, 1985.
26. G. L. Temeles, J. B. Gibbs, J. S. D'Alonzo, I. S. Sigal, and E. M. Scolnick: Yeast and mammalian *ras* proteins have conserved biochemical properties. *Nature* 313:700, 1985.
27. R. A. Weinberg: A molecular basis of cancer. *Sci Am* 249:126, 1983.
28. G. M. Cooper: Cellular transforming genes. *Science* 218:801, 1982.
29. O. T. Avery, C. M. Mac Leod, and M. McCarty: Studies on the chemical nature of the substance inducing transformation of pneumococcal types: Induction of transformation by a sesoxyribonucleic acid fraction isolated from *Pneumococcus* Type III. *J Exp Med* 79:137, 1944.
30. C. Shih, B. Shilo, M. P. Goldfarb, A. Dannenberg, and R. A. Weinberg: Passage of phenotypes of chemically transformed cells via transfection of DNA and chromatin. *Proc Natl Acad Sci USA* 76:5714, 1979.
31. C. Shih, L. C. Padhy, M. J. Murray, and R. A. Winberg: Transforming genes of carcinomas and neuroblastomas introduced into mouse fibroblasts. *Nature* 290:261, 1981.
32. C. S. Cooper, M. Park, D. G. Blain, M. A. Tainsky, K. Huebner, C. M. Croce, and G. F. Vande Woude: Molecular cloning of a new transforming gene from a chemically transformed human cell line. *Nature* 311:29, 1984.
33. C. Shih and R. A. Winberg: Isolation of a transforming sequence from a human bladder carcinoma cell line. *Cell* 29:161, 1982.
34. G. Goubin, D. S. Goldman, J. Luce, P. E. Neiman, and G. M. Cooper: Molecular cloning and nucleotide sequence of a transforming gene detected by transfection of chicken B-cell lymphoma DNA. *Nature* 302:114, 1983.
35. M. Goldfarb, K. Shimizu, M. Prucho, and M.

Wigler: Isolation and preliminary characterization of a human transforming gene from T24 bladder carcinoma cells. *Nature* 296:404, 1982.
36. B. Z. Shilo and R. A. Weinberg: Unique transforming gene in carcinogen-transformed mouse cells. *Nature* 289:607, 1981.
37. L. F. Parada, C. J. Taabin, C. Shih, and R. A. Weinberg: Human EJ bladder carcinoma oncogene is homologue of Harvey sarcoma virus *ras* gene. *Nature* 297:474, 1982.
38. C. J. Der, T. G. Krontiris, and G. M. Cooper: Transforming genes of human bladder and lung carcinoma cell lines are homologous to the *ras* genes of Harvey and Kirsten sarcoma viruses. *Proc Natl Acad Sci USA* 79:3637, 1982.
39. E. Santos, S. R. Tronick, S. A. Aaronson, S. Pulciani, and M. Barbacid: T24 human bladder carcinoma oncogene is an activated form of the normal human homologue of BALB- and Harvey-MSV transforming genes. *Nature* 298:343, 1982.
40. S. Pulciani, E. Santos, A. V. Lauver, L. K. Long, S. A. Aaronson, and M. Barbacid: Oncogenes in solid human tumours. *Nature* 300:539, 1982.
41. K. Shimizu, M. Goldfarb, Y. Suard, M. Perucho, Y. Li, T. Kamata, J. Feramisco, E. Stavnezer, J. Fogh, and M. H. Wigler: Three human transforming genes are related to the viral *ras* ocnogenes. *Proc Natl Acad Sci USA* 80:2112, 1983.
42. E. H. Chang, M. E. Furth, E. M. Scolnick, and D. R. Lowry: Tumorigenic transformation of mammalian cells induced by a normal human gene homologous to the oncogene of Harvey murine sarcoma virus. *Nature* 299:470, 1982.
43. R. G. Greig, T. P. Loestler, D. L. Trainer, S. P. Corwin, L. Miles, T. Kline, R. Sweet, S. Yokoyama, and G. Poste: Tumorigenic and metastatic properties of "normal" and *ras*-transfected NIH/3T3 cells. *Proc Natl Acad Sci USA* 82:3698, 1985.
44. J. M. Bishop: Molecular themes in oncogenesis. *Cell* 64:235, 1991.
45. T. Hunter: Cooperation between oncogenes. *Cell* 64:249, 1991.
46. K. Brown, A. Buchmann, and A. Balmain: Carcinogen-induced mutations in the mouse c-Ha-*ras* gene provide evidence of multiple pathways for tumor progression. *Proc Natl Acad Sci USA* 87:538, 1990.
47. J.-S. Hoffmann, M. Fry, J. J. Kandace, J. Williams, and L. A. Loeb: Codons 12 and 13 of H-*ras* protooncogene interrupt the progression of DNA synthesis catalyzed by DNA polymerase α. *Cancer Res* 53:2895, 1993.
48. T. G. Krontiris, B. Devlin, D. D. Karp, N. J. Robert, and N. Risch: An association between the risk of cancer and mutations in the HRAS1 minisatellite locus. *N Engl J Med* 329:517, 1993.
49. J. B. Cohen, S. D. Broz, and A. D. Levinson: Expression of the H-*ras* proto-oncogene is controlled by alternative splicing. *Cell* 58:461, 1989.
50. M. A. Nelson, B. W. Futscher, T. Kinsella, J. Wymer, and G. T. Bowden: Detection of mutant Ha-*ras* genes in chemically initiated mouse skin epidermis before the development of benign tumors. *Proc Natl Acad Sci USA* 89:6398, 1992.
51. T. R. Fox, A. M. Schumann, P. G. Watanabe, B. L. Yano, V. M. Maher, and J. J. McCormick: Mutational analysis of the H-*ras* oncogene in spontaneous C57BL/6 × C3H/He mouse liver tumors and tumors induced with genotoxic and nongenotoxic hepatocarcinogens. *Cancer Res* 50:4014, 1990.
52. W. L. Cerny, K. A. Mangold, and D. G. Scarpelli: K-*ras* mutation is an early event in pancreatic duct carcinogenesis in the Syrian golden hamster. *Cancer Res* 52:4507, 1992.
53. R. Kumar, S. Sukumar, and M. Barbacid: Activation of *ras* oncogenes preceding the onset of neoplasia. *Science* 248:1101, 1990.
54. W. E. Pierceall, M. L. Kripke, and H. N. Ananthaswamy: N-*ras* mutation in ultraviolet radiation-induced murine skin cancers. *Cancer Res* 52:3946, 1992.
55. M. H. Ricketts and A. D. Levinson: High-level expression of c-H-*ras*1 fails to fully transform rat-1 cells. *Mol Cell Biol* 8:1460, 1988.
56. R. E. Finney and J. M. Bishop: Predisposition to neoplastic transformation caused by gene replacement of H-*ras*1. *Science* 260:1524, 1993.
57. L. Tong, A. M. de Vos, M. V. Milburn, J. Jancarik, S. Noguchi, S. Nishimura, K. Miura, E. Ohtsuka, and S.-H. Kim: Structural differences between a *ras* oncogene protein and the normal protein. *Nature* 337:90, 1989.
58. U. Krengel, L. Schlichting, A. Scherer, R. Schumann, M. Frech, J. John, W. Sabsch, E. F. Pai, and A. Wittinghofer: Three-dimensional structures of H-*ras* p21 mutants: Molecular basis for their inability to function as signal switch molecules. *Cell* 62:539, 1990.
59. T. C. Thompson, J. Southgate, G. Ketchener, and H. Land: Multistage carcinogenesis induced by *ras* and *myc* oncogenes in a reconstituted organ. *Cell* 56:917, 1989.
60. D. A. Greenhalgh, D. J. Welty, A. Player, and S. H. Yuspa: Two oncogenes, v-*fos* and v-*ras*, cooperate to convert normal keratinocytes to squamous cell carcinoma. *Proc Natl Acad Sci USA* 87:643, 1990.
61. J. L. Bos: ras concogenes in human cancer: A review. *Cancer Res* 49:4682, 1989.
62. M. A. Knowles and M. Williamson: Mutation of H-*ras* is infrequent in bladder cancer: Confirmation by single-strand conformation polymorphism analysis, designed restriction fragment length polymorphisms, and direct sequencing. *Cancer Res* 53:133, 1993.
63. T. Enomoto, M. Inoue, A. O. Perantoni, N. Terakawa, O. Tanizawa, and J. M. Rice: K-*ras* activation in neoplasms of the human female reproductive tract. *Cancer Res* 50:6139, 1990.

64. D. Sidransky, T. Tokino, S. R. Hamilton, K. W. Kinzler, B. Levin, P. Frost, and B. Vogelstein: Identification of *ras* oncogene mutations in the stool of patients with curable colorectal tumors. *Science* 256:102, 1992.
65. M. Tada, M. Omata, S. Kawai, H. Saisho, M. Ohto, R. K. Saiki, and J. J. Sninsky: Detection of *ras* gene mutations in pancreatic juice and peripheral blood of patients with pancreatic adenocarcinoma. *Cancer Res* 53:2472, 1993.
66. J. Schlessinger: How receptor tyrosine kinases activate *ras*. *Trends Biochem Sci* 18:8, 1993.
67. C. M. Crews and R. L. Erikson: Extracellular signals and reversible protein phosphorylation: What to mek of it all. *Cell* 74:215, 1993.
68. F. McCormick: How receptors turn *ras* on. *Nature* 363:15, 1993.
69. S. E. Egan and R. A. Weinberg: The pathway to signal achievement. *Nature* 365:781, 1993.
70. P. H. Duesberg, K. Bister, and P. K. Vogt: The RNA of avian acute leukemia virus MC29. *Proc Natl Acad Sci USA* 74:4320, 1977.
71. G. J. Kato and C. V. Dang: Function of the c-Myc oncoprotein. *FASEB J* 6:3065, 1992.
72. G. Klein: The role of gene dosage and genetic transpositions in carcinogenesis. *Nature:* 294:313, 1981.
73. R. Taub, I. Kirsch, C. Morton, G. Lenoir, D. Swan, S. Tronick, S. Aaronson, and P. Leder: Translocation of the c-*myc* gene into the immunoglobulin heavy chain locus in human Burkitt lymphoma and murine plasmacytoma cells. *Proc Natl Acad Sci USA* 79:7837, 1982.
74. R. Dalla-Favera, M. Bregni, J. Erikson, D. Patterson, R. C. Gallo, and C. M. Croce: Human c-*myc* oncogene is located on the region of chromosome 8 that is translocated in Burkitt lymphoma cells. *Proc Natl Acad Sci USA* 79:7824, 1982.
75. S. Crews, R. Barth, L. Hood, J. Prehn, and K. Calame: Mouse c-*myc* oncogene is located on chromosome 15 and translocated to chromosome 12 in plasmacytomas. *Science* 218:1319, 1982.
76. G. L. C. Shen-Ong, E. J. Keath, S. P. Piccoli, and M. D. Cole: Novel *myc* oncogene RNA from abortive immunoglobulin gene recombination in mouse plasmacytomas. *Cell* 31:443, 1982.
77. J. M. Adams, S. Gerondakis, E. Webb, L. M. Corcoran, and S. Cory: Cellular *myc* oncogene is altered by chromosome translocation to an immunoglobulin locus in murine plasmocytomas and is rearranged similarly in human Burkitt lymphomas. *Proc Natl Acad Sci USA* 80:1982, 1983.
78. R. Dalla-Favera, F. Wong-Staal, and R. C. Gallo: *onc* gene amplification in promyelocytic leukaemia cell line HL-60 and in primary leukaemic cells of the same patient. *Nature* 299:61, 1982.
79. K. Alitalo, M. Schwab, C. C. Lin, H. E. Varmus, and J. M. Bishop: Homogeneously staining chromosomal regions contain amplified copies of an abundantly expressed cellular oncogene (c-*myc*) in malignant neuroendocrine cells from a colon carcinoma. *Proc Natl Acad Sci USA* 80:1707, 1983.
80. C. D. Little, M. M. Nau, D. N. Carney, A. F. Gazdar, and J. D. Minna: Amplification and expression of the c-*myc* oncogene in human lung cancer cell lines. *Nature* 306:194, 1983.
81. B. K. Suchy, M. Sarafoff, R. Kerler, and H. M. Rabes: Amplification, rearrangements, and enhanced expression of c-*myc* in chemically induced rat liver tumors in vivo and in vitro. *Cancer Res* 49:6781, 1989.
82. J. Brennan, T. O'Connor, R. W. Makuch, A. M. Simmons, E. Russell, R. I. Linnoila, R. M. Phelps, A. F. Gazdar, D. C. Ihde, and B. E. Johnson: *myc* family DNA amplification in 107 tumors and tumor cell lines from patients with small cell lung cancer treated with different combination chemotherapy regimens. *Cancer Res* 51:1708, 1991.
83. M. F. Melhem, A. I. Meisler, G. G. Finley, W. H. Bryce, M. O. Jones, I. I. Tribby, J. M. Pipas, and R. A. Koski: Distribution of cells expression *myc* proteins in human colorectal epithelium, polyps, and malignant tumors. *Cancer Res* 52:5853, 1992.
84. E. M. J. J. Berns, J. G. M. Klijn, W. L. J. van Putten, I. L. van Staveren, H. Portegen, and J. A. Foekens: c-*myc* amplification is a better prognostic factor than HER2/*neu* amplification in primary breast cancer. *Cancer Res* 52:1107, 1992.
85. D. D. Von Hoff, J. R. McGill, B. J. Forseth, K. K. Davison, T. P. Bradley, D. R. Van Devanter, and G. M. Wahl: Elimination of extrachromosomally amplified MYC genes from human tumor cells reduces their tumorigenicity. *Proc Natl Acad Sci USA* 89:8165, 1992.
86. E. M. Blackwood and R. N. Eisenman: Max: A helix-loop-helix protein that forms a sequence-specific DNA binding complex with Myc. *Science* 251:1211, 1991.
87. G. C. Prendergast, D. Lawe, and E. B. Ziff: Association of Myn, the murine homology of Max, with c-Myc stimulates methylation-sensitive DNA binding and Ras cotransformation. *Cell* 65:395, 1991.
88. L. Kretzner, E. M. Blackwood, and R. N. Eisenman: Myc and Max proteins possess distinct transcriptional activities. *Nature* 359:426, 1992.
89. W. Gu, K. Cechova, V. Tassi, and R. Dalla-Favera: Opposite regulation of gene transcription and cell proliferation by c-Myc and Max. *Proc Natl Acad Sci USA* 90:2935, 1993.
90. B. Amati, M. W. Brooks, N. Levy, T. D. Littlewood, G. I.Evan, and H. Land: Oncogenic activity of the c-Myc protein requires dimerization with Max. *Cell* 72:233, 1993.
91. G. C. Prendergast, D. Lawe, and E. B. Ziff: As-

sociation of Myn, the murine homolog of Max, with c-Myc stimulates methylation-sensitive DNA binding and *ras* cotransformation. *Cell* 65:395, 1991.

92. T. Hunter: A tail of two src's: Mutatis mutandis. *Cell* 49:1, 1987.
93. S. A. Courtneidge: Activation of the pp60c-src kinase by middle T antigen bending or by dephosphorylation. *EMBO J* 4:1471, 1985.
94. J. A. Cooper and B. Howell: The when and how of *src* regulation. *Cell* 73:1051, 1993.
95. C. Bjelfman, F. Hedborg, I. Johansson, M. Nordenskjöld, and S. Pahlman: Expression of the neuronal form of $pp60^{c\text{-}scr}$ in neuroblastoma in relation to clinical stage and prognosis. *Cancer Res* 50:6908, 1990.
96. C. A. Cartwright, A. I. Meisler, and W. Eckhart: Activation of the $pp60^{c\text{-}src}$ protein kinase is an early event in colonic carcinogenesis. *Proc natl Acad Sci USA* 87:558, 1990.
97. T. Pawson and G. D. Gish: SH2 and SH3 domains: From structure to function. *Cell* 71:359, 1992.
98. A. J. Muller, A.-M. Pendergast, K. Parmar, M. H. Havlik, N. Rosenberg, and O. N. Witte: En bloc substitution of the SRC homology region 2 domain activates the transforming potential of the c-Abl protein tyrosine kinase. *Proc Natl Acad Sci USA* 90:3457, 1993.
99. Y. Maki, T. Bos, C. Davis, M. Starbuck, and P. Vogt: Avian sarcoma virus 17 carries the *jun* oncogene. *Proc Natl Acad Sci USA* 84:2848, 1987.
100. J. Schütte, J. D. Minna, and M. J. Birrer: Deregulated expression of human c-*jun* transforms primary rat embryo cells in cooperation with an activated c-Ha-*ras* gene and transforms rat-1a cells as a single gene. *Proc Natl Acad Sci USA* 86:2257, 1989.
101. V. R. Baichwar and R. Tjian: Control of c-*jun* activity by interaction of a cell-specific inhibitor with regulatory domain δ: Differences between v- and c-*jun*. *Cell* 63:815, 1990.
102. J. Schütte, J. Viallet, M. Nau, S. Segal, J. Fedorko, and J. Minna: *jun*-B inhibits and c-*fos* stimulates the transforming and *trans*-activating activities of c-*jun*. *Cell* 59:987, 1989.
103. M. P. Finkel, C. A. Relly, Jr., and B. O. Biskis: Viral etiology of bone cancer. *Front Radiat Theor Oncol* 10:28, 1975.
104. T. Jenuwein and R. Müller: Structure-function analysis of fos protein: A single amino acid change activates the immortalizing potential of v-*fos*. *Cell* 48:647, 1987.
105. K. Macleod, D. Leprince and D. Stehelin: The *ets* gene family. *Trends Biochem Sci* 251, 1992.
106. A. Seth and A. Papas: The c-ets-1 proto-oncogene has oncogenic activity and is positively autoregulated. *Oncogene* 5:1761, 1990.
107. S. P. Goff, E. Gilboa, O. N. Witte, and D. Baltimore: Structure of the Abelson murine leukemia virus genome and the homologous cellular gene: Studies wih clonal viral DNA. *Cell* 22:777, 1980.
108. O. N. Witte: Role of the BCR-ABL oncogene in human leukemia: Fifteenth Richard and Hilda Rosenthal Foundation Award Lecture. *Cancer Res* 53:485, 1993.
109. G. Q. Daley, R. A. Van Etten, and D. Baltimore: Induction of chronic myelogenous leukemia in mice by the $P210^{bcr/abl}$ gene of the Philadelphia chromosome. *Science* 247:824, 1990.
110. J. Willem Voncken, S. Griffiths, M. F. Greaves, P. K. Pattengale, N. Heisterkamp, and J. Groffen: Restricted oncogenicity of BCR/ABL p10 in transgenic mice. *Cancer Res* 52:4534, 1992.
110a. C. L. Sawyers, J. McLaughlin, A. Goga, M. Havlik, and O. Witte: The nuclear tyrosine kinase c-Abl negatively regulates cell growth. *Cell* 77:121, 1994.
111. A. M. Pendergast, L. A. Quilliam, L. D. Cripe, C. H. Bassing, Z. Dai, N. Li, A. Batzer, K. M. Rabun, C. J. Der, J. Schlessinger, and M. L. Gishizky: BCR-ABL-induced oncogenesis is mediated by direct interaction with the SH2 domain of the GRB-2 adaptor protein. *Cell* 75:175, 1993.
112. B. Lüscher, E. Christenson, D. W. Litchfield, E. G. Krebs, and R. N. Eisenman: Myb DNA binding inhibited by phosphorylation at a site deleted during oncogenic activation. *Nature* 344:517, 1990.
113. D. Venturelli, M. T. Mariano, C. Szczylik, M. Valtieri, B. Lange, W. Crist, M. Link, and B. Calabretta: Down-regulated c-*myb* expression inhibits DNA synthesis of T-leukemia cells in most patients. *Cancer Res* 50:7371, 1990.
114. S. Haldar, C. Beatty, Y. Tsujimoto, and C. M. Croce: The *bcl*-2 gene encodes a novel G protein. *Nature* 342:195, 1989.
115. M. Katsumata, R. M. Siegel, D. C. Louie, T. Miyashita, Y. Tsujimoto, P. C. Nowell, M. I. Green, and J. C. Reed: Differential effects of Bcl-2 on T and B cells in transgenic mice. *Proc Natl Acad Sci USA* 89:11376, 1992.
116. Z. N. Oltvai, C. L. Milliman, and S. J. Korsmeyer: Bcl-2 heterodimerizes in vivo with a conserved homolog, Bax, that accelerates programed cell death. *Cell* 74:609, 1993.
117. L. H. Boise, M. González-Garcia, C. E. Postema, L. Ding, T. Lindsten, L. A. Turka, X. Mao, G. Nunez, and C. B. Thompson: *bcl*-x, a *bcl*-2-related gene that functions as a dominant regulator of apoptotic cell death. *Cell* 74:597, 1993.
118. G. Nunez, L. London, D. Hockenbery, M. Alexander, J. P. McKearn, and S. J. Korsmeyer: Deregulated Bcl-2 gene expression selectively prolongs survival of growth factor-deprived hematopoietic cell lines. *J Immunol* 144:3602, 1990.
119. T. E. Allsopp, S. Wyatt, H. F. Paterson, and A. M. Davies: The proto-oncogene *bcl*-2 can selectively rescue neurotropic factor-dependent neurons from apoptosis. *Cell* 73:295, 1993.

120. D. M. Hockenbery, Z. N. Oltvai, X.-M. Yin, C. L. Milliman, and S. J. Korsmeyer: Bcl-2 functions in an antioxidant pathway to prevent apoptosis. *Cell* 75:241, 1993.
121. R. Sen and D. Baltimore: Multiple nuclear factors interact with the immunoglobulin enhancer sequences. *Cell* 46:705, 1986.
122. S. Gosh, A. M. Gifford, L. R. Viviere, P. Tempst, G. P. Nolan, and D. Baltimore: Cloning of the p50 DNA binding subunit of NF-κB: Homology to rel and dorsal. *Cell* 62:1019, 1990.
123. M. Kieran, V. Blank, F. Logeat, J. Vanderkerckhove, F. Lottspeich, O. Le Ball, M. B. Urban, P. Kourisky, P. A. Baeuerie, and A. Israel: The DNA binding subunit of NF-κB is identical to factor KBF1 and homologous to the *rel* oncogene product. *Cell* 62:1007, 1990.
124. T. Lernbecker, U. Müller, and T. Wirth: Distinct NF-κB/rel transcription factors are responsible for tissue-specific and inducible gene activation. *Nature* 365:767, 1993.
125. T. Graf and H. Beug: Role of the v-*erb*A and v-*erb*B oncogenes of avian erythroblastosis virus in erythroid cell transformation. *Cell* 34:7, 1983.
126. K. Damm, C. C. Thompson, and R. M. Evans: Protein encoded by v-*erb*A functions as a thyroid-hormone receptor antagonist. *Nature* 339:593, 1989.
127. C. Desbois, D. Aubert, C. Legrand, B. Pain, and J. Samarut: A novel mechanism of action for v-*Erb*A: Abrogation of the inactivation of transcription factor AP-1 by retinoic acid and thyroid hormone receptors. *Cell* 67:731, 1991.
128. M. Sharif and M. L. Privalsky: v-*erb*A oncogene function in neoplasia correlates with its ability to repress retinoic acid receptor action. *Cell* 66:885, 1991.
129. R. F. Doolittle, M. W. Hunkapiller, L. E. Hood, S. G. Devare, K. C. Robbins, S. A. Aaronson, and H. N. Antoniades: Simian sarcoma virus *onc* gene, v-*sis*, is derived from the gene (or genes) encoding a platelet-derived growth factor. *Science* 221:275, 1983.
130. M. D. Waterfield, G. T. Scrace, N. Whittle, P. Stoobant, A. Johnson, A. Wasteson, B. Westermark, C.-H. Heldin, J. S. Huang, and T. F. Deuel: Platelet-derived growth factor is structurally related to the putative transforming protein $p28^{sis}$ of simian sarcoma virus. *Nature* 304:35, 1983.
131. H. N. Antoniades and M. W. Hunkapiller: Human platelet-derived growth factor (PDGF): Amino terminal amino acid sequence. *Science* 220:963, 1983.
132. S. G. Devare, E. P. Reddy, J. D. Law, K. C. Robbins, and S. A. Aaronson: Nucleotide sequence of the simian sarcoma virus genome: Demonstration that is acquired cellular sequence encode the transforming gene product $p28.^{v\text{-}sis}$ *Proc Natl Acad Sci USA* 80:731, 1983.
133. T. F. Deuel, J. S. Huang, S. S. Huang, P. Stroobant, and M. D. Waterfield: Expression of a platelet-derived growth factor-like protein in simian virus transformed cells. *Science* 221:1348, 1983.
134. J. S. Huang, S. S. Huang, and T. F. Keuel: Transforming protein of simian sarcoma virus stimulates autocrine growth of SSV-transformed cells through PDGF cell-surface receptors. *Cell* 39:79, 1984.
135. T. Hunter: Oncogenes and growth control. *Trends Biochem Sci* July 1985, p. 275.
136. V. B. Lokeshwar, S. S. Huang, and J. S. Huang: Intracellular turnover, novel secretion, and mitogenically active intracellular forms of v-*sis* gene product in simian sarcoma virus-transformed cells. *J Biol Chem* 265:1665, 1990.
137. J. Downward, Y. Yarden, E. Mayes, G. Scrace, N. Totty, P. Stockwell, A. Ullrich, J. Schlessinger, and M. D. Waterfield: Close similarity of epidermal growth factor receptor of v-erb-B oncogene protein sequences. *Nature* 307:521, 1984.
138. Y. Kokai, J. N. Myers, T. Wada, V. I. Brown, C. M. LeVea, J. G. Davis, K. Dobashi, and M. I. Greene: Synergistic interactin of p185c-*neu* and the EGF receptor leads to transformation of rodent fibroblasts. *Cell* 58:287, 1989.
139. C. I. Bargmann and R. A. Winberg: Increased tyrosine kinase activity associated with the protein encoded by the activated neu oncogene. *Proc Natl Acad Sci USA* 85:5394, 1988.
140. Y.-F. Zhai, H. Beittenmiller, B. Wang, M. N. Gould, C. Oakley, W. J. Esselmann, and C. W. Welsch: Increased expression of specific protein tyrosine phosphatases in human breast epithelial cells neoplastically transformed by the *neu* oncogene. *Cancer Res* 53:2272, 1993.
141. L. Bouchard, L. Lamarre, P. J. Tremblay, and P. Jolicoeur: Stochastic appearance of mammary tumors in transgenic mice carrying the MMTV/c-*neu* oncogene. *Cell* 57:931, 1989.
142. C. T. Guy, M. A. Webster, M. Schaller, T. J. Parsons, R. D. Cardiff, and W. J. Muller: Expression of the *neu* protooncogene in the mammary epithelium of transgenic mice induces metastatic disease. *Proc Natl Acad Sci USA* 89:10578, 1992.
143. P. Devilee, M. Van Vliet, N. Kuipers-Dijleshoony, P. L. Pearson, and C. J. Cornelisse: Somatic genetic changes on chromosome 18 in breast carcinomas: Is the DCC gene involved? *Oncogene* 6:311, 1991.
144. M. Streuli, N. X. Krueger, P. D. Ariniello, M. Tang, J. M. Munro, W. A. Blattler, D. A. Adler, C. M. Disteche, and H. Saito: Expression of the receptor-linked protein tyrosine phosphatase LAR: Proteolytic cleavage and shedding of the CAM-like extracellular regions. *EMBO J* 11:897, 1992.
145. D. J. Slamon, G. M. Clark, S. G. Wong, W. J. Levin, A. Ullrich, and W. L. McGuire: Human breast cancer: Correlation of relapse and survival

with amplification of the HER-2/*neu* oncogene. *Science* 235:177, 1987.
146. G. M. Clark and W. L. McGuire: Follow-up study of HER-2/*neu* amplification in primary breast cancer. *Cancer Res* 51:944, 1991.
147. M. F. Press, M. C. Pike, V. R. Chazin, G. Hung, J. A. Udove, M. Markowicz, J. Danyluk, W. Godolphin, M. Sliwkowski, R. Akita, M. C. Paterson, and D. J. Slamon: Her-2/*neu* expression in node-negative breast cancer: Direct tissue quantitation by computerized image analysis and association of overexpression with increased risk of recurrent disease. *Cancer Res* 53:4960, 1993.
148. A Berchuck, A. Kamel, R. Whitaker, B. Kerns, G. Olt, R. Kinney, J. T. Soper, R. Doge, D. L. Clarke-Pearson, P. Marks, S. McKenzie, S. Yin, and R. C. Bast, Jr.: Overexpression of HER-2/*neu* is associated with poor survival in advanced epithelial ovarian cancer. *Cancer Res* 50:4087, 1990.
149. Y. Yonemura, I. Ninomiya, A. Yamaguchi, S. Fushida, H. Kimura, S Ohoyama, I. Miyazaki, Y. Endou, M. Tanaka, and T. Sasaki: Evaluation of immunoreactivity of erbB-2 protein as a marker of poor short term prognosis in gastric cancer. *Cancer Res* 51:1034, 1991.
150. C.-M. Tsai, K.-T. Chang, R.-P. Perng, T. Mitsudomi, M.-H. Chen, C. Kadoyama, and A. F. Gazdar: Correlation of intrinsic chemoresistance of non-small-cell lung cancer cell lines with HER-2/*neu* gene expression but not with *ras* gene mutations. *JNCI* 85:897, 1993.
151. M. F. Roussel, J. R. Downing, C. W. Rettenmier, and C. J. Sherr: A point mutation in the extracellular domain of the human CSF-1 receptor (c-*fms* proto-oncogene product) activates its transforming potential. *Cell* 55:979, 1988.
152. M. Z. Ratajczak, S. M. Luger, K. DeRiel, J. Abraham, B. Calabretta, and A. M. Gewirtz: Role of the KIT protooncogene in normal and malignant human hematopoiesis. *Proc Natl Acad Sci USA* 89:1710, 1992.
153. R. Oskam, F. Coulier, M. Ernst, D. Martin-Sanca, and M. Barbacid: Frequent generation of oncogenes by in vitro recombination of TRK protooncogene sequences. *Proc Natl Acad Sci USA* 85:2964, 1988.
154. A. Nakagawara, M. Arima-Nakagawara, N. J. Scavarda, C. G. Azar, A. B. Cantor, and G. M. Brodeur: Association between high levels of expression of the TRK gene and favorable outcome in human neuroblastomas. *N Engl J Med* 328:847, 1993.
155. S. Giordano, Z. Zhen, E. Medico, G. Guadino, F. Galimi, and P. M. Comoglio: Transfer of motogenic and invasive response to scatter factor/hepatocyte growth factor by transfection of human MET protooncogene. *Proc Natl Acad Sci USA* 90:649, 1993.
156. D. J. Slamon and M. J. Cline: Expression of cellular oncogenes during embryonic and fetal development of the mouse. *Proc Natl Acad Sci USA* 81:7141, 1984.
157. R. Müller, D. J. Slamon, J. M. Tremblay, M. J. Cline, and I. M. Verma: Differential expression of cellular oncogenes during pre- and postnatal development of the mouse. *Nature* 299:640, 1982.
158. A. S. Goustin, C. Betsholtz, S. Pfeifer-Ohlsson, H. Persson, J. Rydnert, M. Bywater, G. Holmgren, C.-H. Heldin, B. Westermark, and R. Ohlsson: Coexpression of the *sis* and *myc* proto-oncogenes in developing human placenta suggests autocrine control of trophoblast growth. *Cell* 41:301, 1985.
159. R. L. Michell, L. Zokas, R. D. Schreiber, and I. M. Verma: Rapid induction of the expression of proto-oncogene *fos* during human monocytic differentiation. *Cell* 40:209, 1985.
160. R. Müller, T. Curran, D. Müller, and L. Guilbert: Induction of c-*fos* during myelomonocytic differentiation and macrophage proliferation. *Nature* 314:546, 1985.
161. E. Sariban, T. Mitchell, and D. Kufe: Expression of the c-*fms* proto-oncogene during human monocytic differentiation. *Nature* 316:64, 1985.
162. P. H. Reitsma, P. G. Rothberg, S. M. Astrin, J. Trial, Z. Bar-Shavit, A. Hall, S. L. Teitelbaum, and A. J. Kahn: Regulation of *myc* gene expression in HL-60 leukaemia cells by a vitamin D metabolite. *Nature* 306:492, 1983.
163. R. Makino, K. Hayashi, and T. Sugimura: c-*myc* transcript is induced in rat liver at a very early stage of regeneration or by cycloheximide treatment. *Nature* 310:697, 1984.
164. M. Goyette, C. J. Petropoulos, P. R. Shank, and N. Fausto: Expression of a cellular oncogene during liver regeneration. *Science* 219:510, 1983.
165. D. Boettiger, S. Anderson, and T. M. Dexter: Effect of *src* infection on long-term marrow cultures: Increased self-renewal of hemopoietic progenitor cells without leukemia. *Cell* 36:763, 1984.
166. S. H. Yuspa, A. E. Kilkenny, J. Stanley, and U. Lichti: Keratinocytes blocked in phorbol ester-responsive early stage or terminal differentiation by sarcoma viruses. *Nature* 314:459, 1985.
167. B. H. Sweet and M. R. Hilleman: The vacuolating virus, SV40. *Proc Exp Biol Med* 105:420, 1960.
168. B. E. Eddy, G. S. Borman, G. E. Grubbs, and R. D. Young: Identification of the oncogenic substance in rhesus monkey cell cultures as simian virus 40. *Virology* 17:65, 1962.
169. H. M. Shein and J. F. Enders: Transformation of human renal cell cultures: I. Morphology and growth characteristics. *Proc Natl Acad Sci USA* 48:1164, 1962.
170. G. Khoury, C.-J. Lai, D. Solomon, M. Israel, and P. Howley: The human papovaviruses and their potential role in human diseases, in H. H. Hiatt, J. D. Watson, and J. A. Winsten, *Origins of Hu-*

man Cancer. Cold Spring Harbor, NY: Cold Spring Harbor Laboratory, 1977, pp. 971–988.

171. C. Prives, E. Gilboa, M. Revel, and E. Winocour: Cell-free translation of simian virus 40 early messenger RNA coding for viral T-antigen. *Proc Natl Acad Sci USA* 74:457, 1977.
172. P. Rigby: The transforming genes of SV40 and polyoma. *Nature* 282:781, 1979.
173. M. R. D. Scott, K.-H. Westphal, and P. W. J. Rigby: Activation of mouse genes in transformed cells. *Cell* 34:557, 1983.
174. M. Kriegler, C. F. Perez, C. Hardy, and M. Botchan: Transformation mediated by the SV40 T antigens: Separation of the overlapping SV40 early genes with a retroviral vector. *Cell* 38:483, 1984.
175. R. L. Brinster, H. Y. Chen, A. Messing, T. van Dyke, A. J. Levine, and R. D. Palmiter: Transgenic mice harboring SV40 T-antigen genes develop characteristic brain tumors. *Cell* 37:367, 1984.
176. R. D. Palmiter, H. Y. Chen, A. Messing, and R. L. Brinster: SV40 enhancer and large-T antigen are instrumental in development of choroid plexus tumors in transgenic mice. *Nature* 316:457, 1985.
177. A Messing, H. Y. Chen, R. D. Palmiter, and R. L. Brinster: Peripheral neuropathies, hepatocellular carcinomas, and islet cell adenomas in transgenic mice. *Nature* 316:461, 1985.
178. D. Hanahan: Heritable formation of pancreatic β-cell tumours in transgenic mice expressing recombinant insulin/simian virus 40 oncogenes. *Nature* 315:115, 1985.
179. J. Feuteun, L. Sompayrac, M. Fluck, and T. Benjamin: Localization of gene functions in polyoma virus DNA. *Proc Natl Acad Sci USA* 73:4169, 1976.
180. Y. Ito, J. R. Brocklehurst, and R. Dulbecco: Virus-specific proteins in the plasma membrane of cells lytically infected or transformed by polyoma virus. *Proc Natl Acad Sci USA* 74:4666, 1977.
181. M. Rassoulzadegan, A. Cowie, A. Carr, N. Glaichenhaus, R. Kamen, and F. Cuzin: The roles of individual polyoma virus early proteins in oncogenic transformation. *Nature* 300:713, 1982.
182. G. Della Valle, R. G. Fenton, and C. Basilico: Polyoma large T antigen regulates the integration of viral DNA sequences into the genome of transformed cells. *Cell* 23:347, 1981.
183. D. J. Donoghue, C. Anderson, T. Hunter, and P. L. Kaplan: Transmission of the polyoma virus middle T gene as the oncogene of a murine retrovirus. *Nature* 308:748, 1984.
184. G. Carmichael, B. S. Schaffhausen, G. Mandel, T. J. Liang, and T. L. Benjamin: Transformation by polyoma virus is drastically reduced by substitution of phenylalanine for tyrosine at residue 315 of middle-sized tumor antigen. *Proc Natl Acad Sci USA* 81:679, 1984.
185. A. A. Schreier and J. Gruber: Viral T-antigen interactions with cellular proto-oncogene and antioncogene products. *JNCI* 82:354, 1990.
186. J. W. Ludlow, J. A. DeCaprio, C.-M. Huang, W.-H. Lee, E. Paucha, and D. M. Livingston: SV40 large T antigen binds preferentially to an underphosphorylated member of the retinoblastoma susceptibility gene product family. *Cell* 56:57, 1989.
187. G. Orth, F. Breitburd, M. Favre, and O. Croissant: Papillomaviruses: Possible role in human cancer, in H. H. Hiatt, J. D. Watson, and J. A. Winsten, eds.: *Origins of Human Cancer*. Cold Spring Harbor, NY: Cold Spring Harbor Laboratory, 1977, pp. 1043–1068.
188. T. d. Kessis, R. J. Slebos, W. G. Nelson, M. B. Kastan, B. S. Plunkett, S. M. Han, A. T. Lorincz, L. Hedrick, and K. R. Cho: Human papillomavirus 16 E6 expression disrupts the p53-mediated cellular response to DNA damage. *Proc Natl Acad Sci USA* 90:3988, 1993.
189. M. H. Schiffman, H. M. Bauer, R. N. Hoover, A. G. Glass, D. M. Cadell, B. B. Rush, D. R. Scott, M. E. Sherman, R. J. Kurman, S. Wacholder, C. K. Stanton, and M. M. Manos: Epidemiologic evidence showing that human papillomavirus infection causes most cervical intraepithelial neoplasis. *JNCI* 85:958, 1993.
190. M. Schenffner, B. A. Werness, J. M. Hulbregtse, A. J. Levine, and P. M. Howley: The E6 oncoprotein encoded by human papillomavirus types 16 and 18 promotes the degradation of p53. *Cell* 63:1129, 1990.
191. P. F. Lamberg, H. Pan, H. C. Pitot, A. Liem, M. Jackson, and A. E. Griep: Epidermal cancer associated with expression of human papillomavirus type 16 E6 and E7 oncogenes in the skin of transgenic mice. *Proc Natl Acad Sci USA* 90:5583, 1993.
192. L. Chen, S. Ashe, M. C. Singhal, D. A. Galloway, I. Hellström, and K. E. Hellström: Metastatic conversion of cells by expression of human papillomavirus type 16 E6 and E7 genes. *Proc Natl Acad Sci USA* 90:6523, 1993.
193. J. J. Trentin, Y. Yabe, and G. Taylor: The quest for human cancer viruses. *Science* 137:835, 1962.
194. J. K. McDougall, L. B. Chen, and P. H. Gallimore: Transformation in vitro by adenovirus type 2—A model system for studying mechanisms of oncogenicity, in H. H. Hiatt, J. D. Watson, and J. A. Winsten, eds.: Cold Spring Harbor, NY: Cold Spring Habor Laboratory, 1977, pp. 1013–1025.
195. H. E. Ruley: Adenovirus early region 1A enables viral and cellular transforming genes to transform primary cells in culture. *Nature* 304:602, 1983.

196. R. Javier, K. Raska, Jr., and T. Shenk: Requirement for the adenovirus type 9 E4 region in production of mammary tumors. *Science* 257:1267, 1992.
197. P. Whyte, N. M. Williamson, and E. Harlow: Cellular targets for transformation by the adenovirus E1A proteins. *Cell* 56:67, 1989.
198. G. Fourel, C. Trepo, L. Bougueleret, B. Henglein, A. Ponzetto, P. Tiollais, and M.-A. Buendia: Frequent activation of N-*myc* genes by hepadnavirus insertion in woodchuck liver tumors. *Nature* 347:294, 1990.
199. H. F. Maguire, J. P. Hoeffler, and A. Siddiqui: HBV X protein alters the DNA binding specificity of CREB and ATF-2 by protein-protein interactions. *Science* 252:842, 1991.
200. B. Roizman, N. Frenkel, E. D. Kieff, and P. G. Spear: The structure and expression of human herpersvirus DNAs in productive infection and intransformed cells, in H. H. Hiatt, J. D. Watson, and J. A. Winsten, eds.: *Origins of Human Cancer.* Cold Spring Harbor, NY: Cold Spring Harbor Laboratory, 1977, pp. 1069–1111.
201. P. Lengyel: Tumor-suppressor genes: News about the interferon connection. *Proc Natl Acad Sci USA* 90:5893, 1993.

8

Tumor Suppressor Genes

HISTORICAL PERSPECTIVES

In Chapter 7, the role of activated oncogenes in causing malignant transformation of cells was discussed. The excitement surrounding this research dominated the scene in cancer cell biology for a number of years. It was a relatively clear and satisfying way to explain cancer. It also unified a number of theories about how chemicals, irradiation, and viruses could cause cancer. They all converged into one theme: damage to DNA causing point mutations, chromosomal rearrangements, translocations, or amplifications, all of which could lead to the activation of cellular protooncogenes that could take over and dominate a cell's behavior, turning it into a cell programmed to survive and proliferate.

Thus, the idea was that cancer genes, once activated, became dominant genes and caused a dominant genetic change in cells. There were only one or two flies in the ointment. Back in 1969, Henry Harris and his colleagues[1] had shown that when malignant cells were fused with nonmalignant cells, most of the hybrid cells were nontumorigenic. If cancer is due to a dominant genetic event, this did not make sense at all; hence this observation was virtually ignored for almost 20 years. Another fly in the ointment was the report of Alfred Knudson, Jr., in 1971, of a hereditary form of the eye tumor retinoblastoma in which some gene carriers acquired bilateral eye tumors, some had unilateral disease, and a small minority had no tumors (reviewed in Ref. 2). Moreover, only three to four tumor loci per affected patient were observed. Since there are more than 1 million cells in a retina, it is a rare cell indeed that actually becomes cancerous, even though all the cells carry the defective gene. This strongly suggested a second genetic event, the inherited mutation by itself not being sufficient, and led to Knudson's "two-hit hypothesis." This hypothesis suggests that in the hereditary disease, one defective gene is inherited as a germ-line mutation and a second mutation, occurring after conception, is necessary to induce a tumor, whereas in the nonhereditary form of the disease, both mutations occur as somatic, postconception events (see Chap. 3).

There were other unsettling findings that did not fit the dominant oncogene theory of cancer. For example, in solid human tumors as opposed to leukemias and lymphomas, the most commonly observed cytogenetic abnormalities were chromosomal deletions. And even when investigators began to be able to detect oncogene mutations and amplifications by sensitive molecular genetic techniques, such mutations could be found in only 15% to 30% of human cancers.[3,4] Thus, a number of investigators began to think more seriously that loss of some inhibitory or regulatory gene function was involved in causing cancer.

A big advance in this theory was made when it was shown that when a single human chromosome 11 from a normal human fibroblast was introduced by microcell transfer into HeLa cells or Wilms' tumor cells, the ability of these cells to induce progressive tumors in nude mice was suppressed (reviewed in Ref. 5). Subsequent studies have shown deletions in specific regions of chromosomes in a number of human cancers, suggesting that the loss of "tumor suppressor" genetic information is a common event in human malignant disease (Table 8-1). The presence of genes to inhibit uncontrolled cell proliferation helps to explain why human beings have only about a 25% chance of developing a full-blown cancer, even though we experience 10^{16} cell mitoses in a lifetime.

The first tumor suppressor gene that was cloned was the *RB* gene that is the defective gene in retinoblastoma. Cavenee et al.[6] used restriction fragment length polymorphisms (RFLPs) to map the defective gene to chromosome 13q14 and showed that a loss of heterozygosity at this locus in the tumor was due to loss of the normal allele from the unaffected parent.[7] This indicated a germ-line mutation, uncovered by the loss of heterozygosity, and helped substantiate the Knudson hypothesis. The *RB* gene was subsequently cloned by Friend et al.[8] It is now known that a variety of other human cancers have inactivated *RB* alleles, including sarcomas, small-cell lung, bladder, and a few breast carcinomas (reviewed in Ref. 9). However, the role of *RB* loss of heterozygosity (LOH) in these tumors is not clear.

A number of other tumor suppressor genes or candidate tumor suppressor genes have been cloned and are being characterized (Table 8-2).[10] This almost certainly is only the tip of the iceberg and many more will likely be discovered as more is learned about cancer cell genetics and the map of the human genome. At this point, however, a number of generalizations can be made (Table 8-3), and a comparison with the activation of oncogenes is instructive.[11] A single mutation can be sufficient to activate an oncogene (e.g., *ras*); a second is not crucial, since there would not necessarily be any particular selective pressure to sustain it. Mutations in *onc* genes are gain-of-function events and lead to increased cell proliferation and decreased cell differentiation. Oncogene mutations do not appear to be inherited through the germ line, since their dominant effects could disrupt normal development. Oncogenes are mutated in a wide variety of human cancers (e.g., *ras*, *myc*). In contrast, tumor suppressor gene inactivations are loss-of-function events, usually requiring a mutational event in

Table 8-1 Evidence of Loss of Genetic Information in Human Cancers[a]

Tumor Type	Chromosome Region(s) Involved
Wilms' tumor-sporadic	11p13, 11p15
Wilms' tumor-familial	Unknown
Retinoblastoma	13q14
Osteogenic sarcoma[b]	13q14, 17p
Soft tissue sarcoma[b]	13q14
Neuroblastoma	1p, 14q, 17
Glioblastoma multiforme (astrocytoma)	10, 17p
Bladder carcinoma	9q, 11p, 17p
Breast carcinoma	1q, 11p, 13q, 17p
Colorectal carcinoma	5q, 17p, 18q
Renal cell carcinoma	3p
Multiple endocrine neoplasia type 1	11q
Multiple endocrine neoplasia type 2	1p, 10, 22
Tumors associated with bilateral acoustic neurofibromatosis	22q
Uveal melanoma	2
Melanoma	1, 6
Myeloid leukemia	5q
Small cell lung cancer	3p, 13q, 17p
Non small cell lung cancer	3p, 11p, 13q, 17p

[a]Data derived from cytogenetic and RFLP analyses.

[b]Second malignancies in familial retinoblastoma patients.

Source: From Stanbridge.[5]

Table 8-2 Some Known or Candidate Tumor Suppressor Genes

Gene	Cancer Types	Product Location	Mode of Action	Hereditary Syndrome
APC	Colon carcinoma	Cytoplasm?	?	Familial adenomatous polyposis
DCC	Colon carcinoma	Membrane	Cell adhesion molecule	—
NF1	Neurofibromas	Cytoplasm	GTPase-activator	Neurofibromatosis type 1
NF2	Schwannomas and meningiomas	Inner membrane?	Links membrane to cytoskeleton?	Neurofibromatosis type 2
p53	Colon cancer; many others	Nucleus	Transcription factor	Li-Fraumeni syndrome
RB	Retinoblastoma	Nucleus	Transcription factor	Retinoblastoma
RET	Thyroid carcinoma; pheochromocytoma	Membrane	Receptor tyrosine kinase	Multiple endocrine neoplasia type 2
VHL	Kidney carcinoma	Membrane?	?	von Hippel-Lindau disease
WT-1	Nephroblastoma	Nucleus	Transcription factor	Wilms' tumor

Source: Reprinted with permission from Marx, Learning how to suppress cancer. *Science* 261:1385. Copyright 1993 by the American Association for the Advancement of Science.

one allele followed by loss or inactivation of the other allele. They are recessive in nature, and the resulting mutations may be inherited through the germ line. One point of similarity is that somatic mutational events can occur both in oncogenes and in tumor suppressor genes and may accumulate over a lifetime.

A point should be made about the terms *dominant* and *recessive*. In the classic Mendelian sense, the terms refer to an inheritance pattern resulting from the interplay between one paternal and one maternal allele in a diploid offspring. In cancer cells, this principle often does not hold. Chromosomal duplications, losses, and rearrangements often occur such that normal ploidy is disrupted. Thus, a cancer cell may often be something other than diploid. It is clear from experimental studies that the balance between oncogene expression and tumor suppressor gene expression is a gene dosage effect.[12] For example, hybrid cell formation between a normal fibroblast and a malignant cell will usually produce a nontumorigenic hybrid if one malignant chromosome set is present but not if there are two malignant sets. Furthermore, hybrids containing two copies of a chromosome bearing a tumor suppressor gene show more stable suppression of the malignant phenotype than cells having only one copy. As noted in Chapter 7, the finding of "dominant" oncogenes is really a cell culture phenomenon, resulting from the neoplastic transformation of cells like mouse 3T3 cells after transfection with an activated oncogene. This sort of transformation event is seldom seen if normal diploid cells are used. Moreover, when malignant cells expressing a known oncogene are fused with normal diploid fibroblasts, malignancy is usually suppressed even though

Table 8-3 Properties of Oncogenes and Tumor Suppressor Genes

Property	Oncogenes	Tumor Suppressor Genes
Mutational events involved in cancer	One	Two
Function of mutation	Gain of function ("dominant")	Loss of function ("recessive")
Germ-line inheritance	No	Yes
Somatic mutations	Yes	Yes
Effect on growth control	Activate cell proliferation	Negatively regulate growth promoting genes
Effects of gene transfection	Transform partly abnormal fibroblasts (e.g., NIH3T3)	Suppress malignant phenotype in malignant cells
Genetic alterations	Point mutations, gene rearrangements, amplification	Deletions, point mutations

the oncogene continues to be expressed.[12] (For this reason the term *tumor suppressor gene* is preferred to "antioncogene".[5]) Thus, the terms *dominant* and *recessive* do not retain the classic Mendelian meaning in cancer.

A word about the mechanisms of action of the tumor suppressor genes is warranted here, even through this is discussed in more detail for each gene below. Some of the suppressor gene products are localized in the cell nucleus and act as transcription factors. Some occur at the cell membrane and act in signal transduction, cell adhesion, or production of a normal extracellular matrix. Others appear to act as conduit for cell membrane–cytoskeleton interactions. At least two (and probably many more yet to be identified) are involved in DNA repair.

PROPERTIES OF INDIVIDUAL TUMOR SUPPRESSOR GENES

RB

Characterization of the RB Protein

The *RB* gene, about 200 kilobases (kb) in length, is located on chromosome 13q14 and has 27 exons coding for a protein of 105 to 110 kDa, depending on the species in which it is produced (reviewed in Ref. 11). It is a nuclear protein and acts to regulate the cell cycle (see below and Chap. 10). Mutations in the *RB* gene have been detected in retinoblastomas, osteosarcomas, bladder, small-cell lung, prostate, breast, and cervical carcinomas and some types of leukemia. In contrast to hereditary retinoblastomas, mutations of *RB* in these other cancers appear to be somatic rather than germ-line because children with the inherited mutant allele may later develop osteosarcomas but only rarely get the other tumors mentioned, even though all the cells in their bodies must bear the mutation. Hence, different cell types respond differently to a germ-line *RB* mutation. Malignant transformation in the tissues, therefore, must require additional mutations, probably because cells have redundant means to provide cell cycle regulation. The frequency of *RB* mutations detected in various tumor types also varies. Such *RB* mutations or deletions are seen in almost all, if not all, retinoblastomas, 60% of small-cell lung carcinoma cell lines, about 33% of bladder and mammary carcinomas, and to a much lesser extent in other tumor types; yet *RB* gene alteration is thought to be important for the carcinogenic process only in retinoblastomas, osteosarcoma, and small-cell lung cancers.[11,13] But this may simply reflect our lack of knowledge of the molecular biology of these tumors.

A number of *RB* gene mutations have been detected in various tumor types, including frameshift and chain termination mutations, deletions of entire exons, and point mutations. Many of these mutations affect domains between amino acids 393–572 and 646–772, which are involved in binding of viral proteins such as SV40 large Tag, adenovirus E1A, or human papillomavirus E7, and cell-cycle regulatory proteins. Some *RB* gene mutations also decrease the ability of the RB protein to be phosphorylated.[14]

There is also an RB-related protein of 107 kDa, called p107, that binds the viral proteins noted above and also appears to play a role in cell-cycle regulation. RB and p107 have about 30% sequence homology and bind similar transcription factors such as E2F (see below), but p107 affects the cell cycle differently than RB protein and the E2F-p107 complex may modulate the activation of promoters that the E2F-RB complex does not.

Interactions of RB Protein with Viral Oncogenic Proteins

The transforming proteins of three oncogenic DNA viruses—SV40, adenovirus (type 5) EIA, and human papillomaviruses type 16 and 18 E7 protein—bind avidly to the p105 RB protein and to the RB-related protein p107 (reviewed in Refs. 11 and 15). Interestingly, SV40 large Tag, adenovirus E1B, and HPV E6 bind p53, another cell-cycle regulatory protein, suggesting that oncogenic DNA viruses have captured this mechanism to work their will on the replicative machinery of the cell in order to make sure that the enzymes for nucleotide synthesis, DNA polymerases, etc., are there to foster their own replication.

SV40 Tag, adenovirus E1A, and HPV E7 contain homologous regions of amino acids that are involved in p105 and p107 binding. If these regions are altered or mutated, binding of these DNA virus oncoproteins to these RB-type proteins is inhibited and their transforming ability is diminished, strongly suggesting that ability to bind these cell-cycle regulatory proteins is de rigueur for their ability to induce a malignant phenotype.

Role of *RB* in Reversing the Malignant Phenotype

Introduction of a wild-type, nonmutated *RB* gene by retroviral- or transfection-mediated gene transfer into a variety of human cancer cells that have an inactivated *RB* gene results in reversion to a more normal phenotype, including reversal of morphologic transformation, growth rate, growth in soft agar, and tumorigenicity in nude mice.[13,16,17] Such reversal has been noted for retinoblastoma, osteosarcoma, bladder, and prostate carcinoma cells. These data demonstrate that normal function of RB is crucial for maintenance of cell growth control.

Requirement of a Functional *RB*-1 Gene in Development

Surprisingly, gene knockout of the *RB*-1 gene by homologous recombination in mouse embryonic stem cells, followed by microinjection of the ES cells into blastocysts and implantation into foster mothers, has shown that the embryos survive until about 14 to 15 days and then die due to massive cell death in the developing central nervous system and lack of hematopoiesis, particularly of erythroid cells.[18–20]

The reason that this is surprising is that if all cell types require *RB* gene expression for regulation of the cell cycle, then how can embryos survive for 14 to 15 days, a time during which a number of cell lineages have already developed, and why is the major defect noted in only two tissues? Another puzzling thing is that heterozygous mice, developed by knocking out only one *RB* allele, survived for up to 11 months; however, some of these animals developed pituitary adenocarcinomas, but none developed retinoblastomas.[19] In these pituitary carcinomas, the remaining wild-type allele was lost, so in this case the two-hit hypothesis for tumor development held up. It should also be noted that transfer of a normal human *RB* minitransgene into the mutant mice corrected the developmental defects.[18] One can only conclude from data of this sort that not all cell lineages rely exclusively on RB for control of cell proliferation and differentiation and that there are species differences in the target cells for neoplastic transformation after abrogation of RB function.

There is, however, a report that expression of SV 40 large Tag, driven by a luteinizing hormone β gene promoter, in transgenic mice produced heritable ocular tumors similar to human retinoblastoma, and an association between SV40 Tag and RB p105 was shown in the tumor tissues.[21] These latter results suggest that if RB function is disrupted in a specific way, similar tissue tropism for a carcinogenic effect of RB knockout can occur across different species.

Cell-Cycle Regulation by RB

Cell-cycle regulation is covered in more detail in Chapter 10, but some points about the role of RB protein in this process are discussed briefly here. The fact that the RB protein plays a key role in cell-cycle regulation and that its activity in this role is determined by its phosphorylation state is now well established.[22–24] RB is a nuclear phosphoprotein (pRB) that is hypophosphorylated through most of the "resting" G_1 phase of the cell cycle. There are at least 10 serine/threonine phosphorylation sites on RB, and it is a substrate for phosphorylation by the cyclin-dependent kinases (CDKs). Both cyclin E, which appears at the G_1 restriction point, and cyclin A, which is newly synthesized at the beginning of S phase, can phosphorylate RB[24] and are probably both involved in maintenance of a hyperphosphorylated state of pRB. Hyperphosphorylated pRB binds less tightly to its "nuclear anchor" binding site, which keeps it in the nucleus, and less tightly to the transcription factor E2F, involved in activating genes progressing through the cell cycle. Hypophosphorylated pRB may dampen this activation signal by "sequestering" E2F so that it cannot bind to its promoter/enhancer DNA sites. The growth inhibitory function of pRB can also be downregulated by binding to viral oncoproteins such as SV40 Tag, E1A, or E7, which bind to hypophosphorylated pRB, or by mutations that alter the ability of pRB to bind to its nuclear anchor protein(s). Any of these three events then—hyperphosphorylation, binding to oncoproteins, or mutations—could have the same end result, namely, the inability of pRB to inhibit cell-cycle progression. Since normal cells do not usually carry SV40 Tag, E1A, or E7 oncoproteins, there is presumably a normal cellular pRB binding protein whose binding is displaced by the viral proteins. At least two such genes that encode RB-binding proteins (RBP-1 and RBP-2) have been cloned.[25] The binding of the viral proteins, in contrast to the normal pRB binding proteins, may then displace pRB from its nuclear anchor, causing it to be lost from the nucleus. Normal cell-cycle progression, as opposed to viral or oncogene-induced cell-cycle progression, is mediated through normal mitogenic signals that turn on cyclins and cyclin-dependent kinases that phosphorylate pRB. Another way to look at it is that binding of pRB to E2F prevents E2F from activating cell-cycle progression genes, and either hyperphosphorylation, binding to viral oncoproteins, or mutation in the pRB-binding "pocket" for E2F can displace E2F to carry out its growth-promoting duty. Of note also is the observation that microinjection of pRB into cycling osteosarcoma cells in culture caused cell cycle arrest and coinjection of pRB with c-Myc but not Ha-Ras, c-Jun, or c-Fos inhibited the ability of pRB to arrest the cell cycle.[26]

Interactions of RB Protein with Transcription Factors and DNA Regulatory Elements

The RB protein pRB has both an indirect and direct role in regulating gene transcription. As noted above, it can bind the transcription factor E2F and prevent it from activating genes involved in cell proliferation. E2F is a transcription factor originally identified by its role as an activator of the adenovirus E2 promoter. Similar promoter sequences have been found in the genes encoding dihydrofolate reductase, thymidylate synthase, DNA polymerase α, ribonucleotide reductase, c-*myc*, N-*myc*, and c-*myb* (reviewed in Ref. 27). All of these genes are involved in turning on cell division, and E2F appears to regulate their expression. One way that these genes can be negatively regulated is by the binding of pRB to E2F. The domains in pRB that are involved in the binding "pocket" for E2F have been identified, and it is this same region of pRB that is involved in the binding of SV40 Tag, E1A, and E7, and that is inactivated by deletion or point mutation in human tumors.[27–29] Add to this the observations that only underphosphorylated pRB binds to E2F, that the E2F-pRB complex becomes prevalent during the G_1 phase of the cell cycle, and that the phosphorylation of pRB occurs close to G_1/S transition and releases it from E2F,[28,30] and you have a compelling argument that pRB is an important negative regulator of E2F transcriptional activation and that the phosphorylation state of pRB is what controls the formation of the E2F-pRB complex. As noted previously, this complex can also be dissociated by competition of viral oncoproteins for binding to pRB or be prevented from forming by mutations in pRB.

In addition to negatively regulating cell proliferation by inactivating E2F, pRB has some more direct actions as a transcriptional regulator. pRB has ability to bind DNA itself, and it has been shown to repress c-*fos* expression and AP-1 transcriptional activity in cycling 3T3 cells. Thus, pRB appears to be able to bind to its own *cis*-acting control element termed RCE.[31] The pRB protein also induces TGF-β1 gene expression in epithelial cells, in which TGF-β1 is a growth-inhibiting factor, and represses TGF-β1 expression in fibroblasts, in which TGF-β1 can act as a growth promoter.[32] Furthermore, pRB has been found to activate expression of TGF-β2 in epithelial cells via an action at ATF-2 transcriptional regulatory elements.[33]

p53

Characterization of p53 and Its Mutations

Originally, p53 was thought to be an oncogenic protein. This 53-kDa protein was first detected as a complex with SV40 Tag in SV40-transformed cells.[34] A similar complex was found between p53 and E1B protein in adenovirus-infected cells. The p53 protein was subsequently found in a variety of transformed mouse cell lines, cultured human tumor cells, and in virally, chemically, or radiation-induced murine tumors.[35–38] Even more indicting was the fact that transfection of the p53 gene was found to immortalize and transform cells[39,40] and to cooperate with *ras* in inducing transformation.[41] It was also noted that the cellular half-life of p53 was increased in SV40-transformed cells, a fact that at the time was thought, for the wrong reasons, to foster the transforming action of p53. This latter observation will resurface again later in the story.

Only gradually, over about 5 years following the observations of the transforming ability of p53, did it become clear that the transforming effects of p53 were due to a mutant protein and that the nonmutated, wild-type p53 negatively controls cellular proliferation and suppresses cell transformation and tumorigenesis.[42–44]

The p53 gene is located on chromosome 17p13 in the human genome. The gene contains 10 coding exons and is expressed in all cells of the body, although at low levels in most tissues. The human protein is 393 amino acids long and contains at least nine potentially phosphorylated serine residues, one of which, serine 316, is phosphorylated by a cyclin-dependent kinase (reviewed in Ref. 11). There is a nuclear translocation domain near the CDK phosphorylation site, suggesting a cell-cycle-dependent signal for nuclear translocation of p53. Nuclear localization is important for p53 to function as a negative regulator of cell proliferation and as a tumor suppressor gene. In some human cancers, a defect in p53 function relates to its sequestration in the cell cytoplasm and inability to be transported to the cell nucleus.

There is high sequence homology for p53 among animal species; for example, there is about 56% amino acid homology all the way from frogs *(Xenopus)* to humans, with 90% to 100% homology in some regions of the protein. Interestingly, these are the regions that are most often found to contain mutations in human cancer, strongly implicating these regions of the protein as important to its regulatory functions.

Mutations of p53 are the most common genetic alterations observed in human cancers (50% to 60%),[11] and there are several "hot spots" for these mutations (Fig. 8-1). Most of the mutations are missense point mutations in carcinomas, whereas in sarcomas, deletions, insertions, and rearrangements are more common and point mutations are rare. Some sarcomas contain an amplification of an oncogene called *mdm2*, whose protein product inactivates p53 (see below). Different mutational hot spots of the p53 gene are seen in different tissues. For example, 53% of liver cancers in areas of high endemic exposure for hepatitis B infection and aflatoxin B1 have mutations in codon 249.

Germ-line mutations of p53 are also observed in some families with a high incidence of cancer. The Li-Fraumeni syndrome is one such case. Many members of these families have missense and nonsense mutations in one p53 allele and tend to get osteosarcomas, adrenocortical carcinomas, breast carcinomas, or brain cancers, often at an early age.[45,46] Curiously, colon carcinoma is not prevalent in these families, even though p53 mutations are often seen in colon cancer, suggesting that germ-line mutations tend to make certain tissues more susceptible to later somatic mutation than other tissues or that some tissues have additional mechanisms for regulating cell proliferation that must be knocked out before p53 mutations become important for the cell's economy.

Mutations in the p53 protein can have at least three phenotypic effects: (1) loss of function, in which a missence mutation abrogates p53's ability to block cell division or reverse a transformed phenotype; (2) gain of function, as demonstrated by the introduction of a mutant p53 gene into cells lacking wild-type (WT) p53, which induces a tumorigenic phenotype; and (3) *trans*-dominant mutation, seen when a mutant p53 allele is introduced into cells bearing a WT p53 allele, resulting in an overriding of the normal inhibitory function of p53. This latter effect is sometimes called a dominant-negative effect. As noted above, the cellular half-life of p53 in transformed cells is often longer than that of WT p53 in normal cells. The reason for that is now clear for cells bearing a p53 mutation. There are conformational differences in the mutant protein that render it less susceptible to degradation.[47] This longer half-life of the mutant form may play a role in the dominant-negative effect.

Mutagenesis of p53

As discussed above, the types of mutations of p53 vary with cell type, as do the hot spots for mutations in different tumor types. Lung tumors

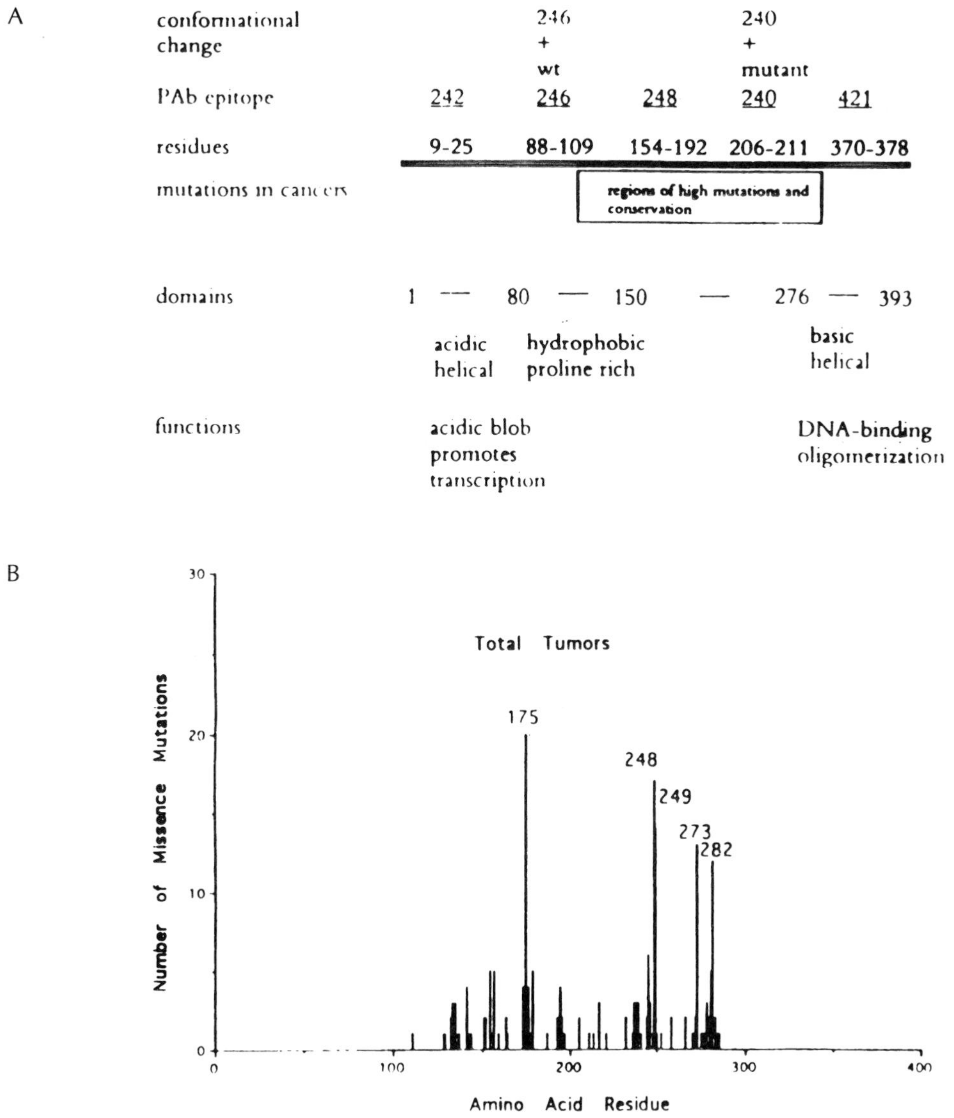

Figure 8-1 (A) A schematic representation of the domains of structure of the p53 protein. (B) The positions of p53 missense mutations in the p53 gene from 191 human cancers. The codon numbers or amino acid residue numbers are indicated on a linear representation of the protein. The height of the line at each codon indicates the number of independent times a mutation occurred at that codon. (From Levine.[11])

contain both base transition and transversion mutations, but colon tumors contain primarily base transitions, often C → T. CpG dinucleotides are frequent sites of mutation, and this raises the question of whether tissue-specific methylation patterns of C in CpG sites could play a role in the types of p53 mutations observed, since methylated C residues in CpG doublets are known to have a higher mutation rate than nonmethylated C. This also raises the issue of the type of carcinogens to which different tissues may be exposed. The instance of liver cancer, as an example, has already been mentioned. More than half of hepatocellular carcinomas (HCC) from high aflatoxin B1 (AFB1) exposure areas have G → T transversions in the third position of codon 249(AGG), which results in replacement of arginine by serine.[48] This can also be shown by exposure of human hepatocytes to AFB1 in culture.[49] On the other hand, HCC tissue taken from patients in areas of low AFB1 exposure do not usually display G → T transversions of the p53 gene, and the observed mutations are found in other regions of the gene.[50] Thus, different carcinogens can apparently mutate p53 in different segments of the gene, but the end result is the same, i.e., inactivation of p53 function.

G → T transversions of p53 occur at high frequency in tobacco-related human cancers, including small-cell and non-small-cell lung cancers, esophageal carcinomas, and squamous cell carcinomas of the head and neck (reviewed in Ref. 51). Benzo[*a*]pyrene (BP), a component of cigarette smoke, produced a high incidence (70%) of G → T transversions in BP-induced murine skin carcinomas, whereas 7,12-dimethylbenz[*a*]anthracene-induced skin tumors had a similar p53 mutation frequency but a low rate of G → T transversions. These data support the concept that different carcinogens attack the p53 gene differently. Furthermore, ultraviolet B radiation–induced mouse skin carcinomas contained a prevalence of C → T transitions.[52]

An additional important fact should be noted here. Most p53 mutations occur in the nontranscribed strand of DNA, and since the nontranscribed strand is more slowly repaired, this raises the potential for these errors to be passed on to daughter cells.[53]

Ability of p53 to Reverse Cellular Transformation and Tumorigenesis

In several diverse cell systems and tumor cell types, introduction of a WT p53 gene into cells growing in culture usually blocks cell proliferation and hangs the cells up at the G_1/S transition point in the cell cycle. Moreover, suppression of the neoplastic phenotype in culture and of tumorigenicity in nude mice is usually observed. Introduction of a mutated p53 gene, on the other hand, does not block cell proliferation or tumorigenicity and may, in fact, enhance them. Such effects have been observed in human colorectal,[54] lung,[55] and prostate[56] carcinoma cells as well as in glioblastomas,[57] osteosarcomas,[58] and acute lymphoblastic leukemia cells.[59]

Role of p53 in Cell-Cycle Progression and in Inducing Apoptosis

It is now clear that WT p53 not only has antiproliferative and antitransforming activity but also possesses the ability to induce programmed cell death (apoptosis) after exposure of cells to DNA-damaging agents such as γ-irradiation or anticancer drugs.[60] The concept that p53 is a growth-regulatory protein fits with its short half-life (5 to 20 minutes in normal mouse cells and 1 to 2 hours in normal human cells), its nuclear location and transcription factor activity (see below), and its increased synthesis in DNA-damaged cells (reviewed in Ref. 61). WT p53 regulates the transcription of a number of cell-replication–associated genes. Growth arrest induced by WT p53 blocks cells prior to or near the restriction point in late G_1 phase and produces a decrease in the mRNA levels for genes involved in DNA replication and cell proliferation, such as histone H3, proliferating cell nuclear antigen (PCNA), DNA polymerase α, and b-*myb*.[62] In order to carry out these gene regulatory events, WT p53 has to assume a certain conformational structure, apparently modulated by its phosphorylation state, and to oligomerize so that it can bind to DNA.[61] Mutant p53 cannot achieve the appropriate conformation and can block WT p53 function by forming oligomers with it. It should also be noted that the "anti-apoptosis" gene *bcl*-2 (see Chap. 10) can be downregulated via a p53-dependent negative response element in the *bcl*-2 gene.[62a] Interestingly, coexpression of the c-*myc* and *bcl*-2 genes can overcome p53-induced apoptosis by altering the intracellular translocation of p53 to the nucleus.[62b]

Not all types of apoptosis, however, are mediated by p53. For example, whereas induction of apoptosis in thymocytes by γ-irradiation or the DNA-damaging drug etoposide is via a p53-dependent pathway, that induced by glucocorticoids in thymocytes is not.[63,64] WT p53 is required for the response to DNA damage; cells having mutant or no p53 fail to respond appropriately. In fact, WT p53 enhances sensitivity to ionizing irradiation[65] and anticancer drugs such as 5-fluorouracil, etoposide, and doxorubicin,[66] whereas p53 mutations increase resistance to ionizing radiation.[67] Thus, the absence or mutation of p53 leads to an increase in cellular resistance to these agents, implying that cancer cells in patients can acquire resistance to chemotherapeutic agents or irradiation through mutations or loss of p53.

These effects may seem somewhat paradoxical, but they are understandable if one thinks of p53 as a protector or "molecular policeman" monitoring the integrity of the genome.[68] When DNA is damaged, p53 accumulates and stops DNA replication and cell division until DNA has time to repair itself. If this is not possible or the DNA repair mechanisms fail, p53 triggers a cell suicide response. Thus, in the case of massive damage to DNA in which DNA repair is not possible, the cell dies. If p53 is mutated or lost, the cell goes on its merry way, replicating its damaged DNA, passing on mutations to daughter cells, and giving cells a survival advantage in the face of DNA damage. Cells that do this are genetically less stable and accumulate mutations and gene rearrangements, leading to the generation of an ever-increasing malignant state. This sort of event could explain, at least in part, the increased rate of mutation ("mutator phenotype") seen in tumor progression (see Chap. 6).

The above findings indicate that WT p53 acts as a "checkpoint" control protein that stops the cell cycle before S phase when DNA damage is present. Thus, p53 is analogous to the RAD9 gene of yeast that inhibits cell-cycle progression following DNA damage.[69] Loss of RAD9 or p53 causes cells to undergo a greater frequency of mutations and gene amplifications.[69–71] For example, when fibroblasts from patients with Li-Fraumeni syndrome are passaged in vitro, they may lose the remaining WT p53 allele; when they do, they have a greatly increased ability to amplify drug-resistance genes in response to a drug called PALA.[70,71] Introduction of a WT p53 gene back into these cells via a retroviral vector restored cell-cycle control and reduced the frequency of gene amplification to background levels.[71] Other factors, however, may also allow gene amplification to occur in tumor cells, since tumor cells with functional p53 can still amplify genes.[70]

A possible therapeutic result may be gained by taking advantage of p53's ability to induce apoptosis in tumor cells. For example, when spheroids of human lung cancer cells grown in culture were treated with a retroviral vector containing a WT p53 gene, apoptosis was induced in the cells.[72]

p53 Binding

Even though a lot is known about the biological actions of p53 (e.g., the ability to induce G_1 arrest, to induce apoptosis following DNA damage, to inhibit tumor cell growth, and to preserve genetic stability), how it does all this is not clear. Three mechanisms have been postulated.[73] In one, p53, acting as an activating transcription factor, is thought to stimulate the expression of genes that cause growth suppression. A second postulated mechanism is via inhibition of transcription of TATA-box–controlled genes, perhaps by binding to TATA-box–binding proteins. A third hypothesis is that p53 directly interferes with DNA replication by interacting with the DNA replication apparatus.

There is some evidence for all of these. Because p53 has been found in association with a wide variety of cellular proteins, its possible actions via these interactions are legion (Table 8-4). While it is unlikely that p53 binds to all these proteins at the same time in any given cell type, it may bind to them when it is in different conformations or phosphorylation states, in different cell types, or in the same cell type under different physiologic conditions.

There is significant evidence that p53 can act

Table 8-4 p53-Associated Proteins

Cellular proteins
CBF, CCAAT binding factor
E6-AP, mediates HPV E6 interaction with p53
HSP70, heat shock protein 70
MDM2, cellular oncoprotein
RPA, replication protein A
SP1, general transcription factor
TBP, TATA binding protein
WT1, Wilms' tumor gene product
Viral oncoproteins
Ad5 E1B, adenovirus type 5 E1B
EBNA-5, Epstein-Barr nuclear antigen 5
HPV16/18 E6, human papillomavirus 16/18 E6
SV40 Tag, simian virus 40 large T antigen

Source: From Pietenpol and Vogelstein.[73]

as a transcription factor (see below). There is also evidence that p53 from human or yeast cells binds to the TATA-box–binding protein TBP, which is part of the general transcriptional regulator complex TFIID[74] (see Chap. 5). One observation was that both WT and mutant p53 bind to TBP, suggesting that functional discrimination between the two lies elsewhere in the regulatory machinery.[74] However, another report states that WT but not mutant p53 binds to human TBP.[75] In any case, it is likely that p53 regulates TATA-mediated transcriptional events in human cells. The evidence that p53 can directly interfere with DNA replication is based on the findings that it binds to SV40 Tag and interferes with Tag's ability to unwind DNA and recruit DNA pol α to the initiation complex and that p53 binds to and inhibits the function of the cellular DNA replication factor called RPA.[76]

As noted in Chapter 7, some of the DNA virus oncoproteins bind to p53. SV40 Tag, adenovirus E1B, and HPV E6 each bind to p53, blocking its ability to bind DNA and to act as a transcription factor. Thus, binding to these oncoproteins is similar to having a mutated p53 in that the tumor suppressor function is abrogated.

Another way that the function of p53 can be inactivated is binding to an oncogene product called MDM-2, the gene for which was originally discovered because it was contained as an amplified gene on a double-minute chromosome in transformed mouse cells (hence the name, derived from murine double-minute gene-2 or *mdm*-2).[77] This gene encodes MDM-2, a 90-kDa cellular phosphoprotein that was later found to copurify with p53 and to be tightly bound to it. When a WT p53-expressing plasmid was transfected into rat embryo fibroblasts that had been cotransfected with the *mdm*-2 gene, the ability of p53 to stimulate transcription of a reporter gene bearing a p53-responsive element was inhibited.[78] Thus, MDM-2 can bind to and block p53-mediated *trans*-activation. MDM-2 does this apparently by binding to the acidic activation domain of p53 and "concealing" it from its regulatory element in a promoter/enhancer DNA sequence.[79] Amplification of the *mdm*-2 gene has now also been observed in a number of human cancers (see below); for example, amplified *mdm*-2 genes are seen in about 30% of soft tissue sarcomas.[79a]

Role of p53 as a Transcriptional Regulator

Wild-type p53 binds to specific DNA sequences, termed p53 response elements, and directly stimulates the transcription of a number of genes. In addition, WT p53 negatively regulates a number of genes, including c-*fos*, c-*jun*, *hsp*70, *RB*, interleukin 6, PCNA, and the multiple drug resistance gene *mdr*1.[74,75] That p53 could act as a transcription factor was first realized from experiments in which the first 75 amino acids of p53 were fused to the DNA-binding domain of the yeast protein GAL-4 and shown to enhance transcription of GAL-4 regulated genes.[80,81] Later it was found that the p53 protein itself could act as a transcriptional activator, a function that mutant p53 has lost.[82] Binding of WT p53 to its response elements appears to require that it form a tetramer, the formation of which may be inhibited by its binding to a mutant form of p53, to viral oncogene proteins, or to MDM-2.

Examples of Genes Regulated by p53

As noted above, p53 can bind to DNA in a sequence-specific manner. The p53 response element contains a 20-bp consensus-binding site consisting of two copies of 10-bp sequences separated by a 13-bp sequence (reviewed in Ref. 83). The consensus sequence is 5′(A/G)(A/G)(A/G)C(A/T)(T/A)G(C/T)(C/T)(C/T)-3′. The p53 gene contains a strong transcriptional activating sequence in its amino terminal domain, and both copies of the 10-bp response element are required for it to bind efficiently. Gene sequences downstream from this response element can be stimulated by p53 binding. Several genes with p53 response elements have been identified. These include creatine kinase, "growth arrest DNA damage" inducible gene (GADD 45), a GLN retroviral element, *mdm*-2, and a gene called WAF1/Cip 1, whose gene product binds to cyclin-dependent protein kinases and inhibits their action.[83]

Why p53 would activate transcription of a gene (*mdm*-2) that produces a protein which inactivates it is one of those biological paradoxes for which nature is famous. It is possible that this represents a feedback loop by which overproduction of p53 could be downregulated, i.e., by production of an antagonist to its action.[83a] It is not clear, though, why a cell would need this escape hatch to start proliferating again, since the cellular $t_{1/2}$ of normal p53 is so short.

The induction of WAF1/Cip1 by p53 appears to be key to its action as a tumor suppressor. The WAF1/Cip 1 gene promoter contains a p53 response element. This gene encodes a 21-kDa protein that binds CDKs and arrests cells in midcycle. Mutant p53 protein does not induce the WAF1/Cip gene. Introduction of the WAF1/Cip gene into cultured human brain, lung, and colon cancer cells suppresses their proliferation.[83]

Some of the protein structural requirements for p53 to bind to DNA have been defined. As noted, the N-terminus contains the transactivating domain of the protein. The C-terminus contains a nuclear localization signal, the DNA-binding domain, and sequences involved in protein–protein interactions. Adjacent to the protein–protein oligomerization domain is a conserved serine, which is a phoshorylation site for casein kinase II. Phosphorylation by CKII can activate DNA-binding by p53.[84] However, the majority of mutations in p53 are in the hydrophobic midregion of the protein, and binding of p53 to other cellular proteins and to DNA are disrupted by these mutations,[85,86] indicating that the correct conformation (i.e., its protein folding state) is a crucial requirement for placing the functional domains in the right configuration.

The p53 gene is expressed at high levels in certain human cancers (see below, under "Clinical Studies"), and in most cases it appears to be a mutant form that is overexpressed. However, in some types of cells, it may be the WT p53 that is overexpressed. This creates a paradox, one that has stirred up some controversy because it forces one to pose the antidogmatic question: Can the untimely expression or overexpression of WT p53 itself favor the survival of cells with a transformed phenotype? This question makes people feel uncomfortable because the obvious answer is a resounding no or "why would anyone ask such a stupid question?" Well, perhaps one should be reminded of the fact that the function of transcription factors is context-dependent, both from the point of view of the cell type and of the genes being regulated. The same transcription factor can activate some genes and repress others, and there is some cell-type specificity to their expression and function (see Chap. 5 for a number of examples). In the case of p53, it may depend on when and how it is expressed in the lifetime of cells that determine its role in the cellular economy. A few examples can be cited. The p53 gene was reported to be produced in acute myelogenous leukemia (AML) blast cells but not at detectable levels in normal myelopoietic cells, and there was a correlation with p53 expression and secondary plating efficiency of the AML blasts.[87] WT p53 is expressed at elevated levels in murine epidermal carcinoma cells compared to normal epidermal cells and benign papillomas.[88] Moreover, treatment of AML blasts in culture with an antisense oligonucleotide to p53 demonstrated selective toxicity for human AML blasts over normal bone marrow cells.[89] While this effect might be due to blocking the translation of a mutant form of p53, p53 from primary AML blasts is usually wild-type in the conserved domains (reviewed in Ref. 89). A possible explanation for this phenomenon is that there is a "rebound" overexpression of p53 after initial lowering of p53 protein. This overexpression could foster apoptosis in the antisense-treated AML cells.

Effects of Altered p53 Expression in Transgenic Mice

In order to investigate the role of p53 in mammalian development and malignant transformation in vivo, Donehower et al.[90] abrogated the function of the p53 gene by a gene knockout approach using homologous recombination in murine embryonic stem cells followed by implantation into a blastocyst. Somewhat surprisingly, mice homozygous for the null allele (i.e., no functional p53 allele was present) developed normally, indicating that a normal p53 gene is not required for the growth and differentiation of normal tissues. However, by about 6 months of age, these animals began to develop a variety of neoplasms, including sarcomas, lymphomas, and embryonal carcinomas. Thus, normal developing cells must have a redundancy of mechanisms to regulate cell proliferation, a conclusion similar to that reached for the *RB* gene knockout experiments described above. Why, then, do these animals develop tumors later in life? One hypothesis is that so long as DNA in developing cells is intact, p53 expression may be of no great importance, but once DNA damage begins to accumulate with the "hard knocks" of living (endogenous DNA damage due to generation of oxygen radicals, spontaneous base deamination, depurination, etc.), p53 function is required to guard the genome from allowing DNA lesions to go unrepaired and passing mutations on to daughter cells. Thus, in the absence of p53, these genetic lesions begin to accumulate and tumorigenesis occurs.

Clinical Studies of p53

The accumulated literature on p53 gene alterations in human cancers is truly prodigious and too vast to cite here. A summary of these data is, however, attempted, and the reader is referred to some recent reviews.[91–93b] Mutations of the p53 gene have been reported for almost every type of human cancer and occur in about 50% of human cancers taken as a whole, even though the frequency varies among cancer types. The catalogue of cancers in which p53 mutations have been detected includes cancers of the bladder, bone, brain, breast, cervix, colon, esophagus, hematopoietic cells, larynx, liver, lung, lymphoid tissue, ovary, pancreas, prostate, skin, stomach, thyroid, and uterine endometrium. Mutations of the p53 gene are very common in colorectal cancers (70%), lung cancers (50%), and breast cancers (40%).

In addition to mutations, p53 is expressed at high levels in a number of tumor types, and its overexpression is an unfavorable prognostic indicator for several cancers, including breast,[94,95] ovarian,[96] endometrial,[97] esophageal,[98] lung,[99] and urinary bladder carcinomas[100] as well as malignant melanoma.[101] While mutation and/or overexpression of p53 appear to be late events in certain human cancers such as colon carcinoma,[102] these events may occur earlier in other cancers. For example, p53 was found to accumulate in about 30% of early dysplastic lung tissues, 68% of lung carcinomas in situ, and 80% of invasive tumors,[103] indicating that abnormalities of p53 expression begin early in lung carcinoma and may provide an early diagnostic marker for this disease. In some cancers, mutations of the p53 gene appear to occur at low frequency, as in prostate and cervical cancer. Nevertheless, their occurrence has a negative implication for survival.[93] In cervical cancer, the loss of p53 function is related to increased degradation of the WT protein due to its binding by the human papillomavirus E6 oncoprotein, which targets it for proteolytic digestion.

The serum of some cancer patients contains antibodies to p53 protein.[93,104] Such antibodies have been found in patients with breast, lung, colon, ovarian, and lymphoreticular cancers. These antibodies react with both mutant and WT p53, so the significance of their presence is not clear. It may be that dying cells release p53, normally only an intracellular protein, into the circulation, causing an immune response to be mounted. Possibly, the titers of these antibodies could be analyzed in prospective clinical trials to test their diagnostic and prognostic significance.

As mentioned above, p53's apparent yin-yang partner in regulating cell proliferation, *mdm*-2, is overexpressed in a number of human cancers. Overexpression of *mdm*-2 in cells increases their tumorigenic potential and overcomes the antiproliferative activity of WT p53. Overexpression of *mdm*-2 can result from gene amplification, as is the case in about 30% of examined human bone and soft tissue sarcomas[79a,105] and in about 10% of malignant gliomas.[106] Overexpression of *mdm*-2 appears to occur by increased gene transcription without amplification in leukemias[107] and breast carcinoma cell lines.[108]

Wilms' Tumor Suppressor Gene *wt*-1

Wilms' tumor is a renal cancer called nephroblastoma that occurs in children; in some cases, there is a genetic predisposition. In about 10% of cases, the tumors are bilateral. It is a rare tumor, occurring in about one in 10,000 children, and is associated with alterations at distinct loci on chromosome 11. Certain distinct clinical syndromes have been associated with chromosome 11 abnormalities (reviewed in Refs. 11 and 109). A deletion in the short arm of chromosome 11 at band 13p is associated with the WAGR syndrome (Wilms' tumor, aniridia, genitourinary malformation, and mental retardation). Genetic mapping of this region in tumor and normal tissue from these patients led to the identification of a gene called *wt*-1, which has mutations and loss of heterozygosity in Wilms' tumor tissue.[110]

Thus, *wt*-1 is a gene locus that has the properties of a tumor suppressor gene that plays an important role in urogenital development.

Other syndromes have also been found in association with chromosome 11 defects, including the Beckwith-Wiedemann syndrome, in which the 11p15 locus is involved. These patients also have a high incidence of Wilms' tumors, implicating a second tumor suppressor gene locus on chromosome 13 called *wt*-2. Both the *wt*-1 and *wt*-2 loci have tumor suppressor activity when introduced into Wilms' tumor cell lines tested for tumorigenicity in nude mice.[111]

Abnormalities of chromosome locus 11p have been observed in several different human cancer types. Loss of heterozygosity (LOH) for DNA markers at 11p has been seen for rhabdomyosarcoma, hepatoblastoma, hepatocellular, bladder, breast, non-small-cell lung, ovarian, and testicular carcinomas as well as Wilms' tumor (reviewed in Ref. 112). Deletions of the short arm of chromosome 11 are associated with hepatitis B virus integration, and human fibroblasts with 11p deletions are sensitive to transformation by HPV and BK viruses. Introduction of human chromosome 11 into BK virus-transformed mouse cells suppresses their tumorigenicity.[112]

The 50-kb *wt*-1 gene at 11p13 has 10 exons and codes for a 3-kb mRNA, but a number of mRNA splice variants have been detected. The WT-1 gene product is a 46-to-49-kDa proline/glutamine-rich protein containing four zinc-finger DNA-binding motifs (reviewed in Ref. 109). These domains have homology to the early growth response (EGR) family of transcription factors, but WT-1 binding to at least one of the EGR response elements (EGR-1) represses transcription rather than stimulating it.[113] In the mouse, WT-1 is first expressed at day 8 of gestation in intermediate mesoderm and subsequently in differentiating mesothelium, spinal cord, brain, and the urogenital ridge. Expression peaks at day 17 of gestation (which is 21 days long in the mouse) and is low in adult tissues, indicating a key role for WT-1 in development. This is borne out in gene knockout experiments in transgenic mice. Homozygous loss of the *wt*-1 gene results in embryonic death about day 11, with abnormal development of the kidneys, gonads, heart, lungs, and mesothelium.[109]

In addition to LOH at the 11p13 locus, point mutations and small deletions of one allele of *wt*-1 have been observed as germ-line defects in some children with a genetic predisposition to develop Wilms' tumor. The tumors from these children have loss of the remaining wild-type allele by chromosomal nondisjunction or recombination events (reviewed in Ref. 114), fulfilling Knudson's two-hit hypothesis. One observed point mutation in the *trans*-activation domain of *wt*-1 found in a Wilms' tumor patient converts the encoded protein from a transcriptional repressor to an activator of the EGR-1 promoter.[114]

In addition to EGR-1 sites, WT-1 can also act as a transcriptional repressor of other growth-related genes, including insulinlike growth factor II and platelet-derived growth factor (PDGF) A chain. However, under some circumstances, WT-1 can activate these genes.[115] It has been found that WT-1 can activate or suppress transcription from similar response elements, depending on how it is bound. It does this by acting through separate functional domains. Suppression of growth-related genes occurs by binding to two independent binding sites 5′ and 3′ relative to the transcription start site. WT-1 functions as a transcriptional activator when it binds only at the 5′ or 3′ site but not both.[116] Amino acids 84 to 179 are required for transcriptional suppression, whereas the domain containing amino acids 180 to 294 mediates transcriptional activation. A second WT-1 DNA binding site has been found that is also involved in transcriptional suppression of growth-related genes, including PDGF-A, Ki-*ras*, EGF receptor, insulin receptor, c-*myc*, and tumor growth factor β3.[115]

How the opposing roles of WT-1 may be modulated in cells is not clear. It may be that certain mutations or deletions disrupt the ability of WT-1 to bind to the DNA sites required for transcriptional repression, leaving only the activation signal or no signal at all in the protein. Another possibility is that interaction with other cellular proteins determines the availability of WT-1 binding sites in a cell-context-dependent manner. For example, WT-1 has been shown to form complexes with p53, and this interaction modulates the ability of these two tumor suppressor proteins to regulate their responsive genes.[117] In the absence of p53, WT-1 is a *trans*-activator rather than a repressor of EGR-1 genes. Furthermore, WT-1 binding to p53 enhances p53's ability to *trans*-activate the muscle creatine kinase promoter. It is also possible that dominant-negative mutations may occur in WT-1 that prevent the normal function of a remaining normal *wt*-1 allele. Such a mutation, involving a deletion of the third zinc finger, has been observed.[117]

Adenomatous Polyposis Coli (*APC*) Gene

Familial adenomatous polyposis (FAP) is an autosomal dominant disease that occurs in one out of 10,000 individuals in the United States, Europe, and Japan and accounts for about 10% of colorectal cancers.[11] These patients develop thousands of colonic polyps during the second to third decade of life, and a small percentage of them become cancerous. However, these people are highly likely to develop colorectal cancer eventually unless they are treated (usually by colectomy). The gene involved in this disorder was found on chromosome 5q21 and cloned.[118,119] This gene, called *APC,* was found to contain point mutations in the germ line of patients with FAP. Frameshift, nonsense, and missense mutations, clustered in the first third of the structural gene, have also been found in these individuals. LOH of this genetic locus has been observed in 35% to 45% of colorectal cancers in patients who do not have FAP, indicating that the *APC* gene has an important tumor suppressor function in colorectal tissue. The *APC* gene has the information for a very large protein of 2843 amino acids, but no clear function for the APC gene product has been assigned. The protein does, however, bind to α- and β-catenins that are associated with and important for the function of the adhesion molecule cadherin,[119a,119b] suggesting a role for APC in cytoskeletal–extracellular matrix (ECM) interactions that control cell growth and differentiation. It should be noted that other tumor suppressor proteins also play a role in cell cytoskeleton-ECM interactions (see below).

Deleted in Colorectal Cancer (*DCC*) Gene

LOH and allelic loss at chromosome 18q is common in colorectal cancer, occurring in more than 70% of carcinomas and about 50% of large adenomas. Since this defect is much less frequent in small, early-stage adenomas, it is thought to contribute to tumor progression more than initiation and to be altered by somatic mutational events. A gene deleted at 18q21 in colorectal cancer has been cloned and called *DCC.*[120] *DCC* mRNA is reduced or absent in more than 85% of the colorectal cancer cell lines studied, and it is found at low levels in several tissue types, including normal colonic mucosa. As opposed to the *APC* gene, which is only found mutated in colorectal but not other human cancers, the *DCC* gene shows LOH or loss of expression in colorectal, gastric, esophageal, pancreatic, and prostatic carcinomas (reviewed in Ref. 121). The DCC protein has significant amino acid–sequence homology with the neural cell adhesion molecule N-CAM, suggesting that it has a role in cell–ECM interactions, the loss of which might be involved in tumor invasion and metastasis (see Chap. 11).

Hereditary Nonpolyposis Colorectal Cancer (*HNPCC*) Gene

The HNPCC syndrome, also known as the Lynch syndrome, occurs in about one of every 200 people and increases the risk of developing colonic, ovarian, uterine, and kidney cancers, often before 50 years of age.[122] Studies of affected families indicated a linkage to a chromosome 2p locus. This predisposition to cancer is inherited in an autosomal dominant manner. It was thought that the *HNPCC* gene would turn out to be a tumor suppressor gene, but—unlike the case with many such genes—both alleles of the affected chromosome 2p locus were retained in HPCC tumors, whereas other tumor suppressor genes are usually lost or inactivated during tumorigenesis (reviewed in Ref. 123). Studies of HNPCC colorectal cancers and a subset of sporadic colorectal cancers with a similar pathologic pattern revealed alterations in microsatellite DNA involving abnormal dinucleotide or trinucleotide repeats (insertions or deletions). These and other data indicated that HNPCC and a subset of sporadic colorectal tumors were related to a heritable defect producing replication errors of microsatellite sequences, so-called RER^+ cells.

The fact that the RER^+ phenotype was reminiscent of some mismatch repair defects in bacteria and yeast was a serendipitous clue that led to characterization of the gene, once it was cloned.[123–125] There are at least three ways in which mismatched nucleotides arise in DNA: (1) deamination of methyl C to T, creating a G–T mispair; (2) misincorporation of a nucleotide during DNA replication, e.g., at an apurinic site; and (3) genetic recombination producing heteroduplexes with mismatched bases (reviewed in Ref. 124). All organisms from *Escherichia coli* to humans have enzyme systems to repair such defects.

The affected locus in HNPCC maps to chromosome 2p22-21 and contains the human homologue (hMSH2) of a bacterial gene MutS that is responsible for mismatch recognition in methyl-directed mismatch repair and of a yeast gene *msh* 2, mutants of which in yeast cause instability of dinucleotide repeat sequences. The mutation rate of $(CA)_n$ repeats was directly measured in RER^+ human tumor cells and shown to be at least 100-fold that of RER^- cells.[125] This increased mutation rate was due to a defect in strand-specific mismatch repair. The finding that $(CA)_n$ repeats are unstable in RER^+ cancer cells and defective in the gene known to stabilize repetitive sequences in *E. coli* and *Saccharomyces cerevisiae* supports the idea of a mutator phenotype since these defects would be expected to accumulate with time and cause genetic instability. The *HNPCC* gene is the first example of a DNA repair defect being associated with a tumor suppressor function. A second such defect to a gene called hMLH1 has been reported (see Chap. 3), and more are likely to be discovered, since DNA repair is so important for maintaining the integrity of the genome.

Neurofibromatosis Genes *NF*-1 and *NF*-2

Neurofibromatosis (von Recklinghausen's disease) varies from a mild form with café-au-lait spots on the skin to a severe form with large disfiguring neurofibromas resulting from the tremendous overproliferation of Schwann cells. The syndrome affects one out of every 3500 people, and in 50% there is an inherited defective gene or genes. One of these, *NF*-1, maps to chromosome 17q11,[126] and like some other tumor suppressor genes such as *RB*, *wt*-1, and p53, can act in a dominant-negative fashion. Inheritance of one defective allele and subsequent loss or inactivation of the remaining normal allele occur in this disease. The *NF*-1 gene encodes a protein with significant sequence homology to GTPase-activating proteins (GAP) that modulate the function of the Ras oncoprotein.[127] Indeed, NF-1 has been shown to bind to human Ras p21 and stimulate GTPase activity.[128] These data suggest that loss of NF-1 GAP activity would keep Ras in its active Ras-GTP state and prolong the signal for cell proliferation. Somewhat curiously, although NF-1 is expressed in all tissues of the body, mutations have so far been found only in neurofibromas and not other cancers. The fact that both GAP and NF-1 regulate the function of Ras suggests that there is redundancy in the regulation of Ras and that tissues susceptible to carcinogenic transformation by loss of NF-1 activity have little regulatory control of Ras by GAP, leaving NF-1 as the key regulator.[129]

A second neurofibromatosis susceptibility gene, *NF*-2, has been cloned; it appears to connect the cell membrane to the internal cytoskeleton, suggesting that its loss of function causes cytoskeletal disorganization that leads to abnormal cell proliferation.[130,131] *NF*-2 maps to chromosome 22q12 and mutations found in tumors (usually vestibular schwannomas or meningiomas) often result in truncated protein products.

Von Hippel-Lindau (VHL) Syndrome and Renal Cell Carcinoma Gene

The VHL syndrome is dominantly inherited and predisposes carriers to develop one or more of three types of cancer: brain hemangioblastomas, pheochromocytomas, or renal cell carcinomas.[129] The *VHL* gene maps to chromosome 3p25; its protein product is a cell surface molecule that, like NF-2 and DCC, appears to be involved in cell surface–ECM interactions and/or signal transduction mechanisms.[132] LOH and translocation of other chromosome 3p markers in renal cell carcinomas strongly suggest that there are other tumor suppressor genes to be found on this chromosome.

Indeed, it is already clear that many other tumor suppressor genes are going to be found in human cancers. Several others have been identified and are awaiting cloning and characterization of their function. These include the *NB*1

1p36 locus in neuroblastoma, the *MLM* 9p21 locus in melanoma, *MEN* 1 at 11q13 in pituitary adenoma, *BCNS* at 9q31 in medulloblastoma, *RCC* at 3p14 in renal cell cancers, and *BRCA*1 at 17q21 in breast and ovarian cancer.[129]

IDENTIFICATION OF TUMOR SUPPRESSOR GENES

One way to identify new tumor suppressor genes is to ask, by way of cell hybridization and chromosome transfer experiments, which human chromosomes can suppress the malignant phenotype. As noted at the beginning of this chapter, the original findings that led to the tumor suppressor gene hypothesis were that the tumorigenic phenotype could be suppressed when malignant cells were fused with normal cells. Although that was a big advance in our knowledge, it was difficult to determine specifically which chromosomes harbored the tumor suppressor gene or genes. In order to get around this difficulty, a technique to transfer single chromosomes into cells has been developed (Fig. 8-2).[132] This technique involves isolating chromosomes from colcemid-treated cells, which prevents mitotic spindle formation, allowing each chromosome to condense as an individual unit within its

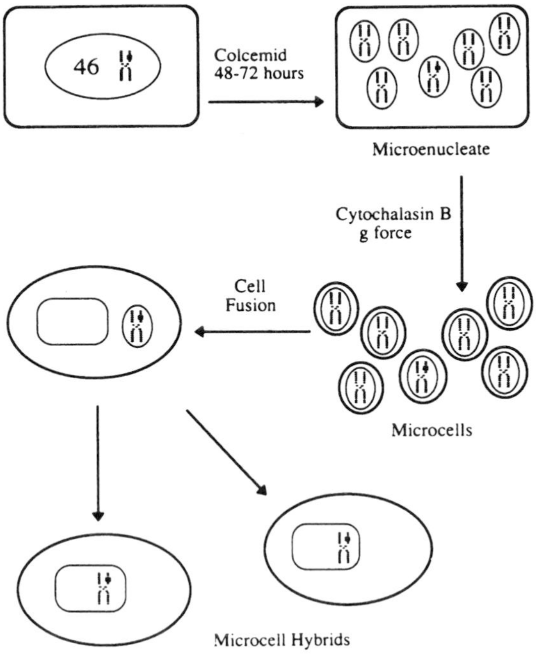

Figure 8-2 Diagram of the microcell transfer procedure. A more detailed explanation of the technique is outlined in the text. The marked chromosome represents the human chromosome containing a dominant selectable gene (i.e., the bacterial *neo* gene or *gpt* gene). (From Anderson and Stanbridge.[133])

Table 8-5 Tumor Suppression Associated with the Transfer of Single Human Chromosomes via Microcell Fusion

Tumor Cell Line	Tumor-Suppressing Chromosome: Expected[a]	Suppressed	Nonsuppressed
Cervical carcinoma (HeLa)	11	11	X
Cervical carcinoma (SiHa)	—	11	NR
Retinoblastoma	13	13	NR
Renal cell carcinoma	3	3	11
Wilms' tumor	11p13	11p15	X, 13
Colorectal carcinoma (COKFu)	5, 17, 18	5, 18	11
Colorectal carcinoma (SW480)	5, 17, 18	4, 17, 18	15
Endometrial carcinoma	—	1, 6, 9	11
Melanoma	6	6	NR
Neuroblastoma (NGP)	1	1p, 17	11
Neuroblastoma (SK-N-MY)	1	1	11
Fibrosarcoma (HT1080)	1	1, 11	2, 7, 12
Fibrosarcoma (HT1080)	1	17	1, 11, 13
Rhabdomyosarcoma (A204)	11	11	NR
Rhabdomyosarcoma (RD)	11	11p, 11q	NR
Bladder carcinoma	13	13	NR
Breast carcinoma	11	11	NR
Prostate carcinoma	13	13	NR

[a]Predicted from cytogenetic and RFLP analyses. NR, not reported.
Source: From Anderson and Stanbridge.[133]

own nuclear membrane. The cells are then enucleated by cytochalasin B treatment and centrifugation, producing microcells that can be fused to recipient cells. In order to identify which microcell contains which chromosome(s), a gene marker, sometimes artificially introduced (e.g., a bacterial drug resistance gene like *neo*), is required. Using this technique, the presence of putative tumor suppressor genes can be located on a given chromosome (Table 8-5). In most instances, transfer of a single copy of a normal chromosome is sufficient to induce growth inhibition in cell culture and/or suppression of tumor growth in nude mice. Specificity is demonstrated by the observation that random chromosomes not carrying a tumor suppressor gene do not suppress cell proliferation or tumor growth ("nonsuppressed" column in Table 8-5). Given the fact that a chromosome has a tumor suppressor function, the next task is to find the gene and to characterize its mechanism, no mean feat indeed.

REFERENCES

1. H. Harris, O. J. Miller, G. Klein, P. Worst, and T. Tachibana: Suppression of malignancy by cell fusion. *Nature* 223:363, 1969.
2. A. G. Knudson, Jr.: Hereditary cancer, oncogenes, and antioncogenes. *Cancer Res* 45:1437, 1985.
3. R. E. Hollingsworth and W.-H. Lee: Tumor suppressor genes: New prospects for cancer research. *JNCI* 83:91, 1991.
4. J. M. Bishop: The molecular genetics of cancer. *Science* 235:305, 1987.
5. E. J. Stanbridge: Human tumor suppressor genes. *Annu Rev Genet* 24:615, 1990.
6. W. K. Cavenee, T. P. Dryja, R. A. Phillips, W. F. Benedict, R. Godbout, B. L. Gallie, A. L. Murphree, L. C. Strong, and R. L. White: Expression of recessive alleles by chromosomal mechanisms in retinoblastoma. *Nature* 305:779, 1983.
7. W. K. Cavenee, M. F. Hansen, M. Nordenskjold, E. Kock, I. Maumenee, J. Squire, R. A. Phillips, and B. L. Gallie: Genetic origin of mutations predisposing to retinoblastoma. *Science* 228:501, 1985.
8. S. H. Friend, R. Bernards, S. Rogelj, R. A. Weinberg, J. M. Rapaport, D. M. Albert, and T. P. Dryja: A human DNA segment with properties of the gene that predisposes to retinoblastoma and osteosarcoma. *Nature* 323:643, 1986.
9. C. J. Marshall: Tumor suppressor genes. *Cell* 64:313, 1991.
10. J. Marx: Learning how to suppress cancer. *Science* 261:1385, 1993.
11. A. J. Levine: The tumor suppressor genes. *Annu Rev Biochem* 62:623, 1993.
12. H. Harris: The analysis of malignancy by cell fusion: The position in 1988. *Cancer Res* 48:3302, 1988.
13. D. W. Goodrich, Y. Chen, P. Scully, and W.-H. Lee: Expression of the retinoblastoma gene product in bladder carcinoma cells associates with a low frequency of tumor formation. *Cancer Res* 52:1968, 1992.
14. D. J. Templeton, S. H. Park, L. Lanier, and R. A. Weinberg: Nonfunctional mutants of the retinoblastoma protein are characterized by defects in phosphorylation, viral oncoprotein association, and nuclear tethering. *Proc Natl Acad Sci USA* 88:3033, 1991.
15. N. Dyson, K. Buchkovich, P. Whyte, and E. Harlow: The cellular 107K protein that binds to adenovirus E1A also associates with the large T antigens of SV40 and JC virus. *Cell* 58:249, 1989.
16. H.-J. Su Huang, J.-K. Yee, J.-Y. Shew, P.-L. Chen, R. Bookstein, T. Friedmann, E. Y.-H. P. Lee, and W.-H. Lee: Suppression of the neoplastic phenotype by replacement of the RB gene in human cancer cells. *Science* 242:1563, 1988.
17. R. Takahashi, T. Hashimoto, H.-J. Xu, S.-X. Hu, T. Matsui, T. Miki, H. Bigo-Marshall, S. A. Aaronson, and W. F. Benedict: The retinoblastoma gene functions as a growth and tumor suppressor in human bladder carcinoma cells. *Proc Natl Acad Sci USA* 88:5257, 1991.
18. E.Y.-H. P. Lee, C.-Y. Chang, H. Nanpin, Y.-C. J. Wang, C.-C. Lai, K. Herrup, W.-H. Lee, and A. Bradley: Mice deficient for Rb are nonviable and show defects in neurogenesis and haematopoiesis. *Nature* 359:288, 1992.
19. T. Jacks, A. Fazeli, E. M. Schmitt, R. T. Bronson, M. A. Goodell, and R. A. Weinberg: Effects of an *Rb* mutation in the mouse. *Nature* 359:295, 1992.
20. A. R. Clarke, E. R. Maandag, M. van Roon, N. M. T. van der Lugt, M. van der Valk, M. L. Hooper, A. Berns, and H. te Riele: Requirement for a functional *Rb*-1 gene in murine development. *Nature* 359:328, 1992.
21. J. J. Windle, D. M. Albert, J. M. O'Brien, D. M. Marcus, C. M. Disteche, R. Bernards, and P. L. Mellon: Retinoblastoma in transgenic mice. *Nature* 343:665, 1990.
22. J. A. DeCaprio, J. W. Ludlow, D. Lynch, Y. Furukawa, J. Griffin, H. Piwnica-Worms, C.-M. Huang, and D. M. Livingston: The product of the retinoblastoma susceptibility gene has properties of a cell cycle regulatory element. *Cell* 58:1085, 1989.

23. D. W. Goodrich, N. P. Wang, Y.-W. Qian, E. Y.-H. P. Lee, and W.-H. Lee: The retinoblastoma gene product regulates progression through the G1 phase of the cell cycle. *Cell* 67:293, 1991.
24. P. W. Hinds, S. Mittnacht, V. Dulic, A. Arnold, S. I. Reed, and R. A. Weinberg: Regulation of retinoblastoma protein functions by ectopic expression of human cyclins. *Cell* 70:993, 1992.
25. D. Defeo-Jones, P. S. Huang, R. E. Jones, K. M. Haskell, G. A. Vuocolo, M. G. Hanobik, H. E. Huber, and A. Oliff: Cloning of cDNAs for cellular proteins that bind to the retinoblastoma gene product. *Nature* 352:251, 1991.
26. D. W. Goodrich and W.-H. Lee: Abrogation by c-*myc* of G1 phase arrest induced by RB protein but not by p53. *Nature* 360:177, 1992.
27. K. Helin, J. A. Lees, M. Vidal, N. Dyson, E. Harlow, and A. Fattaey: A cDNA encoding a pRB-binding protein with properties of the transcription factor E2F. *Cell* 70:337, 1992.
28. S. P. Cellappan, S. Hiebert, M. Mudryj, J. M. Horowitz, and J. R. Nevins: The E2F transcription factor is a cellular target for the RB protein. *Cell* 65:1053, 1991.
29. W. G. Kaelin, Jr., W. Krek, W. R. Sellers, J. A. DeCaprio, F. Ajchenbaum, C. S. Fuchs, T. Chittenden, Y. Li, P. J. Farnham, M. A. Blanar, D. M. Livingston, and E. K. Flemington: Expression cloning of a cDNA encoding a retinoblastoma-binding protein with E2F-like properties. *Cell* 70:351, 1992.
30. S. J. Weintraub, C. A. Prater, and D. C. Dean: Retinoblastoma protein switches the E2F site from positive to negative element. *Nature* 358:259, 1992.
31. P. D. Robbins, J. M. Horowitz, and R. C. Mulligan: Negative regulation of human c-*fos* expression by the retinoblastoma gene product. *Nature* 346:668, 1990.
32. S.-J. Kim, H.-D. Lee, P. D. Robbins, K. Busam, M. B. Sporn, and A. B. Roberts: Regulation of transforming growth factor β1 gene expression by the product of the retinoblastoma-susceptibility gene. *Proc Natl Acad Sci USA* 88:3052, 1991.
33. S.-J. Kim, S. Wagner, F. Liu, M. A. O'Reilly, P. D. Robbins, and M. R. Green: Retinoblastoma gene product activates expression of the human TGF-β2 gene through transcription factor ATF-2. *Nature* 358:331, 1992.
34. D. P. Lane and L. V. Crawford: T antigen is bound to a host protein in SV40 transformed cells. *Nature* 278:261, 1979.
35. W. G. Dippold, G. Jay, A. B. De Leo, G. Khoury, and L. J. Old: p53 transformation-related protein: Detection by monoclonal antibody in mouse and human cells. *Proc Natl Acad Sci USA* 78:1695, 1981.
36. L. V. Crawford, D. C. Pim, E. G. Gurney, P. Goodfellow, and J. Taylor-Papadimitriou: Detection of a common feature in several human tumor cell lines—A 53,000-dalton protein. *Proc Natl Acad Sci USA* 78:41, 1981.
37. H. Jörnvall, J. Luka, G. Klein, and E. Appella: A 53-kilodalton protein common to chemically and virally transformed cells shows extensive sequence similarities between species. *Proc Natl Acad Sci USA* 79:287, 1982.
38. V. Rotter: p53, a transformation-related cellular-encoded protein, can be used as a biochemical marker for the detection of primary mouse tumor cells. *Proc Natl Acad Sci USA* 80:2613, 1983.
39. J. R. Jenkins, K. Rudge, and G. A. Currie: Cellular immortalization by a cDNA clone encoding the transformation-associated phosphoprotein p53. *Nature* 312:651, 1984.
40. D. Eliyahu, A. Raz, P. Gruss, D. Givol, and M. Oren: Participation of p53 cellular tumour antigen in transformation of normal embryonic cells. *Nature* 312:646, 1984.
41. L. F. Parada, H. Land, R. A. Weinberg, D. Wolf, and V. Rotter: Cooperation between gene encoding p53 tumour antigen and *ras* in cellular transformation. *Nature* 312:649, 1984.
42. C. A. Finlay, P. W. Hinds, and A. J. Levine: The p53 protooncogene can act as a suppressor of transformation. *Cell* 57:1083, 1989.
43. D. Eliyahu, D. Michalovitz, S. Eliyahu, O. Pinhasi-Kimhi, and M. Oren: Wild-type p53 can inhibit oncogene-mediated focus formation. *Proc Natl Acad Sci USA* 86:8763, 1989.
44. G. P. Zambetti, D. Olson, M. Labow, and A. J. Levine: A mutant p53 protein is required for maintenance of the transformed phenotype in cells transformed with p53 plus *ras* cDNAs. *Proc Natl Acad Sci USA* 89:3952, 1992.
45. D. Malkin, F. P. Li, L. C. Strong, J. F. Fraumeni, Jr., C. E. Nelson, D. H. Kim, Jayne Kassel, M. A. Gryka, F. Z. Bischoff, M. A. Tainsky, and S. H. Friend: Germ line p53 mutations in a familial syndrome of breast cancer, sarcomas, and other neoplasms. *Science* 250:1233, 1990.
46. T. Frebourg, J. Kassel, K. T. Lam, M. A. Gryka, N. Barbier, T. I. Andersen, A.-L. Borresen, and S. H. Friend: Germ-line mutations of the p53 tumor suppressor gene in patients with high risk for cancer inactivate the p53 protein. *Proc Natl Acad Sci USA* 89:6413, 1992.
47. C. A. Finlay, P. W. Hinds, T.-H. Tan, D. Eliyahu, M. Oren, and A. J. Levine: Activating mutations for transformation by p53 produce a gene product that forms an hsc70-p53 complex with an altered half-life. *Mol Cell Biol* 8:531, 1988.
48. I. Hsu, R. Metcalf, T. Sun, J. Welsh, N. Wang, and C. Harris: Mutational hotspot in the p53 gene in human hepatocellular carcinomas. *Nature* 350:427, 1991.
49. F. Aguilar, S. P. Hussain, and P. Cerutti: Aflatoxin B_1 induces the transversion of G $\rightarrow$ T in codon 249 of the p53 tumor suppressor gene in

human hepatocytes. *Proc Natl Acad Sci USA* 90:8586, 1993.
50. S. Kress, U.-R. Jahn, A. Buchmann, P. Bannasch, and M. Schwarz: p53 mutations in human hepatocellular carcinomas from Germany. *Cancer Res* 52:3220, 1992.
51. B. Ruggeri, M. DiRado, S. Y. Zhang, B. Bauer, T. Goodrow, and A. J. P. Klein-Szanto: Benzo[*a*]pyrene-induced murine skin tumors exhibit frequent and characteristic G to T mutations in the p53 gene. *Proc Natl Acad Sci USA* 90:1013, 1993.
52. S. Kress, C. Sutter, P. T. Strickland, H. Mukhtar, J. Schweizer, and M. Schwarz: Carcinogen-specific mutational patern in the p53 gene in ultraviolet B radiation–induced squamous cell carcinomas of mouse skin. *Cancer Res* 52:6400, 1992.
53. S. Kanjilal, W. E. Pierceall, K. K. Cummings, M. L. Kripke, and H. N. Ananthaswamy: High frequency of p53 mutations in ultraviolet radiation-induced murine skin tumors: Evidence for strand bias and tumor heterogeneity. *Cancer Res* 53:2961, 1993.
54. S. J. Baker, S. Markowitz, E. R. Fearon, J. K. V. Willson, and B. Vogelstein: Suppression of human colorectal carcinoma cell growth by wild-type p53. *Science* 249:912, 1990.
55. T. Takahashi, D. Carbone, T. Takahashi, M. M. Nau, T. Hida, I. Linnoila, R. Ueda, and J. D. Minna: Wild-type but not mutant p53 suppresses the growth of human lung cancer cells bearing multiple genetic lesions. *Cancer Res* 52:2340, 1992.
56. W. B. Isaacs, B. S. Carter, and C. M. Ewing: Wild-type p53 suppresses growth of human prostate cancer cells containing mutant p53 alleles. *Cancer Res* 51:4716, 1991.
57. W. E. Mercer, M. T. Shields, M. Amin, G. J. Sauve, E. Appella, J. W. Romano, and S. J. Ullrich: Negative growth regulation in a glioblastoma tumor cell line that conditionally expresses human wild-type p53. *Proc Natl Acad Sci USA* 87:6166, 1990.
58. L. Diller, J. Kassel, C. E. Nelson, M. A. Gryka, G. Litwak, M. Gebhardt, B. Bressac, M. Ozturk, S. J. Baker, B. Vogelstein, and S. H. Friend: p53 functions as a cell cycle control protein in osteosarcomas. *Mol Cell Biol* 10:5772, 1990.
59. J. Cheng, J.-K. Yee, J. Yeargin, T. Friedmann, and M. Haas: Suppression of acute lymphoblastic leukemia by the human wild-type p53 gene. *Cancer Res* 52:222, 1992.
60. M. B. Kastan, O. Onyekwere, D. Sidransky, B. Vogelstein, and R. W. Craig: Participation of p53 protein in the cellular response to DNA damage. *Cancer Res* 51:6304, 1991.
61. S. J. Ullrich, C. W. Anderson, W. E. Mercer, and E. Appella: The p53 tumor suppressor protein, a modulator of cell proliferation. *J Biol Chem* 267:15259, 1992.
62. D. Lin, M. T. Shields, S. J. Ullrich, E. Appella, and W. E. Mercer: Growth arrest induced by wild-type p53 protein blocks cells prior to or near the restriction point in late G_1 phase. *Proc Natl Acad Sci USA* 89:9210, 1992.
62a. T. Miyashita, M. Harigai, M. Hanada, and J. C. Reed: Identification of a p53-dependent negative response element in the *bcl*-2 gene. *Cancer Res* 54:3131, 1994.
62b. J. J. Ryan, E. Prochownik, C. A. Gottlieb, I. J. Apel, R. Merino, G. Nunez, and M. F. Clarke: c-*myc* and *bcl*-2 modulate p53 function by altering p53 subcellular trafficking during the cell cycle. *Proc Natl Acad Sci USA* 91:5878, 1994.
63. S. W. Lowe, E. M. Schmitt, S. W. Smith, B. A. Osborne, and T. Jacks: p53 is required for radiation-induced apoptosis in mouse thymocytes. *Nature* 362:847, 1993.
64. A. R. Clarke, C. A. Purdie, D. J. Harrison, R. G. Morris, C. C. Bird, M. L. Hooper, and A. H. Wyllie: Thymocyte apoptosis induced by p53-dependent and independent pathways. *Nature* 362:849, 1993.
65. P. M. O'Connor, J. Jackman, D. Jondle, K. Bhatia, I. Magrath, and K. W. Kohn: Role of the p53 tumor suppressor gene in cell cycle arrest and radiosensitivity of Burkitt's lymphoma cell lines. *Cancer Res* 53:4776, 1993.
66. S. W. Lowe, H. E. Ruley, T. Jacks, and D. E. Housman: p53-dependent apoptosis modulates the cytotoxicity of anticancer agents. *Cell* 74:957, 1993.
67. J. M. Lee and A. Bernstein: p53 mutations increase resistance to ionizing radiation. *Proc Natl Acad Sci USA* 90:5742, 1993.
68. D. P. Lane: p53, guardian of the genome. *Nature* 358:15, 1992.
69. T. A. Weinert and L. Hartwell: Characterization of RAD9 of *Saccharomyces cerevisiae* and evidence that its function acts post-translationally in cell cycle arrest after DNA damage. *Mol Cell Biol* 10:6554, 1990.
70. L. R. Livingstone, A. White, J. Sprouse, E. Livanos, T. Jacks, and T. D. Tlsty: Altered cell cycle arrest and gene amplification potential accompany loss of wild-type p53. *Cell* 70:923, 1992.
71. Y. Yin, M. A. Tainsky, F. Z. Bischoff, L. C. Strong, and G. M. Wahl: Wild-type p53 restores cell cycle control and inhibits gene amplification in cells with mutant p53 alleles. *Cell* 70:937, 1992.
72. T. Fujiwara, E. A. Grimm, T. Mukhopadhyay, D. W. Cai, L. B. Owen-Schaub, and J. A. Roth: A retroviral wild-type p53 expression vector penetrates human lung cancer spheroids and inhibits growth by inducing apoptosis. *Cancer Res* 53:4129, 1993.
73. J. A. Pietenpol and B. Vogelstein: No room at the p53 inn. *Nature* 365:17, 1993.
74. D. W. Martin, R. M. Munoz, M. A. Subler, and S. Deb: p53 binds to the TATA-binding protein-TATA complex. *J Biol Chem* 268:13062, 1993.

75. E. Seto, A. Usheva, G. P. Zambetti, J. Momand, N. Horikoshi, R. Weinmann, A. J. Levine, and T. Shenk: Wild-type p53 binds to the TATA-binding protein and represses transcription. *Proc Natl Acad Sci USA* 89:12028, 1992.
76. A. Dutta, J. M. Ruppert, J. C. Aster, and E. Winchester: Inhibition of DNA replication factor RPA by p53. *Nature* 365:79, 1993.
77. L. Cahilly-Snyder, T. Yang-Feng, U. Francke, and D. L. George: Molecular analysis and chromosomal mapping of amplified genes isolated from a transformed mouse 3T3 cell line. *Somatic Cell Mol Genet* 13:235, 1987.
78. J. Momand, G. P. Zambetti, D. C. Olson, D. George, and A. J. Levine: The *mdm*-2 oncogene product forms a complex with the p53 protein and inhibits p53-mediated transactivation. *Cell* 69:1237, 1992.
79. J. D. Oliner, J. A. Pietenpol, S. Thiagalingam, J. Gyuris, K. W. Kinzler, and B. Vogelstein: Oncoprotein MDM2 conceals the activation domain of tumour suppressor p53. *Nature* 362:857, 1993.
79a. C. Cordon-Cardo, E. Latres, M. Drobnjak, M. R. Oliva, D. Pollack, J. M. Woodruff, V. Marechal, J. Chen, M. F. Brennan, and A. J. Levine: Molecular abnormalities of mdm2 and p53 genes in adult soft tissue sarcomas. *Cancer Res* 54:794, 1994.
80. S. Fields and S. K. Jang: Presence of a potent transcription activating sequence in the p53 protein. *Science* 249:1046, 1990.
81. L. Raycroft, H. Wu, and G. Lozano: Transcriptional activation by wild-type but not transforming mutants of the p53 antioncogene. *Science* 249:1049, 1990.
82. S. E. Kern, K. W. Kinzler, A. Bruskin, D. Jarosz, P. Friedman, C. Prives, and B. Vogelstein: Identification of p53 as a sequence-specific DNA-binding protein. *Science* 252:1708, 1991.
83. W. S. El-Deiry, T. Tokino, V. E. Velculescu, D. B. Levy, R. Parsons, J. M. Trent, D. Lin, W. E. Mercer, K. W. Kinzler, and B. Vogelstein: WAF1, a potential mediator of p53 tumor suppression. *Cell* 75:817, 1993.
83a. C.-Y. Chen, J. D. Oliner, Q. Zhan, A. J. Fornace, Jr., B. Vogelstein, and M. B. Kastan: Interactions between p53 and MDM2 in a mammalian cell cycle checkpoint pathway. *Proc Natl Acad Sci USA* 91:2684, 1994.
84. T. R. Hupp, D. W. Meek, C. A. Midgley, and D. P. Lane: Regulation of the specific DNA binding function of p53. *Cell* 71:875, 1992.
85. C. C. Harris: p53: At the crossroads of molecular carcinogenesis and risk assessment. *Science* 262:1980, 1993.
86. R. Srinivasan, J. A. Roth, and S. A. Maxwell: Sequence-specific interaction of a conformational domain of p53 with DNA. *Cancer Res* 53:5361, 1993.
87. L. J. Smith, E. A. McCulloch, and S. Benchimol: Expression of the p53 oncogene in acute myeloblastic leukemia. *J Exp Med* 164:751, 1986.
88. K.-A. Han and M. F. Kulesz-Martin: Altered expression of wild-type p53 tumor suppressor gene during murine epithelial cell transformation. *Cancer Res* 52:749, 1992.
89. E. Bayever, K. M. Haines, P. L. Iversen, R. W. Ruddon, S. J. Pirruccello, C. P. Mountjoy, M. A. Arneson, and L. J. Smith: Selective cytotoxicity to human leukemic myeloblasts produced by oligodeoyribonucleotide phosphorothioates complementary to p53 nucleotide sequences. *Leuk Lymph* 12:223, 1994.
90. L. A. Donehower, M. Harvey, B. L. Slagle, M. J. McArthur, C. A. Montgomery, Jr., J. S. Butel, and A. Bradley: Mice deficient for p53 are developmentally normal but susceptible to spontaneous tumours. *Nature* 356:215, 1992.
91. M. Hollstein, D. Sidransky, B. Vogelstein, and C. C. Harris: p53 mutations in humans cancers. *Science* 253:49, 1991.
92. T. Frebourg and S. H. Friend: The importance of p53 gene alterations in human cancer: Is there more than circumstantial evidence? *JNCI* 85:1554, 1993.
93. C. C. Harris and M. Hollstein: Clinical implications of the p53 tumor-suppressor gene. *N Engl J Med* 329:1318, 1993.
93a. S. Bi, F. Lanza, and J. M. Goldman: The involvement of "tumor suppressor" p 53 in normal and chronic myelogenous leukemia hemopoiesis. *Cancer Res* 54:582, 1994.
93b. M. Drobnjak, E. Latres, D. Pollack, M. Karpeh, M. Dudas, J. M. Woodruff, M. F. Brennan, and C. Cordon-Cardo: Prognostic implications of p53 nuclear overexpression and high proliferation index of Ki-67 in adult soft-tissue sarcomas. *JNCI* 86:549, 1994.
94. A. D. Thor, D. H. Moore II, S. M. Edgerton, E. S. Kawasaki, E. Reihsaus, H. T. Lynch, J. N. Marcus, L. Schwartz, L.-C. Chen, B. H. Mayall, and H. S. Smith: Accumulation of p53 tumor suppressor gene protein: An independent marker of prognosis in breast cancers. *JNCI* 84:845, 1992.
95. J. Isola, T. Visakorpi, K. Holli, and O.-P. Kallioniemi: Association of overexpression of tumor suppressor protein p53 with rapid cell proliferation and poor prognosis in node-negative breast cancer patients. *JNCI* 84:1109, 1992.
96. J. Kupryjanczyk, A. D. Thor, R. Beauchamp, V. Merritt, S. M. Edgerton, D. A. Bell, and D. W. Yandell: p53 gene mutations and protein accumulation in human ovarian cancer. *Proc Natl Acad Sci USA* 90:4961, 1993.
97. M. F. Kohler, A. Berchuck, A. M. Davidoff, P. A. Humphrey, R. K. Dodge, J. D. Iglehart, J. T. Soper, D. L. Clarke-Pearson, R. C. Bast, Jr., and J. R. Marks: Overexpression and mutation of p53 in endometrial carcinoma. *Cancer Res* 52:1622, 1992.

98. W. P. Bennett, M. C. Hollstein, R. A. Metcalf, J. A. Welsh, A. He, S.-M. Zhu, I. Kusters, J. H. Resau, B. F. Trump, D. P. Lane, and C. C. Harris: p53 mutation and protein accumulation during multistage human esophageal carcinogenesis. *Cancer Res* 52:6092, 1992.
99. D. C. Quinlan, A. G. Davidson, C. L. Summers, H. E. Warden, and H. M. Doshi: Accumulation of p53 protein correlates with a poor prognosis in human lung cancer. *Cancer Res* 52:4828, 1992.
100. A. S. Sarkin, G. Dalbagni, C. Cordon-Cardo, Z.-F. Zhang, J. Sheinfeld, W. R. Fair, H. W. Herr, and V. E. Reuter: Nuclear overexpression of p53 protein in transitional cell bladder carcinoma: A marker for disease progression. *JNCI* 85:53, 1993.
101. N. J. Lassam, L. From, and H. J. Kahn: Overexpression of p53 is a later event in the development of malignant melanoma. *Cancer Res* 53:2235, 1993.
102. R. Iggo, K. Gatter, J. Bartek, D. Lane, and A. L. Harris: Increased expression of mutant forms of p53 oncogene in primary lung cancer. *Lancet* 335:675, 1990.
103. W. P. Bennett, T. V. Colby, W. D. Travis, A. Borkowski, R. T. Jones, D. P. Lane, R. A. Metcalf, J. M. Samet, Y. Takeshima, J. R. Gu, K. H. Vähäkangas, Y. Soini, P. Pääkkö, J. A. Welsh, B. F. Trump, and C. C. Harris: p53 protein accumulates frequently in early bronchial neoplasia. *Cancer Res* 53:4817, 1993.
104. S. Labrecque, N. Naor, D. Thomson, and G. Matlashewski: Analysis of the anti-p53 antibody response in cancer patients. *Cancer Res* 53:3468, 1993.
105. J. D. Oliner, K. W. Kinzler, P. S. Meltzer, D. L. George, and B. Vogelstein: Amplification of a gene encoding a p53-associated protein in human sarcomas. *Nature* 358:80, 1992.
106. G. Reifenberger, L. Liu, K. Ichimura, E. E. Schmidt, and V. P. Collins: Amplification and overexpression of the MDM2 gene in a subset of human malignant gliomas without p53 mutations. *Cancer Res* 53:2736, 1993.
107. C. E. Bueso-Ramos, Y. Yang, E. deLeon, P. McCown, S. A. Stass, and M. Albitar: The human MDM-2 oncogene is overexpressed in leukemias. *Blood* 82:2617, 1993.
108. M. S. Sheikh, Z.-M. Shao, A. Hussain, and J. A. Fontana: The p53-binding protein MDM2 gene is differentially expressed in human breast carcinoma. *Cancer Res* 53:3226, 1993.
109. J. A. Kreidberg, H. Sariola, J. M. Loring, M. Maeda, J. Pelletier, D. Housman, and R. Jaenisch: WT-1 is required for early kidney development. *Cell* 74:679, 1993.
110. K. M. Call, T. Glaser, C. Y. Ito, A. J. Buckler, J. Pelletier, D. A. Haber, E. A. Rose, A. Kral, H. Yeger, W. H. Lewis, C. Jones, and D. E. Housman: Isolation and characterization of a zinc finger polypeptide gene at the human chromosome 11 Wilms' tumor locus. *Cell* 60:509, 1990.
111. S. F. Dowdy, C. L. Fasching, D. Araujo, K.-M. Lai, E. Livanos, B. E. Weissman, and E. J. Stanbridge: Suppression of tumorigenicity in Wilms' tumor by the p15.5-p14 region of chromosome 11. *Science* 254:293, 1991.
112. M. Negrini, A. Castagnoli, J. V. Pavan, S. Sabbioni, D. Araujo, A. Corallini, F. Gualandi, P. Rimessi, A. Bonfatti, C. Giunta, A. Sensi, E. J. Stanbridge, and G. Barbanti-Brodano: Suppression of tumorigenicity and anchorage-independent growth of BK virus-transformed mouse cells by human chromosome 11. *Cancer Res* 52:1297, 1992.
113. S. L. Madden, D. M. Cook, J. F. Morris, A. Gashler, V. P. Sukhatme, and F. J. Rauscher, III: Transcriptional repression mediated by the WT1 Wilms' tumor gene product. *Science* 253:1550, 1991.
114. S. Park, G. Tomlinson, P. Nisen, and D. A. Haber: Altered trans-activational properties of a mutated Wt1 gene product in a WAGR-associated Wilms' tumor. *Cancer Res* 53:4757, 1993.
115. Z.-Y. Wang, W.-Q. Qiu, K. T. Enger, and T. F. Deuel: A second transcriptionally active DNA-binding site for the Wilms' tumor gene product, WT1. *Proc Natl Acad Sci USA* 90:8896, 1993.
116. Z.-Y. Wang, Q.-Q. Qiu, and T. F. Deuel: The Wilms' tumor gene product WT1 activates or suppresses transcription through separate functional domains. *J Biol Chem* 268:9172, 1993.
117. S. Maheswaran, S. Park, A. Bernard, J. F. Morris, F. J. Rauscher III, D. E. Hill, and D. A. Haber: Physical and functional interaction between WT1 and p53 proteins. *Proc Natl Acad Sci USA* 90:5100, 1993.
118. K. W. Kinzler, M. C. Nilbert, B. Vogelstein, T. M. Bryan, D. B. Levy, K. J. Smith, A. C. Preisinger, S. R. Hamilton, P. Hedge, A. Markham, M. Carlson, G. Joslyn, J. Groden, R. White, Y. Miki, Y. Miyoshi, I. Nishisho, and Y. Nakamura: Identification of a gene located at chromosome 5q21 that is mutated in colorectal cancers. *Science* 251:1366, 1991.
119. G. Joslyn, M. Carlson, A. Thivers, H. Albertsen, L. Gelbert, W. Samowitz, J. Groden, J. Stevens, L. Spririo, M. Robertson, L. Sargeant, K. Krapcho, E. Wolff, R. Burt, J. P. Hughes, J. Warrington, J. McPherson, J. Wasmuth, D. LePaslier, H. Abderrahim, D. Cohen, M. Leppert, and R. White: Identification of deletion mutations and three new genes at the familial polyposis locus. *Cell* 66:601, 1991.
119a. B. Rubinfeld, B. Souza, I. Albert, O. Müller, S. H. Chamberlain, F. R. Masiarz, S. Munemitsu, and P. Polakis: Association of the APC gene product with β-catenin. *Science* 262:1731, 1993.
119b. L.-K. Su, B. Vogelstein, and K. W. Kinzler: Association of the APC tumor suppressor protein with catenins. *Science* 262:1734, 1993.

120. E. R. Fearon, K. R. Cho, J. M. Nigro, S. E. Kern, J. W. Simons, J. M. Ruppert, S. R. Hamilton, A. C. Preisinger, G. Thomas, K. W. Kinzler, and B. Vogelstein: Identification of a chromosome 18q gene that is altered in colorectal carcinoma. *Science* 247:49, 1990.
121. X. Gao, K. V. Honn, D. Grignon, W. Sakr, and Y. Q. Chen: Frequent loss of expression and loss of heterozygosity of the putative tumor suppressor gene DCC in prostatic carcinomas. *Cancer Res* 53:2723, 1993.
122. H. T. Lynch, T. C. Smyrk, P. Watson, S. J. Lanspa, J. F. Lynch, P. M. Lynch, R. J. Cavalieri, and C. R. Boland: Genetics, natural history, tumor spectrum, and pathology of hereditary nonpolyosis colorectal cancer: An updated review. *Gastroenterology* 104:1535, 1993.
123. F. S. Leach, N. C. Nicolaides, N. Papadopoulos, B. Liu, J. Jen, R. Parsons, P. Peltomäki, P. Sistonen, L. A. Aaltonen, M. Nyström-Lahti, X.-Y. Guan, J. Zhang, P. S. Meltzer, J.-W. Yu, F.-T. Kao, D. J. Chen, K. M. Cerosaletti, R. E. K. Fournier, S. Todd, T. Lewis, R. J. Leach, S. L. Naylor, J. Weissenbach, J.-P. Mecklin, H. Jävinen, G. M. Petersen, S. R. Hamilton, J. Green, J. Jass, P. Watson, H. T. Lynch, J. M. Trent, A. de la Chapelle, K. W. Kinzler, and B. Vogelstein: Mutations of a *mut*S homolog in hereditary nonpolyposis colorectal cancer. *Cell* 75:1215, 1993.
124. R. Fishel, M. K. Lescoe, M. R. S. Rao, N. G. Copeland, N. A. Jenkins, J. Garber, M. Kane, and R. Kolodner: The human mutator gene homolog MSH2 and its association with hereditary nonpolyposis colon cancer. *Cell* 75:1027, 1993.
125. R. Parsons, G.-M. Li, M. J. Longley, W.-H. Fang, N. Papadopoulos, J. Jen, A. de la Chapelle, K. W. Kinzler, B. Vogelstein, and P. Modrich: Hypermutability and mismatch repair deficiency in RER^+ tumor cells. *Cell* 75:1227, 1993.
126. M. R. Wallace, D. A. Marchuk, L. B. Anderson, R. Letcher, H. M. Odeh, A. M. Saulino, J. W. Fountain, A. Brereton, J. Nicholson, A. L. Mitchell, B. H. Brownstein, and F. S. Collins: Type 1 neurofibromatosis gene: Identification of a large transcript disrupted in three NF1 patients. *Science* 249:181, 1990.
127. G. Xu, P. O'Connell, D. Viskochil, R. Cawthon, M. Robertson, M. Culver, D. Dunn, J. Stevens, R. Gesteland, R. White, and R. Weiss: The neurofibromatosis type 1 gene encodes a protein related to GAP. *Cell* 62:599, 1990.
128. G. A. Martin, D. Viskochil, G. Bollag, P. C. McCabe, W. J. Crosier, H. Haubrock, L. Conroy, R. Clark, P. O'Connell, R. M. Cawthon, M. A. Innis, and F. McCormick: The GAP-related domain of the neurofibromatosis type 1 gene product interacts with *ras* p21. *Cell* 63:843, 1990.
129. A. G. Knudson: Antioncogenes and human cancer. *Proc Natl Acad Sci USA* 90:10914, 1993.
130. J. A. Trofatter, M. M. MacCollin, J. L. Rutter, J. R. Murrell, M. P. Duyao, D. M. Parry, R. Eldridge, N. Kley, A. G. Menon, K. Pulaski, V. H. Haase, C. M. Ambrose, D. Munroe, C. Bove, J. L. Haines, R. L. Martuza, M. E. MacDonald, B. R. Seizinger, M. P. Short, A. J. Buckler, and J. F. Gusella: A novel moesin-, ezrin-, radixin-like gene is a candidate for the neurofibromatosis 2 tumor suppressor. *Cell* 72:791, 1993.
131. G. A. Rouleau, P. Merel, M. Luchman, M. Sanson, J. Zucman, C. Marineau, K. Hoang-Xuan, S. Demczuk, C. Desmaze, B. Plougastel, S. M. Pulst, G. Lenoir, E. Bijlsma, R. Fashold, J. Dumanski, P. de Jong, D. Parry, R. Eldridge, A. Aurias, O. Delattre, and G. Thomas: Alteration in a new gene encoding a putative membrane-organizing protein causes neuro-fibromatosis type 2. *Nature* 363:515, 1993.
132. F. Latif, K. Tory, J. Gnarra, M. Yao, F.-M. Duh, M. L. Orcutt, T. Stockhouse, I. Kuzmin, W. Modi, L. Geil, L. Schmidt, F. Zhou, H. Li, M. H. Wei, F. Chen, G. Glenn, P. Choyke, M. M. Walther, Y. Weng, D. S. R. Duan, M. Dean, D. Glavac, F. M. Richards, P. A. Crossey, M. A. Ferguson-Smith, D. LePaslier, I. Chumakov, D. Cohen, A. C. Chinault, E. R. Maher, W. M. Lineham, B. Zbar, and M. I. Lerman. Identification of the von Hippel-Lindau disease tumor suppressor gene. *Science* 260:1317, 1993.
133. M. J. Anderson and E. J. Stanbridge: Tumor suppressor genes studied by cell hybridization and chromosome transfer. *FASEB J* 7:826, 1993.

9

Growth Factors and Signal Transduction Mechanisms

GROWTH FACTORS

Many "factors" that affect the proliferation of eukaryotic cells have been identified (Table 9-1). A growth factor is usually defined as a substance that stimulates cell proliferation and often also promotes cell differentiation of specific target cells. Excluded from this class of agents are substances that are simply nutrients, such as glucose, essential amino acids, vitamins, and key minerals. The impetus for studies of growth factors largely derives from earlier observations that most mammalian cells growing in culture require the presence of animal serum to grow and proliferate. Numerous attempts have been made to isolate growth factors from serum. The isolation and characterization of these factors may help to elucidate the altered growth control of malignant cells because transformed cells usually have a lower growth requirement for serum and some transformed cells appear to produce their own growth factors.

Growth factors are now established as important regulators of embryonic growth and development, cellular proliferation and differentiation, and tumor cell proliferation. While the "modern era" of research on growth factors dates to the late 1940s and early 1950s, with the seminal work of Rita Levi-Montalcini, Stanley Cohen, and Viktor Hamburger, research in this field actually started about 50 years earlier than that (reviewed in Ref. 1), with the work of a young scientist named Thorburn Brailsford Robertson, who came from the University of Adelaide in 1905 to study with Jacob Loeb at the University of California. Loeb was studying the effects of chemicals and salt solutions on the cellular motility of *Paramecium.* This work stimulated Robertson's interest in how biological functions can be modified by chemical substances, and this, in turn, led to his interest in regulation of growth of developing organisms. In a series of papers published in 1916 in *The Journal of Biological Chemistry,* Robertson described the growth-stimulatory effects of an anterior pituitary extract on the growth of juvenile mice from 20 to 60 weeks of age. The active component was extracted from desiccated pituitary glands with alcohol and precipitated in ether. He called this substance "tethelin," from the Greek word for "growing." Interest in tethelin as a pharmaceutical apparently stimulated production of it by a drug manufacturer in Philadelphia, H. L. Mulford's. (The first biotech company!) There appears to be no record of its use in humans.

Other early work on "growth factors" included that of Alexis Carrel at the Rockefeller Institute

Table 9-1 Characteristics of Some Representative Growth Factors

Factor	Original Source	Target Cell	Molecular Weight
Insulin	Beta cells of pancreas	General	6,000
Insulinlike growth factors			
IGF-1	Human plasma	General	7,650
IGF-2	Human plasma	General	7,500
Basic FGF (FGF-2)	Bovine pituitary	Fibroblasts, myoblasts, smooth muscle, chondrocytes, glial cells, vascular endothelium	14,000
NGF	Mouse submaxillary gland, snake venoms, cultured cells	Sympathetic ganglia cells and sensory neurons	26,500
EGF	Mouse submaxillary gland, human urine	Epidermal cells, various epithelial cells, vascular endothelial cells, chondrocytes, fibroblasts, glial cells	6,000
PDGF AA, AB, and BB	Human platelets, placenta	Fibroblasts, glial cells, arterial smooth muscle cells	24,000–31,000
Transforming growth factors			
TGF-α	Various virally transformed cell types and cancer cells	Similar to EGF	≅6,000
TGF-β	Various transformed cell types, normal placenta, kidney, and platelets	Similar to EGF	25,000
Angiogenin	Human colon carcinoma cell line	Capillary endothelium	14,000
Colony-stimulating factors (human)			
GM-CSF	Placenta	Granulocyte/macrophage progenitor cells	22,000
G-CSF	Placenta	Granulocyte progenitor cells	20,000
M-CSF	Urine	Macrophage progenitor cells	45,000
Interleukins			
IL-1	Normal and malignant macrophages, keratinocytes, astrocytes	T lymphocytes, B lymphocytes, fibroblasts, hepatocytes, chondrocytes, hypothalamic cells (fever center)	15,000
IL-2 (TCGF)	T lymphocytes	T lymphocytes (that become T helper cells or cytotoxic T cells)	23,000
IL-3 (Multi-CSF)	T lymphocytes	Eosinophil, mast cell, granulocyte, and macrophage progenitors; T lymphocytes	28,000
Mammary-derived growth factor (MDGF)	Human milk and mammary tumors	Normal mammary epithelial cells, epidermoid carcinoma cells	62,000
Uterine-derived growth factor (UDGF)	Pregnant sheep uterus	Rat mammary and uterine tumor cells	4,000–6,000

and Eric Horning in Melbourne. In 1928, Carrel reported that an unstable protein extracted from embryonic tissue or lymphocytes stimulated growth of animal tissues in culture dishes. He called these "trephones" but did not characterize them further. About the same time, Horning, working with J. M. Byrne and K. C. Richardson, observed that cellular outgrowth from pieces of a mouse sarcoma was stimulated by extracts from embryonic or tumor tissue and that chick embryo intestinal fragments exhibited enhanced cellular proliferation in the presence of embryonic extract plus tethelin, which they obtained from Robertson. Unfortunately, the research trail on these growth-stimulating substances grew cold soon after these reports, with the death of Robertson in 1930 at age 46 and the subsequent lack of encouragement Horning received for this work. It was not until 20 years later that the work of Levi-Montalcini, Cohen,

Hamburger, and their collaborators established this field on a firm footing with the biochemical characterization of growth factors as definitive entities.

A large number of growth-promoting factors have now been found. In general, these are polypeptides of relatively low molecular weight (6000 to 30,000 Da). Specific receptors (high-affinity, saturable) have been identified for many of them, and these receptors are usually cell-surface receptors, a number of which have been shown to possess endogenous protein kinase activity (e.g., insulin, IGFs, EGF, PDGF, FGFs, and TGF-α). Some of these receptors undergo receptor-mediated endocytosis, which may be involved in downregulation of receptor-mediated action, in transference of a receptor-mediated signal to the cell nucleus, or both. New growth factors continue to be discovered, and only some of the better-characterized ones are discussed here. The almost ubiquitous presence of growth factors in a wide variety of tissues leads one to speculate that each tissue may have its own growth-modulating substances. During development and tissue differentiation, these substances probably act locally, either as paracrine- or autocrine-regulatory chemical messengers. They continue to act as needed during adult life for tissue renewal and wound repair. In cancerous tissue, the secretion of certain of these chemical messengers becomes unregulated and cellular proliferation continues unabated.

Some of the factors discussed here are primarily of historical interest, but because they shed light on how the whole field developed, they are included. For example, the discovery that insulin can be a growth-promoting substance for cultured cells and that there are insulinlike substances in human plasma that act similarly led to the discovery of the insulinlike growth factors (IGF-1 and IGF-2). The discovery in 1948 of a peptide factor (NGF) that stimulated nerve outgrowth in chick embryo limb buds was the initial finding that led investigators to look for more such factors and to the serendipitous discovery of epidermal growth factor (EGF). This discovery, in turn, led indirectly to the discovery of the transforming growth factors (TGF-α and -β). Similarly, biochemical characterization of the EGF receptor and elucidation of its amino acid sequence led to the discovery that certain cellular oncogenes code for growth factor receptors or parts thereof (see Chap. 7). Such is the wonderful and unpredictable nature of science.

Insulin

Soon after insulin was isolated, it was found to help support the growth of cells in culture.[2] The idea that insulin might be a growth-stimulatory factor was supported by the observation that many of the mitogenic peptides derived from blood have an insulinlike activity. Whether or not insulin has a physiologically important mitogenic activity in vivo, however, is questionable because supraphysiologic concentrations of insulin are usually needed to stimulate cell proliferation and the mitogenic effect is usually small compared with that of total serum or other mitogenic peptides.[3] Insulin, however, frequently acts synergistically with other growth factors, probably because it is required for optimal uptake and utilization of needed nutrients.

Characterization of the cell surface receptor for insulin has led to a broader understanding of how insulin and insulinlike factors work. The insulin receptor is a tetrameric disulfide-linked complex containing two α subunits of 125,000 MW and two β subunits of 90,000 MW. It has been shown that the insulin receptor has intrinsic protein kinase activity that autophosphorylates its β subunit, as well as certain other substrates, on tyrosine residues.[4–6] Phosphorylation of the insulin receptor is activated by binding the ligand, and it increases the kinase activity of the receptor for other substrates. As we will see, this kind of ligand-receptor interaction leading to activation of an intrinsic receptor kinase activity is a common feature of polypeptide growth factors (Fig. 9-1).[7] Tyrosine kinase receptors interact with a number of signal transduction pathways. This is discussed in more detail later in this chapter.

Insulinlike Growth Factors

The discovery of IGFs came about from a variety of approaches but initially from the identification of insulinlike activities in plasma or serum. At first these activities had various names such as

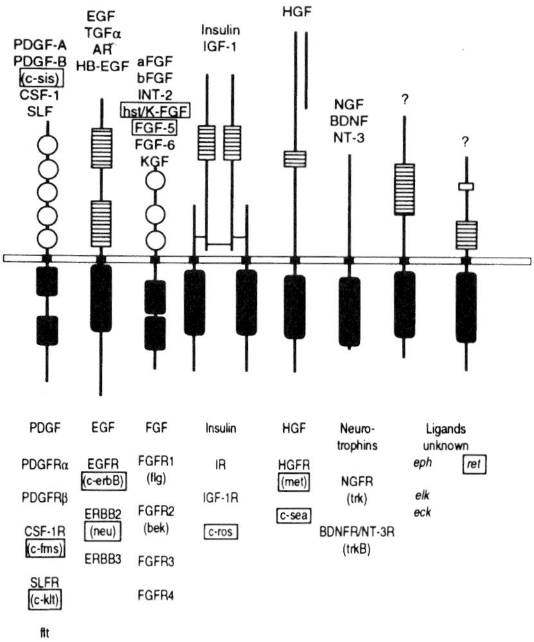

Figure 9-1 Transmembrane tyrosine kinases. Structural features of various tyrosine kinase receptors are shown. Each receptor family is designated by a prototype ligand. Growth factors known to bind to receptors of a given family are listed above, and receptors that constitute each family are listed below. Boxes denote those growth factors or receptors whose genes were initially identified as activated oncogenes. The c-*onc* designation is used to specify cellular homologues of retroviral oncogenes. Open circles illustrate immunoglobulinlike repeats. Dashed boxes indicate cysteine-rich domains. Dotted boxes indicate conserved tyrosine kinase domains. (Reprinted with permission from Aaronson, Growth factors and cancer. *Science* 254:1146. Copyright 1991 by the American Association for the Advancement of Science.)

nonsuppressible insulinlike activities, somatomedins, and *multiplication-stimulating activity.*

A group of polypeptides present in plasma that have insulinlike activity but could not be neutralized by antibodies directed against insulin, called the nonsuppressible insulinlike activities (NSILA), were separated into two fractions: NSILA-P, which was precipitable by ethanol at low pH, and NSILA-S, which was soluble under these conditions. Two peptides of about 7000 MW with mitogenic activity were isolated from the NSILA-S fraction.[8] These have now been called insulinlike growth factors 1 and 2 (IGF-1 and IGF-2).

A number of years ago it was realized that the growth-promoting activity of pituitary growth hormone was mediated through factors present in the serum. These factors, termed *somatomedins,* which have a wide variety of stimulatory actions—including stimulation of sulfate uptake into cartilage and cartilage growth, insulinlike activity on muscle and adipose tissue, and proliferation-promoting activity for cultured cells—have been isolated. Somatomedin C was found to be a basic polypeptide, of about 7000 MW, that stimulates DNA synthesis in cartilage and cultured fibroblasts. Its activity circulates in plasma bound to carrier proteins, which may provide a "reservoir" for the polypeptide and slow its removal from the blood.

Dulak and Temin[9] isolated a small polypeptide from serum, using the same initial steps as for NSILA purification but avoiding the acid-ethanol precipitation step. Further purification produced a fraction that was 6000 to 8000 times more active than whole calf serum in stimulating replication of chick embryo fibroblasts. The purified fraction, which they called multiplication-stimulating activity (MSA), had a minimal amount of insulinlike activity.

It is now known that the insulinlike growth-promoting activities described above are all related and belong to the IGF family of closely related mitogenic polypeptides. These polypeptides can be placed into two groups based on their isoelectric points. The basic group has isoelectric points above pH 7.5 and includes IGF-1, somatomedin C, and basic somatomedin. All of these are now known to be identical. Moreover, IGF-2 and MSA represent the same activity and belong to the neutral-acidic group of insulinlike activities with isoelectric points below pH 7.5. The circulating levels of IGF-1 depend on pituitary growth hormone and IGF-1 mediates its action, whereas plasma levels of IGF-2 do not appear to be so regulated. Because of the high levels of IGF-2 in fetal serum, it is thought to be an important fetal growth factor.[10] IGFs are thought to be generated in the liver, but other cells such as cultured fibroblasts and fetal tissues can also produce them. In addition to growth hormone, such factors as nutritional status and circulating insulin levels can regulate IGF production.

The IGF-1 receptor is structurally similar to the insulin receptor, also having a tetrameric subunit structure with two α subunits of 125,000

MW and two β subunits of 90,000 MW and possessing tyrosine kinase activity, whereas the IGF-2 receptor is a single transmembrane chain with a small intracellular domain, lacking tyrosine kinase activity but possessing mannose 6-phosphate receptor activity (reviewed in Ref. 11).

There is considerable cross-reactivity between the various ligands and receptors in the insulin and IGF family. For example, insulin can bind to both insulin and IGF-1 receptor, and IGF-1 and IGF-2 can bind to the insulin receptor.[12]

IGF-1 and IGF-2 as well as their receptors are present in a wide variety of human cancer types,[13] suggesting that paracrine or autocrine growth factor stimulation of proliferation of tumor cells by IGFs may contribute to their malignant phenotype. IGF-1 and IGF-2 receptors are expressed on breast cancer cells, and IGF-1 and -2 both have mitogenic effects on these cells.[14] The growth-promoting effect of estrogen on breast cancer cells has been postulated to be mediated by IGF-1, and interestingly, the antiestrogen drug tamoxifen has been reported to lower IGF-1 serum levels in breast cancer patients and to inhibit IGF-1 gene expression in the liver and lungs of rats bearing DMBA-induced mammary tumors,[15] suggesting that part of tamoxifen's antitumor action in breast cancer patients may stem from its ability to reduce production of IGF-1. Moreover, the synthetic retinoid fenretinide has also been shown to lower plasma levels of IGF-1 in breast cancer patients, suggesting a rationale for combination therapy with a retinoid and tamoxifen.[16]

Nerve Growth Factor

Nerve growth factor (NGF) was discovered as a result of some experiments designed to test the effects of rapidly growing tissue on nerve development in the limb-bud regions of chick embryos. In these experiments, two mouse sarcomas, S-37 and S-180, were transplanted to chick embryo limb-bud regions. Subsequently, a pronounced sensory innervation of the tumor tissue was observed.[17] Further experiments showed that the nerve-growth promoting effects of the sarcomas were most pronounced for neurons of the sympathetic nervous system and that the effector was a soluble protein.[18] Nerve growth factor, which has been purified from various snake venoms[19] and from mouse submaxillary gland,[20,21] consists of two subunits of about 13,000 MW, each with three intrachain disulfide bonds. When NGF is isolated from submaxillary gland, it is bound in a 7S protein complex of about 140,000 MW. NGF has several structural and functional similarities to insulin.[22] There is significant amino acid sequence homology with insulin, and the majority of the identical amino acid residues are clustered in regions of NGF that align with the insulin A and B chain segments of proinsulin but are separated by the 35 residues needed for the C activation peptide of proinsulin. There are an additional 37 amino acid residues at the carboxyl terminal end of NGF that extend beyond the sequences of its homology with proinsulin, but these residues are similar to the insulin B chain, suggesting a gene duplication event. Both insulin and NGF evoke similar biological responses in their respective target tissues, including increased uptake of glucose and nucleosides and increased RNA, protein, and lipid synthesis. Thus, NGF appears to be a product of a gene that evolved from the same or a similar ancestral gene as proinsulin. In addition to its insulinlike actions, NGF increases nerve fiber outgrowth and induces specific enzymes involved in sympathetic nervous transmission, for example, tyrosine hydroxylase and dopamine β-hydroxylase, in developing nerve cells.[3]

But these are only some of the cold, hard scientific facts. The story about how this all came about is much more interesting.[23] One of the codiscoverers of NGF, Rita Levi-Montalcini, was born in Turin, Italy, where she attended medical school and became board-certified in neurology and psychiatry. She became interested in the regulation of neuronal development and began studying development of the nervous system in chick embryos in Giuseppe Levi's laboratory in Turin. Unfortunately, both she and Levi were barred from academic life in 1939 as a result of Mussolini's anti-Semitic "Manifesto delle razze." Determined to continue her work, she set up a primitive lab in her basement and continued to do research even as the bombs fell during World War II. In some of her early studies, she found that fewer nerve cells grew into an area where the chick limb bud had been eliminated, suggesting that limb bud cells were releasing some trophic factor that stimulates nerve cell growth. A paper published on this topic caught Viktor Hamburger's eye, and he invited her to Washington University in St. Louis in 1946 to continue her studies. Intrigued by Bueker's observation[17] that mouse sarcomas transplanted to areas of the limb bud caused nerve cells to grow, she began to try to purify the trophic factor produced by the sarcoma cells. About that time, she began to collaborate with Stanley Cohen, a biochemist who was then at Washington University, on the purification. Initially, they found that the growth-stimulatory material contained both protein and nucleic acid, so they added snake venom, which is known to contain a diesterase that breaks down nucleic acids, to see if they could determine which component contained the activity. What they found was that even their snake-venom-only controls also had nerve growth activity. This led to the idea that salivary glands might be a source of the active substance. This turned out to be the case, and the material we now know as NGF was ultimately purified from mouse salivary glands, which also proved to be a source for epidermal growth factor (EGF) as well. In 1986, Levi-Montalcini and Cohen won the Nobel Prize for their pioneering studies on growth factors.

It is now known that there is a family of NGF-like factors (reviewed in Ref. 24). This family includes brain-derived neurotrophic factor (BDNF), and neurotrophins NT-3, -4, and -5. The members of this family have about 60% sequence homology and their activity appears to be limited to neuronal tissue. Various neurotrophins have growth-promoting activity for various nervous tissues: e.g., NGF has positive cell survival and differentiation effects on sensory, sympathetic, and cholinergic neurons as well as on PC12 pheochromocytoma cells; BDNF has positive effects on sensory, cholinergic, and dopaminergic neurons and on retinal ganglion cells but not PC12 cells; and NT-3, NT-4, and NT-5 are all growth factors for sensory neurons but have somewhat different effects on other nerve tissue cell types.

A high-affinity receptor for NGF has been identified as the 140-kDa glycoprotein gene product of the *trk*-A protooncogene.[25,26] This receptor, known as gp140$^{trk\text{-A}}$, has tyrosine kinase activity and appears to act through the Ras, Raf-1, MAP kinase signal transduction system.[27] Two Trk-A homologues, Trk-B and Trk-C, are receptors for other members of the neurotrophin family. Specificity for response of nerve tissue cells is due to the fact that there is cell-type specific distribution of the Trk receptors. The transforming version of Trk was first observed in colonic cancer cells as a truncated, chimeric protein fused to tropomyosin and having tyrosine kinase activity.[28]

A second low-affinity receptor for NGF, called gp75^{NGFR}, has also been identified and in some cell types is involved in a crucial way with gp140$^{\mathrm{trk\text{-}A}}$ in producing a response; in other cells, however, gp140$^{\mathrm{trk\text{-}A}}$ by itself is sufficient.[24]

In a study of tissue from 80 untreated neuroblastomas, the absence of gp140$^{\mathrm{trk\text{-}A}}$ mRNA was associated with tumor progression,[29] suggesting that a lack of a response system for NGF fosters unregulated proliferation of malignant nerve cell tissue.

Epidermal Growth Factor

Epidermal growth factor (EGF) was discovered during the course of some experiments on NGF activity in submaxillary gland extracts.[30] When these extracts were injected into newborn animals, precocious eyelid opening and eruption of incisor teeth were observed. These phenomena were found to be caused by stimulation of epidermal growth and by keratinization. The material responsible for these effects was isolated from mouse submaxillary gland and found to be a low-molecular-weight, heat-stable polypeptide.[30] It has been purified from mouse submaxillary gland and human urine. A single polypeptide chain of 53 amino acids, it has three intramolecular disulfide bonds that are required for biological activity. The EGF activity isolated from submaxillary gland at neutral pH is bound to a carrier protein to form a high-molecular-weight (74,000) complex. The complex consists of two molecules of EGF (6000 MW) and two molecules of binding protein (29,300 MW). The binding of EGF to its carrier protein depends on the presence of the carboxyl terminal arginine residue, and because the binding protein has arginine esteropeptidase activity, it has been suggested that biologically active EGF is generated from a precursor protein by the action of the carrier protein peptidase.[31] A similar but not identical carrier protein with arginine esteropeptidase activity has been discovered in association with NGF, and the carrier enzyme is also involved in formation of NGF from its precursor.[32] The EGF activity isolated from human urine is very similar in structure and function to mouse EGF and has significant amino acid sequence homology with it, suggesting that human and mouse EGF evolved from the same gene.[33] It has also been found that the gastrin antisecretory hormone urogastrone in human urine is, in fact, EGF; this suggests an additional biological function of this factor.[34]

The mouse EGF gene has been cloned and is surprisingly large compared with the gene size required to code for the mature EGF polypeptide (reviewed in Ref. 35). Mature EGF is only 53 amino acids long, yet the gene contains sufficient information to encode about 1200 amino acids. Contained within this 1200 amino acid sequence are eight EGF-like sequences, indicating that the gene codes for a large prepro-EGF molecule that is then processed to mature EGF and a family of EGF-like polypeptides in a manner similar to the production of ACTH from pro-opiomelanocortin.

A wide variety of cells—including epidermal cells, corneal cells, fibroblasts, lens cells, glial cells, granulosa cells, vascular endothelial cells, and a large variety of human cancer cells (see below)—have specific cell surface receptors for EGF, indicating the ubiquitous nature of the target cells for this growth factor. Specific receptors for EGF have even been found in liver membrane fractions from certain fish. Thus, EGF is a phylogenetically old protein.

It is now realized that there is a family of EGF-like growth factors that includes TGF-α, amphiregulin (a growth factor first detected in phorbol ester-stimulated MCF-7 cultured breast carcinoma cells), and poxvirus growth factors. All the members of the EGF family are 50 to 60 amino acids in length, have 6 half cystines in the same register (indicating the structural importance of the three intramolecular disulfide bonds), bind with high-affinity to EGF receptors, and produce mitogenic responses in EGF-sensitive cells.[35] The role of the large, eight-EGF sequence-containing prepro-EGF is not clear, but in some cells it exists as a membrane protein and retains EGF-like biological activity. Several other membrane-associated, growth-factor-containing proteins have also been discovered. A number of these have multiple EGF-like repeats (see below, under TGF-α). In some cases, e.g., TGF-α, cleavage of the repeats releases soluble growth factors. In other cases, the membrane-anchored growth-factor-containing polyproteins may be acting as "juxtacrine" growth factors in developmental processes that require cell–cell interaction. This function may not be brought about by secreted, diffusible growth factors.

The finding that a cultured human epidermal carcinoma cell line, A431, has a large number of specific EGF receptors has provided a model system in which to study the receptor and the effects of EGF–receptor interaction. The study of EGF receptors has produced some precedent-setting results that relate to a variety of growth factors.

The EGF receptor (EGFR) is an intrinsic membrane glycoprotein of about 170,000 MW.[36] It is monomeric, unlike the tetrameric insulin and IGF-1 receptors, and contains a core polypeptide of 1186 amino acids and N-linked oligosaccharides, the latter of which make up about 25% of the molecular weight of the receptor. The receptor binds EGF with high affinity ($K_d = 10^{-9}$ to 10^{-10} M) and high specificity. Occupation of the receptor by EGF induces an autophosphorylation of the receptor as well as phosphorylation of certain other cellular substrates (see below). The autophosphorylation site turned out to be a tyrosine[37] rather than a serine or threonine, which was a surprising finding at the time, because the only other known tyrosine kinase was the pp60^{src} product of the *src* oncogene (see Chap. 7). This observation for the EGF receptor came before similar findings for the insulin, IGF, and PDGF receptors and so was indeed precedent-setting.

The extracellular domain of the EGFR binds EGF and EGF-like ligands with high affinity, contains 10 to 11 N-linked oligosaccharides, and has a high content of half-cystine residues that could form up to 25 disulfide bonds, providing a complicated tertiary structure. The cytoplasmic domain contains the tyrosine kinase activity, four sites for tyrosine autophosphorylation, and several sites for serine/threonine phosphorylation that are substrates for phosphorylation by non-receptor kinases such as protein kinase C, suggesting cross-talk between kinase systems and additional control mechanisms for receptor activity. For example, PKC-mediated phosphorylation of threonine 654 attenuates EGF-activated autophosphorylation. There is evidence that binding of EGF to its receptor induces a dimerization that may be related to its functional activation. The ErbB oncoprotein is a truncated version of the EGF receptor and lacks most of the ligand-binding domain.

An important biological regulator in animals, EGF is capable of eliciting a wide variety of physiologic and cellular responses (Table 9-2). It is not clear, however, what its key biological role in vivo is. In cell culture systems, EGF stimulates cell proliferation of a wide variety of cell types. A number of studies have shown that maximal stimulation of DNA synthesis in cultured cells occurs at about 25% saturation of EGF binding sites, indicating that only a fraction of available receptors need be occupied by EGF to trigger a mitogenic response. Morever, cells grown in the presence of EGF continue to proliferate after the cultures become confluent, simulating the loss of density-dependent growth inhibition observed in transformed cells. This suggests that a continued exposure to the growth-stimulating activity of EGF or EGF-like factors could be involved in the excessive proliferation of cells and the loss of feedback inhibition of growth seen in neoplasia. In this regard, it has been observed that certain transformed cell types produce their own EGF-like growth factor that appears to bind to EGF cell surface receptors (see below).

In addition to human epidermoid carcinoma cells, other human cancer cells have receptors for EGF or EGF-like growth factors. Among these are ovarian, cervical, renal, lung, bladder, and breast carcinomas and glioblastomas (reviewed in Ref. 38), but this is not a universal finding; human leukemia and lymphoma cells have low or absent EGF receptors. In human epidermoid A431 carcinoma cells as well as in certain other human cancer cells (e.g., glioblastomas), the gene coding for the EGF receptor is amplified several fold,[39] which explains why some cell types have high receptor content and why they may be so responsive to a proliferative signal provided by EGF-like growth factors. Truncation and deletion mutants of EGFR have also been observed in human cancers.[38]

Table 9-2 Biological Effects of Epidermal Growth Factor

In vivo
Accelerated proliferation/differentiation
Of skin tissue
Of corneal epithelial tissue
Of lung and tracheal epithelia
Potentiation of methylcholanthrene carcinogenesis
Inhibition of gastric acid secretion
Increased activity of ornithine decarboxylase and accumulation of putrescine
Formation of fatty liver
Increase of disulfide group content in skin
Hepatic hypertrophy and hyperplasia
Potentiation of cleft palate
Organ cultures
Accelerated proliferation/differentiation
Of skin tissue
Of corneal epithelial tissue
Of mammary gland epithelial tissue
Induction of ornithine decarboxylase and accumulation of putrescine
Enhanced protein synthesis, RNA synthesis
Inhibition of palate fusion
Cell cultures
Increased transport
Of α-aminoisobutyrate
Of deoxyglucose
Of K^+
Activation of glycolysis
Stimulation of macromolecular synthesis
Of hyaluronic acid
Of RNA
Of protein
Of DNA
Enhanced cell multiplication
Alteration of membrane properties
Stimulated hCG secretion
Increased biogenesis of fibronectin
Enhanced prostaglandin biosynthesis
Alteration of viral growth
Increased squame production by keratinocytes

Source: From Carpenter and Cohen.[30]

Occupancy of EGF receptors by EGF triggers a mitogenic response. The molecular signals involved in this triggering mechanism have been defined and a number of cellular events associated with the EGF response have been identified. Stimulation of the mitogenic cascade involves activation of the Ras-Raf-MAP kinase system (see "Signal Transduction Mechanisms," below). The cellular responses to EGF include both early and late events (for review, see Ref.

40). Within 5 minutes after EGF addition to cultures of responsive cells, increases in protein phosphorylation, ion fluxes, and membrane ruffling occur. Internalization of bound EGF, increased amino acid and glucose transport, onset of RNA synthesis, and increased protein synthesis occur, in that order, between 15 and 60 minutes after initiation of EGF binding to its cell surface receptor. In contrast, induction of DNA synthesis requires exposure of cells to EGF for about 12 hours, and onset of cell division follows within a few hours after that.

The relationship between the early and late cellular responses is not clear. Presumably the pleiotypic response to EGF, as well as to other growth factors, follows some closely regulated temporal cascade of events. Since increased phosphorylation of membrane and other cellular proteins is one of the earliest events observed after EGF binding, it is a likely candidate for the signaling event of the cascade.

Internalization of EGF occurs in cells by means of a process called receptor-mediated endocytosis, which is common to a number of polypeptide hormones and growth factors (Fig. 9-2).[41] After a polypeptide binds to its specific receptor, there is a clustering of ligand-bound receptor into patches on the cell surface. These clusters are then invaginated into organelles called clathrin-coated pits (clathrin is a protein lining the cytoplasmic face of these vesiclelike structures). The pits then pinch off from the cell membrane and form true vesicles. These vesicles become acidified, and under the acidic conditions, the ligand becomes dissociated from the receptor. These vesicles are then shuttled to lysosomes, where they fuse with the lysosomes, and the receptors are degraded therein. In some cases (e.g., transferrin receptors), the endocytotic vesicles appear to become associated with the Golgi complex and receptors may then be recycled to the cell surface. Because this process, in effect, removes receptors from the cell surface, the cells for a time become refractory to further stimulation by exogenous growth factor. This refractory period, called receptor downregulation, appears to be a common pathway for regulation of cellular responses to a number of polypeptides, including EGF, insulin, transferrin, α_2-macroglobulin, immunoglobulins, and low-density lipoprotein (LDL).[42] In the case of EGF binding to cultured human fibroblasts, recovery of complete EGF binding capacity requires several hours and depends on RNA and protein synthesis,[43] indicating that the resynthesis of EGF receptors is required.

A number of growth factors induce a rapid change in ion fluxes across cell membranes as an early event. These events include increased Na^+/H^+ exchange, increased levels of intracellular free Ca^{2+}, and a transient rise in intracellular pH.[44–46] Although these events are commonly induced changes in cells after exposure to a wide variety of mitogenic agents, it is not clear what role they play in triggering the events leading to DNA synthesis.

The role that EGF plays in vivo, particularly in humans, is not clear. EGF has been isolated from human urine, as noted, and thus presumably does have a physiologic effect in adult tissues, probably as a paracrine growth factor involved in tissue renewal and wound repair. EGF is present in human milk, suggesting that it may play a role in milk production and may be a growth-promoting agent for the newborn.[47] Because EGF is acid-stable and at least partially active after oral administration, it may conceivably have such a function. In addition, EGF has been reported to restore spermatogenesis in male mice whose submandibular glands (the major organ or EGF production in mice) have been removed, suggesting a role for EGF in male reproductive function.[48] High-molecular-weight forms of EGF-like activities have been found in the urine of patients with disseminated cancer, and in a number of breast cancer cases there is a correlation with the presence of EGF-like material in the urine and advanced stage of the disease.[49]

The finding that the *erb*B oncogene encodes a protein that has significant sequence homology with the cytoplasmic, protein kinase–containing domain of the EGF receptor (see Chap. 7) has suggested that cancer cells may elevate their ability to respond to endogenously or exogenously produced growth factors, which may be a key to their ability to grow autonomously with minimum regulatory control. Since this truncated receptor protein lacks the ligand-binding domain, it may not be as subject to shutoff by the

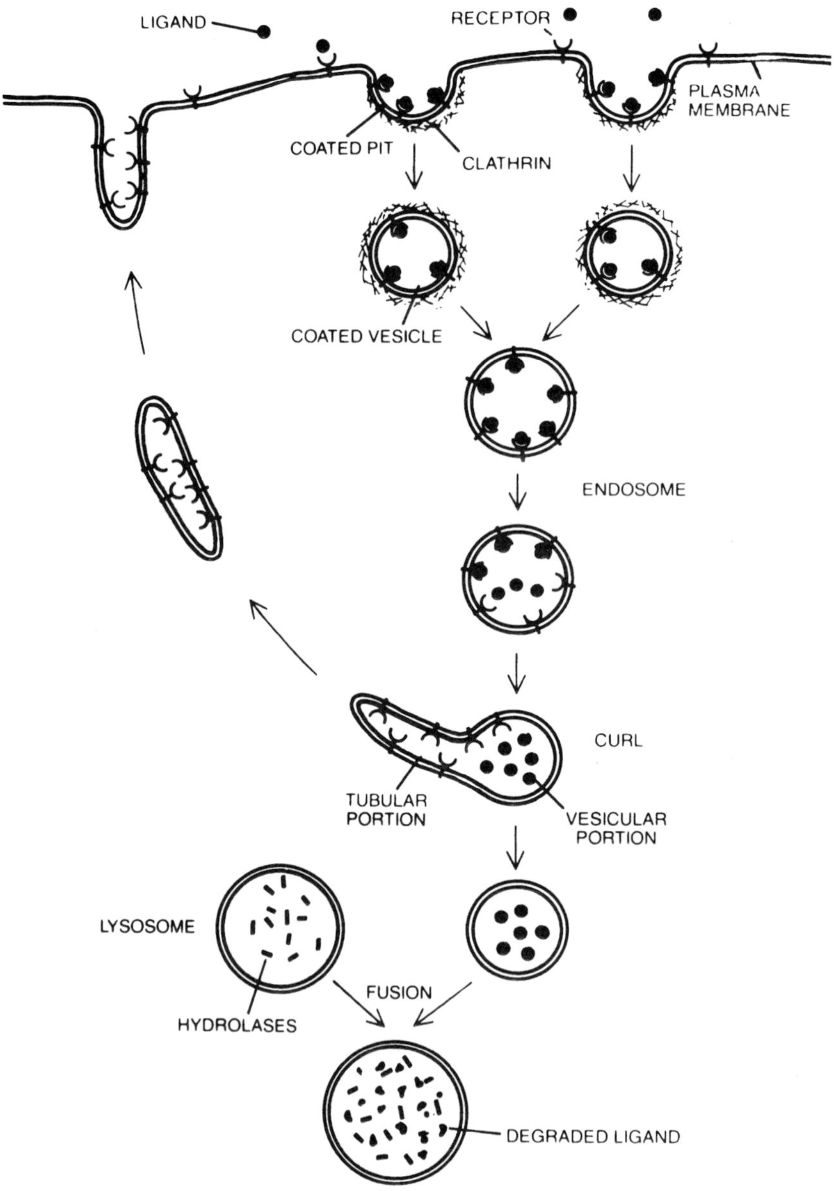

Figure 9-2 Mechanisms involved in receptor-mediated endocytosis. The pathway of receptors and ligands shown here was determined for galactose-terminal glycoproteins but is thought to apply in the case of other ligands and receptors as well. Ligand binds to receptors diffusely and then collects in coated pits, which invaginate and are internalized as coated vesicles, whose fusion gives rise to endosomes and then to a CURL (so designated because it is a "compartment of uncoupling of receptor and ligand"). In the acidic CURL environment, ligand is dissociated from receptors. Ligand accumulates in the vesicular lumen of the CURL, and the receptors are concentrated in the membrane of an attached tubular structure, which then becomes separated from the CURL. The vesicular part moves deeper into the cell and fuses with a lysosome, to which it delivers the ligand for degradation. In some cases (e.g. transferrin receptors), the membranous tubular structure is thought to recycle receptors to the plasma membrane. This is not, however, the case for EGF receptors. (From Dautry-Varsat and Lodish.[41])

downregulation mechanisms normally triggered by growth-factor binding and thus may be constitutively active as a protein kinase.

Fibroblast Growth Factors

Using initiation of DNA synthesis in mouse 3T3 fibroblasts to monitor purification, Gospodarowicz[50] isolated from bovine pituitary a mitogenic polypeptide, which he called fibroblast growth factor (FGF). Distinct from the IGFs, FGF is a potent mitigen for 3T3 cells and other cultured fibroblasts; however, by itself it is only about 30% as potent as whole serum. A combination of FGF and glucocorticoids will replace whole serum as a growth promoter for some cell lines (e.g., 3T3) but not for others (e.g., human diploid fibroblasts). In the latter case, serum is still required for maximal cell proliferation. Although FGF is mitogenic for several cell types, it does not affect cells transformed with polyoma or SV40 virus,[3] presumably because these transformed cells have a lower growth factor requirement.

Originally, two FGF-like polypeptides were purified to homogeneity and characterized[51]: the FGF from bovine pituitary, called basic FGF, and an acidic FGF from bovine brain. Basic FGF has also been isolated from a variety of other tissues, including adrenal gland, corpus luteum, retina, and kidney. There is significant amino acid–sequence homology between the two polypeptides, and they are potent mitogens for diverse cell types including capillary endothelial, vascular smooth muscle, adrenocortical, bone, and ovarian granulosa cells.

It is now clear that the FGFs constitute a large family of growth and differentiation factors for cells of mesenchymal and neuroectodermal origin. It includes the two original members acidic FGF (now called FGF1) and basic FGF (FGF2) as well as seven other current members (there may be more) (reviewed in Refs. 52 and 53). FGF3 (also known as Int-2) is a gene product first observed in MMTV transformed cells and is a 239 amino acid protein with 44% homology to FGF2. FGFs 4, 5, and 6 were found by screening tumors for oncogenes that could transform NIH 3T3 cells. FGF4 was also isolated from Kaposi's sarcoma and is called K-FGF. FGF7 (KGF) is a potent mitogen for keratinocytes and other epithelial cells and differs from the other FGFs in that it does not stimulate the proliferation of fibroblasts or endothelial cells. It is expressed in a restricted number of adult tissues, including kidney, colon, and ileum, but not in brain or lung as the others are. FGF8 is an androgen-induced growth factor cloned from mouse mammary carcinoma cells. FGF9, also called glia-activating factor (GAF), was purified from a human glioma cell line.[54]

The FGF family members have 30% to 70% amino acid sequence identity, with the greatest identity in a "core" region represented by most of the sequence of FGF2, which is considered the prototype FGF. All of the FGFs have the ability to bind heparan sulfate, which is important to their biological function (see below). Except for FGF1 and FGF2, all the FGFs have a N-terminal signal sequence and are secreted from cells. Yet FGF1 and -2 are abundant in the extracellular matrix (ECM), indicating that they must be exported from cells. Binding to the ECM appears to provide a reservoir for FGFs, which can then be mobilized in response to requirements for wound healing, angiogenesis, etc.

FGFs play an important role in embryonic development (reviewed in Ref. 53). For example, FGF1 and FGF2 are produced in early mesoderm and appear to be involved in mesoderm induction. Members of the FGF family appear to regulate differentiation of a variety of cell types during development, and they are expressed in a temporally and spatially regulated way. FGF4 and −5 are expressed in embryonic muscle cell precursors in the mouse. FGF1, −2, and −5 are expressed in nervous system tissue. FGF4 is the first FGF detected in early mouse development, as early as the four-cell stage, continues at least through the early formation of tissue layers (i.e., mesoderm, endoderm, and ectoderm), and is later expressed in a tissue-specific manner. FGF3 and -5 are expressed prior to gastrulation and appear to be restricted to parietal and visceral endoderm, respectively. FGF3, -4, and -5 mRNAs are detectable during mesoderm formation, but in distinct spatial orientations, suggesting a specific role for each FGF in development of specific tissues. FGF1 and -2 are expressed as early as day 10½ in mouse embryos and are detectable in a variety of tissues by day 13½. FGF6 is observed during middle and late gestation, but peak levels are observed on day 15.

Five different FGF receptors (FGFRs) have been cloned and more may be found (reviewed in Refs. 52 and 53). Isoforms of FGFRs are generated by alternate mRNA splicing. FGFRs belong to the tyrosine kinase receptor family and some of them were originally identified by cloning of tyrosine kinases, e.g., the hormone *flg*, chicken *cek*1, and mouse *bek* gene products. The receptors are now numbered FGFR 1–5. FGFR1 is the mammalian *flg* gene product (same as chicken *cek*1); FGFR2 is the murine and human *bek* product (also called K-Sam) and chicken *cek*3; FGFR3 is murine *flg*-2 and chicken *cek*2; FGFR4 and FGFR5 have no other names. As noted above, some FGFRs arise by alternate mRNA splicing. For example, FGFR2 and the receptor for KGF (KGFR) are both derived from the same gene (*bek*) by alternate splicing of one exon.

The FGFRs have several features in common: (1) An extracellular ligand-binding domain with three disulfide bonded loops and an immunoglobulinlike structure; (2) A single transmembrane domain, followed by a relatively long (80 amino acids) juxtamembrane domain; (3) two tyrosine kinase domains separated by a 14 amino acid insert; and (4) a carboxyl-terminal tail of about 50 amino acids, containing tyrosines that may be autophosphorylated upon ligand binding.

There is significant sequence homology among these receptors and overlap in their binding specificity for various FGFs. For instance, one isoform of FGFR1 binds FGF and FGF2 with slightly different affinities. An isoform of FGFR2 binds FGF1, -2, and -4 but not FGF5 or -7, whereas an FGFR2 splice variant binds FGF1 and -7 but not FGF2. FGFR3 can bind FGF1, -2, -4, and -5. FGFR4 binds FGF1 with high affinity and FGF4 and -5 with 10-fold lower affinity. This functional redundancy raises questions about what provides the specificity for response to the FGFs and why different receptors are needed to bind the same ligand. It seems likely that the tissue distribution of the FGFRs and the local production of specific arrays of FGFs provide some selectivity of response in various developing tissues. For example, FGFR1 is highly expressed in developing brain, skin, and growth plates of bones, whereas FGFR2 is highly expressed in choroid plexus, skin, lung, kidney, and brain temporal lobe. FGFR3 is abundant in the intestine, lung, kidney, and bone growth plates. FGFR4 is highly evident in adrenal gland, lung, kidney, and liver. Even though FGFR1 and FGFR2 have similar ligand binding specificities, FGFR1 is predominantly seen in mesenchyme of limb buds and somites, whereas FGFR2 is in highest abundance in epithelial cells of skin and developing internal organs, suggesting alternative and coordinate roles in these two tissue layers as organs develop. FGFR4 is found in muscle cell precursors and may have a prominent role in muscle development.

A feature shared by all FGFs is their high affinity for binding to heparan sulfate and heparan sulfate proteoglycans (HSPGs). A number of studies show an important role for heparan-sulfate-FGF complexes in binding of FGFs to their receptors (Fig. 9-3). HSPG-like molecules may be involved in presentation of FGF to the receptor or in stabilization of high-affinity FGF-receptor complexes, which provides an "anchor" for the interaction with cells. Alternatively, HSPGs may increase the half-life of FGFs by limiting proteolytic degradation and/or HSPGs may act as a reservoir for FGFs by providing long-term storage sites.

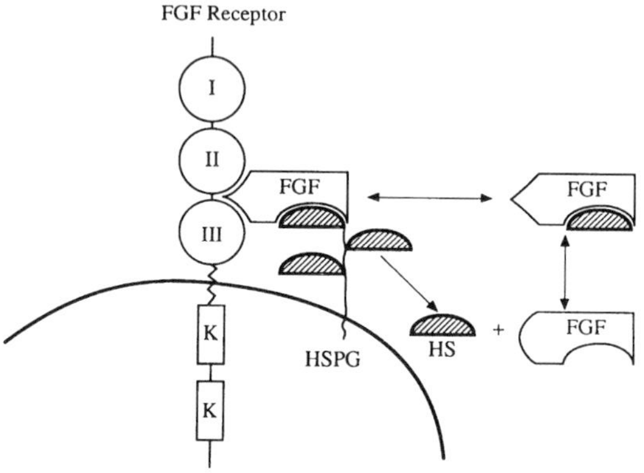

Figure 9-3 A model for heparin-dependent binding of FGF by FGFR. FGF binds to its high-affinity receptor only when bound to either the heparinlike molecule on the cell surface or free soluble heparin. The scheme depicts a possible conformational change of FGF upon heparin binding. HS, heparan sulfate; HSPG, heparan sulfate proteoglycan. I, II, and III are the three disulfide-linked binding domain loops. K = tyrosine kinase domains. (From Givol and Yayon.[52])

FGFs have three possible signal transduction mechanisms: (1) ligand-initiated activation of receptor tyrosine kinase activity that leads to autophosphorylation and phosphorylation of other key cellular proteins such as Raf-1; (2) activation of phospholipase C-γ1 (PLC-γ1) and the phosphoinositol hydrolysis (PI) pathway, leading to mobilization of intracellular calcium and activation of protein kinase C (see below); (3) internal localization of FGF in nuclei, which could lead to DNA binding and direct activation of gene transcription. There is evidence for all three mechanisms. Interestingly, there are high-molecular-weight intracellular forms of FGF1 and -2 that do not appear to get secreted and have nuclear localization sequences. Forms of FGF1 and -2 have also been found in cell nuclei.

Various FGFs are produced by malignant tumors, and constitutive expression of some FGFs induces a transformed phenotype in cultured cells. There are several examples of this. Basic FGF (FGF2) is not produced by normal melanocytes but is made constitutively by human metastatic melanoma cells grown in culture, and transfection of the bFGF gene into normal mouse melanocytes transforms them, although expression of bFGF by itself does not make them tumorigenic in vivo.[55] The ability of some FGFs to induce blood vessel growth has been observed in tumors (tumor angiogenesis). An FGF (K-FGF or FGF4) isolated from Kaposi's sarcoma induces vascularization of this tumor. K-FGF's action is inhibited by a heparin analogue called pentosan polysulfate, which blocks angiogenesis in a Kaposi's sarcoma growing in nude mice and induces tumor regression.[56] Basic FGF has been shown to be an autocrine growth factor for human and rat glioma cells, melanoma cells, and endometrial carcinoma cells (reviewed in Ref. 57). Injection of an antibody to bFGF into tumor-bearing nude mice inhibited tumor growth, partly, at least, because the antibody blocked the potent angiogenic effects of bFGF.[57] Basic FGF is also found in areas of human squamous cell head and neck tumors with a high thymidine labeling index and significant endothelial cell proliferation, indicating a correlation between tumor vascularization, tumor cell proliferation, and bFGF production.[58] By immunostaining for protein and in situ hybridization for mRNA, both acidic and basic FGFs have been detected in about 60% of surgical samples of human pancreatic cancer and their presence correlates with advanced tumor stage and poor prognosis.[59] Presence of bFGF in the cytoplasm of human renal cell carcinomas also correlates with shorter survival time.[60]

Platelet-Derived Growth Factor

The discovery of platelet-derived growth factor (PDGF) arose from an observation by Balk,[61] who found that normal chicken embryo fibroblasts proliferated better in culture medium supplemented with animal serum than in medium containing platelet-poor plasma. He also found that Rous sarcoma virus–transformed cells grew equally well in both, and he speculated that a "wound hormone" was released into serum during the process of clot formation. During the clotting of blood to form serum, a number of things happen, including the conversion of fibrinogen to fibrin and the clumping of platelets followed by platelet factor release. Thus, serum will contain some platelet-derived factors not present in plasma collected in the presence of anticoagulants and then centrifuged to remove the cellular elements and platelets. Other investigators have confirmed the observation that platelet-poor plasma is deficient in growth-promoting activity for cultured cells and have shown that platelet extracts could restore optimal growth to cells cultured in the presence of platelet-poor plasma.[62,63] The growth-promoting factor from human platelets has been purified and characterized.[64–66]

The PDGF family is a group of closely related proteins that consist of $\approx$ 30-kDa dimers of A and B chains: PDGF-AA, PDGF-AB, and PDGF-BB (reviewed in Ref. 67). The A and B chains are 60% homologous with eight conserved cysteines in the same relative positions in each chain. The A chain gene is on chromosome 7 and the B chain gene on chromosome 22. The genes have similar exon/intron organization and presumably arose by gene duplication during evolution.

The primary PDGF A translation product is a 23-kDa polypeptide that dimerizes to a 40-kDa form, which is subsequently processed to a 30-kDa secretory product, PDGF-AA. The PDGF

B chain is translated as a 28-kDa species that dimerizes to a 56-kDa form and then is processed to a 30-kDa secretory product, PDGF-BB. In addition, there are forms of PDGF A and B chains that are retained inside of cells. The primary target cells for PDGF are tissues derived from mesenchyme, although it is also expressed in glial cells and neurons during development.[68,69] Epithelial cells and lymphocytes do not respond to PDGF. Thus, its spectrum of action is less "catholic" than that of EGF. PDGF does not circulate in the blood as a free form; it is present in the α granules of platelets and released locally in an area of tissue damage as platelets aggregate, thus providing a targeted, localized growth hormone for connective tissue in an area undergoing wound healing.

Two distinct 170- to 180-kDa subtypes of PDGF receptors exist, called α and β. The α receptor binds all three PDGF isoforms (AA, AB, and BB), with about equal affinities, whereas the β receptor only binds PDGF-BB with high affinity.[67,70] The β receptor also binds PDGF-AB, but with a 10-fold lower affinity. After ligand binding, the receptor monomers dimerize into αα or ββ homodimers, but there is also some evidence for formation of αβ receptor heterodimers that may have distinct properties. The α and β receptors are structurally and functionally related. Both have extracellular ligand-binding regions made up of five immunoglobulinlike domains, single transmembrane segments, and two cytoplasmic tyrosine kinase domains with an intervening sequence between these TK domains (reviewed in Ref. 67). Thus, PDGF receptors resemble those of colony stimulating factor-1 (CSF-1), the c-*kit* gene product, and the FGF receptor family, suggesting that nature has used related evolutionary schemes to arrive at receptors for various growth factors.

The signal transduction pathways activated by PDGF binding its receptors involves autophosphorylation, which creates interaction with, and sometimes phosphorylation of, a number of potential signal-transducing substrates, including PLC-γ1, RasGAP, PI-3 kinase, $pp60^{c\text{-}src}$, $p62^{c\text{-}yes}$, $p59^{fyn}$, Nck, and GRB2 (reviewed in Refs. 67 and 71). These substrates contain SH2 domains that foster their interaction with activated PDGF receptors. These multiple substrate interactions suggest that activated PDGFR can initiate multiple mitogenic cascades. It is not clear why such redundancy is needed, but different cell types might have a predominance of one mitogenic mechanism over another and thus have one favored way to respond to exogenous signals. Another point, however, is that some of these signal transduction mechanisms are just various components of the same mitogenic cascade. For example, PLC-γ1 and PI-3 kinase appear to be distinct downstream mediators of PDGFR's mitogenic signal and to act in concert with Ras activation.[71] PDGFR activation results in activation of a number of genes involved in cell proliferation, including *myc, fos,* c-*jun, jun*B, and proliferation-"competence" genes such as a gene called JE.[72]

As with EGF, the cellular actions of PDGF can be divided into early and late events. Early events (after 1 to 10 minutes) include tyrosine-specific phosphorylations, stimulation of phosphatidylinositol turnover, and reorganization of actin filaments.[73] Late events (after 30 to 180 minutes) include increased transcription of specific genes, stimulation of IGF-1 binding, and increased amino acid transport. A clear difference between PDGF and EGF or IGF-1 is that the stimulatory signal for cell division provided by PDGF is "locked in" after the cells are exposed to PDGF for only about 30 minutes, and PDGF can then be removed from the growth medium, whereas EGF and IGF-1 must be present continually to stimulate cell division. Moreover, PDGF by itself does not induce a mitogenic response but requires other plasma factors, among which are IGF-1 and EGF. For example, when growth-arrested 3T3 fibroblasts are treated with PDGF for a short period of time and then transferred to platelet-poor plasma, the cells are observed to enter S phase 12 hours after the addition of plasma. This 12-hour lag before the onset of DNA synthesis occurs regardless of whether the plasma is added at the same time as PDGF or up to 13 hours after this. Thus, PDGF and plasma factors appear to control different events in the cell cycle. Scher et al.[74] have postulated that PDGF induces cells to become "competent" to enter the S phase of the cell cycle and that plasma factors allow only competent cells to undergo "progression" through the G_1 phase to enter S phase. Thus, PDGF is thought to prime cells to respond to other growth factors in plasma. As noted earlier, the PDGF-induced competent state must be stable for at least 13 hours after PDGF is removed, since the addition of plasma at any time up to 13 hours will permit progression into S phase.

Among the most potent factors in plasma that stimulate progresion are the insulinlike growth factors. The evidence for this comes from experiments in which plasma from hypophysectomized rats was shown to be 20-fold less potent in permitting PDGF-treated competent 3T3 cells to enter S phase than normal rat plasma. The addition of a low concentration (10^{-9} M) of IGF-1 to cultures of PDGF-treated cells grown in plasma from hypophysectomized rats allowed the cells to enter S phase. Pure IGF-1 without plasma, however, did not do this, indicating that other factors in plasma are also needed for progression to occur. By using this assay system, the competence- and progression-stimulating activities of a variety of growth factors have been tested. Factors that have potent competence activity include PDGF, FGF, Ca^{2+}, and "wounding" (i.e., scraping a clear area through a sheet of confluent cells). Progression-inducing agents are IGF-1, IGF-2, insulin, and EGF. All the mechanisms controlled by these two classes of growth factors are not clear, but they probably relate to the commitment or restriction point at which cells make a commitment to enter S phase. Once this commitment is made, cells enter S and complete the cell cycle; if they are delayed somewhere past this commitment point, they frequently die. The competence-initiating factors may induce the synthesis of a critical initiator protein, whereas progression-inducing agents may promote cell division by stimulating the enzymes necessary for DNA synthesis. However, there appears to be more than one point at which PDGF-treated cells can be arrested before they enter into S phase,[74] which suggests a cascade of events, each of which might have different regulatory signals. The fact that a specific competence-inducing factor appears to be required for the proliferation of fibroblastlike cells suggests that other tissues may have similar requirements for competence activities. The less specific progression-stimulating activities, such as those of the IGFs, may be general growth-promoting agents needed for the growth and development of many tissues in the body.

Of interest is the fact that transformation of fibroblasts with SV40 virus circumvents the need for both competence and progression activities. The SV40-transformed 3T3 cells grow to high density in either serum or platelet-poor plasma and require less serum than nontransformed cells. Thus the requirement for PDGF appears to be lost or greatly diminished during the process of transformation. In addition, transformed human and mouse cell lines that produce tumors in nude mice grow to high density in platelet-poor plasma, whereas cell lines that are not tumorigenic in nude mice grow poorly in platelet-poor plasma.[75] These data imply that transformed cells can produce a PDGF-like growth factor that occupies PDGF receptors and thus does not require exogenous PDGF. It has now been shown that a wide variety of murine and human cell lines transformed with oncogenic viruses or chemical carcinogens produce a PDGF-like substance that competes for binding to PDGF receptors and that antibody to PDGF can inhibit this PDGF-like activity.[76] Moreover, a number of human cancer cell lines have also been shown to secrete PDGF-like factors; these include cells derived from osteosarcomas, glioblastomas, and fibrosarcomas.[77–80] Some human tumor cells coexpress both PDGF-like factors and PDGF receptors, setting the stage for self-stimulation of cell proliferation, given exposure to the right progression factors.

When human melanoma cells are transfected with PDGF-BB cDNA, they produce, in nude mice, actively growing nests of tumor cells with a distinct stroma and abundant blood vessels, suggesting a role for PDGF as an inducer of a vascularized connective tissue stroma on which tumor cells can thrive.[81] Amplification and/or overexpression of PDGF receptors has been observed in human glioblastomas,[82] and activation of PDGF-B gene expression in concert with the PDGF-β receptor gene correlates with the conversion of human hydatidiform made into choriocarcinoma.[83]

A t(5;12)(q33;p13) chromosomal translocation that creates a fusion between the PDGF β receptor gene and a novel *ets*-like oncogene called *tel* has been observed in chronic myelomonocytic leukemia (CMML).[84] This creates a fusion transcript derived from the tyrosine kinase domain of the PDGF-β receptor on chromosome

5 and the *tel* gene on chromosome 12. The PDGF-β receptor-*tel* fusion product would be expected to have high oncogenic potential since it couples a growth factor receptor to an oncogene product that may cause constitutive activation of the tyrosine kinase cascade without the requirement for ligand binding. This translocation event appears to occur before CMML converts to blast crisis (acute myelocytic leukemia, or AML) and thus to be an early mutation in the multistep carcinogenic process of AML.

The first clear link was forged between growth factors and oncogene products when it was discovered that the PDGF B chain has a virtually identical amino acid sequence to the product of the *sis* oncogene first isolated from a sarcoma virus carried by the Woolly monkey (see Chap. 7). This cellular oncogene was apparently picked up during evolution by the simian sarcoma virus and is part of the transforming activity of the virus. The v-*sis* oncogene can activate α and β PDGF receptors and initiate cellular transformation when it binds intracellularly to PDGF receptors.[85]

Transforming Growth Factors

The discovery of transforming growth factors (TGFs) came out of experiments showing that mouse 3T3 cells transformed with murine or feline sarcoma viruses rapidly lost their ability to bind EGF, whereas cells infected with nontransforming RNA viruses maintained normal levels of cell-surface EGF receptors.[86] These initial results suggested that the sarcoma virus genome produced something that altered EGF receptors. Later, however, it was found that murine sarcoma virus–transformed mouse fibroblasts produced a polypeptide growth factor that competed for binding with EGF on cell surfaces.[87,88] This factor was called sarcoma growth factor (SGF), a 6000- to 10,000-MW, heat-stable, trypsin-sensitive polypeptide that stimulated proliferation of transformed and untransformed fibroblasts. It competed with EGF for binding to EGF receptors, but it had a different molecular weight and was immunologically distinct from EGF. It also had the interesting property of being able to promote anchorage-independent growth in cultures of normal fibroblasts and thus to confer on normal cells properties associated with the transformed phenotype. This phenomenon was reversible, so that after the SGF was removed from the growth medium, the cells regained normal growth properties. Thus, SGF appeared to be a growth factor produced specifically by transformed cells and capable of stimulating their proliferation. This was the first observation suggesting that neoplastic cells are capable of autostimulation by producing their own growth factors. In this way, they could presumably escape the negative feedback systems of the normal host that control the production and release of endogenous hormones and growth factors.

Although the EGF-competing activity of SGF was contained in a 6000- to 10,000-MW fraction, the cellular transforming activity appeared to require, in addition, a fraction of higher molecular weight (20,000 to 25,000). The transforming activity of SGF isolated from murine or feline sarcoma virus–transformed cells was subsequently shown to be separable into two fractions: one of about 6000 MW, which competes with EGF for binding to EGF receptors and induces only small colonies of normal rat kidney (NRK) cells in soft agar, and one of about 25,000 MW, which does not compete for EGF binding but is required for production of large colonies of NRK cells in soft agar.[89,90] The former, EGF receptor–binding form has been termed transforming growth factor-α (TGF-α) and the latter is called TGF-β. Like EGF, TGF-α is a potent mitogen and appears to act through the same receptor, but by itself it is only a weak inducer of anchorage-independent cell growth, as is EGF itself. TGF-β is also mitogenic for NRK cells and also, to some extent, for mouse and human fibroblasts, but it does not induce anchorage-independent growth if added alone to test cultures of NRK cells; however, it acts synergistically with either TGF-α or EGF to induce the transformed phenotype and anchorage-independent growth.

A wide variety of RNA and DNA tumor viruses stimulate production of TGF-like substances in cells transformed by them. Oncogenic RNA viruses such as Harvey, Kirsten, and Moloney murine sarcoma viruses as well as Abelson murine leukemia virus and DNA tumor viruses such as SV40 and polyoma have this property.[91] Not all the TGFs produced by cells transformed by these viruses are identical, and thus there are most likely families of TGF-α and TGF-β produced by different types of tumor cells. There is a close relationship between the release of TGF-α and cellular transformation by murine sarcoma viruses, as shown by the use of temperature-sensitive mutants of the transforming viruses.[91,92] When cells are grown at the temperature permissive for cell transformation, TGF-α is produced, but when they are grown at the nonpermissive temperature, the factor is not produced.

TGF-α

As discussed previously, TGF-α belongs to a family of GFs that includes EGF, amphiregulin, and vaccinia virus growth factor. TGF-α is produced as a 160 amino acid proTGF-α form that is cleaved to produce a 50 amino acid soluble form of TGF-α (reviewed in Ref. 93). Although TGF-α was first found in culture fluids of oncogenically transformed cells and is expressed by a wide variety of human cancer cells, its expression is not limited to neoplastic cells. During rodent embryogenesis, it is expressed in maternal decidua and in developing kidney, pharynx, and otic vesicle. TGF-α mRNA and/or protein is also found in adult pituitary, brain, keratinocytes, ovarian theca cells, and macrophages, implying a role in the economy of normal adult tissues as well. TGF-α and EGF have a similar ability to promote proliferation and differentiation of mammalian mesenchymal and epithelial cells. This is not unexpected, since they activate the same cell surface receptors.

In addition to the soluble form of TGF-α, there is a membrane-anchored form (the putative "juxtacrine" form). This is a feature of TGF-α that it has in common with several other membrane anchored proteins bearing EGF-like repeats on their extracellular surface (Fig. 9-4). Some of these membrane-bound glycoprotein forms are cleaved to yield soluble EGF-like GFs, while others—such as the *Drosophila Notch, Delta,* and *Crumbs* gene products as well as the *Caenorhabitidis elegans lin*-12 and *glp*-1—are not known to be cleaved to release soluble factors. While the role of these cell surface EGF-like repeats is not clear, it seems probable that they interact with receptors on the surface of adjacent cells to sustain cell–cell adhesion and cell–cell regulation of proliferation and differentiation.[94]

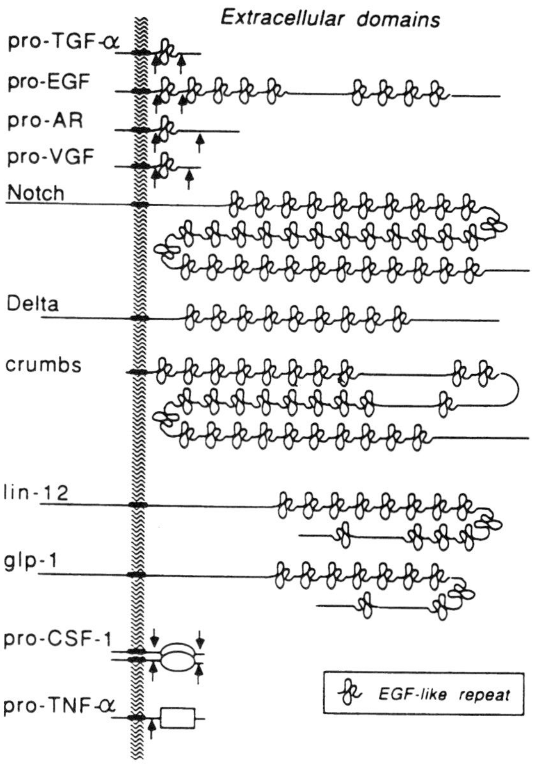

Figure 9-4 Membrane-anchored growth factors and related molecules. Drawings represent the general domain structure of each molecule listed on the *left*. The *shaded vertical line* represents the plasma membrane. *AR,* amphiregulin; *VGF,* vaccinia virus growth factor; *CSF-1,* colony-stimulating factor-1; TNF, tumor necrosis factor. *Arrows,* known sites of cleavage. (From Massagué.[93])

While TGF-α plays a role in normal development, wound healing, ECM production, angiogenesis, and cellular adhesion, it is clear that its production at the wrong time or wrong place or its overproduction can favor neoplastic transformation and/or progression. Transfection of the TGF-α gene into cultured cells can be transforming. Overexpression or inappropriate production of TGF-α has been observed in human lung adenocarcinoma,[95] squamous cell carcinoma,[96] breast carcinoma,[97] endometrial adenocarcinoma,[98] and hepatocellular carcinoma.[99] TGF-α has been detected in the urine of patients with hepatocellular carcinoma[100] and in effusion fluids of patients with a variety of cancers—including ovarian, breast, and lung cancers—often as a bad prognostic sign.[101] In addition, liver carcinomas have been shown to develop in transgenic mice that constitutively overexpress TGF-α.[99,102] This latter observation supports the notion that deregulated expression of TGF-α is a problem for cells. It is expressed in developing liver, repressed in adult liver, and reexpressed in regenerating liver,[103] suggesting that its expression is coupled to cell proliferation and differentiation in a carefully regulated way.

TGF-β

TGF-β is one member, it turns out, of a "superfamily" of growth-modulating factors that have a wide variety of functions (reviewed in Refs. 104 to 106). This superfamily contains five members of the TGF group, bone morphogenesis and developmental patterning factors, and regulators of reproductive function (Fig. 9-5). Among the actions of the TGF family, various cellular actions include an antiproliferative response involving control of *myc* gene expression and prevention of RB hyperphosphorylation via cyclin-dependent kinases, stimulation of chemotaxis of fibroblasts and other cell types, increased ECM formation, stimulation of wound healing and bone formation, increased neuron survival, and possible ability to prevent breast cancer progression. This is a Herculean group of molecules indeed! We will attempt to put some of these actions in perspective here.

TGF-βs have been viewed as growth-stimulatory for mesenchymal cells and growth-inhibitory for epithelial cells, but this is largely based on cell-culture studies and on the positive effects of TGF-βs on wound healing.[104] In vivo, TGF-βs appear to be mostly growth-inhibitory. Thus, the effects on stimulating ECM formation, wound healing, angiogenesis, and bone formation seem paradoxical. The explanation for this appears to lie in the facts that TGF-βs are strong cellular chemoattractants and that they can indirectly cause fibroblast and endothelial cell proliferation by upregulating the expression of PDGF. For example, local administration of TGF-β1 to the chicken chorioallantoic membrane caused increased density of fibroblasts, ingrowth of blood vessels, and increased ECM deposition without marked stimulation of cell

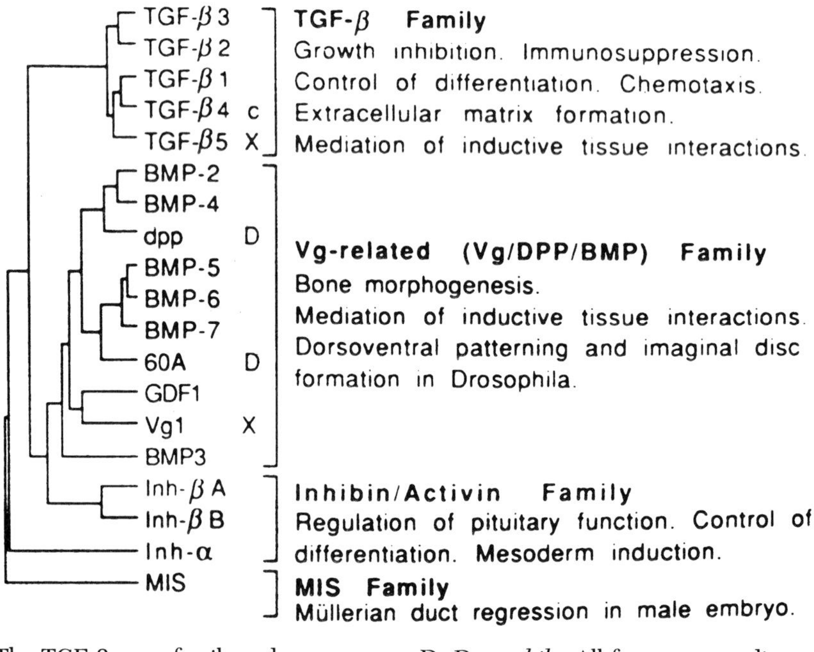

Figure 9-5 The TGF-β superfamily and some representative activities. The hypothetical relationship dendrogram is based on th protein sequence of the bioactive domains of these factors. All factors are human except as indicated: c, chicken; X, *Xenopus;* and D. *Drosophila.* All factors act as dimers. Activins are dimers of βA and/or βB chains, and inhibins are dimers of one α chain and one βA or βB chain. (From Massagué.[105])

proliferation.[104] Yet some cell proliferation was observed at the periphery of the site of administration, where the concentration of TGF-β1 was lower. This latter effect appears to be due to a biphasic dose-response effect on PDGF-AA synthesis, in which stimulation of PDGF-AA synthesis occurs at low concentrations of TGF-β and inhibition occurs at high concentrations. These data support a way in which TGF-β1 could stimulate or inhibit cell proliferation in a given cell type depending on its local concentration. However, the growth-inhibitory effects of TGF-β on epithelial cells and most likely on lymphoid and myeloid cells cannot be thereby explained, because these cells lack PDGF receptors and do not respond to PDGF. Thus, for the mechanism that inhibits cell proliferation of these cell types, we must search elsewhere.

Several explanations for the antiproliferative effects of TGF-βs suggest themselves. TGF-β1 has been shown to block the expression of c-*myc* at the level of transcriptional initiation, and this action is inhibited by SV40 T antigen, human papillomavirus-16 E7, and adenovirus E1A oncoproteins (reviewed in Ref. 104). Since all of these oncoproteins are known to interact with the retinoblastoma protein RB and since the function of pRB is known to be regulated by its phosphorylation state, a clue to TGF-β's antiproliferative effects is provided. To this end, it has been shown, in cultured mink lung epithelial cells used as a model system, that addition of TGF-β1 to the culture medium in mid- to late G_1 phase of the cell cycle prevents phosphorylation of RB and arrests cells in late G_1, suggesting that the TGF-β1–dependent growth inhibitory pathway involves TGF-β's ability to cause RB to remain in its underphosphorylated, growth-suppressive state.[107] It was further shown that TGF-β1 can inhibit the expression and/or function of G_1-phase cyclins, including cyclin E, which regulate the CDK2 phosphokinase that phosphorylates RB and other cell-cycle regulatory proteins (see Chap. 10),[108] thus completing the picture implicating RB in TGF-β's action.

At least nine different TGF-β–binding proteins are known, but only two appear to have true receptor activity in that they both bind TGF-βs and produce a signal-transducing response. These are called TGF-β receptors I and II (reviewed in Ref. 105). TGF-β receptors (TGF-βR) I and II are 53- and 75-kDa glycoproteins, respectively, and are present on essentially all nontransformed cells. Cells bearing mutants of receptors I and II that fail to bind TGF-β lack the typical gene regulatory and growth inhibitory responses of TGF-βs. TGF-βR II has an external ligand-binding domain and a cytoplasmic serine/threonine protein kinase domain rather than a tyrosine kinase domain, as seen for a number of other growth factors, making it somewhat unique among GF receptors. Both TGF-βRI and II appear to be required for full TGF-β action since cell lines bearing type II receptors but lacking type I fail to respond to TGF-β, and fusion of these cells with cells containing type I receptors restores response to TGF-β. Thus, there appears to be some cooperativity between the two receptor types. It may be that type I TGF-βR assists in presentation of ligand to type II R or that it modulates the kinase activity of type II R. There is evidence that TGF-β binds directly to type II receptor; this complex is then recognized by receptor I, which becomes phosphorylated by the kinase activity of receptor II, thus initiating the first step of the signal transduction cascade.[108a] It should be noted that the active forms of TGF-βs are dimers that may induce receptor dimerization after binding, similar to that seen for ligands that bind tyrosine kinase receptors.

Since TGF-βR I and II bind most forms of TGF-β (although they bind TGF-β1 and -β3 better than -β2, for example), the common conundrum arises: What provides the specificity of cellular response? It is not perfectly clear, but there is another important player that may explain some of this differential response of various cells to various TGF-β-like ligands. This is betaglycan (sometimes called TGF-β receptor III). Betaglycan is a membrane-anchored proteoglycan and binds TGF-βs with high affinity. As with the FGFs, this binding to a proteoglycan may act to present ligand to receptors, perhaps in a somewhat cell-specific way, and may also serve as a storage depot for TGF-β.

One question that arises is how do tumor cells escape the antiproliferative action of TGF-β? Some tumor cells lack or have defective TGF-β receptors.[105,108b] Some tumor cells are stimulated rather than inhibited by TGF-β.[109] The growth of some tumor cells may be augmented by the TGF-β-stimulated deposition of ECM and the concomitant angiogenesis.[110] And some tumor cells may be indirectly stimulated by TGF-β via its action to enhance production of PDGF. Nevertheless, the ability of TGF-β to inhibit cancer cell proliferation holds out some therapeutic promise. For example, certain chemopreventive agents such as tamoxifen and retinoic acid analogues can induce the production of TGF-β in stromal fibroblasts, and this may well be the mechanism of action of some of these agents. Indeed, tamoxifen has been shown to have this enhancing effect on the stroma of human breast cancers.[106]

Angiogenesis Factors

It has been known for more than 100 years that solid tumors can become vascularized. It was not appreciated until the 1950s, however, that growing tumors elicit new capillary growth from the host,[111] a process called tumor angiogenesis. The mechanism of this angiogenesis was shown to involve release of some substance(s) from growing tumors that stimulates outgrowth of capillaries from the host's vasculature. This was demonstrated by implanting tumors into the cheek pouch of hamsters in such a way that the normal stromal tissue of the host animal was separated from the tumor tissue by a filter with very small pores (0.45 μ in diameter) that would not allow cells to migrate but would allow large molecules to diffuse between tumor and host tissues.[112,113] In these experiments, the growing tumors elicited the proliferation of new capillaries in the host tissue, indicating the release of a diffusible substance by the tumor that stimulates capillary growth. This factor was called tumor angiogenesis factor (TAF).[114] Folkman and his colleagues showed that tumor cells transplanted into the corneas of rabbits initially grew slowly, but after about a week, small capillaries began to grow outward from the iris toward the tumor and when the capillaries reached the tumor, it began to grow rapidly.[115] Corneal implants of normal adult tissues or of rapidly dividing embryonic tissue did not induce capillary growth. Injection of tissue extracts into the cornea and application of extracts directly onto the chorioallantoic membrane of a fertile chicken egg have been used to demonstrate the presence of TAF. A wide variety of tumors have been examined for TAF activity, and many tumors have been found to contain it. The ability to induce angiogenesis, however, is not restricted to neoplastic cells. Angiogenesis can also be induced by spleen lymphocytes, thymocytes, peritoneal macrophages, and testicular grafts from newborn mice and by leukocyte invasion of the cornea (for review, see Ref. 116). Thus, it is not clear that all the capillary outgrowth in an area surrounding a growing tumor is due to release of a TAF from tumor cells themselves. It may be that the induction of capillary growth by tumors is, in fact, the result of a combination of factors.

A number of factors are known to stimulate vascular endothelial cell proliferation in vitro, including FGFs and TGF-α (see Table 9-3).[117] Some TAF-containing tumor extracts stimulate capillary endothelial cell proliferation in culture if the endothelial cells are grown on a substratum of type I collagen in the presence of platelet-released factors, suggesting that stimulation of capillary endothelial proliferation in vivo depends on the right "microenvironment" and the right concentration of factors. It could be that a combination of competence-inducing and progression-stimulating factors is needed for stimulation, in much the same way as was noted for the actions of PDGF and IGF-1 on fibroblast proliferation.

Angiogenesis, of course, is also a normal process by which new blood vessels are formed—for example, in development of the placenta, in vascularization of developing organs, and in wound healing. Under these conditions, however, angiogenesis is highly regulated, being turned on for specific periods of time and then shut off. It is an unregulated form of angiogenesis that occurs in tumors and in certain other diseases, such as arthritis, diabetic retinopathy, and hemangiomas.

A number of steps are required for angiogenesis to occur: (1) local dissolution of the subendothelial basal lamina of the existing vessels, (2) proliferation of endothelial cells, (3) migration of endothelial cells toward the angiogenic stimulus, and (4) laying down of a basal lamina around the nascent capillary. The different an-

Table 9-3 Angiogenic Polypeptides[a]

	M_r	Subunit	pI	Endothelial Cell Mitogenicity *in Vitro*	Other Angiogenesis-Related Biological Activities
bFGF	18,000	1	9.6	+	Both bFGF and aFGF are mitogenic for a wide variety of cell types; bind to heparan sulfate proteoglycan and copper; stimulate endothelial cells to migrate and form tubes and to increase production of proteases and plasminogen activator; and to act as embryonic inducers.
aFGF	16,400	1	5	+	
VEGF/VPF	45,000	2	8.5	+	Proliferation activity highly specific for vascular endothelial cells; secretory proteins; increase vascular permeability; induce plasminogen activator and plasminogen activator inhibitor in endothelial cells; structurally related to PDGF.
PD-ECGF	45,000	1	5	+	Stimulates endothelial cell DNA synthesis and chemotaxis; proliferation activity not reported; amplifies DNA synthesis activity of FGFs on endothelial cells.
TGF-α	5,500	1	6.8	+	Transforms normal cells into transformed phenotype; binds to EGF receptor.
Angiogenin	14,100	1	9.5	○	Stimulates endothelial cells to form diacylglycerol and to secrete prostacyclin by activating phospholipase C and phospholipase A_2, respectively; has a unique ribonucleolytic activity essential for neovascularization.
TGF-β	25,000	2	4	−	Enhances extracellular matrix production; binds to copper; chemotactic for monocytes
TNF-α	55,000	3	4	−	Induces production of bFGF in endothelial cells and enhances its secretion; chemotactic for monocytes; activates macrophages.

[a]Angiogenic activities of the listed polypeptides were determined by chick embryo chorioallantoic membrane assay, corneal micropocket assay, or hamster cheek pouch assay. These assays cannot currently be used to compare potency quantitatively. However, the FGFs and angiogenin appear to be the most potent. VPF, vascular permeability factor; PD-ECGF, platelet-derived endothelial cell growth factor; TGF, transforming growth factor; PDGF, platelet-derived growth factor, +, stimulatory; ○, no effect; —, inhibitory.

Source: From Folkman and Shing.[117]

giogenesis factors shown in Table 9-3 may interact in different parts of this cascade. For example, FGFs and vascular endothelial growth factor (VEGF) are directly mitogenic for endothelial cells; TGF-β stimulates ECM deposition to help form a basal lamina; angiogenin may help create new "tracks" for vessel formation by ribonucleolytic action.

The first purification of an angiogenesis factor was based on affinity of such factors for heparin, and this led to the identification of basic and acidic FGFs as angiogenesis factors. Since then a number of others have been isolated and characterized (reviewed in Ref. 117). A number of such factors have been shown to be produced and secreted by human tissues. For example, VEGF is produced by human gliomas[118] and epidermoid carcinoma cells.[119] In some cases, angiogenesis factors are found in the urine[120] or effusion fluids[121] of cancer patients, and their presence relates to conversion of hyperplasia to neoplasia[122] and to tumor progression.[123] Indeed, there is much evidence to suggest that neovascularization or conversion to the "angiogenic phenotype" is involved in tumor progression.

Most cancers in humans are of epithelial origin and may grow slowly and remain localized (in situ) for many years before they become invasive and metastatic (see Chap. 11). Evidence suggests that part of this change from in situ carcinoma to invasive malignant cancer involves neovascularization of the tumor. Epithelial cancers do not develop normal vascular beds like normal tissues and depend to a large extent on the diffusion of oxygen and substrates for growth. When tumor cells are too far away from the capillary blood supply for diffusion to provide the needed nutrients, the cells may die. This explains why the cores of large, solid tumors are often necrotic. As long as the tumors remain small, they can obtain sufficient nutrients by diffusion; as they grow and progress to a more malignant cell type, however, this process becomes limiting. At that point, tumors may be stimulated to release angiogenic factors that induce capillary outgrowth from the host's surrounding normal tissues into the tumor. Although this does not result in a full vascularization of the tumors, it does provide nutrients for their growth. This process of angiogenesis is believed to be part of the process involved in converting in situ carcinomas to aggressive malignant tumors. Thus, blocking the process could inhibit or significantly slow this conversion. This concept has led to a search for antiangiogenic agents.

A number of inhibitors of angiogenesis have been discovered. Langer et al.[124,125] reported that extracts of bovine or shark cartilage inhibited the vascularization and growth of rabbit V_2 carcinoma and mouse B16 melanoma in vivo. This activity did not appear to be related to an antiprotease or anticollagenase action or to a direct inhibition of tumor cell growth. Folkman et al.[126] reported that heparin or a heparin fragment in the presence of cortisone inhibited angiogenesis, caused tumor regression, and inhibited metastasis of a number of murine tumor types in vivo. The heparin fragment was identified as a hexasaccharide. This stimulated a search for steroidlike agents that would have this antiangiogenic activity. Using the chick chorioallantoic membrane assay, it has been found that a number of compounds with the steroid nucleus have this inhibitory activity.[127] Interestingly, steroids that lack glucocorticoid or mineralocorticoid activity also inhibit angiogenesis, indicating that the two types of activity are separable. Agents with different structural alterations of the pregnane nucleus have been examined, and a structure-activity relationship has been established. For example, when the 11-, 17-, and 21-hydroxyl groups of the steroid ring are absent, as in progesterone, antiangiogenic activity is lost. On the other hand, 17α-hydroxyprogesterone is as potent as hydrocortisone in inhibiting angiogenesis in the presence of heparin; but corticosterone, which lacks the hydroxyl group in the 17 position, is only about 20% as potent as hydrocortisone. Estrone and testosterone are weakly active. In addition, a synthetic pentasaccharide representing a sulfated glycosaminoglycan fragment of heparin was shown to be as potent as heparin itself in the assay. It is noteworthy that neither steroids nor the pentasaccharide alone have antiangiogenic activity.

The mechanism of this steroid-plus-heparin antiangiogenic action remains obscure, but it may relate to inhibition of the formation of the basement membrane around the new endothelial cells.[127] Even though the mechanism to explain these results is not clear, there is the intriguing possibility that tumor growth and invasion in vivo could be inhibited by agents such as the "angiostatic" agents. That this may indeed work is suggested by studies showing that heparin-steroid conjugates retarded the growth of Lewis lung carcinomas in mice.[128] Furthermore, combination therapy with the antiangiogenic agents tetrahydrocortisol, β-cyclodextrin tetradecasulfate, and the tetracycline analogue minocycline plus cytotoxic drugs such as cyclophosphamide reduced in vivo growth of Lewis lung tumors and the number and size of metastases from the primary tumors.[129]

Other inhibitors of angiogenesis include the antiestrogens tamoxifen, nafoxidine, and clomiphene (in the chick chorioallantoic membrane assay),[130] the ECM protein thrombospondin and peptides derived from it (in the rat cornea in vivo assay),[131] and genistein, an isoflavonoid analogue of a natural product found in soy products (in an in vitro endothelial cell proliferation assay).[132] Since soy products are widely used in the diets of Japanese people, who as a population have less breast, prostate, and colon cancer than western populations, these latter authors speculate that a diet rich in soy may act as a chemopreventive agent by this mechanism.[132]

Hematopoietic Growth Factors

The hematopoietic growth factors (Table 9-4) include erythropoietin, which stimulates red blood cell formation; the granulocyte, macrophage, and granulocyte-macrophage colony stimulating factors (G-CSF, M-CSF and GM-CSF); the interleukins (of which there are at least 13 and still counting), which act on various stem cell populations in hematopoiesis; and various factors such as stem cell factor (SCF) and leukemia inhibitory factor (reviewed in Ref. 133).

The first hematopoietic GF to be discovered was erythropoietin (EPO) in 1906. By the mid

Table 9-4 The Hemopoietic Regulators

Regulator (Abbreviation)	Responding Hemopoietic Cells
Erythropoietin (Epo)	E, Meg
Granulocyte—macrophage colony stimulating factor (GM-CSF)	G, M, Eo, Meg, E
Granulocyte colony stimulating factor (G-CSF)	G, M
Macrophage colony stimulating factor (M-CSF)	M, G
Multipotential colony stimulating factor (Multi-CSF/IL-3)	G, M, Eo, Meg, Mast, E, Stem
Interleukin-1 (IL-1)	T, Stem
Interleukin-2 (IL-2)	T, B
Interleukin-4 (IL-4)	B, T, G, M, Mast
Interleukin-5 (IL-5)	Eo, B
Interleukin-6 (IL-6)	B, G, Stem, Meg
Interleukin-7 (IL-7)	B, T
Interleukin-9 (IL-9)	T, Meg, Mast
Interleukin-10 (IL-10)	T
Interleukin-11 (IL-11)	Meg, B
Interleukin-12 (IL-12)	NK
Megakaryocyte colony stimulating factor (Meg-CSF)	Meg
Stem cell factor (SCF)	Stem, G, E, Meg, Mast
Leukemia inhibitory factor (LIF)	Meg
Oncostatin M (OSM)	?
Macrophage inflammatory protein α (MIP-1α)	Stem

Abbreviations: G, granulocytes; M, macrophages; Eo, eosinophils; E, erythroid cells; Meg, megakaryocytes; Stem, stem cells; Mast, mast cells; T, T lymphocytes; B, B lymphocytes; NK, natural killer cells.
Source: From Metcalf.[133]

1960s, semisolid culture techniques that could support the growth of blood cell colonies from normal bone marrow became available, and it soon became clear that soluble substances released into the conditioned culture medium of such cell types were necessary to support the growth of these colonies. In the 1970s, several of these "colony stimulating factors" began to be purified and characterized; by 1983 EPO, GM-CSF, G-CSF, M-CSF, and interleukins-1, -2, and -3 had all been purified. Between 1984 and 1986, human cDNAs for EPO, GM-CSF, and G-CSF became available, allowing their development for clinical use. These are considered by many observers to be the first clinically and commercially successful products of the new age of genetic engineering.

The colony stimulating factors are a subset of regulatory polypeptides of the "cytokine" family that are involved in the proliferation and differentiation of granulocytes and monocyte/macrophages. The term *CSF* has stuck, and subsets of CSFs, based on their ability to stimulate particular pathways of hematopoietic cell differentiation, have been identified (Fig. 9-6).[134]

The CSFs are glycoproteins with 15,000- to 21,000-Da polypeptide chains and variable amounts of carbohydrate. Except for M-CSF, which is a homodimer, they consist of a single polypeptide chain. They are produced by multiple cell types, including fibroblasts, placenta, endothelial cells, lymphocytes, and bone marrow stromal cells. Blood levels of CSFs are normally low, but their production can be elevated rapidly in response to infection.

In vitro, some cell-type specificity can be demonstrated: GM-GSF and IL-3 stimulate formation of granulocyte and macrophage colonies; G-CSF favors granulocyte colony formation; and M-CSF fosters macrophage colony growth. In addition to stimulating progenitor cell proliferation and cellular commitment to a particular differentiation pathway, the CSFs are also necessary to maintain functional activity of mature cells, e.g., chemotaxis, phagocytosis, and production and release of cytotoxic factors.[135]

Interestingly, receptors for multiple CSFs are present on many hematopoietic progenitor cell types. There is a redundancy in the signaling process for hematopoietic cell proliferation, as if nature had built in multiple mechanisms to protect the host from invading organisms and other stresses. For example, granulocyte-macrophage progenitor cells and their maturing progeny express membrane receptors for GM-, G-, M-, and multi-CSF. In addition, CSF-occupied receptors can initiate multiple functions in responding cells, implying that broad signaling cascades are initiated by receptor occupancy. As one might expect from these observations, there are numerous potential interactions among CSFs. For instance, combinations of two CSFs can produce additive or synergistic responses. Because CSF receptor levels are low (a few hundred per cell),[135] occupancy of more than one type of CSF receptor may be required for an optimal proliferative response. Alternatively, progenitor cells may have multiple receptor types, so that they are able to respond either to their normal, most appropriate ligand and secondarily to another less optimal ligand that may be turned on by a different stress and/or different CSF-producing cell type, so that the host can respond to any of a number of emergencies. In general, more mature, committed, or single-lineage cells can respond to stimulation by single growth factors, whereas less mature stem cells often require combined signaling from multiple factors.[133] It should be noted that CSFs were the first growth factors to show clearly that GFs can both stimulate cell proliferation of developing stem cells and also induce a differentiation pathway in cell-lineage progenitor cells. Some hematopoietic growth factors have a fairly limited range of target cells and others have a wide spectrum of target cells. For example, IL-3 and IL-6 act on hematopoietic precursor cell types (Fig. 9-6), and leukemia inhibitory factor (LIF) acts on megakaryocytes, osteoblasts, neuronal tissue, hepatocytes, and adipocytes.[133]

Other ligand-receptor interactions have also been observed. Binding of GM-CSF to its receptor downregulates expression of G-CSF receptors. GM-CSF and IL-3 compete for binding to the same receptor. Macrophages are induced by IL-3 and M-CSF to produce G-CSF and also by GM-CSF to produce M-CSF. Obviously, there is a great deal of cross-talk between cells in the hematopoietic system, and it must require some finely tuned regulation, the mechanisms for which are only vaguely understood.

Despite all the redundancies in the system, it is clear that in populations of bipotential progenitor cells, G-CSF fosters development of the cells in the granulocyte lineage and M-CSF fosters development of cells in the monocyte/macrophage lineage. This is borne out in vivo in that injection of G-CSF into mice induces a greater increase in peripheral blood granulocytes than other blood cells, whereas GM-CSF induces a rise in both macrophages and granulocytes (reviewed in Ref. 135). IL-3 administration elicits a rise in granulocytes, macrophages, eosinophils, and megakaryocytes, as might be expected from its broad target-cell specificity.

Clinically, the CSFs have been used in AIDS, aplastic anemia, congenital or cyclic neutropenia, and cancer. The latter use has been to restore bone marrow function after chemotherapy,

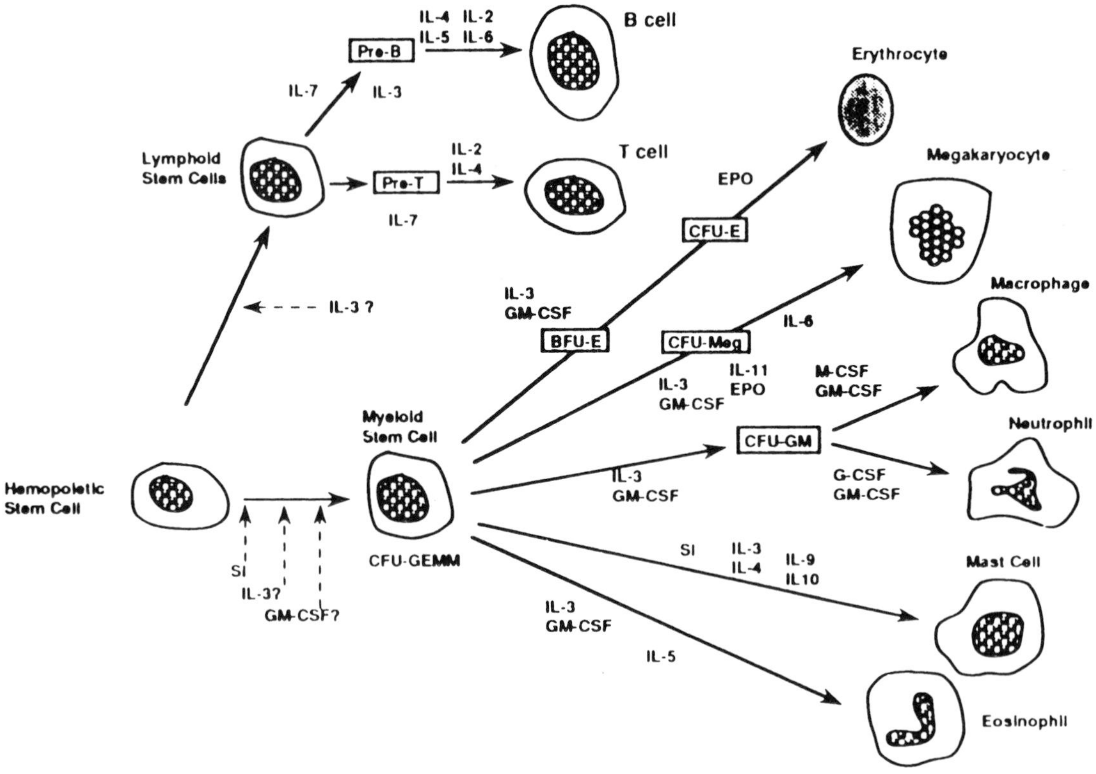

Figure 9-6 Hemopoiesis and cytokines. (From Miyajima et al.[134])

often accompanied by bone marrow transplantation. Both G-CSF and GM-CSF have been shown to replenish peripheral blood neutrophils after high-dose chemotherapy followed by autologous bone marrow transplantation.[136,137] Positive benefits include decreasing the frequency of infections and shortening the stay in the hospital.

An intriguing sidelight of CSF therapy was the observation of a dramatic rise in the number of progenitor cells in the peripheral blood.[138] Usually, these cells are largely restricted to the bone marrow. This raises the possibility of harvesting stem cells from the peripheral blood rather than the marrow. Advantages of this are the ability to recover many more stem cells and decreased trauma to the patient. A number of cytokines are currently being evaluated to attain maximal mobilization of stem cells into the peripheral blood. Clinical trials are being done to evaluate the use of peripheral stem cells rather than bone marrow cells, and the results are encouraging.

The receptors for hematopoietic GFs have several common features (reviewed in Refs. 133 and 139). They have highly related α chains with low-affinity binding sites, which—when dimerized with β chains that provide some GF-specificity—produce a high-affinity receptor. Once receptor dimerization occurs, association with one of a family of cytoplasmic tyrosine kinases such as Tyk2 or JAK2 induces tyrosine phosphorylation on cellular substrates involved in the signal transduction cascade. Thus, though the CSF family of receptors are not themselves tyrosine kinases, once activated, they become receptor tyrosine kinases in disguise. A model for how these interactions occur is shown in Figure 9-7, where the α binding component of the cytokine binds to the α chain of the receptor, providing a stable "matrix" for sequestration of one β chain of the receptor, followed by dimerization of the receptor via a second β chain binding site on the growth factor. This leads to activation of a cytoplasmic tyrosine kinase of the JAK family. Since different cytokine receptors are capable of generating qualitatively different signals in the same cells, it is likely that the receptors have some signaling specificity through their β subunits either for recruiting different sets of JAKs, for attracting different substrates, or both.

Hepatocyte Growth Factor/Scatter Factor (HGF/SF)

Hepatocyte growth factor (HGF) and scatter factor (SF) were originally thought to be distinct cytokines that stimulated proliferation of cultured hepatocytes and promoted motility of ep-

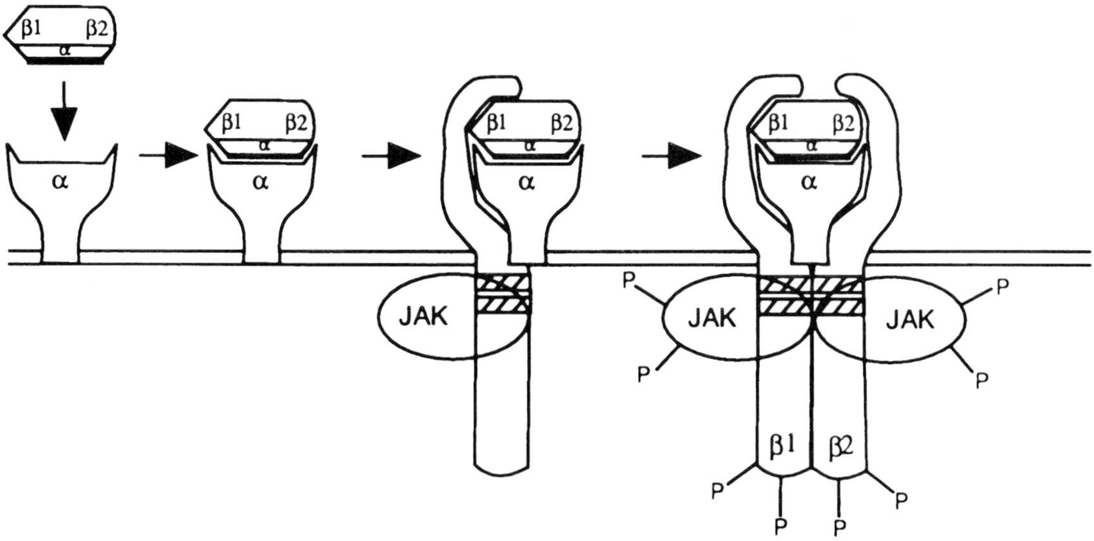

Figure 9-7 Stepwise formation of a generic cytokine receptor complex. Hatched boxes in β components represent conserved box 1 and box 2 sequences. (From Stahl and Yancopoulos.[139])

ithelial cells, respectively. HGF was first identified in the serum of partially hepatectomized rats as a potent mitogen for cultured rat hepatocytes and later also found in human plasma and serum, rat liver, and rat platelets (reviewed in Ref. 140). HGF has been cloned and sequenced.[141,142] Scatter factor (SF) was originally found as a secretory product of fibroblasts that dissociates epithelial cell colonies into individual cells and stimulates migration of epithelial cells (reviewed in Ref. 143). Purified SF also promotes invasiveness of cultured human carcinoma cells into collagen matrices, suggesting a role of SF in metastasis. Once both SF and HGF were cloned, it became clear that they were the same molecule.

HGF/SF is a disulfide-linked heterodimer of 55- to 65-kDa and 32- to 36-kDa subunits and is expressed in several tissues of mesodermal origin, including vascular smooth muscle cells. It has angiogenic properties[140] as well as the ability to induce morphologic changes and induce anchorage-independent growth in HGF/SF transfected epithelial cells.[143]

The receptor for HGF/SF has been found to be identical to the c-*met* protooncogene, which is another of the receptor tyrosine kinase family.[144] Activation of this receptor, which is found on a variety of cells including keratinocytes, melanocytes, endothelial cells, and other epithelial cells, triggers autophosphorylation and produces stimulation of the phosphatidyl inositol hydrolysis pathway and activation of Ras by shifting the equilibrium toward the active GTP-bound state.[145] By now, the reader is no doubt beginning to recognize a commonality in all this growth-factor-receptor business.

Miscellaneous Growth Factors

More modulators of cell proliferation and differentiation are being identified as they are looked for in normal and tumor tissues and cell lines derived from different tissues. Some of these are bone-cell-derived growth factors (BDGF),[146] uterine-derived growth factor (UDGF),[147] mammary gland–derived growth factor (MDGF),[148] melanocyte growth factor,[149] lung cancer,[150] ovarian cancer,[151] and Wilms' tumor[152]–derived growth factors as well as a family of estrogen-inducible growth factors called estromedins, found in a variety of tissues including uterus, kidney, and pituitary gland.[153] Whether all of these growth factors are in fact distinct chemical entities or are members of already identified growth factor families is yet to be determined.

The ubiquity of growth and differentiation factors leads one to predict that they will be found in all tissues in the body. These factors most likely act by paracrine or autocrine mechanisms to induce cell renewal or tissue repair of damage, but they may, under some circumstances, be released from the tissue and act on other organs through an endocrine mechanism like other known hormones. When cells undergo malignant transformation, they may continue to produce these factors, much as their normal proliferating stem cell counterparts do. Although the normal stem cells stop proliferating at some point and stop making these factors, tumor cells may continue to make them until they undergo enough genetic drift to become growth factor–independent and capable of autonomous growth. What missing signal, or missing signal receptor, is necessary to convert the proliferating cancer cell type into a differentiating cell type is not known for most human cancers, but clearly this is an important area for future research in cancer biology. Candidates for negative growth regulators have been found in a number of cell types, including lymphocytes, granulocytes, liver, mammary gland, epidermis, and fibroblasts. One of these growth regulatory substances, Oncostatin M, is a 28- to 36-kDa polypeptide cytokine that is produced by activated T lymphocytes and phorbol ester–treated monocytes; it inhibits the ability of melanoma cells and other cancer cell lines to grow in vitro. Paradoxically, it is also a potent mitogen for AIDS-derived Kaposi's sarcoma cells in culture.[154] These kinds of data indicate the complexity of growth factor–stimulated events and demonstrate the cellular "context"-dependency of their actions.

A variety of growth-regulatory factors for mammary gland epithelial cells have been reported (reviewed in Ref. 155). Some of these factors have been detected in milk, some in a conditioned medium of cultured mammary cells, some in mammary tissue extracts, and some even in neoplastic cells. Some appear to be produced constitutively and others are induced by antiestrogens. One regulatory factor, called mammastatin, is produced by normal mammary cells[156] and can be detected in serum of women at the onset of menstruation and in rat mammary gland in late pregnancy. This suggests mammastatin or similar factors as candidates for differentiation-inducing agents that may provide the protective effect of early pregnancy for breast cancer.

SIGNAL TRANSDUCTION MECHANISMS

In general, there are three types of signal transduction mechanisms for growth modulating substances: (1) receptors coupled to tyrosine kinase activity; (2) receptors coupled to guanine nucleotide–binding proteins, which, in turn, may activate or inhibit adenylate cyclase, activate phosphoinositide hydrolysis, leading to protein kinase C activation and intracellular Ca^{2+} release, or modulate cell membrane ion channels; and (3) intracellular receptors such as those for steroid hormones, thyroid hormone, and retinoic acid, all of which have DNA-binding domains as well as ligand-binding domains and can interact directly with DNA to modulate gene transcription. All of these receptor-mediated signal transduction mechanisms are potential sites for upregulation or deregulation in cancer cells—for example, by oncogene activation or overexpression or by tumor suppressor gene inactivation.

Tyrosine Kinases

The tyrosine kinase–coupled receptors mentioned above are one potential target for carcinogenic alteration. Activation of these receptors can lead to phosphorylation of a number of key substrates. Many growth factor receptors mediate their cellular effects by intrinsic tyrosine kinase activity, which, in turn, may phosphorylate other substrates involved in mitogenesis. As noted in Chapter 7, a number of transforming oncogene products have growth factor or growth factor receptor–like activities that work via a tyrosine kinase–activating mechanism. For example, the v-*src* gene product is itself a cell membrane associated tyrosine kinase. The v-*sis* oncogene product is virtually homologous to the B-chain of platelet-derived growth factor (PDGF). The v-*erb* product is a truncated form of the epidermal growth factor (EGF) receptor. The *fms* gene product is analogous to the receptor for colony stimulating factor CSF-1. The *met* and *trk* protooncogene products turn out to be receptors for hepatocyte growth factor (HGF) and nerve growth factor (NGF), respectively.

Some of the key substrates for receptor-tyrosine kinase–coupled activity include (1) phospholipase C (PLCγ), which, in turn, activates phosphatidyl inositol hydrolysis, releasing the second messengers diacylglycerol (DAG) and inositol trisphosphate ($InsP_3$) that activate protein kinase C (PKC) and mobilize intracellular calcium release (a number of tumor promoters also activate PKC); (2) the GTPase activating protein GAP that modulates Ras protooncogene protein function; (3) Src-like tyrosine kinases; (4) PI-3 kinase that associates with and may modulate the transforming activity of polyoma middle T antigen and the v-*src* and v-*abl* gene products; (5) the *raf* protooncogene product, which is itself a serine/threonine protein kinase.

Thus, activation of protein kinases is a key mechanism in regulating signals for cell proliferation. The substrates of these kinases include transcription regulatory factors such as those linked to mitogenic signaling pathways, e.g., proteins encoded by the *jun, fos, myc, myb, rel,* and *ets* protooncogenes.

The central role of tyrosine phosphorylation in cell proliferative signaling mechanisms also provides a target for chemotherapy. Tyrosine analogues that block tyrosine phosphorylation by acting selectively on tyrosine kinases may provide such agents. For example, if one could selectively block the tyrosine phosphorylating activity of an overexpressed or inappropriately activated *src, fms,* or *erb*B oncogene, one might be able to shut off a proliferative signal in certain kinds of cancer cells. Similarly, if one could design a peptide that would mimic an overexpressed protein kinase substrate such as Src or Raf, one might be able to specifically block their activation. Tyrphostins might be examples of such agents. They are a group of natural tyrosine analogues that block phosphorylation of tyrosine residues. They have been shown to block proliferation of cultured cells.[157]

Other protein kinases may also be targets for

Table 9-5 Mammalian, Drosophila, and Yeast Protein Kinases[a]

Mammals

Protein-serine/threonine kinases

- Cyclic nucleotide-regulated
  - cAMP-dependent protein kinases (C_α, C_β)
  - cGMP-dependent protein kinase
- Calmodulin-regulated
  - Phosphorylase kinase (distinct liver and muscle forms?)
  - Myosin light-chain kinases (skeletal, smooth muscle)
  - Type II-calmodulin dependent protein kinase
    - (brain α, β, β′ subunits; liver α, α′ subunits; muscle β, β′ subunits)
  - Calmodulin-dependent protein kinases I and III
- Diacylglycerol-regulated
  - Protein kinase Cs (α, β and β′, γ, δ [RP14])
- Others
  - Casein kinases I and II
  - Nuclear protein kinases N1 and N2
  - Protease-activated kinases I and II
  - Glycogen synthase kinases 3 and 4
  - Heme-regulated protein kinase
  - Double-stranded RNA regulated protein kinase
  - Double-stranded DNA regulated protein kinase
  - S6 kinase
  - β-adrenergic receptor kinase
  - Rhodopsin kinase
  - Histone H1 kinase
  - Hydroxymethylglutaryl-CoA reductase kinase
  - Pyruvate dehydrogenase kinase
  - Branched chain ketoacid dehydrogenase kinase
  - Polypeptide-dependent protein kinase
  - Polyamine-stimulated protein kinase
  - c-*mos*, c-*raf*, A-*raf*, *pks*, *pim*-1 proteins
  - CDC-R (PSK-J3), CDC2Hs, PSK H1, PSK-C3

Protein-tyrosine kinases

- *src* gene family
  - pp60$^{\text{c-arc}}$ (fibroblast, neuronal forms)
  - pp62$^{\text{c-yes}}$, pp56^{lck}
  - *fgr, hck, fyn, lyn* proteins
- *abl* gene family
  - p150$^{\text{c-abl}}$ (type I and type II N-terminus)
  - *arg* protein
- *fps* gene family
  - p98$^{\text{c-fps}}$
  - NCP94
  - c-*fps*-related proteins (TKR11 and TKR16)
- Growth factor receptors
  - EGF receptor family
    - EGF receptor (c-*erb*B protein)
    - *neu* protein (*erb*B2 protein)
  - Insulin receptor family
    - Insulin receptor
    - IGF-1 receptor
    - c-*ros, met, trk* proteins
  - PDGF receptor family
    - PDGF receptor
    - CSF-1 receptor (c-*fms* protein)
    - c-*kit* protein
  - c-*sea,ret* proteins
- Others
  - p75 (liver)
  - p120 (brain

Drosophila

Protein-serine/threonine kinases

- Cyclic nucleotide-regulated
  - cAMP-dependent protein kinase-related (C0, C1, C2)
  - cGMP-dependent protein kinase-related (GO [2 genes], G1)
- Diacylglycerol-regulated
  - Protein kinase C
- Others
  - Casein kinase II
  - *raf* protein

Protein-tyrosine kinases

- D*src*64B protein
- D*src*28C protein
- D*ash* protein
- *fps*-related protein
- EGF receptor (types I, II, III *N*-terminus)
- Insulin receptor
- *sevenless* protein

Yeast

- Cyclic nucleotide-regulated
  - cAMP-dependent protein kinase-related
    - TPK1, TPK2, TPK3, SRA3 *(S. cerevisiae)*
- Others
  - CDC28 *(S. cerevisiae)* (≈cdc2$^+$ in *S. pombe*)
  - CDC7 *(S. cerevisiae)*
  - KIN28 *(S. cerevisiae)*
  - wee1$^+$ *(S. pombe)*
  - nim1$^+$ *(S. pombe)*
  - STE7, STE11 *(S. cerevisiae)*
  - KIN1, KIN2 *(S. cerevisiae)*
  - SNF1 *(S. cerevisiae)*
  - ran1$^+$ *(S. pombe)*

[a] Protein kinases are listed under protein-serine/threonine kinase and protein-tyrosine kinase headings in subfamily groups. Protein kinases included in this table have either been characterized as distinct by complete or partial protein purification, or have been identified as unique based on nucleotide sequencing. The reader should be aware, however, that many of the protein kinases whose existence is deduced from sequences of cDNA clones have not yet been proven to be protein kinases. Conversely, until the complete amino acid sequences are available for all of the protein kinases which have been identified based on their enzymatic activity, one cannot be certain that they are distinct proteins.

Source: From Hunter.[158]

chemotherapeutic intervention. In recent years, protein kinases have been discovered at a rapid pace. The catalytic domains of these enzymes share a significant amount of sequence homology, yet the cellular localization, substrate specificity, and ligands that activate them may vary among cell types. While there may not be "1001 kinases" in mammalian cells, the number discovered is well over 100 and may well go a lot higher (Table 9-5).[158] In view of the importance of phosphorylation-dephosphorylation reactions in cellular regulatory mechanisms, this provides a rich pool of targets indeed.

Protein Phosphatases

Let us not forget the catalysts for the other half of this reaction—the phosphatases. Although it has been known for a long time that protein phosphatases play a regulatory role in certain cellular metabolic functions (e.g., in the activation-inactivation steps for glycogen synthase and phosphorylase), it has only recently been demonstrated that phosphatases play a role in the activity of various receptors and in the function of certain cell-cycle regulating genes (reviewed in Refs. 159 and 160). For example, expression of a truncated, abnormal protein tyrosine phosphatase in baby hamster kidney (BHK) cells produces multinucleated cells, possibly by dephosphorylating the cyclin-dependent kinase $p34^{cdc2}$.[160] Activation of $p34^{cdc2}$ requires dephosphorylation of a tyrosine residue, and this activation drives the cell from G_2 into M phase. The truncated phosphatase apparently interferes with the normal synchrony between nuclear formation and cell division.

Protein tyrosine phosphatases (PTPases), it is now known, are a diverse family of enzymes that exist in cell membranes. Some of them are associated with receptors that have tyrosine kinase activity. Phosphatases are also in other intracellular locations. The aberrant phosphorylation state of tyrosine in certain key proteins such as c-Src or c-Raf that can lead to cellular transformation could theoretically come about due to deregulation of a protein kinase or under expression of a protein phosphatase. For example, cells treated with vanidate, a PTPase inhibitor, had increased protein phosphotyrosine levels and a transformed phenotype.[161] Further evidence that PTPases are involved in cancer is the observation that receptor-linked PTPase γ (one of the PTPase isozymes) is located on chromosome 3, which is deleted in renal cell and lung carcinomas, suggesting that the PTPase γ gene may act as a tumor suppressor gene. Thus, one could predict that a high level of expression of specific PTPases may be able to reverse the malignant phenotype, and one can think of strategies, then, to transfect these genes into tumor cells or deliver inducers of the enzymes to tumor cells.

G Protein–Linked Receptors

As noted above, guanine nucleotide–binding protein-coupled receptors are a diverse set of ligand-activated receptors that regulate adenylyl cyclase, ion channels, certain protein kinases, and other signal transduction mechanisms. They all share a common general structure (Fig. 9-8)[162] with an external ligand-binding domain, membrane-spanning domains, and an intracellular domain that interacts with various G-protein complexes and contains sites for phosphorylation. The ligands that interact with such receptors include α and β adrenergic agonists and antagonists, angiotensin, serotonin, bombesin, bradykinin, acetylcholine (muscarinic type), vasopressin, and vasoactive intestinal polypeptide (VIP).

G proteins are heterotrimers composed of an α (39 to 46 kDa), β (37 kDa), and γ (8 kDa) subunits (reviewed in Ref. 163). The β and γ subunits form a tightly associated complex ($G_{\beta\gamma}$) that functions as a unit and forms trimers with an α (G_α) subunit. The α subunit has a high-affinity binding site for GTP or GDP. The GDP-bound form of α binds tightly to the $\beta\gamma$ complex and is inactive. When GTP is bound it displaces GDP, dissociates α from $\beta\gamma$, and elicits the regulatory function of α. While it has been thought that the activated G_α subunit alone regulates ion channels, adenylyl cyclase, phospholipase $C\beta$, and other enzymes, it is now apparent that the $G_{\beta\gamma}$ dimer also plays a role in modulating the activity of these effector systems.[164]

The G_α subunits possess intrinsic GTPase activity. Modification of G_α by cholera toxin activates G_α proteins by inhibiting their GTPase activity, and binding of pertussis toxin blocks

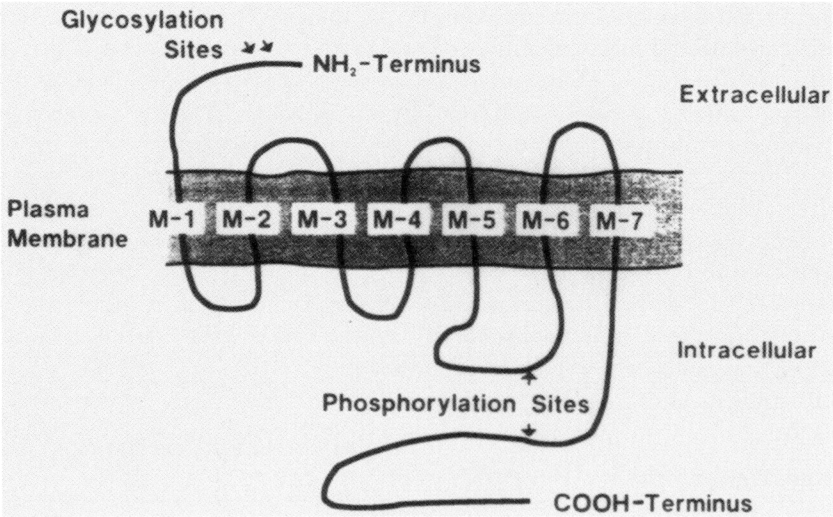

Figure 9-8 Schematic representation of the membrane organization of plasma membrane receptors (such as adrenergic receptors, muscarinic acetylcholine receptors, substance K receptors, or opsins) that are linked to G proteins. An extracellular amino terminal region with sites of glycosylation on asparagine residues is followed by seven membrane-spanning domains (M1 to M7) interspersed with three intracellular and three extracellular loops and then an intracellular carboxyl terminus. The consensus sequences expected at sites for phosphorylation are found in the third intracellular loop and carboxyl terminal regions. (From Taylor and Insel.[162])

receptor-mediated activation of G proteins. Thus, these two toxins, which ADP-ribosylate different sites on G_α subunits, are often used as tools to investigate the role of G proteins in various physiologic systems. G proteins are anchored in the plasma membrane of cells by lipid modifications of the subunits; γ subunits are prenylated and some α subunits are myristolated.

There are at least 21 distinct G_α subunits (from 17 genes), and these make up four major subfamilies based on their amino acid sequence homology: G_s, G_i, G_q, G_{12}.[163] G_s family α subunits, when activated by ligand binding of G protein–coupled receptors, stimulate the activity of adenylyl cyclase and open calcium channels. G_i subunits, when activated, mediate K^+ channel opening but inhibit Ca^{2+} channels and adenylyl cyclase. There are at least four distinct β and six γ subunits, but probably many more await discovery.

Activation of adenylyl cyclase to produce cyclic AMP and its subsequent activation of cAMP-dependent protein kinases is one of the major "second messenger" signal transduction systems in cells from bacteria to humans.

There are six subfamilies of adenylyl cyclases in mammalian cells, and they share common structural motifs (reviewed in Ref. 165), with a short amino terminal and two ≈40-kDa cytoplasmic domains held in place by two hydrophobic transmembrane domains called M1 and M2 (Fig. 9-9). When G protein–coupled receptors are activated by ligands, a stimulatory or inhibitory effect on adenylyl cyclase may occur, depending on the type of G protein activated (Fig. 9-9). Adenylyl cyclase activity may also be modulated by calcium-calmodulin, 3′-AMP, and perhaps by other kinases. This activation is self-limiting because the activated G_α subunit has GTPase activity, and the hydrolysis of GTP leaves GDP in the binding site and favors reassociation of G_α and $G_{\beta\gamma}$, thus deactivating the G protein–adenylyl cyclase complex.

The formation of cAMP, in turn, activates protein kinases (PKA) that can have a profound effect on cellular metabolism (Table 9-6).[166] While many of the cellular actions of cAMP are due to activation of PKA, other actions are attributed to a direct action of gene transcription via binding to cAMP-binding proteins (CREB) that act as transcription factors (see Chap. 5).

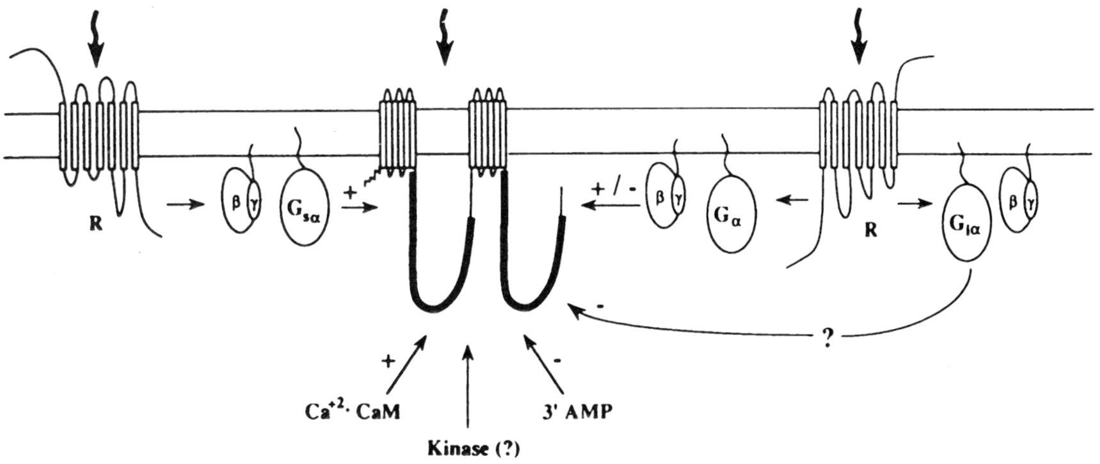

Figure 9-9 Regulation of mammalian adenylyl cyclases. R, G protein–coupled receptors; G_s and G_i, G proteins that stimulate and inhibit adenylyl cyclase, respectively; Gα, any G protein α subunit; CaM, calmodulin; 3′-AMP, a P-site inhibitor. (From Tang and Gilman.[165])

Because cAMP appears to play a key role in cell proliferation and differentiation, a number of investigators have speculated that alterations in the cyclic nucleotide-generating or response systems may be altered during malignant transformation. A considerable amount of work has been done to show this, and there does appear to be some correlation, but the exact role of cyclic nucleotides in this process has not yet been defined. A number of studies have shown that transformed fibroblasts regain a number of the characteristics of untransformed cells after treatment with cAMP analogues. These characteristics include a more flattened morphology, an increased adhesion to the substratum, a decreased agglutinability by lectins, and a decreased rate of cell proliferation. However, not all transformed cells respond to cAMP treatment in this way. The response is determined by the cell of origin of the transformed cell line. Fibroblastic cells, in general, tend to respond to cAMP in the previously described manner, whereas epithelial cells often do not. Experiments with various clones of rat kidney cells, for example, showed that in a fibroblastic clone, intracellular levels of cAMP rose as the cells reached confluency, but this was not the case in an epithelial clone.[167] Murine sarcoma virus–transformed fibroblastic clones of rat kidney cells did not have elevated intracellular cAMP at confluency, and they responded to treatment with exogenous cAMP analogues by exhibiting a slower growth rate and a flattened cell morphology. Neither the growth nor the morphology of the epithelial clone was affected by exogenous cAMP.

It appears that certain parts of the response system for cAMP differ in certain kinds of cells. In support of this idea, experiments with the S49 lymphoma cell line showed that the proliferation of these cells was inhibited by cAMP but was not inhibited in a mutant S49 cell line defective in cAMP-dependent protein kinase.[168] Similar results have been obtained for mutant Chinese hamster ovary cells with a variant cAMP-dependent protein kinase. There may also be alterations in the cAMP-binding regulatory subunit of cAMP-dependent protein kinase or in the translocation step involved in the nuclear uptake of protein kinase, a step that appears to be required for response to cAMP in some cells. Decreased binding of cAMP and an altered cAMP-binding protein have been demonstrated in a cAMP-unresponsive line of Walker 256 carcinosarcoma cells compared with the responsive parent line.[169] The nuclear translocation of cAMP-binding proteins and protein kinase was also markedly diminished in the unresponsive tumor cells after treatment with dibutyryl cAMP.[170]

It is clear that cAMP affects the proliferation rate of some normal and transformed cultured cells and that cAMP levels are lower in some transformed cell lines. It is not clear if changes

Table 9-6 Effects of cAMP on Various Cellular Processes

Enzyme or Process Affected	Tissues or Organism	Change in Activity or Rate[a]
Protein kinase[b]	Several	+
Phosphorylase	Several	+
Glycogen synthetase	Several	−
Phosphofructokinase	Liver fluke	+
Lipolysis	Adipose	+
Clearing factor lipase	Adipose	−
Amino acid uptake	Adipose	−
Amino acid uptake	Liver and uterus	+
Synthesis of several enzymes	Liver	+
Net protein synthesis	Liver	−
Gluconeogenesis	Liver	+
Ketogenesis	Liver	+
Steroidogenesis	Several	+
Water permeability	Epithelial	+
Ion permeability	Epithelial	+
Calcium resorption	Bone	+
Renin production	Kidney	+
Discharge frequency	Cerebellar Purkinje	−
Membrane potential	Smooth muscle	+
Tension	Smooth muscle	−
Contractility	Cardiac muscle	+
HC1 secretion	Gastric mucosa	+
Fluid secretion	Insect salivary glands	+
Amylase release	Parotid gland	+
Insulin release	Pancreas	+
Thyroid hormone release	Thyroid	+
Calcitonin release	Thyroid	+
Research of other hormones	Anterior pituitary	+
Histamine release	Mast cells	−
Melanin granule dispersion	Melanocytes	+
Aggregation	Platelets	−
Aggregation	Cellular slime molds	+
Messenger RNA synthesis	Bacteria	+
Synthesis of several enzymes	Bacteria	+
Proliferation	Thymocytes	+
Cell growth	Tumor cells	−

[a]+, increase; −, decrease.

[b]Stimulation of protein kinase is known to mediate the effects of cAMP on several systems, such as the glycogen synthetase and phosphorylase systems, and may be involved in many or even most of the other effects of cAMP.

Source: From Sutherland.[166]

in cAMP levels or the cAMP-response system are *responsible* for the appearance of the transformed phenotype and, more important, for the loss of normal growth control. In some cells, alterations in intracellular cAMP appear to be more closely related to the morphologic characteristics of the transformed phenotype than to growth control; in fact, the two events are clearly dissociable in certain cell types. Nevertheless, it is clear that induction of cAMP in several types of cultured neoplastic cells induces a more differentiated, less transformed phenotype.

Several lines of evidence implicate G protein–coupled receptors in malignant transformation.[171] Overexpression of acetylcholine or serotonin receptors in NIH 3T3 cells causes ligand-dependent transformation. Bombesinlike peptides are secreted by some small-cell lung carcinoma cells and stimulate their growth, and antibodies to bombesin inhibit tumor cell proliferation. Some pituitary, adrenal cortical, and ovarian tumors have point mutations in G proteins coupled to adenylyl cyclase that could lead to constitutive overproduction of cAMP.

The α_{1B}-adrenergic receptor is a member of the G protein–coupled receptor superfamily and activates PI hydrolysis, a signaling pathway that is activated by a number of growth factors and that plays a crucial role in mitogenesis (see below). Mutation of three amino acid residues in the 3rd intracellular loop (see Fig. 9-8) increases the binding affinity of norepinephrine and its ability to stimulate PI hydrolysis by two to three orders of magnitude.[172] Moreover, this activating mutation renders the receptor constitutively active, stimulating PI turnover even in the absence of ligand. When the wild-type gene for the α_{1B} receptor is transfected into rat or NIH 3T3 fibroblasts, the cells express high levels of this receptor, become transformed in response to norepinephrine, and form tumors when injected into nude mice. When the mutated gene is transfected into fibroblasts, the cells spontaneously form transformed foci in the absence of ligand and have an enhanced ability to form tumors in nude mice. Thus, the α_{1B} adrenergic receptor gene acts like a protooncogene and, when activated or overexpressed, is a transforming oncogene. These data suggest that other G protein–coupled receptors of this type can act as oncogenes in certain cell types. This further suggests a host of strategies for chemotherapeutic interdiction of this system—for example, the design of specific antagonists of the G protein–coupled receptors that may be activated or overexpressed in tumor cells.

There is also evidence that alteration of G-protein subunits themselves can cause alterations in fibroblast growth characteristics. For example, transfection and overexpression of a mutated G protein α_{i2}-subunit gene, a gene shown to be involved in proliferation of fibroblasts and differentiation of myeloid cells, in fibroblasts produces increased cell proliferation and anchorage-independent growth, indicating a role for this G-protein subunit in regulation of fibroblast cell proliferation and in transformation events.[173]

Phosphatidylinositol (PI) Hydrolysis, Protein Kinase C (PKC) Activation, and Calcium Mobilization

It has been known for some time that mitogenic effects of growth-stimulatory agents are accompanied by a transient increase in monovalent cation influx into cells and a transient increase in intracellular pH in a wide variety of cell types, including human cells. In addition, divalent cations, especially calcium, are not only important for cell adhesion and cell attachment to the substratum but also affect cell behavior. Extracellular calcium concentration has been shown to be an important determinant of cell proliferation and differentiation in cell cultures of murine or human epidermal keratinocytes.[174] In the presence of 0.02 to 0.09 m*M* Ca^{2+}, keratinocytes proliferate rapidly in monolayer cultures, but when the extracellular Ca^{2+} concentration is raised to 1.2 m*M*, cell-to-cell contact occurs and proliferation ceases; the cells also express skin differentiation markers, become stratified, and produce keratin. Carcinogen-transformed mouse skin cells, however, are resistant to the induction of differentiation by high Ca^{2+}. Calcium ions are also essential for the fusion of chick myoblasts (undifferentiated muscle cells) into myotubes in muscle differentiation.[175] The precise molecular mechanisms involved in these Ca^{2+}-regulated effects are not known, but some are modulated by a 17,000-MW Ca^{2+}-binding protein called calmodulin that is present in all cells. A calmodulin–calcium complex is involved in the activation of several enzyme systems, including protein kinases, phosphodiesterase, and adenylyl cyclase, which controls cellular levels of 3′,5′-cyclic AMP.

A number of hormones and cyclic AMP induce a transient cellular uptake of Ca^{2+} and/or the mobilization of Ca^{2+} from intracellular stores such as the rough endoplasmic reticulum (RER) and mitochondria. Elevated intracellular Ca^{2+} can have a number of metabolic effects, including the activation of a Ca^{2+}-dependent protein kinase, protein kinase C. Activation of protein kinase C occurs by the action of a number of hormones and mitogenic agents, including the tumor-promoting phorbol esters (see Chap. 6). The substrates of protein kinase C include cytoskeletal proteins, cell surface receptors, and 40,000- 60,000-, or 80,000-MW proteins of as yet unknown function in various cell types. It is not clear which of these might be involved in modulating cell proliferation and differentiation, but it is likely that these effects involve the phosphorylation/dephosphorylation of cytoskeletal elements.

It is known that Ca^{2+} mobilization from intracellular stores occurs by means of occupation of cell surface receptors by various pharmacologically active ligands. Subsequent to ligand binding, cell membrane phospholipid turnover is stimulated, leading to generation of diacylglycerol and inositol trisphosphate (the so-called PI effect). Diacylglycerol operates within the cell membrane to activate protein kinase C, and inositol trisphosphate is released into the cytoplasm to function as a second messenger by mobilizing intracellular Ca^{2+}. The idea that phospholipid metabolites constitute another class of second messengers produced in cells by a wide variety of hormones and growth factors came originally from the observation by Hokin and Hokin[176] of stimulated incorporation of ^{32}P into phospholipids in hormonally activated tissues. These investigators demonstrated increased phospholipid turnover in pancreatic tissue exposed to acetylcholine. Later it became clear that catabolism of inositol lipids is stimulated in many different tissues by many different external signals (Table 9-7). The link of inositol trisphosphate ($InsP_3$) generation to Ca^{2+} mobilization is now widely recognized and explains one pathway leading from receptor occupation to Ca^{2+} mobilization, protein kinase C activation, and subsequent cellular responses (Fig. 9-10).[177]

The action of a number of growth factors and mitogens is also known to increase formation of diacylglycerol and $InsP_3$ in cells. The subsequent stimulatory events include (1) activation of protein kinase C, (2) release of intracellular Ca^{2+}, and (3) a transient increase in Na^+/H^+ exchange that increases the intracellular pH. These signals are similar in many mitogen-stimulated cells and are almost certainly involved in regulation of cell proliferation as well as of other metabolic events. Oncogenes such as the *src* and *ros* oncogene products can also stimulate $InsP_3$ formation, apparently by activating inositol lipid kinases (see Chap. 7). It is intriguing to speculate that certain oncogenic products may transform cells by causing a prolonged, uncontrolled stimulation of inositol lipid turnover, producing a continual supply of the second signals, diacylglycerol and $InsP_3$.

Like many other signal transduction systems, the PI system has turned out to be much more complicated than originally thought. For example, it is now apparent that activation of PKC may occur by a number of alternative routes and that some activations of PKC may be transient, leading to early or rapid cellular responses, while other activities may be more sustained, leading to late cellular responses such as cell proliferation and differentiation events (Fig. 9-11).[178] Phospholipases C, A_2, and D all appear to be involved in PKC activation, with PLC being

Table 9-7 Summary of the Tissues That Respond to External Stimuli by a Change in the Hydrolysis of Inositol Phospholipids and Formation of Inositol Trisphosphate

Tissue	Stimulus
Rabbit iris smooth muscle	Acetylcholine
Brain	Acetylcholine
Parotid	Acetylcholine, substance P
Pancreas	Acetylcholine, cerulein
Avian salt gland	Acetylcholine
Liver	Adrenaline, ATP, vasopressin, angiotensin, platelet-activating factor
Blowfly salivary gland	Serotonin
Blood platelets	ADP, thrombin, platelet-activating factor
Sympathetic ganglion	Vasopressin
GH_3 pituitary tumor cells	Thyrotropin-releasing hormone
Adrenal cortex	Angiotensin
Pancreatic islets	Glucose
Neutrophils	*f*-Methionyl-leucyl-phenylalanine
HL-60 leukemic cells	*f*-Methionyl-leucyl-phenylalanine
Leukocytes	Pseudomonal leukocidin
T-lymphoblastoid cells	Phytohemagglutinin
Leukemic basophils	Antigen
Swiss 3T3 cells	PDGF, bombesin, vasopressin
Neuroblastoma-glioma hybrid NG108-15	Bradykinin
Photoreceptors	Photons
Sea urchin eggs	Spermatozoa

Source: From Berridge and Irvine.[177]

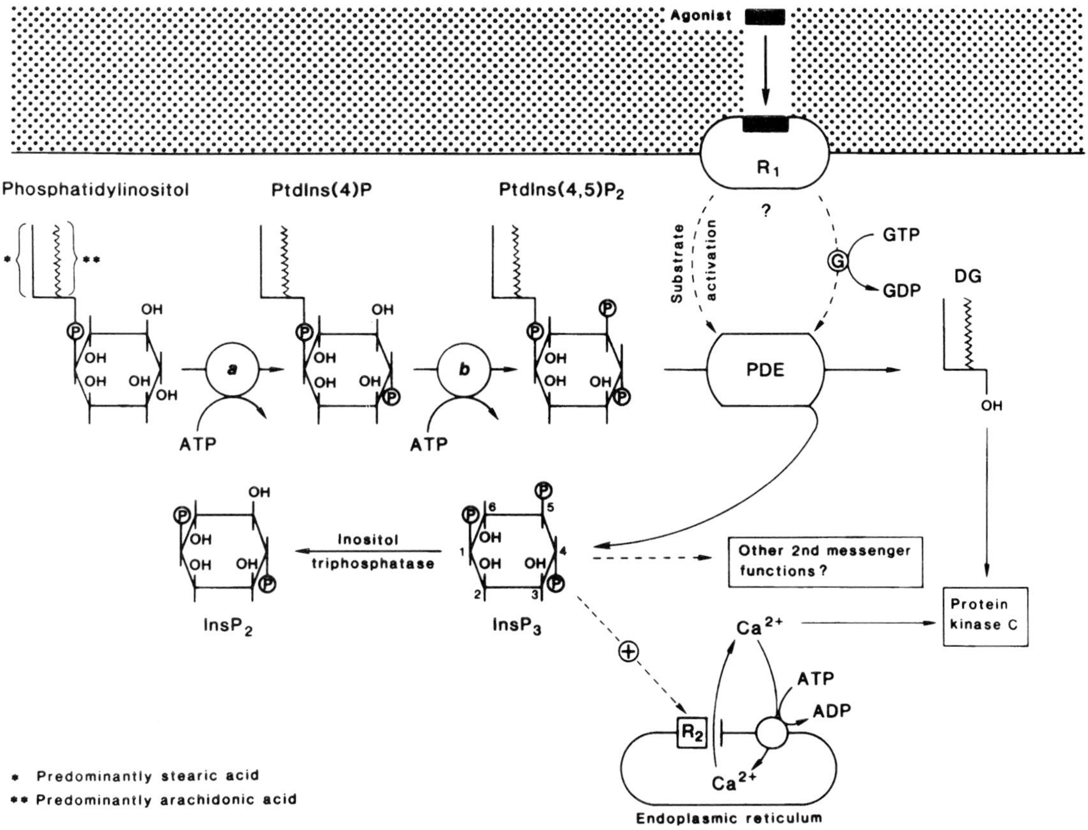

Figure 9-10 The proposed role of inositol trisphosphate as an intracellular second messenger. Agonists bind to external receptors (R_1) to stimulate the hydrolysis of PtdIns(4.5)P_2 by a phosphodiesterase (PDE) to form diacylglycerol (DG) and $InsP_3$. The latter may have a number of second messenger functions, one of which is to bind to a specific receptor (R_2) on the endoplasmic reticulum to release calcium. The action of $InsP_3$ is curtailed by an inositol trisphosphatase that removes a phosphate from the 5-position to form $InsP_2$. Diacylglycerol (DG) and Ca^{++} are required for activation of protein kinase C. *a* = phosphatidylinositol kinase; *b* = PtdIns(4)P kinase; PDE = PtdIns(4,5)P_2 phosphodiesterase. (Modified from Berridge and Irvine.[177])

more involved in the early response pathway and PLA_2 and PLD in the late response pathway. The reaction products of phosphatidyl choline hydrolyis by PLA_2, i.e., the *cis*-unsaturated fatty acids and lysophosphatidylcholine, can both enhance PKC activation. The reader will recall that the tumor-promoting phorbol esters mimic DAG for activation of PKC (Chap. 6) and, being more stable than DAG, favor the sustained, late response pathway involved in cell proliferation and differentiation.

In addition, other inositol phosphates such as Ins(1,3,4,5,6)P_5 and $InsP_6$, which actually compose the bulk of inositol phosphate content of mammalian cells, appear to have a role in cell signaling processes.[179] PKC also is a family of enzymes consisting of three major classes: the "classic" PKCs α, β, and γ, the "novel" PKC isoforms δ, ϵ, η, and θ, and the "atypical" isoforms ζ, λ, and ι (reviewed in Refs. 180 and 181). These isoforms are differentially expressed in cells and have differences in their cofactor requirements and their capacity for activation by DAG and nonsaturated fatty acids (Table 9-8). Thus, it is likely that they have different physiologic roles. All members of the PKC family have a highly conserved catalytic domain that includes an ATP binding site and a variable regulatory domain. The regulatory domain of the classical PKC group has a Ca^{2+}-binding domain that accounts for their activation by Ca^{2+}. Both the classic and novel PKCs have a tandem repeat

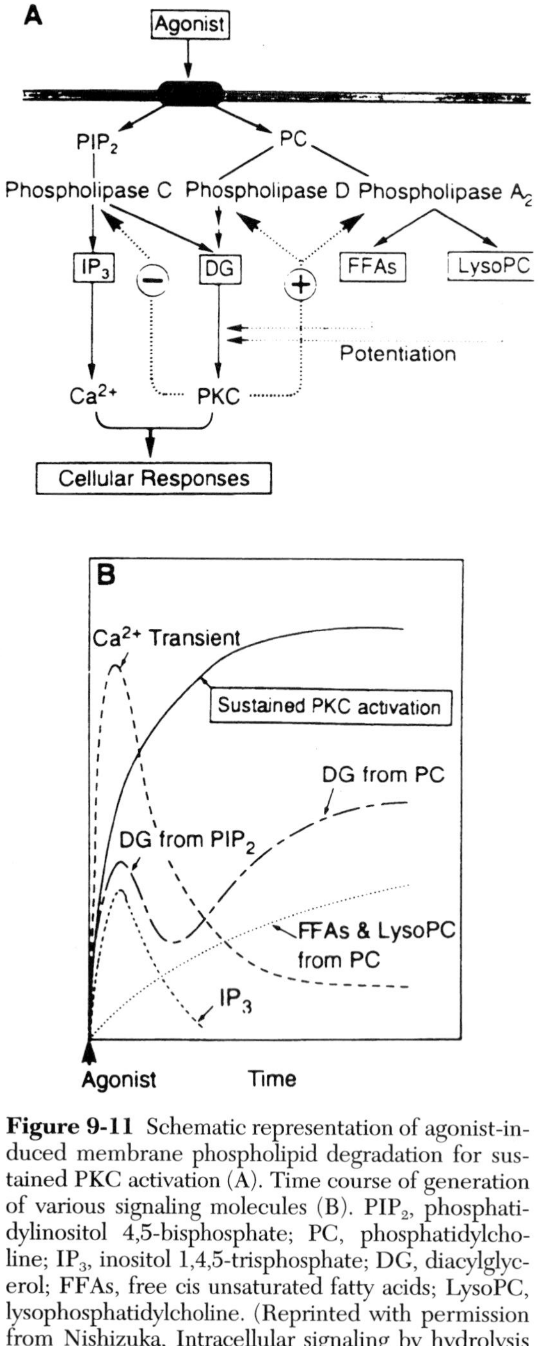

Figure 9-11 Schematic representation of agonist-induced membrane phospholipid degradation for sustained PKC activation (A). Time course of generation of various signaling molecules (B). PIP_2, phosphatidylinositol 4,5-bisphosphate; PC, phosphatidylcholine; IP_3, inositol 1,4,5-trisphosphate; DG, diacylglycerol; FFAs, free cis unsaturated fatty acids; LysoPC, lysophosphatidylcholine. (Reprinted with permission from Nishizuka, Intracellular signaling by hydrolysis of phospholipids and activation of protein kinase C. *Science* 258:607. Copyright 1992 by the American Association for the Advancement of Science.)

of a cysteine-rich zinc-finger domain that binds phospholipid. The atypical PKCs have only a single copy of this domain, and this is thought to be the reason that these PKCs do not respond to phorbol esters.[179] They are neither activated by phorbol esters during short-term exposure nor downregulated by them during long-term exposure, as the classic PKCs are. The novel PKCs are insensitive to Ca^{2+} but respond to DAG and phorbol esters. The atypical PKCs do not respond to Ca^{2+}, DAG, or phorbol esters but require phosphatidyl serine and are activated by *cis*-unsaturated fatty acids.[181]

In view of the apparent central role of protein kinase C in cell proliferation and the observations that overexpression of two isoforms of PKC, PKCβ and PKCγ, in rodent cells correlates with their tumorigenicity in nude mice and that expression of mutant forms of PKCα increases metastatic potential of tumor cells, it seems logical that inhibitors of overexpressed or mutant forms of PKC may have some therapeutic value. Some known PKC inhibitors have been examined for their effects on tumor cell growth in vitro and in vivo. Two such inhibitors, (4E)-*N,N*-dimethyl-D-*erythro*-sphingenine(DMS) and (4E)-*N,N,N*-trimethyl-D-*erythro*-sphingenine (TMS) have been shown to inhibit the growth of human gastric cancer cells in culture and in nude mice.[182] It has also been observed that PKCβ is expressed in invasive but not noninvasive human gastric cancers and that inhibitors of PKC, threo-dihydrosphingosine and staurosporine inhibit the invasiveness of cultured gastric carcinoma cells.[183]

Nuclear Receptors

Most of the nuclear receptors act as transcriptional regulators that have DNA binding domains and that directly affect gene transcription. These include the steroid hormone, thyroid hormone, and retinoic acid receptors. Binding of ligand in these cases leads to gene activation or repression and so in this sense, the activated receptor is itself the "second messenger." Since these were discussed in Chapter 5, they will not be discussed again here. One subset of nuclear receptors, the retinoic acid receptor pathway, is discussed in greater detail in Chapter 15 because

Table 9-8 PKC Subspecies in Mammalian Tissues

Group	Subspecies	Apparent molecular mass (kDa)	Activators[b]	Tissue Expression
cPKC	α	76,799	Ca^{++}, DAG, PS, FFA, LysoPC	Universal
	βI	76,790	Ca^{++}, DAG, PS, FFA, LysoPC	Some tissues
	βII	76,933	Ca^{++}, DAG, PS, FFA, LysoPC	Many tissues
	γ	78,366	Ca^{++}, DAG, PS, FFA, LysoPC	Brain only
nPKC	δ	77,517	DAG, PS	Universal
	ϵ	83,474	DAG, PS, FFA	Brain and others
	η(L)[a]	77,972	?	Lung, skin, heart
	θ[a]	81,571	?	Skeletal muscle (mainly)
aPKC	ζ	67,740	PS, FFA	Universal
	λ[a]	67,200	?	Ovary, testis and others

[a]The detailed enzymological properties of the η(L)-, θ-, and λ-subspecies have not yet been clarified.
[b]The activators for each subspecies are determined with calf thymus H1 histone and bovine myelin basic protein as model phosphate acceptors.
Abbreviations: DAG, diacylglycerol; PS, phosphatidylserine; FFA, *cis*-unsaturated fatty acid; LysoPC, lysophosphatidylcholine.
Source: From Asaoka, et al.[181]

of the importance of retinoids in the chemoprevention of cancer.

MAP Kinase System Revisited

The important role of the mitogen-activated protein kinase (MAPK) cascade as a signal transduction system for growth factors, tumor promoters, and oncogene-activated cell proliferation was discussed in Chapter 7. The reader should be reminded that this is one of the major signal transduction systems in cells and has important interconnections with other cell signal systems. Like all of these systems, it gets more complicated the deeper one looks.

The reader will remember that the activity of the transcription factor AP-1 (Jun/Fos) is rapidly increased in response to a wide variety of environmental stimuli and that part of this is due to increased transcription of the *jun* and *fos* genes. An additional level of regulation, however, is due to the posttranslational modification of the Jun and Fos proteins.

The MAPK system includes the extracellular signal-regulated MAP kinases 1 (ERK1) and 2 (ERK2) that phosphorylate a transcription factor called TCF (ternary complex factor), increasing its transcriptional activation function to turn on c-*fos* when it is complexed with serum response factor (SRF). SRF-TCF activates the serum response element (SRE), a *cis*-regulatory element that is involved in c-*fos* activation.

Activation of c-*jun* also occurs in response to many mitogenic stimuli, but its response is not identical to that of c-*fos*; for example, c-*jun* is activated by ultraviolet (UV) light but c-*fos* transcription is only mildly increased by UV light (reviewed in Ref. 184). The transcriptional activity of c-Jun protein is enhanced by phosphorylation of serine residues Ser-63 and Ser-73 in its transactivation domain. Phosphorylation of these two residues is increased in response to UV light, receptor tyrosine kinases, Src family tyrosine kinases, and Ha-Ras.

A c-Jun N-terminal kinase called JNK1 has been identified that phosphorylates Ser-63 and -73 of c-Jun.[184] Both UV light and Ha-Ras activate JNK1, which, it turns out, is a distant cousin of the MAP kinase family. This family contains the proline-directed MAPKs as well as the cyclin-dependent protein kinases. JNK1 appears to be a component of a novel signal transduction pathway, and since it is activated by oncogene proteins and UV irradiation, it may be an important factor in tumor promotion.[184]

Nitric Oxide

Nitric oxide (NO) is a stable free radical gas that surprisingly has been found to be a major biological messenger in several mammalian organ systems. One would not intuitively predict that a gas would be an important signal transduction molecule in an animal organism. But both NO and more recently carbon monoxide have been shown to do just that (reviewed in Ref. 185). NO has turned out to be a major "messenger molecule" regulating immune function, blood vessel dilation, and neurotransmission in both the central and peripheral nervous systems (Fig. 9-12).

NO was first observed to be an important signaling mechanism in macrophages as a mediator of tumoricidal and bactericidal actions. A role in regulating blood vessel dilation was discovered from the ability of acetylcholine to relax vascular smooth muscle via a factor released from the endothelium. This endothelial-derived relaxing factor turned out to be NO. Later, it was found that several cell types could generate NO endogenously by an enzyme called NO synthase. Its activity can be induced by Ca^{2+}, via an action of calmodulin, in neurons and endothelial cells, but in macrophages NO synthase activity is Ca^{2+}-independent. Neuronal NO synthase (NOS) is phosphorylated by cAMP-dependent protein kinase, PKC, and Ca^{2+}/calmodulin protein kinase. Phosphorylation appears to regulate NOS activity, at least that affected by PKC, phosphorylation by which reduces NOS catalytic activity.

NO has a number of cellular actions. It combines with superoxide anion to yield peroxynitrite that decomposes to hydroxide free radical and NO_2 free radical. Since NO synthase inhibitors block macrophage cytotoxic actions, NO generation appears to be crucial to the tumor cell–killing effect of macrophages. In endothelial cells, NO activates guanylyl cyclase to generate cGMP, which stimulates cGMP-dependent protein kinase and results in smooth muscle relaxation. In the brain, neurotransmitters such as glutamate or *N*-methyl-D-aspartate (NMDA) induce NOS activity, suggesting a role for NO in the function of these neurotransmitters. In some cell types, e.g., brain and red blood cells, NO enhances ADP ribosylation of proteins such as glycerol-3-phosphate dehydrogenase (GAPDH), suggesting a regulatory role of NO in glycolysis.

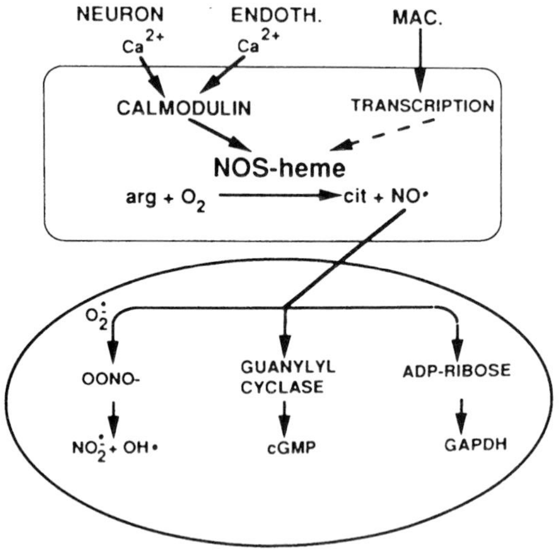

Figure 9-12 NO synthesis and mechanisms of action as an intercellular messenger in macrophages, blood vessels, and neurons. MAC., macrophage; cit, citrulline; arg, arginine; NOS, NO synthase; GAPDH., glyceraldehyde-3-phosphate dehydrogenase. (From Lowenstein and Snyder.[184])

REFERENCES

1. A. W. Burgess: Reflections on biochemistry. *Trends Biochem Sci* 14:117, 1989.
2. G. O. Gey and W. Thalhimer: Observations on the effects of insulin introduced into the medium of tissue cultures. *JAMA* 82:1609, 1924.
3. D. Gospodarowicz and J. S. Moran: Growth factors in mammalian cell culture. *Annu Rev Biochem* 45:531, 1976.
4. M. Kasuga, F. A. Karlsson, and C. R. Kahn: Insulin stimulates the phosphorylation of the 95,000-dalton subunit of its own receptor. *Science* 215:185, 1982.
5. L. Petruzzelli, R. Herrera, and O. M. Rosen: Insulin receptor is an insulin-dependent tyrosine kinase: Copurification of insulin-binding activity and protein kinase activity to homogeneity from human placenta. *Proc Natl Acad Sci USA* 81:3327, 1984.
6. K-T. Yu and M. P. Czech: Tyrosine phosphorylation of the insulin receptor β subunit activates the receptor-associated tyrosine kinase activity. *J Biol Chem* 259:5277, 1984.
7. S. A. Aaronson: Growth factors and cancer. *Science* 254:1146, 1991.
8. E. Rinderknecht and R. E. Humbel: The amino acid sequence of human insulin-like growth factor I and its structural homology with proinsulin. *J Biol Chem* 253:2769, 1978.
9. N. C. Dulak and H. M. Temin: A partially purified polypeptide fraction from rat liver cell conditioned medium with multiplication-stimulating activity for embryo fibroblasts. *J Cell Physiol* 81:153, 1973.
10. J. Zapf and V. R. Froesch: Insulin-like growth factors/somatomedins: Structure, secretion, biological actions and physiological roles. *Hormone Res* 24:121, 1986.
11. M. P. Czech: Signal transmission by the insulin-like growth factors. *Cell* 59:235, 1989.

12. J. E. Fradkin, R. C. Eastman, M. A. Lesniak, and J. Roth: Specificity spillover at the hormone receptor-exploring its role in human disease. *N Engl J Med* 320:640, 1989.
13. W. H. Daughaday: Editorial: The possible autocrine/paracrine and endocrine roles of insulin-like growth factors of human tumors. *Endocrinology*127:1, 1990.
14. K. J. Cullen, D. Yee, W. S. Sly, J. Perdue, B. Hampton, M. E. Lippman, and N. Rosen: Insulin-like growth factor receptor expression and function in human breast cancer. *Cancer Res* 50:48, 1990.
15. H. T. Huynh, E. Tetenes, L. Wallace, and M. Pollak: In vivo inhibition of insulin-like growth factor I gene expression by tamoxifen. *Cancer Res* 53:1727, 1993.
16. R. Torrisi, F. Pensa, M. A. Orengo, E. Catsafados, P. Ponzani, F. Boccardo, A. Costa, and A. Decensi: The synthetic retinoid fenretinide lowers plasma insulin-like growth factor I levels in breast cancer patients. *Cancer Res* 53:4769, 1993.
17. E. D. Bueker: Implantation of tumours in the hind limb field of the embryonic chick and the developmental response of the lumbosacral nervous system. *Anat Rec* 102:369, 1948.
18. S. R. Cohen, R. Levi-Montalcini, and V. Hamburger: A nerve growth stimulating factor isolated from sarcomas 37 and 180. *Proc Natl Acad Sci USA* 40:1014, 1954.
19. R. A. Hogue-Angeletti, W. A. Frazier, J. W. Jacobs, H. D. Niall, and R. A. Bradshaw: Purification, characterization, and partial amino acid sequence of nerve growth factor from cobra venom. *Biochemistry* 15:26, 1976.
20. S. Varon, J. Nomura, and E. M. Shooter: The isolation of the mouse nerve growth factor protein in a high molecular weight form. *Biochemistry* 6:2202, 1967.
21. V. Bocchini and P. U. Angeletti: The nerve growth factor: Purification as a 30,000-molecular-weight protein. *Proc Natl Acad Sci USA* 64:787, 1969.
22. W. A. Frazier, R. H. Angeletti, and R. A. Bradshaw: Nerve growth factor and insulin. *Science* 176:482, 1972.
23. M. Holloway: Finding the good in the bad. *Sci Am* 1:31, 1993.
24. R. A. Bradshaw, T. L. Blundell, R. Lapatto, N. Q. McDonald, and J. Murray-Rust: Nerve growth factor revisited. *Trends Biochem Sci* 18:48, 1993.
25. D. R. Kaplan, B. L. Hempstead, D. Martin-Zanca, M. V. Chao, and L. F. Parada: The trk proto-oncogene product: A signal transducing receptor for nerve growth factor. *Science* 252:554, 1991.
26. C. Cordon-Cardo, P. Tapley, S. Jing, V. Nanduri, E. O'Rourke, F. Lamballe, K. Kovary, R. Klein, K. R. Jones, L. F. Reichardt, and M. Barbacid: The *trk* tyrosine protein kinase mediates the mitogenic properties of nerve growth factor and neurotrophin-3. *Cell* 66:173, 1991.
27. K. W. Wood, C. Sarnecki, T. M. Roberts, and J. Bienis: *ras* mediates nerve growth factor receptor modulation of three signal-transducing protein kinases: MAP kinase, Raf-1, and RSK. *Cell* 68:1041, 1992.
28. D. Martin-Zanca, S. H. Hughes, and M. Barbacid: A human oncogene formed by the fusion of truncated tropomyosin and protein tyrosine kinase sequences. *Nature* 319:743, 1986.
29. T. Suzuki, E. Bogenmann, H. Shimada, D. Stram, and R. C. Seeger: Lack of high-affinity nerve growth factor receptors in aggressive neuroblastomas. *JNCI* 85:377, 1993.
30. G. Carpenter and S. Cohen: Epidermal growth factor. *Annu Rev Biochem* 48:193, 1979.
31. J. M. Taylor, W. M. Mitchell, and S. Cohen: Characterization of the binding protein for epidermal growth factor. *J Biol Chem* 249:2188, 1974.
32. P. Frey, R. Forand, T. Maciag, and E. M. Shooter: The biosynthetic precursor of epidermal growth factor and the mechanism of its processing. *Proc Natl Acad Sci USA* 76:6294, 1979.
33. R. H. Starkey, S. Cohen, and D. N. Orth: Epidermal growth factor: Identification of a new hormone in human urine. *Science* 189:800, 1975.
34. H. Gregory: Isolation and structure of urogastrone and its relationship to epidermal growth factor. *Nature* 257:325, 1975.
35. G. Carpenter and S. Cohen: Epidermal growth factor. *J Biol Chem* 265:7709, 1990.
36. S. Cohen, H. Ushiro, C. Stosbeck, and M. Chinkers: A native 170,000 epidermal growth factor receptor-kinase complex from shed plasma membrane residues. *J Biol Chem* 257:1523, 1982.
37. H. Ushiro and S. Cohen: Identification of phosphotyrosine as a product of epidermal growth factor-activated protein kinase in A-431 cell membranes. *J Biol Chem* 255:8363, 1980.
38. I. E. Garcia de Palazzo, G. P. Adams, P. Sundareshan, A. J. Wong, J. R. Testa, D. D. Bigner, and L. M. Weiner: Expression of mutated epidermal growth factor receptor by non-small cell lung carcinomas. *Cancer Res* 53:3217, 1993.
39. T. A. Libermann, H. R. Nusbaum, N. Razon, R. Kris, I. Lax, H. Soreq, N. Whittle, M. D. Waterfield, A. Ullrich, and J. Schlessinger: Amplification, enhanced expression and possible rearrangement of EGF receptor gene in primary brain tumours of glial origin. *Nature* 313:144, 1985.
40. C. F. Fox, P. S. Linsley, and M. Wrann: Receptor remodeling and regulation in the action of epidermal growth factor. *Fed Proc* 41:2988, 1982.

41. A. Dautry-Varsat and H. F. Lodish: How receptors bring proteins and particles into cells. *Sci Am* 250:52, 1984.
42. J. L. Goldstein, R. G. W. Anderson, and M. S. Brown: Coated pits, coated vesicles, and receptor-mediated endocytosis. *Nature* 279:679, 1979.
43. G. Carpenter and S. Cohen: ^{125}I-labeled human epidermal growth factor: Binding, internalization, and degradation in human fibroblasts. *J Cell Biol* 71:159, 1976.
44. W. H. Moolenaar, R. Y. Tsien, P. T. van der Saag, and S. W. de Laat: Na^+/H^+ exchange and cytoplasmic pH in the action of growth factors in human fibroblasts. *Nature* 304:645, 1983.
45. P. Rothenberg, L. Glaser, P. Schlesinger, and D. Cassel: Activation of Na^+/H^+ exchange by epidermal growth factor elevates intracellular pH in A431 cells. *J Biol Chem* 258:12644, 1983.
46. C. P. Burns and E. Rozengurt: Extracellular Na^+ and initiation of DNA synthesis: Role of intracellular pH and K^+. *J Cell Biol* 98:1082, 1984.
47. G. Carpenter: Epidermal growth factor is a major factor-promoting agent in human milk. *Science* 210:198, 1980.
48. O. Tsutsumi, H. Kurachi, and T. Oka: A physiological role of epidermal growth factor in male reproductive function. *Science* 233:975, 1986.
49. K. Eckert, A. Granetzny, J. Fisher, E. Nexo, and R. Grosse: A M_r 43,000 epidermal growth factor-related protein purified from the urine of breast cancer patients. *Cancer Res* 50:642, 1990.
50. D. Gospodarowicz: Purification of a fibroblast growth factor from bovine pituitary. *J Biol Chem* 250:2515, 1975.
51. F. Esch, A. Baird, N. Ling, N. Ueno, F. Hill, L. Denoroy, R. Klepper, D. Gospodarowicz, P. Böhlen, and R. Guillemin: Primary structure of bovine pituitary basic fibroblast growth factor (FGF) and comparison with amino-terminal sequence of bovine brain acidic FGF. *Proc Natl Acad Sci USA* 82:6507, 1985.
52. D. Givol and A. Yayon: Complexity of FGF receptors: Genetic basis for structural diversity and functional specificity. *FASEB J* 6:3362, 1992.
53. K. Miller and A. Rizzino: Developmental regulation and signal transduction pathways of fibroblast growth factors and their receptors, in M. Nielsin-Hamilton, ed.: *Growth Factors and Signal Transduction Pathways in Development.* New York: Wiley-Liss, 1994.
54. M. Miyamoto, K-I. Naruo, C. Seko, S. Matsumoto, T. Kondo, and T. Kurokawa: Molecular cloning of a novel cytokine cDNA encoding the ninth member of the fibroblast growth factor family, which has a unique secretion property. *Mol Cell Biol* 13:4251, 1993.
55. G. P. Dotto, G. Moellmann, S. Ghosh, M. Edwards, and R. Halaban: Transformation of murine melanocytes by basic fibroblast growth factor cDNA and oncogenes and selective suppression of the transformed phenotype in a reconstituted cutaneous environment. *J Cell Biol* 109:3115, 1989.
56. A. Wellstein, G. Zugmaier, J. A. Califano III, F. Kern, S. Paik, and M. E. Lippman: Tumor growth dependent on Kaposi's sarcoma–derived fibroblast growth factor inhibited by pentosan polysulfate. *JNCI* 83:716, 1991.
57. J. L. Gross, W. F. Herblin, B. A. Dusak, P. Czerniak, M. D. Diamond, T. Sun, K. Eidsvoog, D. L. Dexter, and A. Yayon: Effects of modulation of basic fibroblast growth factor on tumor growth in vivo. *JNCI* 85:121, 1993.
58. S. Schultz-Hector and S. Haghayegh: β-fibroblast growth factor expression in human and murine squamous cell carcinomas and its relationship to regional endothelial cell proliferation. *Cancer Res* 53:1444, 1993.
59. Y. Yamanaka, H. Friess, M. Buchler, H. G. Beger, E. Uchida, M. Onda, M. S. Kobrin, and M. Korc: Overexpression of acidic and basic fibroblast growth factors in human pancreatic cancer correlates with advanced tumor stage. *Cancer Res* 53:5289, 1993.
60. D. M. Nanus, B. J. Schmitz-Dräger, R. J. Motzer, A. C. Lee, V. Vlamis, C. Cordon-Cardo, A. P. Albino, and V. E. Reuter: Expression of basic fibroblast growth factor in primary human renal tumors: Correlation with poor survival. *JNCI* 85:1597, 1993.
61. S. D. Balk: Calcium as a regulator of the proliferation of normal, but not of transformed, chicken fibroblasts in a plasma-containing medium. *Proc Natl Acad Sci USA* 68:271, 1971.
62. R. Ross, B. Glomset, B. Kariya, and L. Harker: A platelet-dependent serum factor that stimulates the proliferation of arterial smooth muscle cells in vitro. *Proc Natl Acad Sci USA* 71:1207, 1974.
63. N. Kohler and A. Lipton: Platelets as a source of fibroblast growth-promoting activity. *Exp Cell Res* 87:297, 1974.
64. H. N. Antoniades, C. D. Scher, and C. D. Stiles: Purification of human platelet-derived growth factor. *Proc Natl Acad Sci USA* 76:1809, 1979.
65. C.-H. Heldin, B. Westermark, and A. Wasteson: Platelet-derived growth factor: Purification and partial characterization. *Proc Natl Acad Sci USA* 76:3722, 1979.
66. E. W. Raines and R. Ross: Platelet-derived growth factor: I. High yield purification and evidence for multiple forms. *J Biol Chem* 257:5154, 1982.
67. B. Westermark and C.-H. Heldin: Platelet-derived growth factor in autocrine transformation. *Cancer Res* 51:5087, 1991.
68. M. Sasahara, J. W. U. Fries, E. W. Raines, A. M. Gown, L. E. Westrum, M. P. Frosch, D. T. Bonthron, R. Ross, and T. Collins: PDGF B-chain in neurons of the central nervous system, posterior pituitary, and in a transgenic model. *Cell* 64:217, 1991.

69. H-J. Yeh, K. G. Ruit, Y-X. Wang, W. C. Parks, W. D. Snider, and T. F. Deuel: PDGF A-chain gene is expressed by mammalian neurons during development and in maturity. *Cell* 64:209, 1991.
70. N. E. Olashaw, W. Kusmik, T. O. Daniel, and W. J. Pledger: Biochemical and functional discrimination of platelet-derived growth factor α and β receptors in BALB/c-3T3 cells. *J Biol Chem* 266:10234, 1991.
71. M. Valius and A. Kazlauskas: Phospholipase C-γ1 and phosphatidylinositol 3 kinase are the downstream mediators of the PDGF receptor's mitogenic signal. *Cell* 73:321, 1993.
72. B. J. Rollins, E. D. Morrison, and C. D. Stiles: Cloning and expression of JE, a gene inducible by platelet-derived growth factor and whose product has cytokine-like properties. *Proc Natl Acad Sci USA* 85:3738, 1988.
73. C. D. Stiles: The molecular biology of platelet-derived growth factor. *Cell* 33:653, 1983.
74. C. D. Scher, R. C. Shepard, H. N. Antoniades, and C. D. Stiles: Platelet-derived growth factor and the regulation of the mammalian fibroblast cell cycle. *Biochim Biophys Acta* 560:217, 1979.
75. C. D. Scher, W. J. Pledger, P. Martin, H. N. Antoniades, and C. D. Stiles: Transforming viruses directly reduce the cellular growth requirement for a platelet derived growth factor. *J Cell Physiol* 97:371, 1978.
76. D. F. Bowen-Pope, A. Vogel, and R. Ross: Production of platelet-derived growth factor-like molecules and reduced expression of platelet-derived growth factor receptors accompany transformation by a wide spectrum of agents. *Proc Natl Acad Sci USA* 81:2396, 1984.
77. D. T. Graves, A. J. Owen, and H. N. Antoniades: Evidence that a human osteosarcoma cell line which secretes a mitogen similar to platelet-derived growth factor required growth factors present in platelet-poor plasma. *Cancer Res* 43:83, 1983.
78. M. Nistér, C.-H. Heldin, A. Wasteson, and B. Westermark: A glioma-derived analog to platelet-derived growth factor: Demonstration of a receptor competing activity and immunological crossreactivity. *Proc Natl Acad Sci USA* 81:926, 1984.
79. C. Betsholtz, B. Westermark, B. Ek, and C.-H. Heldin: Coexpression of a PDGF-like growth factor and PDGF receptors in a human osteosarcoma cell line: Implications for autocrine receptor activation. *Cell* 39:447, 1984.
80. P. Pantaziz, G. G. Pelicci, R. Dalla-Favera, and H. N. Antoniades: Synthesis and secretion of proteins resembling platelet-derived growth factor by human glioblastoma and fibrosarcoma cells in culture. *Proc Natl Acad Sci USA* 82:2404, 1985.
81. K. Forsberg, I. Valyi-Nagy, C.-H. Heldin, M. Herlyn, and B. Westermark: Platelet-derived growth factor (PDGF) in oncogenesis: Development of a vascular connective tissue stroma in xenotransplanted human melanoma producing PDGF-BB. *Proc Natl Acad Sci USA* 90:393, 1993.
82. T. P. Fleming, A. Saxena, W. C. Clark, J. T. Robertson, E. H. Oldfield, S. A. Aaronson, and I. U. Ali: Amplification and/or overexpression of platelet-derived growth factor receptors and epidermal growth factor receptor in human glial tumors. *Cancer Res* 52:4550, 1992.
83. L. Holmgren, F. Flam, E. Larsson, and R. Ohlsson: Successive activation of the platelet-derived growth factor β receptor and platelet-derived growth factor B genes correlates with the genesis of human choriocarcinoma. *Cancer Res* 53:2927, 1993.
84. T. R. Golub, G. F. Barker, M. Lovett, and D. G. Gilliland: Fusion of PDGF receptor β to a novel ets-like gene, tel, in chronic myelomonocytic leukemia with t(5;12) chromosomal translocation. *Cell* 77:307, 1994.
85. B. E. Bejcek, R. M. Hoffman, D. Lipps, D. Y. Li, C. A. Mitchell, P. W. Majerus, and T. F. Deuel: The vis-sis oncogene product but not platelet-derived growth factor (PDGF) a homodimer activate PDGF α and β receptors intracellularly and initiate cellular transformation. *J Biol Chem* 267:3289, 1992.
86. G. J. Todaro, J. E. DeLarco, and S. Cohen: Transformation by murine and feline sarcoma viruses specifically blocks binding of epidermal growth factor to cells. *Nature* 264:26, 1976.
87. J. E. DeLarco and G. J. Todaro: Growth factors from murine sarcoma virus-transformed cells. *Proc Natl Acad Sci USA* 75:4001, 1978.
88. A. B. Roberts, L. C. Lamb, D. L. Newton, M. B. Sporn, J. E. DeLarco, and G. J. Todaro: Transforming growth factors: Isolation of polypeptides from virally and chemically transformed cells by acid/ethanol extraction. *Proc Natl Acad Sci USA* 77:3494, 1980.
89. M. A. Anzano, A. B. Roberts, J. M. Smith, M. B. Sporn, and J. E. DeLarco: Sarcoma growth factor from conditioned medium of virally transformed cells is composed of both type α and type β transforming growth factors. *Proc Natl Acad Sci USA* 80:6264, 1983.
90. J. Massagué: Type β transforming growth factor from feline sarcoma virus-transformed rat cells: Isolation and biological properties. *J Biol Chem* 259:9756, 1984.
91. B. Ozanne, T. Wheeler, and P. L. Kaplan: Cells transformed by RNA and DNA tumor viruses produce transforming factors. *Fed Proc* 41:3004, 1982.
92. M. B. Sporn and A. B. Roberts: Autocrine growth factors and cancer. *Nature* 313:745, 1985.
93. J. Massagué: A model for membrane-anchored growth factors. *J Biol Chem* 265:21393, 1990.
94. S. T. Wong, L. F. Winchell, B. K. McCune, H.

S. Earp, J. Teixido, J. Massagué, B. Herman, and D. C. Lee: The TGF-α precursor expressed on the cell surface binds to the EGF receptor on adjacent cells, leading to signal transduction. *Cell* 56:495, 1989.
95. K.-I. Imanishi, K. Yamaguchi, M. Kuranami, E. Kyo, T. Hozumi, and K. Abe: Inhibition of growth of human lung adenocarcinoma cell lines by anti-transforming growth factor-α monoclonal antibody. *JNCI* 81:220, 1989.
96. M. Reiss, E. B. Stash, V. F. Vellucci, and Z.-l. Zhou: Activation of the autocrine transforming growth factor α pathway in human squamous carcinoma cells. *Cancer Res* 51:6254, 1991.
97. S. E. Bates, N. E. Davidson, E. M. Valverius, C. E. Freter, R. B. Dickson, J. P. Tam, J. E. Kudlow, M. E. Lippman, and D. S. Salomon: Expression of transforming growth factor α and its messenger ribonucleic acid by estrogen and its possible functional significance. *Mol Endocrinol* 2:543, 1988.
98. Y. Gong, G. Ballejo, L. C. Murphy, and L. J. Murphy: Differential effects of estrogen and antiestrogen on transforming growth factor gene expression in endometrial adenocarcinoma cells. *Cancer Res* 52:1704, 1992.
99. H. Takagi, R. Sharp, C. Hammermeister, T. Goodrow, M. O. Bradley, N. Fausto, and G. Merlino: Molecular and genetic analysis of liver oncogenesis in transforming growth factor α transgenic mice. *Cancer Res* 52:5171, 1992.
100. Y. C. Yeh, J. F. Tsai, L. Y. Chuang, H. W. Yeh, J. H. Tsai, D. L. Florine, and J. P. Tam: Elevation of transforming growth factor α and α-fetoprotein levels in patients with hepatocellular carcinoma. *Cancer Res* 47:896, 1987.
101. C. L. Arteaga, A. R. Hanauske, G. M. Clark, C. K. Osborne, P. Hazarika, R. L. Pardue, F. Tio, and D. D. Von Hoff: Immunoreactive α transforming growth factor activity in effusions from cancer patients as a marker of tumor burden and patient prognosis. *Cancer Res* 48:5023, 1988.
102. G. H. Lee, G. Merlino, and N. Fausto: Development of liver tumors in transforming growth factor α transgenic mice. *Cancer Res* 52:5162, 1992.
103. J. E. Mead and N. Fausto: Transforming growth factor α may be a physiological regulatory of liver regeneration by means of an autocrine mechanism. *Proc Natl Acad Sci USA* 86:1558, 1989.
104. H. L. Moses, E. Y. Yang, and J. A. Pietenpol: TGF-β stimulation and inhibition of cell proliferation: New mechanistic insights. *Cell* 63:245, 1990.
105. J. Massagué: Receptors for the TGF-β family. *Cell* 69:1067, 1992.
106. M. B. Sporn and A. B. Roberts: Transforming growth factor-β: Recent progress and new challenges. *J Cell Biol* 119:1017, 1992.
107. M. Laiho, J. A. DeCaprio, J. W. Ludlow, D. M. Livingston, and J. Massagué: Growth inhibition by TGF-β linked to suppression of retinoblastoma protein phosphorylation. *Cell* 62:175, 1990.
108. Y. Geng and R. A. Weinberg: Transforming growth factor β effects on expression of G_1 cyclins and cyclin-dependent protein kinases. *Proc Natl Acad Sci USA* 90:10315, 1993.
108a. J. L. Wrana, L. Attisano, R. Wieser, F. Ventura, and J. Massagué: Mechanism of activation of the TGF-β receptor. *Nature* 370:341, 1994.
108b. K. Park, S.-J. Kim, Y.-J. Bang, J.-G. Park, N. K. Kim, A. B. Roberts, and M. B. Sporn: Genetic changes in transforming growth factor β (TGF-β) type II receptor gene in human gastric cancer cells: Correlation with sensitivity to growth inhibition by TGF-β. *Proc Natl Acad Sci USA* 91:8772, 1994.
109. P. Schroy, J. Rifkin, R. J. Coffey, S. Winawer, and E. Friedman: Role of transforming growth factor β_1 in induction of colon carcinoma differentiation by hexamethylene bisacetamide. *Cancer Res* 50:261, 1990.
110. A. Bassols and J. Massagué: Transforming growth factor β regulates the expression and structure of extracellular matrix chondroitin/dermatan sulfate proteoglycans. *J Biol Chem* 263:3039, 1988.
111. R. W. Merwin and G. H. Algire: The role of graft and host vessels in the vascularization of grafts of normal and neoplastic tissue. *JNCI* 17:23, 1956.
112. R. L. Ehrmann and M. Knoth: Choriocarcinoma: Transfilter stimulation of vasoproliferation in the hamster cheek pouch—studied by light and electron microscopy. *JNCI* 41:1229, 1968.
113. M. Greenblatt and P. Shubik: Tumor angiogenesis: Transfilter diffusion studies in the hamster by the transparent chamber technique *JNCI* 41:111, 1968.
114. J. Folkman, E. Merler, C. Abernathy, and G. Williams: Isolation of a tumor factor responsible for angiogenesis. *J Exp Med* 113:275, 1971.
115. M. A. Gimbrone, Jr., R. S. Cotran, S. B. Leapman, and J. Folkman: Tumor growth and neovascularization: An experimental model using the rabbit cornea. *JNCI* 52:413, 1974.
116. P. M. Gullino: Angiogenesis and oncogenesis. *JNCI* 61:639, 1978.
117. J. Folkman and Y. Shing: Angiogenesis. *J Biol Chem* 267:10931, 1992.
118. K. H. Plate, G. Breier, H. A. Weich, and W. Risau: Vascular endothelial growth factor is a potential tumour angiogenesis factor in human gliomas in vivo: *Nature* 359:845, 1992.
119. Y. Myoken, Y. Kayada, T. Okamoto, M. Kan, G. H. Sato, and J. D. Sato: Vascular endothelial cell growth factor (VEGF) produced by A-431 human epidermoid carcinoma cells and identifica-

tion of VEGF membrane binding sites. *Proc Natl Acad Sci USA* 88:5819, 1991.

120. M. Nguyen, H. Watanabe, A. E. Budson, J. P. Richie, and J. Folkman: Elevated levels of the angiogenic peptide basic fibroblast growth factor in urine of bladder cancer patients. *JNCI* 85:241, 1993.
121. K.-T. Yeo, H. H. Wang, J. A. Nagy, T. M. Sioussat, S. R. Ledbetter, A. J. Hoogewerf, Y. Zhou, E. M. Masse, D. R. Senger, H. E. Dvorak, and T.-K. Yeo: Vascular permeability factor (vascular endothelial growth factor) in guinea pig and human factor and inflammatory effusions. *Cancer Res* 53:2912, 1993.
122. J. Folkman, K. Watson, D. Ingber, and D. Hanahan: Induction of angiogenesis during the transition from hyperplasia to neoplasia. *Nature* 339:58, 1989.
123. M. Maxwell, S. P. Naber, H. J. Wolfe, E. T. Hedley-Whyte, T. Galanopoulos, J. Neville-Golden, and H. N. Antoniades: Expression of angiogenic growth factor genes in primary human astrocytomas may contribute to their growth and progression. *Cancer Res* 51:1345, 1991.
124. R. Langer, H. Conn, J. Vacanti, C. Haudenshild, and J. Folkman: Control of tumor growth in animals by infusion of angiogenesis inhibitor. *Proc Natl Acad Sci USA* 77:4331, 1980.
125. A. Lee and R. Langer: Shark cartilage contains inhibitors of tumor angiogenesis. *Science* 221:1185, 1983.
126. J. Folkman, R. Langer, R. J. Linhardt, C. Haudenschild, and S. Taylor: Angiogenesis inhibition and tumor regression caused by heparin or a heparin fragment in the presence of cortisone. *Science* 221:719, 1983.
127. R. Crum, S. Szabo, and J. Folkman: A new class of steroids inhibits angiogenesis in the presence of heparin or a heparin fragment. *Science* 230:1375, 1985.
128. P. E. Thorpe, E. J. Debyshire, S. P. Andrade, N. Press, P. P. Knowles, S. King, G. J. Watson, Y.-C. Yang, and M. Rao-Betté: Heparin-steroid conjugates: New angiogenesis inhibitors with antitumor activity in mice. *Cancer Res* 53:3000, 1993.
129. B. A. Teicher, E. A. Sotomayor, and Z. D. Huang: Antiangiogenic agents potentiate cytotoxic cancer therapies against primary and metastatic disease. *Cancer Res* 52:6702, 1992.
130. A. Gagliardi and D. C. Collins: Inhibition of angiogenesis by antiestrogens. *Cancer Res* 53:533, 1993.
131. S. S. Tolsma, O. V. Volpert, D. J. Good, W. A. Frazier, P. J. Polverini, and N. Bouck: Peptides derived from two separate domains of the matrix protein thrombospondin-1 have anti-angiogenic activity. *J Cell Biol* 122:497, 1993.
132. T. Fotsis, M. Pepper, H. Adlercreutz, G. Fleischmann, T. Hase, R. Montesano, and L. Schweigerer: Genistein, a dietary-derived inhibitor of in vitro angiogenesis. *Proc Natl Acad Sci USA* 90:2690, 1993.
133. D. Metcalf: Hemopoietic regulators. *Trends Biochem Sci* 17:286, 1992.
134. A. Miyajima, T. Kitamura, N. Harada, T. Yokota, and K.-I. Arai: Cytokine receptors and signal transduction. *Annu Rev Immunol* 10:295, 1992.
135. D. Metcalf: Control of granulocytes and macrophages: Molecular, cellular, and clinical aspects. *Science* 254:529, 1991.
136. W. P. Hammond IV, T. H. Price, L. M. Souza, and D. C. Dale: Treatment of cyclic neutropenia with granulocyte colony-stimulating factor. *N Engl J Med* 320:1306, 1989.
137. J. Vose, P. Bierman, A. Kessinger, P. Coccia, J.Anderson, F. Oldham, C. Epstein, and J. Armitage: The use of recombinant human granulocyte-macrophage colony stimulating factor for the treatment of delayed engraftment following high dose therapy and autologous hematopoietic stem cell transplantation for lymphoid malignancies. *Bone Marrow Transpl* 7:139, 1991.
138. U. Duhrsen, J.-L. Vileval, J. Boyd, G. Kannourakis, G. Morstyn, and D. Metcalf: Effects of recombinant human granulocyte colony-stimulating factor on hematopoietic progenitor cells in cancer patients. *Blood* 72:2074, 1988.
139. N. Stahl and G. D. Yancopoulos: The alphas, betas, and kinases of cytokine receptor complexes. *Cell* 74:587, 1993.
140. F. Bussolino, M. F. DiRenzo, M. Ziche, E. Bocchietto, M. Olivero, L. Naldini, G. Gaudino, L. Tamagnone, A. Coffer, and P. M. Comoglio: Hepatocyte growth factor is a potent angiogenic factor which stimulates endothelial cell motility and growth. *J Cell Biol* 119:629, 1992.
141. T. Nakamura, T. Nishizawa, M. Hagiya, T. Keki, M. Shimonishi, A. Sugimura, K. Tashiro, and S. Shimizu: Molecular cloning and expression of human hepatocyte growth factor. *Nature* 342:440, 1989.
142. K. H. Miyazawa, D. Tsubouchi, K. Naka, M. Takahashi, N. Okigakj, H. Arakaki, S. Nakayama, O. Hirono, K. Sakiyama, E. Takahashi, Y. Gohda, Daikuhara, and N. Kitamura: Molecular cloning and sequence analysis of cDNA for human hepatocyte growth factor. *Biochem Biophys Res Commun* 163:967, 1989.
143. Y. Uehara and N. Kitamura: Expression of a human hepatocyte growth factor/scatter factor cDNA in MDCK epithelial cells influences cell morphology, motility, and anchorage-independent growth. *J Cell Biol* 117:889, 1992.
144. D. P. Bottaro, J. S. Rubin, D. L. Faletto, A. M.-L. Chan, T. E. Kmiecik, G. F. Vande Woude, and S. A. Aaronson: Identification of the hepatocyte growth factor receptor as the c-met protooncogene product. *Science* 251:802, 1991.
145. A. Graziani, D. Gramaglia, P. dalla Zonca, and

P. M. Comoglio: Hepatocyte growth factor/scatter factor stimulates the *ras*-guanine nucleotide exchange. *J Biol Chem* 268:9165, 1993.
146. M. R. Urist, R. J. DeLange, and G. A. M. Finerman: Bone cell differentiation and growth factors. *Science* 220:680, 1983.
147. T. Ikeda and D. A. Sirbasku: Purification and properties of a mammary-uterine-pituitary tumor cell growth factor from pregnant sheep uterus. *J Biol Chem* 259:4049, 1984.
148. M. Bano, D. S. Salomon, and W. R. Kidwell: Purification of a mammary-derived growth factor from human milk and human mammary tumors. *J Biol Chem* 260:5745, 1985.
149. M. Eisinger, O. Marko, S.-I. Ogata, and L. J. Old: Growth regulation of human melanocytes: Mitogenic factors in extracts of melanoma, astrocytoma, and fibroblast cell lines. *Science* 229:984, 1985.
150. J. M. Siegfried: Detection of human lung epithelial cell growth factors produced by a lung carcinoma cell line: Use in culture of primary solid lung tumors. *Cancer Res* 47:2903, 1987.
151. G. B. Mills, C. May, M. McGill, C. M. Roifman, and A. Mellors: A putative new growth factor in ascitic fluid from ovarian cancer patients: Identification, characterization, and mechanism of action. *Cancer Res* 48:1066, 1988.
152. C. R. Burrow and P. D. Wilson: A putative Wilms' tumor-secreted growth factor activity required for primary culture of human nephroblasts. *Proc Natl Acad Sci USA* 90:6066, 1993.
153. T. Ikeda, D. Danielpour, and D. A. Sirbasku: Isolation and properties of endocrine and autocrine type mammary tumor cell growth factors (estromedins). *Prog Cancer Res Ther* 31:171, 1984.
154. S. A. Miles, O. Martinez-Maza, Ahmad Rezai, L. Magpantay, T. Kishimoto, S. Nakamura, S. F. Radka, and P. S. Linsley: Oncostatin M as a potent mitogen for AIDS—Kaposi's sarcoma-derived cells. *Science* 255:1432, 1992.
155. F. R. Miller, D. McEachern, and B. E. Miller: Growth regulation of mouse mammary tumor cells in collagen gel cultures by diffusible factors produced by normal mammary gland epithelium and stromal fibroblasts. *Cancer Res* 49:6091, 1989.
156. P. R. Ervin, Jr., M. S. Kaminski, R. L. Cody, and M. S. Wicha: Production of mammastatin, a tissue-specific growth inhibitor, by normal human mammary cells. *Science* 244:1585, 1989.
157. P. Yaish, A. Gazit, C. Gilon, and A. Levitzki: Blocking of EGF-dependent cell proliferation by EGF receptor kinase inhibitors. *Science* 242:933, 1988.
158. T. Hunter: A thousand and one protein kinases. *Cell* 50:823, 1987.
159. T. Hunter: Protein-tyrosine phosphatases: The other side of the coin. *Cell* 58:1013, 1989.
160. E. H. Fischer, H. Charbonneau, and N. K. Tonks: Protein tyrosine phosphatases: A diverse family of intracellular and transmembrane enzymes. *Science* 253:401, 1991.
161. J. K. Klarlund: Transformation of cells by an inhibitor of phosphatases acting on phosphotyrosine in proteins. *Cell* 41:707, 1985.
162. P. Taylor and P. A. Insel: Molecular basis of drug action, in W. B. Pratt and P. Taylor, eds.: *Principles of Drug Action,* New York: Churchill Livingstone, 1990, pp. 103–200.
163. J. R. Hepler and G. G. Gilman: G proteins. *Trends Biochem Sci* 17:383, 1992.
164. D. E. Clapham and E. J. Neer: New roles for G-protein $\beta\gamma$-dimers in transmembrane signalling. *Nature* 365:403, 1993.
165. W.-J. Tang and A. G. Gilman: Adenylyl cyclases. *Cell* 70:869, 1992.
166. E. W. Sutherland: Studies on the mechanism of hormone action. *Science* 177:401, 1972.
167. I. Pastan and M. Willingham: Cellular transformation and the "morphologic phenotype" of transformed cells. *Nature* 274:645, 1978.
168. J. Hochman, P. A. Insel, H. R. Bourne, P. Coffino, and G. M. Tomkins: A structural gene mutation affecting the regulatory subunit of cyclic AMP-dependent protein kinase in mouse lymphoma cells. *Proc Natl Acad Sci USA* 72:5051, 1975.
169. Y. S. Cho-Chung, T. Clair, P. N. Yi, and C. Parkinson: Comparative studies on cyclic AMP binding and protein kinase in cyclic AMP-responsive and -unresponsive Walker 256 mammary carcinomas. *J Biol Chem* 252:6335, 1977.
170. Y. S. Cho-Chung, T. Clair, and P. Huffman: Loss of nuclear cyclic AMP binding in cyclic AMP-unresponsive Walker 256 mammary carcinoma. *J Biol Chem* 252:6349, 1977.
171. S. A. Aaronson: Growth factors and cancer. *Science* 254:1146, 1991.
172. L. F. Allen, R. J. Lefkowitz, M. G. Caron, and S. Cotecchia: G-protein-coupled receptor genes as protooncogenes: Constitutively activating mutation of the α_{1b}-adrenergic receptor enhances mitogenesis and tumorigenicity. *Proc Natl Acad Sci USA* 88:11354, 1991.
173. S. Hermouet, J. J. Merendino, Jr., J. S. Gutkin, and A. M. Spiegel: Activating and inactivating mutations of the α subunit of G_{i2} protein have opposite effects on proliferation of NIH 3T3 cells. *Proc Natl Acad Sci USA* 88:10455, 1991.
174. J. R. Stanley and S. H. Yuspa: Specific epidermal protein markers are modulated during calcium-induced terminal differentiation. *J Cell Biol* 96:1809, 1983.
175. A. E. Warner, S. C. Guthrie, and N. B. Gilula: Antibodies to gap-junction protein selectively disrupt junctional communication in the early amphibian embryo. *Nature* 311:127, 1984.
176. M. R. Hokin and L. E. Hokin: Enzyme secretion and the incorporation of P^{32} into phospholipids of pancreas slices. *J Biol Chem* 203:967, 1953.

177. M. J. Berridge and R. F. Irvine: Inositol trisphosphate, a novel second messenger in cellular signal transduction. *Nature* 312:315, 1984.
178. Y. Nishizuka: Intracellular signaling by hydrolysis of phospholipids and activation of protein kinase C. *Science* 258:607, 1992.
179. F. S. Menniti, K. G. Oliver, J. W. Putney, Jr., and S. B. Shears: Inositol phosphates and cell signaling: New views of $InsP_5$ and $InsP_6$. *Trends Biochem Sci* 17:53, 1993.
180. L. A. Selbie, C. Schmitz-Peiffer, Y. Sheng, and T. J. Biden: Molecular cloning and characterization of PKC, an atypical isoform of protein kinase C derived from insulin-secreting cells. *J Biol Chem* 268:24296, 1993.
181. Y. Asaoka, S.-i. Nakamura, K. Yoshida, and Y. Nishizuka: Protein kinase C, calcium and phospholipid degradation. *Trends Biochem Sci* 17:414, 1992.
182. K. Endo, Y. Igarashi, M. Nisar, Q. Zhou, and S. I. Hakomori: Cell membrane signaling as target in cancer therapy: Inhibitory effect of *N,N*-dimethyl and *N,N,N*-trimethyl sphingosine derivatives on in vitro and in vivo growth of human tumor cells in nude mice. *Cancer Res* 51:1613, 1991.
183. G. K. Schwartz, J. Jiang, D. Kelsen, and A. P. Albino: Protein kinase C: A novel target for inhibiting gastric cancer cell invasion. *JNCI* 85:402, 1993.
184. B. Dérijard, M. Hibi, I.-H. Wu, T. Barrett, B. Su, T. Deng, M. Karin, and R. J. Davis: JNK1: A protein kinase stimulated by UV light and ha-*ras* that binds and phosphorylates the c-Jun activation domain. *Cell* 76:1025, 1994.
185. C. J. Lowenstein and S. H. Snyder: Nitric oxide, a novel biologic messenger. *Cell* 70:705, 1992.

10

Cell-Cycle Regulation and Apoptosis

CELL-CYCLE REGULATION

Introduction and Overview

It is clear that nature must control the amount of cell proliferation that occurs. If it did not, we would be subject to the cinematic science fiction horror of *The Blob*. In this film, a meteor brings a blob of space protoplasm to earth; the blob grows until it consumes some of the inhabitants of a small town, but Steve McQueen (in his first film) and his pals figure out how to stop its growth—by freezing it. If even a single organism were allowed to proliferate in an uncontrolled fashion, it could potentially take over the universe. As incredible as this sounds, consider for a moment a calculation by O'Farrell[1] on what would happen if the simple bacterium *Escherichia coli* were allowed to proliferate at its maximum rate. A single *E. coli* cell weighs 10^{-12} g. If allowed to grow in medium that contains everything it needs, it divides once every 20 minutes. At this rate of cell division, one *E. coli* would pass through 72 cell doublings in 24 hours. This would produce in one day an increase in mass from 10^{-12} g to 4000 metric tons (an amplification of 4×10^{21} fold, i.e., $2^{72} = 10^{72} \times \log 2 = 10^{21.6}$). In 2 days, the mass of *E. coli* growing in this uncontrolled, exponential manner would exceed by many times the mass of the entire earth. This extreme example points out the stark reality that living organisms must have built-in mechanisms to regulate their proliferation. Indeed, even bacteria spend more time not growing than growing. Of course, several environmental factors come into play, not the least of which is the availability of nutrients, water, necessary trace elements, etc. Nevertheless, as metazoan organisms evolved, they developed ways to regulate their proliferation rate, in addition to a simple response to nutrients, so that cells would stop growing in order to differentiate into multifunctional tissues. Thus, regulation of cell proliferation became carefully coupled to tissue differentiation processes, and multicellular organisms have "learned" to regulate their cell cycles in precise spatial and temporal ways during development. Cancer is a situation in which these carefully monitored regulatory processes go crazy in the locale of a particular tissue.

The development of our knowledge about cell-cycle regulation is itself a fascinating story and takes us through a tale of fundamental discoveries in two kinds of yeast, sea urchins, clams, fruit flies, frogs, mice, and humans. This story serves as a wonderful example of why fundamental research should be supported for its own sake, even though its primary aim at the time may simply be the pursuit of knowledge.

The story of the factors involved in cell-cycle regulation goes back many years. Distinct phases of a division cycle (i.e., G_1, S, G_2, and M) were defined in the 1950s when tritium labeling and cell synchronization techniques became available to score mitoses and measure the time between one mitotic wave and another in cycling cells (reviewed in Ref. 2). This method consists of labeling cells with a pulse of tritiated thymidine, taking an aliquot of cells at various times after the labeling, fixing the cells, and counting the percentage of mitoses that are labeled in autoradiograms from each time point. The percentage of labeled mitoses will rise from zero to a peak as the cells that were in S phase at the time of the pulse go through mitosis. Following the peak, there will be a "trough" as the cells in G_1 at the time of label go through M. A second cycle will then show a similar peak of labeled M-phase cells as the first "wave" comes back through the cell cycle, but the peak will be lower because of dilution of the [^{3}H] thymidine as another round of DNA synthesis occurs and because of some spread of cell-cycle times as synchrony is diminished (Fig. 10-1). In this way, the classic cell cycle of $G_1 \rightarrow S \rightarrow G_2 \rightarrow M$ was established.

It became clear from early studies of yeast mutants that certain genetically controlled factors play a key role in regulating the cell cycle. In the 1970s, Hartwell and his colleagues identified mutants of the budding yeast *Saccharomyces cerevisiae* that had defects at specific

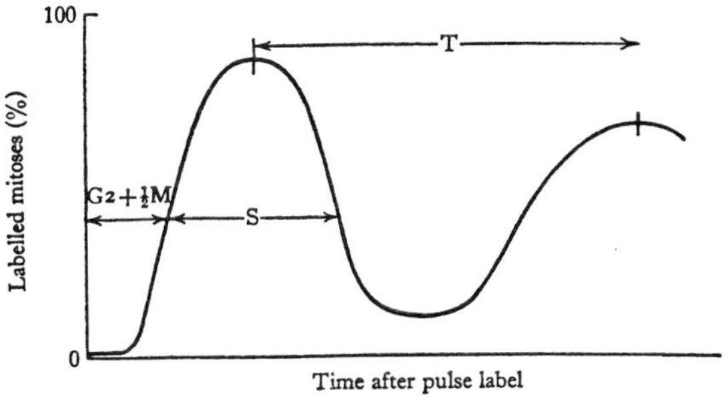

Figure 10-1 Diagram of labeled mitoses (metaphases) in successive cell samples after a pulse of tritiated thymidine. (From Mitchison.[2])

stages of the cell cycle. Temperature-sensitive mutations that were defective in initiation of DNA replication, DNA elongation, DNA ligation, tubulin assembly, spindle elongation, chromatin assembly, sister chromatid separation, nuclear division, and cytokinesis were identified (reviewed in Ref. 3). This led to the concept of "checkpoints" in the cell cycle that sense the completion of one event before allowing the cell to proceed to the next event. (This is discussed in more detail below.) Hartwell et al. identified a series of genes called cdc genes, the mutation of which produced defects in cell cycle progression. The mutated genes were postulated to be the ones involved in the checkpoints. One gene, called *cdc*28, appeared to control entry into mitosis. A similar gene, *cdc*2, was later discovered in the fusion yeast *Saccharomyces pombe* by Nurse and Bisset.[4] That these genes had the same function was shown by the observation that the *cdc*28 gene could substitute for *cdc*2 mutants of *S. pombe* in allowing the cell cycle to proceed.[5]

Another piece fell into place when it was realized that a cytoplasmic factor, called maturation promoting factor (MPF), in unfertilized frog (*Xenopus*) eggs induced immature oocytes to undergo mitotic division and was conserved in oocytes of distantly related species such as starfish. MPF was bound to be a protein complex containing a factor identical to the *cdc*2 gene product of *S. pombe* (reviewed in Ref. 6). It was also later shown that a homologous human gene could substitute for a defective *cdc*2 gene in *S. pombe*,[7] showing the evolutionary linkage for this key cell-cycle gene.[8]

Another key to the puzzle was discovered when Hunt and his students in the physiology course at Woods Hole Marine Biological Laboratory were looking for a simple experiment to study the development of sea urchin eggs (reviewed in Refs. 8 and 9). They labeled the eggs continuously with [^{35}S] methionine after fertilization and analyzed, by SDS-gel electrophoresis, the pattern of labeled proteins over time. They found that one protein of about 55 kDa was strongly labeled after fertilization but that its presence seemed to oscillate with the division cycle in that it built up during interphase and was lost at about the time the egg divided. Hunt and his colleagues made similar observations in a second species of sea urchins and in the clam *Spisula solidissima,* and they named this protein "cyclin."[10]

What remained was to link all these pieces together. The linkage of MPF to *cdc*2 and then to cyclin came from several approaches (reviewed in Ref. 9). Purification of MPF from frog eggs showed that it contained CDC2, the gene product of *cdc*2, and a cyclin called cyclin B. CDC2 was shown to have a protein kinase activity that phosphorylated histone H1 and that oscillated in activity with cell-cycle phase. Reconstitution experiments using frog egg cytoplasm depleted of endogenous transcripts showed that the presence of a cyclin whose synthesis varied with the cell cycle was tied to progression of cell division. Identification of cyclin homologues in the yeast *S. cerevisiae* (called CLN 1, 2, and 3), whose levels vary with various phases of the cell cycle and which bind to the CDC2 homologue CDC28 and regulate its kinase activity (as well as similar data from other organisms), provides a picture of the cell cycle, as shown in Figure 10-2.

Unfortunately or fortunately, depending on your point of view, the story has become much more complicated, even in yeast. Genetic studies in *S. pombe* have identified a negative regulator of CDC2, called *wee* 1, which is a tyrosine kinase that phosphoryates a specific tyrosine on CDC2, tyrosine 15, leading to inactivation of CDC2. CDC2 is reactivated by a phosphatase, called CDC25 in *S. pombe,* which removes the phosphate at tyrosine 15. A second phosphorylation step, phosphorylation of threonine 167, also appears to be required for activation of CDC2 kinase activity. Thus, the phosphorylation state of CDC2 is important for its regulation. Similar activation/inactivation events for CDC2-like kinases have been described in *Drosophila* and mammalian cells.Indeed CDC2-like kinases are key cell-cycle regulators in all cell types examined, making them the "mother" of all cell-cycle kinases.

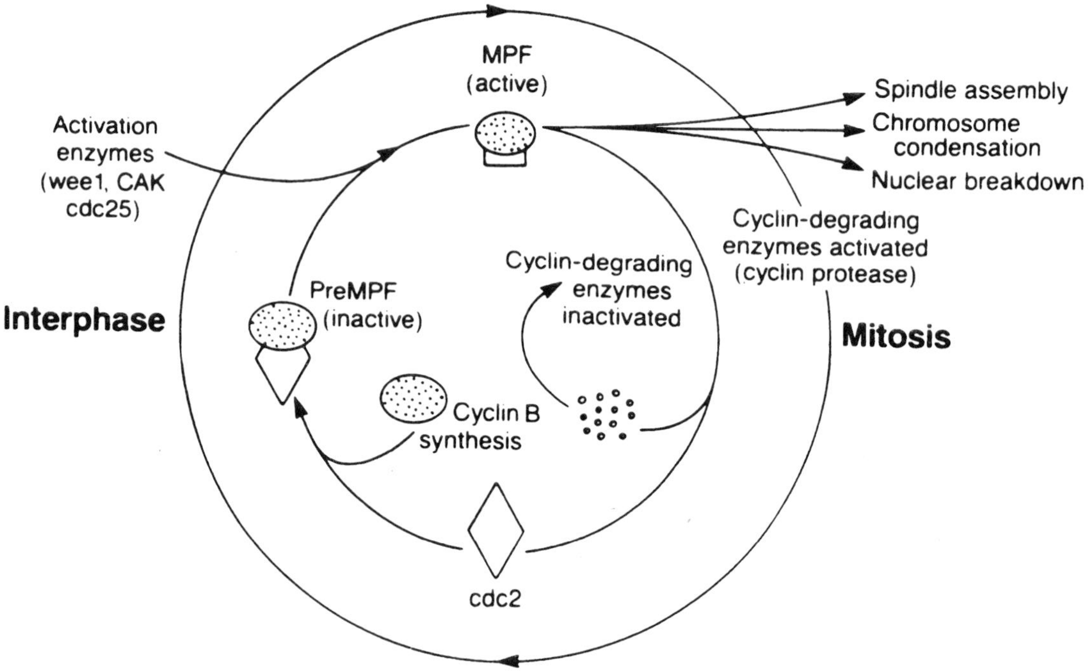

Figure 10-2 The cyclin/cdc2 cycle. During interphase, cyclin B accumulates and associates with cdc2 to form pre-MPF. This is then sequentially phosphorylated by wee1 (a cyclin-cdc2-specific protein tyrosine kinase) on Try15 of cdc2, then by CAK (cdc2-activating kinase) on Thr161; cdc25, a protein tyrosine phosphatase then dephosphorylates Tyr15, leaving active MPF. This triggers mitosis and activates cyclin protease. As cyclin is broken down, MPF disperses and the cyclin-degrading enzymes are inactivated. Thus, cyclin begins to accumulate once more. (From Kirschner.[9])

Cyclin-Dependent Protein Kinases (CDKs)

Cyclin-dependent protein kinases, of which CDC2 is only one, it turns out, are crucial regulators of the timing and coordination of eukaryotic cell-cycle events. Transient activation of members of this family of kinases occurs at specific cell-cycle phases. In budding yeast, G_1 cyclins encoded by the CLN genes (see above) interact with and are necessary for the activation of the CDC2 kinase (also called $p34^{cdc2}$), driving the cell cycle through a regulatory point called START and committing cells to enter S phase. START is analogous to the G_1 restriction point in mammalian cells.[11]

In human cells, cell-cycle events are regulated by several CDKs. Two of these, CDC2 (also called CDK1) and CDK2, are 60% to 65% homologous to yeast CDC2/CDC28 but appear to be functionally distinct. Human CDC2 is involved mainly in control of mitosis, whereas CDK2 is involved in G_1 and G_1/S transition points. CDK2 and CDC2 are 34-kDa proteins containing a conserved kinase catalytic core found in all eukaryotic protein kinases. However, they are inactive as monomers and their activation requires binding to cyclins as well as phosphorylation by a CDK-activating kinase on threonine-160 (for CDK2) or -161 (for CDC2). Negative regulation occurs by phosphorylation on tyrosine 15 and threonine 14. Removal of the inhibitory phosphates from cyclin B-CDC2 complexes at the G_2/M transition activates the CDC2 kinase activity and triggers entry into mitosis. The crystal structure of human CDK2 indicates that it is like other members of the highly conserved family of protein kinases, but it has a unique helix-loop segment that interferes with ATP and protein substrate binding.[12] That may explain why its activation events are different from those of other kinases.

Other CDKs have been discovered, including CDKs 3, 4, 5, and 6, and there are probably several others based on the finding of several related cDNAs (reviewed in Ref. 13). They all have more than 40% sequence homology to CDC2 in the protein kinase domain. It is not as yet clear what all these CDKs do, but a number of them have been shown to substitute for CDC28 in yeast mutants and/or to bind certain cyclins. For example, CDKs 1, 2, and 3 can all substitute for CDC28 in yeast and carry out START function; CDK1 binds A and B cyclins; CDK2 binds A, D, and E cyclins; and CDKs 4, 5, and 6 bind D cyclins.[13]

It is clear that the CDKs must bind to a cyclin to become active protein kinases and that different cyclins bind to different CDKs. CDC2, in complex with B-type cyclins, is the major protein kinase acting at the S/G2 and G2/M transitions. CDK2, on the other hand, in complex with cyclin A, is the kinase active at the beginning of DNA replication.

Cyclins

The originally discovered cyclins, cyclins A and B, identified in sea urchins, act at different phases of the cell cycle. Although both cyclins A and B interact with $p34^{cdc2}$ (CDC2) to induce maturation and mitosis in *Xenopus* oocytes, the synthesis and destruction of cyclin A occurs earlier in the cell cycle than that of cyclin B. Cyclin A is first detected near the G_1/S transition and cyclin B is first synthesized during S phase and accumulates in complexes with $p34^{cdc2}$ as cells approach the G_2-to-M transition. Cyclin B is then abruptly degraded during mitosis. Thus, cyclins A and B regulate S and M phase but do not appear to play a role in G_1 control points such as the restriction point (R point) where key factors have accumulated to commit cells to enter S phase.

Three more recently discovered mammalian cyclins, C, D1, and E, are the cyclins that regulate the key G_1 and G_1/S transition points (reviewed in Ref. 14). Unlike cyclins A and B, cyclins C, D1, and E are synthesized during the G_1 phase in mammalian cells. Cyclin C levels change only slightly during the cell cycle but peak in early G_1. Cyclin E peaks at the G_1/S transition, suggesting that it controls entry into S. Three distinct cyclin D forms, D1, 2, and 3, have been discovered and are differentially expressed in different mouse cell lineages. These D cyclins all have human counterparts. Cyclin D levels are growth factor–dependent in mammalian cells: when resting cells are stimulated by growth factors, D-type cyclin levels rise earlier than cyclin E levels, implying that they act earlier in G_1 than E cyclins. Cyclin D levels drop rapidly when growth factors are removed from the medium of cultured cells. All of these cyclins (C, D, and E) form complexes with, and regulate the activity of, various CDKs, and these complexes control the various G_1, G_1/S_1, and G_2/M transition points (Fig. 10-3).

In some cell types, different D-type cyclins are synthesized at different times in G_1, sometimes in response to different growth stimuli. When cyclin D1 function is deregulated, as it appears to be in human parathyroid adenomas in which a chromosome 11 inversion places the D1 gene next to the parathyroid hormone promoter,[15] it may be acting as an oncogene. Similarly, translocations, amplifications, retroviral insertions, and inappropriate transcriptional activation (the latter noted in some lymphomas, squamous cell tumors, and breast carcinomas) imply that D1 gene deregulation may be a less than rare oncogene-activation event in human cancer.[14] Overexpression of cyclin D1, driven by the MMTV promoter, induces mammary hyperplasia and carcinoma in transgenic mice that have received the gene construct,[15a] suggesting that cyclin D1 overexpression plays a role in breast cancer.

Cyclins A, D, and E can form complexes with the retinoblastoma protein RB and its analogue p107, but in distinct ways.[16] Coexpression of cyclin A or E with RB induces hyperphosphorylation of RB and overrides the ability of RB to suppress exit from G_1. Cyclins B and C do not

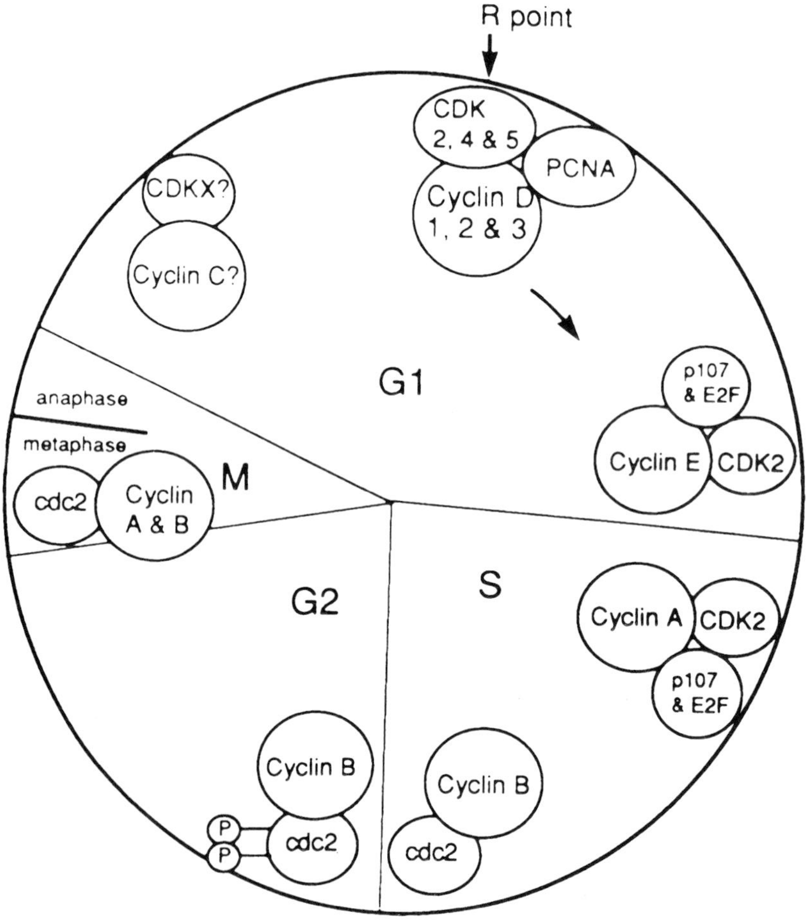

Figure 10-3 Who binds whom and when in the cell cycle. CDKX binds cyclin C and has kinase activity, but the CDK-type has not yet been determined. The R point is the restriction point, at which the cell cycle can proceed in a serum-independent manner. The cyclin D complex formed around this point is detected through the G1–S boundary and into S phase. Cyclins G and F are not represented on this figure as their function is unknown. (From Pines.[13])

have this activity. Remember that hypophosphorylated RB interacts with the transcription factor E2F during G_1 to repress the expression of target genes, whereas hyperphosphorylation of RB prevents its interaction with E2F, releasing it to promote expression of genes involved in entry of cells into S phase (Chap. 8). Likewise, introduction of cyclin D2 together with RB into RB-negative osteosarcoma cells hyperphosphorylates RB and overrides its ability to suppress exit from G_1.[17] Cyclin D1 did not have this effect. Thus, all D-type cyclins do not function equivalently. Those cyclins that bind to and induce RB phosphorylation do so by activating $p34^{cdc2}$ kinase activity. Interestingly, cell cycle–dependent expression of cyclin D1 is dependent on the presence of a functional RB protein, suggesting a regulatory loop between RB and cyclin D1.[17a]

A variety of viral oncogenic proteins also get into the act here. The adenovirus E1A protein binds a cyclin A–$p34^{cdc2}$ complex, via its interaction with p107 and RB,[18] and may act to "strip" RB or p107 from the cyclin–CDK complex, aiding in CDK activation. Introduction of a constitutively acting c-*myc* gene into BALB/3T3 mouse fibroblasts activated cyclin A expression and produced a growth factor–independent association of cyclin A–CDK2 with the transcription factor E2F, which correlated with an increase in E2F transcriptional activity.[19] In this model system, *myc*-transformed cells reduced cyclin D1 expression in early G_1. In addition, both the *src* gene product $p60^{c\text{-}src}$ [20] and SV 40 large T antigen[21] are phorphorylated by $p34^{cdc2}$, suggesting that this phosphorylation event is involved in the effects of these oncogenic proteins on DNA replication and cell proliferation.

Interestingly, negative growth regulators also interact with the cyclin-CDK system. For example, TGF-β1, which inhibits proliferation of epithelial cells by interfering with G_1/S transition, reduced the stable assembly of cyclin E-CDK2 complexes in mink lung epithelial cells and prevented the activation of CDK2 kinase activity and the phosphorylation of RB. This was one of the first pieces of data suggesting that the mammalian G_1 cyclin-dependent kinases are targets for negative regulators of the cell cycle.[22]

A whole raft of CDK inhibitors have now been discovered (reviewed in Refs. 23 and 24). These have been called CDIs or CKIs. They include Cdi1, which interacts with CDC2, CDK2, and CDK3 but not CDK4.[25] In HeLa cells, Cdi1 is expressed at the G_1/S transition, and its overexpression delays progression through the cell cycle, apparently via its action as a phosphatase. A p53-inducible 21-kDa protein called Cip1 (also WAF1) binds tightly to CDKs and inhibits the phosphorylation of RB by cyclin A–CDK2, cyclin E–CDK2, cyclin D1–CDK4, or cyclin D2–CDK4 complexes.[26] A 16-kDa protein (p16) that binds CDK4 and inhibits cyclin D–CDK4 catalytic activity has been found in human cells.[27]

Three other putative CDIs have been found: INK4, PIC1, and CIP2/CD11 (reviewed in Ref. 23). How all these CDIs might act in regulating cell-cycle progression is illustrated in Figure 10-4. Key questions are: How do all these negative regulators work and what regulates their expression? One idea is that CDIs are barriers or "dams" that act at different control points in the cell cycle.[24] The concept is that as cyclins are synthesized, they bind a CDI and form an inactive complex with a partner CDK. Once the cyclin–CDK complex concentration exceeds that of the CDI (which is thought to bind stoichiometrically), the cyclin–CDK complex can become catalytically active. Of course, CDI levels could also be regulated themselves by selective expression and/or degradation during distinct cell-cycle phases.

It turns out that at least one of these CDIs is most likely a potent tumor suppressor. The p16 inhibitor of CDK4 is located on the short arm of chromosome 9, in a region frequently homozygously deleted in a wide variety of human cancers, including melanoma and cell lines derived from lung, breast, brain, bone, skin, bladder, kidney, ovary, and lymphocytic cancers.[28,29] Because of the ubiquitous nature of this deletion in human cancers, the gene coding for p16 has been called the multiple tumor suppressor 1 (MTS1) gene.

Well, one might now ask, given the controlling role of cyclin and cyclin-dependent kinases in the regulation of cell-cycle progression and cell division, what are the cellular substrates for these key kinases? Several have been proposed, based on the ability of CDKs to phosphorylate substrates in vitro (Table 10-1).[30] Several of the potential substrates for $p34^{cdc2}$ have the consen-

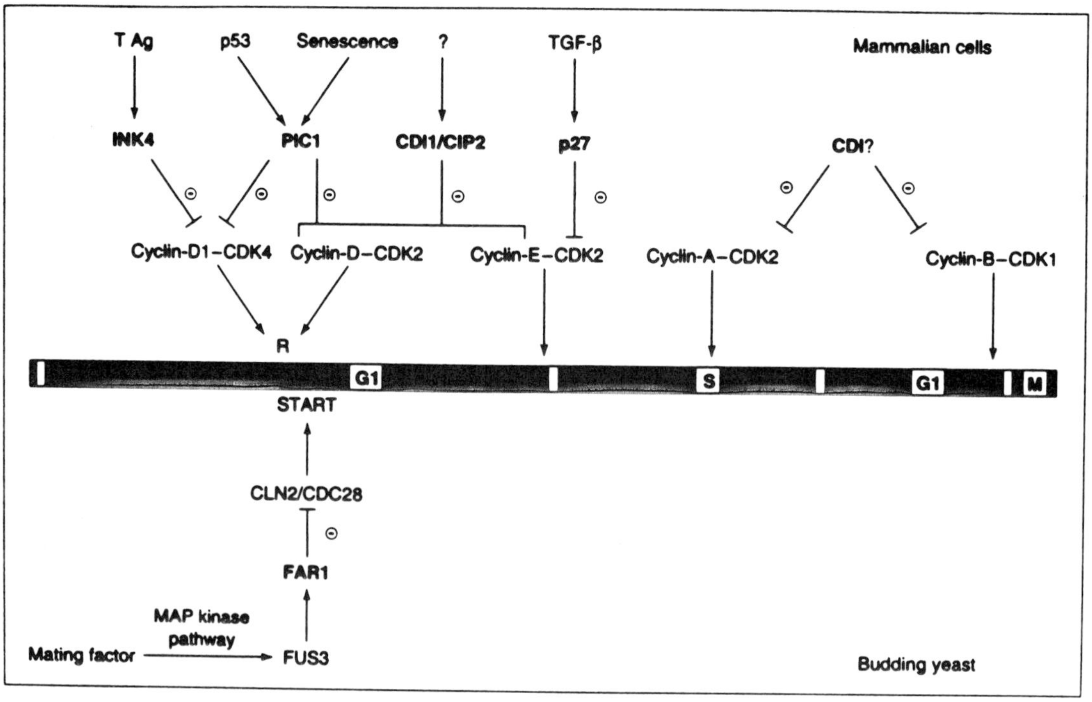

Figure 10-4 The potential sites of action of CDK inhibitors in the cell cycle. The newly isolated CDIs and the cyclin–CDK complexes with which they have been found to interact are illustrated. The cyclin–CDK complexes are assigned to major control points in the cell cycle according to the prevailing weight of evidence. (From Pines.[23])

sus sequence Lys-Ser/Thr-Pro-X-Lys/Arg. Proteins that are substrates for p34^{cdc2} contain a common motif that predicts a β-turn, allowing them to bind DNA in the minor groove. This motif is found frequently in nuclear proteins involved in transcriptional regulation, such as zinc fingers or helix-turn-helix containing proteins, and it has been predicted that phosphorylation of this motif would alter protein structure and inhibit DNA binding (reviewed in Ref. 30). One role of p34^{cdc2}-mediated phosphorylation of DNA-binding proteins could be to remove them from DNA and allow chromosomes to become more highly condensed during mitosis.

Table 10-1 Potential p34^{cdc2} Substrates[a]

Substrate	Sequence	Same Site During M Phase in Vivo as in Vitro	Possible Role in M Phase
Lamin B	P-L-S-P-T-R	Yes	Nuclear lamina disassembly
Histone H1	K/R-S/*T*-P-X-K	Yes	Chromosome condensation
pp60$^{c\text{-}src}$	Q-*T*-P-N-K R-*T*-P-S-R T-S-P-Q-R	Yes	Cytoskeletal rearrangements
NO38, nucleolin	*T*-P-X-K	Yes	Nucleolar disassembly
RNA polymerase II	S-P-T-*S*-P-S-Y	Unknown	Transcription inhibition
Cyclin B	Unknown	Unknown	Regulation of p34^{cdc2} activity
EF-1β, EF-1γ	Unknown	Unknown	Translation inhibition
SV40 large T	H-S-*T*-P-P-K-K-K-R-K-V	Unknown	Unknown
Consensus	S/*T*-P-X-Z		

[a]The phosphorylated residue(s) is underlined. X is a polar amino acid; Z is generally a basic amino acid.
Source: From Moreno and Nurse.[30]

Checkpoints in the Cell Cycle

The fidelity of the process of cell division requires that an accurate copy of a completed genome be passed on to each daughter cell. In other words, earlier events in the cell cycle such as completion of DNA replication must be accomplished for later events such as mitosis and cytokinesis to occur. Eukaryotic cells have developed "sensors" to monitor these events and to make sure that they occur in correct sequence and that one event is completed before the next one is carried out. These sensors have been called "checkpoints."

Checkpoints are feedback controls that monitor and regulate various steps in cell-cycle progression. These feedback controls have three components: a sensor that monitors the completion of a later, downstream event; a biochemical signal that is generated by the sensor; and a response element in the cell-cycle "engine" that halts or delays the checkpoint event.[31]

Figure 10-5 illustrates the division cycle in eukaryotic cells and some of the checkpoints that cells sense as they move through the cycle. There are checkpoints at the decision to enter S phase (START), at the decision to enter mitosis (G_2/M checkpoint), and at the decision to exit from mitosis. In most cell cycles, inhibition of events downstream from START halts progression of the cell-cycle engine. For example, lack of completion of DNA replication prevents cells from entering mitosis. Embryos differ from somatic cells in the stringency of this control, however. For instance, fertilized frog eggs continue to activate and inactivate MPF even when treated with drugs that block DNA synthesis or spindle assembly. This loose stringency may be allowed for several cell cycles in frog embryos (which do not have a START checkpoint), but once the embryos have generated more than 100 nuclei, inhibition of DNA synthesis blocks entry into mitosis. At this time unreplicated DNA is sensed by a feedback control that prevents activation of MPF (reviewed in Ref. 31). Thus, control of activation of MPF is one of the biochemical mechanisms for checkpoint control for entry into M phase. Understanding the events involved in checkpoint controls evolved from the pioneering work of Hartwell and his colleagues with the budding yeast *S. cerevisiae* (reviewed

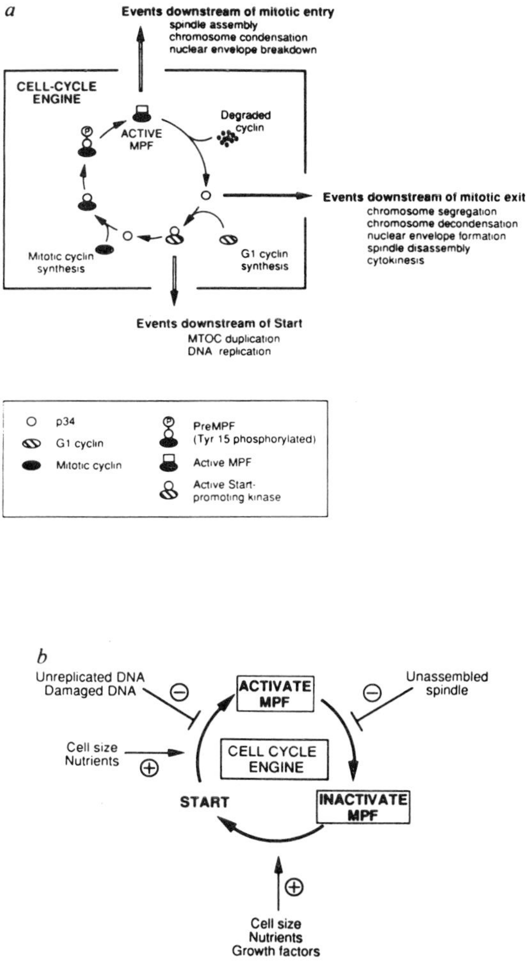

Figure 10-5 The cell-cycle engine and its regulation. (A) The cyclin and p34 components of the cell-cycle engine are represented at different points during the cell cycle, as are the downstream events that they induce. MTOC, microtubule organizing centre (the centrosome in animals and the spindle pole body in fungi). It is not known whether the degradation of G1 cyclins is regulated during the cell cycle. (B) Feedback controls inhibit the ability of the cell-cycle engine to pass key checkpoints, while increasing cell size stimulates passage through the checkpoints. (From Murray.[31])

in Ref. 3). They identified, by mutational analysis, several genes involved in a cell-cycle control. Among these are the *rad* 9, *rad* 17, and *rad* 24 genes, which are involved in arresting the cell cycle at G_2/M in response to damaged DNA. Mutations in the *rad* 9 gene allow irradiated cells with DNA damage to proceed through cell division, whereas irradiated wild-type cells stop in G_2 until DNA damage is repaired. Other eukaryotic cells, including mammalian cells, have a similar cell-cycle checkpoint. For example, a temperature-sensitive mutation (i.e., one that loses function at a higher temperature) in a mammalian gene called RCC1 (repressor of chromosome condensation) causes cultured cells to cease DNA replication and enter M prematurely at the nonpermissive temperature. The RCC1 gene product is a chromatin-associated protein and interacts with a G protein called Ran, which may be involved in stabilizing-DNA replication forks.[31] Budding yeast, fission yeast, and *Drosophila* have proteins with functions similar to RCC1.

Mutations in other components of the cell-cycle engine can have a deleterious effect on the operation of feedback controls at G_2/M (Fig. 10-6). A number of these involve regulation of the phosphorylation of $p34^{cdc2}$. Overexpression of CDC25, the phosphatase that removes an inhibiting tyrosine phosphate, expression of *cdc*2-3w dominant alleles of *cdc*2, mutation of tyrosine 15, or inactivation of tyrosine kinases Wee 1, and Mik 1, which catalyze the inhibiting phosphorylation of Tyr 15, all abolish the ability of unreplicated DNA to prevent entry into M in both fission yeast and frog embryos.

Biochemical mechanisms involved in feedback control over exit from mitosis impinge on the ability of cells to exit from M in response to completion of spindle assembly. In the budding yeast, the *bub* (budding inhibited by benomyl) and *mad* (mitotic arrest deficient) genes are involved in this checkpoint. These genes appear to activate the machinery that degrades the mitotic cyclins, which in turn allows exit from M. Mutations in these genes prevent cells from maintaining high levels of MPF in the presence of microtubule depolymerizing agents. The wild-type genes maintain MPF levels in the presence of such agents, thus blocking inappropriate exit from M. Of interest is the fact that one of these genes, *mad* 2, encodes the α-subunit of a prenyl transferase, which, the reader may recall, is involved in the prenylation of G proteins such as Ras, a key step in their ability to become anchored in cell membranes. The fact that both Mad 2 and RCC1, which are involved in two different checkpoint feedback controls, act on G proteins suggests a broad role for G proteins as checkpoint mediators.

A key question remains, and that is how the checkpoint mechanism detects the presence of unreplicated or incompletely replicated DNA. One model suggests that cells sense the presence of replication structures by the complex of proteins associated with replication forks during the elongation phase of replication.[32] When replication is proceeding, the cell mechanism senses a specific conformation at the replication fork complex that prevents initiation of mitotic events. Once DNA synthesis is complete, the replication fork complex dissipates, acting as a hallmark for completion of this step and allowing cells to proceed to M phase.

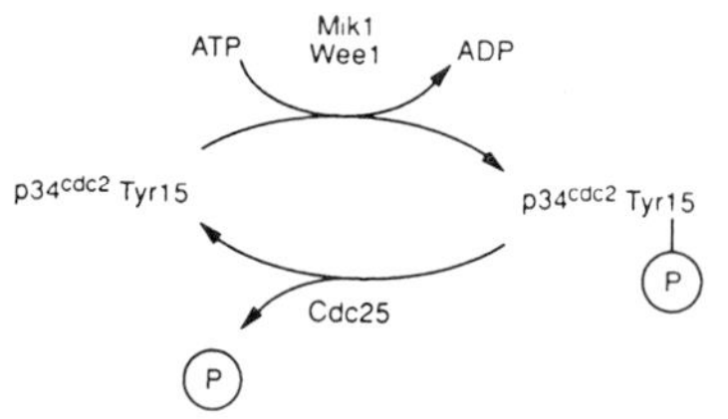

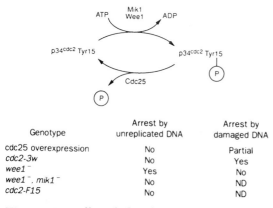

Genotype	Arrest by unreplicated DNA	Arrest by damaged DNA
cdc25 overexpression	No	Partial
cdc2-3w	No	Yes
wee1⁻	Yes	No
wee1⁻, mik1⁻	No	ND
cdc2-F15	No	ND

Figure 10-6 Effect of cdc2 phosphorylation on feedback controls in fission yeast. The effect of different genetic manipulations in fission yeast on the ability of unreplicated or damaged DNA to arrest the cell cycle in G2. ND, not done. (From Murray.[31])

Checkpoint Controls and Cancer

Although both yeast and mammalian cells have G_1/S and G_2/M checkpoints, currently the G_1/S checkpoint is best understood in mammalian cells and the G_2/M control is best understood in yeast.[33] One gene for which there is strong evidence of its function at the G_1/S checkpoint in mammalian cells is p53. High levels of p53 block cell-cycle progression at the G_1/S interface, and tumor cells lacking p53, or having dominant-negative mutant forms, lack the G_1/S delay that occurs after exposure to ionizing radiation.[34]

In addition, cells taken from patients with the repair-defective syndrome ataxia telangiectasia lack the irradiation-induced increase in p53 observed in normal cells.[34] AT cells also fail to arrest at G_1/S in response to irradiation, suggesting that AT cells are defective in responses to damaged DNA. Loss of the G_1/S checkpoint in mammalian cells that correlates with loss or mutation of p53 also allows gene amplification to occur.[35] Fibroblasts derived from Li-Fraumeni syndrome patients, who have a mutated p53 gene, also have a high frequency of chromosomal abnormalities. All the data point to an important role of p53 as a feedback control that keeps cells from entering S phase when they contain damaged DNA.

Since one of the hallmarks of cancer cells is genetic instability, it seems highly probable that cancer cells have weak or deficient checkpoint controls of their cell cycle. This deficiency may actually be taken advantage of therapeutically. Data from cultured cell lines indicate that, as cells become transformed, their cell-cycle controls becomed weaker (reviewed in Ref. 31). If tumor cells in vivo also have weaker feedback controls than normal cells, they would not be expected to survive DNA-damaging agents as well as normal cells. This is because once cells proceed past the checkpoints, they must continue through to completion of cell division or they undergo programmed cell death (apoptosis). Tumor cells, having weak checkpoint controls, proceed along their merry way in the face of anticancer drug damage to their DNA replicating machinery, whereas normal cells will have a greater tendency to stop and repair themselves. Thus, treatment of cancer cells with checkpoint inhibitors may increase their susceptibility to DNA-damaging agents and increase specificity, since normal cells, with stronger checkpoints, could survive such treatment.

APOPTOSIS

The study of cell death is a lively field, as the tremendous spate of recent publications and scientific meetings covering the subject give testimony. But it was not always so. Evidence for the existence of two morphologically distinct types of cell death was obtained by Kerr[36] in 1965 from histochemical studies of ischemic injury to rat liver. Some cell death occurred with the typical changes seen in tissue necrosis: Clumping of chromatin into ill-defined masses, swelling of organelles, flocculent densities in the matrix of mitochondria, membrane disintegration, and infiltration of inflammatory cells. Cells in some areas of the damaged liver, however, died a different death. They contained chromatin compacted into sharply delineated masses, condensation of the cytoplasm, and outcropping of cytoplasmic "blebs" or protuberances that became pinched off (apoptotic bodies) and released, to be devoured by tissue phagocytic cells. No inflammatory reaction, however, was noted around cells dying by this second mechanism. Kerr's further studies with Currie and Wyllie[37] showed that this second mechanism of cell death occurs as tissues undergo remodeling during developmant and in that sense is a "physiologic cell death." The original term for this phenomenon, "shrinkage necrosis," did not seem an appropriate one for this process, so they searched for another. A colleague of theirs at the University of Aberdeen, professor James Carmack of the Department of Greek, suggested the term *apoptosis* (pronounced ap′ ə tōsis, with the second "p" silent), meaning "falling off" of petals from a flower or leaves from a tree. The term has stuck ever since. The concept of apoptosis, however, was largely ignored until the mid- to late 1980s, when the discovery of the *ced* genes in the roundworm *Caenorhabditis elegans* and of the *bcl*-2 gene in B lymphocytes (see below) put the field on a solid genetic basis.

A number of the cellular morphologic and biochemical changes that occur during apoptosis have been worked out (reviewed in Ref. 38). These include morphologic changes that can be observed both at the light and electron microscopic levels (Figs. 10-7 and 10-8).

Biochemical Events in Apoptosis

One of the most extensively studied biochemical events in apoptosis is the double-strand cleavage of nuclear DNA that occurs at linker strands between nucleosomes, producing fragments that are multiples of about 185 base pairs. These fragments can be observed as characteristic apoptotic DNA "ladders" by agarose gel electrophoresis (Fig. 10-9). This cleavage appears to involve both DNase I and DNase II endonucleases and possibly another 18-kDa nuclease called

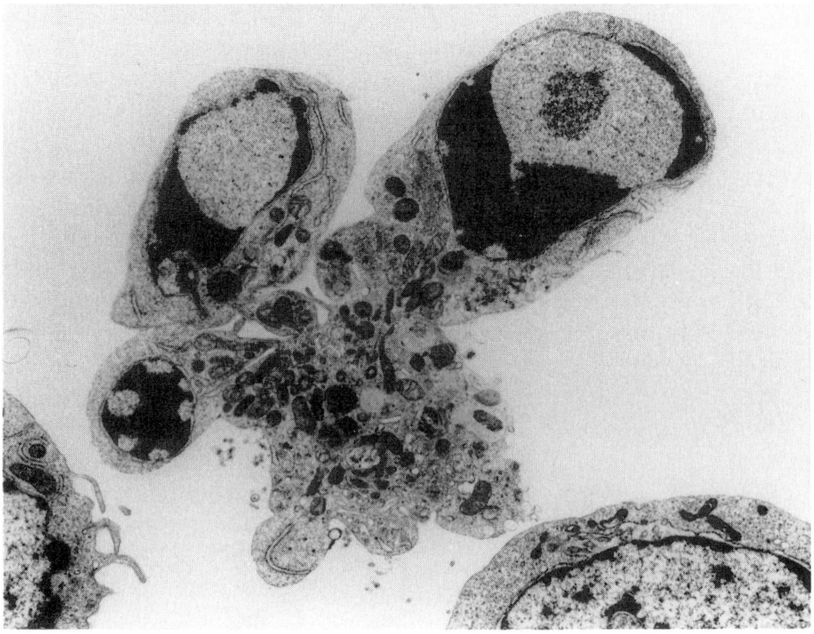

Figure 10-7 Apoptosis of murine NS-1 cell occurring spontaneously in culture. Note the discrete nuclear fragments with characteristic segregation of compacted chromatin, the crowding of organelles, and the marked convolution of the cellular surface. (From Kerr and Harmon.[38])

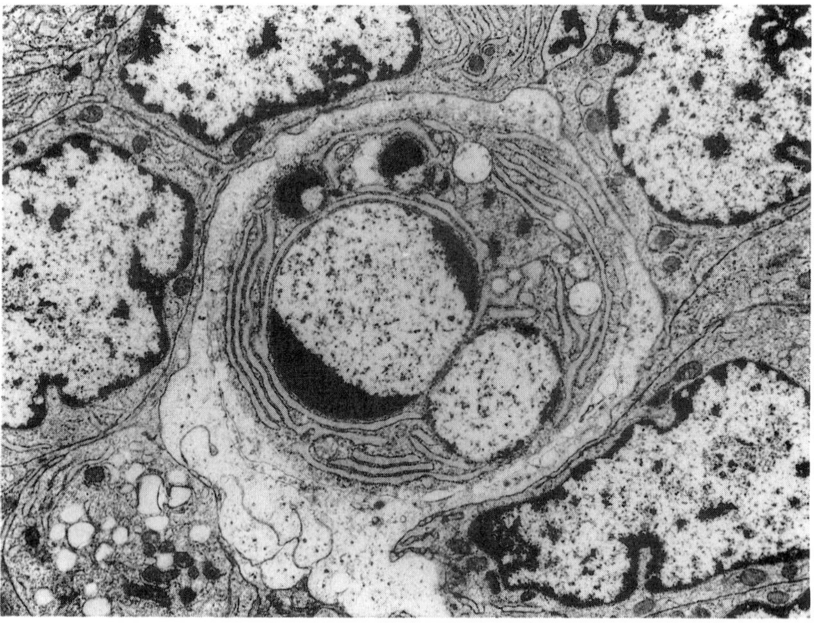

Figure 10-8 Apoptotic body containing well-preserved rough endoplasmic reticulum and four nuclear fragments, which has been phagocytosed by an intraepithelial macrophage in the rat ventral prostate 2 days after castration. (From Kerr and Harmon.[38])

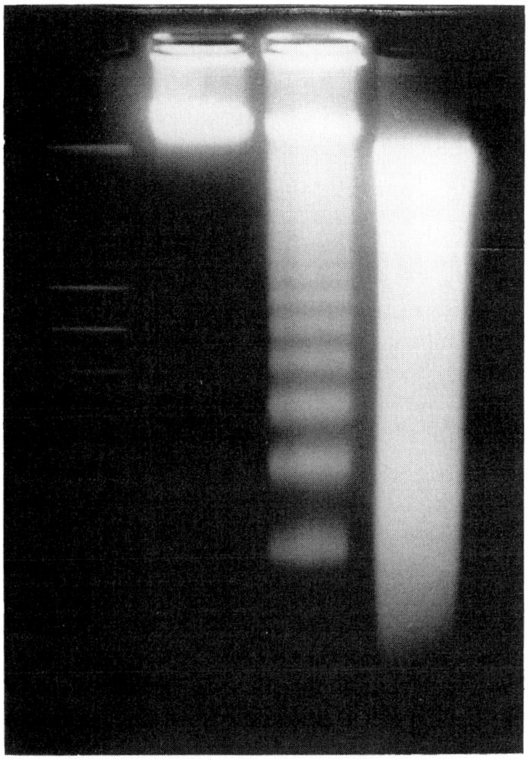

Figure 10-9 Agarose gel electrophoresis of DNA extracted from cultures of murine P-815 cells. Ethidium bromide stain; photographed under UV illumination. (Lane *1*) Molecular weight markers (DRIgest III); (lane *2*) log phase control culture; (lane *3*) culture showing extensive, morphologically typical apoptosis 8 hr after heating at 44°C for 30 min; (lane *4*) culture showing massive necrosis with karyolysis 72 hr after repeated freezing and thawing. (From Kerr and Harmon.[38])

NUC-18, isolated from thymocytes (reviewed in Ref. 39). However, the exact endonucleolytic steps and the enzymes involved in DNA cleavage during apoptosis have not been clearly defined. Since DNase I is Ca^{2+}- and Mg^{2+}-dependent for its activity and since apoptosis in some (but not all) cell types involves Ca^{2+} influx, DNase I has been implicated. Regardless of the exact identity of the endonuclease(s) involved, however, one thing is clear: the cleavage occurs at linker regions between histone octamer–containing nucleosomes rather than at specific restriction endonuclease sites, indicating that the cleavage sites are topologically recognized rather than recognized by DNA sequence.

Other biochemical changes that appear to correlate with the induction of the apoptotic cascade of events (reviewed in Refs. 38 to 41) include Ca^{2+} ion influx into cells, induction of transglutaminase, which catalyzes the formation of ε (γ-glutamyl) lysine cross-links between proteins, disruption of microtubules, induction of calmodulin expression, loss of cell membrane phospholipid asymmetry resulting in exposure of phosphatidylserine on the cell surface, activation of a cell surface receptor called Fas/Apo 1, activation of protein kinase C (some cells only), and induction of a neutral sphingomyelinase in the plasma membrane that releases phosphocholine and ceramide.[42]

Most of these biochemical events are not clearly linked in a cause-and-effect way with apoptosis, and some of them occur during apoptosis in some cell types but not in others. Thus, it is not clear which, if any, of them are universal mechanisms of apoptosis. A brief review of the involvement of these events will suffice here.

Apoptosis can be triggered by a number of agents or stimuli, including events triggered in tissue differentiation during development; removal of growth factors or hormones required for cell survival; exposure to TGF-β, tumor necrosis factor, or glucorticoids in cell types sensitive to their negative regulation; and exposure to DNA-damaging anticancer drugs or environmental toxins. Sustained increases in intracellular free Ca^{++} precede apoptosis induced by a number of agents or conditions, and apoptosis is delayed or inhibited when Ca^{++} is depleted from the cellular growth medium. In addition, Ca^{++} ionophores such as A23187, which carries Ca^{++} into cells, induces apoptosis in some cell types. However, Ca^{++} influx does not accompany apoptosis in all cell types and thus may not be an essential requirement, but rather an after effect of cell membrane perturbation.

Induction of transglutaminase accompanies apoptosis in several cell types, but again it is not an invariant biochemical indicator of apoptosis.[40] There is evidence that this enzyme causes protein–protein cross-linking to produce a protein net or meshwork that holds the cell intact and prevents leakage of cellular contents until the apoptotic bodies bud off and are consumed by phagocytes.[38]

Loss of microtubular structures and reorganization of the cytoskeleton occurs in apoptotic cells. Vinca alkaloids such as vincristine, which cause microtubule dissolution, can trigger apoptosis. However, cytochalasin B, an inhibitor of actin polymerization, inhibits formation of apoptotic bodies but not the other features of apoptosis.[39]

Increased levels of the mRNA encoding the calcium-binding protein calmodulin have been observed in apoptotic cells, and calmodulin antagonists inhibit apoptosis in some cell types, suggesting a way in which Ca^{++} might be mediating some of the cellular effects, e.g., through calmodulin-stimulated kinases.

Evidence for the effects of protein kinase C (PKC) activators and inhibitors is equivocal. Phorbol esters, which stimulates PKC, inhibit apoptosis in some cell types, but inhibition of PKC inhibits apoptosis in others.[42]

Several changes in cell surface membranes accompany apoptosis and may be related to the way in which phagocytes recognize apoptotic cells to cause a controlled cell degradation, thus preventing the marked inflammatory response that accompanies tissue necrosis. Part of this recognition by phagocytes of apoptotic cells appears to involve the rearrangement and exposure of phospholipids noted above and by exposure of desialylated (or unsialylated) oligosaccharides, which may act as lectins to attract phagocytic cells to the surface of the dying cells (reviewed in Ref. 39).

Another cell surface component that is involved in mediating apoptosis is the cell surface receptor called APO-1/Fas (reviewed in Refs. 41 and 43). APO-1 was first identified as an antibody receptor for a monoclonal antibody that triggers apoptosis in human leukemia cells in culture. It is the human homologue of Fas, the murine thymocyte surface antigen involved in regulation of T-lymphocyte proliferation and differentiation. Fas also has homology to the receptor for tumor necrosis factor and has been implicated in cell killing by cytotoxic lymphocytes, which induce apoptosis in their target cells.[44]

Genes Implicated in the Control of Apoptosis

The field of apoptosis was put on a solid genetic footing by the discovery of Horvitz and colleagues of genes in *C. elegans* that induce cell death. This creature is a beautiful model for studies of cell biology. It is a microscopic, transparent roundworm that has only has 1090 cells, and the lineage of each cell can be traced during development. About 131 of the embryonic cells of *C. elegans* undergo apoptosis or "programmed cell death," as it is sometimes called. (Strictly speaking, programmed cell death and apoptosis are not synonymous, because the former is a type of cell death that is a normal process of tissue remodeling and differentiation during development, and the latter can be induced by a wide variety of cytotoxic drugs and physical stimuli; yet the morphologic and biochemical processes appear to be very similar if not identical).

Horvitz and coworkers have identified a series of genes in *C. elegans,* called the *ced* genes, which regulate programmed cell death during embryonic development.[45] Two of these genes, *ced*-3 and *ced*-4, are "suicide" genes that induce cell death, and another, *ced*-9 is an "antideath" gene that protects cells from apoptosis.

A big breakthrough came when it was discovered that *ced*-9 was about 25% identical to a gene called *bcl*-2, identified by Korsmeyer and his colleagues, in human B-cell lymphomas.[46] The *bcl*-2 gene is activated by a translocation commonly seen in follicular lymphomas, and its overexpression protects the lymphoma cells from cytotoxic drugs. Together with c-*myc*, it can immortalize cells in culture.[47] Expression of the *bcl*-2 gene prevents apoptosis rather than stimulating cell proliferation, it was later found, and it is also involved in the normal differentiation pathway of lymphocytes together with its partner *bax* (see below). The ability of *ced*-9 to inhibit apoptosis in *C. elegans* resembles the ability of the *bcl*-2 gene to do the same in mammalian cells (reviewed in Ref. 40). Furthermore, expression of the human *bcl*-2 gene in *C. elegans* overcomes the cell death mediated by *ced*-3 and *ced*-4. Thus, *bcl*-2 and *ced*-9 appear to have the same function in organisms as diverse as roundworms and humans.

It is not exactly clear what that function is, but in mammalian cells the *bcl*-2 gene product acts in concert with another gene product called Bax. Bax is a 21-kDa protein that has significant amino acid sequence homology to Bcl-2 and can form homodimers as well as heterodimers with the Bcl-2 protein. Bax counteracts the action of Bcl-2 when it pairs with it. Thus, the ratio of Bax to Bcl-2 appears crucial to whether cells live or die when challenged by a loss of growth factors or the presence of cytotoxic agents. When Bax predominates, it binds all the Bcl-2 and may form Bax-Bax dimers as well. If Bcl-2 is in excess, it binds all the Bax and forms Bcl-2 homodimers, making cells resistant to apoptosis. During T lymphocyte differentiation, Bcl-2 levels are high in immature, rapidly proliferating T cells and drop precisely at the point when they receive a signal to kill off self-reactive cells.

Gene knockout experiments have shown that Bcl-2 is not the only negative controller of cell death.[48] Cell death becomes excessive in the thymus and kidney, for example, during mouse development when Bcl-2 is knocked out, but the nervous system appears to develop normally. Thus, even though Bcl-2 has been shown to suppress apoptosis in a variety of cell types, other genes must be involved in protection from apoptosis in some tissues of the body. Two candidates are gene products called Bcl-X_L and Bcl-X_S, identified by Boise et al.[49] These are also members of the Bcl-2 family and appear to result from differential mRNA splicing of a single gene transcript. They have opposite effects in transfected cells; Bcl-X_L is similar in size to Bcl-2 and, like Bax, acts to suppress mitosis; Bcl-X_S has a 63 amino acid deletion, does not form dimers with Bcl-2, but also allows apoptosis to proceed, perhaps by competing with Bcl-2 for its substrates or regulators.[43]

The function of Bcl-2 in preventing apoptosis is still obscure. One postulate is that Bcl-2 blocks the formation or effects of oxygen radicals that may be generated during apoptosis. Another is that Bcl-2 associates with a member of the Ras family to overcome or circumvent a cell-death signal.

Other genes involved as players on the apoptosis stage include c-*myc*, p53, c-*fos*, the gene for interleukin-1β–converting enzyme (ICE), and *crmA*, a cytokine-response modifier gene encoded by cowpox virus. In some cells, c-*myc* overexpression induces apoptosis that can be inhibited by Bcl-2. Since expression of c-*myc* is correlated with cell proliferation in many cell types (see Chap. 7), this sounds like a paradox. The paradox can be resolved by thinking about what turns the c-*myc* gene on in cells, i.e., mitogens such as growth factors. When growth factors are removed or are in short supply in a cell's environment, the cell is stuck with lots of c-Myc, and lots of signals that say "divide!" On the other hand, some key signals are missing due to growth factor deprivation, so the cell now gets a countersignal that says "die peacefully" without the fuss of bringing in all those inflammatory cells as pallbearers.

Expression of the p53 gene is also involved in apoptotic events in many cell types. As noted in Chapter 8, p53 is a protector of the genome in the sense that it stops cells with damaged DNA from dividing; in cells whose DNA is too badly damaged to be repaired before the next division cycle, it triggers apoptosis. Like c-Myc, however, p53 is not an indispensable requirement for all types of apoptosis, even though it is that for apoptosis in DNA-damaged cells.[39]

Various oncogene products can suppress apoptosis. These include adenovirus E1b, Ras, and v-Abl.[43] However, continuous expression of c-Fos, like that of c-Myc, is associated with apoptosis in skin, bone, and atretic follicles of the corpus luteum.[50]

Overexpression of the ICE gene, which shares sequence homology with the *ced*-3 gene of *C. elegans,* induces apoptosis in rat fibroblasts, which can be suppressed by Bcl-2 or the cowpox virus *crm* A gene.[51] It is of interest that microinjection of *crm* A into chicken dorsal root ganglia neurons prevents cell death due to deprivation of nerve growth factor, suggesting that ICE participates in neuronal death in vertebrates.[51]

Role of Apoptosis in Normal Development

Apoptosis plays a vital role in the embryonic development of multicellular organisms, including higher vertebrates. It is required for tissue remodeling and development in several tissues (reviewed in Ref. 38). This includes deletion of interdigital webs, palatal fusion, development of the intestinal mucosa and retina, and regression of larval organs (in amphibia). During adult life, apoptosis occurs continuously in proliferating cell populations such as intestinal crypt cells, differentiating spermatogonia, and T lymphocytes. It is also involved in ovarian follicular atresia, involution of postlactating breast tissue, and probably in the normal aging process. Indeed, apoptosis most likely occurs in all tissues of the body where cell turnover is a normal event.

Apoptosis in Cancer

Apoptosis occurs in most if not all solid cancers. Ischemia, infiltration of cytotoxic lymphocytes, and release of tumor necrosis factor may all play a role in this. It would be therapeutically advantageous to tip the balance in favor of apoptosis over mitosis in tumors. It is clear that a number of anticancer drugs induce apoptosis in cancer cells. The problem is that they also usually do this in normal proliferating cells. To manipulate the genes involved in inducing apoptosis in tumor cells *selectively* would be a great advantage. Understanding how those genes work, even in so primitive an animal as *C. elegans,* may go a long way toward achieving this goal.

REFERENCES

1. P. H. O'Farrell: Cell cycle control: Many ways to skin a cat. *Trends Cell Biol* 2:159, 1992.
2. J. M. Mitchison: *The Biology of Cell Cycle.* New York: Cambridge University Press, 1971.
3. L. H. Hartwell and T. A. Weinert: Checkpoints: Controls that ensure the order of cell cycle events. *Science* 246:629, 1989.
4. P. Nurse and Y. Bisset: Gene required in G1 for commitment to cell cycle and in G2 for control of mitosis in fission yeast. *Nature* 292:558, 1981.
5. D. H. Beach, B. Durkacz, and P. Nurse: Functionally homologous cell cycle control genes in budding and fission yeast. *Nature* 300:706, 1982.
6. A. W. Murray and M. W. Kirschner: Dominoes and clocks: The union of two views of the cell cycle. *Science* 246:614, 1989.
7. M. G. Lee and P. Nurse: Complementation used to clone a human homologue of the fission yeast cell cycle control gene cdc2. *Nature* 327:31, 1987.
8. J. Pines: Cyclins: Wheels within wheels. *Cell Growth Diff* 2:305, 1991.
9. M. Kirschner: The cell cycle then and now. *Trends Biochem Sci* 17:281, 1992.
10. T. Evans, E. T. Rosenthal, J. Youngblom, D. Distel, and T. Hunt: Cyclin: A protein specified by maternal mRNA in sea urchin eggs that is destroyed at each cleavage division. *Cell* 33:389, 1983.
11. A. B. Pardee: G_1 events and regulation of cell proliferation. *Science* 246:603, 1989.
12. H. L. DeBondt, J. Rosenblatt, J. Jancarik, H. D. Jones, D. O. Morgan, and S.-H. Kim: Crystal structure of cyclin-dependent kinase 2. *Nature* 363:595, 1993.
13. J. Pines: Cyclins and cyclin-dependent kinases: Take your partners. *Trends Biochem Sci* 18:195, 1993.
14. C. J. Sherr: Mammalian G_1 cyclins. *Cell* 73:1059, 1993.
15. T. Motokura, T. Bloom, H. G. Kim, H. Juppner, J. V. Ruderman, H. M. Kronenberg, and A. Arnold: A novel cyclin encoded by a bcl1-linked candidate oncogene. *Nature* 350:512, 1991.

15a. T. C. Wang, R. D. Cardiff, L. Zukerberg, E. Lees, A. Arnold, and E. V. Schmidt: Mammary hyperplasia and carcinoma in MMTV-cyclin D1 transgenic mice. *Nature* 369:669, 1994.

16. S. F. Dowdy, P. H. Hinds, K. Louie, S. I. Reed, A Arnold, and R. A. Weinberg: Physical interactions of the retinoblastoma protein with human D cyclins. *Cell* 73:499, 1993.
17. M. E. Ewen, H. K. Sluss, C. J. Sherr, H. Matsushime, J.-Y. Kato, and D. M. Livingston: Functional interactions of the retinoblastoma protein with mammalian D-type cyclins. *Cell* 73:487, 1993.

17a. H. Müller, J. Lukas, A. Schneider, P. Warthoe, J. Bartek, M. Eilers, and M. Strauss: Cyclin D1 expression is regulated by the retinoblastoma protein. *Proc Natl Acad Sci* USA 91:2945, 1994.

18. B. Faha, M. E. Ewen, L.-H. Tsai, D. M. Livingston, and E. Harlow: Interaction between human cyclin A and adenovirus E1A-associated p107 protein. *Science* 255:87, 1992.
19. P. Jansen-Dürr, A. Meichle, P. Steiner, M. Pagano, K. Finke, J. Botz, J. Wessbecher, G. Draetta, and M. Eilers: Differential modulation of cyclin gene expression by MYC. *Proc Natl Acad Sci USA* 90:3685, 1993.
20. D. O. Morgan, J. M. Kaplan, J. M. Bishop, and H. E. Varmus: Mitosis-specific phosphorylation of $p60^{c\text{-}src}$ by $p34^{cdc2}$-associated protein kinase. *Cell* 57:775, 1989.
21. D. McVey, L. Brizuela, I. Mohr, D. R. Marshak, Y. Gluzman, and D. Beach: Phosphorylation of large tumour antigen by cdc2 stimulates SV40 DNA replication. *Nature* 341:503, 1989.
22. A. Koff, M. Ohtsuki, K. Polyak, J. M. Roberts, and J. Massagué: Negative regulation of G1 in mammalian cells: Inhibition of cyclin E-dependent kinase by TGF-β. *Science* 260:536, 1993.
23. J. Pines: Arresting developments in cell-cycle control. *Trends Biochem Sci* 19:143, 1994.
24. K. Nasmyth and T. Hunt: Dams and sluices. *Nature* 366:634, 1993.
25. J. Gyuris, E. Golemis, H. Chertkov, and R. Brent: Cdi1, a human G1 and S phase protein phosphatase that associates with Cdk2. *Cell* 75:791, 1993.
26. J. W. Harper, G. R. Adami, N. Wei, K. Keyomarsi,

and S. J. Elledge: The p21 Cdk-interacting protein Cip1 is a potent inhibitor of G1 cyclin-dependent kinases. *Cell* 75:805, 1993.
27. M. Serrano, G. J. Hannon, and D. Beach: A new regulatory motif in cell-cycle control causing specific inhibition of cyclin D/CDK4. *Nature* 366:704, 1993.
28. A. Kamb, N. A. Gruis, J. Weaver-Feldhaus, Q. Liu, K. Harshman, S. V. Tavtigian, E. Stockert, R. S. Day, III, B. E. Johnson, and M. H. Skolnick: A cell cycle regulator potentially involved in genesis of many tumor types. *Science* 264:436, 1994.
29. T. Noborl, K. Miura, D. J. Wu, A. Lois, K. Takabayashi, and D. A. Carson: Deletions of the cyclin-dependent kinase-4 inhibitor gene in multiple human cancers. *Nature* 368:753, 1994.
30. S. Moreno and P. Nurse: Substrates for p34cdc2: In vivo veritas? *Cell* 61:549, 1990.
31. A. W. Murray: Creative blocks: Cell-cycle checkpoints and feedback controls. *Nature* 359:599, 1992.
32. J. J. Li and R. J. Deshaies: Exercising self-restraint: Discouraging illicit acts of S and M in eukaryotes. *Cell* 74:223, 1993.
33. L. Hartwell: Defects in a cell cycle checkpoint may be responsible for the genomic instability of cancer cells. *Cell* 71:543, 1992.
34. M. B. Kastan, Q. Zhan, W. S. El-Deiry, F. Carrier, T. Jacks, W. V. Walsh, B. S. Plunkett, B. Vogelstein, and A. J. Fornace, Jr.: A mammalian cell cycle checkpoint pathway utilizing p53 and GADD45 is defective in ataxia-telangiectasia. *Cell* 71:587, 1992.
35. L. Livingstone, A. White, J. Sprouse, E. Livanos, T. Jacks, and T. D. Tlsty: Altered cell cycle arrest and gene amplification potential accompany loss of wild-type p53. *Cell* 70:923, 1992.
36. J. F. R. Kerr: A histochemical study of hypertrophy and ischaemic injury of rat liver with special reference to changes in lysosomes. *J Pathol Bacteriol* 90:419, 1965.
37. J. F. R. Kerr, A. H. Wyllie, and A. R. Currie: Apoptosis: A basic biological phenomenon with wide-ranging implications in tissue kinetics. *Br J Cancer* 26:239, 1972.
38. J. F. R. Kerr and B. V. Harmon: Apoptosis, in L. D. Tomei and F. O. Cope, eds.: *The Molecular Basis of Cell Death.* New York: Cold Spring Harbor Laboratory Press. 1991:5–29.
39. S. J. Martin, D. R. Green, and T. G. Cotter: Dicing with death: Dissecting the components of the apoptosis machinery. *Trends Biochem Sci* 19:26, 1994.
40. D. L. Vaux: Toward an understanding of the molecular mechanisms of physiological cell death. *Proc Natl Acad Sci USA* 90:786, 1993.
41. M. K. L. Collins and A. L. Rivas: The control of apoptosis in mammalian cells. *Trends Biochem Sci* 18:307, 1993.
42. W. D. Jarvis, R. N. Kolesnick, F. A. Fornari, R. S. Traylor, D. A. Gewirtz, and S. Grant: Induction of apoptotic DNA damage and cell death by activation of the sphingomyelin pathway. *Proc Natl Acad Sci USA* 91:73, 1994.
43. G. T. Williams and C. A. Smith: Molecular regulation of apoptosis: Genetic controls on cell death. *Cell* 74:777, 1993.
44. E. Rouvier, M.-F. Luciani, and P. Golstein: Fas involvement in Ca(2+)-independent T cell-mediated cytotoxicity. *J Exp Med* 177:195, 1993.
45. M. O. Hengartner, R. E. Ellis, and H. R. Horvitz: *Caenorhabditis elegans* gene ced-9 protects cells from programmed cell death. *Nature* 356:494, 1992.
46. S. J. Korsmeyer: Bcl-2: A repressor of lymphocyte death. *Immunol Today* 13:285, 1992.
47. R. P. Bissonnette, F. Echeverri, A. Mahboubi, and D. R. Green: Apoptotic cell death induced by c-*myc* is inhibited by *bcl*-2. *Nature* 359:552, 1992.
48. C. L. Sentman, J. R. Shutter, D. Hockenbery, O. Kanagawa, and S. J. Korsmeyer: Bcl-2 inhibits multiple forms of apoptosis but not negative selection in thymocytes. *Cell* 67:879, 1991.
49. L. H. Boise, M. González-Garcia, C. E. Postema, L. Ding, T. Lindsten, L. A. Turka, X. Mao, G. Nunez, and C. B. Thompson: *bcl*-x, a bcl-2-related gene that functions as a dominant regulator of apoptotic cell death. *Cell* 74:597, 1993.
50. R. J. Smeyne, M. Vendrell, M. Hayward, S. J. Baker, G. G. Miao, K. Schilling, L. M. Robertson, T. Curran, and J. I. Morgan: Continuous c-*fos* expression precedes programmed cell death in vivo. *Nature* 363:166, 1993.
51. V. Gagliardini, P.-A. Fernandez, R. K. K. Lee, H. C. A. Drexler, R. J. Rotello, M. C. Fishman, and J. Yuan: Prevention of vertebrate neuronal death by the crmA gene. *Science* 263:826, 1994.

11

Biology of Tumor Metastasis

PHASES OF TUMOR GROWTH IN VIVO

In humans, the earliest detectable malignant lesions are often referred to as in situ cancers (Fig. 11-1). These are small tumors (usually only a few millimeters in diameter) that are localized in tissues. They are usually detected only if they can be directly visualized, for instance, as in the case of carcinoma in situ of the uterine cervix, urinary bladder, or skin. At this stage, the tumor is usually avascular, lacking its own network of blood vessels to supply oxygen and nutrients. The latter are provided primarily by diffusion, and this limitation results in slow growth of the tumor. As a result, these lesions may remain dormant for several years. The critical events that trigger the conversion of a dormant tumor into a more rapidly growing invasive neoplasm are not well understood, but this conversion is associated with the vascularization of tumors, stimulated by tumor angiogenesis factors (see Chap. 9). The vascularized tumor begins to grow more rapidly. It compresses surrounding tissue, invades through basement membranes, and metastasizes. Metastasis occurs early for some tumors (e.g., melanoma, small-cell carcinoma of the lung) and late for others (e.g., some thyroid carcinomas). Metastatic potential is determined by the invasiveness of a given tumor, and for some tumors it reflects the size of the primary tumor and the duration of tumor progression (number of population doublings). Vascularization appears to contribute to tumor progression, since the increased supply of nutrients and the resulting increased number of proliferating cells favor the propagation of more aggressively growing cells and the appearance of new subclones of cells with a more malignant phenotype.

The progression of growth of a human solid tumor is shown schematically in Figure 11-2. Because the ordinate (number of cancer cells) is a log scale, the magnitude of the changes that occur after vascularization of the tumor is somewhat deceiving. The dormant phase of growth, which may go on for several years, achieves a diameter of only a few millimeters (10^6 cells, 1-mg mass). Once vascularization occurs, the growth becomes more rapid, and a clinically detectable tumor (10^9 cells, 1-g mass) may be achieved within a few months or years, depending on the cell type. A tumor of this size would be about 1 cm in diameter and be just within the realm of detection by sensitive diagnostic methods. In other words, the patient will have a tumor burden of about 1 billion cells before it can be diagnosed clinically. Within another few years, a tumor that is not treated could approach a potentially lethal tumor burden (10^{12} cells, 1-kg mass). If one thinks of tumor growth as represented in Figure 11-2 in terms of the number

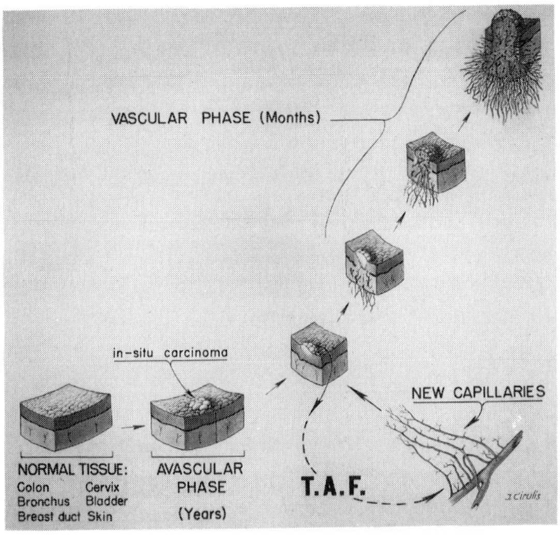

Figure 11-1 Growth phases of a carcinoma. A tumor originating in normal epithelium is separated from the vascular system by the basement membrane separating the epithelium from the underlying connective tissue stroma. Such in situ carcinomas can remain dormant for years. During tumor progression, more aggressively growing cancer cells evolve. These cells may evoke a response from the host that involves invasion of the tumor by immune lymphocytes, macrophages, and polymorphonuclear leukocytes. Vascularization of the tumor can occur by direct invasion through the basement membrane and into small blood vessels and by release of tumor angiogenesis factors by the tumor and/or by lymphocytes and macrophages in the tumor bed. Once the tumor becomes vascularized, it grows more rapidly and can metastasize. (From Folkman.[1])

of cancer cell doublings, by the time a cancer is detected clinically it will have already gone through approximately two-thirds of its lifetime, with about 30 cell population doublings.[2] If tumor growth is unchecked, five more population doublings would produce a cancer of about 32 g that would be about 4 cm in diameter if it were all in one solid sphere. By five more doublings, a tumor of 1-kg mass (10^{12} cells) would be reached. The growth curve depicted in Figure 11-2 resembles a typical Gompertzian growth curve for cells growing in culture, in that it has a lag phase of relatively slow cell proliferation, a logarithmic growth phase of rapid cell doubling, and a phase of slow growth, eventually reaching a steady state of cell proliferation and cell loss. Similar growth kinetics are seen for some tumors transplanted into experimental animals.[3] For many human cancers, however, the growth kinetics depicted in Figure 11-2 are only an approximation. For this kind of growth kinetics to occur, there would have to be virtually no cell loss as a tumor progressed through its logarithmic phase of growth. This is most likely not true for many human tumors, because a significant amount of cell loss due to cell death and/or exfoliation from the tumor usually occurs during growth. Limitation of nutrients in the central, less well vascularized areas of tumors and host defense mechanisms also play a role in tumor cell loss. In addition, some human tumors have short mass doubling times (e.g., small-cell carcinoma of the lung, embryonal carcinoma of the testis), whereas others have much longer mass doubling times (e.g., carcinomas of the breast and colon). Some tumors are better vascularized than others. Some will become lethal more rapidly because they invade surrounding tissues and metastasize very early in their development.

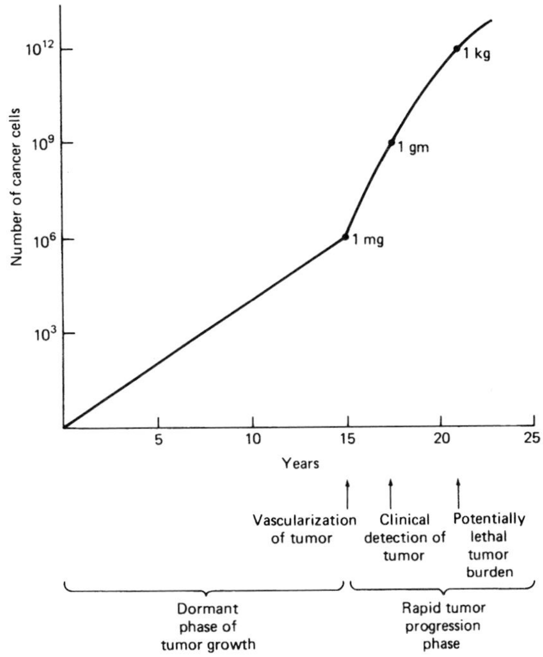

Figure 11-2 Theoretical growth curve of a human carcinoma. The dormant phase of growth may occur over many years, leading eventually to a more aggressive, invasive tumor. This is followed by a more rapid growth phase. Tumors are generally not clinically detectable until they reach a diameter of about 1 cm ($\cong 10^{9}$ cells). A tumor burden of 10^{12} cells is approaching lethality.

Some will grow in areas limited by fixed anatomic boundaries and be compressed or contained, and others will grow in critical areas and be lethal before they get very large because of their early compromising of critical functions (e.g., brain tumors). Nor do these kinetics hold for most leukemias or a number of lymphomas. For example, the onset and course of acute lymphocytic leukemia in children is much more rapid, occurring frequently in early childhood and becoming lethal within 6 months to 1 year if untreated. Similarly, the growth kinetics of Burkitt's lymphoma reflect a much more rapid growth. The mass doubling time of Burkitt's lymphoma is 1 to 2 days, as opposed to about 50 days for a typical breast carcinoma.[4] One point is clear, however, and that is that most human tumors are relatively far advanced by the time of diagnosis, and about 50% of patients have metastatic spread by the time their cancer is clinically detected.[2]

INVASION AND METASTASIS: THE HALLMARKS OF MALIGNANT NEOPLASIA

The steps involved in the invasion and metastatic spread of cancer cells are illustrated in Figure 11-3. Tumor cells can spread by direct extension into a body cavity, such as the pleural or peritoneal space, or the cerebrospinal fluid. In these cases, tumor cells released into the body space can seed out onto tissue surfaces and develop new growths where they become embedded. Examples of cancers that spread in this way are lung cancers that enter the pleural cavity, ovarian cancers that shed cells into the peritoneal cavity, and brain tumors that shed cells into the cerebrospinal fluid. Tumor cells metastasize by invading blood vessels or lymphatic channels. Although it has frequently been said that carcinomas metastasize primarily through the lymphatic system and sarcomas through the blood vessels, this distinction is somewhat arbitrary, since the

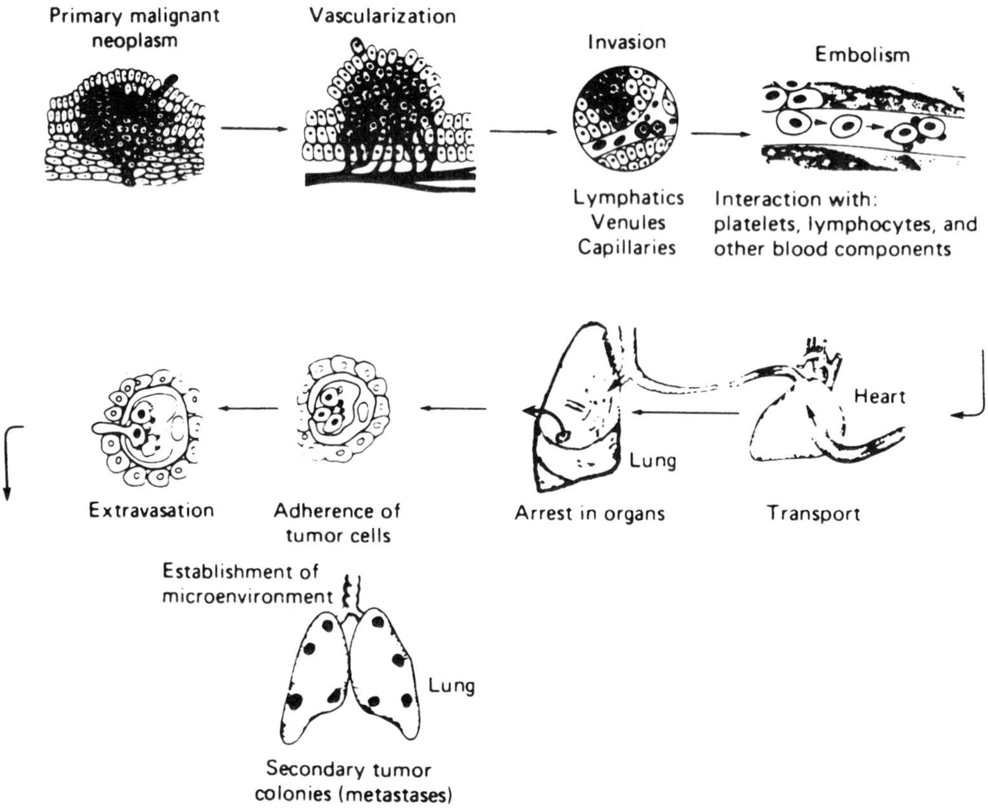

Figure 11-3 Steps involved in cancer metastasis. (From Fidler.[5])

blood and lymph systems communicate freely, and it has been shown that cancer cells that invade lymphatic channels enter the bloodstream and vice versa.[6] Capillaries, venules, and lymph vessels offer little resistance to penetration by tumor cells because of their thin walls and relatively "loose" intercellular junctions. Arteries and arterioles, on the other hand, are surrounded by dense connective tissue sheaths made up of collagen and elastic fibers and hence are rarely invaded by tumor cells.

The mechanisms for invasion of tumor cells through tissue barriers and into blood and lymphatic vessels are not well understood, but they appear to involve both mechanical and enzymatic processes. As a tumor grows, the pressure exerted on surrounding tissue tend to force cells between intercellular spaces. It is unlikely that this process, in itself, could explain the penetration of cancer cells through tissue barriers such as basement membranes. For this to occur, the release of certain degradative enzymes appears to be necessary. Indeed, as discussed below, tumors are known to contain and secrete a variety of proteolytic enzymes that may be involved in this step. Some of the lytic enzymes found in high concentration in tumor fluids are listed in Table 11-1. The enzyme activities released by growing tumors destroy surrounding cells and degrade tissue barriers, allowing tumor cells to penetrate.

After tumor cells invade the lymphatic or vascular vessels, they may form a local embolus by interaction with other tumor cells and blood cells and by stimulating fibrin deposition. Individual cells or clumps of cells are then shed from these sites and spread to distant organs by the lymph or blood vessels. Tumor cells that enter the lymphatic system travel to regional lymph nodes, in which some tumor cells may be trapped and produce a metastatic growth. However, all the tumor cells are not necessarily trapped or "filtered out" in the first few lymph nodes draining an area of tissue containing a tumor. Fisher and Fisher[8] found that some tumor cells are not trapped in the first groups of lymph nodes and can thus reach the circulatory system early in the course of tumor cell dissemination. The presence of tumor cells in the blood does not invariably mean that distant metastases will form. The vast majority of circulating tumor cells shed from solid tumors do not survive in the blood, and only about 0.1% live long enough to form a distant metastasis.[9]

During circulation in the vascular system, tumor cells can undergo a variety of interactions, including aggregation with platelets, lymphocytes, and neutrophils, leading to the formation of emboli that can become lodged in the capillary bed of a distant organ. These clumps of cells adhere to the capillary endothelium and elicit the formation of a fibrin matrix that appears to favor the survival of the cancer cells. A number of years ago, Wood and coworkers[10] showed that adherence of cancer cells to capillary endothelium and subsequent thrombus formation are involved in metastasis. They injected V_2 carcinoma cells (a type of rabbit carcinoma) stained with the dye Trypan Blue into small arteries in a rabbit's ear in which a chamber had been inserted to visualize the capillary endothelium, even though the blood was flowing briskly. Within 30 minutes, a thrombus formed around the cancer cells. By 24 hours, the cancer cells began to di-

Table 11-1 Relative Activities of Some Lytic Enzymes in Tumor Fluids[a]

	Relative Activities					
Fluid Assayed	Dipeptidase	Arginase	Acid Phosphatase	β-Glucuronidase	Aryl Sulfatase	Cathepsin
Normal mouse plasma	0.2	0.1	6	1	2	5
Normal mouse intraperitoneal fluid	1.3	0.2	1	1	1	12
Peripheral tumor fluid	20	2	21	3	5	15
Central tumor fluid	24	2	43	9	11	63

[a]Activities are based on rates of hydrolysis of substrates in a standardized assay. Tumor data include interstitial fluid from unicentric transplants of sarcoma 37, MCIM rhabdomyosarcoma, methylcholanthrene-induced sarcoma, the Harding-Passey melanoma, and mouse mammary carcinomas.

Source: Data from Sylvén.[7]

vide. By 48 hours, invasion of the endothelium was apparent, and by 72 hours, a tumor metastasis was established. Subsequent experiments indicated that fibrinolytic agents that lysed tumor-cell-containing thrombi reduced metastasis formation.

It has been experimentally demonstrated that both the size of a circulating tumor-cell-containing embolus and its "deformability" during passage through capillary beds are related to the formation of metastatic foci in a given tissue[11,12]; however, these factors do not solely determine the site of localization of a metastasis. Substantial evidence indicates that there is a tissue "tropism" for some circulating cancer cells. For example, the distribution of metastasis after an injection of certain kinds of tumor cells reflects a preference for certain tissue sites rather than just the first capillary system encountered by the tumor cells. This suggests that there are specific adhesive interactions between circulating tumor cells and cells of given host tissues.

The adhesion of tumor cells to capillary endothelium in susceptible organs appears to damage the vessel walls and to lead to the accumulation of neutrophils that may penetrate the spaces between endothelial cells and open up a channel through which tumor cells can also penetrate.[13] Moreover, platelets that aggregate at the site of the thrombus release mediators, such as histamine, that promote capillary permeability, allowing the migration of tumor cells through the endothelium. The role of platelets in this process is implied from several lines of evidence.[14,15] Many murine tumors aggregate platelets in vitro and in vivo. Addition of fibroblasts to a tumor cell inoculum enhances platelet aggregation and the number of metastases, whereas induction of thrombocytopenia in the host animal or treatment with aspirin, at doses that decrease platelet aggregation, decrease tumor metastases. Aggregation of platelets and release of their contents can be induced by a number of factors, including collagen, thrombin, and arachidonic acid. Platelets accumulate in areas of endothelial cell regeneration following trauma, and platelet-released factors have a mitogenic effect on a number of different cell types, including endothelial cells (see Chap. 9). Elastase and collagenase are released from platelets, and this alters the connective tissue of the vessel wall.

Platelet aggregation also produces an increase in serum thrombin, which, in turn, increases the amount of fibrin deposited on the endothelial wall. This deposition of fibrin stimulates the release of plasminogen activator from neutrophils, macrophages, and other cells to induce fibrinolysis through plasmin, thus generating more proteolytic activity in the area of the tumor thrombus. Once tumor cells migrate through the vascular wall, they quickly establish themselves in the new environment and begin to proliferate. This is fostered by the release of angiogenesis factors from tumor cells or host lymphocytes and macrophages that promotes vascularization of the nidus of tumor cells. In the presence of platelets or platelet-released factors, the mitogenic activity of angiogenesis factors for endothelial cells growing on a collagen substratum is greatly enhanced.[15] Thus, the local aggregation of platelets in the area of a tumor-cell-containing thrombus activates a whole cascade of events that can promote the extravasation and new growth of tumor cells at a metastatic site.

Somewhat paradoxically, the presence of immune lymphocytes that recognize tumor cells may enhance the colonization of metastatic sites. For example, the metastatic potential of a mixture of intravenously injected murine melanoma cells and immune lymphocytes is enhanced if the ratio of immune lymphocytes to melanoma cells is about 1000:1, but if the ratio is 5000:1, a reduction in metastases is observed.[16] The lower number of sensitized lymphocytes may favor the formation of tumor thrombi that are necessary for extravasation; the higher number leads to killing of the majority of tumor cells. In other words, as far as the tumor is concerned, "a little immunity is a good thing." The ability of cancer cells to take advantage of the host's own inflammatory response mechanisms, and at the same time avoid destruction by the host's immunologic defense system, gives the cancer cells a tremendous selective advantage.

Metastasis Is at Least Partly a Selective Process

Clinical observations on the pathogenesis of cancer metastases have shown that there is a tendency for primary tumors arising in a given organ to metastasize to particular distant sites (Table 11-2). Some of this is explainable by the nature

Table 11-2 Most Frequent Sites of Metastasis for Some Human Carcinomas

Site of Primary Tumor	Most Common Sites of Metastasis
Breast	Axillary lymph nodes, opposite breast through lymphatics, lung, pleura, liver, bone, brain, adrenal, spleen, and ovary
Colon	Regional lymph nodes, liver, lung, by direct extension to urinary bladder or stomach
Kidney	Lung, liver, and bone
Lung	Regional lymph nodes, pleura, diaphragm (by direct extension), liver, bone, brain, kidney, adrenal, thyroid, and spleen
Ovary	Peritoneum, regional lymph nodes, lung, and liver
Prostate	Bones of spine and pelvis, regional lymph nodes
Stomach	Regional lymph nodes, liver, lung, and bone
Testis	Regional lymph nodes, lung, and liver
Urinary bladder	By direct extension to rectum, colon, prostate, ureter, vagina, bone; regional lymph nodes; bone; lung; peritoneum; pleura; liver; and brain
Uterine endometrium	Regional lymph nodes, lung, liver, and ovary

Source: Data from Anderson[17] and Rubin.[18]

of the venous and lymphatic drainage of a given tissue and by the presence of natural channels created by fascial planes and nerve sheaths. In addition, as already mentioned, tumors impinging on body cavities, such as the pleural or peritoneal space, can spread by shedding tumor cells directly into the cavity. For example, many carcinomas—such as those of the breast, stomach, colon, and lung—metastasize most frequently to regional lymph nodes, but they also metastasize to certain distant organs more frequently than to others. In general, the lungs and the liver are the most common sites of visceral metastases because of their large bulk and abundant blood supply. However, a tendency of some carcinomas to metastasize to certain other tissues is also seen. Adenocarcinomas of the breast often metastasize to lungs, liver, bones, adrenals, and ovaries. Stomach carcinomas spread to liver, lungs, and bone, and, by direct extension, into the peritoneal cavity. Lung cancers frequently metastasize to brain. Carcinomas of the prostate have a predilection for metastasis to bones of the spine. In this latter case, the fact that venules draining the vertebral column anastomose with those draining the prostate explain how some tumor cells could directly reach the bones of the spine. Such clinical observations led Paget[19] to propose the "seed and soil" hypothesis, which states that certain cancer cells ("seeds") will grow readily in certain tissues ("soil") but not in others.

Studies in experimental animals have also suggested a propensity for certain types of cancer cells to metastasize to certain organs. Murine thymomas, plasmacytomas, melanomas, fibrosarcomas, and histiocytomas have been demonstrated to have some organ specificity to the pattern of their metastatic spread (for review, see Ref. 5). Thus, both clinical and experimental animal studies indicate that the metastatic spread of cancer cells is not a random event but reflects properties of individual host tissues and, possibly, of the circulating malignant cells themselves.

Fidler[20] has developed a model system with which to determine whether tumor cells themselves can choose the site of metastatic spread and whether cells of a given malignant neoplasm have the same or different metastatic potential (Fig. 11-4). He injected a B16 mouse melanoma line (B16-F0) intravenously into syngeneic C57 BL/6 mice and 2 to 3 weeks later removed colonies of melanoma cells growing in the lungs. The melanoma cells were dissected free of lung tissue and established in tissue culture as a line called B16-F1. Cells from this cell culture line were then injected into C57 BL/6 mice, and 3 weeks later the lung colonies produced by these cells were again removed and cultured to yield B16-F2. This procedure was continued until a line of cells, designated B16-F10, that produced significantly more lung tumors per input cell than the B16-F1 line after either intravenous or intracardiac (left ventricle) injection, was obtained.[22] The increased number of lung colonies per input cell for the F10 line was not due to nonspecific trapping in the first capillary bed encountered, since injection into the left ventricle of the same number of cells (radioactively labeled with [^{125}I]iododeoxyuridine) of the F1 and F10 cell lines resulted in equivalent numbers of cells localized in the lung 2 minutes later, but by 2 weeks postinjection, eightfold more metastatic colonies were found in the lungs of animals injected with F10 cells than in those that received F1 cells. Since cells injected into the left ventricle had to pass through the capillary beds of

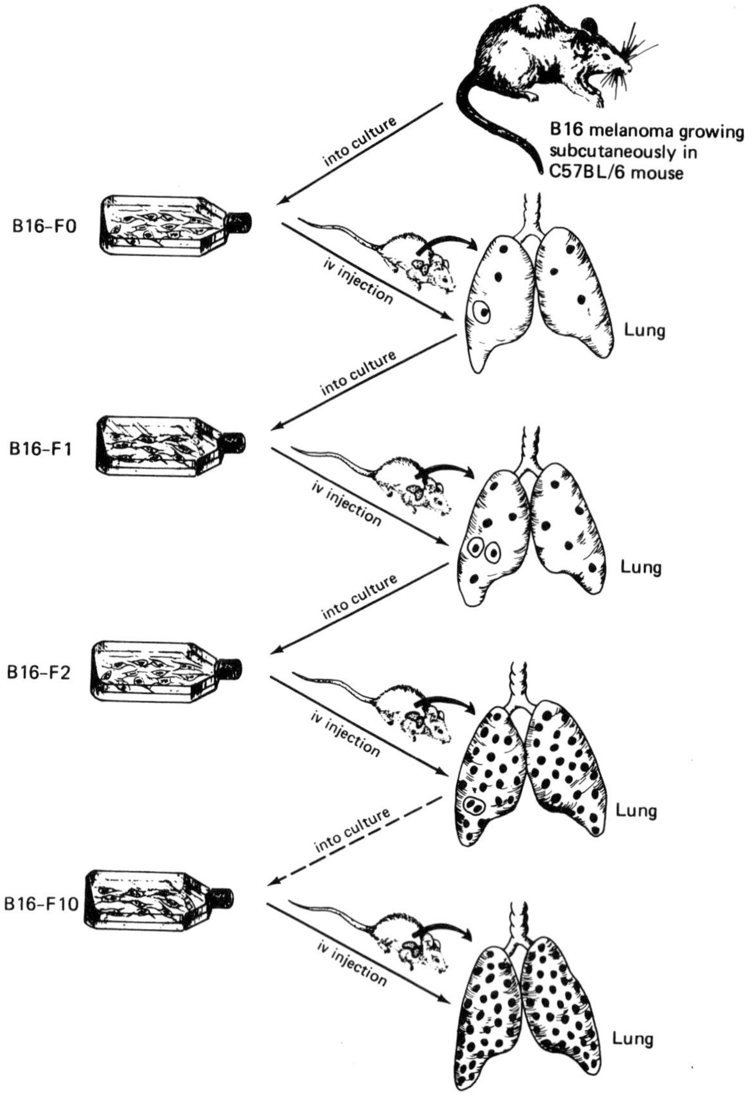

Figure 11-4 Selection of a highly metastatic line of mouse B16 melanoma. C57 BL/6 mice were injected intravenously with the cultured parent line (F0) of B16 melanoma cells. About 2 weeks later, lung colonies were isolated and placed into cell culture dishes (F1 line). When these cells grew to a sufficient number, they were again injected intravenously into mice, and after a few weeks, lung colonies were again isolated and placed in culture. This procedure was carried out several times until a line of highly metastatic, "lung-seeking," B16 cells (F10 line) was isolated. (From Fidler.[21])

other tissues before they reached the lung, the ability of F10 cells to seed out and grow in the lung selectively must be due to something other than nonspecific trapping. Moreover, the observation that B16-F10 formed primarily lung colonies after left ventricular injection whereas the F1 line produced both pulmonary and extrapulmonary colonies, indicated that the invasion and growth of the F10 cells were organ-site specific. Nicolson and his colleagues[23,24] have also selected B16 melanoma lines that selectively colonize brain, adrenal, or ovary. Taken together, these data suggest that at least part of the organ-site specificity of cancer metastases is determined by the cancer cells themselves.

Fidler and Kripke,[25] in experiments in which the parent B16-F1 line was cloned in cell culture prior to injection into syngeneic mice, determined that individual cells of a given cancer have different metastatic potentials. The original uncloned F1 line produced similar numbers of metastases in different animals, but the cloned sublines differed markedly, with some clones producing a very high number of lung metastases and some very few. This indicates that a highly metastatic population of cells preexists in the parent melanoma line. Similar marked heterogeneity in metastatic potential has been found in cloned lines of an ultraviolet-induced fibrosarcoma of the C3H mouse, a murine mammary tumor, a methylcholanthrene-induced fibrosarcoma, a murine sarcoma virus-induced fibrosarcoma, and a transformed rat epithelial cell line (for review, see Ref. 26). Thus, the data indicate that primary malignant tumors contain subpopulations of cells with vastly differing metastatic potential. This adds another important parameter to the list of phenotypic characteristics for which malignant neoplasms are heterogeneous. Both human and animal neoplasms also display heterogeneity in their growth rate, metabolic characteristics, immunogenicity, and sensitivity to irradiation and cytotoxic drugs (for review, see Ref. 26). The development of resistance to therapy by human cancers probably reflects this heterogeneity. Moreover, because it is the metastatic cells that ultimately prove fatal to the patient, directing studies toward defining the specific biochemical characteristics of metastatic cells would seem to be one of the more promising ways to design anticancer therapies that will cure patients of their disease.

Biochemical Characteristics of Metastatic Tumor Cells

There is strong evidence that the development of metastatic clones of cells within a tumor cell population results from genetic instability and selection of cells that have some selective advantage for invasion and metastasis rather than from an adoptive process by which all the cells in a tumor population gradually acquire the ability to metastasize.[27] Moreover, when highly metastatic clones of cells are isolated from a heterogeneous tumor cell population, they are genetically unstable and can, with continual passage in vitro or in vivo, produce more metastatic as well as less metastatic subpopulations of cells.[28] In other words, the metastatic phenotype is itself unstable. Highly metastatic clones, for example, have been shown to have about a fivefold greater rate of spontaneous mutation to ouabain and 6-thioguanine resistance than poorly metastatic clones of several rodent tumors.[28] Moreover, in one study it was shown that the B16-F1 and B16-F10 mouse melanoma lines do not differ as much in their inherent metastatic potential as in the rate at which they generate metastatic cells during passage in culture;[29] this finding supports the concept that cells with a higher metastatic potential are those with the greatest genetic instability and thus the greatest adaptability to a new environment.

An implication of these findings is that even though tumors may be of monoclonal origin, and metastasis may initially arise from one clone of cells[30,31], diversity and heterogeneity are likely to arise quickly within a metastatic tumor population. This is discouraging because it decreases the possibility of finding consistent phenotypic characteristics that would lend themselves to a single chemotherapeutic or immunotherapeutic approach to attacking the most dangerous cells in a cancer patient—namely, those with the highest metastatic potential. Nevertheless, metastatic cells must share certain biochemical characteristics, among which are the ability to invade through basement membranes, to evade the host's immune defenses, to attach to endothelial surfaces, to extravasate into the tissue parenchyma at a distant site (probably by means of similar mechanisms involved in initial invasion), and to elicit development of a vascular network in their new home. Thus, determination of which biochemical parameters are associated with the ability to do these things could still provide a common point for therapeutic attack. Some of the candidates for this commonality are discussed next.

Relationship of Cancer Metastasis to Normal Tissue Invasion Events

It is important to keep in mind that malignant cells use some of the same tissue degradative and invasive mechanisms that are utilized by normal processes such as cell migration and tissue remodeling in embryonic development, wound healing, and trophoblast invasion of the uterine wall during normal pregnancy. The last of these events can be used as an example.

If a successful implantation of the blastocyst into the uterine wall is to occur in pregnancy, trophoblast cells must cross the basement membranes of the uterine epithelium and vasculature. Several lytic enzymes, proteolytic enzyme regulators, and growth factors are involved in this process (reviewed in Ref. 32). For example, the production of protease of the urokinase-type plasminogen activator (u-PA) by murine trophoblasts coincides with invasion of the mouse blastocyst into the uterine wall. Similarly, human trophoblast cells express u-PA receptors and bind u-PA. Metalloproteases such as stromelysin and the 92-kDa form of type IV collagenase are also produced by trophoblast cells, as are tissue inhibitors of metalloproteases (TIMPS).

The difference between normal trophoblast invasion of the uterine wall and cancer cell metastasis, however, is that the former is a tightly regulated process with stringent termination signals. For example, production of TIMP at the time of termination of implantation shuts off metalloprotease activity. Expression of TGF-β is activated at this time, and it induced differention of cytotrophoblasts into syncytiotrophoblasts, which are noninvasive. TGF-β also induces expression of TIMP and an inhibitor of u-PA. An intriguing possibility is that there is a feedback loop here, in that production of u-PA by the invading trophoblast could release TGF-β from the extracellular matrix of the uterine epithelium, which, in turn, activates the production of u-PA inhibitors and TIMP to terminate the invasive process. Interestingly, TGF-β actually appears to stimulate growth of advanced, metastatic melanoma cells instead of inhibiting their proliferation, as is seen for normal melanocytes or early melanoma lesions, suggesting that metastatic cells inappropriately respond to negative signals.

Role of Lytic Enzymes in the Metastasis Cascade

In order for cancer cells to carry out a successful metastasis, a group of cells within the primary tumor must invade through the host tissue cells and extracellular matrix (ECM), enter the circulation, arrest at a distant vascular bed, extravasate into the target organ's ECM and interstitium, proliferate as a new colony, and induce a new blood supply (reviewed in Ref. 33). A number of these steps require the release of lytic enzymes.

The invasive and metastatic potential of cancer cells has been correlated in a number of studies with the activity of various proteases, including serine proteases such as plasmin (activated by plasminogen activator), thiol proteases such as the cathepsins, and metalloproteases such as type IV collagenase. These proteolytic activates do not go unabated in tissues, even tumor tissues, because there are a number of tissue protease inhibitors that keep them in check under normal conditions. Proteases, as noted above, are needed for a number of natural processes such as normal tissue repair, tissue remodeling during development, and implantation of the blastocyst and growth of the placenta during normal pregnancy. In these instances, as opposed to highly malignant tumors, the proteases and antiproteases are kept in a tightly regulated balance, the mecahnisms for which are not entirely clear but probably involve the local release of growth factors, feedback from the ECM, etc. For example, metalloproteases are induced by interleukin-1, EGF, and PDGF[34], whereas TGF-β has been shown to induce the production of plasminogen activator inhibitor type 1 and to decrease the degradation of the ECM by human fibrosarcoma cells in culture.[35] Thus, normally, there is a stringently regulated process that controls the release of proteases and then inactivates them once they have done their job. Tumor cells of the metastatic variety have lost or do not respond to this control mechanism.

Another point that is clear is that individual proteases do not act alone in the metastatic process but rather are part of a cascade of lytic activity. For instance, plasminogen activator activates plasmin, which, in turn, can activate type IV collagenase. Of the plasminogen activators, the urokinase type (u-PA) has been most closely linked to the metastatic phenotype (reviewed in Ref. 33). Several studies also support an important role for type IV collagenase in tumor metastasis. Moreover, benign proliferative lesions of the breast, benign polyps of the colon, as well as normal colon and gastric mucosa have low levels of a 72-kDa form of type IV collagenase, but their invasive counterparts express high levels of this enzyme. Also, type IV collagenolytic activity can be inhibited by retinoic acid, and this correlates with loss of the invasive phenotype in cultured human melanoma cells.[36]

Another important concept for understanding the biology of tumor metastasis is the interaction of cancer cells with the surrounding stroma on which they grow. There is cross-talk among the cancer cells, the ECM, and the supporting stroma. As an epithelial tumor grows and breaches the ECM, the tumor cells come into contact with the fibroblasts and other mesenchymal cells in the supporting stroma. Via production and secretion of various growth factors and cytokines and interaction among tumor cells, stromal cells, and ECM components, the process of invasion and metastasis goes on. This is also part of the process by which tumors become vascularized. They secrete tumor angiogenesis factors that induce the growth of vascular endothelial channels through the stroma and ECM to reach the tumor; this appears to be the time in the life cycle of a malignant neoplasm when it undergoes a spurt of growth and becomes more aggressive.[37]

Plasminogen activators (PAs) of either the u-PA or tissue (t-PA) type are neutral serine proteases whose primary proteolytic activity is to convert the zymogen plasminogen into plasmin, which is a "nondiscriminate" protease that degrades a number of ECM components including fibronectin, laminin, and type IV collagen and activates other matrix metalloproteases. This suggests that PAs have a pivotal role in activating a hydrolytic cascade that is capable of attacking the ECM (reviewed in Ref. 38). The u-PA type of PA has been most closely linked to the metastatic phenotype, and antibodies to u-PA can block human hepatoma cell invasion in the chick chorioallantoic membrane assay as well as mouse melanoma B16F10 metastasis following tail-vein injection.[33] High levels of u-Pa and its inhibitor PAI-1 have been found in cytosolic extracts of human breast carcinomas and they correlate with poor prognosis.[39]

The cathepsins are a family of cysteine proteases that also appear to be involved in the metastatic process. Cathepsin B activity is elevated in a variety of human and animal tumors and is found at higher levels in metastatic as opposed to nonmetastatic B16 melanoma cells.[40] Cathepsin L is expressed at higher levels in a wide variety of human cancers than in their normal counterpart tissues.[41]

Cathepsin B mRNA levels are elevated in human colorectal carcinoma tissue in a tumor stage–specific way.[42] The increased cathepsin B gene expression was found at a time when colorectal tumors were in the process of invading the bowel wall. Cathepsin D levels in breast cancer tissue have been correlated with poor prognosis,[43] but examination of breast cancer cell lines did not show a correlation with their in vitro metastatic potential, suggesting that the high tumor tissue levels of cathepsin D are due to the stromal components of the tumor, such as infiltrating inflammatory cells, rather than the cancer cells themselves.[44]

Chambon and colleagues[45] have isolated a gene of the ECM-degrading metalloprotease family, called stromelysin-3 (ST3), which is expressed in stromal cells of invasive breast carcinomas but not in less advanced in situ breast carcinomas. Furthermore, the fact that the gene is expressed at high levels in the stromal cells of invasive breast carcinomas but not in carcinoma cells themselves suggests that release of a factor from the carcinoma cells induces the expression of the stromelysin-3 gene and that this event is related to tumor progression. Thus, the products of the stromelysin gene family are an important potential target for breast cancer therapy.

The ST3 gene was expressed in all of a series of invasive breast carcinomas, in a number of their metastases, and in some in situ breast carcinomas.[46] ST3 mRNA and protein were found in the fibroblastic cells surrounding the neoplastic cells in a pattern similar to that of u-PA, suggesting that these degradative enzymes cooperate during tumor progression. This pattern of expression differed from that of type IV collagenase, which was found expressed in fibroblasts throughout the stroma of primary breast carcinomas and in fibroadenomas, the latter of which did not contain ST3 transcripts.[46] A correlation has also been found between ST3 mRNA levels and local invasiveness of head and neck carcinomas.[47]

There is a lot of information linking type IV collagenase to tumor progression and metastasis.

Type IV collagenase comes in two varieties: a 72-kDa form originally purified from a metastatic mouse tumor and a 92-kDa form identified originally in neutrophils (reviewed in Ref. 48). Both enzymes cleave type IV collagen at a single site, and they also degrade elastin and gelatin. As it turns out, the 72- and 92-kDa type IV colalgenases are two of at least 11 members of the matrix metalloprotease (MMP) family that also includes interstitial collagenase (MMP-1), stromelysin-1 (MMP-3), stromelysin-3 (MMP-11), as well as the 72-kDa collagenase (MMP-2) and the 92-kDa collagenase (MMP-9). The 72-kDa and/or the 92-kDa collagenase have been found at elevated levels in melanoma[48] and in cancers of human skin,[49] colon,[50] breast,[51] lung,[52] prostate,[53] and bladder.[54] In most of these cases, the elevated levels correlated with a higher tumor grade and invasion. A factor produced by mouse and human cancer cell lines stimulates production of MMP-1, MMP-2, and MMP-3 by human fibroblasts, indicating a way in which human cancer cells could induce these lytic enzymes in stromal tissues with which they come into contact.[55] MMP-2 levels are significantly elevated in the serum of patients with metastatic lung cancer; in those with high levels, response to chemotherapy was poor.[56]

Some direct evidence for a role of MMP-9 in tumor metastasis comes from experiments in which MMP-9 expression was induced in tumorigenic but nonmetastatic rat embryo cells by transfection with an MMP-9 expression vector.[56a] The MMP-9–transfected cells became metastatic, and cells that lost MMP-9 expression with continual passage in culture also lost their ability to metastasize.

Collagenases are not unique to tumor cells, being produced also by inflammatory cells and by normal involuting epithelial duct cells. However, there is evidence to indicate that transplanted animal tumors can produce collagenases in vivo as well as in vitro. Thus, both tumor cells themselves and infiltrating inflammatory cells may contribute to the destruction of the basement membrane in the area of invasive tumor growth. Collagenase activity of malignant tumor tissue is consistently higher than that of corresponding benign tumors, and within a series of cell lines of variable metastatic potential, there is a quantitative relationship between the amount of type IV collagenase activity and the degree of metastasis.[57]

Tissues in the body have mechanisms to limit the amount of basement membrane turnover. During wounding, tissue repair, and various inflammatory conditions, lytic enzymes similar to those produced by invasive tumor cells are released. Various kinds of protease and collagenase inhibitors circulate in the blood and are found in normal tissues. For example, α-2-macroglobulin is a potent circulating inhibitor of collagenase. The extracellular matrix also contains a family of cationic proteins that inhibit collagenase activities. Treatment with collagenase inhibitors has been shown to retard invasion of human tumors in nude mice[58] and of tumor cell invasiveness in an in vitro assay.[59] It is likely, therefore, that in the area of an invasive tumor, the normal balance between collagenases and anticollagenase activity is lost, perhaps because the tumor locally overwhelms the available anticollagenase activity, because invasive tumor cells secrete an anticollagenase inhibitor, or because in the area where tumor cells breach the basement membrane, extracellular matrix components are not laid down in a normal fashion.

An important family of metalloprotease inhibitors found in tissues are the tissue inhibitors of metalloprotease (TIMPs). Two of these have been identified: TIMP-1 and TIMP-2. In some animal models, administration of TIMP-1 inhibits metastasis, and transfection of antisense TIMP-1 RNA induces oncogenicity in murine 3T3 cells (reviewed in Ref. 60). Addition of TIMP-2 to the cell culture medium has been shown to block the invasiveness of human fibrosarcoma cells in an in vitro assay.[61] These and other data, taken together, suggest a therapeutic approach to block tumor cell metastasis by administration of metalloprotease inhibitors or inducers of inhibitor activity.

Other proteases and protease inhibitors have been found in tumor tissue, including tumor-associated trypsinogen-2 (TAT-2) and a corresponding inhibitor (TATI).[62] Undoubtedly, more tumor-associated proteases and tissue protease inhibitors will be found in the future; this whole area is a fertile ground for learning more about the biology of tumor metastasis. Perhaps it will also lead to the discovery of new anticancer drugs.

Role of Cell Surface Components in Metastasis

As discussed in Chapter 4, malignant transformation of mammalian cells is accompanied by multiple changes in the cell surface or plasma membrane of cells. These alterations are morphologic, functional, and immunologic in nature. For instance, changes in membrane fluidity, cell surface ionic charge, lectin-binding affinity, cell permeability and transport mechanisms, intercellular communication, cell surface–associated enzymes and receptors, turnover and shedding of cell surface components, activity of cell surface–associated proteases, cellular adhesion to extracellular matrix, and reactivity with antibodies have all been observed in one or more types of malignantly transformed cells.

While the biochemical basis for all these changes is often not clear, several generalizations can be made. One fact that is clear is that compositional changes in cell surface glycoproteins, glycolipids, and mucins occur, and many of these may help explain the differences between cancer and normal cells noted above. It is becoming increasingly clear that many changes occurring on the cell surface of cancer cells are related to changes in cell surface carbohydrates. One of the commonly observed alterations in cell surface carbohydrates is a shift to higher-molecular-weight glycans caused by one or more of the following: increased branching on the trimannosyl core of asparagine-linked oligosaccharides, increased polylactosaminoglycan chain formation, and increased sialylation (reviewed in Ref. 63). These changes in carbohydrates have also been associated with reduced cellular adhesion to the extracellular matrix and with increased invasiveness and metastatic potential of tumor cells.

Recently, the enzymatic basis for some of these changes has been found. There are several glycosyltransferase activities in mammalian cells that are responsible for the synthesis and processing of oligosaccharides on cell surface glycoproteins and glycolipids. One important family of these are the *N*-acetyl-glucosaminyl transferases. Six of these have been identified: GlcNAc transferases I through VI. The activities of certain of these are altered during transformation of cells by oncogenic viruses or in cancer tissues. Examples include an increase in GlcNAc transferase V in BHK cells transformed with polyomaviruses or Rous sarcoma virus, increases in GlcNAc transferase III in hepatomas, and increases of both GlcNAc III and V in *ras*-transformed NIH 3T3 cells (reviewed in Ref. 63). In addition, alterations in the levels of oligosaccharide-elongating enzymes have been found in virally transformed cells. These changes in enzyme activities appear to explain the increased branching of N-linked oligosaccharides, the elongated polylactosaminoglycan chains, and the increased sialylation of carbohydrates often observed in cancer cells. This, in turn, explains, at least in part, the altered antigenicity of tumor cells and the fact that many monoclonal antibodies made against animal and human tumor cells recognize a carbohydrate-determined epitope. As yet, many of the genes for these carbohydrate synthesis and processing enzymes have not been cloned and their regulatory mechanisms are yet to be clarified. Nor have specific inhibitors been found. Yet these enzymes clearly represent a future target for the development of anticancer drugs. A role of plasma membrane components in organ colonization by metastatic cells has been demonstrated by experiments in which plasma membrane vesicles shed from B16-F10 cells were fused with B16-F1 cells. The results were that the B16-F1 cells with fused membranes from F10 cells were converted from a low to a high lung-colony-forming cell type.[64]

A further indication of the role of cell surface glycoproteins in tissue-specific arrest of cells in vivo was obtained by incubating B16-F1 or B16-F10 cells in culture with tunicamycin or swainsonine, drugs that block formation and maturation of asparagine-linked oligosaccharides, respectively, before injecting them intravenously into mice.[65,66] The drug-treated cells remained viable but formed significantly fewer lung colonies. Furthermore, the tunicamycin-treated cells did not adhere to endothelial cell monolayers in culture, suggesting that cell surface glycoproteins containing asparagine-linked (N-linked) oligosaccharides are required for the tumor–host-cell interactions involved in tissue arrest and metastatic colony formation.

Several lines of evidence indicate a relationship between cell surface sialic acid content and metastatic potential of tumor cells. For 10 cell lines derived from a polyomavirus-induced rat renal sarcoma, the ability of the cell lines to metastasize spontaneously from subcutaneous sites in syngeneic hosts correlated with the degree of sialylation of cell surface glycoconjugates and the platelet-aggregating activity of these glycoconjugates.[67] Moreover, the ability of a wide variety of rodent tumor cell lines—including rat renal sarcoma, rat mammary adenocarcinomas, chemically and virally transformed mouse lines, and B16 melanoma lines—to metastasize from subcutaneous sites correlates with the total sialic acid content and the degree of sialylation of galactose and *N*-acetylgalactosamine residues of cell surface glycoconjugates.[68] The fact that these latter two sugars are found on O-linked (serine or threonine) oligosaccharides of cell surface glycoproteins and on the O-linked GAG chains of proteoglycans suggests that these molecules are also important for tumor cell attachment and metastasis.

Differences in lectin-binding characteristics between low and high metastatic cell lines also suggest an important role for cell surface carbohydrates contained in glycoproteins or glycolipids. In a study of DBA-2 mouse T-cell lymphoma sublines with variable metastatic behavior, it was shown that low metastatic tumor lines expressed receptor sites for soybean agglutinin and *Vicia villosa* lectin, whereas in metastatic lines the respective lectin-binding sites were blocked by sialic acid (shown by treating the cells with neuraminidase), strongly implying that differences in sialylation patterns on the cell surface, which are involved in the masking or unmasking of terminal sugars, influence metastatic potential of tumor cells.[69] Supporting this concept are the findings of Dennis et al.,[70] who have demonstrated that a wheat germ agglutinin–(WGA) resistant mouse tumor cell line (MDW40), derived from a highly metastatic line called MDAY-D2 by exposure to cell-killing concentrations of WGA, loses its metastatic phenotype. Although the cell glycoproteins of both MDW40 and MDAY-D2 have as a major class of their N-linked oligosaccharides $Man_{5\text{-}9}$-containing forms, the nonmetastatic mutant cell type contains a unique triantennary class of N-linked oligosaccharides that lacks sialic acid and galactose, indicating the presence of incompletely processed N-linked oligosaccharide chains. Interestingly, when WGA-resistant cells reacquire the metastatic phenotype, they regain the oligosaccharide composition of the metastatic parent cell type.

Role of Extracellular Matrix (ECM) Components and the Basement Membranes in Tumor Metastasis

As noted in Chapter 5, the basement membrane of epithelial tissues consists of type IV collagen, fibronectin, laminin, entactin, heparan-sulfate–containing proteoglycans, and, in some tissues, type V collagen. Invading tumor cells encounter basement membranes in a variety of ways. They have to breach this barrier to invade the underlying stroma. To invade adjacent tissues, the tumor cells must also locally disrupt the basement membrane of that tissue. To gain access to the blood vasculature, they must invade through the basement membrane of capillary endothelium. Finally, when an embolus containing tumor cells lodges in a distant capillary bed, the tumor cells have to attach to endothelial basement membranes and invade it once again. Disruption of the basement membrane, on which an in situ carcinoma develops (step 1 in metastasis), could occur by release of degradative enzymes, as discussed previously, or abnormalities in the quantity or quality of basement membrane components laid down by tumor cells, or a combination of these mechanisms. The latter mechanism could result from decreased synthesis or decreased assembly of extracellular matrix components.

Laminin is one of the key ECM components of the basement membranes and one that tumor cells are capable of producing. Four categories of laminin production by human cancer cells in culture can be defined[71]: (1) "laminin producers/secretors," which produce a considerable amount of complete laminin (a 950-kDa trimer made up of one A and two B subunits)[71a] and shed about 25% of that produced into the culture medium; (2) "high laminin secretors," which shed over 60% of the synthesized 950 kDa laminin molecule from their surface; (3) "laminin A-subunit–deficient cells," characterized by cancer cells that produce the B but not the A subunits; and (4) "low laminin producers" which produce only trace quantities of the laminin A and B subunits. Seven of 10 human cell lines tested were either unable to biosynthesize one or both laminin subunits or to retain laminin in a cell-associated matrix. This situation contrasts with most nonmalignant human cell types tested in that normal cells produced complete laminin and shed very little into the culture medium, suggesting that normal epithelial cells deposit the laminin they produce into a more stable extracellular matrix than their cancerous counterparts do. Moreover, because epithelial cells are a principal biosynthetic source of the basal lamina to which they attach, the aberrant basement membranes frequently observed in human carcinomas may arise at least in part as a result of the impaired ability of malignant cells to synthesize or to deposit basal lamina components.

Fibronectin is another important ECM attachment factor for some cell types, such as fibroblasts, and it has been postulated that inability of transformed cells to produce or deposit fibronectin on their cell surface contributes to their invasive and metastatic potential. However, no simple, direct relationship between fibronectin production and metastatic potential can be made (see below).

A comment should be made here on the differences between tumor cell invasion into capillary endothelial basement membranes and that into lymphatic channels. Lymphatic capillaries lack a "tight" basement membrane containing type IV collagen and laminin.[72] Thus, a tumor cell that has already invaded into the stromal interstitial space does not have to cross another basement membrane to enter the lymphatic circulation. Infiltration into lymphatic channels, then, is a path of lesser resistance for invading epithelial cancer cells and probably accounts for the fact that lymphatic spread is usually the first type of metastasis observed clinically in patients with carcinomas, the most common kind of human cancer.

At the other end of the circuit—namely, the place where metastatic cells lodge in the capillary beds of other organs and invade—several distinct steps involving the basement membrane have been postulated: (1) adhesion to endothelium, (2) retraction of endothelial cells, (3) migration onto the endothelial basement membrane, (4) breaching of the basement membrane, and (5) locomotion or invasion into the interstitial space of the target tissue.[73] Some investigators favor the idea that tumor cells themselves can attach directly to the endothelial basement membrane by means of attachment factors of their own or receptors for host-tissue attachment factors; indeed, there is evidence to support this idea. Presumably this could occur in open spaces between endothelial cell–cell contact or after distortion of endothelial cell–cell boundaries induced by the tumor cell embolus. Attachment of normal cells to underlying matrix occurs through the cell surface integrins that bind the glycoproteins fibronectin or laminin (see Chap. 5), and tumor cells also take advantage of these attachment mechanisms. These attachment factors may be synthesized by the tumor cells themselves, or the cells may use factors already present in the matrix.

Fibronectin can mediate attachment of certain kinds of tumor cells to collagen, endothelial-derived matrix, or plastic culture dishes in vitro, and this can be blocked with antifibronectin antibodies. Thus, some investigators have postulated that fibronectin is an important attachment factor for metastatic tumor cells.[74] Other data, however, suggest that fibronectin production or attachment is not crucial to metastasis. For example, when parental mammary adenocarcinoma cells and their metastasis-derived clones were examined for ability to produce or release fibronectin, no difference in these characteristics was found between these different types of tumor cells.[75]

Laminin is another attachment factor that may play a key role in metastatic tumor cell binding to basement membranes. Several types of laminin-binding proteins have been identified on cell surface membranes. These include a high-affinity 67-kDa receptor, galactoside binding lectins, galactosyltransferase, sulfatides, and integrins, the family of cell surface receptors that bind various ECM components including fibronectin, vitronectin, thrombospondin, collagen, and von Willebrand factor as well as laminin.

The interactions of tumor cells with the ECM are important in determining the invasiveness and metastatic potential of cancer cells, and the ability of cancer cells to attach to laminin correlates with their metastatic potential (reviewed in Ref. 76). The 67-kDa high-affinity laminin receptor has been particularly associated with a cancer cell's metastatic capability, in that highly metastatic cells express higher levels of laminin receptors on their surface than do less metastatic or benign tumor cells of the same tissue type. A number of examples can be cited: the number of laminin receptors on breast carcinoma cells correlates with the extent of lymph node metastases and poor prognosis in patients[77]; the number of 67-kDa laminin receptors also correlates with the degree of invasiveness and metastasis of colon carcinoma cells in patients with that disease.[76]

A number of studies indicate that metastatic spread of tumors in experimental animal models can be influenced by substances that interfere with tumor cell binding to adhesion molecules in basement membranes. For example, coinjection of tumor cells with high metastatic potential together with antibodies against laminin reduces metastases.[78] Proteolytic fragments of laminin that bind to the 67-kDa receptor on the surface of tumor cells have been shown to inhibit metastasis of melanoma cells (reviewed in Ref. 79). Also pretreatment of murine melanoma cells with the synthetic peptide Arg-Gly-Asp-Ser (RGDS), which binds the laminin receptor, inhibits their metastasis after injection into syngeneic mice and results in an increased rate of clearance of melanoma cells from the pulmonary microcirculation.[80] However, a 20 amino acid synthetic polypeptide that represents the laminin binding domain of the 67-kDa receptor[81] or a 19 amino acid polypeptide representing a sequence of the lamina A chain,[82] when injected into mice prior to injection of melanoma cells, increases attachment of the melanoma cells to subendothelial matrix and enhances lung metastasis. In this latter experiment, lamininlike peptides apparently foster tumor cell attachment by, in effect, acting like laminin itself. In the case of the RGDS peptide, the peptide binds to the tumor cell surface and prevents efficient attachment to laminin, whereas the laminin peptide fosters formation of the cell-receptor–laminin complex.

Tissue Adhesion Properties of Metastatic Cells

The organ-site-specific localization of metastatic tumor cells has been suggested to result from specific adhesion of metastatic cells to the endothelial cells or endothelial basal lamina of the tissues for which they have a tropism. This specific adhesive quality appears to relate to the composition of cell surface glycoproteins, which form the attachments between cells as well as between cells and the basal lamina.

Organ-specific adhesion of metastatic tumor cells to cryostat sections of specific tissues has been shown. B16-F10 melanoma cells adhere much more to lung than to liver, brain, heart, or testis, and a murine reticulum cell sarcoma (M5076), which metastasizes specifically to liver in vivo, adheres to liver cryostat sections much more than to lung, brain, heart, or testis.[83] This specific adhesion process was inhibited by first treating the tumor cells with neuraminidase + β-galactosidase or with tunicamycin to block addition of asparagine-linked oligosaccharides to glycoproteins, indicating the importance of glycoproteins for the observed tissue-specific adhesion. Cryostat sections have also been used to select organ-specific metastatic cells: unselected B16 cells with high metastatic potential for lung over other tissues in vivo have been obtained by repeated incubations with mouse lung tissue sections.[84] In contrast, B16 cells selected on cryostat sections of mouse brain did not show a significant change in their metastatic organ site tropism in vivo. This indicates either a difference in the adhesive qualities of the tissues or a difference in the mechanisms by which metastatic cells attach.

Adhesion studies have also been carried out between specific types of endothelial cells and metastatic tumor cells. Teratoma cells with ovary-seeking properties in vivo have been shown to adhere preferentially to mouse ovary endothelial cells over brain endothelial cells, whereas glioma cells adhered preferentially to brain endothelial cells.[85]

Other attachment factors are also important in the metastatic process. Some of them act to inhibit the metastatic potential of tumor cells by favoring *intercellular* adhesiveness and limiting cell detachment. Although cell–cell adhesion is a complex process involving at least four families of adhesion molecules (integrins, immunoglobulins, selectins, and cadherins, see Chap. 4), a significant amount of data implicate the Ca^{2+}-dependent E-cadherin as a critically important adhesion factor to maintain epithelial integrity (reviewed in Refs. 86 and 87). E-Cadherin plays a key role in the normal development of epithelial tissues, and antibodies to it disturb developmental processes in the early embryo. Loss or aberrant expression of E-cadherin has been implicated in the invasive and metastatic potential of tumor cells. *ras* transformed, invasive Madin-Darby canine kidney (MDCK) cells lack E-cadherin expression, but if the E-cadherin cDNA is transfected into these cells, they lose their invasiveness.[88] Similarly, noninvasive clones of *ras*-transformed MDCK cells were rendered invasive by transfection of a plasmid encoding E-cadherin–specific antisense RNA. Moreover, human cancer cell lines from bladder, breast, lung, and pancreas carcinomas were noninvasive by an in vitro assay if they expressed E-cadherin and invasive if they did not.[89] The former could be rendered invasive if treated with monoclonal antibodies to E-cadherin, and the latter could be made noninvasive by transfection with E-cadherin cDNA.

In human bladder cancers, decreased expression of E-cadherin was observed in only 5 of 24 superficial tumors but in 19 of 25 invasive cancers, and a correlation of low expression with increased stage, grade, and poor survival was found.[86]

Other members of the cell–cell adhesion molecule (CAM) family have also been shown to be involved in tumor cell metastasis. For example, expression of NCAM-B has been found to down-regulate matrix metalloproteases 1 and 9, indicating that expression of NCAM-B on the cell surface can regulate the turnover of the surrounding extracellular matrix.[90]

Some CAMS expressed in tissues may, however, foster the ability of tumor cells to seed out in vascular beds in a tissue-specific way. For instance, a 90-kDa lung-specific melanoma cell–binding molecule called Lu-ECAM is expressed on the endothelia of pleural and subpleural capillaries and venules and fosters the attachment of "lung-seeking" melanoma cells.[91] Pretreatment with antibodies to Lu-ECAM inhibited colonization of lungs in mice by lung-seeking B16-F10 cells but had no effect on liver metastatic colonies produced by liver-seeking B16-F10 cells or on lung metastases produced by other lung-metastatic cell lines.[91]

Taken together, these data indicate the importance of cell adhesion to the ECM and of cell–cell adhesion molecules (CAMs) in the expression of the metastatic phenotype. Strategies, then, to increase the expression of normal CAMS in tumor tissue might be thought of as ways to modulate this phenotype.

Ability of Metastatic Tumor Cells to Escape the Host's Immune Response

Cell surface antigens representing the major histocompatibility complex (MHC) antigens of the mouse (H-2 genes) and human (HLA genes) play a role in immune surveillance, tumorigenicity, and metastatic potential in both mouse and human cancer (see Chap. 13). Cytolytic lymphocytes recognize cell surface alterations of neoplastic cells associated with MHC antigens. Experiments in mice have demonstrated that metastatic properties of certain mouse tumors are correlated with the expression of class I MHC antigens. Using cloned cell lines of differing metastatic capability derived from the 3-methylcholanthrene-induced fibrosarcoma T10 of mice, a correlation was observed between the in vivo metastatic potential and expression of the H-2D and H-2K antigens.[92] Metastatic clones expressed only $H\text{-}2D^b$ and $H\text{-}2D^k$ MHC antigens, but lacked $H\text{-}2K^b$ and $H\text{-}2K^k$ expression. Nonmetastatic clones had $H\text{-}2D^b$ on their cell surface but not $H\text{-}2D^k$, suggesting that the D^k antigen contributes to the metastatic potential of these clones. Furthermore, when genes coding for the H-2K region were transfected into the metastatic cloned T10 cells, these cells expressed K^b and K^k antigens on their surface and lost their metastatic ability in vivo, even though they remained locally tumorigenic. These results strongly imply that the MHC system is involved in immune surveillance that limits the viability of circulating, metastatically potent tumor cells. This contention is supported by the fact that the H-2K gene-transfected cells were more immunogenic and more susceptible to killing by cytolytic lymphocytes than their H-2K-negative counterparts.[92]

Chemotactic Factors in Cancer Cell Migration

As noted in Chapter 5, cellular migration occurs normally throughout the life of multicellular organisms. In early embryonic development, migration of neural crest cells, hematopoietic cells, and germ cells occurs, enabling the embryonic progenitor cells to reach their destination for organogenesis. Examples of migratory cells in adult life are the motility of spermatozoa during fertilization, movement of cells during wound healing and tissue repair, and migration of leukocytes and macrophages in the inflammatory process. In these instances, movement of cells is under the influence of several regulatory signals, including cell-to-cell contact, the nature of the extracellular matrix, and chemotactic factors that regulate cellular motility and directionality of cell movement. In cancer cell metastasis, similar kinds of mechanisms come into play, but in an unregulated fashion.

Chemotactic factors for both leukocytes and nonleukocytic cells have been identified (reviewed in Ref. 93). Sources of chemotactic factors include native types I, II, and III collagen; collagenolytic breakdown products; lymphocyte-derived chemotactic factors; and complement-derived peptides. Chemotactic factors generated from the fifth complement component (C5) are chemotactic for both leukocytic and nonleukocytic cell types. A C5-derived fibroblast chemotactic factor has a molecular weight of about 80 kDa and is clearly distinguishable from smaller leukocyte and tumor cell chemotactic factors generated from C5.[93,94]

In the early 1970s, Hayashi and coworkers described chemotactic responses in several tumor cell lines.[95,96] They demonstrated that metastatic tumors developed at skin sites injected with chemotactic factors and proposed that this mechanism was similar to that of leukocyte migration and localization at sites of inflammation. Later, Ward and his colleagues,[94,97] as well as others (reviewed in Ref. 93), observed chemotactic responses for a number of tumor cell types. Potent tumor cell chemotactic factors were shown to be generated from intact C5 as well as from the C5a fragment of C5, the latter of which is chemotactic for leukocytes. The active component was shown to be generated from C5 or C5a by proteolytic cleavage to a 6000 MW peptide. Since both C5a and lysosomal proteases that can generate tumor cell chemotactic factor from C5a are present in inflammatory exudates, it was predicted that inflammatory sites would favor the generation of tumor cell chemotactic factor in vivo, and this was subsequently shown to be correct in animal models.[93] Interestingly, when inflammatory reactions were generated in vivo, or when preformed tumor-cell chemotactic factors were injected intraperitoneally, an increased number of metastases formed in tumor-bearing treated animals compared with tumor-bearing control animals.

C5-derived chemotactic factor has also been observed in human neoplastic effusion fluids (peritoneal and pleural effusions as well as cerebrospinal fluids). Human tumor cells were also found to undergo chemotactic-factor–stimulated motility when incubated with effusion fluid from the same patient or with C5-derived factor generated in vitro.[93] Thus, these findings suggest that tumor-cell chemotactic factors can modulate the metastatic potential of tumor cells in vivo and may explain the tendency of tumors to metastasize to areas of tissue injury.

Role of Oncogenes in Tumor Metastasis

In Chapter 7, the role of oncogene activation in tumorigenesis and tumor progression was described. Activation of a number of oncogenes has been associated with the invasive, metastatic phenotype in different tumor types, but whether this is a direct cause of the induction of metastasis or whether it is a reflection on the increased survival potential that goes along with oncogene activation in cells that have a selective advantage to survive in a new tissue environment is not clear. Another possibility is that activation of cellular oncogenes is involved in producing the genetic instability that leads to metastasis. Such an observation was made by transfection of the v-Ha-*ras* gene into rat mammary carcinoma cells induced by DMBA.[98] In this case, the v-Ha-*ras*-transfected cells became genetically unstable, as demonstrated by the acquisition of additional chromosomal abnormalities, and developed more distant metastases than mammary tumor cells transfected with a control plasmid. Coexpression of v-*fos* in a *src*-transformed rat cell line produced cells with a greater invasive capacity in in vitro assays and a higher metastatic potential in vivo.[99] Transfection of a c-*erbB*-2 gene, activated by mutation, into low metastatic potential mouse colonic carcinoma cells significantly enhanced lung metastasis.[100] In humans, the detection of ErbB2-positive cells in the bone marrow correlated with the incidence of metastasis.[101] In patients with overt metastases, the incidence of metastatic ErbB2-positive cells in the bone marrow was 68% in breast cancer patients and 28% in colorectal cancer patients and correlated with clinical stage of tumor progression. These data suggest that ErbB2 expression is a marker for cells that exhibit the metastatic phenotype and that have a selective advantage for survival during the metastatic process. Overexpression of c-*myc*, c-*erb*, c-Ki-*ras*, and *hst* oncogenes has been observed in metastatic gastric cancers,[102] and *mdm2* gene amplification has been seen in metastatic osteosarcomas.[103] This activation or overexpression of cellular oncogenes is a common phenomenon in metastatic cancers. There may also be tissue-associated, oncogene-related growth factors that differentially stimulate the growth of tumor cells in specific tissues, as has been for a lung-derived growth factor that stimulates the proliferation of lung-seeking metastatic cells in mouse model systems.[104]

Angiogenesis in Invasive and Metastatic Cancer

There is abundant evidence that tumors can induce angiogenesis through release of a variety of factors (Chap. 9) and that vascularized tumors have a higher propensity to metastasize than nonvascularized (carcinoma in situ) or poorly vascularized tumors (reviewed in Ref. 33). The process of angiogenesis involves three functions: motility, proteolysis, and cell proliferation. Motility is required for endothelial cell chemotaxis toward the angiogenic stimulus as well as alignment of an endothelial "sprout" or outgrowth. Proteolysis is required for penetration of the endothelial cell buds into the extracellular matrix. Proliferation of endothelial cells is needed to form the new capillaries that feed into the tissue. This same process is utilized by metastatic cancer cells and most likely involves the same mechanisms. In addition, tumor cells can produce their own angiogenesis factors that foster their vascularization once they set up housekeeping in a new target tissue. Thus, antiangiogenesis agents may be good candidates for new anticancer drugs.

Cloning "Metastasis Genes" or "Metastasis Suppressor Genes"

Clearly, it is of utmost importance to know which cellular genes are involved in the expression of the metastatic phenotype and to learn how they are regulated. Theoretically, an approach similar to the one used to isolate transforming genes from human and animal cancer cells could be used here (see Chap. 7). Therefore, if one selected for metastatic clones of cancer cells, created a library of genes from such cells, and tested for metastatic potential of the transfected cells, one should be able to determine if specific segments of DNA are associated with the metastatic phenotype. Individual segments could then be sequenced and the amino acid sequence of putative protein products predicted. By then comparing the amino acid sequences with those of known proteins in the computerized databases, one could determine if the "metastasis proteins" are related to known cellular proteins. In addition, labeled cDNA probes representing the cloned "metastasis genes" could be prepared and their transcription product (mRNA) looked for in metastatic and nonmetastatic tumor cells. Also, synthetic peptides could be made from the putative amino acid sequence, monoclonal antibodies prepared, and the presence of such proteins looked for in metastatic versus nonmetastatic cells.

An approach like this has been used to identify genes associated with the metastatic phenotype of transformed murine fibroblasts. Bernstein and Weinberg[105] isolated DNA from a human metastatic tumor and transfected it into Ha-*ras* oncogene (isolated from the EJ human bladder carcinoma cell line)–transformed NIH 3T3 fibroblasts. The Ha-*ras*-transformed cells, without the metastatic cell DNA, formed nonmetastasizing fibrosarcomas in mice; however, one of the clones of cells resulting from transfection with DNA from human metastatic cells yielded a lung metastasis after it was subcutaneously injected into mice. DNA was isolated from this metastasis, and a second round of transfection of 3T3 cells was carried out. When these cells were then injected into mice, a much higher incidence of metastasis was noted than with the initially transfected cells. Four of the resulting metastases contained DNA fragments in common. These fragments were not homologous to the *myc* oncogene or any of the *ras* oncogenes. The enticing finding, however, was that the metastatic phenotype could be transferred from cell to cell by discrete DNA segments, implying that specific genes are closely associated with the ability of cells to metastasize. The nature of genes that confer metastatic potential are only beginning to be known. They may be related to one of the known cellular oncogenes or to an as yet undiscovered family of genes.

A gene called pGM21 was found to be associated with high metastatic potential in rat mammary adenocarcinomas. This gene was fished out by constructing a cDNA library from substraction hybridization of mRNA from a poorly metastatic cell line with the cDNA from a highly metastatic cell line.[106] Sequencing of one of the cDNAs so isolated contained a 45-nucleotide segment homologous to human elongation factor-1 subunit α.

In addition, one could look for metastasis *suppressor* genes by cell fusion experiments similar to those described in Chapter 8, that is, by examining the metastatic potential of hybrid cells prepared by fusion of high metastatic and low metastatic tumor cells and asking which chromosomes were found in hybrid cells that were nonmetastatic. Such experiments have been done. Ichikawa et al.[107] fused highly metastatic rat mammary carcinoma cells, transfected with v-Ha-*ras*, with nonmetastatic parent mammary carcinoma cells. Several hybrid clones of cells were isolated that grew as primary tumors but were nonmetastatic. Interestingly, these cells continued to express v-Ha-*ras*. These data strongly suggested the presence of a metastatic suppressor gene that could overcome the ability of high levels of v-Ha-*ras* to foster metastasis.

A candidate for such a suppressor gene, at least for some cancer cell types, has been found. A metastasis suppressor gene, called *nm*23, was identified by mRNA subtraction experiments comparing the content of mRNA found in metastatic versus nonmetastatic murine melanoma cells.[108] The levels of *nm*23 mRNA were 10-fold lower in melanoma cell lines of high metastatic potential compared to those with low potential. Subsequently, a similar gene has been found in human cells and low levels of its expression have been correlated with metastasis and poor patient prognoses in breast,[109] hepatocellular,[110] and ovarian[111] carcinomas and malignant melanoma.[112] However, in human colonic tissue, *nm*23 mRNA levels were increased in colonic carcinoma cells compared to normal colonic mucosa,[113] suggesting that *nm*23 gene expression is controlled differently in different tissues.

Through an intriguing bit of comparative genetic sleuthing, the function of the *nm*23 gene has been found. There is a gene in the fruit fly *Drosophila* that, when mutated, causes morphologically deformed wing disks in the larval stage. This gene, called *awd* for "abnormal wing disks" gene, has been cloned and sequenced. After the *nm*23 gene sequence was determined, a gene database search revealed that it was 78% homologous to the *awd* gene. A further clue came when the cDNA clones for nucleoside diphosphate kinase (NDP kinase) were isolated from the slime mold *Dictyostelium* and from a *Myxococcus* microorganism. These cDNA clones were found to be highly homologous to the *nm*23/*awd* gene, and the *awd* gene product was subsequently shown to be a NDP kinase.[114]

NDP kinases are a ubiquitous family of enzymes that catalyze the transfer of the terminal phosphate group of 5′-triphosphate nucleotide donors to diphosphate nucleotide acceptors, e.g., GDP to GTP via ATP. These kinases participate in functions that could affect tumor cell proliferation and metastasis by an action on G-protein coupled signal transduction mechanisms that regulate microtubule assembly, since GTP is required for this function. The NDP kinase coded for by the *awd* gene is associated with microtubules in *Drosophila* larvae.[114] What role this might have in tumor metastasis is speculative at this point, but since microtubules are important for cell locomotion and for response to external signals mediated by the ECM, loss of regulatory mechanisms mediated by NDP kinases could result in loss of normal matrix-cell interactions. This idea is supported by evidence that *nm*23 gene transfection into murine melanoma cells or human breast carcinoma cells inhibits their motility in response to serum, PDGF, or IGF-1.[115] Another interesting observation is that the human purine-binding transcription factor gene PuF has an identical sequence to the *nm*23-H2 gene, a member of the *nm*23 gene family.[116] Since PuF encodes a transcription factor that regulates c-*myc* expression, this suggests a direct link between *nm23* and expression of c-*myc*.

Another putative metastasis suppressor gene has been identified in human mammary epithelial cells.[117] The product of this gene, called "maspin," is related to the serpin family of protease inhibitors (*ser*ine *p*rotease *in*hibitors).[117a] Maspin is expressed in normal mammary epithelial cells but not in most mammary carcinoma cell lines. Transfection of the maspin gene into a human mammary carcinoma cell line did not alter the cells' growth properties in vitro but reduced the ability of the transfected cells to induce tumors and metastasize in nude mice. These cells also had a reduced ability to invade through a basement membrane matrix in vitro. Maspin expression was also reduced or lost in advanced breast cancer specimens from patients, suggesting that maspin is a tumor metastasis suppressor in vivo.

A novel candidate metastasis-associated gene, called *mta*1, is expressed in highly metastatic mammary adenocarcinoma cells of both rat and human origin.[117b] The Mta1 protein is 80 kDa and contains several potential phosphorylation sites. Sequencing of the cDNA indicates it is a novel gene whose function is as yet unknown.

Therapeutic Considerations in the Treatment of Metastasis

As previously discussed, metastatic cells have greater genetic instability than nonmetastatic tumor cells. Thus, as one might suspect, even if individual metastases arise from a single clone of cells, for which there is evidence from both human and animal tumors,[118] cellular heterogeneity is likely to arise within a few population doublings. For example, the metastatic properties of tumor cell clones isolated from individual metastatic lung foci of B16 melanoma cells have been examined for their metastatic characteristics.[119] Initially, the cells of a given metastatic focus displayed "intralesional clonal homogeneity," but as the metastatic lesions grew and progressed, they became populated with cells with heterogeneous phenotypes.

Because, in most cases of cancer, the metastatic lesions kill the patient, it is extremely important to know the best way to treat these lesions. Unfortunately, though metastatic lesions may initially show a homogeneous sensitivity to cancer chemotherapeutic agents, they quickly develop a marked clonal heterogeneity with respect to their chemosensitivity to drugs, such that cells within one metastatic site may vary from cells in other metastatic sites as well as from the primary tumor in their drug responsiveness. Such heterogeneity has been seen in animal[118,120] and human[121] neoplasms. An approach that is still in experimental phases is to use activated macrophages or cytolytic lymphocytes that are capable of attacking tumor cells of various types. This promising approach is discussed in detail in Chapter 13.

A number of potentially antimetastatic drugs have also been investigated. Based on a search for compounds that could inhibit cell motility as a potential inhibitory mechanism of tumor metastasis, the Merck compound L651582 was tested for its antitumor activity.[122] The rationale for this was that certain tumors secrete an autocrine motility factor that stimulates cell locomotion via G-protein–coupled inositol trisphosphate (IP3) generation. L651582, originally developed to inhibit coccidiosis microorganisms, was shown to inhibit proliferation and clonogenic growth of human melanoma, breast, and ovarian carcinomas and *ras*-transformed rat fibroblasts.[123] The drug also significantly prolonged the survival time of nude mice transplanted intraperitoneally with a metastatic human ovarian carcinoma.

Inhibitors of the enzyme glucosylceramide synthase also may provide an approach to inhibit tumor metastasis. There is evidence that carbohydrate residues of cell surface glucosphingolipids (GSLs) play a role in metastasis by being involved in cell attachment. Moreover, GSL composition has been found to differ in cell lines with high versus low metastatic potential. An inhibitor of glucosylceramide synthase is D-threo-1-phenyl-2-decanoylamino-3-morpholino-1-propanol (D-PDMP). Treatment of murine Lewis lung carcinoma cells in culture with D-PDMP produced a time-dependent decrease in levels of all cellular GSLs, reduced the cells' invasive capacity for reconstituted basement membranes, and diminished the lung colonizing ability of treated cells compared to untreated cells after tail-vein injection of cells into mice.[123]

The invasive capacity of human bladder carcinoma cells through an artificial basement membrane has been reported to be reduced by staurosporine, a microbial alkaloid isolated from a *Streptomyces* species and a potent inhibitor of protein kinase C. A rationale for this is that total PKC activity was found to be twofold higher in an invasive bladder carcinoma line than in a non-invasive one.[124] The drug, at concentrations producing minimal cell toxicity, inhibited the penetration of the invasive cell line in a dose-dependent manner by decreasing cell motility rather than tumor cell attachment.[124]

Other potentially clinically useful inhibitors of cancer cell metastasis include the angiogenesis inhibitor O-(chloroacetyl-carbamoyl)fumagillol (TNP-470), which has been shown to inhibit the in vivo metastasis of murine B16 melanoma cells and of M5076 reticulum cell sarcoma cells,[125] and a urokinase-plasminogen activator-IgG fusion protein that displaces u-PA activity from tumor cell surfaces and blocks the metastatic potential of human prostate carcinoma cells in nude mice.[126]

REFERENCES

1. J. Folkman: The vascularization of tumors. *Sci Am* 234:58, 1976.
2. V. T. DeVita, Jr., R. C. Young, and G. P. Canellos: Combination versus single agent chemotherapy: A review of the basis for selection of drug treatment of cancer. *Cancer* 35:98, 1975.
3. F. M. Schabel, Jr.: The use of tumor growth kinetics in planning "curative" chemotherapy of advanced solid tumors. *Cancer Res* 29:2384, 1969.
4. G. G. Steel: Cytokinetics of neoplasia, in J. F. Holland and E. Frei III, eds.: *Cancer Medicine.* Philadelphia: Lea & Febiger, 1973, pp. 125–140.
5. I. J. Fidler: Tumor heterogeneity and the biology of cancer invasion and metastasis. *Cancer Res* 38:2651, 1978.
6. B. Fisher and E. R. Fisher: The interrelationship of hematogenous and lymphatic tumor cell dissemination. *Surg Gynecol Obstet* 122:791, 1966.
7. B. Sylvén: Biochemical factors accompanying growth and invasion, in R. W. Wissler, T. L. Dao, and S. Wood, Jr., eds.: *Endogenous Factors Influencing Host-Tumor Balance.* Chicago: University of Chicago Press, 1967, pp. 267–276.

7a. L. A. Liotta, K. Tryggvason, S. Garbisa, I. Hart, C. M. Foltz, and S. Shafie: Metastatic potential correlates with enzymatic degradation of basement membrane collage. *Nature* 284:67, 1980.

8. E. R. Fisher and B. Fisher: Recent observations on the concept of metastasis. *Arch Pathol* 83:321, 1967.
9. I. J. Fidler: Metastasis: Quantitative analysis of distribution and fate of tumor emboli labeled with ^{125}I-5-iodo-2′-deoxyuridine. *JNCI* 45:775, 1970.
10. S. Wood, Jr., R. R. Robinson, and B. Marzocchi: Factors influencing the spread of cancer: Locomotion of normal and malignant cells in vivo, in R. W. Wissler, T. L. Dao, and S. Wood, Jr., eds.: *Endogenous Factors Influencing Host-Tumor Balance.* Chicago: University of Chicago Press, 1967, pp. 223–237.
11. I. J. Fidler: The relationship of embolic homogeneity, number, size, and viability to the incidence of experimental metastasis. *Eur J Cancer* 9:223, 1973.
12. I. Zeidman: The fate of circulating tumor cells: I. Passage of cells through capillaries. *Cancer Res* 21:38, 1961.
13. B. A. Warren: Environment of the blood-borne tumor embolus adherent to vessel wall. *J Med* 4:150, 1973.
14. G. J. Gasic, T. B. Gasic, N. Galanti, T. Johnson, and S. Murphy: Platelet-tumor cell interactions in mice: The role of platelets in the spread of malignant disease. *Int J Cancer* 11:704, 1973.
15. A. M. Schor, S. L. Schor, and S. Kumar: Importance of a collagen substratum for stimulation of capillary endothelial cell proliferation by tumor angiogenesis factor. *Int J Cancer* 24:225, 1979.
16. I. J. Fidler: Immune stimulation-inhibition of experimental cancer metastasis. *Cancer Res* 34:491, 1974.
17. W. A. D. Anderson, ed.: *Pathology,* 4th ed. St. Louis, MO: C. V. Mosby, 1961.
18. P. Rubin, ed.: *Clinical Oncology,* 4th ed. Rochester, NY: American Cancer Society, 1974.
19. S. Paget: The distribution of secondary growth in cancer of the breast. *Lancet* 1:571, 1889.
20. I. J. Fidler: Selection of successive tumor lines for metastasis. *Nature New Biol* 242:148, 1973.
21. I. J. Fidler: General considerations for studies of experimental cancer metastasis, in H. Busch, ed.: *Methods in Cancer Research.* Vol. XV. New York: Academic Press, 1978, pp. 399–439.
22. I. J. Fidler and G. L. Nicolson: Organ selectivity for survival and growth of B16 melanoma variant tumor lines. *JNCI* 57:1199, 1976.
23. K. W. Brunson, G. Beattie, and G. L. Nicolson: Selection and altered properties of brain-colonizing metastatic melanoma. *Nature* 272:543, 1978.
24. G. L. Nicolson and K. W. Brunson: Organ specificity of B16 melanomas: In vivo selection for organ preference of blood-borne metastasis. *Gann Monogr Cancer Res* 20:15, 1977.
25. I. J. Fidler and M. L. Kripke: Metastasis results from preexisting variant cells within a malignant tumor. *Science* 197:893, 1977.

26. G. Poste and I. J. Fidler: The pathogenesis of cancer metatheses. *Nature* 283:139, 1980.
27. G. L. Nicolson and S. E. Custead: Tumor metastasis is not due to adaptation of cells to a new organ environment. *Science* 215:176, 1982.
28. I. J. Fidler and I. R. Hart: Biological diversity in metastatic neoplasms: Origins and implications. *Science* 217:998, 1982.
29. R. P. Hill, A. F. Chambers, V. Ling, and J. F. Harris: Dynamic heterogeneity: Rapid generation of metastatic variants in mouse B16 melanoma cells. *Science* 224:998, 1984.
30. J. E. Talmadge, S. R. Wolman, and I. J. Fidler: Evidence for the clonal origin of spontaneous metastases. *Science* 217:361, 1982.
31. N. Wang, S. H. Yu, I. E. Liener, R. P. Hebbel, J. W. Eaton, and C. F. McKhann: Characterization of high- and low-metastatic clones derived from a methylcholanthrene-induced fibrosarcoma. *Cancer Res* 42:1046, 1982.
32. S. Strickland and W. G. Richards: Invasion of the trophoblasts. *Cell* 71:355, 1992.
33. L. A. Liotta, P. S. Steeg, and W. G. Stetler-Stevenson: Cancer metastasis and angiogenesis: An imbalance of positive and negative regulation. *Cell* 64:327, 1991.
34. L. M. Matrisian: Metalloproteinases and their inhibitors in matrix remodeling. *Trends Genet* 6:121, 1990.
35. J. F. Cajot, J. Bamat, G. E. Bergonzelli, E. K. O. Kruithof, R. L. Medcalf, J. Testuz, and B. Sordat: Plasminogen-activator inhibitor type 1 is a potent natural inhibitor of extracellular matrix degradation by fibrosarcoma and colon carcinoma cells. *Proc Natl Acad Sci USA* 87:6939, 1990.
36. M. Nakajima, D. Lotan, M. M. Baig, R. M. Carralero, W. R. Wood, M. J. C. Hendrix, and R. Lotan: Inhibition of retinoic acid of type IV collagenolysis and invasion through reconstituted basement membrane by metastatic rat mammary adenocarcinoma cells. *Cancer Res* 49:1698, 1989.
37. J. Folkman, K. Watson, D. Ingber, and D. Hanahan: Induction of angiogenesis during the transition from hyperplasia to neoplasia. *Nature* 339:58, 1989.
38. A. M. P. Montgomery, Y. A. DeClerck, K. E. Langley, R. A. Reisfeld, and B. M. Mueller: Melanoma-mediated dissolution of extracellular matrix: Contribution of urokinase-dependent and metalloproteinase-dependent proteolytic pathways. *Cancer Res* 53:693, 1993.
39. J. Grondahl-Hansen, I. J. Christensen, C. Rosenquist, N. Brummer, H. T. Mouridsen, K. Dano, and M. Blichert-Toft: High levels of urokinase-type plasminogen activator and its inhibitor PAI-1 in cytosolic extracts of breast carcinomas are associated with poor prognosis. *Cancer Res* 53:2513, 1993.
40. J. Rozhin, A. P. Gomez, G. H. Ziegler, K. K. Nelson, Y. S. Chang, D. Fong, J. M. Onoda, K. V. Honn, and B. F. Sloane: Cathepsin B to cysteine proteinase inhibitor balance in metastatic cell subpopulations isolated from murine tumors. *Cancer Res* 50:6278, 1990.
41. S. S. Chauhan, L. J. Goldstein, and M. M. Gottesman: Expression of cathepsin L in human tumors. *Cancer Res* 51:1478, 1991.
42. M. J. Murnane, K. Sheahan, M. Ozdemirli, and S. Shuja: Stage-specific increases in cathepsin B messenger RNA content in human colorectal carcinoma. *Cancer Res* 51:1137, 1991.
43. S. M. Thorpe, H. Rochefort, M. Garcia, G. Freiss, I. J. Christensen, S. Khalaf, F. Paolucci, B. Pau, B. B. Rasmussen, and C. Rose: Association between high concentrations of M_r 52,000 cathepsin D and poor prognosis in primary human breast cancer. *Cancer Res* 49:6008, 1989.
44. M. D. Johnson, J. A. Torri, M. E. Lippman, and R. B. Dickson: The role of cathepsin D in the invasiveness of human breast cancer cells. *Cancer Res* 53:873, 1993.
45. P. Basset, J. P. Bellocq, C. Wolf, I. Stoll, P. Hutin, J. M. Limacher, O. L. Podhajcer, M. P. Chenard, M. C. Rio, and P. Chambon: A novel metalloproteinase gene specifically expressed in stromal cells of breast carcinomas. *Nature* 348:699, 1990.
46. C. Wolf, N. Rouyer, Y. Lutz, C. Adida, M. Loriot, J.-P. Bellocq, P. Chambon, and P. Basset: Stromelysin 3 belongs to a subgroup of proteinases expressed in breast carcinoma fibroblastic cells and possibly implicated in tumor progression. *Proc Natl Acad Sci USA* 90:1843, 1993.
47. D. Muller, C. Wolf, J. Abecassis, R. Millon, A. Engelmann, G. Bronner, N. Rouyer, M.-C. Rio, M. Eber, G. Methlin, P. Chambon, and P. Basset: Increased stromelysin 3 gene expression is associated with increased local invasiveness in head and neck squamous cell carcinomas. *Cancer Res* 53:165, 1993.
48. R. E. B. Seftor, E. A. Seftor, W. G. Stetler-Stevenson, and M. J. C. Hendrix: The 72 kDa type IV collagenase is modulated via differential expression of $\alpha_v\beta_3$ and $\alpha_5\beta_1$ integrins during human melanoma cell invasion. *Cancer Res* 53:3411, 1993.
49. C. Pyke, E. Ralfkiaer, P. Huhtala, T. Hurskainen, K. Dano, and K. Tryggvason: Localization of messenger RNA for M_r 72,000 and 92,000 type IV collagenases in human skin cancers by in situ hybridization. *Cancer Res* 52:1336, 1992.
50. A. T. Levy, V. Cioce, M. E. Sobel, S. Garbisa, W. F. Grigioni, L. A. Liotta, and W. G. Stetler-Stevenson: Increased expression of the M_r 72,000 type IV collagenase in human colonic adenocarcinoma. *Cancer Res* 51:439, 1991.
51. S. Zucker, R. M. Lysik, M. H. Zarrabi, and U. Moll: M_r 92,000 type IV collagenase is increased in plasma of patients with colon cancer and breast cancer. *Cancer Res* 53:140, 1993.

52. H. Ura, R. D. Bonfil, R. Reich, R. Reddel, A. Pfeifer, C. C. Harris, and A. J. P. Klein-Szanto: Expression type IV collagenase and procollagen genes and its correlation with the tumorigenic, invasive, and metastatic abilities of oncogene-transformed human bronchial epithelial cells. *Cancer Res* 49:4615, 1989.

53. M. E. Stearns and M. Wang: Type IV collagenase (M_r 72,000) expression in human prostate: Benign and malignant tissue. *Cancer Res* 53:878, 1993.

54. B. Davies, J. Waxman, H. Wasan, P. Abel, G. Williams, T. Krausz, D. Neal, D. Thomas, A. Hanby, and F. Balkwill: Levels of matrix metalloproteases in bladder cancer correlate with tumor grade and invasion. *Cancer Res* 53:5365, 1993.

55. H. Kataoka, R. DeCastro, S. Zucker, and C. Biswas: Tumor cell-derived collagenase-stimulatory factor increases expression of interstitial collagenase, stromelysin, and 72-kDa gelatinase. *Cancer Res* 53:3154, 1993.

56. S. Garbisa, G. Scagliotti, L. Masiero, C. DiFrancesco, C. Caenozzo, M. Onisto, M. Micela, W. G. Stetler-Stevenson, and L. A. Liotta. Correlation of serum metalloproteinase levels with lung cancer metastasis and response to therapy. *Cancer Res* 52:4548, 1992.

56a. E. J. Bernhard, S. B. Gruber, and R. J. Muschel: Direct evidence linking expression of matrix metalloproteinase 9 (92 kDa gelatinase/collagenase) to the metastatic phenotype in transformed rat embryo cells. *Proc Nat Acad Sci USA* 91:4293, 1994.

57. L. A. Liotta, C. N. Rao, and S. H. Barsky: Tumor invasion and the extracellular matrix. *Lab Invest* 49:636, 1983.

58. D. P. DeVore, D. P. Houchens, A. A. Ovejera, G. S. Dill, and T. B. Hutson: Collagenase inhibitors retarding invasion of a human tumor in nude mice. *Exp Cell Biol* 48:367, 1980.

59. U. P. Torgeirsson, L. A. Liotta, T. Kalebric, and I. M. K. Margulies: Effect of natural protease inhibitors and a chemoattractant on tumor cell invasion in vitro. *JNCI* 69:1049, 1982.

60. L. A. Liotta, P. S. Steeg, and W. G. Stetler-Stevenson: Cancer metastasis and angiogenesis: An imbalance of positive and negative regulation. *Cell* 64:327, 1991.

61. A. Albini, A. Melchiori, L. Santi, L. A. Liotta, P. D. Brown, and W. G. Stetler-Stevenson: Tumor cell invasion inhibited by TIMP-2. *JNCI* 83:775, 1991.

62. E. Koivunen, A. Ristimaki, O. Itkonen, S. Osman, M. Vuento, and U.-H. Stenman: Tumor-associated trypsin participates in cancer cell-mediated degradation of extracellular matrix. *Cancer Res* 51:2107, 1991.

63. E. W. Easton, J. G. M. Bolscher, and D. H. v. d. Eijnden: Enzymatic amplification involving glycosyltransferases forms the basis for the increased size of asparagine-linked glycans at the surface of NIH 3T3 cells expressing the N-*ras* proto-oncogene. *J Biol Chem* 266:21674, 1991.

64. G. Poste and G. L. Nicholson: Arrest and metastasis of blood-borne tumor cells are modified by fusion of plasma membrane vesicles from highly metastatic cells. *Proc Natl Acad Sci USA* 77:399, 1980.

65. T. Irimura, R. Gonzalez, and G. Nicolson: Effects of tunicamycin on B16 metastatic melanoma cell surface glycoproteins and blood-borne arrest and survival properties. *Cancer Res* 41:3411, 1981.

66. M. J. Humphries, K. Matsumoto, S. L. White, and K. Olden: Oligosaccharide modification by swainsonine treatment inhibits pulmonary colonization by B16-F10 murine melanoma cells. *Proc Natl Acad Sci USA* 83:1752, 1986.

67. E. Pearlstein, P. L. Salk, G. Yogeeswaran, and S. Karpatkin: Correlation between spontaneous metastatic potential, platelet-aggregating activity of cell surface extracts, and cell surface sialylation in 10 metastatic-variant derivatives of a rat renal sarcoma cell line. *Proc Natl Acad Sci USA* 77:4336, 1980.

68. G. Yogeeswaran and P. L. Salk: Metastatic potential is positively correlated with cell surface sialylation of culture murine tumor cell lines. *Science* 212:1514, 1981.

69. P. Altevogt, M. Fogel, R. Cheinsong-Popov, J. Dennis, P. Robinson, and V. Schirrmacher: Different patterns of lectin binding and cell surface sialylation detected on related high- and low-metastatic tumor lines. *Cancer Res* 43:5138, 1983.

70. J. W. Dennis, J. P. Carver, and H. Schachter: Asparagine-linked oligosaccharides in murine tumor cells: Comparison of a WGA-resistant (WGA^r) nonmetastatic mutant and a related WGA-sensitive (WGA^s) metastatic line. *J Cell Biol* 99:1034, 1984.

71. G. P. Frenette, T. E. Carey, J. Varani, D. R. Schwartz, S. E. G. Fligiel, R. W. Ruddon, and B. P. Peters: Biosynthesis and secretion of laminin and laminin-associated glycoproteins by nonmalignant and malignant human keratinocytes: Comparison of cell lines from primary and secondary tumors in the same patient. *Cancer Res* 48:5193, 1988.

71a. B. P. Peters, R. J. Hartle, R. F. Krzesicki, T. G. Kroll, F. Perini, J. E. Balun, I. J. Goldstein, and R. W. Ruddon: The biosynthesis, processing, and secretion of laminin by human choriocarcinoma cells. *J Biol Chem* 260:14732, 1985.

72. S. H. Barsky, A. Baker, G. P. Siegel, S. Togo, and L. A. Liotta: Use of anti-basement membrane antibodies to distinguish blood vessel capillaries from lymphatic capillaries. *Am J Surg Pathol* 7:667, 1983.

73. G. L. Nicolson, T. Irimura, M. Nakajima, and J. Estrada: Metastatic cell attachment to and invasion of vascular endothelium and its underlying basal lamina using endothelial cell monolayers, in G. L. Nicholson and L. Milas, eds.: *Cancer Invasion and Metastasis: Biologic and Therapeutic Aspects.* New York: Raven Press, 1984, pp. 145–167.
74. R. H. Kramer, R. Gonzalz, and G. L. Nicolson: Metastatic tumor cells adhere preferentially to the extracellular matrix underlying vascular endothelial cells. *Int J Cancer* 26:639, 1980.
75. A. Neri, E. Rouslahti, and G. L. Nicolson: Distribution of fibronectin on clonal cell lines of a rat mammary adenocarcinoma growing in vitro and in vivo at primary and metastatic sites. *Cancer Res* 41:5082, 1981.
76. V. Cioce, V. Castronovo, B. M. Shmookler, S. Garbisa, W. F. Grigioni, L. A. Liotta, and M. E. Sobel. Increased expression of the laminin receptor in human colon cancer. *JNCI* 83:29, 1991.
77. S. Martignone, S. Menard, R. Bufalino, N. Cascinelli, R. Pellegrini, E.Tagliabue, S. Andreola, F. Rilke, and M. I. Colnaghi: Prognostic significance of the 67-kilodalton laminin receptor expression in human breast carcinomas. *JNCI* 85:398, 1993.
78. V. P. Terranova, L. A. Liotta, R. G. Russo, et al.: Role of laminin in the attachment and metastasis of murine tumor cells. *Cancer Res* 42:2265, 1982.
79. J. B. McCarthy, A. P. N. Skubitz, S. L. Palm, and L. T. Furcht: Metastasis inhibition of different tumor types by purified laminin fragments and a heparin-binding fragment of fibronectin. *JNCI* 80:108, 1988.
80. M. J. Humphries, K. Olden, and K. M. Yamada: A synthetic peptide from fibronectin inhibits experimental metastasis of murine melanoma cells. *Science* 233:467, 1986.
81. G. Taraboletti, D. Belotti, R. Giavazzi, M. E. Sobel, and V. Castronovo: Enhancement of metastatic potential of murine and human melanoma cells by laminin receptor peptide G: Attachment of cancer cells to subendothelial matrix as a pathway for hematogenous metastasis. *JNCI* 85:235, 1993.
82. T. Kanemoto, R. Reich, L. Royce, D. Greatorex, S. H. Adler, N. Shiraishi, G. R. Martin, Y. Yamada, and H. K. Kleinman: Identification of an amino acid sequence from the laminin A chain that stimulates metastasis and collagenase IV production. *Proc Natl Acad Sci USA* 87:2279, 1990.
83. P. A. Netland and B. R. Zetter: Organ-specific adhesion of metastatic tumor cells in vitro. *Science* 224:1113, 1984.
84. P. A. Netland and B. R. Zetter: Metastatic potential of B16 melanoma cells after in vitro selection for organ-specific adherence. *J Cell Biol* 101:720, 1985.
85. G. Poste and G. L. Nicolson: Arrest and metastasis of blood-borne tumor cells are modified by fusion of plasma membrane vesicles from highly metastatic cells. *Proc Natl Acad Sci USA* 77:399, 1980.
86. P. P. Bringuier, R. Umbas, H. E. Schaafsma, H. F. M. Karthaus, F. M. J. Debruyne, and J. A. Schalken: Decreased E-cadherin immunoactivity correlates with poor survival in patients with bladder tumors. *Cancer Res* 53:3241, 1993.
87. Y. Doki, H. Shiozaki, H. Tahara, M. Inoue, H. Oka, K. Iihara, T. Kadowaki, M. Takeichi, and T. Mori: Correlation between E-cadherin expression and invasiveness in vitro in a human esophageal cancer cell line. *Cancer Res* 53:3421, 1993.
88. K. Vleminckx, L. Vakaet, Jr., M. Mareel, W. Fiers, and F. V. Roy: Genetic mainpulation of E-cadherin expression by epithelial tumor cells reveals an invasion suppressor role. *Cell* 66:107, 1991.
89. U. H. Frixen, J. Behrens, M. Sachs, G. Eberle, B. Voss, A. Warda, D. Lochner, and W. Birchmeier: E-Cadherin-mediated cell-cell adhesion prevents invasiveness of human carcinoma cells. *J Cell Biol* 113:173, 1991.
90. K. Edvardsen, W. Chen, G. Rucklidge, F. S. Walsh, B. Obrink and E. Bock: Transmembrane neural cell-adhesion molecule (NCAM), but not glycosyl-phosphatidylinositol-anchored NCAM, down-regulates secretion of matrix metalloproteinases. *Proc Natl Acad Sci USA* 90:11463, 1993.
91. R. C. Johnson, D. Zhu, H. G. Augustin-Voss, and B. U. Pauli: Lung endothelial dipeptidyl peptidase IV is an adhesion molecule for lung-metastatic rat breast and prostate carcinoma cells. *J Cell Biol* 121:1423, 1993.
92. R. Wallich, N. Bulbuc, G. J. Hammerling, S. Katzav, S. Segal, and M. Feldman: Abrogation of metastatic properties of tumour cells by de novo expression of H-2K antigens following H-2 gene transfection. *Nature* 315:301, 1985.
93. J. Varani and F. W. Orr: Chemotaxis in non-leukocytic cells, in P. A. Ward, ed.: *Handbook of Inflammation.* Vol. 4. Amsterdam: Elsevier, 1983, pp. 211–244.
94. A. G. Romualdez and P. A. Ward: A unique complement-derived chemotactic factor for tumor cells. *Proc Natl Acad Sci USA* 72:4128, 1975.
95. H. Hayashi, K. Yoshida, T. Ozaki, and K. Ushijima: Chemotactic factor associated with invasion of cancer. *Nature* 226:174, 1970.
96. K. Ushijima, H. Nishi, A. Ishikura, and H. Hayashi: Characterization of two different factors chemotactic for cancer cells from tumor tissue. *Virchows Arch* B21:119, 1976.
97. A. G. Romualdez, P. A. Ward, and T. Torikata:

Relationship between the C5 peptides chemotactic for leukocyte and tumor cells. *J Immunol* 117:1762, 1976.

98. T. Ichikawa, N. Kyprianou, and J. T. Isaacs: Genetic instability and the acquisition of metastatic ability by rat mammary cancer cells following v-H-*ras* oncogene transfection. *Cancer Res* 50:6349, 1990.
99. S. Taniguchi, M. Tatsuka, K. Nakamatsu, M. Inoue, H. Sadano, H. Okazaki, H. Iwamoto, and T. Baba: High invasiveness associated with augmentation of motility in a *fos*-transferred highly metastatic rat 3Y1 cell line. *Cancer Res* 49:6738, 1989.
100. K. Yusa, Y. Sugimoto, T. Yamori, T. Yamamoto, K. Toyoshima, and T. Tsuruo: Low metastatic potential of clone from murine colon adenocarcinoma 26 increased by transfection of activated c-*erb*B-2 gene. *JNCI* 82:1633, 1990.
101. K. Pantel, G. Schlimok, S. Braun, D. Kutter, F. Lindemann, G. Schaller, I. Funke, J. R. Izbicki, and G. Riethmuller: Differential expression of proliferation-associated molecules in individual micrometastatic carcinoma cells. *JNCI* 85:1419, 1993.
102. G. N. Ranzani, N. S. Pellegata, C. Previdere, A. Saragoni, A. Vio, M. Maltoni, and D. Amadori: Heterogeneous protooncogene amplification correlates with tumor progression and presence of metastases in gastric cancer patients. *Cancer Res* 50:7811, 1990.
103. M. Ladanyi, C. Cha, R. Lewis, S. S. C. Jhanwar, A. G. Huvos, and J. H. Healey: MDM2 gene amplification in metastatic osteosarcoma. *Cancer Res* 53:16, 1993.
104. P. G. Cavanaugh and G. L. Nicolson: Purification and some properties of a lung-derived growth factor that differentially stimulates the growth of tumor cells metastatic to the lung. *Cancer Res* 49:3928, 1989.
105. S. C. Bernstein and R. A. Weinberg: Expression of the metastatic phenotype of cells transfected with human metastatic tumor DNA. *Proc Natl Acad Sci USA* 82:1726, 1985.
106. S. M. Phillips, A. J. Bendall, and I. A. Ramshaw: Isolation of gene associated with high metastatic potential in rat mammary adenocarcinomas. *JNCI* 82:199, 1990.
107. T. Ichikawa, Y. Ichikawa, and J. T. Isaacs: Genetic factors and suppression of metastatic ability of v-Ha-ras-transfected rat mammary cancer cells. *Proc Natl Acad Sci USA* 89:1607, 1992.
108. P. S. Steeg, G. Bevilacqua, L. Kooper, U. P. Thorgeirsson, J. E. Talmadge, L. A. Liotta, and M. E. Sobel: Evidence for a novel gene associated with low tumor metastatic potential. *JNCI* 80:200, 1988.
109. C. Hennessy, J. A. Henry, F. E. B. May, B. R. Westley, B. Angus, and T. W. J. Lennard: Expression of the antimetastatic gene nm23 in human breast cancer: An association with good prognosis. *JNCI* 83:281, 1991.
110. T. Nakayama, A. Ohtsuru, K. Nakao, M. Shima, K. Nakata, K. Watanabe, N. Ishii, N. Kimura, and S. Nagataki: Expression in human hepatocellular carcinoma of nucleoside diphosphate kinase, a homologue of the nm23 gene product. *JNCI* 84:1349, 1992.
111. M. Mandai, I. Konishi, M. Koshiyama, T. Mori, S. Arao, H. Tashiro, H. Okamura, H. Nomura, H. Hiai, and M. Fukumoto: Expression of metastasis-related nm23-H1 and nm23-H2 genes in ovarian carcinomas: Correlation with clinicopathology, EGFR, c-erbB-2, and c-erbB-3 genes, and sex steroid receptor expression. *Cancer Res* 54:1825, 1994.
112. V. A. Florenes, S. Aamdal, O. Myklebost, G. M. Maelandsmo, O. S. Bruland, and O. Fodstad: Levels of nm23 messenger RNA in metastatic malignant melanomas: Inverse correlation to disease progression. *Cancer Res* 52:6088, 1992.
113. M. Haut, P. S. Steeg, J. K. V. Willson, and S. D. Markowitz: Induction of nm23 gene expression in human colonic neoplasms and equal expression in colon tumors of high and low metastatic potential. *JNCI* 83:712, 1991.
114. J. Biggs, E. Hersperger, P. S. Steeg, L. A. Liotta, and A. Shearn: A *Drosophila* gene that is homologous to a mammalian gene associated with tumor metastasis codes for a nucleoside diphosphate kinase. *Cell* 63:933, 1990.
115. J. D. Kantor, B. McCormick, P. S. Steeg, and B. R. Zetter: Inhibition of cell motility after nm23 transfection of human and murine tumor cells. *Cancer Res* 53:1971, 1993.
116. E. H. Postel, S. J. Berberich, S. J. Flint, and C. A. Ferrone: Human c-myc transcription factor PuF identified as nm23-H2 nucleoside diphosphate kinase, a candidate suppressor of tumor metastasis. *Science* 261:478, 1993.
117. Z. Zou, A. Anisowicz, M. J. C. Hendrix, A. Thor, M. Neveu, S. Sheng, K. Rafidi, E. Seftor, and R. Sager: Maspin, a serpin with tumor-suppressing activity in human mammary epithelial cells. *Science* 263:526, 1994.

117a. J. Potempa, E. Korzus, and J. Travis: The serpin superfamily of proteinase inhibitors: Structure, function, and regulation. *J Biol Chem* 269:15957, 1994.

117b. Y. Toh, S. D. Pencil, and G. L. Nicolson: A novel candidate metastasis-associated gene *mta*1, differentially expressed in highly metastatic mammary adenocarcinoma cell lines. *J Biol Chem* 269:22958, 1994.

118. J. E. Talmadge, K. Benedict, J. Madsen, and I. J. Fidler: Development of biological diversity and susceptibility to chemotherapy in murine cancer metastases. *Cancer Res* 44:3801, 1984.
119. G. Poste, J. Tzeng, J. Doll, R. Greig, D. Rieman, and I. Zeidman: Evolution of tumor cell hetero-

geneity during progressive growth and individual lung metastases. *Proc Natl Acad Sci USA* 79:6574, 1982.
120. T. Tsuruo and I. J. Fidler: Differences in drug sensitivity among tumor cells from parental tumors, selected variants, and spontaneous metastases. *Cancer Res* 41:3058, 1981.
121. N. Tanigawa, Y. Mizuno, T. Hashimura, K. Honda, K. Satomura, Y. Hikasa, O. Niwa, T. Sugahara, O. Yoshida, D. H. Kern, and D. L. Morton: Comparison of drug sensitivity among tumor cells within a tumor, between primary tumor and metastases, and between different metastases in the human tumor colony-forming assay. *Cancer Res* 44:2309, 1984.
122. E. C. Kohn, and L. A. Liotta: L651582: A novel antiproliferative and antimetastasis agent. *JNCI* 82:54, 1990.
123. J.-I. Inokuchi, M. Jimbo, K. Momosaki, H. Shimeno, A. Nagamatsu, and N. S. Radin: Inhibition of experimental metastasis of murine Lewis lung carcinoma by an inhibitor of glucosylceramide synthase and its possible mechanism of action. *Cancer Res* 50:6731, 1990.
124. G. K. Schwartz, S. M. Redwood, T. Ohnuma, J. F. Holland, M. J. Droller, and B. C.-S. Liu: Inhibition of invasion of invasive human bladder carcinoma cells by protein kinase C inhibitor staurosporine. *JNCI* 82:1753, 1990.
125. M. Yamaoka, T. Yamamoto, T. Masaki, S. Ikeyama, K. Sudo, and T. Fujita: Inhibition of tumor growth and metastasis of rodent tumors by the angiogenesis inhibitor *O*-(chloroacetyl-carbamoyl)fumagillol (TNP-470`M-1470). *Cancer Res* 53:4262, 1993.
126. C. W. Crowley, R. L. Cohen, B. K. Lucas, G. Liu, M. A. Shuman, and A. D. Levinson: Prevention of metastasis by inhibition of the urokinase receptor. *Proc Natl Acad Sci USA* 90:5021, 1993.

12

Host–Tumor Interactions: Tumor Effects on the Host

A malignant tumor growing in vivo produces a number of effects on the host (Table 12-1). The end result of one or more of these effects is what ultimately proves fatal. The effects of a growing cancer in a patient may include fever, anorexia (loss of appetite), weight loss and cachexia (body wasting), infection, anemia, and various hormonal and neurologic symptoms. These may occur out of proportion to the size of the tumor. A relatively small tumor may cause many symptoms, whereas another tumor may produce few symptoms and remain occult until it is far advanced and has metastasized. Malignant tumors affect host functions by compression, invasion, and destruction of normal tissues and also by the elaboration of substances that circulate in the bloodstream. The effects of tumor-produced factors are called collectively the *paraneoplastic syndromes.* Approximately 15% of patients hospitalized with advanced malignancy will have clinically apparent systemic effects in organ systems distant from the primary neoplasm, even though there is no evidence of metastasis to the affected organ.[1] From 50% to 75% of cancer patients eventually experience a paraneoplastic syndrome. A common form of paraneoplastic syndrome is related to ectopic hormone production by growing tumors.

PAIN

Pain is frequently associated with malignant disease. Cancer-induced pain is often unrelenting and difficult to treat, requiring the use of addicting narcotic drugs. It is one of the most difficult problems for the patient, the physician, and the cancer patient's family. The cause of pain in a cancer patient may be destruction of tissue by the tumor, infection, stretching of internal organs, pressure, or obstruction. An example of pain that results from tissue destruction is the bone pain from invasive or metastatic cancers that cause periosteal irritation, pressure in the medullary space of bone, or fractures.

Infection can result from decreased immunity or from an obstruction that decreases drainage from a tissue. Malignant neoplasms originating in mucous membranes, such as those of the oral cavity, vagina, or rectum, may ulcerate early in the course of disease and produce inflammation and infection at the site of ulceration. A painful infection that often occurs in cancer patients with decreased immunity is herpes zoster, which produces a "nerve pain," or causalgia, following the distribution of the affected nerve.

Growth of tumors in areas that have minimal room for expansion results in pain due to pressure. A typical example is a brain tumor, either

Table 12-1 Effects of a Malignant Neoplasm on Host Functions

Effect	Manifestations	Causes
Pain	Pressure, colic, headache, etc., depending on location of tumor	Tissue destruction by tumor, obstruction, pressure on organs, and infection
Cachexia	Muscle wasting, loss of body fat, and generalized weakness	Anorexia, nausea, malabsorption, involvement of digestive organs, metabolic demands of tumor and host's defense system, TNF, lipolytic factor
Anemia	Pallor, weakness, and fatigue	Invasion of bone marrow and crowding out of normal hematopoietic cells, hemolysis, hemorrhage, and decreased erythrocyte production and survival
Leukopenia	Infections	Crowding out of normal marrow cells, chemotherapy, radiotherapy, and trapping of cells in spleen
Thrombocytopenia	Petechia, purpura, and internal hemorrhaging	Crowding out of marrow, chemotherapy, radiation therapy, immune destruction, and hypercoagulability of blood
Blood hyper-coagulability	Thrombophlebitis, disseminated intravascular coagulation, hemorrhage, renal failure, and shock	Release of thromboplastinlike substances by tumor
Fever	Sweating, malaise, and confusion	Infection, pyrogenic substances released by tumor
Hormone release	Examples include Cushing's syndrome (ACTH), water retention (ADH), gynecomastia or precocious puberty (hCG), hypercalcemia (PTH), and hypoglycemia (insulinlike substances)	Ectopic production by tumor, tumor involving organ producing hormone
Hypercalcemia	Lethargy, weakness, nausea, confusion, coma, decreased gastrointestinal tract motility, cardiac arrhythmias, and kidney stones	Tumor invasion or metastasis to bone, ectopic PTH release, and tumor production of osteolytic substances, including prostaglandins, osteoclast-stimulating substance, and parathyroid hormone-related protein (PTHrP)
Neurologic syndromes	Muscle weakness, decreased reflexes, and muscle atrophy	Cachexia, "toxohormones"(?)
Dermatologic involvement	Hyperpigmentation, erythema, petechiae, purpura, hirsuitism, pruritus, and infection (e.g., herpes zoster)	Hormone release by tumor, thrombocytopenia, invasion of skin by tumor, allergic reactions, and decreased immunity

primary or metastatic, enlarging in the cranial vault. Pain due to a brain tumor results from the pressure on blood vessels and membranous septa and may be "referred pain" in the sense that it will occur over the distribution of one of the cranial nerves rather than at the primary site of the tumor. Neoplasms growing in a nasal sinus or an eye produce pain by similar mechanisms.

Involvement of visceral organs may produce pain when the cancerous growth causes stretching of an organ, pressure on an adjacent structure, or obstruction. However, the slow enlargement of a tumor may not cause significant pain, and for this reason may cancers of internal organs, such as the stomach, colon, pancreas, and liver, remain occult until they are far advanced. Pain from visceral organs is frequently referred pain because the stimuli are carried over sympathetic nerves and may enter the spinal cord at an area distant from the site of the tumor. For example, pain from the esophagus is sometimes referred to the shoulder, and stomach pain may be felt more in the chest than in the abdomen. Obstruction of a hollow organ, such as the stomach, small intestine, bile duct, or colon, may produce cramps and colicky pain that may become severe if complete obstruction occurs.

Some cancers produce a characteristic pattern of pain. For example, advanced breast cancer is painful if an inflammatory reaction distends the breast or if an extensive infiltration of the chest wall occurs. Swelling of involved lymph nodes in the axilla may cause severe shoulder or arm pain. Metastases to the liver, ovary, or other visceral organs may produce referred pain, and metastatic involvement of bone may produce pain in the affected part. Metastases to the brain can produce headache. In lung cancer, pain is a late syndrome and indicates local invasion or distant metastases. It may be manifest as severe shoulder or arm pain or bone pain due to periosteal stretching. Extensive skeletal pain may indicate multiple myeloma or advanced prostatic carcinoma. Diffuse bone pain and joint discomfort due to leukemia may simulate arthritic disease.

NUTRITIONAL EFFECTS

Nutritional disorders and malnutrition are frequently the most disabling effects of cancer. These effects are manifest by weight loss, hypermetabolism in body tissues, and ultimately body wasting or cachexia. This tissue wasting affects predominantly muscle and fat but probably involves all organs with the exception of the heart, liver, and brain.[2] The symptoms of malnutrition are not strictly correlated with the size of the tumor or the rate of tumor growth, because some patients with widespread tumors have minimal symptoms until very late in their disease. Furthermore, the amount of cachexia is usually out of proportion to the expected metabolic demands based on the size of a tumor. Ultimately, almost all cancer patients will experience cachexia. The cachexia seen in cancer patients is caused by a variety of factors including reduced appetite (anorexia), decreased digestive functions, metabolic demands of the tumor and of the host's defense systems, and factors ("toxohormones") released by tumors.

Many patients with advanced cancer have anorexia. Nausea resulting from tumor-released factor or from the side effects of therapy (many anticancer drugs produce nausea and vomiting) contributes to the loss of appetite. In addition, abnormalities in taste sensation that make certain foods less palatable have been reported.[3] Certain patients with liver involvement may have a postprandial hyperglycemia that signals the "glucostat" in the central nervous system to reduce appetite.[4] Liver dysfunction and wasting of tissues also produce elevated circulating levels of amino acids and fatty acids that may decrease appetite, probably by a central nervous system–mediated mechanism.[2,4]

Although anorexia is a major contributing factor to the cachexia of advanced cancer, it cannot fully explain the progressive weight loss associated with malignant disease. In certain patients, malabsorption of nutrients from the gastrointestinal tract may occur if the GI tract is obstructed or ulcerated. Direct involvement of the liver, bile duct, or pancreas can reduce the production of bile salts or digestive enzymes needed for digestion of foodstuffs. But even in patients who have no evidence of malabsorption or direct involvement of the digestive system, weight loss frequently occurs. Thus, there must be something related to the cancerous growth itself that contributes to weight loss and cachexia. Observations in experimental animals support this.[5,6] After the inoculation of a tumor into an animal, an initial increase in carcass weight may occur for several days because of fluid retention. Thereafter, the weight of the animal's carcass progressively decreases as the tumor grows. Normal tissues other than the liver, brain, and heart lose weight and have a decreased nitrogen content. With continued tumor growth, the animals go into a negative nitrogen balance, yet the tumor continues to gain weight. Thus, the tumor and the host's tissues appear to comprise two separate metabolic compartments.[5]

A tumor continues to grow and sequester nutrients even in the face of starvation of the host organism; host tissues are, in fact, dismantled, by depletion of protein, fat, and carbohydrate, to feed the tumor. Mider et al.[7] have proposed the idea that malignant tumors are nitrogen "traps" in that the tumor continues to have a positive nitrogen balance in the face of protein loss from the host animal's tissues, and the nitrogen sequestered by the tumor does not become available to the host.

In tumor-bearing animals that are force-fed, tumor growth is stimulated, but no sustained weight gain of the animal occurs.[6] However, it has been found that the relative growth of tumor and host tissues depends on the type of nutrient supplied.[8] If mammary carcinoma–bearing rats are fed carbohydrate alone, neither host nor tumor tissue growth is stimulated. Amino acids alone, given in amounts adequate to improve host nitrogen balance, stimulate tumor growth. Adequate amino acids and carbohydrates given together induce optimal tumor growth and maintain host tissues. In contrast, when a diet that is isocaloric with the carbohydrate–amino acid diet but consists of both fat and amino acids is fed intravenously to the animals, host tissue is maintained and no stimulation of tumor growth is observed. These data suggest that the host's normal tissues and the mammary carcinoma have different primary mechanisms for energy metabolism. If these tumors have a high rate of anaerobic glycolysis, for example they would be expected to consume large amounts of glucose or gluconeogenic precursors (at the expense of the host) during active growth. Thus, when animals are fed amino acids alone or amino acids plus glucose, tumor growth is stimulated. However, when calories are provided as a mixture of a nongluconeogenic substrate (i.e., fat) and amino acids, the host tissues, which can use nongluconeogenic substrates more efficiently than the tumor, regain some advantage for maintenance of their own growth. These kinds of results hold out the hope that appropriate alimentation of cancer patients could maintain their nutritional balance without stimulating tumor growth. In a study of patients with small-cell lung cancer, however, only short-term gains in weight and caloric intake were achieved when these patients were placed on total parenteral nutrition for 4 weeks; no long-term differences in nutritional status were observed, nor was there any significant improvement in response rates to chemotherapy or in overall survival.[9]

Experiments in tumor-bearing animals indicate a number of mechanisms by which a tumor growing in a host can cause increased use of energy and body wasting even in the face of a normal caloric intake. A study of sarcoma-bearing rats, before the onset of cachexia, has shown that there is an increased rate of glucose turnover, secondary to both an increased rate of gluconeogenesis and an increased rate of recycling of glucose, mostly from normal tissues to tumor.[10] The tumor-bearing animals had significantly lower plasma glucose and higher blood lactic acid levels compared with non-tumor-bearing controls, but this abnormality was not due to changes in serum insulin or glucagon levels. The sarcoma-bearing rats also had a higher rate of glucose production than controls, most likely because of increased glucose synthesis from lactate in the host's liver and gluconeogenesis from amino acid precursors. The explanation for these findings is that the tumor has a high rate of glucose utilization because of its high rate of glycolysis. This increased consumption of glucose by the tumor forces the host to increase endogeneous production of glucose to maintain glucose levels. To do this, the host increases the rate of both glucose recycling (from lactate) and gluconeogenesis. These are energy-inefficient processes, and along with the high metabolic demands of the tumor itself, place metabolic trains on the host that, even if normal caloric intake is maintained, can lead to weight loss and eventually to cachexia.

Other animal studies support these conclusions. For example, Cameron and Ord[11] found that rats bearing Morris hepatomas relied on gluconeogenesis to maintain glucose levels, mobilized more liver glycogen, catabolized more of their muscle proteins, and had higher blood lactate levels than normal control animals. Lindmark et al.[12] found that sarcoma-bearing mice had an increased fat oxidation and loss of body lipids as well as a significantly higher energy expenditure in relation to their food intake compared with pair-fed controls.

Studies in cancer patients also bear out the findings with tumor-bearing animals. Noncachectic lung cancer patients, under conditions of constant caloric and nitrogen intake, had a significantly higher turnover rate of total body protein, as measured by continuous infusion of [^{14}C]lysine, than did control individuals.[13] In addition, muscle catabolism rates, determined by 3-methylhistidine/creatinine excretion rates, were elevated in the lung cancer patients. Glucose production rates were also higher in the patient group, but serum levels of ACTH, insulin, and glucagon were normal, as were 24-hour urinary cortisol levels. Thus, abnormalities in the levels of these hormones could not explain the increased protein turnover, glucose production, and muscle catabolism noted in these patients. Increased synthesis and breakdown of whole-body protein has also been observed in children with newly diagnosed leukemia or lymphoma.[14] In another study, comparing malnourished cancer patients with malnourished patients without cancer, it was found that the cancer patients had a doubling of glucose turnover, indicating that cancer patients have an increased glucose drain, which could account for a loss of about 0.9 kg of body fat per 30-day period.[15] In cachectic patients with colorectal cancer compared with a group of age-matched normal subjects, elevated rates of glucose production and recycling by means of lactate have been observed.[16]

In addition to the great nutrient demands of growing tumor tissue because of high growth fraction and their inefficient use of substrates (e.g., high rate of glycolysis), the demands placed on the body's defense systems may expend a significant amount of energy. Maintaining cells with a high turnover rate, such as cells of the bone marrow involved in the production of granulocytes, lymphocytes, and monocytes, requires a tremendous amount of energy. Even under normal circumstances, the bone marrow produces several billion cells a day, and the increased demands placed on that system by a chronic illness such as cancer may be very significant. Decreased protein synthesis and loss of protein in the urine may also account for nitrogen loss in some patients. For example, patients with advanced cancer often have low albumin levels in their blood, and this is at least partly due to decreased albumin synthesis in the liver.[2] Many cancer patients also have significant proteinuria[17] and lose protein that way. Thus, the greatly increased demands on the cancer patient's energy stores produces a generalized hypermetabolic state with an increased rate of turnover of normal tissue components.

The production and release of certain substances from tumors have sometimes been invoked as the mecahnism of cachexia in cancer patients. Nakahara and Fukuoka[18] observed that the decreased hepatic catalase activity often seen in tumor-bearing animals and patients[19] could be produced by injection of a water-soluble, thermostable, and ethanol-precipitable material that was extracted from human gastric or rectal carcinoma. They called this *toxohormone.* This substance appears to be a polypeptide, and when it is injected into normal animals it produces, in addition to decreased liver catalase activity, decreased levels of plasma iron, liver ferritin, and diphosphopyridine nucleotides. It also causes involution of the thymus, an enlarged liver and spleen, and an increased level of liver protoporphyrin. In addition, substances that uncouple oxidative phosphorylation in normal liver mitochondria have been found in the rat Novikoff hepatoma[20] and in the serum of sarcoma-bearing rats.[21] A cytotoxic "diffusible polypeptide" that inhibits the growth of a variety of cells has been found in ascites or pleural fluids from tumor-bearing animals and cancer patients.[22] The implication of these data is that tumors secrete substances that can have a far-reaching effect on host functions and that these substances may contribute significantly to the aberrant metabolism and cachexia observed in tumor-bearing animals. It remains to be proven, however, that such substances cause these phenomena in the cancer patient.

Another factor that may be involved in the metabolic changes seen in cancer patients is a pyrogen released by leukocytes called *endogenous pyrogen* (now known to be the same as interleukin-1). It has been observed that human leukocytic pyrogen, isolated from monocytes, stimulates skeletal muscle protein degradation in vitro.[23] Thus, it has been speculated that fever and increased protein turnover, both of which are commonly seen in cancer patients, are mediated by the same factor.[14]

One of the cachexia-inducing factors present in the serum of cancer patients is tumor necrosis factor-α (TNF-α), also known as cachectin. Although TNF-α levels are elevated in the serum of some cancer patients,[24] there is often no correlation with elevated serum TNF-α levels and weight loss (reviewed in Ref. 25). Thus, other cachectic factors are likely to play a role in this phenomenon. One such factor may be a lipolytic factor identified in extracts of the murine colon adenocarcinoma MAC16, which when injected into mice bearing a non–cachexia-inducing tumor induced weight loss without a significant reduction in food intake.[25] A similar factor was found in the serum of cancer patients with cachexia. In this same MAC16 model system, a diet rich in fish oil containing ω-3 fatty acids significantly reduced weight loss.[26] This was associated with a maintenance of total body fat and muscle mass.

HEMATOLOGIC EFFECTS

Hematologic complications occur frequently in patients with cancer, particularly in those with disseminated cancer. Depression of the hematopoietic tissues of the bone marrow can occur as a result of direct invasion by cancer cells or the abnormal expansion of the immature bone marrow cell compartment that occurs in leukemia. Metastases to the bone marrow occur in patients with carcinomas of the breast, prostate, lung, adrenal, thyroid, and kidney; malignant melanoma; neuroblastoma; Hodgkin's disease; reticulum cell sarcoma; and other lymphomas. Bone marrow depression also occurs as a complication of chemotherapy or radiotherapy. As in the case of the nutritional abnormalities of cancer patients, indirect effects of cancer growing elsewhere than in the bone marrow can also be observed, possibly because of the remote effects of tumor-released products.

Erythropoiesis

The most common hematologic effect of cancer is anemia. Anemia can develop in cancer patients by a number of mechanisms, including malnutrition, blood loss due to neoplastic invasion of tissues, autoimmune hemolysis, and decreased red blood cell production and survival.[27] The decreased nutritional status of cancer patients can lead to a secondary anemia resulting from decreased levels of folic acid and other essential nutrients. Folic acid deficiency is common in patients who have difficult eating, such as those with head and neck cancers, and in cachectic patients. Malabsorption of folate has been reported in patients with intestinal lymphoma, reticulum cell sarcoma, and leukemia and also in patients who have had bowel resections or extensive radiation to the abdomen. Decreased vitamin B_{12} absorption, coupled with anemia, is a common finding in patients with gastric carcinoma. Anemia secondary to ulcerating lesions is commonly seen in cancers of the gastrointestinal tract, head and neck, urinary bladder, and uterus. This blood loss may be occult and leads in time to an iron-deficiency anemia. Iron-deficiency anemia can also occur in cancer patients as a result of malabsorption or hemolysis. An immune-type hemolytic anemia resulting from production of antibodies directed against the patient's own erythrocytes is seen in certain types of cancer, particularly those involving the lymphatic and reticuloendothelial system (e.g., chronic lymphocytic leukemia, Hodgkin's disease, and other malignant lymphomas). The mechanism of this apparent autoimmunity is not well understood. A number of cancer patients have a decreased total plasma iron-binding capacity and an increased rate of removal of iron from the plasma. In many of these patients, however, erythropoiesis is not decreased and, in fact, may be higher than normal, but the rate of erythrocyte destruction is elevated.[28] The decreased survival of erythrocytes appears to be due to their damage or destruction as they pass through the tumor mass[29] or to production of a hemolytic substance by the tumor[30] or to a combination of these factors.

The opposite of anemia, namely, a marked increase in circulating erythrocytes (erythrocytosis), occurs in some cancer patients, most notably those with renal tumors, cerebellar hemangioblastomas, and hepatocellular carcinomas, and less frequently in patients with uterine fibroma, pheochromocytoma, adrenal adenoma, ovarian carcinoma, and carcinoma of the lung.[27] This erythrocytosis appears to be caused by the ectopic production of erythropoietin by the tumors.

Leukopoiesis

Leukopenia can occur in cancer patients as part of the bone marrow depression seen with marrow metastases. The mechanism is thought to be a "crowding out" of the hematopoietic cells of the marrow. A decrease in circulating white blood cells is also sometimes observed in cancer patients with tumor-involved enlarged spleens, presumably as a result of increased trapping of these cells by the spleen. The most common cause of leukopenia in cancer patients, however, is associated with the bone-marrow-depressing effects of cancer chemotherapeutic agents and radiation therapy.

Increased circulating leukocytes (leukocytosis) is, of course, seen in leukemias, in which the nature of the disease process itself results in an overproduction of immature leukocytes, but it is also observed in patients with other cancers, usually metastatic ones. The stimulus for this leukocytosis is not understood, but may involve the release of colony-stimulating factors (CSFs) or other leukocyte-stimulating growth factors by tumors that favor leukopoiesis similar to the erythropoietic effect noted previously. Because many cancer patients have chronic infections of one sort or another, the leukocytosis seen in many patients may be secondary to the infectious process.

Platelets

Depression of the number of circulating platelets (thrombocytopenia) is the most common platelet abnormality observed in cancer patients. A decreased platelet count produces bleeding due to the accompanying blood coagulation defect. The most common cause of thrombocytopenia in cancer patients is impaired production of megakaryocytes, the platelet progenitor cell, in the bone marrow. The mechanism for this is similar to that described for leukopenia (see above)—that is, marrow infiltration by tumor and increased sequestration of platelets in the spleen. An immune type of thrombocytopenia, similar to that described for immune hemolysis and resulting in increased platelet destruction, has also been described.[31] Increased platelet consumption also leads to a lowered platelet count; blood hypercoagulability, observed in some cancer patients, is responsible. In some cases, hypercoagulability is widespread and leads to a phenomenon known as disseminated intravascular coagulation (DIC), characterized by an extensive activation of the blood coagulation system within the circulatory system and the deposition of fibrin clots in small blood vessels. It has been postulated that DIC is due to the release of thromboplastinlike substances by the tumor.[32,33] Some studies indicate that a low-grade DIC occurs in many cancer patients.[34] The intravascular coagulation that typifies this syndrome consumes clotting factors, particularly factors I (fibrinogen), II, V, VII, and XIII, and platelets, thus decreasing the platelet count. Secondarily, the increased fibrin deposition activates the plasminogen system, eliciting an increased activity of the major fibrinolysin, plasmin, which, in turn, leads to the presence of fibrin degradation products in the blood. The most common clinical manifestation of DIC is bleeding because of the consumption of coagulation factors and platelets. The bleeding may be intermittent or continuous and tends to occur in the urinary tract, gastrointestinal tract, lungs, and skin. Organ failure (usually renal) secondary to localized obstruction of the microcirculation and shock due to generalized obstruction of the microcirculation are other clinical effects of DIC.

A few cancer patients may display thrombocytosis, presumably as the result of stimulation of megakaryocyte proliferation in the marrow; however, the mechanism for this is not known.

Activation of the coagulation pathway in cancer patients may be nonspecific in that it results from local obstruction of a duct or bronchus followed by thrombus formation, an inflammatory response to the tumor, necrosis of tumor tissue, or shedding of mucins or other tumor cell surface components into the bloodstream.[35] It can also be specific in that thrombin generation can occur through tumor production of a cytokine that activates host endothelial cells or monocytes or by induction of thrombin formation by tumor cell–derived procoagulants. Some tumors tend to produce procoagulants that induce fibrin formation. These include small-cell lung, renal cell, and ovarian carcinomas and malignant melanomas. Others release urokinase (u-PA), which converts plasminogen to plasmin and activates the coagulation pathway by this mechanism.[35,36] These latter tumors include non-small-cell lung, colon, prostate, and breast carcinomas. There is also a correlation between a high propensity to form deep vein thromboses and the subsequent development of invasive cancer.[37]

FEVER AND INFECTION

Fever is a common manifestation of malignant neoplastic disease. It is due to the systemic effects of the malignant process itself or to infection, although the latter is the most frequent cause. The incidence of fever is high in patients with advanced disease; about 70% of cancer patients experience febrile episodes during hospitalization.[38,39] The incidence of fever is related to the type of cancer. Patients with acute leukemia have elevated temperatures during about 50% of their hospitalization; patients with Hodgkin's disease have elevated temperatures about 26% of the time. Other diseases with a high incidence of fever are lymphosarcoma and reticulum cell sarcoma. Fever is also common in patients with widely disseminated "solid" cancers, such as metastatic carcinomas of lung, kidney, pancreas, and gastrointestinal tract, and particularly in those patients with liver involvement.[38] The presence of fever often correlates with a poor prognosis. The mechanism of tumor-induced fever is not well established, but there is evidence for a fever-inducing (pyrogenic) substance in the urine of febrile patients with Hodgkin's disease[40] and in tissue culture fluids of lymphoid cell cultures derived from lymphoma patients.[41]

Patients with cancer have an increased incidence of infection with such pathogenic organisms as staphylococci, streptococci, and pneumococci, which frequently cause infections in patients without cancer. In addition, cancer patients become infected with "opportunistic" microorganisms that are normally held in check by host defense mechanisms. These latter organisms include bacteria of the *Pseudomonas* genus, fungi, such as *Candida albicans,* and viruses such as herpes zoster. Infection is present at the time of death in 70% to 80% of patients with leukemia or lymphoma and in 15% to 40% of patients with metastatic carcinoma.[38]

The type of infection depends to some extent on the nature of the impairment of the host's defense mechanisms (Table 12-2).[42] For example, patients with marked granulocytopenia most frequently develop infections of the respiratory tract and gastrointestinal tract with organisms such as *Escherichia coli, Klebsiella pneumonia, Pseudomonas aeruginosa,* or *Staphylococcus aureus.* Patients whose primary deficiency is in cell-mediated immunity tend to develop infections with *Listeria, Salmonella,* and *Mycobacterium* species, for example. Patients with a primary deficiency in humoral immunity are at increased risk for infections by *Streptococcus pneumoniae* and *Hemophilus influenzae.*

Lung, skin, gastrointestinal tract, and urinary tract are the most common sites of infection in cancer patients. The site of infection is often determined by the location of the primary tumor or its metastases. For example, pneumonia and abscess formation in the lung occur peripherally to tumors blocking major bronchi, and urinary

Table 12-2 Factors Predisposing to Infection among Patients with Cancer and Organisms Commonly Involved

Granulocytopenia (e.g., acute leukemia)
- Usually with associated damage to body barriers (especially alimentary canal mucosa, respiratory tract ciliary function, and integument)
  - Common organisms
    - Gram-negative bacilli: *Pseudomonas aeruginosa, Klebsiella pneumoniae,* and *Escherichia coli*
    - Gram-positive cocci: *Staphylococcus aureus* and *Staphylococcus epidermidis*
    - Yeasts: *Candida* species, *Torulopsis glabrata*
    - Fungi: *Asperigillus* species, *Mucor*

Cellular immune defiency (e.g., lymphoma)
- Common organisms
  - Bacteria: *Listeria monocytogenes, Salmonella* species, *Mycobacterium* species, *Nocardia asteroides,* and *Legionella pneumophilia*
  - Viruses: varicella zoster, herpes simplex, and cytomegalovirus (CMV)
  - Fungi: *Cryptococcus neoformans, Histoplasma capsulatum,* and *Coccidioides immitis*
  - Protozoa: *Pneumocystis carinii* and *Toxoplasma gondii*
  - Helminths: *Strongyloides stercoralis*

Humoral immune dysfunction (e.g., multiple myeloma)
- Common organisms: *Streptococcus pneumoniae* and *Hemophilus influenzae*

Obstruction to natural passages (e.g., solid tumors)
- Common sites: respiratory tract, biliary tract, and urinary tract
- Common organisms: locally colonizing forms

Central nervous system dysfunction (e.g., brain tumors)
- Common sites: pneumonitis and urinary tract infection
- Common organisms: locally colonizing forms

Infections associated with medical procedures
- Procedures: intravascular catheters, urinary catheters, and respiratory assist devices
- Common organisms: locally colonizing forms

Source: From Pizzo and Schimpff.[42]

tract infections occur in patients whose tumors obstruct the ureter or urinary bladder. The increased susceptibility of cancer patients to infection, however, is most often related to depression of normal host defense mechanisms induced by the disease process itself or by chemotherapy or irradiation. Depression of granulocytes, the humoral immune response, and cell-mediated immunity all occur in cancer patients as a result of the disease process (e.g., bone marrow involvement) or immunosuppressive therapy. Granulocytopenia is responsible for many of the bacterial infections seen in cancer patients, especially those with hematologic malignancies or metastatic involvement of the marrow. Patients with chronic lymphocytic leukemia often have low levels of immunoglobulins, and bacterial infections occur in a high percentage of these patients. Cell-mediated immunity plays the key role in the body's defense against certain bacteria (e.g., *Mycobacterium tuberculosis*), fungi (e.g., *Cryptococcus*), and viruses (e.g., herpes), and infections with these organisms are frequently seen in patients with malignant lymphoma (e.g., Hodgkin's disease).

Before the widespread use of antibiotics, pneumococci, streptococci, and staphylococci were major causes of infection in cancer patients. More recently the biggest problem has been with such Gram-negative bacilli as *Pseudomonas aeruginosa, Klebsiella* species, and *E. coli.* About 25% of such infections are caused by *P. aeruginosa,* and this organism accounts for about 35% to 50% of the fatal septicemias in patients with acute leukemia.[38] The lung is the most common site of infection with *Pseudomonas* organisms. Yeast infections, particularly with *Candida* and *Aspergillus,* have become increasingly prevalent in hospitalized cancer patients, especially those with hematologic malignancies or lymphomas. *Candida* infections are the most common; they usually occur in the mouth and gastrointestinal tract, but the infection may become disseminated. The incidence of viral infections is also high in cancer patients. For example, herpes zoster, which is rare in the general population over 50 years of age, occurs in 3% to 15% of patients with lymphoma, multiple myeloma, or chronic lymphocytic leukemia.[38] Because of the weakened condition of patients with advanced cancer and their markedly compromised defense mechanisms, their infections are extremely difficult to control and are often the ultimate cause of death. In patients with hematologic malignancies and lymphomas, infection is the most common lethal event.

HORMONAL EFFECTS

In Chapter 4, the production of ectopic hormones was discussed. Since the biochemical form of the hormone released by the tumor may have low biological activity (e.g., "big" ACTH secreted by lung tumors), the incidence of clinical syndromes related to ectopic hormones underestimates the actual frequency of this phenomenon. Nevertheless, a number of instances of clinical syndromes related to ectopic hormone production have been clearly defined. In addition, the clinical symptoms exhibited by patients with ectopic humoral syndromes are often more complex than would be expected from the overproduction of a single hormone. This is probably because the complex host–tumor interactions may modify or mask the hormonal effects. For example, in a patient with severe infection and cachexia, abnormal physiologic events resulting from ectopic hormone production may be attributed to other causes. Also, some tumors produce multiple hormones, some of which may have different actions.[43,44] A clinical rule of thumb, however, is that "if a patient has overproduction of a hormone, look for a tumor, and if a patient has a tumor, look for evidence of hormone overproduction."[45]

HYPERCALCEMIA

Elevated plasma calcium occurs in 10% to 20% of patients with disseminated cancer[46] and is the most frequent paraneoplastic syndrome. This condition affects multiple organ systems, and the resulting events may be more immediately life-threatening than the cancer itself. Manifestations of hypercalcemia include central nervous system effects (e.g., confusion, psychotic behavior, coma), gastrointestinal effects (e.g., decreased motility, ulceration), kidney stones and salt loss by the kidney, and cardiac arrythmias and cardiovascular system collapse. A number of mechanisms may operate in cancer patients to produce hypercalcemia. An obvious one is the direct invasion of bone by metastatic tumor cells. This may lead to direct destruction of bone, with release of calcium into the bloodstream. Hypercalcemia has been observed, however, in patients in whom there is no evidence of bony metastasis, and this may account for at least 15% of cases of malignant hypercalcemia.[47] Carcinomas of the breast, lung, kidney, thyroid, ovary, and colon are tumors that commonly have bony metastases and produce hypercalcemia. In a number of cases, however, there is no obvious correlation between clinically detectable bone involvement and the degree of hypercalcemia.[48] The ectopic secretion of parathyroid hormone (PTH), which increases calcium resorption from bone, has been noted for certain tumor types, particularly those of the lung and kidney. There is, however, a group of hypercalcemic patients without evident bone metastases and in whom there is no increased level of PTH.[49]

Other factors or conditions thought to play a role in malignant hypercalcemia are prostaglandins, particularly PGE_2 and its metabolites, colony-stimulating factors (CSFs), transforming growth factors (TGFs), interleukin-1, TNF-α, TNF-β and certain osteolytic polypeptides not yet completely characterized. In addition, alterations in calcium absorption from the gastrointestinal tract have been invoked as a possible mechanism (reviewed in Refs. 50 and 51).

Extracts of certain patients' tumors contain an activity that is osteolytic in bone tissue culture systems, suggesting that certain tumors themselves produce osteolytic substances. Both animal and clinical studies suggest that one of these substances is a prostaglandin. For example, $HSDM_1$ fibrosarcoma-bearing mice[52] and VX_2 carcinoma-bearing rabbits[53] develop hypercalcemia without evidence of bony metastases, and high concentrations of prostaglandin E_2 (PGE_2) are found in the tumor masses and venous drainage from the tumors. Treatment of rats bearing the Walker 256 carcinosarcoma with indomethacin or aspirin, drugs that block prostaglandin synthesis, prevents tumor growth in bone.[54] In a study by Voelkel et al.[55] of VX_2 carcinoma-bearing rabbits, it was found that PGE_2 and a metabolite of PGE_2, called PGE_2-M, were elevated in the plasma in parallel with the acute-phase reactant ceruloplasmin, and the elevation of PGE_2-M levels preceded by 7 to 10 days the development of hypercalcemia. Administration of indomethacin to these animals completely prevented the hypercalcemia and largely prevented the elevated ceruloplasmin levels, suggesting that PGE_2 production by the tumor is involved both in the resorption of calcium from bone and the increased synthesis or release of acute-phase reactants by the liver. Moreover, PGE_2 has potent bone-resorbing activity in vitro.[52,56] Several clinical studies also intimate that prostaglandins have a role in malignant hypercalcemia. Seyberth et al.[57] have reported that, in hypercalcemic patients with a variety of solid tumors, the urinary excretion of PGE_2-M was increased and that both serum calcium concentrations and PGE_2-M excretion fell after treatment with indomethacin or aspirin. In studies by Robertson et al.,[58,59] about 35% of hypercalcemic patients had undetectable PTH but elevated plasma PGE. These investigators also found that elevated plasma PGE levels served to predict a therapeutic response of hypercalcemia to indomethacin. Breast cancers have been reported to contain more prostaglandins than uninvolved adjacent breast tissue or benign tumors from other patients.[60] Hypercalcemia or bone metastases, or both, were present in patients whose breast cancers contained large amounts of PGE. A high percentage of human breast cancers have also been reported to produce osteolytic activity in vitro, and in a significant number of these cultures the osteolytic activity was inhibited by aspirin.[61] These data strongly suggest that prostaglandins, particularly those of the PGE family, have a significant role in causing malignancy-induced hypercalcemia. As in the case of the proteolytic activities found in and around tumors, however, it is not clear how much of the prostaglandin is produced by the tumor cells themselves and how much is released by host cells perhaps involved in the body's reaction to an invading malignant tumor.[62]

Hypercalcemia has also been observed in association with human cancers that produce a CSF-type factor. Two patients with squamous cell carcinomas were found to produce CSF and to have marked hypercalcemia.[63] When these tumors were transferred into nude mice, they produced granulocytosis and hypercalcemia in the animals, which promptly returned to normal when the tumors were excised. Treatment of the mice with prednisone or with indomethacin had no effect on serum calcium levels, and neither PTH nor PGE levels were elevated in the serum. Clear evidence of bone resorption was present in the tumor-bearing mice. These results suggest a paraneoplastic syndrome of granulocytosis and hypercalcemia associated with tumor production of CSF.

TGFs may also be involved in the induction of malignant hypercalcemia in tumor-bearing animals and cancer patients. Both conditioned medium from cultured rat Leydig cell tumors and extracts of Leydig cell tumors growing in vivo produce a bone-resorbing factor with the biochemical characteristics of a TGF-α-like substance (e.g., competes for binding of EGF with its receptors, stimulates colony growth in soft agar).[64] That TGF-α can cause bone resorption has been confirmed by experiments in which pure TGF-α was added to fetal rat long bones previously labeled by incorporation of $^{45}Ca^{2+}$ (bone-resorbing activity was measured by release of $^{45}Ca^{2+}$).[65] Moreover, addition of human TGF-α and TGF-β to organ cultures of mouse calvaria has been shown to stimulate production of PGE_2 and to increase bone resorption, both of which were inhibited by indomethacin.[66] These data suggest that TGFs stimulate bone resorption by a mechanism that involves enhanced local production of PGE_2.

The most important factor that causes hypercalcemia in human cancer patients appears to be parathyroid hormone–related protein (PTHrP). PTHrP was isolated from a cell line established from a lung cancer of a patient with hypercalcemia and the gene subsequently cloned.[67] The PTHrP gene has six exons that give rise to three mRNAs by alternative splicing. These transcripts encode three related proteins with a common sequence of 139 amino acids which differ only in the C-terminus, giving rise to three polypeptides 139, 141, and 173 amino acids in length (reviewed in Ref. 68). The first 13 amino-terminal amino acids are 70% homologous to PTH, and in a variety of biological assays a polypeptide containing the first 34 amino acids of PTHrP induces bone resorption, hypercalcemia, phosphate transport, and cAMP production in bone and kidney cells.[68] Immunohistochemical staining of several human cancers, including lung, bladder, and colon carcinomas as well as osteogenic sarcomas, revealed the presence of PTHrP.[51] Immunoreactive PTHrP was found in 21 of 22 tumor tissues from patients with hypercalcemia but in only 5 of 34 cancers from patients with normal calcium levels.[69] PTHrP mRNA levels detected by Northern blot analysis were significantly higher in breast cancer tissues from women who subsequently developed bone metastases, suggesting that PTHrP gene expression is associated with the onset of such metastases.[68]

Additional evidence for the role of PTHrP in humoral hypercalcemia of malignancy (HHM) comes from studies showing that injection of a neutralizing antibody to PTHrP into hypercalcemic nude mice bearing a PTHrP-producing human squamous cell lung carcinoma resulted in a decrease in serum calcium levels, a decrease in bone resorption, and an increase in bone formation.[70] These effects were similar to those seen when the tumors were surgically excised, indicating that the source of the PTHrP was the human cancerous tissue.

NEUROLOGIC EFFECTS

Patients with neoplastic disease may have disorders of the nervous or muscular systems that are related to the presence of a malignant tumor, but not directly due to invasion or metastases of these systems. A term used to describe these effects is *carcinomatous neuromyopathy*. These disorders have been reported to occur in about 7% of cancer patients, most often in those with carcinomas of the lung, ovary, stomach, prostate, breast, colon, and cervix, in that order.[71] Although some of the neuromuscular effects can be explained on the basis of cachexia, many cannot. The most common abnormality is muscle wasting and weakness with decreased tendon reflexes, out of proportion to the degree of cachexia.[71] This is seen more often in patients with extensive local or metastatic disease. Actual degeneration and demyelinization of peripheral nerves is observed in some patients. Other examples of neuromuscular syndromes reported in cancer patients are inflammatory degenerative diseases (polymyositis) and a myasthenia-gravis–

like syndrome. The cause of these effects is unknown, but they may be due to the release of toxic, biologically active substances by malignant neoplasms.

DERMATOLOGIC EFFECTS

Cutaneous manifestations are frequent concomitants of certain forms of cancer.[72] Examples of this include hyperpigmentation of the skin, reddening or flushing (erythema), bleeding into the skin (purpura), abnormal hair growth (hirsutism), and itching (pruritus). Skin pigmentation can occur as a result of ectopic release of ACTH or melanocyte-stimulating hormone (MSH) or as a result of excess corticosteroid release by adrenal gland invasion. An unusual dark hyperpigmentation of the skin associated with skin hyperplasia, called acanthosis nigricans, occurs in the axillae, lower back, neck, groin, and antecubital spaces of some patients with carcinomas of the gastrointestinal tract, uterus, prostate, ovary, and kidney and with lymphomas. Erythema tends to occur more frequently in patients with lymphomas, leukemias, and carcinomas of the gastrointestinal tract, breast, lung, and cervix. Bleeding into the skin may take the form of small, pinpoint purplish red spots (petechiae) or larger, more diffuse areas of hemorrhage (purpura). These occur in patients with thrombocytopenia resulting from marrow involvement or as a result of chemotherapy or radiation therapy. Hirsutism occurs with some ovarian cancers and adrenal tumors that secrete male hormones. Pruritus is common in Hodgkin's disease and other lymphomas, and also occurs occasionally in patients with various carcinomas (e.g., of the pancreas, stomach, brain).

In some cases, the causes of the dermatologic effects of cancer are clearly defined—for example, in the case of the known biological effects of excess hormonal production, or the results of decreased platelet count—but in others (e.g., pruritus, acanthosis nigricans, erythema), the cause is not known.

Kaposi's sarcoma is a malignancy of endothelial cell origin that manifests itself by vascular, raised or flat, pink, brown, or blue lesions that can arise anywhere on the skin or in the mucosa of the mouth or gastrointestinal tract. The lesions can resemble bruises or nevi and occasionally they appear first in enlarged lymph nodes. Although formerly this was a rare tumor, usually seen in elderly men of Mediterranean descent, it occurs with increased frequency in AIDS patients.

REFERENCES

1. T. C. Hall: The paraneoplastic syndromes, in P. Rubin, ed.: *Clinical Oncology,* 4th ed. Rochester: American Cancer Society, 1974, pp. 119–128.
2. C. Waterhouse: Nutritional disorders in neoplastic disease. *J Chronic Dis* 16:637, 1963.
3. W. DeWys: Metastases and disseminated cancer, in P. Rubin, ed.: *Clinical Oncology,* 4th ed. Rochester: American Cancer Society, 1974, pp. 508–525.
4. W. DeWys: Working conference on anorexia and cachexia of neoplastic disease. *Cancer Res* 30:2816, 1970.
5. S. D. Morrison: Partition of energy expenditure between host and tumor. *Cancer Res* 31:98, 1971.
6. A. Theologides: Pathogenesis of cachexia in cancer: A review and a hypothesis. *Cancer* 29:484, 1972.
7. G. B. Mider, H. Tesluk, and J. J. Morton: Effect of Walker carcinoma 256 on food intake, body weight, and nitrogen metabolism in growing rats. *Acta Univ Int Cancer* 6:409, 1948.
8. G. P. Buzby, J. L. Mullen, T. P. Stein, E. E. Miller, C. L. Hobbs, and E. F. Rosato: Host-tumor interaction and nutrient supply. *Cancer* 45:2940, 1980.
9. W. K. Evans, R. Makruck, G. H. Clamon, R. Feld, R. S. Weiner, E. Moran, R. Blum, F. A. Shepherd, K. N. Jeejeebhoy, and W. D. DeWys: Limited impact of total parenteral nutrition on nutritional states during treatment for small cell lung cancer. *Cancer Res* 45:3347, 1985.
10. M. E. Burt, S. F. Lowry, C. Gorshboth, and M. E. Brennan: Metabolic alterations in a noncachectic animal tumor system. *Cancer* 47:2138, 1981.
11. I. L. Cameron and V. A. Ord: Parenteral level of glucose intake on glucose homeostasis, tumor growth, gluconeogenesis, and body consumption in normal and tumor-bearing rats. *Cancer Res* 43:5228, 1983.
12. L. Lindmark, S. Edstrom, L. Ekman, I. Karlberg, and K. Lundholm: Energy metabolism in nongrowing mice with sarcoma. *Cancer Res* 43:3649, 1983.
13. D. Heber, R. T. Chlebowski, D. E. Ishibashi, J. N. Herrold, and J. B. Block: Abnormalities in glucose and protein metabolism in noncachectic lung cancer patients. *Cancer Res* 43:4815, 1982.
14. C. L. Kien and B. M. Camitta: Increased whole-body protein turnover in sick children with newly

diagnosed leukemia or lymphoma. *Cancer Res* 43:5586, 1983.
15. E. Edén, S. Edström, K. Bennegard, T. Scherstén, and K. Lundholm: Glucose flux in relation to energy expediture in malnourished patients with and without cancer during periods of fasting and feeding. *Cancer Res* 44:1718, 1984.
16. C. P. Holroyde, C. L. Skutches, G. Boden, and G. A. Reichard: Glucose metabolism in cachectic patients with colorectal cancer. *Cancer Res* 44:5910, 1984.
17. D. Rudman, R. K. Chawla, D. W. Nixon, W. R. Vogler, J. W. Keller, and R. C. MacDonnell: A system of cancer-related urinary glycoproteins: Biochemical properties and clinical applications. *Trans Assoc Am Physicians* 90:286, 1977.
18. W. Nakahara: Toxohormone, in H. Busch, ed.: *Methods in Cancer Research.* Vol. II, New York: Academic Press, 1967, pp. 203–237.
19. T. Ohnuma: Hepatic catalase, in J. F. Holland and E. Frei III, eds.: *Cancer Medicine.* Philadelphia: Lea & Febiger, 1973, pp. 1044–1046.
20. T. M. Devlin and M. P. Pruss: Oxidative phosphorylation in transplanted Novikoff hepatoma of the rat. *Fed Proc* 17:211, 1958.
21. G. Nanni and A. Casu: In vitro uncoupling of oxidative phosphorylation in normal liver mitochondria by serum of sarcoma bearing rats. *Experientia* 17:402, 1961.
22. B. Sylvén and B. Holmberg: On the structure and biological effects of a newly discovered cytotoxic polypeptide in tumor fluid. *Eur J Cancer* 1:199, 1965.
23. V. Baracos, H. P. Rodemann, C. A. Dinarello, and A. L. Goldberg: Stimulation of muscle protein degradation and prostaglandin E_2 release by leukocytic pyrogen (interleukin-1): A mechanism for the increased degradation of muscle proteins during fever. *N. Engl J Med* 308:553, 1983.
24. A. Oliff: The role of tumor necrosis factor (cachectin) in cachexia. *Cell* 54:141, 1988.
25. S. A. Beck, H. D. Mulligan, and M. J. Tisdale: Lipolytic factors associated with murine and human cancer cachexia. *JNCI* 82:1922, 1990.
26. M. J. Tisdale and J. K. Dhesi: Inhibition of weight loss by ω-3 fatty acids in an experimental cachexia model. *Cancer Res* 50:5022, 1990.
27. W. B. Kremer and J. Laszlo: Hematologic effects of cancer, in J. F. Holland and E. Frei III, eds.: *Cancer Medicine.* Philadelphia: Lea & Febiger, 1973, pp. 1085–1099.
28. G. A. Hyman: Anemia in malignant neoplastic disease. *J Chronic Dis* 16:645, 1963.
29. V. E. Price, R. E. Greenfield, W. R. Sterling, and R. C. MacCardle: Studies on the anemia of tumor-bearing animals: III. Localization of erythrocyte iron within the tumor. *JNCI* 22:877, 1959.
30. K. Oh-Uti, S. Inoue, and S. Minato: Purification of an anemia-inducing factor from human placenta and its application to diagnosis of malignant neoplasms. *Cancer Res* 40:1686, 1980.
31. S. Ebbe, B. Wittels, and W. Damesbek: Autoimmune thrombocytopenic purpura ("ITP") type with chronic lymphocytic leukemia. *Blood* 19:23, 1962.
32. R. B. Davis, A. Theologides, and B. J. Kennedy: Comparative studies of blood coagulation and platelet aggregation in patients with cancer and nonmalignant diseases. *Ann Intern Med* 71:67, 1969.
33. S. P. Miller, J. Sanchez-Avalos, T. Stafanski, and L. Zuckerman: Coagulation disorders in cancer: I. Clinical and laboratory studies. *Cancer* 20:1452, 1967.
34. S. D. Peck and C. W. Reiquam: Disseminated intravascular coagulation in cancer patients: Supportive evidence. *Cancer* 31:1114, 1973.
35. L. R. Zacharski, M. Z. Wojtukiewicz, V. Constantini, D. L. Ornstein, and V. A. Memoli: Pathways of coagulation/fibrinolysis activation in malignancy. *Semin Thromb Hemost* 18:104, 1992.
36. L. R. Zacharski, V. A. Memoli, D. L. Ornstein, S. M. Rousseau, W. Kisiel, and B. J. Kudryk: Tumor cell procoagulant and urokinase expression in carcinoma of the ovary. *JNCI* 85:1225, 1993.
37. P. Prandoni, A. W. A. Lensing, H. R. Buller, A. Cogo, M. H. Prins, A. M. Cattelan, S. Cuppini, F. Noventa, and J. W. Ten Cate: Deep-vein thrombosis and the incidence of subsequent symptomatic cancer. *N Engl J Med* 327:1128, 1992.
38. G. P. Bodey: Infections in patients with cancer, in J. F. Holland and E. Frei III, eds.: *Cancer Medicine.* Philadelphia: Lea & Febiger, 1973, pp. 1135–1165.
39. R. T. Silver: Fever and host resistance in neoplastic diseases. *J Chronic Dis* 16:677, 1963.
40. J. E. Sokal and K. Shimaoka: Pyrogen in the urine of febrile patients with Hodgkin's disease. *Nature* 215:1183, 1967.
41. P. Bodey: Pyrogen production in vitro by lymphoid tissue in malignant lymphoma. *J Clin Invest* 51:11a, 1972.
42. P. A. Pizzo and S. C. Schimpff: Strategies for the prevention of infection in the myelosuppressed or immunosuppressed cancer patient. *Cancer Treat Rep* 67:223, 1983.
43. L. H. Rees, G. A. Bloomfield, G. M. Rees, B. Corrin, L. M. Franks, and J. G. Ratcliffe: Multiple hormones in a bronchial tumor. *J Clin Endocrinol Metab* 38:1090, 1974.
44. Y. Hirata, S. Matsukura, H. Imura, T. Yakura, S. Ihjima, C. Nagase, and M. Itoh: Two cases of multiple hormone-producing small cell carcinoma of the lung: Coexistence of tumor ADH, ACTH, and β-MSH. *Cancer* 38:2575, 1976.
45. G. W. Liddle and J. H. Ball: Manifestations of cancer mediated by ectopic hormones, in J. F. Holland and E. Frei III, eds.: *Cancer Medicine.* Philadelphia: Lea & Febiger, 1973, pp. 1046–1957.
46. R. G. Josse, D. R. Wilson, J. N. M. Heershe, J. R. F. Mills, and T. M. Murray: Hypercalcemia with

ovarian carcinoma: Evidence of a pathological role for prostaglandins. *Cancer* 48:1233, 1981.

47. D. Chopra and E. P. Clerkin: Hypercalcemia and malignant disease. *Med Clin North Am* 59:441, 1975.
48. A. Besarab and J. F. Caro: Mechanisms of hypercalcemia in malignancy. *Cancer* 41:2276, 1978.
49. D. Powell, F. R. Singer, T. M. Murroy, C. Minkin, and J. T. Potts, Jr.: Nonparathyroid humoral hypercalcemia in patients with neoplastic disease. *N Engl J Med* 289:176, 1973.
50. S. H. Doppelt, D. M. Slovik, R. M. Neer, J. Nolan, R. M. Zusman, and J. T. Potts, Jr.: Gut-mediated hypercalcemia in rabbits bearing VX_2 carcinoma: New Mechanisms for tumor-induced hypercalcemia. *Proc Natl Acad Sci USA* 79:640, 1982.
51. S. Kramer, F. H. Reynolds, M. Castillo, D. M. Valenzuela, M. Thorikay, and J. M. Sorvillo: Immunological identification and distribution of parathyroid hormone-like protein polypeptides in normal and malignant tissues. *Endocrinology* 128:1927, 1991.
52. A. H. Tashjian, Jr., E. F. Voelkel, L. Levine, and P. Goldhaber: Evidence that the bone resorption-stimulating factor produced by mouse fibrosarcoma cells is prostaglandin E_2: A new model for hypercalcemia of cancer. *J Exp Med* 136:1329, 1972.
53. E. F. Voelkel, A. H. Tashjian, Jr., R. Franklin, E. Wasserman, and L. Levine: The VX_2 carcinoma model in the rabbit. *Metabolism* 24:973, 1975.
54. T. J. Powles, S. A. Clark, D. M. Easty, G. C. Easty, and A. M. Neville: The inhibition by aspirin and indomethacin of osteolytic tumour deposits and hypercalcemia in rats with Walker tumour, and its possible application to human breast cancer. *Br J Cancer* 28:316, 1973.
55. E. F. Voelkel, L. Levine, C. A. Alper, and A. H. Tashjian, Jr.: Acute phase reactants ceruloplasmin and haptoglobulin and their relationship to plasma prostaglandins in rabbits bearing the VX_2 carcinoma. *J Exp Med* 147:1078, 1978.
56. D. C. Klein and L. G. Raisz: Prostaglandins: Stimulators of bone resorption in tissue culture. *Endocrinology* 86:1436, 1970.
57. H. W. Seyberth, G. V. Segre, J. L. Morgan, B. J. Sweetman, J. T. Potts, Jr., and J. A. Oates: Prostaglandins as mediators of hypercalcemia associated with certain types of cancer. *N Engl J Med* 293:1278, 1975.
58. R. P. Robertson, D. J. Baylink, J. J. Marini, and H. W. Adkinson: Elevated prostaglandins and suppressed parathyroid hormone associated with hypercalcemia and renal cell carcinoma. *J Clin Endocrinol Metab* 41:164, 1975.
59. R. P. Robertson, D. J. Baylink, S. A. Metz, and K. B. Cummings: Plasma prostaglandin E in patients with cancer with and without hypercalcemia. *J Clin Endocrinol Metab* 43:1330, 1976.
60. A. Bennett, J. S. Simpson, A. M. McDonald, and I. F. Stamford: Breast cancer, prostaglandins, and bone metastases. *Lancet* 2:1218, 1975.
61. T. J. Powels, M. Dowsett, G. C. Easty, D. M. Easty, and A. M. Neville: Breast cancer osteolysis, bone metastases, and anti-osteolytic effect of aspirin. *Lancet* 1:608, 1976.
62. G. Eilon and G. P. Mundy: Direct resorption of bone by human breast cancer cells in vitro. *Nature* 276:726, 1978.
63. Y. Kondo, K. Sato, H. Ohkawa, Y. Ueyama, T. Okabe, N. Sato, S. Asano, M. Mori, N. Ohsawa, and K. Kosaka: Association of hypercalcemia with tumors producing colony-stimulating factor(s). *Cancer Res* 43:2368, 1983.
64. K. J. Ibbotson, S. M. D'Souza, K. W. Ng, C. K. Osborne, M. Niall, T. J. Martin, and G. R. Mundy: Tumor-derived growth factor increases bone resorption in a tumor associated with humoral hypercalcemia of malignancy. *Science* 221:1292, 1983.
65. K. J. Ibbotson, D. R. Twardzik, S. M. D'Souza, A. R. Hargeaves, G. J. Todaro, and A. R. Mundy: Stimulation of bone resorption in vitro by synthetic transforming growth factor-alpha. *Science* 228:1007, 1985.
66. A. H. Tashjian, Jr., E. F. Voelkel, M. Lazzaro, F. R. Singer, A. B. Roberts, R. Derynck, M. E. Winkler, and L. Levine: α and β human transforming growth factors stimulate prostaglandin production and bone resorption in cultured mouse calvaria. *Proc Natl Acad Sci USA* 82:4535, 1985.
67. L. J. Suva, G. A. Winslow, R. E. H. Wettenhall, R. G. Hammonds, J. M. Moseley, H. Diefenbach-Jagger, C. P. Rodda, B. E. Kemp, H. Rodriguez, E. Y. Chen, P. J. Hudson, T. J. Martin, and W. I. Wood: A parathyroid hormone-related protein implicated in malignant hypercalcemia: Cloning and expression. *Science* 237:893, 1987.
68. Z. Bouizar, F. Spyratos, S. Deytieux, M.-C. de Vernejoul, and A. Jullienne: Polymerase chain reaction analysis of parathyroid hormone-related protein gene expression in breast cancer patients and occurrence of bone metastases. *Cancer Res* 53:5076, 1993.
69. T. Tsuchihashi, K. Yamaguchi, Y. Miyake, K. Otsubo, K. Nagasaki, S. Honda, Y. Akiyama, Y. Maeda, M. Koike, O. Alper, S. Kumura, and K. Abe: Parathyroid hormone–related protein in tumor tissues obtained from patients with humoral hypercalcemia of malignancy. *JNCI* 82:40, 1990.
70. S. C. Kukreja, T. J. Rosol, S. A. Wimbiscus, D. H. Shevrin, V. Grill, E. I. Barengolts, and T. J. Martin: Tumor resection and antibodies to parathyroid hormone-related protein cause similar changes on bone histomorphometry in hypercalcemia of cancer. *Endocrinology* 127:305, 1990.
71. E. P. Richardson, Jr.: Neurological effects of cancer, in J. F. Holland and E. Frei III, eds.: *Cancer Medicine.* Philadelphia: Lei & Febiger, 1973, pp. 1057–1067.
72. W. L. Dobes, Jr., and R. R. Kierland: Dermatologic effects of cancer, in J. F. Holland and E. Frei III, eds.: *Cancer Medicine.* Philadelphia: Lea & Febiger, 1973, pp. 1067–1074.

13

Tumor Immunology: Immunologic Approaches to the Diagnosis and Treatment of Cancer

All the cells in the body have antigenic determinants on their cell surfaces that reflect the expression of the major histocompatibility complex (MHC) genes of that organism. In humans, this gene complex, called the HLA complex, is located on chromosome 6. The letters HLA stand for human leukocyte antigens, reflecting the cells in which the expression of these genes was initially determined. These genes determine recognition of self from nonself and are involved in the rejection of transplanted tissue from a foreign host and in other aspects of the immune response system. When a normal cell becomes transformed into a malignant cell, it undergoes biochemical changes that often result in the production of new cellular antigens. These new antigens may be recognized by the host organism as foreign. Although new antigenic determinants may be present in other parts of a cancer cell, the ones that are most important in cancer cell recognition are most likely located on the cell surface, where they are "perceived" by interactions with host cells or shed into the bloodstream where they are recognized as foreign. New antigens are found in tumors induced by chemicals, viruses, or irradiation in experimental animals. For reasons that will be discussed later, the antigenicity of tumors that arise spontaneously in animals (i.e., those that arise in certain animal species without experimental induction or interference) and of many human tumors is low—so low, in fact, that tumor cells escape detection by the host or are able to circumvent a relatively weak reaction by the host.

HISTORICAL PERSPECTIVES

The history of tumor immunology has been an up-and-down affair for almost a century. There have been times of extreme optimism, almost hubris, about the importance of the immune system in moderating or even rejecting tumor cell growth. These highs have been followed by lows in which all but a few stalwarts gave up on the idea that the immune system was capable of mounting any meaningful response against cancer cells growing in the body. This early checkered history of tumor immunology has been reviewed by Scott.[1]

The crux of the problem with the early studies of tumor rejection lay in the difficulty of differentiating true tumor rejection from rejection of a foreign tissue by a genetically incompatible host. Even today this is a problem, because experimental tumors are often transplanted again and again and may change over time to the extent that they may no longer be perfectly "syngeneic" with their host animal strain.[1] Thus, the very term *syngeneic* is in a sense a misnomer, because if a tumor that arises in a strain of mice, for example, is passaged to other inbred mice to maintain histocompatibility and then subsequently shown to evoke a rejection response, it is by strict definition not syngeneic with the host animal. Moreover, if an individual animal within an inbred strain becomes mutated in a histocompatibility locus, that individual would not be syngeneic with other members of the strain. Perhaps the term *autochthonous* is a better one to describe the relationship between a tumor and its own host.[1]

A few historical benchmarks can be pointed out. As early as 1910, Peyton Rouse recognized that engrafted tumors could be rejected, but he raised the question whether this was "simply one expression of a resistance to the growth of engrafted tissues in general." Leo Loeb, utilizing one of the first inbred strains of mice—the Japanese waltzing mouse—is credited with carrying out the first successful series of tumor transplantations in 1904. He was able to obtain virtually 100% successful "takes" in the waltzing mice and no takes in unrelated white mice.

Even though it is now known that a wide range of malignant tumors in experimental animals or in humans have tumor-associated cell surface antigens, this was not generally accepted until the 1950s. Prior to the 1940s, when little was known about the existence, let alone the immense complexity, of transplantation rejection antigens (the histocompatibility antigens), it was thought that the small group of then-known transplantable tumors (e.g., Ehrlich carcinoma, Jensen sarcoma, Walker 256 carcinosarcoma, Sarcoma 180, and Sarcoma 37) could be transplanted from one animal to another animal of different genotype because they had no incompatibility for the host and thus were not rejected.[2] Even then, however, it was appreciated that the vast majority of animal tumors were rejected by allogeneic hosts (same species, different genotype). This led to the belief that tumor cells, with the exception of the few transplantable ones, carried potent antigenic determinants that caused their rejection by the host. And in fact, experiments in the 1930s and 1940s showed that preimmunization of animals with arrested tumor cells increased the host animals' resistance to tumor transplantation. This led to the idea that an immunologic cure for cancer was possible. However, after highly inbred strains of mice with known genotypes became available in the 1940s, it soon became apparent that most of the tumor rejection phenomena studied previously were due to histocompatibility differences between the tissues of mice of different genotype. This led to the conclusion that tumor cells themselves had no distinguishing immunologic features. This low point was followed by another wave of enthusiasm that started with the experiments by Foley[3] in 1953. He induced sarcomas with methylcholanthrene (MC) in inbred C3H mice, grafted them into other C3H mice, and ligated the tumors to induce tumor necrosis. Subsequent challenge of these mice with the same sarcoma frequently led to rejection of the second tumor graft, whereas the tumors grew in control animals that had not received the initial graft. Mammary carcinomas arising spontaneously in the same strain of mice were not rejected when transplanted into the mice pregrafted with MC-induced sarcomas. These experiments indicated that tumor rejection could, in fact, be due to antigenic determinants of the tumor itself rather than simply to differences in histocompatibility antigens. Prehn and Main[4] confirmed these findings by showing that skin grafts from one mouse to another of the same inbred strain were not rejected, but that MC-induced sarcomas were rejected by host animals in which the tumors were grafted and then ligated. Klein et al.[5] subsequently demonstrated that secondary tumor transplants were rejected by the same animal (autochthonous tumor host) following surgical removal of the primary tumor and antigenic stimulation by injections of tumor cells previously inactivated by irradiation. These experiments excluded the possibility that histocompatibility differences were responsible for tumor rejection in inbred strains. These findings were subsequently confirmed in several laboratories, and the principle is now well established. It is important to keep in mind that tumor-associated antigens induce only a *relative* degree of host resistance, depending on the tumor burden in the animal. Tumor rejection occurs if the size of the tumor challenge is within a certain threshold, but resistance is overwhelmed by a larger tumor cell burden.[2] This has important therapeutic implications for the potential use of immunotherapy in cancer treatment (see below).

With further experimentation in tumor immunology, the potential role of the immune system in modulating tumor growth loomed larger and larger. In the early 1970s, the theory of "immune surveillance" became popular.[6] In brief, this theory states that tumor cells contain aberrant cell surface antigens that a host's immune system can recognize and react to as soon as the concentration of these foreign antigens reaches a certain threshold level. This goes on all the time, and in most young, healthy animals or people, it prevents the growth of aberrant cells. However, as the individual ages, the immune surveillance mechanism becomes defective and the probability that tumor cells will escape rejection increases. This theory has had several proponents (as well as detractors[7]), and it has led to the idea that nonspecific stimulation of the immune system by such bacterial agents as the *Mycobacterium bovis* strain *Bacille* Calmette-Guérin (BCG) and *Corynebacterium parvum (C. parvum)* can increase the body's ability to reject tumor growth. The initial enthusiasm for this, however, has waned as more extensive clinical trials have been carried out. More recently, other immune system modulating agents, such as interferon and interleukin-2, have received considerable attention and aroused a certain amount of excitement (and also some cynicism). Such has been the ebb and flow of research in tumor immunology. It does seem fair to say, however, that expanding knowledge of the body's immune system should lead to better ways to manipulate it to tip the balance in favor of the host.

Studies on cancer induction in experimental animals have shown that, in general, tumors induced with chemicals, irradiation, or physical agents (e.g., implantation of plastic films) have unique antigenic determinants. Even two different tumors induced in the same animal by the same agent are antigenically distinct. On the other hand, virally induced tumors contain new antigens with a common (or cross-reactive) specificity for all tumors induced by the same virus, regardless of tumor cell type or animal species. Tumors initiated by viruses may, however, carry weaker, specific tumor-associated antigens. The relative antigenic strengths and cross-reactivities of animal tumors induced by various types of agents are indicated in Table 13-1. Several types of cell surface antigens can arise on tumor cells, depending on the nature of the carcinogenic agent. Those antigens that are involved in tumor transplant rejection are called tumor-associated transplantation antigens (TATA). Virus-associated antigens in virally induced tumors may be intracellular (e.g., nuclear T antigen: see Chap. 7) or on the surface (e.g., viral envelope pro-

Table 13-1 Relative Antigenic Strengths of Certain Tumors

		Cross-Reactivity of Tumors with		
Etiologic Agents	Relative Antigenic Strength[a]	Other Primary Tumors in the Same Individual Induced by the Same Agent	Tumors Induced by the Same Agent in Other Individuals	Tumors Induced by Other Agents in Other Individuals
DNA viruses	++	++	++	−
RNA viruses	+	+	+	−[b]
Chemical carcinogens	++	−(±)	−(±)	−
Radiation	−(±)	−(±)	−(±)	
"Spontaneous"	−(±)			

[a]Even when the relative antigenic strength is high (++), the absolute antigenic strength compared to that of tissue alloantigens may be very low.

[b]Cross-reactivity may occur between tumors induced by closely related viruses.

Source: Modified from Reif.[8]

teins), the latter type probably being those involved in tumor transplant rejection. In addition, certain embryonic or fetal antigens may reappear on various kinds of tumors, including those that arise spontaneously.

As noted above, chemically induced tumors possess TATA unique for each neoplasm.[4,5,9] However, evidence for additional cross-reacting TATA from chemically induced tumors has also been obtained. For example, cross-protection has occasionally been observed in mice preimmunized against one chemically induced sarcoma and then challenged with a different chemically induced tumor.[10,11] These studies are frequently complicated by the fact that chemically induced mouse sarcomas often express antigens of the endogenous MuLV on the cell surface, and some reports have indicated that immunity induced to MuLV can protect against the development of chemically induced sarcomas in mice.[12] Moreover, the MuLV envelope glycoprotein gp70 has been detected on the surface of many MC-induced murine sarcoma cells, and those chemically produced sarcomas expressing gp70 on the cell surface induce serum antibodies against viral envelope antigens (VEA) gp70 and another VEA, p15E, whereas tumors lacking gp70 on the cell surface do not.[13] Thus, the presence of MuLV VEA on the tumor cells' surface appears to account for a major portion of the common cross-reactivity seen between chemically induced mouse sarcomas. In some instances, however, the cross-reactivity that occurs cannot be accounted for by expression of MuLV antigens.[11]

Cell surface antigens are found on the surface of cells in tumors induced by RNA or DNA oncogenic viruses. Similar antigens are found on the surface of cells lytically infected with but not transformed by these viruses. Thus, the question: Which if any of these are related to the TATA of tumors induced by these viruses? The C-type RNA tumor viruses express several classes of virus-associated envelope antigens. In most instances, cells infected with C-type RNA viruses express additional surface antigens qualitatively distinct from viral structural proteins, and immunity to virus particles can occur independently from immunity to the TATA of virally induced tumors,[14] indicating that at least certain TATA are not typical viral structural proteins. A tumor-associated transplantation antigen has been solubilized and partially purified from plasma membranes of murine leukemia cells induced with the Rauscher RNA murine leukemia virus,[15] and its biochemical and immunologic characterization indicated that the major TATA could be separated from the major viral structural proteins gp70 and p30 as well as from histocompatibility antigens. This supports the idea that the TATA of virally induced tumors are distinct from the major structural antigens of the tumor-inducing virus. A key question that remains to be answered is whether the TATA are coded for directly by the viral genome or insertion of the viral genome into the host's genetic material somehow induces the expression of a new or repressed host gene.

The DNA-tumorigenic viruses also confer new antigenicities on the cells they infect or transform. A specific cellular immune response to tumors induced by these viruses also occurs. In the case of SV40 virus, for example, the same region of DNA that codes for SV40-specified nuclear T antigen is also involved in the synthesis of TATA, and genetic as well as immunologic evidence suggest that the antigenic sites for T and TATA are located on the same protein, but at different sites.[16] This suggests that, at least in the case of SV40, the TATA are coded directly by the viral genome. In virally induced tumors, however, it is conceivable that virally coded proteins, including some structural proteins, could augment the immunogenicity of tumor cells that also carry on their surface additional distinct TATA resulting from the transformation process. This could account for the presence of tumor-type–specific rejection as well as cross-reactivity between tumors induced by different oncogenic agents.

The phenomenon of some shared and some distinct tumor-associated antigens is also observed in human cancer. Tumor-associated antigens have been reported for a wide variety of human neoplasms (reviewed in Refs. 1 and 17). Several methods have been used to define human tumor immunity. Antisera to human cancer cells or extracts of human cancer cells have been prepared by inoculating other species. In a number of instances, these antisera seemed to be tumor specific. However, when these antisera were more thoroughly tested, they were found to be directed against normal cell components present in low concentration in many normal cell types or in normal tissues not initially tested. A second approach has been to screen sera from cancer patients for antibodies to selected cell lines derived from other patients' tumors of the same histologic type. The problem with this approach is similar to that encountered in animal studies before inbred strains were available, namely, the contributions of the histocompatibility system to the observed immune reactions. A third method is aimed at estimating the cell-mediated immune response (see below) more directly by determining the ability of tumor extracts to affect the function (e.g., migration in capillary tubes) of leukocytes from patients with the same or different tumors. With this approach, the so-called leukocyte migration inhibition (LMI) assay, it has been found that some human cancers possess common antigens that appear to elicit a cell-mediated immune response that cannot be explained by simple allogeneic histocompatibility reactions.[18] A fourth approach is to study reactions of cancer patients' sera with surface antigens of their own as well as other patients' cultured cancer cells. With this method, three classes of surface antigens have been defined in a study of human melanoma, renal cancer, and astrocytoma.[19–21] In the case of melanoma, the three classes are as follows: Class I includes unique melanoma antigens expressed only on the individual patient's own (autologous) melanoma cells and not on any other cell tested. Class II are shared melanoma antigens that are expressed not only on autologous melanoma cells, but also on other patients' (allogeneic) melanoma cell lines. These two classes of antigens are not detected on autologous, allogeneic, or xenogeneic (other species) normal cells or on nonmelanoma tumor cells. Class III are antigens that are not restricted to melanoma cells and that appear to be present on autologous normal cells, as well as on allogeneic and xenogeneic normal and tumor cells. Thus, these data indicate that a variety of antigenic determinants are present on malignant melanoma cells, some of which are unique to a given tumor, some of which are tumor-type specific, and some of which are present, at least to some extent, on a variety of normal cells and other types of tumor cells. It is of interest that individual patients produce antibodies to one or more of these classes of antigen. Thus, they may develop immunity to their own tumor and also to other patients' tumors. Similar data have been obtained for human renal and brain tumors.[19–21] These findings indicate the complexity of the human immune response to cancer and have significant implications for regimen designs in immunotherapy.

Mouse monoclonal antibodies (mAbs) to human melanoma cells have been developed, and with these reagents a greater refinement of the original classification reported by Shiku et al.[19] has been generated. Using the mAbs, four categories of melanoma-associated antigens can be distinguished: (1) antigens found on all cells of melanocyte origin, whether normal or malignant melanocytes; (2) antigens found on adult but not fetal or newborn melanocytes and on a subset of melanomas; (3) antigens found on fetal and newborn but not on adult melanocytes and on a subset of melanomas; and (4) antigens found on a subset of melanomas but not on fetal, newborn, or adult normal melanocytes. Thus, a number of the originally detected antigenic determinants appear to be differentiation-stage–specific antigens. The mouse monoclonal antibodies to melanoma cells that have been developed detect 13 distinct types of cell surface antigens on cultured human melanocytes and melanoma tissue samples.[21] Five of the melanoma surface antigens have been well characterized biochemically: (1) a GD3 disialoganglioside, (2) a chondroitin sulfate proteoglycan, (3) HLA (human lymphocyte antigen) class II antigens, (4) a 130,000-MW glycoprotein (gp130), and (5) a 95,000-MW glycoprotein (gp95). Of the 13 antigen types, 3 have been related to the stage of melanocyte differentiation, and 6 are additional markers for subsets of cultured melanoma cells.

Of the biochemically identified antigens, the GD3 disialoganglioside is a predominant form found on cultured melanoma cells, a variety of normal cells of neuroectodermal origin, and a high percentage of melanomas, astrocytomas, and sarcomas. The chondroitin sulfate proteoglycan is expressed on the surface of most melanoma and astrocytoma cell lines and on cultured normal melanocytes, but not on other tumor types. The HLA class II antigen is detected on all melanomas tested and on a variety of other tumor types but not on normal melanocytes. Expression of the glycoprotein gp130 is generally restricted to a subset of melanomas and some normal cells, and gp95 is detectable on a small number of melanomas and is not seen in normal tissues or other tumor types. It is of interest that primary melanomas in patients have lower HLA class II antigen expression than metastatic melanomas, and in primary melanomas HLA class II–positive cells appear to accumulate at sites of tumor invasion,[21] suggesting that expression of altered HLA class II antigens is a late event correlating with tumor progression to a more invasive, metastatic cell type. It could be that the more aggressive tumor cells have adapted to the patient's immune defense system by altering their major antigenic recognition components so that they can escape or inhibit the hosts' immune defenses.

MECHANISMS OF THE IMMUNE RESPONSE TO CANCER

In order to understand the immune response to cancer cells, we need to review the characteristics and functions of the cellular components involved in this response.

Antigen-Presenting Cells (APCs)

In order for cells involved in the immune response to react to foreign organisms or cells, the nonself antigens present in these organisms or cells must be "presented" in a way that the immune cells can "see" them. This requires that the foreign antigens be processed into smaller bits of information and be presented to immune cells as part of a complex with cell surface MHC molecules. There are two types of MHC molecules: class I, expressed on all cells, and class II, expressed on macrophages, dendritic cells, B cells, and occasionally on other cell types. All three of these cell types can present antigens to CD4 (cell surface marker)-bearing T cells (helper T cells; see below), which initiate most immune responses (reviewed in Refs. 22 and 23).

Antigen-derived peptides are generated in APC cells by one of two routes. Antigenic peptides present in the cytosol—for example, from viral-infected cells or from engulfed tumor-associated antigens—are derived by proteolytic degradation and transported into the endoplasmic reticulum, where they bind to nascent MHC class I molecules.[22] In this way, T lymphocytes bearing the CD8 cell surface marker ($CD8^+$ cells), become stimulated to become cytotoxic T cells (CTLs). $CD8^+$ cells recognize processed antigens presented on MHC class I cell surface molecules. Antigenic peptides generated during endocytosis of engulfed antigenic molecules can bind to MHC class II molecules that are targeted to the endocytic pathway on their way to being cycled back to the cell surface. Antigens presented in conjunction with MHC class II stimulate $CD4^+$ T cells, which become activated T-helper cells.

Macrophages scavenge dead and dying cells, engulf and kill many types of bacteria, present antigens to T cells, and are themselves effector cells in cell-mediated immune reactions; when activated, they can kill tumor cells. Phagocytic dendritic cells are present in many tissues, where they can pick up and process antigens and then migrate to lymphoid tissues. Those that differentiate in lymphoid tissues acquire the ability to present antigens to T cells, and they are highly effective in activating $CD4^+$ cells.[23] They do not appear to have a cytotoxic effector function themselves. B cells have been implicated as "presenters" of soluble protein antigens to $CD4^+$ cells. The interaction of T and B cells is discussed in more detail below.

Although intact antigenic proteins must be processed to generate antigenic peptides and copresented with MHC class I or II molecules, soluble peptides can also bind directly to empty class I or class II molecules present on the cell surface of APC cells. Such empty MHC molecules are potentially important targets for synthetic immune system—stimulatory peptides that could be antitumor vaccines (see below).

T Lymphocytes and T-Cell Activation

The lymphocytic stem cell produced in the bone marrow has two pathways of differentiation. One pathway requires the thymus gland and leads to the generation of cells called thymus-dependent, or T lymphocytes, that are involved in cell-mediated or delayed immune reactions. Precursor T cells (prothymocytes) migrate to the thymus gland, where they are processed into functionally competent cells and are then released into the circulation, from which they populate the peripheral lymphoid tissue. The second pathway produces "bursal-equivalent" or B lymphocytes. The term *bursal equivalent* derives from the fact that this class of cells was first clearly delineated in chickens that have a distinct bursa in which these cells are produced. In higher animals and humans, the equivalent B-lymphocyte–producing tissues appear to be the lymph tissue of the gastrointestinal tract and certain areas of the spleen. Both T and B lymphocytes are present in lymph nodes and spleen, although their relative concentrations vary within these organs. Both types of cells circulate in the blood, but about 70% of circulating lymphocytes are T cells.

A number of subpopulations of human T cells have been defined on a functional basis and on the basis of their cell surface marker antigens. One of these cell populations carries a distinct surface marker called CD4. These $CD4^+$ cells constitute about 55% to 65% of peripheral T cells and are the $T_{helper/inducer}$ cells of the immune response system. These cells can respond directly to antigen (although interaction with APCs is usually involved in the response, as indicated before) or to such lectins as Con A or phytohemagglutinin by undergoing a burst of cell proliferation. These cells provide helper function to other T cells, to B lymphocytes, and to macrophages mediated at least in part by the release of various lymphokines. The $T_{helper/inducer}$ cells do not themselves have a cytolytic effector function, but they play a role in stimulating the generation of cytolytic $CD8^+$ T cells. It is the $CD4^+$ helper/inducer cells that are the target for the AIDS virus, and their loss, as observed in this disease, is devastating to the immune system.

Activated T lymphocytes also release a chemotactic factor for macrophages and polymorphonuclear leukocytes (PMNs). This factor promotes the sequestration of these phagocytic cells in the area of a cell-mediated immune reaction, and PMNs appear to aid in this reaction by the release of proteases and other lytic enzymes that promote destruction of target cells and aid in clearing the cellular debris left by target cell killing.

$T_{suppressor}$ cells also have the CD8 surface marker. The cells suppress the proliferative response of T cells to alloantigens and the production of immunoglobulin by B lymphocytes, possibly by the release of a suppressor factor. Normally, these cells function to modulate the immune response system and prevent overresponse to an antigenic stimulus. However, excessive $T_{suppressor}$ cell activity can produce generalized immunosuppression and decrease the immune response to a number of foreign antigens, including those present on tumor cells. On the other hand, a loss of $T_{suppressor}$ cells has been observed in cases of excessive immune response such as occurs in certain autoimmune diseases, including systemic lupus erythematosus, hemolytic anemia, and inflammatory bowel disease. Thus, the balance of T_{helper} and $T_{suppressor}$ activities regulates the immune response and determines the outcomes of antigenic stimulation of the host.

In tumor-bearing animals, populations of specific suppressor T cells develop and their appearance favors tumor growth. For example, when 3-methylcholanthrene-induced sarcomas are injected into normal A/J mice, the tumors grow and kill the animals within 40 days, but if tumor cells are transplanted into mice previously made immune to this sarcoma, the tumors are rejected in about 2 weeks.[24,25] If thymus or spleen cells from tumor-bearing mice are infused into preimmunized mice at the same time that the mice are injected with tumor cells, the tumor is not rejected and tumor growth is enhanced. The explanation for this is that the thymus and spleen of tumor-bearing animals contain a large population of $T_{suppressor}$ cells that counteract the immune response of the host to the tumor. Interestingly, if the tumor is removed from tumor-bearing host animals, the suppressor cell activity in the thymus and spleen that was directed against the tumor-induced immune response disappears, suggesting that a large tumor burden with significant shedding of tumor antigen may overtax the immune system and tip the balance in favor of the suppressor arm of the immune response. There is evidence that the suppression of cytolytic T cell activity against tumor cells comes about by an autoimmune response against the specific clone of T cells that is cytolytic for the tumor. This autoimmune reaction is mediated by $T_{suppressor}$ cells that recognize surface markers of a specific "idiotype" on cytolytic T cells and that specifically attack T cells bearing that idiotype.[26] This is called autologous anti-idiotypic immunity.

Since CD4$^+$ T cells become the helper cells that control the activation of B cells, macrophages, and in some instances CD8$^+$ cells, the activation of CD4$^+$ cells is a keystone of the immune response system. CD4$^+$ cells that have not previously been activated ("naive" CD4$^+$ cells) require two distinct signals to proliferate and differentiate into activated effector cells.[23] The receptor for the first activation event is the T-cell receptor (TCR) that recognizes antigen in the form of fragments bound to MHC class II molecules.[27] The T-cell receptor is a multisubunit complex that has separate antigen-binding and signal transduction subunits.[28] Binding of an antigen–MHC II complex by the TCR leads to a signal transduction pathway involving cytoplasmic protein tyrosine kinases and phosphatases, followed by activation of the phosphatidyl inositol pathway, activation of Ras, and activation of ser/thr kinases and phosphatases. These events induce CD4$^+$-associated helper activity, including production of cytokines such as IL-2.

The second activation event is not carried out via a TCR-mediated event but involves a costimulatory signal. The costimulatory signal involves cell–cell interaction between a cell surface component of APC cells called B7 and one on naive T cells called CD28 (Fig. 13-1). If costimulation does not occur, i.e., if an APC cell does not have the correct B7 components (these are called nonprofessional APCs), no immune stimulatory response occurs and IL-2 production does not occur. In fact, inactivation or immune anergy occurs. The B7 molecule is expressed on macrophages and dendritic cells, the prototypical professional APCs, and also on activated T and B cells. A second B7 surface component called B7-2 has been found in mice in which the first B7 gene, B7-1, was knocked out.[29] This second B7-2 component was unmasked in the gene-knockout mice and also identified by monoclonal antibody[30] and gene cloning[31] experiments. B7-2 was shown to interact with CD28 on T cells as well as a second cell surface "counterreceptor"

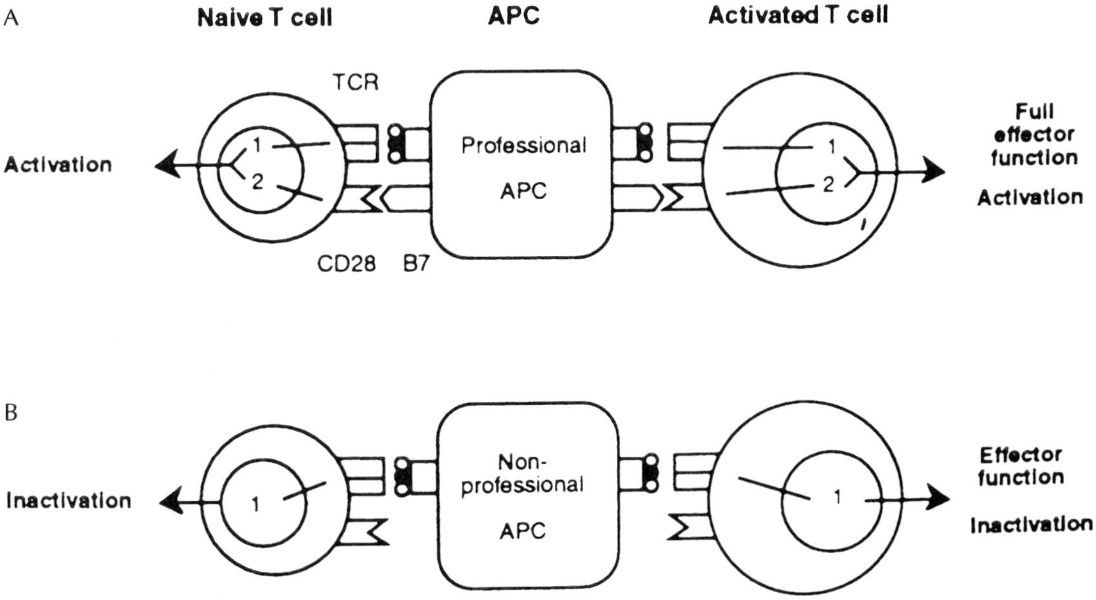

Figure 13-1 The outcome of specific antigen recognition by T cells (signal 1) is determined by costimulation delivered by professional APCs (signal 2). (A) Signal 1 is triggered by ligation of the T-cell receptor (TCR) to specific peptide–MHC complexes. Signal 2 provides costimulation by interaction of CD28 on T cells with B7 on antigen presenting cells (APCs). (B) Signal 1 in the absence of signal 2 results in inactivation or "anergy." (Reprinted with permission from Lanzavecchia, Identifying strategies for immune intervention. *Science* 260:937. Copyright 1993 by the American Association for the Advancement of Science.)

component on T cells, called CTLA4, thus providing a second T cell activation pathway.

The three main effector functions of activated T cells are to (1) stimulate B cells to produce antibodies, (2) induce inflammatory reactions, and (3) kill abnormal cells, e.g., virus-infected cells and tumor cells. They do this by producing a number of cytokines for local activation of neighboring cells and by direct cell–cell contact. Effector T cells also carry "homing devices" that allow them to migrate to relevant tissue sites. By using homing receptors that recognize lymph node or blood vessel adhesion molecules, lymphocytes can find their way to target organs. For example, activated T and B cells express integrin α4β1, which has a high affinity for vascular cell adhesion molecule 1 (VCAM-1) expressed on the luminal surface of inflamed blood vessels (reviewed in Refs. 32 and 33). Unstimulated lymphocytes, in general, have lymphoid tissue homing receptors, and activated lymphocytes usually lack lymphoid organ receptors but carry receptors that allow them to home to inflammatory sites, tumor tissues, etc.

When activated, naive T cells produce IL-2 only, which stimulates clonal expansion of T cells. Once this occurs, they become competent to produce a wide variety of cytokines (also called lymphokines). Which ones they produce depends on the local environment and their interaction with APCs. For example, $CD4^+$ cells can differentiate down one of two helper-cell pathways and become T-helper cell type 1 (T_H1)

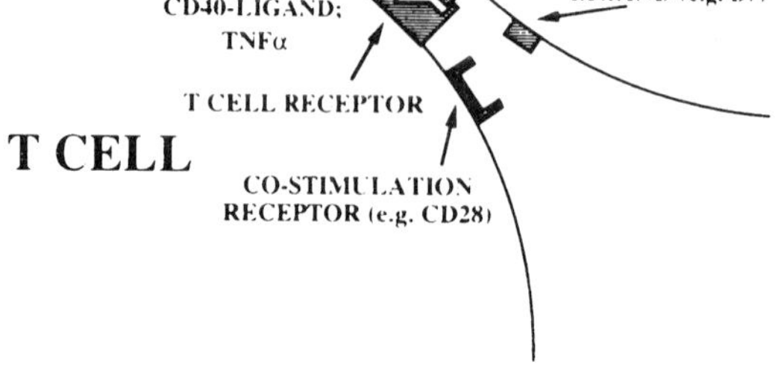

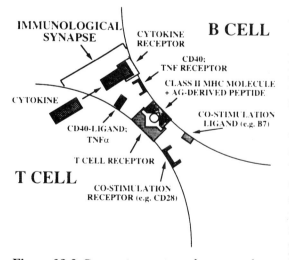

Figure 13-2 Cognate interactions; the immunological synapse. (From Paul and Seder.[34])

or T-helper cell type 2 (T_H2) cells. T_H1 cells produce IL-2, interferon γ (IFN-2), and tumor necrosis factor β (TNFβ); induce inflammatory responses; and act as helpers in the production of opsonizing antibodies that bind to Fc receptors on phagocytes (reviewed in Refs. 22 and 34). T_H2 cells produce IL-4, IL-5, IL-6, IL-10, and IL-13 and help B cells in the production of nonopsonizing antibodies of the IgG, IgA, and IgE class. Local production of IL-4, IL-10, and TNF-β favor the development of the T_H2 pathway, whereas IFN-α and IL-12 favor the T_H1 pathway. Depending on the type of eliciting immune stimulus, one or the other of these pathways is favored in the local inflammatory or tissue-altered environment.

B Lymphocytes and B-Cell Activation

The B cells also originate in the bone marrow but mature differently from T cells. They are the precursors of antibody-forming plasma cells. When B cells are stimulated by antigen either directly (a few antigens do not require interaction with T-helper cells to elicit antibody formation) or indirectly by T-cell interactions, B cells specifically activated by the antigenic stimulus proliferate and differentiate into immunoglobulin-producing plasma cells.

Activated T-helper cells interact with antigen-stimulated B cells in a manner analogous to that of T-cell–APC interactions. Antigen, processed by B cells and presented on the cell surface together with MHC class II antigens, interacts with receptors on T cells. Contact with $CD4^+$ helper cells stimulates B cells to mature, multiply, and differentiate into antibody-secreting plasma cells as well as into a clone of memory B cells. Lymphokines secreted by $CD4^+$ cells aid in this maturation process.

The interaction of $CD4^+$ helper cells and B cells is complex (Fig. 13-2). To achieve B-cell activation, T_H cells produce a set of soluble cytokines that act at various stages in the growth and differentiation of B cells. The interaction of a T_H cell component called CD40 with a CD40 receptor on B cells activates B cells and makes them competent to respond to soluble cytokines produced by T cells. The signal generated through the CD40 ligand-receptor interaction also plays a role in antibody switching from the more primitive IgM class to other "more sophisticated" immunoglobulins such as IgG and IgE. Cytokines produced by T_H2 cells (e.g., IL-4, IL-5, IL-6, and IL-10) are released locally and act locally in a tight network of T and B cells that has been called the "immunological synapse."[34]

Individual T_H-derived cytokines have distinct actions in stimulating B cells (reviewed in Ref. 34). IL-4 acts as a costimulant of B-cell proliferation together with anti-IgM antibodies or with the T_H cell surface CD40 ligand. IL-4 also moderates Ig class switching from IgM to IgG4 and IgE. IL-5 acts on B cells to induce a high rate of secretion of Ig molecules and to act along with IL-4 in Ig gene-switching events. IL-6 and IL-10 both strongly promote Ig secretion from differentiated B cells. IL-13 can also cause Ig class switching to IgE production.

Initially, the activated B cell looks like a primitive blast cell and produces mainly IgM-type immunoglobulin. As it matures to a plasma cell, it produces mainly IgG-type immunoglobulin. Antibody directed against the tumor antigen is released from the expanded clone of plasma cells that are specifically producing it. These antibodies can induce tumor cell killing by means of antibody-mediated, complement-dependent cell lysis. This mechanism of cell killing, however, appears to play a minor role in the immune reaction against cancer. A more active cytotoxic reaction in which antibodies participate is the so-called antibody-dependent, cell-mediated cytotoxity (ADCC) reaction. Antibody released by plasma cells adheres to antigens on the tumor cell surface and this attracts cells that have receptors for the Fc portion of IgG. Cells with such Fc receptors on their surface include macrophages, T lymphocytes, and natural killer cells. This mechanism for cell killing is in addition to the direct killing effect of cytolytic T cells. As discussed below, soluble antitumor antibodies or antigen-antibody complexes, when present in high concentrations, may actually block cell-mediated cytotoxicity by binding recognition sites on cytotoxic cells.

Natural Killer Cells

In animals and humans, another population of cells is cytotoxic for tumor cells. These cells appear to belong to the lymphocyte class, but they lack surface markers that clearly place them in a specific category. They are nonphagocytic and nonadherent, they appear to possess Fc receptors after activation, and they have a low density of certain T-cell markers. Thus, these cells are thought to be derived from clones of immature pre-T lymphocytes. Their ability to kill tumor cells does not depend on prior immunization of the host and does not appear to depend on the generation of antitumor antibodies: hence, these cells have been called natural killer (NK) cells.[35] They will kill tumor cells from both syngeneic and heterologous animals species and, in this respect, are rather indiscriminate killer cells. In some animal systems, populations of NK cells have been shown to recognize certain tumor antigens and to have cytotoxic activity against normal thymus cells, macrophages, and bone marrow cells.[36] It has been postulated that NK cells can recognize several different types of specificities on cells that may have some cross-reactive antigenic determinants. This could explain the broad cytotoxic specificity that NK cells possess. It is of interest that NK cells are stimulated to become active killer cells by IFN-γ, which is released by T cells as well as by activated NK cells. The latter mechanism would provide a positive feedback system even in animals lacking a functional thymus gland (e.g., nude mice). The stimulation of NK cells may partly explain the antitumor activity of interferon when injected into animals or cancer patients.

It has also been postulated that NK cells play a role in immune surveillance, particularly for virus-induced tumors, and some NK cell reactivity against the major envelope glycoprotein of endogeneous C-type viruses has been detected in mice.[37] Mice with a genetic deficiency in NK cell production have a decreased resistance to the growth of transplanted tumors, and a disease in humans (Chediak-Higashi syndrome), characterized by a 500-fold impairment of NK cell function, is associated with a high incidence of lymphoproliferative disorders, some of which are malignant.[38] The fact that nude, athymic mice that lack normal T-cell mediated immune defense mechanisms but have NK cells can still reject some heterologous tumor transplants suggests that NK cell activity does have an important immune surveillance effect against tumor cells. This effect is apparently easily overwhelmed, however, since heterologous tumor transplants generally grow much more easily in nude mice than in mice with a normal thymus gland.

While NK cells do not carry the typical T cell markers, they may be generated from a common progenitor cell. For example, NK cells have been generated from CD34$^+$ hematopoietic progenitor cells cultured with IL-2. As noted above, NK cells are activated by interferons and by IL-12 (reviewed in Ref. 39). When activated, NK cells undergo a morphologic change characterized by the acquisition of intracellular granules that contain lytic enzymes. The broad cell-killing specificity of NK cells appears to be mediated by the absence of normal MHC class I molecules on the surface of cells, which makes them susceptible to NK cell–mediated cytotoxicity. NK cells also produce a number of cytokines, including IFN-γ, G-CSF, GM-CSF, IL-1, and TGFβ.[39] NK cell depletion promotes metastasis in experimental animal systems, suggesting that NK cells play an important role in tumor cell surveillance.[39]

Cell-Mediated Cytotoxicity

Tumor cells can be attacked in the body by a variety of mechanisms. These include (1) the activation of macrophages by IFN-γ to produce tumor necrosis factor (TNF-α) and oxygen intermediates such as nitric oxide (NO), which may induce target cell killing; (2) activation of NK cells by IL-2 to become active tumor cell killers; (3) the production of antibodies to tumor-associated antigens that results in "coating" the tumor cells and targeting them for activated macrophages or cytotoxic T lymphocytes (CTLs), the so-called antibody-dependent cellular cytoxicity (ADCC) reactions; (4) antitumor antibody-dependent, complement-mediated tumor cell killing (this appears to be a minor mechanism), and (5) activation of CD8$^+$ T lymphocytes to become CTLs by an MHC-mediated cell-killing mechanism. The most important of these reactions for killing tumor cells in vivo are the cell-mediated ones involving CTLs, macrophages, and NK cells.

The mechanisms of cell-mediated cytotoxicity involve direct cell–cell contact between a killer cell and a target tumor cell (Fig. 13-3).[40] There is a cell–cell contact release of lytic enzymes from CTL cells that attack the tumor cells (reviewed in Ref. 41). This is initiated by T-cell receptor interaction with antigenic peptide plus MHC on the target cell. This triggers a Ca^{++}-dependent pathway leading to polarization and exocytosis of granules containing proteases ("granzymes") and the assembly of channels called perforins from pore-forming proteins (PFPs) in the plasma membrane of the target cell but not the activated T cell. The perforin channels allow the uptake of a variety of proteases, particularly serine proteases called granzymes A, B, and C. These serine proteases may be involved in activation of endonucleases that trigger the DNA fragmentation characteristic of apoptosis. Both perforin assembly and granzyme A appear to be necessary for cell-mediated cell killing because transfection of one or the other gene into inactive killer cells induced only minimal cell lysis, whereas transfection of both genes produced an active killer cell.[41] Furthermore, mice that underwent gene knockout of the perforin gene do not generate potent NK cells or CD8$^+$CTL against virally infected cells.[41,41a]

Another mechanism of cell-mediated cytotoxicity not dependent on perforin or granzymes involves activation of the cell surface receptor called Fas or APO-1, discussed in the apoptosis section of Chapter 10. Persistent stimulation induces T cells to express Fas and the Ca^{2+}-independent component of T-cell mediated cytoxicity seems to work via the Fas pathway.

IMMUNE SURVEILLANCE

The concept of immune surveillance against cancer is an old one and has had its proponents and detractors over the years. The original theory postulated that transformed malignant cells arise frequently in normal individuals and that these cells are usually eliminated by immune response mechanisms. One of the strong arguments against this theory is the inability of spontaneously arising tumors in animals to protect syngeneic hosts from subsequent challenge with the same or other tumors (for review, see Ref. 42). In contrast with these findings, immune protection against chemically or virally induced tumors has often been observed. These findings thus

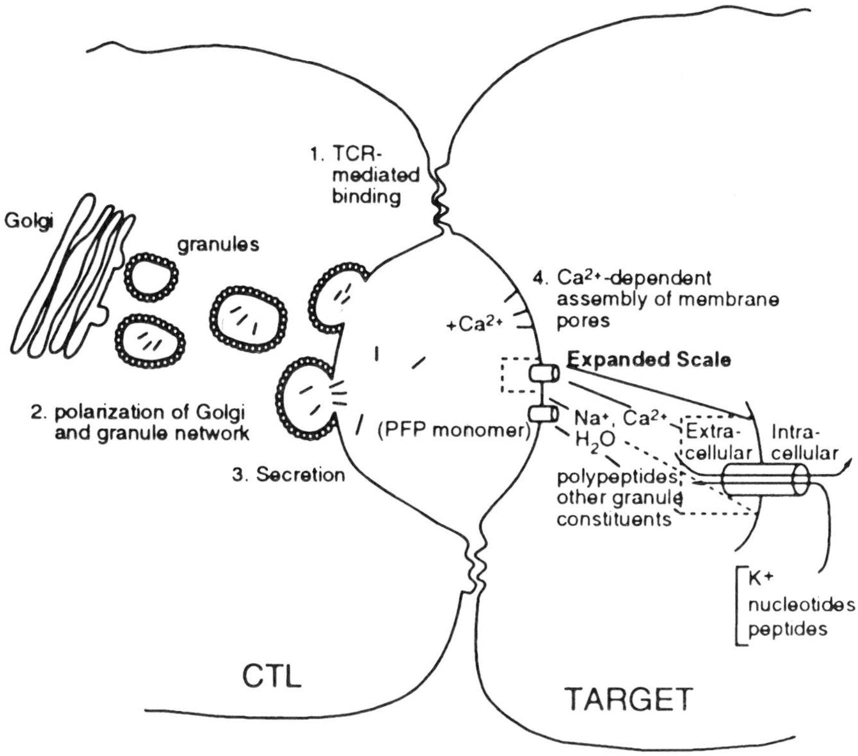

Figure 13-3 CTL-mediated cell killing. TCR, T-cell receptor; PFP, pore-forming proteins. (From Young and Cohn.[40])

pose a dilemma. Which of these observations holds for human cancer? The answer is that probably they both do. There is good evidence that virally induced neoplasms produce a specific immune reaction in animals and that immune protection against tumor growth can be achieved by preimmunization or vaccination with virally induced tumors or certain related viruses. The protection of mice against polyomavirus-induced tumors by preimmunization with tumor grafts and the protection of chickens against a virally induced leukemia (Marek's disease) by vaccination with nonpathogenic turkey herpesvirus are cases in point.[42] Similar results are inferred for humans from the fact that the vast majority of individuals infected with Epstein-Barr virus do not get lymphoma. There is suggestive evidence that cell-mediated immune responses restrain the proliferation of EBV-transformed B lymphocytes in another disease caused by EBV—namely, infectious mononucleosis.[42] This response may be overwhelmed or circumvented in patients who get Burkitt's lymphoma.

One of the reasons that spontaneously arising tumors in animals, and most cancers in humans, are not very immunogenic is that they arise slowly and pass through many biochemical and immunologic alterations before they can be detected. This differs from the relatively rapid appearance of experimentally induced tumors in laboratory animals. During tumor progression, clones of cancer cells evolve that evoke less immunogenic response and that have increased ability to avoid the host's immune response. Thus, spontaneous animal tumors and human cancers have "selected themselves for nonrejectability."[42]

ROLE OF MHC SYSTEM IN THE IMMUNOLOGIC RESPONSE TO CANCER CELLS—WAYS TUMORS AVOID THE HOST'S IMMUNE RESPONSE

The major histocompatibility complex genes (of mice, humans, and other species that have been characterized) code for three general classes.

Class I and class II are cell surface glycoproteins that are important for cellular interactions in the immune system. The MHC components are also called the HLA system in humans. Class I molecules include the self-recognition system antigens, or so-called transplantation antigens, and certain differentiation antigens of hematopoietic cells. Class I transplantation antigens are expressed on all somatic cells and were originally detected as recognition antigens in rejection of foreign tissue grafts, hence the name.[43] Class II molecules (so-called I_a antigens in the mouse) are found primarily on cells of the immune system, namely, B and T lymphocytes and macrophages, and are involved in regulating antibody responses to certain foreign antigens. As noted earlier, activation of T and B lymphocytes by a foreign antigen requires presentation of the foreign antigen in conjunction with MHC cell surface markers. A third type of gene, called class III, encodes components of the complement cascade and is linked to the MHC complex. The role of the MHC complex in tumor rejection has been most thoroughly studied in the mouse, and for this reason these data are discussed in some detail (for review, see Ref. 44).

In the mouse there are about 32 class I genes that map either to the H-2 or Tla loci of the MHC complex. Each genetically homogenous (inbred) strain of mice expresses a distinct set of MHC antigens on its cells. The products of H-2 genes include the highly polymorphic H-2K, H-2D, and H-2L histocompatibility antigens. The H-2 antigens are cell surface glycoproteins that help present the "target" antigen for signaling the attack on tumor cells or virally infected cells by cytolytic T lymphocytes. The cytolytic function of CTLs requires coordinated recognition of foreign antigen and self MHC components. Thus, the H-2 surface markers are said to provide MHC-restricted recognition of foreign antigens by CTL cells. Some CTLs recognize nonself class I antigens themselves as foreign targets, and this "allorecognition" is the basis of transplantation rejection between histoincompatible strains.

Many tumors (but apparently not all) of both mice and humans express cell surface neoantigens that are recognized in association with self MHC markers and become the target of an immune response by the mechanisms described above. As tumor cells evolve through their genetic instability, they appear to be able to attenuate or alter their recognition by the host as foreign. This may be partly because of a loss of expression of specific class I antigens or the expression of altered class I antigens on their cell surface.[45] Loss or attenuated expression of H-2 antigens has been observed in virally or chemically induced murine tumors. For example, a virally induced AKR T-cell leukemia that does not express H-2K antigen on its surface is resistant to killing by T cells and grows to a lethal tumor when transplanted into immunocompetent mice, whereas a leukemic cell line that expresses H-2K is rejected in immunocompetent mice. When the cloned H-2K gene is inserted into H-2K–negative AKR cells, the tumor is rejected.[46] Similarly, the tumorigenicity of various adenoviruses appears to be related to the loss of expression of class I MHC antigens on the surface of adenovirus-transformed cells.[47] Although loss of H-2 gene expression is most likely due to alterations of regulation of H-2 gene transcription in tumor cells, in at least one case (adenovirus-2 infected HeLa cells), loss of cell surface class I MHC antigens is due to impaired transport of class I glycoproteins to the cell surface because of deficient oligosaccharide processing.[48] In this case, the class I antigens remained in the high mannose form (see Chap. 4) and accumulated in the perinuclear region of the HeLa cells. A potentially important therapeutic effect has been noted by Hayashi et al,[49] who observed that α plus β interferon induced reexpression of class I MHC gene products on MHC class I-deficient, adenovirus-12–transformed mouse cells; this made these cells less tumorigenic in immunocompetent syngeneic hosts.

In human tumors also, decreased expression, or lack thereof, of MHC class I genes may be important for recognition and rejection of tumor cells. A number of human tumors of epithelial origin either do not express or display markedly reduced levels of class I antigens on their surface. These include choriocarcinomas, mammary carcinomas, epidermal carcinomas, small-cell lung carcinomas, neuroblastomas, and colorectal carcinomas. For example, in a comparison of MHC class I and II antigens, expressed on colon carcinomas from 25 patients, normal colonic mucosa 5 to 10 cm distal from each tumor, and colonic mucosa from individuals without cancer, significantly reduced expression of MHC class I antigens was found on the colon cancer tissue as compared with the adjacent normal mucosa and that taken from patients without cancer.[50]

Similarly, in a study of 52 tumor tissue samples (comprising 29 ovarian, 15 lung, 1 breast, and 4 colon carcinomas as well as 1 carcinoid and 2 malignant mesenchymal tumors), 20 had undetectable levels of one or more HLA alleles. There was a correlation of HLA loss with tumor grade, with the most poorly differentiated tumors having the greatest losses.[51]

The production of immunosuppressive factors has been observed for some cancer cells. For instance, cultured human HT29 colonic carcinoma cells produce a soluble factor of molecular weight 56,000 that suppresses mitogen-induced activation of T lymphocytes and blocks IL-2 production.[52] Mice bearing a MCA-38 colonic carcinoma produce a factor that impairs the cytotoxic function of $CD8^+$ T cells, decreases their expression of TNF-α and granzyme B genes, and decreases their ability to mediate an antitumor response in vivo.[53] These effects are apparently due to the factor's ability to alter assembly of the T-cell receptor so that it could no longer function in its signal-transduction pathway interactions.

Another mechanism by which tumor cells could alter the ability of the host to recognize them and to respond to their presence is by expressing altered or unique MHC markers. Many of the early reports of unique class I antigens on tumor cells reflected histocompatibility differences between the animal strain from which the tumor was derived and the host into which it was transplanted. Nevertheless, some clear examples of altered MHC antigens on tumor cells exist. One of these is the ultraviolet (UV)-light-induced murine fibrosarcoma 1591. This tumor has on its cell surface novel class I antigens that are different from those normally expressed on C3H mouse cells, the strain in which the tumor arose (reviewed in Ref. 44). One of these novel antigens appears to be derived from the H-2K gene, possibly by multiple gene recombination events that may have occured during tumorigenesis. Interestingly, the expression of these novel H-2 antigens renders the 1591 tumor cells highly immunogenic, yet they generally grow and progress to lethal tumors in the C3H host. This has led to the idea that UV-induced immunosuppressive events are involved in the host's response to these tumor cells and leads to escape of the tumor cells from the host's defenses. The evidence to support this is as follows (reviewed in Refs. 44 and 54): (1) UV-irradiated mice do not reject the growth of UV-induced tumors but do have the ability to reject non-UV induced tumors; (2) suppression of the immune response to UV-induced tumors can be transferred by blood elements to recipient non-UV-irradiated mice, suggesting that a population of specific suppressor cells can modulate the immune reaction to a specific type of tumor cell. Whether this mechanism can be applied in a general way to other tumors, and, most importantly, to "spontaneously" arising animal and human tumors, is not yet clear, but this could be of great therapeutic importance, since theoretically it might be possible to remove selectively or to inhibit the suppressor cells that abolish the immune response to a tumor without blocking the host's ability to maintain immunosuppressive capability for other kinds of immune stimuli.

Expression of certain types of MHC antigens actually favors growth and metastasis of some types of tumors. Although this may seem paradoxical at first, based on the preceding discussion, the explanation most likely lies in the fact that altered or novel MHC antigens may protect tumor cells from killing by CTL cells that depend on MHC-associated recognition of the tumor cells. As indicated in Chapter 11, for example, the metastatic potential of 3-methylcholanthrene-induced T10 fibrosarcomas of mice appears to depend on expression of H-$2D^k$ antigens on the surface.[55] These cells, however, lack expression of H-$2k^b$ and H-$2K^k$ antigens, and their metastatic potential is decreased when these H-2K genes are transfected into the T10 cells. Thus, an altered pattern of MHC gene expression appears to be a determining factor in establishing the malignant characteristics of tumor cells.

The knowledge gained in the extensive studies of the mouse MHC system has not yet been taken advantage of in human cancer. However, the ability to modulate the immune system is clearly an important therapeutic approach. Patients who have large tumor loads or who are immunosuppressed for some reason are not able to mount an effective immune response to their cancers, and as a rule, do not have a very favorable response to therapy.

There are also other ways by which tumors avoid the host's immune response. Variation in the cell surface properties of cancer cell clones appears to occur during tumor progression. These changes are reflected in the degree of immunogenicity of tumor cells and, because of the selection process described earlier, tumor cells with characteristics of "self-recognition" by the host would be favored. Thus, one way tumor cells can avoid rejection is by antigenically mimicking the host's own tissues; a second way is by "jamming" the immune response network. The latter can result from shedding tumor cell surface antigens that react directly with antitumor antibody-bearing cells or with circulating antitumor antibody to form antigen–antibody immune complexes. These complexes circulate and bind to killer cells that have Fc receptors for the antitumor antibody, thereby eliminating antibody-dependent cell-mediated killing of tumor cells. It has often been observed that cancer patients, particularly those with advanced disease, have a poor response to antigens that evoke cutaneous delayed (cell-mediated) hypersensitivity reactions and a diminished lymphocyte response to lectins and antigens in vitro.[56–58] In some patients, a complete lack of response ("anergy") to challenge with antigenic agents that produce delayed cutaneous hypersensitivity reactions is observed. This is associated with a poor prognosis. Conversion from an anergic to a reactive state during treatment is associated with improved survival compared with patients who remain immunoincompetent.[56] Tumors can also subvert the host's immune system in other ways. For example, the release of thromboplastinlike activity that evokes the deposition of fibrin around tumor cell clusters could prevent activated immune cells from attacking the tumor. Generation of low-molecular-weight fibrin degradation products by tumor-released proteases is another mechanism by which tumor cells could escape the immune response system, since some fibrin degradation products have been shown to inhibit antigenic stimulation of lymphocytes.[59] Of course, direct invasion of the bone marrow and lymphoid tissues by tumor, as noted, can produce a lymphopenia and a neutropenia that indirectly leads to subversion of the immune response by decreasing the production of cells involved in that response.

IMMUNOTHERAPY

Rationale for Immunotherapy

In an immunosuppressed patient, the best way to improve immune status is to remove or destroy the bulk of the tumor. This by itself frequently leads to improved immune responsiveness. It is when a tumor is small that therapy aimed at stimulating the patient's immune response to a tumor is most likely to succeed. Various methods to stimulate a patient's immune system nonspecifically have been tried with minimal clinical success (e.g., with the bacterial antigens BCG and *C. parvum*). Another approach is the use of monoclonal antibodies or monoclonal antibody–antitumor agent complexes directed against specific tumor antigenic determinants. Therapy aimed at overcoming a patient's immune unresponsiveness to a tumor could also be useful and could potentially be accomplished by tumor antigen modification to render tumor cells less like normal cells and thereby less recognized as self. This could conceivably be done in vitro by (1) chemical modification of tumor cell membranes by attachment of strongly antigenic chemical haptens, (2) infection of tumor cells with nonpathogenic viruses that produce altered cell surface antigens, or (3) hybridization of tumor cells to more antigenic cells.[42] Reinjection of such in vitro–altered cells (obtained originally by surgery or biopsy) back into the same patient could evoke a heightened immune response that would include antigenic recognition of the original tumor.

Other promising methods of immunotherapy involve the use of various lymphokines and other immune system stimulatory factors, sometimes lumped together under the rubric "biological response modifiers," and the use of the patient's own immune cells, stimulated by factors such as interleukin-2, in a procedure called "adoptive immunotherapy." These treatment modalities are discussed below.

Monoclonal Antibodies

A remarkable advance in the preparation of antibodies that recognize specific antigenic determinants has been provided by the work of Köhler and Milstein.[60] They showed that fusion of an antigenically stimulated lymphocyte with a malignant myeloma cell, whose whole protein-synthetic machinery is geared to make immunoglobulins, results in the formation of a hybrid cell (hybridoma) that produces large amounts of antibody against the antigenic determinant originally recognized by the stimulated lymphocyte (Fig. 13-4). In an immunized animal, different clones of lymphocytes that recognize different antigenic determinants on the injected antigen are produced. The fusion of a population of antibody-producing lymphocytes obtained from the spleen of an immunized animal with myeloma cells will result in clones of hybridomas, each producing antibodies highly specific for a given antigenic determinant recognized by the injected animal as foreign. These highly specific antibodies arising from discrete hybridoma clones are called monoclonal antibodies.

The hybridoma clones are obtained by fusing cells in the presence of Sendai virus or polyethylene glycol to make cell membranes soft and sticky, such that two cells in contact will fuse to form a heterokaryon containing nuclei from both cell types. With continued passage in culture, the nuclei ultimately fuse to form a true hybrid cell. The myeloma cell line that is usually used in these fusions is one that has been selected by development of resistance to 8-azaguanine for a deficiency in an enzyme involved in nucleic acid synthesis, hypoxanthine-guanine phosphoribosyl transferase (HGPRT). The cultures are then grown in a medium (HAT medium) in which only hybrid cells will survive. The fused cells are then cloned to select for hybridomas making specific antibodies. Once it is determined which clones make antibody, these clones can be grown in large cultures or injected back into the peritoneal cavity of animals who become living incubators for the proliferation of antibody-producing cells. Incredible amounts of specific, high-titer antibody can be obtained from the ascites fluid of relatively few animals. Thus, one can obtain a panel of highly specific antibodies directed at different antigenic determinants of a single antigen in a relatively inexpensive manner. Another advantage of this technique is that it is not necessary to have pure antigen to obtain specific antibodies. Specific antibodies have been obtained, for example, by the injection of whole tumor cells or cell homogenates into mice, followed by hybridoma formation and cloning. In this way, specific antibodies directed against animal and human cancer cells have been obtained.

The potential use of monoclonal antibodies (Mabs) that recognize tumor-associated antigenic determinants is far-reaching. Monoclonal antibodies directed against tumor cell surface components can inhibit tumor cell proliferation in culture and in whole animals.[61,62] Administration of monoclonal antibodies directly to patients has been tried with limited success so far. Monoclonal antibodies by themselves will most likely be of therapeutic benefit in clearing circulating tumor cells from the blood; in diminishing the amount of circulating tumor antigen, which could have a blocking effect on subsequent immunotherapy; and in clearing the bone marrow of tumor cells so that it can be used for autologous transplantation in patients after they have been treated with bone-marrow suppressive therapy.[63–65] In the latter case, samples of the patient's marrow are removed, treated with antibody directed against the type of tumor antigen present on the surface of their cells (determined by "typing" of the patient's tumor cells with a panel of monoclonal antibodies prepared against the particular type of tumor that they have), and stored frozen until after the patient is treated. Then the patient's own bone marrow is injected in order to restore normal bone marrow function.

Monoclonal antibodies can also be complexed with antitumor drugs, toxins, or radionuclides (reviewed in Ref. 66). The rationale of this approach is to target toxic substances directly to the tumor cells and spare normal cells. This approach has been tried experimentally with some success. For example, conjugates of antitumor

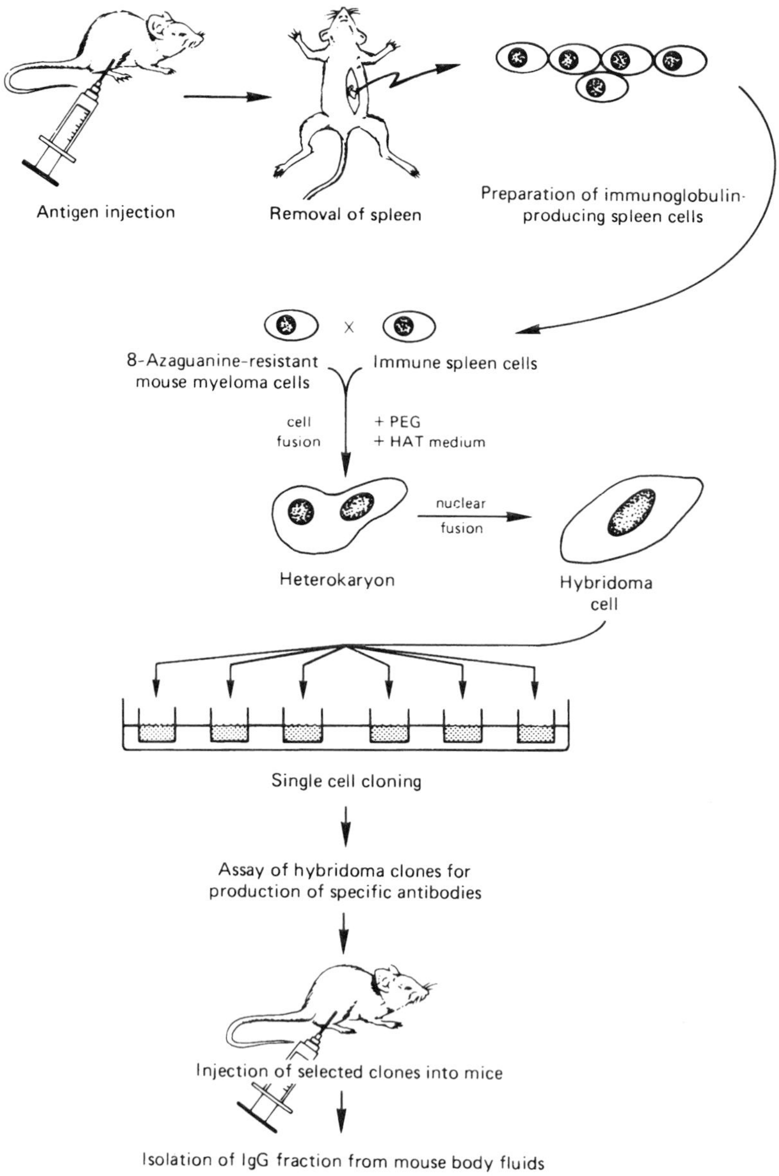

Figure 13-4 Production of monoclonal antibodies. Antigen or whole cells are injected into a mouse, and production of antibody is screened after immunization by examining serum samples. The spleens of antibody-producing mice are removed and single-cell suspensions of spleen cells are prepared. These cells are fused to 8-azaguanine resistant ($HGPRT^-$) mouse myeloma cells in the presence of polyethylene glycol (PEG) and grown in medium containing hypoxanthine, aminopterin, and thymidine (HAT medium). Hybridoma cells are then cloned and screened for antibody production. Large amounts of specific antibody can be obtained by injecting the hybridoma cells back into syngeneic mice and harvesting their blood and ascites fluid.

antibody and the anticancer drug daunomycin have been shown to inhibit the growth of lung carcinomas and hepatomas in animals with more efficacy than the drug or the antibody alone or than an unconjugated mixture of drug and antibody.[67,68] Doxorubicin—conjugated with a monoclonal antibody directed to a human melanoma-associated proteoglycan—has been shown to suppress growth of human melanomas growing in nude mice.[69]

Similarly, immunoconjugates prepared between doxorubicin and a human-mouse chimeric antibody that binds a tumor-associated antigen closely related to the Lewis blood group Y(Le^y) antigen, which is abundantly expressed on human carcinoma cell lines, induced complete regressions of xenografted human lung, breast, and colonic carcinomas growing subcutaneously in athymic mice, even in mice that had metastases.[70] Antitumor effects against xenografted human squamous cell and breast carcinomas in athymic mice have also been observed with a conjugate between doxorubicin and an anti-EGF receptor monoclonal antibody.[71]

Coupling of bacterial toxins such as diphtheria toxin or plant toxins such as abrin and ricin, all of which are potent inhibitors of protein synthesis and extremely toxic for mammalian cells, has also been tried as an approach to more selective tumor cell killing. For example, conjugates of ricin A chain to monoclonal antibodies against T-cell leukemias or Burkitt's lymphoma cells are specifically toxic for the malignant cells bearing the tumor-associated antigen[72–74] and can be shown to selectively kill tumor cells mixed with normal bone marrow.[74] Such toxin–monoclonal antibody conjugates may be useful in clearing tumor cells from the bone marrow prior to autologous marrow transplantation, as indicated above. Whether such drug-antibody or toxin-antibody conjugates will provide an advantage for systemic cancer therapy, however, remains an open question.

A number of clinical trials with Mab-toxin conjugates have been initiated (reviewed in Ref. 66). Although some responses have been observed, a number of problems with this approach have emerged. Some of these problems are common to all monoclonal antibody therapy. They include: (1) the difficulty of Mabs being able to penetrate into solid tumor masses in such a way as to deliver pharmacologically effective concentrations of drug or toxin, (2) the heterogeneity of tumors such that in any given tumor cell population only a minority of cells may have the targeted tumor antigen, (3) formation of anti-IgG antibodies by the patient, (4) the possible toxicity of antigen-antibody complexes (e.g., renal failure), and (5) the blocking effect of high circulating tumor antigen which could neutralize the antibody conjugate. A higher rate of success will likely be achieved by the use of monoclonal antibody–radionuclide complexes, because it may be possible to deliver sufficient radioactivity to the area of the tumor regardless of whether the complex penetrates all the way into the interior of the tumor or whether there is heterogeneity with respect to display of surface antigen. Moreover, such monoclonal antibody–radionuclide conjugates can be used to localize tumors diagnostically. Iodine-125 (^{125}I), iodine-131 (^{131}I), and technetium-99m (^{99m}Tc) have been coupled to monoclonal antibodies and been used successfully to radiolocalize human tumor xenografts growing in nude mice. A promising clinical result was observed in patients with B-cell non-Hodgkin's lymphomas treated with an ^{131}I-labeled Mab specific to a B-cell surface marker called CD20.[75] Six of nine treated patients had tumor responses, including patients with bulky or drug-resistant diseases; four had complete remission and remained disease free for at least 8 months. Thus, this approach using radiolabeled antibodies holds diagnostic as well as therapeutic potential for the treatment of cancer.

An advantage of radiolabeled monoclonal antibody conjugates for diagnosis and therapy is that, regardless of tumor-cell antigenic heterogeneity, as long as a significant number of the cells in the tumor bear the antigens recognized by the antibody, the Mab should be taken up by the tumor. The localized radioisotope should then be effective. For diagnostic purposes, localized radioisotope of appropriate energy can be detected by external radioimaging devices. For a therapeutic goal, the radiolabeled Mabs can kill cells that are millimeters away and thus be effective even against antigen-negative cells close to the antigen-expressing cells.

For a radiolabeled monoclonal antibody conjugate to be effective, it must posses the following characteristics: (1) the radionuclide must be tightly linked to the antibody, so that significant amounts of free radioactivity are not released from the antibody; (2) the chelating agent and the coupling reaction cannot compromise antibody specificity or negatively effect the pharmacokinetics and biodistribution of the Mab; and (3) the Mab conjugate should not cross-react with normal tissues. Several β emitters have been used for Mab conjugates, including iodine-

131 (^{131}I), yttrium-90 (^{90}Y), rhenium-186 (^{186}Re), and copper-67 (^{67}Cu).[66,76]

Future directions in isotopic labeled monoclonal antibody research will focus on the development of better localizing Mabs, the use of α emitters that have higher energy and better localized cell-killing effects, and the development of genetically engineered or human Mabs that are less likely to generate an anti-antibody immune response than the currently used murine antibodies. In addition, other cell targets will be used to direct Mabs to tumors. These include tumor cell surface mucins, for which there is evidence for some tumor type specificity,[77] and cell surface cytokine receptors such as the inducible α chain of the interleukin-2 receptor (IL-2Rα) expressed on the surface of abnormal T lymphocytes.[76] The latter target has been utilized in a clinical study of patients with adult T-cell leukemia. A Mab targeted to the IL-2Rα receptor has been shown to induce remission in about one-third of patients without significant toxicity.[76] This selectivity is presumably due to the fact that resting T cells do not express IL-2Rα, whereas abnormal T cells do. Similar approaches can be visualized for the development of anti-receptor Mabs targeted to receptors for tumor-derived paracrine or autocrine growth factors.

To circumvent the problems with murine Mabs, genetically engineered antibodies combining mouse variable or hypervariable regions (the specific antigen-binding region) with human constant and variable framework regions have been produced. This is done by transfecting hybrid mouse-human immunoglobulin genes into lymphoid cells that can express a chimeric protein from such genes. In this way, human-mouse Mabs to human colorectal, mammary, pancreatic, and B- and T-cell lymphomas have been made (reviewed in Ref. 76). Although such "humanized" Mabs appear to be less immunogenic, the presence of human allotypes on the hybrid immunoglobulin may still provide a "foreign" determinant. Early clinical trials suggest, nevertheless, that the hybrid Mabs are less immunogenic and have a longer plasma half-life.

An ultimate goal is the development of totally human Mabs against cancer cells. A number of approaches have been tried. Human B-cell lymphoid cell lines immortalized by Epstein-Barr virus have been employed as the antibody-producing cell type, but these cells produce low amounts of the expected IgM-type antibodies. The production of human antibodies in immunodeficient SCID mice by reconstituting their bone marrow and lymphoid tissue with human counterpart cells has been attempted and may work, but clinical trials have not yet been done with such Mabs to prove this.

Another approach bypasses hybridoma technology altogether. In this procedure, immunoglobulin variable-region genes from human B cells are cloned by the polymerase chain reaction technique and then expressed in *E. coli* and screened for ability to bind antigen. By this technique, large libraries of immunoglobulin genes can be generated and screened for antigen-binding specificity. Another advantage of this system is that such clones of Ig genes can be genetically manipulated to generate immunoglobulins of desired antigen specificity.

Modification of Tumor Cell Antigenicity

Since many tumors are apparently not immunogenic enough to stimulate an effective immune response against them, increasing their antigenicity could be a way to induce a more effective host immune response. For example, as noted above, one way might be to remove tumor cells from a patient, modify them with chemicals or viruses to make them more immunogenic, and then inject them back into the patient after sterilizing them by x-irradiation or cytotoxic drugs.

Tumor cell immunogenicity could also be increased by modulating expression of MHC gene products on tumor cells. Since class I MHC antigens are necessary for the presentation and recognition of tumor cell neoantigens by cytolytic T lymphocytes, their masking or absence on tumor cell surfaces may be key to their ability to escape an immune response of the host. One way to alter this is to introduce, by DNA-mediated gene transfer, for example, the genes for the missing MHC class I molecules. This has been done in adenovirus-12–transformed cells that lack expression of an H-2 class I gene product and that produce lethal tumors in syngeneic mice.[78] The induced expression of a single type of class I gene was sufficient to block the in vivo tumorigenicity of these cells. Thus, an approach to cancer therapy may be to increase or modulate expression of class I genes in tumor cells. Interferon γ (IFN-γ) can increase expression of class I antigens in certain cells.[78] Other such modulators may also be found.

Another method to boost a patient's response to a tumor may be to transfect cytokine genes into surgically removed tumor cells followed by reinplantation or to deliver such genes by tumor-targeted gene therapy in vivo so that their recognition by immune cells is increased. Several studies in murine tumor models have demonstrated increased antitumor responses in vivo to tumor cells transduced with cytokine genes, including IFN-γ, IL-2, IL-4, IL-6, IL-7, TNFα, and G-CSF (reviewed in Ref. 79). Clinical trials have been carried out using TNFα gene-transduced tumor infiltrative lymphocytes (TIL, see below) or melanoma cells transduced with TNFα or IL-2 genes.[80]

Immune System Modulators

Modulators of the immune system include both naturally occurring materials isolated from microorganisms, synthetic or semisynthetic polypeptides, synthetic compounds, and cytokines produced by cells of the immune system in the body that are involved in the cascade of events leading to an immune response against cancer cells. A partial list of these "biological response modifiers," or biomodulators, is shown in Table 13-2.

Microorganisms

Bacille Calmette-Guérin (BCG), a member of the *Mycobacterium* family, or extracts thereof (e.g., BCG-Mer), have been tried therapeutically in a number of ways. This includes adjuvant therapy for metastatic disease, intralesional in-

Table 13-2 Examples of Biological Response Modifiers

Microorganisms: whole, extracts, and cell wall skeletons
- Bacille Calmette-Guérin (BCG)
- Methanol extractable residue (MER from BCG)
- *Corynebacterium parvum* (*C. parvum* or more correctly *Propionibacterium acnes*)
- *Brucella abortus*
- Bru-Pel (*Brucella abortus* extract)
- *Pseudomonas aeruginosa* (Pseudogen)
- *Bordetella pertussis*
- *Norcardia rubra* (also cell wall extract (N-CWS)
- Picibanil (OK-432, *Streptococcus pyogenes*)
- *Klebsiella* spp.
- *Micrococcus* spp.
- *Lactobacillus casei* (LC 9018)

Chemically identified compounds from natural sources
- Peptidoglycan (gram-negative bacteria)
- WSA (disaccharide bound to a peptidoglycan, *Micrococcus smegmatis*)
- Lipopolysaccharide (LPS, from *Escherichia coli, Salmonella typhimurium*)
- Monophosphoryl lipid A (detoxified LPS)
- Biostim (glycoprotein, *Klebsiella pneumoniae*)

Polysaccharides
- Krestin (PSK)
- Glucan (β1-3 polyglucose, *Saccaromyces cerevisiae*)
- Lentinan (B1-3 polyglucose, *Lentinus edodes*)
- Pustulon
- Levan
- Mannozym

Polypeptides
- Bestatin (*Streptomyces olivorecticuli*)
- Tuftsin (part of Fc portion of leukokinin)
- Muramyldipeptide (MDP, BCG, also a large number of analogues)
- Muramyltripeptide (MTP-PE, a lipophilic analogue of MDP, also a large number of similar analogues)
- FK-565
- Complement C5a analogues

Synthetic compounds
- Levamisole
- Isoprinosine
- Pyrimidinols (ABPP)
- Anthraquinone (Tilorone)
- Azimexone BM 12.531
- Cimexone
- Alkyl-lysophospholipids (a large number of analogues)
- Sodium diethylthiocarbamate
- Thiabendazole
- Methylfurylbutyrolactones (Nafocare B)

Polyribonucleotides
- poly IC
- poly ICLC
- poly AU
- Ampligen

Vaccines/inducers of specific immune response
- Viral oncolysates
- Chemically modified tumor vaccines
- Hapten-modified tumor vaccines
- Normal tumor vaccine with or without adjuvant
- Xenogenized tumor cell vaccine
- Immune RNA
- Transfer factor

Source: Adapted from Talmadge and Clark.[81]

jection for cutaneous melanoma, topical instillation for urinary bladder cancer, and in combination with tumor cell vaccines or with chemotherapy (reviewed in Ref. 81). Other than intralesional use for cutaneous melanoma and instillation therapy for bladder cancer, BCG has not shown a great deal of activity and its current use is limited.

Among other microorganisms tested, *C. parvum* has probably been tested clinically the most, and it has shown some activity intraperitoneally in refractory ovarian cancer. It is thought that these bacterial preparations act primarily to augment the activity of cytotoxic lymphocytes, macrophages, and NK cells. There is evidence that bacterial extracts can induce the production of cytokines, such as IL-1 and TNF (discussed below). This is most likely the mechanism of action of products such as lipopolysaccharide (LPS) and other polysaccharides.[81,82]

Polypeptides

Polypeptides useful as immune system modulators are most often synthetic peptides modeled after naturally occurring ones, e.g., bestatin, tuftsin, muramyldipeptide, and complement component C5a analogues.

Bestatin, originally isolated from a *Streptomyces* species, has been used in clinical trials in Japan and Europe.[81] It has been shown to activate NK cells and macrophages and to be of some benefit as an adjuvant to chemotherapy in acute myelogenous leukemia. Tuftsin is a tetrapeptide that acts as an activator of macrophages and may have some clinical use as an adjuvant to chemotherapy. It stimulates tumor cell killing by macrophages and cytotoxic T cells. FK-565 is a synthetic acyltripeptide analogue of an immunoactive peptide isolated from a *Streptomyces* species and has been shown to inhibit metastasis in a murine model.[83] It also apparently acts by activating macrophages. Analogues of complement component C5a have been synthesized and tested for their ability to stimulate macrophages to produce IL-1 and IL-6, which in turn stimulate other cells in the immune cascade.[84]

The above are just some examples of synthetic peptide analogues that have immune modulatory activity. A growing number of these are being synthesized and tested for activity. Theoretically at least, such peptides could be designed so that they would resist degradation by proteases, have high affinity for cytokine receptors, be orally absorbable, and have the appropriate pharmacokinetics and biodistribution properties that would make them ideal therapeutic agents to augment a patient's immune system. This would circumvent many of the problems of treating patients with the naturally occurring immunomodulatory polypepetides, many of which are more difficult to use because of their pharmacologic properties (e.g., poor oral absorption, rapid destruction by proteases in the body) or their propensity to induce antipolypeptide antibodies, thus limiting their continued use in a given patient.

Other Synthetic Compounds

Examples of other synthetic compounds with immunomodulatory activity are listed in Table 13-2. Of these, levamisole has been shown to have the most clinical utility so far. In patients with stage B2 or C colon cancer, levamisole plus 5-fluorouracil, as adjuvants to surgical resection, have been shown to provide a longer survival than surgery alone.[85] Alkyl-lysophospholipids can induce interferon production by cells, but they have significant systemic toxicity in animals. Azimexone and cimexone are inducers of colony stimulating factors and foster the regeneration of bone marrow stem cells. The pyrimidinedione compound 2-amino-5-bromo-6-phenyl-4(3H)-pyrimidinedione (ABPP) is an interferon inducer that has shown some immune system–stimulating activity in tumor-bearing animals.

Other than levamisole, these synthetic compounds have not demonstrated great clinical utility; however, they may well represent only the first generation of such drugs, which could act similarly to endogenous activators of the immune system and would be easier from a pharmacologic standpoint to administer to patients.

Polyribonucleotides

These agents were developed for clinical use because of their ability to induce interferon production in vivo. They are synthetic double-stranded polymers of nucleic acid bases. In addition to inducing interferon, these agents appear to have other immune system and tumoricidal effects. Poly IC, poly ICLC, poly AU, and ampligen have also been tested clinically (reviewed in Ref. 81). Poly IC is a poor inducer of interferon in humans, probably because it is rapidly degraded by serum ribonucleases. Poly IC complexed with poly-L-lysine (poly ICLC) is less sensitive to ribonucleases and is a potent inducer of interferon production in mice. In phase I clinical trials of poly ICLC, some responses were observed in patients with leukemia, myeloma, and renal cell carcinoma. Phase II clinical trials of poly ICLC have also been carried out. Overall response rates to all these interferon inducers have been low, and it is likely that their usefulness will be limited to augmentation of other treatment modalities.

Cytokines

The term *cytokines,* used in its broadest sense, defines a large group of secreted polypeptides released by living cells that act nonenzymatically in picomolar to nanomolar concentrations to regulate cellular functions.[86] These cellular functions include regulation of immune cell activity (interferons and interleukins), hematopoiesis (colony stimulating factors or CSFs), and regulation of proliferation and differentiation of a wide variety of cell types (peptide growth factors such as EGF, FGF, TGFα, and TGFβ, etc.). Here we will discuss the cytokines that affect the immune system. The other peptide growth factors are discussed in Chapter 9.

INTERFERONS

Interferon was discovered in 1957 by two virologists, A. Isaacs and J. Lindenmann, who were looking for a substance that blocks viral infection of cells.[87] Their research was prompted by the clinical observation that patients seldom come down with two virally induced diseases at the same time. They showed that the medium removed from influenza virus–infected chicken cells grown in culture, when added to other cultures of chicken cells, prevented infection of the second cultures by a different virus. They named this interfering substance interferon. Since that time, numerous studies aimed at isolating and characterizing interferon have been carried out.

Human interferons are classified into three general groups: IFN-α, -β, and -γ.[88] IFN-α is produced by leukocytes and lymphoblastoid cells stimulated by viruses or by microbial cell components. The α interferons are members of a multigene family; they are acid-stable M_r = 16- to 25-kDa polypeptides. IFN-β is a 20-kDa glycoprotein produced by fibroblasts in culture after exposure to various microorganisms, microbial components, or high-molecular-weight polyanions (e.g., poly IC). Two IFN-β genes have been identified. IFN-γ, so-called immune interferon, is produced by T lymphocytes in response to antigenic or mitogenic stimulation. There are at least two species of IFN-γ, a 20-kDa and a 25-kDa glycoprotein, and they are acid-labile. The antiviral action of the interferons appears to involve interference with viral nucleic acid and protein synthesis via an increased rate of viral RNA degradation and decreased rate of peptide chain initiation.

The antitumor effects of interferons appear to result from a stimulation of NK cells and macrophages and from a direct cytotoxic effect on tumor cells. The evidence that multiple actions are involved comes from studies in tumor-bearing animals before and after depletion of NK cells. For example, in a study of the growth of tumors produced by Moloney sarcoma virus–transformed cells in mice, it was found that a mixture of IFN-α and -β markedly stimulated NK cell activity at the site of the tumor and inhibited tumor growth; however, when NK cells were depleted by in vivo treatment with an antibody to NK cells, tumor growth was still inhibited.[89] When the tumor load was very high, however, NK cell depletion did reduce the antitumor effect of IFN, suggesting that IFN has dual actions: direct inhibition of tumor cell multiplication and stimulation of NK cell activity, the latter of which plays a more apparent role in IFN action when the tumor burden is high.

IFN-γ, isolated initially from activated T lymphocytes and later produced by recombinant DNA techniques, is a potent activator of tumoricidal macrophages.[90] It also has marked macrophage migration inhibitory factor activity (MIF). Thus, IFN-γ may be responsible for the immunomodulating activities previously ascribed to the lymphokines macrophage activating factor (MAF) and MIF. One of the mechanisms by which IFN-γ also increases the cellular killing effect of macrophages mediated by IgG antibodies involves induction of F_c receptors on the macrophage surface.[91]

The antitumor immune modulating effect of IFN in vivo and the relatively low host toxicity of IFN led to a number of clinical trials. The greatest therapeutic usefulness of IFN has been in the treatment of a relatively rare form of leukemia called hairy cell leukemia because of the spiked appearance of the cell surface. Most of the earlier clinical trials used IFN-α purified from human leukocytes and later obtained by recombinant DNA techniques. Currently, various recombinant forms of IFN-α, -β, and -γ are in clinical trials for various malignancies and for AIDS-related Kaposi's sarcoma.

An intriguing approach to the activation of tumoricidal macrophages involves the use of liposomes to deliver macrophage-activating factors directly to these phagocytic cells.[92] Liposomes containing phosphatidylcholine and phosphatidylserine are selectively taken up by phagocytic cells, including reticuloendothelial cells in the liver, spleen, lymph nodes, and bone marrow as well as by circulating monocytes. Incorporation of agents that activate macrophages, such as crude macrophage activating factor (MAF), IFN-γ, and the low-molecular weight (M_r = 459) synthetic compound muramyl dipeptide (*N*-acetylmuramyl-L-alanyl-D-isoglutamine; MDP) into liposomes provides a delivery system for these factors to macrophages. Once the carrier liposomes are engulfed, the macrophages become tumoricidal against target cells in vivo. The macrophages thus activated recognize and lyse neoplastic cells in vitro by a mechanism that requires cell-to-cell contact but apparently is independent of MHC antigens.[92] Intravenous administration of liposomes containing MAF plus MDP to nude mice, who were previously injected with B16 melanoma cells and who had spontaneous metastases in the lungs and lymph nodes by the time of liposome treatment, produced 9/18 250-day survivors compared to 2/18 250-day survivors in the liposome-only control group.[92] These data indicate that activation of macrophages in vivo is possible and hold out the possibility that metastatic tumor sites can also be recognized and destroyed by such activated macrophages.

INTERLEUKINS

The interleukins belong to a family of polypeptide growth and differentiation factors called lymphokines. These are factors, produced by lymphocytes or macrophages, that stimulate the proliferation, differentiation, and function of T lymphocytes, B lymphocytes, and certain other cells involved in the immune response (Table 13-3). Initially discovered as soluble factors present in the growth medium of cultured lymphocytes, several such activities have been now identified (reviewed in Ref. 93). These activities were usually defined by their role in simulating an in vitro immune reaction, i.e., promoting the activation and/or proliferation of immune system cells. Accordingly, the following kinds of activities were identified: T-cell mitogenesis factor (TMF), or T-cell stimulating factor (TSF), which fostered T-cell proliferation in response to added plant lectins; killer/helper factors (KHFs), which stimulated in vitro generation of antigen-specific cytolytic T-cells; B-cell helper factors, which could replace T cells in fostering differentiation of B cells into antibody-producing cells; and T-cell growth factor (TCGF), which was produced by mitogen-stimulated lymphocytes and promoted proliferation of antigen-activated T cells. By the use of cloned lymphocyte populations, more extensive purification of lymphokine-containing media, and specific monoclonal antibodies, it was possible to catalogue these factors more definitively. It became clear that several of the previously described activities could be attributed to two distinct polypeptides. The renaming of these factors was adopted at the Second International Lymphokine Workshop held in Ermattingen, Switzerland, in 1979. One of these is interleukin-1 (IL-1) and the other interleukin-2 (IL-2). The term *interleukin* was chosen because it indicates the basic property of these secreted mediators, i.e., to serve as intercellular signals between leukocytes. Several additional interleukins have now been identified (Table 13-3), and some of their cellular functions have been discussed above, under "Mechanisms of the Immune Response to Cancer."

IL-1, -2, -3, and -4 have been introduced into clinical trial for a variety of malignant diseases.

Table 13-3 Characteristics of Some Human Interleukins

Factor	Molecular Weight	Cellular Source	Hematopoietic Activities
IL-1 alpha	17,500	Macrophages, epithelial cells	Affects early hematopoietic progenitors, making them more sensitive to "later" acting factors (hemopoietin 1 activity); induces CSF production by accessory cells; causes multiple effects on lymphoid and other nonhematopoietic cells
IL-1 beta	17,500	Endothelial cells, fibroblasts, other cell types	
IL-2	17,200	Activated T cells	Is a cofactor for growth and differentiation of T and B cells; augments lymphocyte-activated killer activity; induces production of other lymphokines
IL-3	25,000	Activated T cells	Stimulates early growth of granulocyte, monocyte, erythroid, and megakaryocyte progenitor cells; supports mast cell growth; induces acute nonlymphocytic leukemia blasts to proliferate
IL-4	20,000	T cells	Promotes secretion of IgG; induces B-cell growth; synergizes with other growth factors to promote colony growth; enhances growth of mast cell lines
IL-5	18,000	Activated T cells	Induces proliferation and differentiation of eosinophil progenitors
IL-6	21,000	Fibroblasts, macrophages, some T-cell and tumor cell lines	Induces differentiation of B cells; enhances Ig secretion by B cells; synergizes with other growth factors to promote colony growth
IL-7	17,500	Marrow stromal cells	Supports growth of pre-B cells
IL-8	10,000	Activated T cells and monocytes; endothelial cells	Activates neutrophils; attenuates inflammatory events at blood vessel endothelium

Source: From Laver and Moore.[94]

IL-1α and -1β have been employed to reverse bone marrow suppression due to chemotherapy or radiotherapy and to augment immunotherapy. IL-2 has been used in adoptive immunotherapy to stimulate clonal expansion of lymphokine-activated killer (LAK) cells and tumor-infiltrative lymphocytes (TIL) (see below). IL-3 has been utilized to stimulate bone marrow recovery in bone marrow or peripheral stem cell transplantation. IL-4 has been introduced as an immune system stimulator in various cancer treatment regimens.

TUMOR NECROSIS FACTOR

In the late 1800s it was observed that a few cancer patients had regression of their tumors after a full-blown systemic bacterial infection. This led a handful of investigators, including William Coley in the United States, to attempt to treat cancers by infecting patients with certain bacteria (reviewed in Ref. 95). The infectious process proved difficult to control, although a few positive responses were noted. There was sufficient encouragement, however, from this approach to prompt Coley in 1893 to try a mixture of killed bacteria (*Streptococcus pyogenes* and *Serratia marcescens*). When he injected his "Coley's toxins" directly into tumors, he got, in a few cases, some remarkable remissions of the tumor. This approach was taken by a number of investigators and continued on an experimental basis for a number of years. In fact, Coley's toxins were considered to be the only systemic therapy for cancer until the 1930s. With the advent of radiotherapy, chemotherapy, and improved cancer surgery, this treatment was abandoned and forgotten until the early 1980s.

It has been known for a long time that filtrates of certain bacteria such as the mycobacterium BCG and the corynebacterium *C. parvum* can induce a hemorrhagic necrosis of tumors when injected into tumor-bearing mice. Later work showed that an extract of gram-negative bacteria, known as endotoxin or bacterial pyrogen, could induce similar effects. Endotoxin was subsequently identified as a lipopolysaccharide (LPS). Because LPS is quite toxic, it never found its way into clinical use, but *BCG* and *C. parvum* have received extensive clinical trials and were not found to be very effective. The effectiveness of BCG and LPS against certain murine tumors in vivo, however, stimulated continued interest in finding the mechanism of this effect. The evidence pointed to the fact that these agents acted by modulating the host's immune response, probably by stimulating macrophages.

Attempts to define the antitumor mechanism of BCG and LPS led Old and his colleagues to test the serum of mice injected with BCG, LPS, or both for antitumor activity. The serum of BCG- plus LPS-treated mice produced hemorrhagic necrosis of mouse sarcomas in vivo. This effect required both BCG and LPS. Moreover, serum from BCG- and LPS-treated mice was also cytotoxic for a line of transformed mouse fibroblasts (L cells) in culture, but BCG and LPS added directly to the cultures were not, indicating that these agents were eliciting the release of some cytotoxic factors into the serum. It was also found that the in vivo necrotizing factor and the factor cytotoxic to cultured cells were one and the same. The factor was named tumor necrosis factor (TNF).[96]

TNF has also been shown to be identical to a cachexia-inducing factor called "cachectin."[97] This factor, also produced by macrophages, suppresses lipogenic enzymes such as glycero-3-phosphate dehydrogenase. It has been postulated that the cachexia associated with cancer may involve endogenous release of TNF. In addition to macrophages and certain lymphoid cell lines, there is evidence that TNF can also be produced by NK cells and endothelial cells. What role release from these latter cell types might play in vivo is not yet clear. A variety of agents can modulate cellular release of TNF. TNF expression by producer cells is increased by IFN-γ and indomethacin and decreased by glucocorticoids and PGE_2.

The normal biological function of TNF also is a little vague. Since bacterial and parasitic infections stimulate TNF release by immune system cells, it is thought that the normal function of TNF is to help fight off these infectious processes. The role of TNF in increasing phagocytotic activity of PMNs, fat mobilization from adipocytes, and fever induction are most likely involved in TNF's infection-fighting action.

It is likely that TNF is one of a family of tumor cell cytotoxic factors produced in the body. Another biological factor with TNF activity is lymphotoxin (LT), which is produced by mitogen-stimulated T lymphocytes and is thought to play a role in T cell–mediated tumor cell killing.[95] It is now known that the genes for TNF and lymphotoxin are related but not identical. There is about 30% sequence homology at the amino acid level. These gene products have been termed TNF-α (has cachetin activity) and TNF-β (has lymphotoxin activity). TNF-α is produced primarily by monocyte/macrophage cells, although it is also produced by T and B lymphocytes, LAK cells, NK cells, neutrophils, astrocytes, endothelial cells, smooth muscle cells, and a variety of tumor cell lines, whereas TNF-β is produced almost exclusively by lymphoid cells (reviewed in Ref. 98). TNF-α and -β are considered members of the "inflammatory cytokines" and are released at sites of inflammation. They also act as immunostimulants and mediators of host resistance to infectious agents and malignant cells. Overproduction of TNF-α during infection can lead to the septic shock syndrome.

The TNF-α and TNF-β genes are single-copy genes and closely linked within the MHC locus on chromosome 6 in humans. TNF-α is a trimer made up of 17,000-Da polypeptide chains. Native TNF-β is also a trimer; it consists of 25,000-Da subunits, each with one N-linked carbohydrate chain. TNF receptors of 55-kDa and 75-kDa have been identified and most cell types examined have both types. Each receptor type binds TNF-α and -β with similar high affinity, and they appear to transduce signals via G-protein–mediated activation of protein kinases. However, the two TNF receptors may mediate different sets of biological activity, and they have a somewhat different cellular distribution: the 55-kDa receptor is expressed on a wide range of cells and appears to be the receptor involved in direct tumor cell killing, whereas the 75-kDa receptor is expressed primarily on lymphoid and myeloid cells and may be more involved in the immune system effects of TNF.[99] The TNFs have a large number of biological activities and can induce a large number of genes in multiple target cells. These genes include transcription factors, cytokines, growth factors, adhesion molecules, inflammatory mediators, and acute-phase proteins.[98]

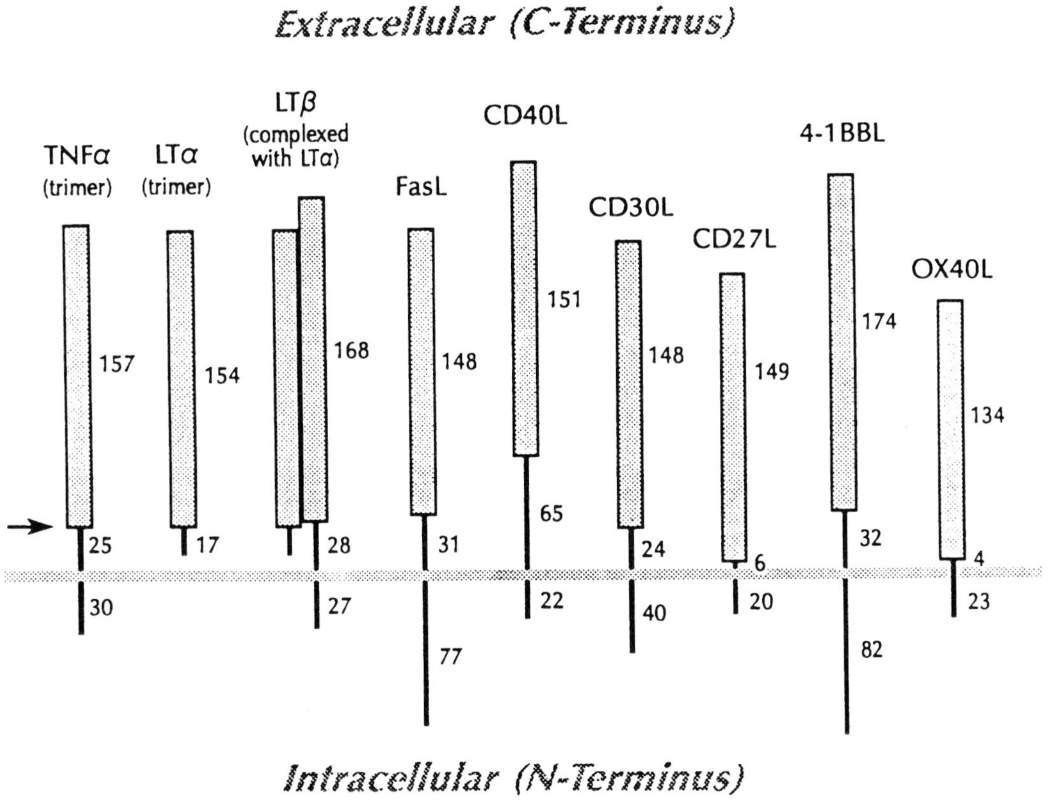

Figure 13-5 The TNF family of cytokines. The homologous C-terminal domains are indicated by open boxes. Extracellular and cytoplasmic domains, which lack sequence homology, are indicated by closed bars. The number of amino acids in each domain is shown. LTα is shown in both secreted and alternative membrane-associated forms, complexed with LTβ. The arrow indicates the proteolytic cleavage site in pro-TNFα that allows for the release of soluble form. Only TNFα, LTα, and LTβ have been shown to form oligomers. (From Smith et al.[99a])

It is now known that the TNF family comprises a large number of T cell cytokines (Fig. 13-5), including TNF-α and TNF-β (LT-α); T-B cell recognition factors CD30, CD40, and CD27; and the apoptosis-inducing ligand Fas.[99a] All of these ligands except LT-α are membrane-bound and appear to produce their effects by contacting receptors on adjacent cells while they stay in the membrane bound form, although some TNF-α (and perhaps other of these ligands) is released by proteases and circulates as a soluble form. There is also a superfamily of TNF receptors (Fig. 13-6), one of which, surprisingly, is the low-affinity nerve growth factor receptor (NGFR). This suggests some evolutionary relatedness that may or may not have anything to do with a commonality of function, but NGFRs are expressed at high levels on follicular dendritic cells of lymphoid germinal centers and TNF receptors are expressed on glial cells.

The TNF receptor family has sequence homology confined to the extracellular region. The canonical motif of all of them is the cysteine-rich repeats, each containing about 6 cysteines and 40 amino acids, but with considerable variation in length and number of repeats.[99a] The cytoplasmic domains are all rather small and variable in sequence homology, suggesting that they may trigger different signal transduction mechanisms. None have obvious catalytic activity sequences (e.g., of a tyrosine kinase–like activity, etc.). Interaction with ligand, however, induces receptor oligomerization, as seen for other cytokines and growth factors.

The tumor cell–killing effect of TNF may be both direct and indirect. There is evidence that

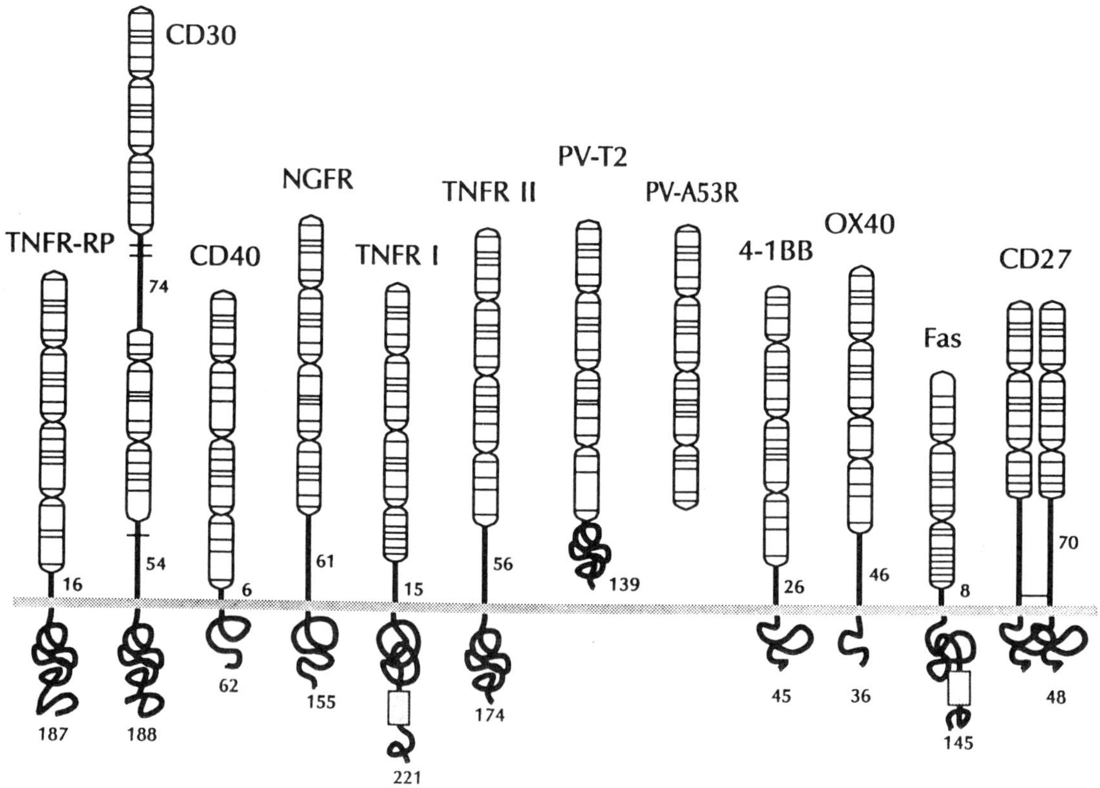

Figure 13-6 The TNF receptor superfamily. Homologous do mains are shown as open boxes and cysteine residues by horizontal lines. Number of amino acids in the (nonhomologous) extracellular linker and cytoplasmic domains are indicated. TNFR-RP is a predicted family member encoded by a transcribed sequence from human chromosome 12p. OX40 is a rat T cell activaton antigen with no reported cognate. In laboratory strains of vaccinia virus, the A53R open reading frame is interrupted by a premature termination codon. (From Smith et al.[99a])

cells which bind TNF can be killed by induction of a cytolytic cascade of events, including generation of oxygen radicals, DNA fragmentation, and apoptosis.[100] Why TNF tends to kill tumor cells preferentially over normal cells is not clear, but it may be related to a higher number of TNF receptors on tumor cells, more active internalization of TNF into tumor cells, ability of tumor cells' lysosomes to more effectively process TNF into a cytotoxic form, or lack of some protective factor in tumor cells. There is evidence that TNF has to be internalized and processed in lysosomes to be cytotoxic, since resistant cells do not degrade TNF as avidly as sensitive cells and inhibitors of lysosomal enzymes reduce TNF cytotoxicity.[100] Indirect killing could occur by TNF-induced activation of macrophages and cytotoxic T cells at the site of the tumor.

The clinical usefulness of TNF has been relatively limited.[101] However, some responses have been observed, and in a study of patients with metastatic renal cell carcinoma who responded to IL-2 therapy, the serum levels of TNF correlated with response: 48 hours after IL-2 infusion, both TNF serum levels and biological activity were higher in the responders than in the nonresponders.[102]

Although TNF-α has some promising antitumor effects, the systemic toxicity observed after intravenous infusion limits its utility. Therefore, efforts have been made to deliver TNF more directly to tumors. One such approach is to encapsulate TNF in liposomes. Liposome-encapsulated recombinant human TNF-α has been reported to retain full immunostimulatory effects but have much reduced toxic effects compared to free TNF-α when injected intravenously into normal rats.[103] Moreover, a new gene therapy approach incorporates the TNF-α gene into TIL cells in order to deliver TNF-producing cells directly to the tumor site (see Chap. 14).

ADOPTIVE IMMUNOTHERAPY

Adoptive immunotherapy is "a treatment approach in which cells with antitumor reactivity are administered to a tumor-bearing host and mediate either directly or indirectly the regression of established tumor."[104] Several types of approaches have been tried. One idea is to take tumor cells from a patient (removed by biopsy or surgery), inactivate the cells by x-irradiation, and then use these cells plus bacterial derived adjuvants such as *BCG* or *C. parvum* to attempt to induce an immune response in the patient to his or her own tumor cells.[105] Another idea is to take the patient's own lymphocytes and activate them in vitro with appropriate activating factors and then inject them back into the patient. The latter approach has been taken by Rosenberg and colleagues,[104] who have obtained, by leukapheresis, large quantities of patients' leukocytes, from which lymphocytes are separated. The lymphocytes are then incubated with recombinant interleukin-2, which stimulates a population of lymphocytes that, when activated, can lyse fresh, noncultured, NK-cell–resistant tumor cells but not normal cells.[106] These cells have been termed lymphokine-activated killer (LAK) cells. In earlier animal studies, Rosenberg et al. demonstrated that the adoptive transfer of LAK cells, incubated in culture with IL-2 followed by additional treatment in vivo with IL-2, induced the regression of pulmonary and hepatic metastases from a wide variety of murine tumors, including melanomas, sarcomas, a colon adenocarcinoma, and a urinary bladder carcinoma (reviewed in Ref. 104). In these model systems, the in vivo administration of both IL-2-activated LAK cells plus subsequent IL-2 was required for tumor regression. Expansion of the LAK cell population in vivo has been shown to occur after injection of more IL-2.

This rationale was then utilized in clinical trials. In vitro IL-2–activated LAK cells have been infused intravenously, followed by intravenous infusion of IL-2 every 8 hours for several days, depending on patient tolerance. With the high doses of IL-2 used in the earlier studies, patients experienced pulmonary edema with dyspnea and severe pulmonary complications. Other side effects include malaise, fever, chills, nausea, vomiting, diarrhea, erythema, pruritus, and fluid retension.[104] These toxic effects are reversible when IL-2 treatment is stopped. Autologous LAK-cell infusion by itself has minimal toxicity. Partial or complete responses have been seen in 13% to 57% of patients with advanced cancer, depending on the tumor type.[107] Best results were seen in patients with renal cell cancer, melanoma, non-Hodgkin's lymphoma, and colorectal cancer. Though it remains to be seen whether this or other forms of adoptive immunotherapy can cure patients with advanced disease, this approach may be the first generation of approaches to adoptive immunotherapy that may someday be an important addition to the armamentarium of the clinical oncologist.

Although lymphocytes circulating in the peripheral blood of cancer patients can in certain cases be activated to become cytotoxic tumor killer cells, these cells are apparently a relatively small percentage of circulating lymphocytes, and it requires high doses of IL-2 in vivo to keep them active. As noted above, these high doses of IL-2 have significant host toxicity.

Techniques have now been developed to isolate T lymphocytes directly from a patient's tumor, and these tumor-infiltrative lymphocytes (TIL) can also be clonally expanded in culture by adding IL-2. TIL cells bear the CD8-cell surface marker. They can kill the host's tumor cells in culture in a major histocompatibility complex (MHC) class I-restricted manner and are 50 to 100 times more potent than LAK cells in reducing lung metastases in mouse model systems.[107] In order for TIL cells to be effective in vivo, IL-2 must also be injected, but it requires lower doses of IL-2 than for LAK cell therapy. However, adjuvant treatment with cyclophosphamide or irradiation is required for TIL therapy to work. The reason for this is not clear, but it may be that cyclophosphamide-induced inhibition of T suppressor cells and/or the reduction of total tumor burden (with subsequent improved access of TIL to tumor tissue) are necessary for this approach to be effective.

Clinical trials have been initiated with TIL + IL-2 with or without cyclophosphamide. Results are better in regimens with cyclophosphamide and overall about 38% of patients with melanoma have had at least a partial response.[107]

Studies with indium-111–labeled TIL have shown that up to 0.015% of injected TIL localize in each gram of tumor tissue. This localization effect has suggested another approach, i.e., using these cells to deliver cytotoxic products directly to tumors. This is discussed under "Gene Therapy" in Chapter 14. LAK cells or TILs could also be used to deliver chemotherapeutic drugs to tumors. For instance, LAK cells loaded with 4′-deoxy-4′-iododoxorubicin produced a significantly higher reduction in lung metastases in tumor-bearing mice than did treatment with the drug alone.[108]

Another approach to adoptive immunotherapy is to use tumor-target–specific cytotoxic T lymphocytes (CTLs). Crowley et al.[109] have generated melanoma-specific CTLs by coculturing human peripheral blood lymphocytes with irradiated, allogeneic melanoma cells that express a restricting HLA-A region antigen, at a lymphocyte:tumor cell rate of 20:1, in the presence of IL-2. CTLs generated in this way were able to kill human melanoma cells bearing the restricting HLA-A phenotype in vitro and in a human-xenograft nude mouse model. As few as 2.5×10^6 T-cells were effective in vivo and were able to produce, in combination with IL-2, 96% and 88% disease-free animals with hepatic metastases when the CTLs were injected 3 and 7 days, respectively, after generation of metastases. IL-2 by itself was ineffective.

The potential clinical advantages of CTL adoptive immunotherapy are as follows[109]: (1) CTLs are MHC-restricted, can be targeted to specific HLA-A expressing cells, and can kill autologous tumor cells as well as allogeneic tumor cells as long as they express the HLA-A region antigen. (2) Thus, allogenic tumors matched by HLA-A region can substitute for autologous tumor in the stimulating tumor to generate specific CTLs. (3) CTLs can be generated from peripheral lymphocytes or lymph node cells from surgical specimens, allowing generation of CTLs from patients with stage I or II disease, whereas TIL cells may only be available from larger metastatic sites. (4) The amount of tumor needed for restimulation of CTLs in culture is small, and if autologous tumor is unavailable for follow-up stimulation, allogenic HLA-A matched allogeneic tumors could be used.

TUMOR VACCINES

For those cancers that are initiated or promoted by infectious viruses, it is conceivable that a vaccine derived from inactivated viruses or a preparation of viral antigens would prevent the onset of the disease. For example, there is a strong correlation between hepatitis B virus (HBV) infection and hepatocellular carcinomas. In a number of countries where hepatitis B infection is widespread, there is a high incidence of liver cancer. In China and 20 other countries where infection is endemic and liver cancer rates are high, an active immunization program against HBV is under way. Although HBV vaccination is not expected to eliminate liver cancer entirely, because some liver tumors are HBV-negative, it is predicted that the rate of liver cancer in endemic areas might be reduced by as much as 80%. This, however, might take 20 to 30 years to be evident, since the infant population receiving immunization now would not be expected to develop liver carcinomas until age 20 to 40.

Other cancers with a likely viral etiology include adult T-cell leukemia associated with HTLV-1 infection endemic in certain parts of Japan and the Caribbean, cervical carcinoma associated with human papillomavirus (HPV), and certain B-cell lymphomas (e.g., Burkitt's) and nasopharyngeal carcinoma associated with Epstein-Barr virus (EBV). However, effective vaccines have not yet been developed or clinically tested for these latter viruses, and it remains to be seen whether these diseases could be prevented by such an approach.

Vaccination of patients who already have cancer is entirely another matter and will likely be much more difficult to accomplish successfully. As noted above, several attempts have been made to elicit tumor immunity by inoculation of sterilized tumor cells or tumor-associated antigen–containing cell lysates, sometimes with adjuvants such as BCG, or by injection of anti-idiotypic antibodies directed against tumor-associated antigens (reviewed in refs. 110, 111).

Such approaches have in certain instances shown some promising results, but overall these therapies have met with limited success.

Another idea is to construct recombinant viruses that express tumor-associated antigens so that infected tumor cells will display the target antigen together with host MHC antigens and immunogenic viral proteins (via the normal cellular mechanism for antigen processing). This has achieved some success in experimental animal studies. For instance, vaccinia virus recombinants expressing a breast cancer–associated mucin called H23 epithelial tumor antigen (ETA) have been utilized to inoculate Fisher rats that were subsequently challenged with *ras*-transformed tumorigenic cells bearing H23 ETA.[110] Vaccination of rats with the recombinant virus expressing H23 ETA prior to challenge with the tumorigenic cells prevented tumor development in 82% of animals, whereas all animals inoculated with tumorigenic cells without prior vaccination developed tumors. These and other data encourage the idea that if an appropriate immune stimulus can be provided to an immunocompetent patient, at least a partial rejection reaction may be mounted in that patient. It is important to point out that this is unlikely to be effective in patients with a large tumor burden or in immunocompromised patients. Nevertheless, as an adjuvant to other treatment modalities such as surgery, radiation, or chemotherapy, immunotherapy may provide the extra needed "punch" to completely eliminate the residual tumor or silent metastases present in a patient, leading to more cures than are currently achievable by standard therapy.

Several cancer vaccines are under development and some of these have reached clinical trial. Various approaches include (reviewed in ref. 112): (1) Antigenic stimulation of the host to ganglioside antigens GM2, GD2, and GD3 (see Chap. 4), which are modified on several types of tumor cells including melanoma; (2) immunization of patients with mucins such as sialylated Thomsen-Friedenreich (TN) antigen or mucins derived from specific tumor types[113,114]; (3) stimulation of a immune response to carcinoembryonic antigen (CEA)–bearing tumors by means of a recombinant (attenuated) vaccinia virus expressing CEA[115]; (4) use of synthetic peptides representing the immunoglobulin epitope of B-cell malignancies; (5) transfection of the B7 "co-stimulation" receptor for immune recognition into tumor cells, followed by their reintroduction into the host; (6) fusion of GM-CSF with variable regions of immunoglobulins on malignant B cells to provide a "one-two punch" of immune reaction to a tumor antigen and a cytokine to stimulate macrophage production[116]; and (7) fusion of cancer cells with activated B lymphocytes to increase CTL recognition of tumor cells.[117]

REFERENCES

1. O. C. A. Scott: Tumor transplantation and tumor immunity: A personal view. *Cancer Res* 51:757, 1991.
2. G. Klein: Experimental studies in tumor immunology. *Fed Proc* 28:1739, 1969.
3. E. J. Foley: Antigenic properties of methylcholanthrene-induced tumors in mice of the strain of origin. *Cancer Res* 13:835, 1953.
4. R. T. Prehn and J. M. Main: Immunity to methylcholanthrene-induced sarcomas. *JNCI* 18:769, 1957.
5. G. Klein, H. O. Sjogren, E. Klein, and K. E. Hellstrom: Demonstration of resistance against methylcholanthrene-induced sarcomas in the primary autochthonous host. *Cancer Res* 20:1561, 1960.
6. F. M. Burnet: Immunological surveillance in neoplasia. *Transplant Rev* 7:3, 1971.
7. R. T. Prehn: The immune reaction as a stimulator of tumor growth. *Science* 176:170, 1972.
8. Å. R. Reif: Evidence for organ specificity of defenses against tumors, in H. Waters, ed.: *The Handbook of Cancer Immunology*. New York: Garland STPM Press, 1978, pp. 173–240.
9. L. J. Old and E. A. Boyse: Immunology of experimental tumors. *Annu Rev Med* 15:167, 1964.
10. M. S. Leffell and J. H. Coggin, Jr.: Common transplantation antigens on methylcholanthrene-induced murine sarcomas detected by three assays of tumor rejection. *Cancer Res* 37:4112, 1977.
11. K. E. Hellstrom, I. Hellstrom, and J. P. Brown: Unique and common tumor-specific transplantation antigens of chemically induced mouse sarcomas. *Int J Cancer* 21:317, 1978.
12. C. E. Whitmore and R. J. Huebner: Inhibition of chemical carcinogenesis by mixed vaccines. *Science* 177:60, 1972.
13. J. P. Brown, J. M. Klitzman, I. Hellstrom, R. C. Nowinski, and K. E. Hellstrom: Antibody response of mice to chemically induced tumors. *Proc Natl Acad Sci USA* 75:955, 1978.
14. E. M. Fenyo and G. Klein: Independence of

Moloney virus–induced cell-surface antigen and membrane-associated virion antigens in immunoselected lymphoma sublines. *Nature* 260:355, 1976.

15. M. J. Rogers, L. W. Law, and E. Appella: Separation of the tumor rejection antigen (TSTA) from the major viral structural proteins associated with the membrane of an R-MuLV-induced leukemia, in R. W. Ruddon, ed.: *Biological Markers of Neoplasia: Basic and Applied Aspects.* New York: Elsevier North Holland, 1978, pp. 53–62.
16. S. S. Tevethia, R. Greenfield, J. Pretell, and M. J. Tevethia: Viral origin of antigenic markers in simian papovavirus SV40-transformed cells, in R. W. Ruddon, ed.: *Biological Markers of Neoplasia: Basic and Applied Aspects.* New York: Elsevier North Holland, 1978, pp. 67–71.
17. H. F. Oettgen and K. E. Hellstrom: *Tumor Immunology,* in J. F. Holland and E. Frei III, eds.: *Cancer Medicine.* Philadelphia: Lea & Febiger, 1973, pp. 951–990.
18. J. L. McCoy, L. F. Jerome, J. H. Dean, E. Perlin, R. K. Oldham, D. H. Char, M. H. Cohen, E. L. Felix, and R. B. Herberman: Inhibition of leukocyte migration by tumor-associated antigens in soluble extracts of human malignant melanoma. *JNCI* 55:19, 1975.
19. H. Shiku, T. Takahashi, T. E. Carey, L. A. Resnick, H. F. Oettgen, and L. J. Old: Cell surface antigens of human cancer, in R. W. Ruddon, ed.: *Biological Markers of Neoplasia: Basic and Applied Aspects.* New York: Elsevier North Holland, 1978, pp. 73–88.
20. C. L. Finstad, C. Cordon-Cardo, N. H. Bander, W. F. Whitmore, M. R. Melamed, and L. J. Old: Specificity analysis of mouse monoclonal antibodies defining cell surface antigens of human renal cancer. *Proc Natl Acad Sci USA* 82:2955, 1985.
21. F. X. Real, A. N. Houghton, A. P. Albino, C. Cordon-Cardo, M. R. Melamed, H. F. Oettgen, and L. J. Old: Surface antigens of melanomas and melanocytes defined by mouse monoclonal antibodies: Specificity analysis and comparison of antigen expression in cultured cells and tissues. *Cancer Res* 45:4401, 1985.
22. A. Lanzavecchia: Identifying strategies for immune intervention. *Science* 260:937, 1993.
23. C. A. Janeway, Jr., and K. Bottomly: Signals and signs for lymphocyte responses. *Cell* 76:275, 1994.
24. S. Fujimoto, M. Greene, and A. H. Sehan: Regulation of the immune response to tumor antigens: I. Immunosuppressor T cells in tumor-bearing hosts. *J Immunol* 116:791, 1976.
25. S. Fujimoto, M. Greene, and A. H. Sehan: Regulation of the immune response to tumor antigens: II. Nature of immunosuppressor T cells in tumor-bearing hosts. *J Immunol* 116:800, 1976.
26. P. M. Flood, M. L. Kripke, D. A. Rowley, and H. Schreiber: Suppression of tumor rejection by autologous anti-idiotypic immunity. *Proc Natl Acad Sci USA* 77:2209, 1980.
27. R. N. Germain: MHC-dependent antigen processing and peptide presentation: Providing ligands for T lymphocyte activation. *Cell* 76:287, 1994.
28. A. Weiss and D. R. Littman: Signal transduction by lymphocyte antigen receptors. *Cell* 76:263, 1994.
29. G. J. Freeman, F. Borriello, R. J. Hodes, H. Reiser, K. S. Hathcock, G. Laszio, A. J. McKnight, J. Kim, L. Du, D. B. Lombard, G. S. Gray, L. M. Nadler, and A. H. Sharpe: Uncovering of functional alternative CTLA-4 counter-receptor in B7-deficient mice. *Science* 262:907, 1993.
30. K. S. Hathcock, G. Laszio, H. B. Dickler, J. Bradshaw, P. Linsley, and R. J. Hodes: Identification of an alternative CTLA-4 ligand costimulatory for T cell activation. *Science* 262:905, 1993.
31. G. J. Freeman, J. G. Bribben, V. A. Boussiotis, J. W. Ng, V. A. Restivo, Jr., L. A. Lombard, G. S. Gray, and L. M. Nadler: Cloning of B7-2: A CTLA-4 counter-receptor that costimulates human T cell proliferation. *Science* 262:909, 1993.
32. I. L. Weissman: Developmental switches in the immune system. *Cell* 76:207, 1994.
33. T. A. Springer: Traffic signals for lymphocyte recirculation and leukocyte emigration: The multistep paradigm. *Cell* 76:301, 1994.
34. W. E. Paul and R. A. Seder: Lymphocyte responses and cytokines. *Cell* 76:241, 1994.
35. R. B. Herberman and H. T. Holden: Natural cell-mediated immunity. *Adv Cancer Res* 27:305, 1978.
36. M. E. Nunn and R. B. Herberman: natural cytotoxicity of mouse, rat, and human lymphocytes against heterologous target cells. *JNCI* 62:765, 1979.
37. J. C. Lee and J. N. Ihle: Characterization of the blastogenic and cytotoxic responses of normal mice to ecotropic C-type gp 71. *J Immunol* 118:928, 1977.
38. K. Karre, G. O. Klein, R. Kiessling, G. Klein, and J. C. Roder: Low natural in vivo resistance to syngeneic leukaemias in natural killer–deficient mice. *Nature* 284:624, 1980.
39. W. J. Murphy, C. W. Reynolds, P. Tiberghien, and D. L. Longo: Natural killer cells and bone marrow transplantation. *JNCI* 85:1475, 1993.
40. J. Ding-E Young and Z. A. Cohn: Cell-mediated killing: A common mechanism? *Cell* 46:641, 1986.
41a. D. Kägi, B. Ledermann, K. Bürki, P. Seiler, B. Odermatt, K. J. Olsen, E. R. Podack, R. M. Zinkernagel, and H. Hengartner: Cytotoxicity mediated by T cells and natural killer cells is greatly impaired in perforin-deficient mice. *Nature* 369:31, 1994.

41. P. C. Doherty: Cell-mediated cytotoxicity. *Cell* 75:607, 1993.
42. G. Klein: Immune and non-immune control of neoplastic development: Contrasting effects of host and tumor evolution. *Cancer* 45:2488, 1980.
43. G. D. Snell: Studies in histocompatibility. *Science* 213:172, 1981.
44. R. S. Goodenow, J. M. Vogel, and R. L. Linsk: Histocompatibility antigens on murine tumors. *Science* 230:777, 1985.
45. K. Karre, H. G. Ljunggren, G. Piontek, and R. Kiessling: Selective rejection of H-2 deficient lymphoma variants suggests alternative immune defense strategy. *Nature* 319:675, 1986.
46. K. Hui, F. Grosveld, and H. Festenstein: Rejection of transplantable AKR leukaemia cells following MHC-DNA-mediated cell transformation. *Nature* 311:750, 1984.
47. R. Bernards, P. I. Schrier, A. Houweling, J. L. Bos, A. J. van der Eb, M. Zijlotra, and C. J. M. Melief: Tumorigenicity of cells transformed by adenovirus type 12 by evasion of T-cell immunity. *Nature* 305:776, 1983.
48. M. Andersson, S. Paabo, T. Nilsson, and P. A. Peterson: Impaired intracellular transport of class I MHC antigens as a possible means for adenoviruses to evade immune surveillance. *Cell* 43:215, 1985.
49. H. Hayashi, K. Tanaka, F. Jay, G. Khoury, and G. Jay: Modulation of the tumorigenicity of human adenovirus-12-transformed cells by interferon. *Cell* 43:263, 1985.
50. C. J. McDougall, S. S. Ngoi, I. S. Goldman, T. Godwin, J. Felix, J. J. DeCosse, and B. Rigas: Reduced expression of HLA class I and II antigens in colon cancer. *Cancer Res* 50:8023, 1990.
51. Z. Vegh, P. Wang, F. Vanky, and E. Klein: Selectively downregulated expression of major histocompatibility complex class I alleles in human solid tumors. *Cancer Res* 53:2416, 1993.
52. E. C. Ebert, A. I. Roberts, D. Devereux, and H. Nagase: Selective immunosuppressive action of a factor produced by colon cancer cells. *Cancer Res* 50:6158, 1990.
53. H. Mizoguchi, J. J. O'Shea, D. L. Longo, C. M. Loeffler, D. W. McVicar, and A. C. Ochoa: Alterations in signal transduction molecules in T lymphocytes from tumor-bearing mice. *Science* 258:1795, 1992.
54. M. L. Kripke: Immunologic mechanisms in UV radiation carcinogenesis. *Adv Cancer Res* 34:69, 1981.
55. R. Wallich, N. Bulbuc, G. J. Hammerling, S. Katzav, S. Segal, and M. Feldman: Abrogation of metastatic properties of tumor cells by de novo expression of H-2K antigens following H-2 gene transfection. *Nature* 315:301, 1985.
56. A. E. Giuliano, D. Rangel, S. H. Golub, E. C. Holmes, and D. L. Morton: Serum-mediated immunosuppression in lung cancer. *Cancer* 43:917, 1979.
57. C. C. Ting, S. C. Tsai, and M. J. Rogers: Host control of tumor growth. *Science* 197:571, 1977.
58. I. Hellstrom, G. A. Warner, K. E. Hellstrom, and H. O. Sjogren: Sequential studies on cell-mediated tumor immunity and blocking serum activity in ten patients with malignant melanoma. *Int J Cancer* 11:280, 1973.
59. G. Girmann, H. Pees, G. Schwarze, and P. G. Scheurlen: Immunosuppression by micromolecular fibrinogen degradation products in cancer. *Nature* 259:399, 1976.
60. G. Köhler and C. Milstein: Continuous cultures of fused cells secreting antibody of predefined specificity. *Nature* 256:495, 1975.
61. H. P. Vollmers, B. A. Imhof, I. Wieland, A. Hiesel, and W. Birchmeier: Monoclonal antibodies: NORM-1 and NORM-2 induce more normal behavior of tumor cells in vitro and reduce tumor growth in vivo. *Cell* 40:547, 1985.
62. H. Masui, T. Kawamoto, J. D. Sato, B. Wolf, G. Sato, and J. Mendelsohn: Growth inhibition of human tumor cells in athymic mice by anti-epidermal growth factor receptor monoclonal antibodies. *Cancer Res* 44:1002, 1984.
63. L. M. Nadler, P. Stashenko, R. Hardy, W. D. Kaplan, L. N. Button, D. W. Kufe, K. N. Antman, and S. F. Schlossman: Serotherapy of a patient with a monoclonal antibody directed against a human lymphoma-associated antigen. *Cancer Res* 40:3147, 1980.
64. R. C. Bast Jr., P. DeFabritiis, J. Lipton, R. Gelber, C. Mauer, L. Nadler, S. Sallan, and J. Ritz: Elimination of malignant clonogenic cells from human bone marrow using multiple monoclonal antibodies and complement. *Cancer Res* 45:499, 1985.
65. J. J. Vredenburgh, Jr., and E. D. Ball: Elimination of small cell carcinoma of the lung from human bone marrow by monoclonal antibodies and immunomagnetic beads. *Cancer Res* 50:7216, 1990.
66. D. M. Goldenberg: New developments in monoclonal antibodies for cancer detection and therapy. *CA* 44:43, 1994.
67. E. Hurwitz, B. Schechter, R. Arnon, and M. Sela: Binding of anti-tumor immunoglobulins and their daunomycin conjugates to the tumor and its metastasis: In vitro and in vivo studies with Lewis lung carcinoma. *Int J Cancer* 24:461, 1979.
68. Y. Tsukada, W. K.-D. Bischof, N. Hirai, H. Hirai, E. Hurwitz, and M. Sela: Effect of a conjugate of daunomycin and antibodies to rat α-fetoprotein on the growth of α-fetoprotein-producing tumor cells. *Proc Natl Acad Sci USA* 79:621, 1982.
69. H. M. Yang, and R. A. Reisfeld: Doxorubicin conjugated with a monoclonal antibody directed to a human melanoma-associated proteoglycan suppresses the growth of established tumor

xenografts in nude mice. *Proc Natl Acad Sci USA* 85:1189, 1988.
70. P. A. Trail, D. Willner, S. J. Lasch, A. J. Henderson, S. Hofstead, A. M. Casazza, R. A. Firestone, I. Hellstrom, and K. E. Hellstrom: Cure of xenografted human carcinomas by BR96-doxorubicin immunoconjugates. *Science* 261:212, 1993.
71. J. Baselga, L. Norton, H. Masui, A. Pandiella, K. Coplan, W. H. Miller, Jr., and J. Mendelsohn: Antitumor effects of doxorubicin in combination with anti-epidermal growth factor receptor monoclonal antibodies. *JNCI* 85:1327, 1993.
72. V. Raso, J. Ritz, M. Basala, and S. F. Schlossman: Monoclonal antibody-ricin A chain conjugate selectively cytotoxic for cells bearing the common acute lymphoblastic leukemia antigen. *Cancer Res* 42:457, 1982.
73. B. K. Seon: Specific killing of human T-leukemia cells by immunotoxins prepared with ricin A chain and monoclonal anti-human T-cell leukemia antibodies. *Cancer Res* 44:259, 1984.
74. M. Bregni, P. DeFabritiis, V. Raso, J. Greenberger, J. Lipton, L. Nadler, L. Rothstein, J. Ritz, and R. C. Bast, Jr.: Elimination of clonogenic tumor cells from human bone marrow using a combination of monoclonal antibody: Ricin A chain conjugates. *Cancer Res* 46:1208, 1986.
75. M. S. Kaminski, K. R. Zasadny, I. R. Francis, A. W. Milik, C. W. Ross, S. D. Moon, S. M. Crawford, J. M. Burgess, N. A. Petry, G. M. Butchko, S. D. Glenn, and R. L. Wahl: Radioimmunotherapy of B-cell lymphoma with [^{131}I]anti-B1(anti-CD20) antibody. *N Engl J Med* 329:459, 1993.
76. T. A. Waldmann: Monoclonal antibodies in diagnosis and therapy. *Science* 252:1657, 1991.
77. R. S. Metzgar, M. A. Hollingsworth, and B. Kaufman: Pancreatic Mucins, in V. L. Go, E. Dimagno, J. Gardner, I. Lebenthal, H. Reber, and G. Scheele, eds.: *The Pancreas.* Vol. 2. New York: Raven Press, 1993, pp. 351–367.
78. K. Tanaka, K. J. Isselbacher, G. Khoury, and G. Jay: Reversal of oncogenesis by the expression of a major histocompatibility complex class I gene. *Science* 228:26, 1985.
79. M. Ogasawara and S. A. Rosenberg: Enhanced expression of HLA molecules and stimulation of autologous human tumor infiltrating lymphocytes following transduction of melanoma cells with γ-interferon genes. *Cancer Res* 53:3561, 1993.
80. S. A. Rosenberg: The immunotherapy and gene therapy of cancer. *J Clin Oncol* 10:180, 1992.
81. J. E. Talmadge and J. Clark: Biological response modifiers: Preclinical and clinical results, in H. M. Pinedo, D. L. Longo, and B. A. Chabner, eds.: *Cancer Chemotherapy and Biological Response Modifiers.* Annual 9. Amsterdam: Elsevier, 1987, pp. 454–472.
82. R. Englehardt, A. Mackensen, and C. Galanos: Phase I trial of intravenously administered endotoxin *(Salmonella abortus equi)* in cancer patients. *Cancer Res* 51:2524, 1991.
83. N. Inamura, K. Nakahara, T. Kino, T. Gotoh, I. Kawamura, H. Aoki, H. Imanaka, and S. Sone: Activation of tumoricidal properties in macrophages and inhibition of experimentally induced murine metastases by a new synthetic acyltripeptide, FK-565. *J Biol Response Modif* 4:408, 1985.
84. E. L. Morgan, S. D. Sanderson, W. Scholz, D. J. Noonan, W. O. Weigle, and T. E. Hugli: Identification and characterization of the effector region within human C5a responsible for stimulation of interleukin-6 synthesis. *J Immunol* 148:3937, 1992.
85. C. G. Moertel, T. R. Fleming, J. S. MacDonald, D. G. Haller, J. A. Laurie, P. J. Goodman, J. S. Ungerleider, W. A. Emerson, D. C. Tormey, J. H. Glick, M. H. Veeder, and J. A. Mailliard: Levamisole and fluorouracil for adjuvant therapy of resected colon carcinoma. *N Engl J Med* 322:352, 1990.
86. C. Nathan and M. Sporn: Cytokines in context. *J Cell Biol* 113:981, 1991.
87. A. Isaacs and J. Lindenmann: Virus interference: I. The interferon. *Proc R Soc Lond [Ser B]* 147:258, 1957.
88. K. C. Zoon and R. Wetzel: Comparative structures of mammalian interferons, in P. E. Came and W. A. Carter, eds., *Interferons and Their Applications.* New York: Springer-Verlag, 1984.
89. K. L. Fresa and D. M. Murasko: Role of natural killer cells in the mechanism of the antitumor effect of interferon on Moloney sarcoma virus–transformed cells. *Cancer Res* 46:81, 1986.
90. L. Varesio, E. Blasi, G. B. Thurman, J. E. Talmadge, R. H. Wiltrout, and R. B. Herberman: Potent activation of mouse macrophage by recombinant interferon-γ. *Cancer Res* 44:4465, 1984.
91. Y. Akiyama, M. D. Lubeck, Z. Steplewski, and H. Koprowski: Induction of mouse IgG2a- and IgG3-dependent cellular cytotoxicity in human monocytic cells (U937) by immune interferon. *Cancer Res* 44:5127, 1984.
92. I. J. Fidler: Macrophages and metastasis—A biological approach to cancer therapy. *Cancer Res* 45:4714, 1985.
93. S. B. Mizel: The interleukins. *FASEB J* 3:2379, 1989.
94. J. Laver and M. A. S. Moore: Clinical use of recombinant human hematopoietic growth factors. *JNCI* 81:1370, 1989.
95. L. J. Old: Tumor necrosis factor (TNF). *Science* 230:630, 1985.
96. E. A. Carswell, L. J. Old, R. L. Kassel, S. Green, N. Fiore, and B. Williamson: An endotoxin-induced serum factor that causes necrosis of tumors. *Proc Natl Acad Sci USA* 72:3666, 1975.
97. B. Beutler and A. Cerami: Cachectin and tumor

necrosis factor as two sides of the same biological coin. *Nature* 320:584, 1986.

98. J. Vilcek and T. H. Lee: Tumor necrosis factor: New insights into the molecular mechanism of its multiple actions. *J Biol Chem* 266:7313, 1991.
99. F. Balkwill: Tumor necrosis factor: Improving on the formula. *Nature* 361:206, 1993.
99a. C. A. Smith, T. Farrah, and R. G. Goodwin: The TNF receptor superfamily of cellular and viral proteins: Activation, costimulation, and death. *Cell* 76:959, 1994.
100. J. W. Larrick and S. C. Wright: Cytotoxic mechanism of tumor necrosis factor-α. *FASEB J* 4:3215, 1990.
101. J. K. McIntosh, J. J. Mule, W. D. Travis, and S. A. Rosenberg: Studies of effects of recombinant human tumor necrosis factor on autochthonous tumor and transplanted normal tissue in mice. *Cancer Res* 50:2463, 1990.
102. J.-Y. Blay, M. C. Favrot, S. Negrier, V. Combaret, S. Chouaib, A. Mercatello, P. Kaemmerlen, C. R. Franks, and T. Philip: Correlation between clinical response to interleukin-2 therapy and sustained production of tumor necrosis factor. *Cancer Res* 50:2371, 1990.
103. R. J. Debs, H. J. Fuchs, R. Philip, E. N. Brunette, N. Duzgunes, J. E. Shellito, D. Liggitt, and J. R. Patton: Immunomodulatory and toxic effects of free and liposome-encapsulated tumor necrosis factor α in rats. *Cancer Res* 50:375, 1990.
104. S. A. Rosenberg, M. T. Lotze, L. M. Muul, S. Leitman, A. E. Chang, S. E. Ettinghausen, Y. L. Matory, J. M. Skibber, E. Shiloni, J. T. Vetto, C. A. Seipp, C. Simpson, and C. M. Reichert: Observations on the systemic administration of autologous lymphokine-activated killer cells and recombinant interleukin-2 to patients with metastatic cancer. *N Engl J Med* 313:1485, 1985.
105. M. G. Hanna, J. S. Brandhorst, and L. C. Peters: Active specific immunotherapy of residual micrometastases: An evaluation of sources, doses, and ratios of BCG with tumor cells. *Cancer Immunol Immunother* 7:165, 1979.
106. J. J. Mule, S. Shu, S. L. Schwarz, and S. A. Rosenberg: Adoptive immunotherapy of established pulmonary metastases with LAK cells and recombinant interleukin-2. *Science* 225:1487, 1984.
107. S. A. Rosenberg: Immunotherapy and gene therapy of cancer. *Cancer Res.* 51(suppl):5074s, 1991.
108. S. Mandruzzato, A. Rosato, V. Bronte, P. Zanovello, N. Amboldi, D. Ballinari, and D. Collavo: Adoptive transfer of lymphokine-activated killer cells loaded ith 4′-deoxy-4′-iododoxorubicin: Therapeutic effect in mice bearing lung metastases. *Cancer Res* 54:1016, 1994.
109. N. J. Crowley, C. E. Vervaert, and H. F. Seigler: Human xenograft-nude mouse of adoptive immunotherapy with human melanoma-specific cytotoxic T-cells. *Cancer Res* 52:394, 1992.
110. F. K. Stevenson: Tumor vaccines. *FASEB J* 5:2250, 1991.
111. M. Hareuveni, C. Gautier, M.-P. Kieny, K. Wreschner, P. Chambon, and R. Lathe: Vaccination against tumor cells expressing breast cancer epithelial tumor antigen. *Proc Natl Acad Sci USA* 87:9498, 1990.
112. J. Cohen: Cancer vaccines get a shot in the arm. *Science* 262:841, 1993.
113. P. O'Boyle, R. Samore, S. Adluri, A. Cohen, N. Kemeny, S. Welt, K. O. Lloyd, H. F. Oettgen, L. J. Old, and P. O. Livingston: Immunization of colorectal cancer patients with modified ovine submaxillary gland mucin and adjuvants induces IgM and IgG antibodies to sialylated Tn. *Cancer Res* 52:5663, 1992.
114. A. Rughetti, V. Turchi, C. A. Ghetti, G. Scambia, P. B. Panici, G. Roncucci, S. Marcuso, L. Frati, and M. Nuti: Human B-cell immune response to the polymorphic epithelial mucin. *Cancer Res* 53:2457, 1993.
115. J. Kantor, K. Irvine, S. Abrams, P. Snoy, R. Olsen, J. Greiner, H. Kaufman, D. Eggensperger, and J. Schlom: Immunogenicity and safety of a recombinant vaccinia virus vaccine expressing the carcinoembryonic antigen gene in a nonhuman primate. *Cancer Res* 52:6917, 1992.
116. M.-H. Tao and R. Levy: Idiotype/granulocyte-macrophage colony-stimulating factor fusion protein as a vaccine for B-cell lymphoma. *Nature* 362:755, 1993.
117. Y. Guo, M. Wu, H. Chen, X. Wang, G. Liu, G. Li, J. Ma, M.-S. Sy: Effective tumor vaccine generated by fusion of hepatoma cells with activated B cells. *Science* 263:518, 1994.

14

Gene-Targeted Therapy for Cancer

Mechanisms for directly replacing or targeting genes or their transcripts include (1) gene therapy, i.e., replacement of a defective or deleted gene in cells; (2) shutting off or inhibiting the expression of a gene transcript with "antisense" oligonucleotides; (3) modulating gene expression by a third oligonucleotide strand that forms a "triple helix" with double-helical DNA; (4) gene knockout by genetic recombination; and (5) inactivation of RNA by RNA enzymes called ribozymes.

GENE THERAPY

The knowledge that led to the feasibility of transferring new genes to mammalian cells was initially generated in the early 1970s from the work on RNA and DNA tumor viruses. This research and that that followed on recombinant DNA, showing that genes could be cloned and inserted into viral vectors, led to the idea that new genes could be inserted into cells by putting these into a virus that could then be used to infect a cell. The development of retriviral vectors for gene transfer came about in the early 1980s; by 1987, the first human gene therapy protocol was proposed, and the first true human gene therapy experiment was carried out in September 1990 (reviewed in Ref. 1).

Initially, gene therapy was thought of as a way to treat, or perhaps cure, human genetic diseases, of which about 4000 are known. Some of these are listed in Table 14-1. For most of the human genetic diseases, the nature of the genetic mutation is not known. However, in those for which it is known, gene therapy is a potential therapeutic reality. Some of the genetic diseases for which clinical trials of replacement gene therapy have been carried out include adenosine deaminase (ADA) deficiency, familial hypercholesterolemia, hemophilia, and cystic fibrosis. However, gene therapy as an approach to treatment has progressed more rapidly for cancer than for any other disease. Current clinical therapeutic protocols for cancer include various genetic manipulations for the treatment of malignant melanoma, ovarian cancer, lymphoma, neuroblastoma, renal cell carcinoma, non-small-cell lung carcinoma, and brain cancer.[1]

Gene Transfer Systems

Gene transfer methods include both physical/chemical techniques and the use of viral vectors as well as combinations of the two.

Table 14-1 Examples of Genetic Diseases That Are Potential Candidates for Gene Therapy

Defective Gene	Disease
Adenosine deaminase	Severe combined immunodeficiency
α1-Antitrypsin	Emphysema
Arginosuccinate synthetase	Citrullinemia
CD-18	Leukocyte adhesion deficiency
Cystic fibrosis transmembrane regulator	Cystic fibrosis
Factor IX	Hemophilia B
Factor VIII	Hemophilia A
α-L-Fucosidase	Fucosidosis
Glucocerebrosidase	Gaucher's disease
β-Glucuronidase	Mucopolysaccharidosis type VII
β-Globin	Thalassemia
β-Globin	Sickle cell anemia
α-L-Iduronidase	Mucopolysaccharidosis type I
Low-density lipoprotein receptor	Familial hypercholesterolemia
Ornithine transcarbamylase	Hyperammonemia
Purine nucleoside phosphorylase	Severe combined immunodeficiency
Sphingomylinase	Niemann-Pick disease

Source: From Morgan and Anderson.[1]

Physical/Chemical Methods

These include (1) coprecipitation of DNA with calcium phosphate and calcium phosphate-mediated cellular uptake, (2) use of polycation- or lipid-DNA complexes, (3) encapsulation of DNA into liposomes or erythrocyte ghosts, (4) microinjection of DNA into cells directly or via high-velocity tungsten "microprojectiles," (5) exposure of cells to rapid pulses of high-voltage current (electrocorporation), and (6) receptor-mediated DNA transfer. All of these methods have worked in vitro. A general characteristic of these methods is the integration into the genome of multiple tandem repeats. These integrated sequences, however, may be unstable, and the overall efficiency of these physical/chemical transfection methods is reported to be only about 1% of treated cells.[2]

Most gene therapy protocols have required the ex vivo transfer of genetic material into a patient's lymphocytes, hematopoietic stem cells, or tumor cells and then reinjection of these cells back into the patients. In order to get the most efficient gene transfer, the initial emphasis has been on viral vector transfer systems, but physical/chemical methods have the advantage of not cotransferring unwanted and potentially dangerous viral genetic material. Of the physical/chemical methods, receptor-mediated and liposome-mediated gene transfer appear to hold the most promise for clinical utility.[1]

Receptor-mediated gene transfer is achieved by linking DNA to a polypeptide ligand via a polylysine linker. Such ligands as transferrin and asialoglycoproteins have been used to target DNA to hepatocytes because receptors for these molecules are found in high abundance on liver cells. Ligand-DNA constructs can be injected in vivo and appear to be taken up in reasonable yield by the liver, where binding and receptor internalization occur (reviewed in Ref. 1). One problem with this approach is that these DNA-ligand–receptor complexes are taken up by endocytotic vesicles and transported to lysosomes in cells, where they may be degraded. Attempts to circumvent the destruction of the genetic material once it enters cells have focused on coinfection of the cells with adenovirus or the use of fusogenic peptides derived from the influenza hemagglutinin (HA) gene product to disrupt endosome functions (reviewed in Ref. 3). Although this method appears promising, in vivo expression of genes transferred by this mechanism appears to be only transient.[3]

Liposome/DNA complexes also have the advantage of not transferring superfluous viral genetic information, and they have been shown to work in vivo with minimal toxicity. The liposomal carriers used for this gene transfer method are constructed with hydrolyzable bonds that can be degraded once fusion with cell membranes occurs; however, this method lacks the tissue-targeting possibilities of receptor-mediated gene transfer and also suffers from the fact that stable, long-term expression of genes thus transferred has not been achieved.

A combination of physical/chemical and viral transfer systems has been attempted by coupling DNA via polylysine to an antibody to the adenovirus hexon protein. The complex is then bound to adenoviruses used to infect cells. Once internalized, the complex is degraded and DNA is released. This works in vitro, but it remains to be shown that it works effectively in vivo.

Viral Vector Gene Transfer Systems

Several viral vectors have been used to transfer DNA to mammalian cells. These include papovaviruses such as SV40; adenovirus; adeno-associated virus; vaccinia virus; herpesviruses such as herpes simplex virus (HSV) and Epstein-Barr virus (EBV); retroviruses of avian, murine, and human origin; and lenteviruses (reviewed in Refs. 1, 3, and 3a). Each of these have advantages and disadvantages. Of the viral systems, retroviruses, adenoviruses, adeno-associated viruses, and herpesviruses have received the most attention for human gene therapy.

RETROVIRAL VECTORS

Most of the early human gene therapy trials were carried out using ex vivo transduction of disabled murine retroviral vectors.[1] (Note: transfer of a gene via a replication-deficient viral vector is termed *transduction,* in distinction from *infection,* which is used for replication-competent viruses.[1]) The advantage of retrovirus-mediated gene transfer is that it can transduce a high percentage of target cells and become stably integrated into the host cell's genome. Disadvantages include the fact that cells must be dividing for integration to occur and the danger of cotransferring contaminating infectious and potentially transforming viruses. Moreover, retroviral entry into cells depends on the presence of the appropriate viral receptor on the target cell surface. Also, retroviral constructs are relatively labile compared to other viruses, which is a disadvantage for potential in vivo use.[3]

That cotransfer of a contaminating infectious and transforming retrovirus along with the disabled, gene-carrying retrovirus is a real potential risk was demonstrated by a study of rhesus monkeys undergoing gene transfer via a retroviral vector in an autologous bone marrow transplantation protocol (reviewed in Ref. 1). It turned out that there was a contaminating replication-competent retrovirus in the preparation used to carry out the gene transfer, and 3 of 8 monkeys in this protocol study developed T-cell lymphomas within 6 months of receiving the contaminating virus-infected bone marrow cells.

ADENOVIRUS VECTORS

Adenoviruses are large double-stranded DNA-containing (35 kilobases) viruses that enter cells via receptor-mediated endocytosis. The viral coat is then degraded in endosomes, and the adenovirus genome is translocated to the cell nucleus, where it remains as an extrachromosomal element. The advantages of adenoviruses for gene therapy are that they are usually minor pathogens for humans and are not known to be cancer-causing. Also, they can infect or transduce nonreplicating cells, a feat that retroviruses do not efficiently accomplish. The main disadvantage is that they do not integrate with the target cell genome and the viral vector DNA is lost with subsequent cell division. Adenoviruses are being studied as vectors to deliver genes to airway tissue (nasopharyngeal and lung epithelium) for treatment of diseases such as cystic fibrosis that affect lung tissue, because adenoviruses efficiently infect airway epithelial cells.

ADENO-ASSOCIATED VIRUSES (AAVS) AS VECTORS

AAVs are relatively small, very stable viruses that can infect human cells. Integration of AAV genomes into host-cell DNA occurs but is less efficient than that of retroviruses. Interestingly, integration of wild-type AAV DNA occurs primarily within a small locus of human chromosome 19 in infected cells. Since this region of chromosome 19 has been implicated in chromosomal rearrangements observed in chronic B-cell leukemias, this is of some concern. However, the therapeutic gene-carrying vectors of AAV do not seem to have the same propensity to integrate into chromosome 19 as the parent wild-type virus.[3]

HERPESVIRUS VECTORS

Herpesviruses can infect neural tissue and thus carry a thymidine kinase (TK), which can selectively (compared to noninfected cells) activate certain nucleoside antimetabolite drugs like acyclovir or gancyclovir. Thus, herpesviruses have been thought of as potential vectors to deliver genes to brain tumors. However, it has proven difficult to engineer herpesviruses that are devoid of the ability to replicate or to kill cells that they have transduced. Nevertheless, advantage has been taken of the ability of herpes simplex virus (HSV) TK to activate gancyclovir by utilizing stereotaxic injection of retrovirally transduced, HSV-TK–producing cells into the central nervous system of glioma-bearing rats.[4] This protocol inhibited glioma growth in 11 of 14 rats so treated. Similar results were obtained for response to gancyclovir in HSV-TK–transduced murine sarcomas.[5] Such results have prompted the development of protocols for the treatment of inoperable brain tumors in humans.

Gene Therapy for Human Genetic Diseases

Adenosine Deaminase (ADA) Deficiency

The treatment of a 4-year-old girl with ADA deficiency in September 1990 was the first true clinical protocol for the use of gene therapy.[6] ADA deficiency is a very rare genetic disease in which afflicted children lack ADA, which is a key enzyme in the purine salvage pathway. Its action is to deaminate deoxyadenosine into deoxyinosine; when it is absent, deoxyadenosine accumulates to levels that are toxic to T lymphocytes, producing a severe immune deficiency syndrome.

ADA deficiency was chosen for the first gene therapy trial for a genetic disease because the ADA gene is a relatively small one that can easily be inserted into a retroviral vector and because it appeared that only about 5% to 10% of normal ADA levels are needed to have a functional immune system.[1] The target cells, T lymphocytes, were isolated from the blood of the young patient, expanded in culture in the presence of IL-2, and transduced with a retroviral vector containing the ADA gene and a neomycin resistance marker gene, Neo^R. Expression of the ADA gene was driven by the retroviral LTR, and the Neo^R gene was expressed from an SV 40 promoter. Once the transduced cells were expanded in the culture, they were collected and transfused back into the patient.

Several important things were learned from this first clinical trial[1]: (1) it was found that lymphocytes from immunodeficient individuals could be isolated, grown in culture, transduced with a new gene, and returned to the patient with minimal toxicity to the patient; (2) the transduced lymphocytes were able to express ADA in vivo and the amount of ADA in the T cells rose to 25% of normal; (3) the total lymphocyte count in the patient's blood rose to normal levels; and (4) the number of ADA gene–transduced T cells and the level of ADA in them remained nearly constant for 6½ months, suggesting that they have some growth advantage in vivo. Follow-up clinical protocols for ADA deficiency include attempts to isolate and transduce $CD34^+$ hematopoietic stem cells along with T lymphocytes. The rationale for this is that $CD34^+$ stem cells provide a potential lifetime source of ADA, since they are self-renewing cells and their progeny continue to provide hematopoietic cells for the life of an individual.

Cystic Fibrosis

A number of approaches have been tried to correct the genetic defects in cystic fibrosis (CF), which is an autosomal recessively inherited disease in which defective salt and water transport leads to pathologies in several organs, including lung, pancreas, and liver. The disease state is related to one or more mutations in a gene that codes for a chloride ion channel called the CF transmembrane conductance regulator (CFTR). Lung pathology is the most threatening part of the disease, leading to the accumulation of thick mucus secretions, chronic infection, and eventual airway obstruction. With the current improved treatment of pulmonary complications, most patients now live into adulthood and many develop severe extrapulmonary complications such as hepatic biliary fibrosis and cirrhosis.

Studies in animals have demonstrated the feasibility of attempting gene replacement therapy in humans. Retrovirally mediated transfer of the CFTR gene into cultured pancreatic cells derived from a patient with CF restored cAMP-stimulated chloride conductance.[7] AAV vectors containing the CFTR gene have been used to transduce airway cells in rabbits' lungs after introduction by a fiberoptic bronchoscope; this led to expression of CFTR mRNA and protein for up to 6 months after vector administration.[8] Liposomes containing a CFTR expression plasmid have also been used to introduce the CFTR gene by tracheal instillation into alveoli deep in the lung of CFTR-deficient mice, leading to correction of the ion conductance defects found in the tracheas of these mice.[9] Recombinant adenoviruses expressing the human CFTR gene have been administered by retrograde infusion through the bile duct into the liver of rats and shown by in situ hybridization and immunohistochemistry to continue to express the human gene for up to 21 days, suggesting a way to treat the liver complications of CF in humans.[10] An E1-deficient adenovirus encoding CFTR was administered into the nasal airway of three patients with CF and shown to correct the chloride transport defect in exfoliated or biopsied nasal epithelial cells.[11] Taken together, these data set the stage for clinical trials of gene therapy for CF.

α1-Antitrypsin (α1-AT) Deficiency

α1-antitrypsin is an antiprotease that blocks the protein-digesting action of the enzyme elastase. The unchecked action of elastase is destructive to lung tissue and individuals who have a genetic defect in this gene produce an α1-antitrypsin that is ineffective in protecting the lung. This leads to pulmonary damage and emphysema. It is estimated that about 40,000 of the 2 million emphysema sufferers in the United States have an α1-AT deficiency that contributed to their disease.

A replication-deficient adenoviral vector containing an adenovirus major late promoter and a recombinant human α1-AT gene has been used to transfect rat respiratory tract cells in vitro and in vivo. After intratracheal administration of the α1-AT gene vector, human α1-AT mRNA and protein were detected in rat lung tissue for at least a week,[12] indicating the potential for human clinical trials.

Duchenne Muscular Dystrophy (DMD)

DMD is an X-linked recessive disease caused by the defective expression of dystrophin. The disease afflicts about 1 in 3500 boys in the United States each year and is the most severe of the muscular dystrophies, causing progressive muscular wasting and death by about age 20 from respiratory failure. A mouse model of DMD has been developed and is useful to test gene therapy for this disease. The feasibility of gene therapy for DMD has been tested in the mouse model called the *mdx* mouse. A dystrophin cDNA-expression vector was used to create transgenic *mdx* mice, in whom the expression of dystrophin in muscle tissue was found to be about 50 times the control level in untreated *mdx* mice.[13] Even better, the transgene mice had normal muscle morphology and function. The problem, of course, for gene therapy of the human disease is that one cannot easily visualize using transgenic technologies in humans. Because the dystrophin gene is large, about 14 kilobases (kb), it is not easy to visualize efficient delivery via a viral vector. One way around that might be to use a 7-kb "minigene," which is the size of the gene in some mildly affected patients with a partial gene deletion.[14] Another problem is that the important target tissue is muscle, which is essentially a nondividing tissue after birth. Even though adenovirus and AAVs can transduce nondividing cells, gene expression via these vectors is only transient in muscle cells. An alternative approach may be to transplant dystrophin gene–transduced myoblasts, which can divide and mature into differentiated muscle cells.[14] Phase I clinical trials have revealed no deleterious effects of myoblast transplantation into muscle tissue of DMD patients, and in some cases dystrophin gene expression was detected in some muscle fibers (reviewed in Ref. 14).

Hunter Syndrome

Hunter syndrome (mucopolysaccharidosis type II) is an X-linked recessive inborn error of metabolism resulting from a mutational defect in lysosomal iduronate-2-sulfatase (IDS). This defect results in the accumulation of the glycosaminoglycans heparan sulfate and dermatan sulfate. Clinical features include coarse facial features, joint contractures, hepatosplenomegaly, and obstructive airway disease. Mental retardation is a feature of severe disease. Patients typically die of cardiac and respiratory complications in early adulthood.

To explore the possibility of gene therapy for this disease, a full-length IDS cDNA construct was inserted into a defective Moloney murine leukemia retroviral vector and transduced into lymphoblastoid cell lines from patients with Hunter syndrome. These cells were then cocultured with fibroblasts from Hunter syndrome patients.[15] Interestingly, not only the lymphoblasts but also the cocultured fibroblasts expressed high levels of IDS. Moreover, no accumulation of glycosaminoglycans was observed in the fibroblasts, indicating that intercellular "metabolic cross-correction" can occur between cell types.

Hepatic Deficiencies and Hemophilia

Adenovirus particles, chemically conjugated to asialoorosomucoid poly L-lysine and bound ionically to DNA have been found to be efficient vectors to deliver genes to hepatocytes, indicating a potential way to target liver cells for gene therapy. Such a system has been employed to deliver canine clotting factor IX to mouse hepatocytes in culture, resulting in high expression of factor IX.[16] A retroviral factor IX–containing vector has also been used to correct hemophilia in vivo in hemophiliac dogs with factor IX deficiency.[17]

Gene Therapy for Cancer

Gene therapy protocols for cancer have focused on the genetic modification of tumor-specific lymphocytes or of tumor cells themselves in attempts to boost the host's immune response to a tumor. Rosenberg and colleagues (reviewed in Ref. 18) carried out the first attempt to deliver engineered genes to human tumors in vivo by inserting the Neo^R gene into tumor-infiltrating lymphocytes (TIL) derived from the same patient. These early experiments were designed to see if the Neo^R gene-bearing TIL could "home" to the patient's tumor in vivo. The idea was that if this worked, other genes could be inserted into TIL cells to deliver immunostimulants or tumor necrosis factors, etc. A replication-defective Moloney murine leukemia retrovirus bearing the Neo^R gene was used to transduce the TIL taken from the patients and expanded in culture. Gene-modified cells could be detected in the circulation for 189 days and in biopsied tumor deposits for up to 64 days after infusion of the Neo^R-bearing TIL. These findings led to the first attempts to deliver therapeutically active molecules to human tumors. The first gene selected was the gene encoding tumor necrosis factor (TNF). The hypothesis was that the TIL would home to the tumor and deliver local doses of TNF, producing a local heightening of the immune response as well as a direct tumor cell-killing effect without the systemic toxicity associated with injection of TNF into the circulation. The TIL transduced with the TNF gene have been shown to secrete TNF in vitro. In vivo, the use of IL-2 in conjunction with the TNF gene-modified TIL will probably be needed to maintain the TIL population.

Based on a number of experimental studies, the delivery of other immune system–modulating cytokines to tumor-bearing host animals and cancer patients is also a feasible approach. Additional cytokine genes under study for experimental use in TIL cells include interferon-γ (IFN-γ), IL-2, and IL-6. In addition, genetic modification of tumor cells themselves to increase their immunogenicity is under study. One idea is to insert cytokine genes into a patient's tumor cells obtained at biopsy and modify them by insertion of cytokine genes such as TNF, IL-2, IL-4, IFN-γ, or GM-CSF and then to reinfuse them in order to stimulate immune cells to reject the tumor. Another idea is to modify the expression of MHC genes in tumor cells to make them more immunogenic. Both approaches have shown promise. For example, when tumor cells from a spontaneously arising mouse renal cell tumor were transduced with the murine IL-4 cDNA in a bovine papillomavirus vector and injected back into mice, they were rejected in a $CD8^+$ T cell–dependent manner.[19] Moreover, established parental untransfected tumors were also destroyed by the immune response generated by injection of the IL-4 genetically modified tumor cells. In a similar study, mouse mammary adenocarcinoma cells, genetically engineered to secrete IL-2, triggered a local inflammatory reaction in vivo that led to rejection of established parental adenocarcinomas.[20] Of note was the fact that heightened systemic immunity selectively inhibited subsequently transplanted parental mammary adenocarcinomas and cured established carcinomas and their lung metastases. But if the parental tumor mass was large, the effectiveness of the immune rejection response was lessened. This should remind the reader of one of the problems with immunotherapy: when the tumor burden is large, the immune response's ability to inhibit its growth is significantly diminished (see Chap. 13).

Efficient transfer of cytokines such as GM-CSF, IL-2, and IFN-α into human renal, ovarian, and pancreatic tumor explants or cell lines has also been achieved,[21,22] setting the stage for clinical trials of such genetically engineered human tumor cells.

Foreign MHC gene constructs have been introduced into malignant tumor cells in efforts to stimulate tumor rejection. In vivo transfer of a cDNA encoding the mouse H-2K^s MHC gene was achieved by intratumoral injection of a modified Moloney murine leukemia retroviral vector.[23] The transfected tumor cells (CT26 mouse colon carcinoma or MCA 106 fibrosarcoma) expressed the H-2K^s gene and induced a cytotoxic T-cell response to H-2k^s as well as, most notably, to other antigens present on *unmodified* tumor cells. Direct transfer of human HLA-B7 has been achieved in human melanoma cells from patients with advanced HLA-B7–negative melanomas by direct injection of DNA-liposome complexes into melanoma tumor nodules.[24] Plasmid DNA was detected in biopsies of treated tumor nodules 3 to 7 days after injection but was not found in the serum. Immune responses to HLA-B7–bearing autologous tumor cells could be demonstrated in in vitro assays. One of five patients had a regression of tumors at the injected site and at distant sites.

Other potentially important approaches for gene therapy of human cancer include delivery of the multidrug-resistance (MDR1) gene to normal bone marrow cells to protect them from the cytotoxic effects of chemotherapy[25] and the introduction of functional tumor suppressor genes into cancers in which they have been inactivated or deleted. Successful transfer of fragments of chromosome 11p15 has been carried out in vitro and was found to arrest growth of cultured human rhabdomyosarcoma cells,[26] suggesting the feasibility of this approach in the treatment of human cancers. Since the 11p15 locus shows loss of heterozygosity (LOH) in rhabdomyosarcomas, Wilms' tumor, and a variety of other human cancers including bladder, lung, ovary, liver, and breast, the antitumor effect of this chromosomal region suggests reexpression of a lost tumor suppressor gene function.

Caveats to and Ethics of Gene Therapy

If gene therapy is to become widely used in the treatment of human disease, a number of technical advances will have to be achieved. It will be necessary to develop gene transfer vehicles or vectors that can be directly injected into patients. Currently, as noted above, most of the gene therapy protocols have been based on ex vivo transduction of viral vectors into human cells in culture and then injecting them back into the patient. This means that sufficient target cells have to be obtained from the patient and that it must be possible to grow them in cell culture outside the body—not always an easy thing to accomplish. Second, the vectors to be used for in vivo gene transfer must be able to seek out and enter the desired target tissues.

If, for example, the goal is to target liver cells, the desired therapeutic effect is not likely to be achieved if all the gene-bearing vectors are taken up by the spleen or the kidney. Third, even if the gene vector reaches and enters the cells of interest, it must integrate safely into chromosome sites that will not result in the knockout of key genes needed for normal function. If the goal is to replace a defective gene, the inserted gene should be able to cause homologous recombination at the defective gene locus. Such recombination might not be necessary, however, if the goal is to replace a tumor suppressor gene that has had both alleles deleted or inactivated. Finally, the introduced gene must be placed under the control of regulatory elements that respond to physiologic stimuli unless no functional disruption would ensue from their being constitutively expressed. For example, if the goal is to carry out gene therapy for diabetes by replacing a defective gene for insulin production, then the expression of engineered insulin genes must be able to sense and respond to the rise and fall in blood glucose levels.

Safety issues of gene therapy must, of course, also be addressed. The incident of the rhesus monkeys who inadvertently received a contaminating infectious retrovirus in a gene transfer experiment was noted above. Thus, any viral vector system for gene therapy must be meticulously checked for replication-competent viruses.

The concept of gene therapy also raises several ethical issues. The idea that it could be used to engineer a race of supermen or to produce 7-foot basketball players or Albert Einsteins at will is an abhorrent thought to most people. That the Nietzschean *Übermensch* concept could be an endpoint of genetic science gone haywire is not a comforting thought. The public must be informed that the replacement of defective genes in somatic cells after birth does not tamper with the germ line and has as its goal the treatment of disease. This is, after all, not very different from transplanting a kidney, liver, or bone marrow into a patient who would die without it.

Direct genetic manipulation of the germ line is quite another matter. The fact that this is not yet a scientific reality for humans does not mean that it could not be done, since it has been done in mice and a few domestic animals. There may be some therapeutic advantage to correcting a genetic defect at the early embryo stage, but the possibility that any genetic manipulation could alter germ-line cells and be passed on to the descendants of such individuals would have to be carefully examined. There are national guidelines for the use of gene therapy, and any such protocols must be approved by the Recombinant DNA Advisory Committee of the National Institutes of Health (a committee whose acronym, RAC, appropriately describes, in the view of some, the intense rigor one must go through to get a gene therapy protocol approved) and by the Food and Drug Administration. The two review mechanisms are meant to be complementary.[27] The review by RAC is designed to ensure broad public discussion of the scientific value as well as the ethical and social concerns. The FDA focuses on the safety and efficacy issues of any proposed therapy, from its first use in humans throughout the time of commercial distribution for wide clinical use.

ANTISENSE OLIGONUCLEOTIDES

Another potential approach to regulating gene expression in cancer cells is by introducing an antisense piece of DNA or RNA into such cells. The molecular trick here can be achieved either by DNA transfer of a coding sequence which has the reverse (complementary) sequence of the normal gene and which has a promoter on the "wrong end," i.e., such that an RNA transcription product will be complementary to the normal mRNA transcript and be able to hybridize with it, or by introducing a smaller piece of DNA (an oligodeoxyribonucleotide) that can do the same thing. It has been shown that antisense transcripts form RNA-RNA duplexes with the normal transcript[28] and that insertion of "flipped gene" constructions of the thymidine kinase (TK) gene can inhibit expression of TK four- to fivefold in intact mouse L cells.[29] Such an approach might be used to inhibit the expression of cancer-associated genes if an antisense gene or oligodeoxyribonucleotide could be directed to cancer cells.

Oligonucleotides complementary to a mRNA can be chemically synthesized. Oligodeoxyribonucleotides are the ones usually used because their chemical synthesis is easier than that of ribonucleotides. The recognition of an mRNA by an antisense oligodeoxyribonucleotide (asODN) is based on Watson-Crick hydrogen bonding between complementary bases, e.g., AUCG in an mRNA binds to TAGC in an asODN. There are at least two mechanisms by which asODNs block mRNA function: by targeting an mRNA-asODN hybrid duplex for cleavage by RNAse H or by binding to the 5′-untranslated region of an mRNA, which inhibits its binding or movement on the 40 S ribosomal subunit.[30] The first mechanism most likely pertains when an asODN is targeted to the coding region of the mRNA and the second when the asODN is directed to the 5′-untranslated region of the mRNA.

A minimum length of 11 to 15 bases is required for specific base sequence recognition of an mRNA: 11 is sufficient if the targeted sequence is rich in GC bases because of the abilities of G-C base pairs to form three hydrogen bonds; whereas at least 15 bases is required for good binding to AT-rich sequences because of the lower binding affinity afforded by two hydrogen bonds. AsODNs of 12 to 20 nucleotides (12 to 20-mers) have often been used to produce biological effects in cells or experimental animals.

The chemical backbones of the asODNs are usually modified so as to render them less susceptible to ribonucleases than the naturally occurring phosphodiesters. For example, oligophosphorothioates, oligomethylphosphonates, and oligo[α]-deoxynucleotides have all been used to produce an antisense effect.[30]

Phosphorothioate asODNs are taken up into cells and as much as 10% of the total cell-associated material reaches the cell nucleus by 24 hours of incubation of cultured cells with [^{35}S]-labeled asODN.[31] Injection of a 27-mer labeled phosphorothiate into mice, rats, or rabbits indicated an initial plasma half-life of 15 to 25 minutes followed by a second half-life of 20 to 40 hours.[31] Repeated injections of the 27-mer provided steady-state plasma concentrations in 6 to 9 days and chronic treatment of experimental animals did not show any untoward toxicity. A phase I clinical trial of a phosphorothioate asODN in patients with acute myelogenous leukemia or myelodysplastic syndrome also showed that phosphorothioate asODNs have minimal toxicity in patients.[32]

A number of examples of the potential usefulness of asODN therapy can be cited. Selective inhibition of human chronic myelogenous leukemia (CML) cells in culture by an 18-mer asODN to *bcr-abl* has been observed.[33] Exposure of MCF-7 human breast carcinoma cells to a *myc* asODN in culture resulted in a 95% inhibition of estrogen-induced c-Myc protein production and a 75% decrease in estrogen-stimulated cell proliferation.[34] Inhibition of proliferation of human colon carcinoma cells that express c-*myb* by an asODN to c-*myb* has been seen.[35] Retrovirus-mediated gene transfer has been used to introduce antisense RNA complementary to human creatine kinase mRNA into human histiocytic lymphoma cells, and this blocked translation of creatine kinase in these cells.[36] Antisense oligodeoxynucleotides specific for the *bcl*-2 gene in B-cell lymphomas has been shown to induce cell death of human pre-B-cell acute lymphocytic leukemia cells.[37] Antisense oligodeoxynucleotides to the N-*myc* gene have been shown to block N-*myc* expression and inhibit proliferation of human neuroblastoma cells.[38] Transfection of a recombinant plasmid containing a genomic segment of the K-*ras* protooncogene in the antisense orientation by electrocorporation in a human lung cancer cell line inhibited expression of K-*ras* and proliferation of the cells.[39]

Such a therapeutic approach remains to be proven in patients, but if in vivo stability and tissue-specific targeting problems can be solved, such therapy would be very attractive because it could produce tumor-specific cell killing with minimal toxicity to the patient.[40] Some in vivo studies support this possibility. For example, systemic treatment of SCID mice bearing human CML or melanoma cells with asODNs to *bcr-abl* or *myb*, respectively, produced increased survival compared to untreated control animals.[40a,40b] Human leukemia-bearing SCID mice treated in vivo with a c-*myb* asODN survived 3½ times longer and had significantly less tissue infiltration of leukemia cells than control animals treated with a sense ODN.[41] Ablation of HTLV-1 *tax*-induced fibroblastic tumors in mice was observed after treatment with an asODN to NF-κB.[42] Similarly, in vivo treatment of mice bearing a fibrosarcoma transfected with a dexamethasone-inducible gene coding for an antisense RNA to the p65 protein of NF-κB produced a significant reduction in tumor growth when the animals were injected with dexamethasone.[43] A human lung carcinoma growing in athymic mice was inhibited from growing after intratracheal instillation of a retroviral asODN K-*ras* construct.[44] Since activation of the *ras* oncogene in human cancers occurs by single base mutations, it is of interest that selective inhibition of expression of a Ha-*ras* gene containing a G → T point mutation in codon 12 by a phosphorothioate asODN has been reported.[45] Thus, it appears that single base changes in an mRNA can be recognized with an appropriate asODN.

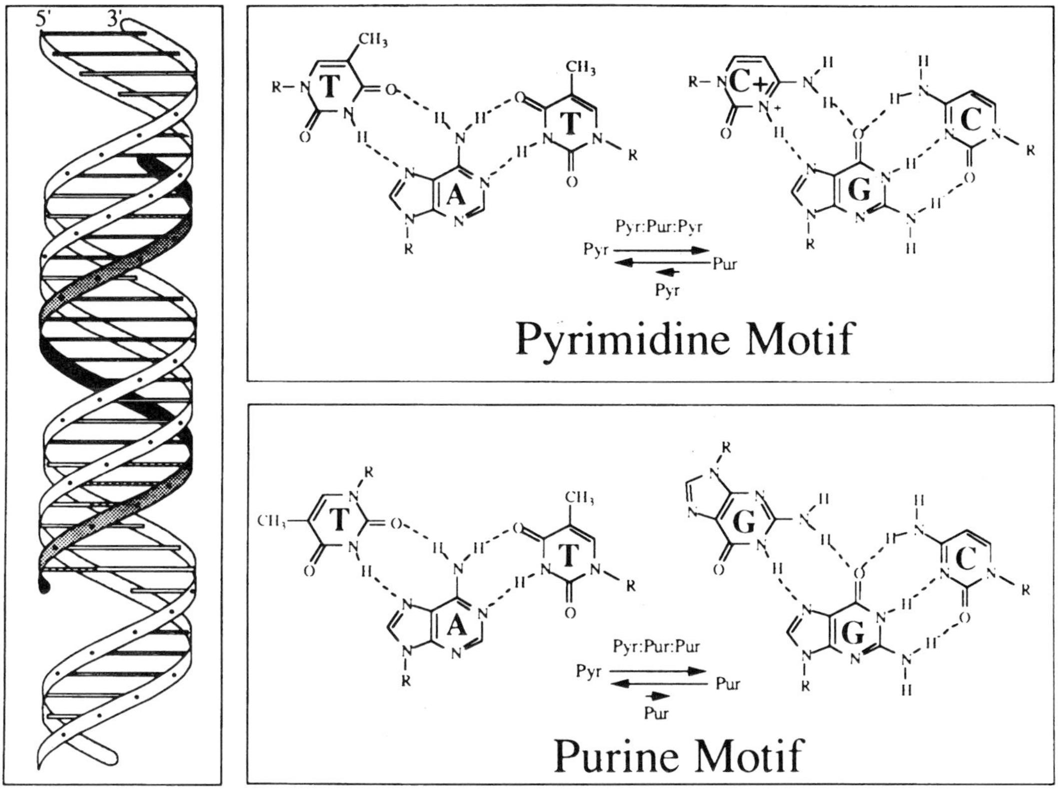

Figure 14-1 Recognition motifs in oligonucleotide-directed triple-helix formation. Both motifs involve binding of oligomers to the DNA major groove, with recognition of consecutive purine bases being generally limited to homopurine sequences. The pyrimidine motif involves oligomer binding parallel to the purine target strand through T-A-T and C-G-C base triplets (stabilized by cytosine protonation). The purine motif involves oligomer binding antiparallel to the purine target strand through T-A-T (or A-A-T) and G-G-C triplets. (From Maher.[46])

TRIPLE-HELICAL DNA: THE ANTIGENE STRATEGY

The antisense approach to inhibitory gene expression is aimed at blocking translation of mRNA by binding to mRNA and either causing its degradation or blocking its ability to function. Another strategy is to target oligonucleotides to DNA as a way to inhibit gene transcription. This is called the "antigene" approach.[30] The molecular basis for this approach is the ability of thymine and protonated cytosine to form a type of hydrogen bond called Hoogsteen hydrogen bonds with A-T and G-C Watson-Crick base pairs, respectively, producing a "triple helix." Triple helices may form by interactions between two pyrimidines and one purine (the "pyrimidine motif") or by recognition of a Watson-Crick double helix by a purine oligonucleotide ("purine motif") (Fig. 14-1).[46] Both motifs involve binding of oligonucleotides in the major groove of DNA.

Studies on the potential for oligonucleotides to bind to DNA and interfere with its function began in the early 1970s,[47] but the formation of a triple-stranded nucleic acid structure was first reported in 1957.[48] Subsequently, the field has developed to the point where it is possible to think about applying this antigene strategy to the treatment of human disease (reviewed in Refs 46 and 49).

To recognize a double-helical DNA sequence, a triple helix-forming oligonucleotide should have a minimum length of 8 to 14 bases,[46] and in human cells a minimum length of 17 bases appears to be required.[30] Triple helices in the pyrimidine motif require cytosine protonation and are stabilized by mild acid pH (pH 5 to 7).

Thus, adenosine-rich homopurine regions appear to form more stable complexes with homopyrimidine oligomers near neutral pH than do guanosine-rich targets. Purine motif complexes do not require acid pH for formation but do require guanosine-rich targets.[46] Triple-helix formation is about two orders of magnitude slower than double-helix formation. In vitro studies have shown the dissociation constant of triple-helical complexes to be in the range of 1×10^{-8} M, and oligonucleotide dissociation rates from a triple helix are slow.[46]

Efficient triple-helix formation requires a homopurine-homopyrimidine sequence in target DNA. It remains a challenge to make oligonucleotides that will bind mixed purine-pyrimidine sequences that occur more often in natural DNA. One approach has been to use linkers of nucleoside phosphoramidites to cross over between DNA strands (Figure 14-2).[50]

Several in vitro experimental systems have been utilized to show the efficacy of triple-helix formation in blocking DNA function (Fig. 14-3). These include blocking sites for restriction endonucleases, DNA methylases, transcription factor binding, and RNA polymerase binding. Some studies have also shown effects of triple-helix–forming oligonucleotides in intact cells. For example, Grigoriev et al.[51] have shown that a psoralen-oligonucleotide conjugate forms a triple helix with DNA in cells and that a triple-helical cross-link can be induced after UV irradiation. This oligonucleotide was designed to target the IL-2Rα promoter site that binds the transcription factor NF-κB and was shown to inhibit gene transcription from this promoter. An octathymidylate oligonucleotide bound to an acridine intercalating agent has been shown to selectively block SV40 DNA replication in CV-1 cells.[52] A G-rich 27-mer oligonucleotide binding to a G-rich sequence in the c-*myc* gene has been shown to lower c-*myc* RNA levels in HeLa cells 2.5 hours after addition to HeLa cell cultures. A similar study using a G-rich 28-mer, targeted to a G-rich element in the IL-2Rα gene, inhibited the PHA response for stimulating IL-2Rα level in lymphocytes.[53] Monoclonal antibodies to triple-helix DNA have been used to show that triple helices can in fact form in the nuclei of cells (reviewed in Ref. 50).

GENE TARGETING BY HOMOLOGOUS RECOMBINATION

Gene targeting is a method to generate offspring (usually done so far in mice) that have a specific gene alteration (reviewed in Refs. 54 to 56). This is carried about by recombinant DNA technology to create a targeting vector containing the altered gene of interest with appropriate homology to the gene that is to be targeted. The modified DNA is then introduced into a pluripotent embryonic stem cell (ES cell) derived from a mouse embryo. The targeting DNA sequence usually contains a drug resistance gene such as Neo^R to allow for selection of ES cells that contain the targeting gene. At a certain frequency, depending on the organism, homologous DNA recombination occurs by a double reciprocal crossover event leading to insertion of the targeting gene into the targeted sequence (Fig. 14-4).[57] Both homologous and nonhomologous types of somatic recombination events occur. Cells most likely evolved these processes to repair DNA damaged by exposure to chemicals or radiation. For example, homologous recombination, by using the genetic sequence infor-

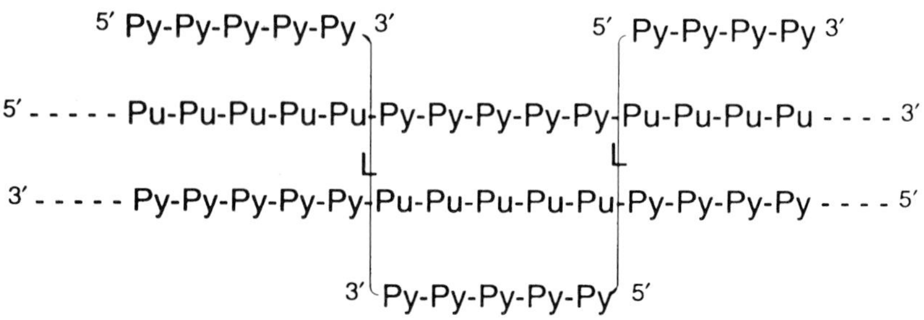

Figure 14-2 Triple helical DNA formed by using nucleoside phosphoramidites to cross over between DNA strands. (from Thuong and Hélène.[50])

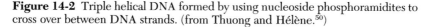

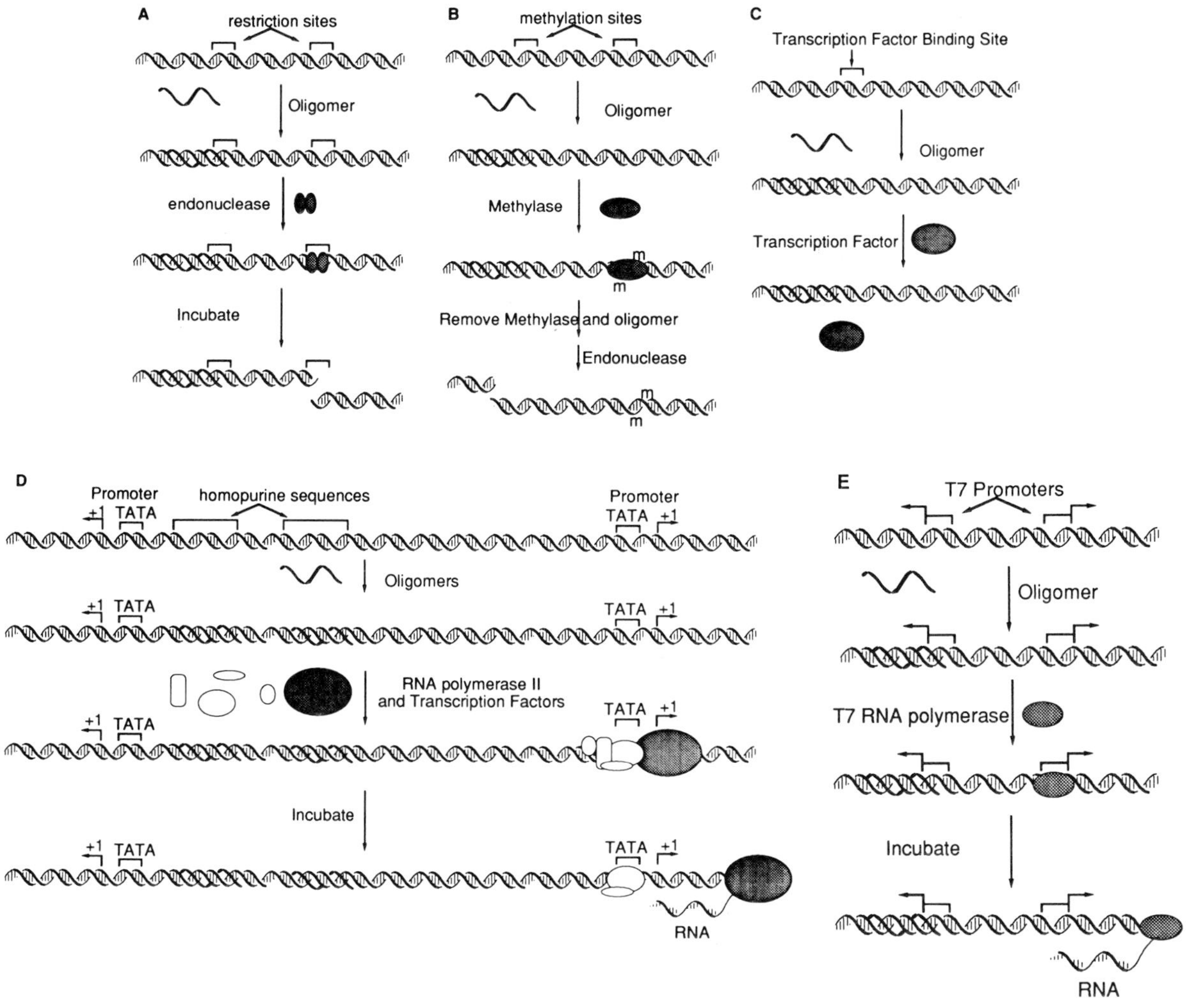

Figure 14-3 Experimental systems documenting site specific inhibition of DNA binding proteins by triple-helix formation. (A) Schematic illustration of site-specific restriction inhibition by triple-helix formation. The restriction site at left overlaps a homopurine target site for triple-helix formation. Cleavage of an internal control restriction site (right) demonstrates that oligonucleotide inhibition is site-specific, and does not affect enzyme activity *per se*. (B) Schematic illustration of site-specific inhibition of restriction methylation by triple-helix formation. When followed by triple-helix destabilization and restriction endonuclease digestion, this procedure directs enzymatic DNA cleavage uniquely to the site of triple-helix formation. (C) Schematic illustration of site-specific Spl inhibition. As a control, oligomers of the same nucleotide composition in scrambled sequence were shown not to affect Spl binding. (D) Schematic illustration depicting inhibition of RNA polymerase II transcription initiation by triple-helical complexes upstream of basal transcription factor binding sites. Transcription from promoters bearing multiple triple-helical complexes is inhibited even when such complexes do not interfere with known protein binding sites. A linked promoter (right) is not affected. (E) Schematic illustration of repression of a bacteriophage RNA polymerase by binding of an oligonucleotide repressor to a homopurine operator sequence that overlaps the promoter. An internal control promoter lacking the operator sequence (right) is not inhibited by triple-helix formation. (From Maher.[46])

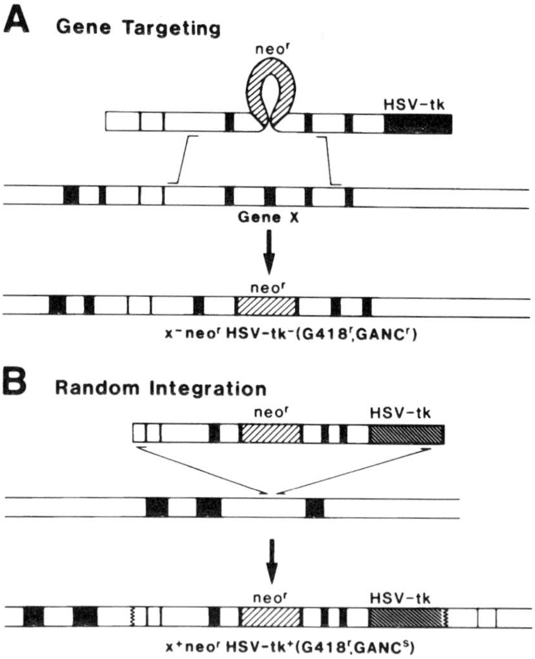

Figure 14-4 (A) Homologous recombination (gene targeting). The targeting vector, a neomycin resistance gene (*neo^R*) flanked by gene x sequences is shown pairing with a chromosomal copy of gene x. Homologous recombination between the targeting vector and genomic x DNA results in the specific disruption of the chromosomal copy of gene x (that is, x^- but now *neo^R*). (B) Nonhomologous recombination (random integration). In nonhomologous recombination the targeting vector is randomly inserted into the chromosomal DNA (here arbitrarily shown to be inserted into gene y). The cells are still x^+ because the targeting vector was inserted somewhere else in the genome. Most nonhomologous or random insertions of the targeting vector occur through the ends of the targeting vector, presumably at an existing double-strand break in the chromosomal DNA. The solid boxes represent exons, the open boxes introns. (From Capecchi.[57])

mation of the undamaged allele of a pair of alleles, could be utilized to repair the damaged allele.

ES cells containing the marker gene sequence are selected in culture by exposing the cells to a neomycin analogue called G418. Cells containing the targeting sequence are microinjected into normal blastocysts, which are then returned to pseudopregnant females for development into embryos and newborn animals. Some of the chimeric animals thus produced will contain the chimeric cells in their gonads and thus be able to transmit the targeting mutant gene to their offspring. Finally, interbreeding of heterozygous siblings produces animals homozygous for the desired mutation.

One of the problems with this technique is that a high percentage of G418-resistant cells will be ones that result from nonhomologous random integration of the targeting vector rather than from homologous recombination. There are methods to enrich the ES cell population for cells that have undergone homologous recombination (reviewed in Ref. 58). For example, if a promoterless Neo^R construct is used, Neo^R will not usually be expressed unless it is integrated into the right position to be driven by the targeted gene's promoter. Also, the number of random integrations may be reduced when a polyadenylation sequence is not included in the targeting vector, because when the targeting vector is correctly inserted, the targeted gene provides this sequence. This favors selection of the cells with homologous recombination because random integrations produce an mRNA that is rapidly degraded. (It should be noted that when the targeting gene is inserted properly, the resulting mRNA and protein will be a fusion product between the targeted gene and the Neo^R gene, but it retains sufficient function to provide drug resistance.)

The efficiency of homologous recombination increases with the length of the genomic sequence of the targeting vector, with peak efficiency between 10 and 14 kb. It is clear that the efficiency of homologous recombination is significantly increased when isogenic DNA sequences are used in the targeting vector; i.e., if the targeted gene is the mouse HGPRT gene, the targeting gene should contain mouse HGPRT sequences. Otherwise, additional unwanted mutations in the targeted gene may occur.

Most gene targeting studies have used electrocorporation to introduce the targeting vector into ES cells. Further identification of the ES cells (after G418 selection) that contain the desired sequence can be carried out by the polymerase chain reaction (PCR) to amplify the mRNA of the expected gene transcript, followed by sequence analysis to make sure it is the right sequence.

Most reports claim that one germ-line chimera is obtained from about 50 blastocysts that have been injected with gene-altered ES cells and subsequently implanted into foster mothers. However, about half of the implanted blastocysts result in live births and about half of these are chimeric at least for some phenotypic characteristic. Thus, generation of a germ-line chimera is difficult compared to obtaining a chimera of other tissues.

In most gene targeting studies, the goal is to knock out the gene coding for a certain function in order to ask the question: what developmental or functional effects does that gene have? The integration of the foreign DNA by homologous recombination disrupts and inactivates the target gene (Fig. 14-4). Usually, one of the two alleles is knocked out in vitro and the second is lost by interbreeding of the offspring to produce homozygous mice that are said to be null for the lost gene. Additional methods, however, have been developed to knock out both alleles in the ES cells in vitro (reviewed in Ref. 58).

Gene knockout experiments have produced some surprising results. Sometimes knockout of a gene predicted to be critical for cell function has no obvious effect on the normal development of animals. Sometimes a gene knockout affects tissues in which its expression or function was not predicted. And sometimes disruption of a gene reveals other hereditary mechanisms of gene regulation such as imprinting. Examples of some gene knockout experiments are shown in Table 14-2.

Some of the surprises include the following. Knockout of the c-*myb* gene causes severe depletion of erythroid and myeloid blood cell precursors and deaths on day 15 of gestation, yet megakaryocytes, which are thought to be derived from the same pluripotent stem cells, are not effected.[59] Disruption of the insulinlike growth factor II (IGF-II) gene led to the discovery that this gene is paternally imprinted in mice.[60] Even though it is known that the p53 gene plays a key role in regulation of cell proliferation, p53-gene null mice appear normal at birth.[61] However, a high percentage of mice develop tumors by 5 months of age, supporting p53's role as a tumor suppressor gene. Inactivation of the *RB* gene causes embryonic death of mice late in their gestation period (day 16),[62] which was not predicted because it was thought that RB was a key cell-cycle regulator that would be important much earlier in development. Knockout of the ubiquitously expressed c-*src* gene produces osteopetrosis and not much else.[63]

These sorts of data favor the concept of functional redundancy in the genome. That is, more than one gene is able to carry out a key function in growth, development, and differentiation. This makes some sense, after all, in that multiorgan higher animal species have presumably developed mechanisms through gene duplication events, etc., to protect themselves from being annihilated by single hits to their genome. Certainly there are genes that are members of multigene families with overlapping functions. The c-*src* gene family is a case in point. Nevertheless, homologous recombination experiments have provided a great amount of information about the functions of genes in growth and development, and these kinds of genetic manipulations are beginning to enable investigators to generate animal models for human diseases, including cancer.

RIBOZYMES

The central dogma of biology until about 1981 contained the concept that DNA is transcribed into RNA, which, in turn, is translated into proteins that do the work of the cell. In this scheme, RNA molecules were thought to be somewhat passive carriers of information and to have no catalytic activity of their own. In 1981, that all changed, although the scientific world at large did not realize it for several years thereafter. At that time, Thomas Cech et al.[64] published a paper describing an RNA molecule from the ciliated protozoan *Tetrahymena thermophila*,which could catalyze the self-splicing of introns from ribosomal RNA. They called this a *ribozyme*. Soon thereafter, Sidney Altman and his colleagues discovered that some RNAs could act as true catalysts by cleaving molecules other than themselves. In 1989, Cech and Altman shared the Nobel Prize in Chemistry for this work.

It is now known that ribozymes, like their protein enzyme counterparts, contain a catalytic center, form specific enzyme-substrate complexes, and can attain levels of catalytic rate enhancement similar to those of protein enzymes

Table 14-2 Mutant Mice Produced by Gene Targeting in ES Cells

Gene Target	Lethality	Comments[a]
p53		+/− Slight effect −/− Tumor susceptibility
Rb	Embryonic	+/− Brain tumors 6–12 mo. −/− RBC maturation defect
Cystic fibrosis transmembrane conductance regulator	Postnatal	+/− Normal −/− Cystic fibrosis
N-*myc*	Embryonic	+/− Normal −/− Embryonic lethal
N-*myc* (leaky)	At birth	+/− Normal −/− Underdeveloped lung
c-*fos*	60% at birth	+/− Normal −/− Osteopetrosis, lymphopenia, gametogenesis, behavior
c-*src*		+/− Normal −/− Osteopetrosis
c-*myb*	Embryonic	+/− Normal −/− Anemia starts day 14
Hox-1.5	Postnatal	+/− Normal −/− Segment disruption
Hox-1.6	Postnatal	+/− Normal −/− Segment disruption
Hox-3.1		+/− Usually normal −/− Segment ID change
En-2		+/− Normal −/− Cerebellar change
Wnt-1 (*int*-1)	Postnatal	+/− Normal −/− Midbrain missing
Calcium calmodulin kinase II α isoform		+/− Not well reported −/− LTP, behavior, learning
DNA methyl-transferase	Embryonic	+/− Normal −/− Retarded development
LIF		+/− Normal −/− Sterile females, slower growth
IGF-II		+/− Paternal dominant −/− Small size
NGF receptor (low-affinity p75)		+/− Normal −/− Fewer sensory nerves
LDL receptor-related protein	Embryonic	+/− Normal −/− Embryos do not implant
Tenascin		+/− Normal −/− Normal

(reviewed in Ref. 65). They require divalent cations both for structural stability and catalytic activity (reviewed in Ref. 66). Rate constants calculated for the enzymatic cleavage events are in the range of $k_c = 350 \text{ min}^{-1}$, which is 10^{11} times that estimated for the uncatalyzed hydrolysis of RNA and comparable to catalytic rates achieved by protein enzymes.[65] Ribozymes have the additional advantage of base-pairing specificity to direct them to a target site (Fig. 14-5),[67] rendering ribozymes as close to "catalytic perfection"[65] as an enzyme could probably be. There is some evolutionary significance to all this also, because RNA may have been the first informational molecule, and its ability to carry both genetic information and functional properties may have facilitated the evolution of cells from the primodial slime.

Ribozymes are best known for their ability to find another strand of RNA and specifically

Table 14-2 *(continued)*

Gene Target	Lethality	Comments[a]
P_0 (myelin component)		+/− Normal −/− Hypomyelination, progressive defects in motor control
MyoD		+/− Elevated Myf-5 −/− Further elevated Myf-5
Myf-5	At birth	+/− Normal −/− Defective rib cage
Apolipoprotein E		+/− Normal −/− Atherosclerosis
Apolipoprotein A		+/− Normal, intermediate −/− Normal but less HDL
Genes affecting immune system		
c-*abl*		+/− Normal −/− Lymphopenia and other
fyn		−/− Thymocyte signaling changes
$p56^{lck}$		+/− Normal −/− No mature T cells
IL-2		+/− Normal −/− Immune response defects
IL-4		+/− Not reported −/− No IgE, less IgG1
Lyt-2		+/− Normal −/− No $CD8^+$ T cells
RAG-1		+/− Not reported −/− No mature B or T cells
RAG-2		+/− Normal −/− No mature B or T cells
CD4		−/− No CD4, normal CD8
T cell receptor (α gene)		+/− Small if any effect −/− No mature αβ T cells
MHC class II $A^b\beta$		+/− Normal −/− Few CD4 T cells
β2-microglobulin		+/− Intermediate −/− No CD8 T cells
IgD (δ heavy chain)		+/− Normal B cells −/− Not yet bred
IgM μ chain		+/− Normal −/− No B cells
λ5		+/− Normal −/− Fewer B cells

[a] +/−, Heterozygous mutants; −/−, homozygous mutants.
Source: From Wilder and Rizzino.[58]

cleave the phosphodiester backbone, which they do by a divalent cation-dependent nucleophilic displacement reaction, generating a 5′-OH and a 3′-cyclophosphate at the cleavage site (Fig. 14-6). Most of these cleavage reactions are Mg^{2+}-dependent, but Mn^{2+}, Ca^{2+}, and Co^{2+} may also act as metal cofactors.[66] The natural affinity of RNA for Mg^{2+} may help explain why many ribozymes depend on this divalent cation for structure and function. It is generally less well known that ribozymes can also catalyze the cleavage of DNA, the polymerization RNA monomers, the replication of RNA strands, the opening of 2′-3′ cyclic phosphate rings, and perhaps even act in a way analogous to tRNA in the peptidyl transfer reaction of protein synthesis.[66]

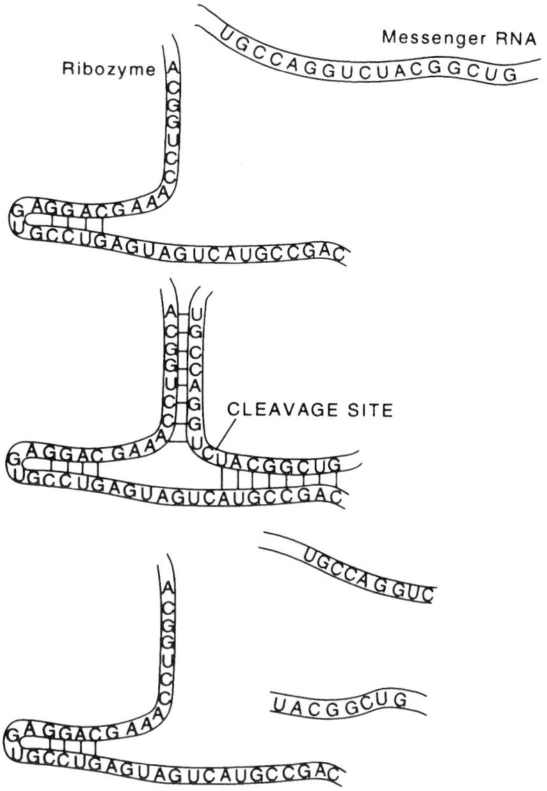

Figure 14-5 Slice of life. A ribozyme bonds to a specific RNA sequence, and then cuts it at a predetermined cleavage point, destroying the sequence. (From Barinaga.[67])

Once the functional activity and specificity of ribozymes became evident, it was not long before the powerful potential therapeutic uses of such molecules became apparent. Ribozymes have now been designed to attack viruses such as HIV[68,69] and oncogenes such as *ras*[69] and to shut off expression of cellular enzymes such as O^6-methylguanine-DNA methyltransferase.[70] It should be noted that the base-pairing specificity of ribozymes with their target RNA also provides an antisense effect (see above), and this, in effect, gives ribozymes a two-edged sword advantage: they can be designed to target a specific mRNA and then to cleave it with their own catalytic cleavage activity. Indeed, such specificity has been taken advantage of to design an antisense ribozyme that can recognize and destroy a mutated form of the *ras* oncogene that differs from its normal cellular *ras* gene counterpart by a single nucleotide.[69]

The key, of course, is to design ribozymes that not only have exquisite specificity but that also can get to the target site inside of cells. The two most commonly engineered ribozymes—the "hammerhead" (depicted in Fig. 14-5) and the "hairpin" (which is larger and can also cleave phosphorothioate linkages)[66]—have to escape cleavage themselves and get into cells, and perhaps in some cases into the nuclei of cells, to be effective. In some cases, a synthetic gene can be designed and placed in a plasmid with the appropriate initiation and termination codons to transfect cells and express the ribozyme of interest.[71,72] Such "directed evolution"[71] offers a way to design highly selective and efficient ribozymes and guide them into target cells.

All of this "Star Wars" biology, however, is not without some problems. The ribozymes themselves are large molecules (up to 100 nucleotides or longer, although shorter ones may work[66]), much larger than the 15- to 25-nucleotide-long oligodeoxynucleotides needed to deliver an antisense activity to cells. Thus, uptake and delivery to the target site is a problem. Second, the ability to discriminate between the target RNA and normal RNAs is crucial and when ribozymes stick around long enough in cells (they have cellular half-lives of up to 72 hours), they may cleave normal cellular RNAs. Nevertheless, ribozymes that can attack and cleave HIV RNA have been designed and clinical trials for this therapeutic approach are being undertaken.

The future therapeutic potential for ribozymes, provided that design and delivery problems can be solved, seems unlimited. Because RNA molecules have informational as well as functional content, they can be designed with a very selective target in mind and amplified many times by standard nucleic acid chemistry techniques. They may also be designed to have very novel functions, functions nature has not thought of yet and for which natural protein enzymes are unknown. For example, ribozymes have been found to catalyze disulfide exchange reactions,[66] and they might be able to regulate the redox potential of cells, protecting them from oxidative damage. Also, it may be possible to target them very specifically in order to carry out in vivo gene knockout experiments, replace a defective function, or provide cell suicide functions to kill cancer cells.

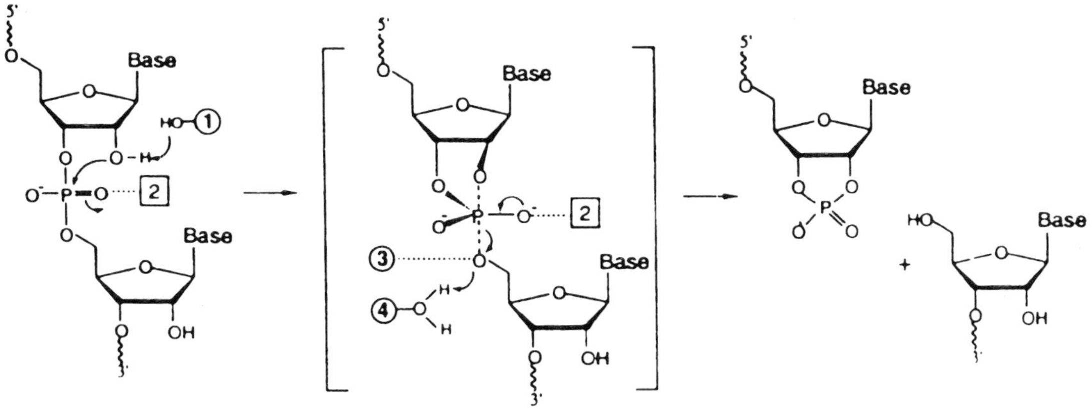

Figure 14-6 One mechanistic class of ribozyme phosphodiester cleavage. Circles represent divalent cations and several ways that they may participate in catalysis; squares represent sites directly supported by experimental data. Although most of the ions are shown participating in the transition state of the reaction, they may also contribute to ground-state stabilization and substrate binding (represented by M2, the square labeled 2). This mechanism is observed in metal-catalyzed RNA hydrolysis and RNA cleavage by the hammerhead, hairpin, hepatitis delta, Neurospora, and tRNA ribozymes. Although chirality is not shown, the attack of the 2′-OH group proceeds by an in-line S_N2 nucleophilic displacement with inversion of the configuration at phosphorus. The M2 square represents a divalent cation that promotes catalysis by the hammerhead ribozyme. In addition to coordinating the nonbridging phosphate oxygen, M2 may also stabilize the oxyanion nucleophile. Although four discrete ions are represented for clarity, a single ion could perform several functions at once (such as M2 and M4, or M1 and M2). (Modified from Pyle.[66])

REFERENCES

1. R. A. Morgan and W. F. Anderson: Human gene therapy. *Annu Rev Biochem* 62:191, 1993.
2. T. Friedmann: Progress toward human gene therapy. *Science* 244:1275, 1989.
3. R. C. Mulligan: The basic science of gene therapy. *Science* 260:926, 1993.

3a. S. S. Lee, L. C. Eisenlohr, P. A. McCue, M. J. Mastrangelo, and E. C. Lattime: Intravesical gene therapy: In vivo gene transfer using recombinant vaccinia virus vectors. *Cancer Res* 54:3225, 1994.

4. K. W. Culver, Z. Ram, S. Wallbridge, H. Ishii, E. H. Oldfield, R. M. Blaese: In vivo gene transfer with retroviral vector-producer cells for treatment of experimental brain tumors. *Science* 256:1550, 1992.
5. F. L. Moolten and J. M. Wells: Curability of tumors bearing herpes thymidine kinase genes transferred by retroviral vectors. *JCNI* 82:297, 1990.
6. R. M. Blaese and W. F. Anderson: *Hum Gene Ther* 1:327, 1990.
7. M. L. Drumm, H. A. Pope, W. H. Cliff, J. M. Rommens, S. A. Marvin, L.-C. Tsui, F. S. Collins, R. A. Frizzell, and J. M. Wilson: Correction of the cystic fibrosis defect in vitro by retrovirus-mediated gene transfer. *Cell* 62:1227, 1990.
8. T. R. Flotte, S. A. Afione, C. Conrad, S. A. McGrath, R. Solow, H. Oka, P. L. Zeitlin, W. B. Guggino, and B. J. Carter: Stable in vivo expression of the cystic fibrosis transmembrane conductance regulator with an adeno-associated virus vector. *Proc Natl Acad Sci USA* 90:10613, 1993.
9. S. C. Hyde, D. R. Gill, C. F. Higgins, A. E. O. Trezise, L. J. MacVinish, A. W. Cuthbert, R. Ratcliff, M. J. Evans, and W. H. Colledge: Correction of the ion transport defect in cystic fibrosis transgenic mice by gene therapy. *Nature* 362:250, 1993.
10. Y. Yang, S. E. Raper, J. A. Cohn, J. F. Engelhardt, and J. M. Wilson: An approach for treating the hepatobiliary disease of cystic fibrosis by somatic gene transfer. *Proc Natl Acad Sci USA* 90:4601, 1993.
11. J. Zabner, L. A. Couture, R. J. Gregory, S. M. Graham, A. E. Smith, and M. J. Welsh: Adenovirus-mediated gene transfer transiently corrects the chloride transport defect in nasal epithelia of patients with cystic fibrosis. *Cell* 75:207, 1993.
12. M. A. Rosenfeld, W. Siegfried, K. Yoshimura, K. Yoneyama, M. Fukayama, L. E. Stier, P. K. Pääkkö, P. Giliardi, L. D. Stratford-Perricaudet, M. Perricaudet, S. Jallat, A. Pavirani, J.-P. Lecocq, and R. G. Crystal: Adenovirus-mediated transfer of a recombinant α1-antitrypsin gene to the lung epithelium in vivo. *Science* 252:431, 1991.
13. G. A. Cox, N. M. Cole, K. Matsumura, S. F. Phelps, S. D. Hauschka, K. P. Campbell, J. A. Faulkner, and J. S. Chamberlain: Overexpression of dystrophin in transgenic mdx mice eliminates dystrophic symptoms without toxicity. *Nature* 364:725, 1993.

14. H. M. Blau: Muscling in on gene therapy. *Nature* 364:673, 1993.
15. S. E. Braun, E. L. Aronovich, R. A. Anderson, P. L. Crotty, R. S. McIvor, and C. B. Whitley: Metabolic correction and cross-correction of mucopolysaccharidosis type II (Hunter syndrome) by retroviral-mediated gene transfer and expression of human iduronate-2-sulfatase. *Proc Natl Acad Sci USA* 90:11830, 1993.
16. R. J. Cristiano, L. C. Smith, M. A. Kay, B. R. Brinkley, and S. L. C. Woo: Hepatic gene therapy: Efficient gene delivery and expression of primary hepatocytes utilizing a conjugated adenovirus-DNA complex. *Proc Natl Acad Sci USA* 90:11548, 1993.
17. M. A. Kay, S. Rothenberg, C. N. Landen, D. A. Bellinger, F. Leland, C. Toman, M. Finegold, A. R. Thompson, M. S. Read, K.-M. Brinkhous, and S. L. C. Woo: In vivo gene therapy of hemophilia B: Sustained partial correction in factor IX-deficient dogs. *Science* 262:117, 1993.
18. S. A. Rosenberg: Immunotherapy and gene therapy of cancer. *Cancer Res* 51:5074s, 1991.
19. P. T. Golumbek, A. J. Lazenby, H. I. Levitsky, L. M. Jaffee, H. Karasuyama, M. Baker, and D. M. Pardoll: Treatment of established renal cancer by tumor cells engineered to secrete interleukin-4. *Science* 254:713, 1991.
20. F. Cavallo, F. Di Pierro, M. Giovarelli, A. Gulino, A. Vacca, A. Stoppacciaro, M. Forni, A. Modesti, and G. Forni: Protective and curative potential of vaccination with interleukin-2-gene-transfected cells from a spontaneous mouse mammary adenocarcinoma. *Cancer Res* 53:5067, 1993.
21. E. M. Jaffee, G. Dranoff, L. K. Cohen, K. M. Hauda, S. Clift, F. F. Marshall, R. C. Mulligan, and D. M. Pardoll: High efficiency gene transfer into primary tumor explants without cell selection. *Cancer Res* 53:2221, 1993.
22. A. Belldegrun, C.-L. Tso, T. Sakata, T. Duckett, M. J. Brunda, S. H. Barsky, J. Chai, R. Kaboo, R. S. Lavey, W. H. McBride, and J. B. deKernion: Human renal carcinoma line transfected with interleukin-2 and/or interferon α gene(s): Implications for live cancer vaccines. *JNCI* 85:207, 1993.
23. G. E. Plautz, Z.-Y. Yang, B.-Y. Wu, X. Gao, L. Huang, and G. J. Nabel: Immunotherapy of malignancy by in vivo gene transfer into tumors. *Proc Natl Acad Sci USA* 90:4645, 1993.
24. G. J. Nabel, E. G. Nabel, Z.-Y. Yang, B. A. Fox, G. E. Plautz, X. Gao, L. Huang, S. Shu, D. Gordon, and A. E. Chang: Direct gene transfer with DNA-liposome complexes in melanoma: Expression, biologic activity, and lack of toxicity in humans. *Proc Natl Acad Sci USA* 90:11307, 1993.
25. J. R. McLachlin, M. A. Eglitis, K. Ueda, P. W. Kantoff, I. H. Pastan, W. F. Anderson, and M. M. Gottesman: Expression of a human complementary DNA for the multidrug resistance gene in murine hematopoietic precursor cells with the use of retroviral gene transfer. *JNCI* 82:1260, 1990.
26. M. Koi, L. A. Johnson, L. M. Kalikin, P. F. R. Little, Y. Nakamura, A. P. Feinberg: Tumor cell growth arrest caused by subchromosomal transferable DNA fragments from chromosome 11. *Science* 260:361, 1993.
27. D. A. Kessler, J. P. Siegel, P. D. Noguchi, K. C. Zoon, K. L. Feiden, and J. Woodcock: Regulation of somatic-cell therapy and gene therapy by the food and drug administration. *N Engl J Med* 329:1169, 1993.
28. J. Tomizawa, T. Itoh, G. Selzer, and T. Som: Inhibition of ColE1 primer formation by a plasmid-specified small RNA. *Proc Natl Acad Sci USA* 78:1421, 1981.
29. J. G. Izant and H. Weintraub: Inhibition of thymidine kniasegene expression by antisense RNA: A molecular approach to genetic analysis. *Cell* 36:1007, 1984.
30. C. Hélène: Rational design of sequence-specific oncogene inhibitors based on antisense and antigene oligonucleotides. *Eur J Cancer* 27:1466, 1991.
31. P. Iversen: In vivo studies with phosphorothioate oligonucleotides: Pharmacokinetics prologue. *Anti-Cancer Drug Des* 6:531, 1991.
32. E. Bayever, P. L. Iversen, M. R. Bishop, J. G. Sharp, H. K. Tewary, M. A. Arneson, S. J. Pirruccello, R. W. Ruddon, A. Kessinger, G. Zon, and J. O. Armitage: Systemic administration of a phosphorothioate oligonucleotide with a sequence complementary to p53 for acute myelogenous leukemia and myelodysplastic syndrome: Initial results of a phase I trial. *Antisense Res Devel* 3:383, 1993.
33. C. Szczylik, T. Skorski, N. C. Nicolaides, L. Manzella, L. Malaguarnera, D. Ventrurelli, A. M. Gewirtz, and B. Calabretta: Selective inhibition of leukemia cell proliferation by BCR-ABL antisense oligodeoxynucleotides. *Science* 253:562, 1991.
34. P. H. Watson, R. T. Pon, and R. P. C. Shiu: Inhibition of c-*myc* expression of phosphorothioate antisense oligonucleotide identifies a critical role for c-*myc* in the growth of human breast cancer. *Cancer Res* 51:3996, 1991.
35. C. Melani, L. Rivoltini, G. Parmiani, B. Calabretta, and M. P. Colombo: Inhibition of proliferation by c-*myb* antisense oligonucleotides in colon adenocarcinoma cell lines that express c-*myb*. *Cancer Res* 51:2897, 1991.
36. J. L. C. Ch'ng, R. C. Mulligan, P. Schimmel, and E. W. Holmes: Antisense RNA complementary to 3′coding and noncoding sequences of creatine kinase is a potent inhibitor of translation in vivo. *Proc Natl Acad Sci USA* 86:10006, 1989.
37. J. C. Reed, C. Stein, C. Subasinghe, S. Haldar, C. M. Croce, S. Yum, and J. Cohen. Antisense-mediated inhibition of BCL2 protooncogene expression and leukemic cell growth and survival: Comparisons of phosphodiester and phosphorothioate oligonucleotides. *Cancer Res* 50:6565, 1990.

38. A. Rosolen, L. Whitesell, N. Ikegaki, R. H. Kennett, and L. M. Neckers: Antisense inhibition of single copy N-*myc* expression results in decreased cell growth without reduction of c-*myc* protein in a neuroepithelioma cell line. *Cancer Res* 50:6316, 1990.
39. T. Mukhopadhyay, M. Tainsky, A. C. Cavendar, and J. A. Roth: Specific inhibition for K-*ras* expression and tumorigenicity of lung cancer cells by antisense RNA. *Cancer Res* 51:1744, 1991.
40. J. E. Karp and S. Broder: New directions in molecular medicine. *Cancer Res* 54:653, 1994.
40a. T. Skorski, M. Nieborowska-Skorska, N. C. Nicolaides, C. Szczylik, P. Iversen, R. V. Iozzo, G. Zon, and B. Calabretta: Suppression of Philadelphia[1] leukemia cell growth in mice by BCR-ABL antisense oligodeoxynucleotide. *Proc Natl Acad Sci USA* 91:4504, 1994.
40b. N. Hijiya, J. Zhang, M. Z. Ratajczak, J. A. Kant, K. DeRiel, M. Herlyn, G. Zon, and A. M. Gewirtz: Biologic and therapeutic significance of MYB expression in human melanoma: *Proc Natl Acad Sci USA* 91:4499, 1994.
41. M. Z. Ratajczak, J. A. Kant, S. M. Luger, N. Hijiya, J. Zhang, G. Zon, and A. M. Gewirtz: In vivo treatment of human leukemia in a SCID mouse model with c-*myb* antisense oligodeoxynucleotides. *Proc Natl Acad Sci USA* 89:11823, 1992.
42. I. Kitajima, T. Shinohara, J. Bilakovics, D. A. Brown, X. Xu, and M. Nerenberg: Ablation of transplanted HTLV-1 *tax*-transformed tumors in mice by antisense inhibition of NF-κB. *Science* 258:1792, 1992.
43. K. A. Higgins, J. R. Perez, T. A. Coleman, K. Dorshkind, W. A. McComas, U. M. Sarmiento, C. A. Rosen, and R. Narayanan: Antisense inhibition of the p65 subunit of NF-κB blocks tumorigenicity and causes tumor regression. *Proc Natl Acad Sci USA* 90:9901, 1993.
44. R. N. Georges, T. Mukhopadhyay, Y. Zhang, N. Yen, and J. A. Roth: Prevention of orthotopic human lung cancer growth by intratracheal instillation of a retroviral antisense K-*ras* construct. *Cancer Res* 53:1743, 1993.
45. B. P. Monia, J. F. Johnston, D. J. Ecker, M. A. Zounes, W. F. Lima, and S. M. Frier: Selective inhibition of mutant Ha-ras mRNA expression by antisense oligonucleotides. *J Biol Chem* 267:19954, 1992.
46. L. J. Maher, III: DNA triple-helix formation: An approach to artificial gene repressors? *BioEssays* 14:807, 1992.
47. A. M. Belikova, N. I. Grineva, D. G. Knoore, and S. D. Mysina: Specific cleavage of DNA on guanosine residues by 2′,3′-*O*-[4-(*N*-2-chlorethyl-*N*-methylamino)benzylidene] uridine and its 5′-methyl-phosphate. *Dokl Acad Nauk SSSR* 212:876, 1973.
48. G. Felsenfeld, D. R. Davies, and A. Rich: Formation of a three-stranded polynucleotide molecule. *J Am Chem Soc* 79:2023, 1957.
49. P. B. Dervan: In R. H. Sarma and M. E. Sarma, eds.: *Human Genome Initiative and DNA Recombination.* New York: Adenine Press, 1990, pp. 37–49.
50. N. T. Thuong, and C. Hélène: Sequence-specific recognition and modification of double-helical DNA by oligonucleotides. *Angew Chem Int Ed Engl* 32:666, 1993.
51. M. Grigoriev, D. Praseuth, A. L. Guieysse, P. Robin, N. T. Thuong, C. Hélène, and A. Harel-Bellan: Inhibition of gene expression by triple helix-directed DNA cross-linking at specific sites. *Proc Natl Acad Sci USA* 90:3501, 1993.
52. F. Birg, D. Praseuth, A. Zerial, N. T. Thuong, U. Asseline, T. Le Doan, and C. Hélène: Inhibition of simian virus 40 DNA replication in CV-1 cells by an oligodeoxynucleotide covalently linked to an intercalating agent. *Nucleic Acids Res* 18:2901, 1990.
53. F. M. Orson, D. W. Thomas, W. M. McShan, D. J. Kessler, and M. E. Hogan: Oligonucleotide inhibition of IL2Rα mRNA transcription by promoter region collinear triplex formation in lymphocytes. *Nucleic Acids Res* 19:3435, 1991.
54. M. R. Capecchi: The new mouse genetics: Altering the genome by gene targeting. *Trends Genet* 5:70, 1989.
55. T. C. Doetschman: Gene targeting in embryonic stem cells. *Biotechnology* 16:89, 1991.
56. E. J. Robertston: Using embryonic stem cells to introduce mutations into the mouse germ line. *Biol Reprod* 44:238, 1991.
57. M. Cappechi: How efficient can you get? *Nature* 348:109. 1990.
58. P. J. Wilder and A. Rizzino: Mouse genetics in the 21st century: using gene targeting to create a cornucopia of mouse mutants possessing precise genetic modifications. *Cytotechnology* 11:79, 1993.
59. M. L. Mucenski, K. McLain, A. B. Kier, S. H. Swerdlow, C. M. Schreiner, T. A. Miller, D. W. Pietryga, W. J. Scott, Jr., and S. S. Potter: A functional c-*myb* gene is required for normal murine fetal hepatic hematopoiesis. *Cell* 65:677, 1991.
60. T. M. DeChiara, E. J. Robertson, and A. Efstratiadis: Parental imprinting of the mouse insulin-like growth factor II gene. *Cell* 64:849, 1991.
61. L. A. Donehower, M. Harvey, B. L. Slagle, M. J. McArthur, C. A. Montgomery, Jr., J. S. Butel, and A. Bradley: Mice deficient for p53 are developmentally normal but susceptible to spontaneous tumors. *Nature* 356:215, 1992.
62. A. R. Clarke, E. R. Maandag, M. van Roon, N. M. T. van der Lugt, M. van der Valk, M. L. Hooper, A. Berns and H. te Riele: Requirement for a functional Rb-1 gene in murine development. *Nature* 359:328, 1992.
63. P. Soriano, C. Montgomery, R. Geske, and A. Bradley: Targeted disruption of the c-src proto-oncogene leads to osteopetrosis in mice. *Cell* 64:693, 1991.
64. T. R. Cech, A. J. Zaug, and P. J. Grabowski: In

vitro splicing of the ribosomal RNA precursor of tetrahymena: Involvement of a guanosine nucleotide in the excision of the intervening sequence. *Cell* 27:487, 1981.
65. T. R. Cech, D. Herschlag, J. A. Piccirilli, and A. M. Pyles: RNA catalysis by a group I ribozyme. *J Biol Chem* 267:17479, 1992.
66. A. M. Pyle: Ribozymes: A distinct class of metalloenzymes. *Science* 261:709, 1993.
67. M. Barinaga: Ribozymes: Killing the messenger. *Science* 262:1512, 1993.
68. B. Dropulic, D. A. Elkins, J. J. Rossi, and N. Sarver: Ribozymes: Use as anti-HIV therapeutic molecules. *Antisense Res Devel* 3:87, 1993.
69. M. Kashani-Sabet, T. Funato, V. A. Florenes, O. Fodstad, and K. J. Scanlon: Supression of the neoplastic phenotype in vivo by an anti-*ras* ribozyme. *Cancer Res* 54:900, 1994.
70. P. M. Potter, L. C. Harris, J. S. Remack, C. C. Edwards, and T. P. Brent: Ribozyme-mediated modulation of human O^6-methylguanine-DNA methyltransferase expression. *Cancer Res* 53:1731, 1993.
71. S. Altman: RNA enzyme-directed gene therapy. *Proc Natl Acad Sci USA* 90:10898, 1993.
72. B. A. Sullenger and T. R. Cech: Tethering ribozymes to a retroviral packaging signal for destruction of viral RNA. *Science* 262:1566, 1993.

15

Cancer Prevention

As a sage once said, "An ounce of prevention is worth a pound of cure." Clearly, the best way to cure cancer is to prevent it. As noted in Chapter 2, cancer epidemiologists generally agree that about one-third of all human cancers are related to cigarette smoking. The most obvious thing, then, for smokers to know is that their risk of cancer is severalfold higher than that of the nonsmoking population. The next most commonly cited relationship to cancer is diet (that is, excluding exposure to the ultraviolet rays of sunlight as a cause of skin cancer). Although there is less agreement among cancer epidemiologists on the role of diet in cancer causation (10% to 70% of cancers—see Chap. 2), there is sufficient evidence to lead the National Cancer Institute and the American Cancer Society to recommend a diet low in fat, high in natural fiber, and rich in fruits and vegetables so as to help prevent or slow the onset of cancer.

DIET AND CANCER PREVENTION

A number of studies support the hypothesis that a diet high in fat and low in carbohydrates, fruits, and fruit and vegetable fiber increases the risk of colorectal[1] and other types of cancer.[2] A considerable amount of evidence from both animal experiments and human epidemiological studies suggests that high-fat and high-caloric diets increase the risk of cancer.[3] Conversely, caloric restriction has been shown to reduce the incidence of chemically induced tumor development in the skin[4] and pancreas[5] of experimental animals.

There are a number of potential cancer-preventive ingredients in foods (Fig. 15-1)[6] that can act at various stages in tumor initiation and promotion (Fig. 15-2). This information—as well as additional data on naturally occurring food constituents and vitamins (see below)—has led to the idea that diet modification can be a way to prevent cancer.[7] One such nutritional intervention trial has been carried out in a region in China where the mortality rates of gastric and esophageal cancer are among the highest in the world.[8] Dietary supplementation with vitamin A (retinol) and zinc reduced the incidence of gastric cancer, and study participants who received supplements of β-carotene, vitamin E, and selenium had a reduced incidence of esophageal cancer. Overall cancer mortality was reduced over a 5.25-year period for those receiving β-carotene, vitamin E, and selenium.

CHEMOPREVENTION

The observations that certain chemicals, some of them natural dietary constituents and some not,

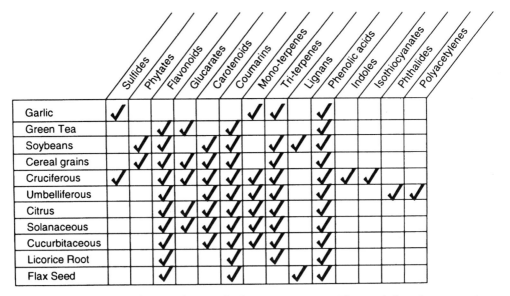

	Sulfides	Phytates	Flavonoids	Glucarates	Carotenoids	Coumarins	Mono-terpenes	Tri-terpenes	Lignans	Phenolic acids	Indoles	Isothiocyanates	Phthalides	Polyacetylenes
Garlic	✓						✓	✓		✓				
Green Tea			✓	✓		✓				✓				
Soybeans		✓	✓		✓	✓		✓	✓	✓				
Cereal grains		✓	✓	✓	✓	✓		✓		✓				
Cruciferous	✓		✓	✓	✓	✓	✓	✓		✓	✓	✓		
Umbelliferous			✓		✓	✓	✓	✓		✓			✓	✓
Citrus			✓	✓	✓	✓	✓	✓		✓				
Solanaceous			✓	✓	✓	✓	✓	✓		✓				
Cucurbitaceous			✓		✓	✓	✓	✓		✓				
Licorice Root			✓			✓		✓		✓				
Flax Seed			✓			✓			✓	✓				

Figure 15-1 Qualitative distribution of major food plant phytochemicals. Fourteen classes of phytochemicals are known or believed to possess cancer-preventive properties. They are believed to appear in greatest abundance in the foods and ingredients included in this diagram. (From Caragay.[6])

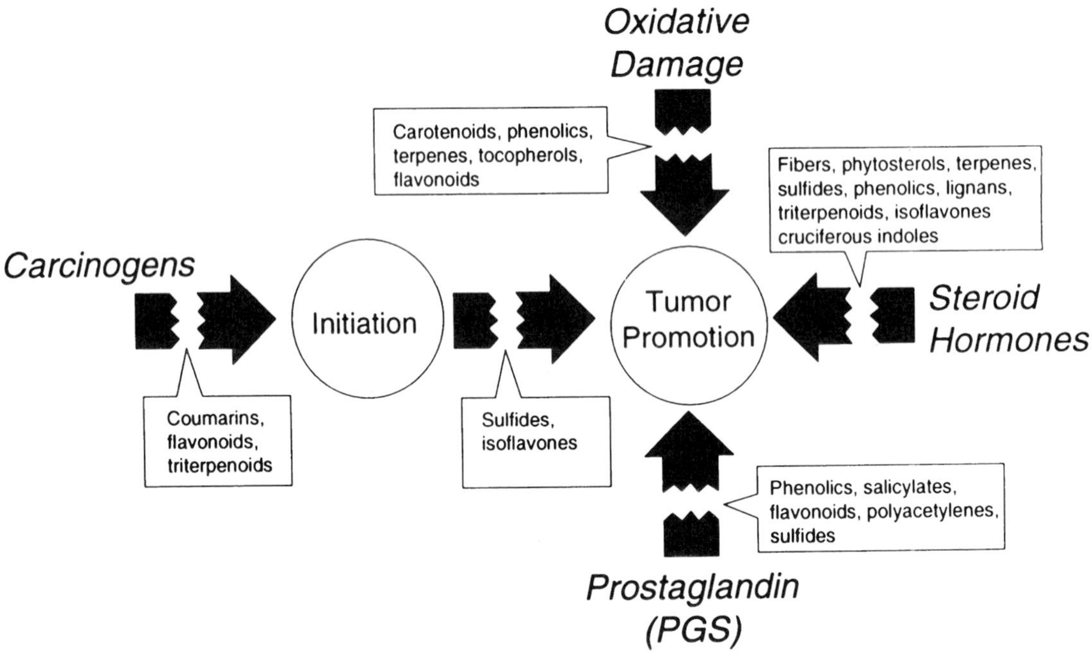

Figure 15-2 Dietary phytochemicals can affect metabolic pathways associated with breast cancer. Certain phytochemicals are known or believed to block specific pathways that lead to the development of breast cancer. (From Caragay.[6])

can decrease tumorigenesis in animals and, by epidemiologic implication, humans as well has led to the idea that the intake of certain chemicals can either prevent cancer or slow its progression. Numerous chemicals have now been tested in experimental animal studies and a number of clinical trials are ongoing.[9]

The idea that carcinogenesis could be prevented by the administration of agents that inhibit carcinogenesis is an attractive one. Theoretically, prevention of carcinogenesis could be accomplished by blockade of the initiation or the promotion-progression phases. Blockade of initiation events could be brought about by agents that decrease the metabolic activation of chemicals to the ultimate carcinogen, increase the detoxification of chemical carcinogens, or prevent the binding of carcinogens to their cellular targets.

Simply put, chemoprevention is the prevention of cancer with drugs. In this context, the word *drugs* is used in its broadest sense to include dietary supplements, vitamins, hormones, antihormones, etc., as well as "real" drugs such as aspirin, oltipraz, and other synthetic agents used for therapeutic purposes.

Cancer chemopreventive agents can be classified as antimutagens/carcinogen blocking agents, antiproliferatives, or antioxidants (Table 15-1); however, there is some overlap among agents in these categories, i.e., they may have more than one mechanism of action. Chemopreventive agents in clinical trial or being considered for clinical trial and considerations relating to their use are shown in Table 15-2. The need for biomarkers to track the success or lack thereof in such clinical trials should be noted here. These studies take years to complete and require a large study population to obtain meaningful data. Thus, it is crucial to have some biochemical or histopathologic means to track the progress of the study population. Unfortunately, there are no sure-fire markers to do this for most types of cancer. For cancers on external surfaces or areas that can be visualized by endoscopic techniques, changes in the degree of dysplasia can be observed (e.g., oral cavity, uterine cervix, and urinary bladder). However, for many other cancers, this is not possible. Hence, noninvasive biochemical tests are needed to track these cases. A few such tests are available—e.g., prostate-specific antigens (PSA) for prostate cancer—but in most "internal" cancers such tests are not specific or sensitive enough to detect early metaplastic or neoplastic changes. Markers of exposure to certain specific types of agents may be useful—e.g., levels or aflatoxin-N7-guanine adducts in the urine.[10] Markers of oxidative damage such as 8-hydroxyguanine can also be found in the urine.[11] The limitation of these tests is that the inciting type of carcinogen has to be known or highly suspected ahead of time.

Table 15-1 Pharmacologic and Chemical Structural Classification of Promising Chemopreventive Agents

Antimutagens/carcinogen blocking agents
Phase II metabolic enzyme inducers: *N*-acetyl-L-cysteine, *S*-allyl-L-cysteine, oltipraz, phenhexyl isothiocyanate
Polyphenols: ellagic acid
Other: curcumin, DHEA,[a] fluasterone (16-fluoro-DHEA)
Antiproliferatives
Retinoids/carotenoids: β-carotene, 4-HPR, 13-*cis*-retinoic acid, vitamin A
Antihormones: finasteride, tamoxifen
Antiinflammatories: aspirin, carbenoxolone, curcumin, 18β-glycyrrhetinic acid, ibuprofen, piroxicam, sulindac
G6PDH inhibitors: DHEA, fluasterone
ODC inhibitors: *N*-acetyl-L-cysteine, aspirin, carbenoxolone, curcumin. DFMO, 18β-glycyrrhetinic acid, 4-HPR, ibuprofen, piroxicam, 13-*cis*-retinoic acid, sulindac, vitamin A
Protein kinase C inhibitors: carbenoxolone, 18β-glycyrrhetinic acid, 4-HPR, tamoxifen
Other: calcium
Antioxidants
Antiinflammatories: see under "Antiproliferatives"
Antioxidants: *N*-acetyl-L-cysteine, β-carotene, curcumin, ellagic acid, fumaric acid
Phase II metabolic enzyme inducers: see under "Antimutagens/Carcinogen Blocking agents"
Thiols: *N*-acetyl-L-cysteine, *S*-allyl-L-cysteine, oltipraz
Retinol
Vitamin C
Vitamin E

[a]DHEA, dehydroepiandrosterone; G6PDH, glucose-6-phosphate dehydrogenase; ODC, ornithine decarboxylase; 4-HPR, all-*trans*-N-(4-hydroxyphenyl)-retinamide; DFMO, 2-difluoromethylornithine.
Source: Modified from Kelloff et al.[9]

Antimutagens/Carcinogen Blocking Agents

Isothiocyanates

Organic isothiocyanates (R—N=C=S) are widely distributed in plants and show up in the human food chain in a variety of "feed it to

Table 15-2 Aspects of Chemoprevention at Major Cancer Sites

	Colorectal	Prostate	Lung	Breast	Bladder	Oral	Cervix
Agents classified by mechanism	Antiinflammatories (e.g., sulindac, piroxicam, aspirin, ibuprofen) Antiproliferatives (e.g., DFMO, calcium, curcumin, 18β-glycyrrhetinic acid)	Testosterone 5α-reductase inhibitors (e.g., finasteride) Retinoids (e.g., 4-HPR) Antiproliferatives (e.g., DFMO) *ras*-farnesylation inhibitors (*e.g.*, *d*-limonene)	Retinoids/carotenoids (e.g., vitamin A, 13-*cis*-retinoic acid, β-carotene) Dithiolthiones and other organosulfur compounds (e.g., oltipraz, *N*-acetyl-L-cysteine)	Antiestrogens (e.g., tamoxifen, toremifene) Retinoids (e.g., 4-HPR) *ras*-farnesylation inhibitors (e.g., perillyl alcohol, *d*-limonene)	Antiinflammatories (e.g., sulindac, piroxicam, aspirin, ibuprofen) Antiproliferatives (e.g., DFMO) Retinoids (e.g., 4-HPR)	Retinoids/carotenoids (e.g., vitamin A, 4-HPR, 13-*cis*-retinoic acid, β-carotene) Antiinflammatories (e.g., carbenoxolone)	Retinoids (e.g., vitamin A, 4-HPR) Antiproliferatives (e.g., DFMO) Folic acid
Intermediate biomarkers	Adenomas, proliferative indices, aberrant crypts, Lewis blood group antigens, sialyl-Tn antigen	PIN, PSA, PAP,[a] cytokeratins (loss of 50–64 kDa), vimentin, nucleolar prominence, DNA content	Cellular atypia/dysplasia in sputum, bronchial atypical metaplasia/dysplasia, PCNA, blood group antigens, p53	Atypical hyperplasia, DCIS, LCIS	TIS, dysplasia, DNA content, F- and G-actins, integrins, loss of heterozygosity (e.g., 9q), blood group antigens, RB	Dysplastic leukoplakia, keratin expression, GGT	CIN
Clinical cohorts (phase II)	Patients with previous adenomas, patients with adenomas <1 cm in diameter	Patients with PIN without prostatic adenocarcinoma; patients scheduled for radical prostatectomy	Patients with recently resected stage 1 lung or laryngeal cancer	Patients scheduled for breast cancer surgery	Patients with previously resected TIS or T_a, T_1 disease without TIS	Patients with dysplastic leukoplakia	HPV-negative patients with CIN III
Clinical cohorts (phase III)	Subjects at high risk (e.g., family history of adenomas or colorectal cancer, previously treated breast or endometrial cancer)	Patients with elevated serum PSA; subjects ≥60 years of age	Patients with previous lung, head or neck cancers; subjects at high risk (e.g., smokers, occupational exposure to asbestos)	Patients with previously treated breast cancer	Subjects at high risk (e.g., occupational exposure to aromatic amines)	Patients with previously treated head and neck cancers; subjects at high risk (e.g., smokers, tobacco chewers)	Patients with CIN

[a]PAP, prostatic alkaline phosphatase; GGT, γ-glutamyl transpeptidase; DFMO, 2-difluoromethylornithine; CIN, cervical intraepithelial neoplasia; CIS, carcinoma *in situ*; DCIS, ductal carcinoma *in situ*; 4-HPR, all-*trans*-*N*-(4-hydroxyphenyl)-retinamide; HPV, human papilloma virus; IEN, intraepithelial neoplasia; LCIS, lobular carcinoma *in situ*; PCNA, proliferating cell nuclear antigen; PIN, prostatic intraepithelial neoplasia; PSA, prostate specific antigen; TIS, transitional cell carcinoma *in situ*.

Source: From Kelloff et al.[9]

Mikey"–type vegetables such as broccoli, Brussels sprouts, cabbage, cauliflower, and kale. In spite of a former President's abhorrence, broccoli and other cruciferous vegetables are not only good for building character but they may also decrease the risk of getting cancer. Allyl isothiocyanates are found in mustard seeds and are responsible for the pungent flavor and odor of mustard. In addition to their characteristic flavors and odors, isothiocyanates have a variety of pharmacologic properties such as antibacterial, antifungal, and antiprotozoal actions. A keen interest has developed in them because a number of studies have shown that they have anticarcinogenic activity (reviewed in Ref. 12).

In rodents, organic isothiocyanates block the production of tumors by a wide variety of carcinogens, primarily by blocking their activation to ultimate carcinogens. The carcinogenic actions of polycyclic aromatic hydrocarbons, azo dyes, ethionine, *N*-2-fluorenylacetamide, and nitrosamines have been shown to be decreased by isothiocyanates.[12] Active agents in this class include α-naphthyl, β-naphthyl, phenyl, benzyl, phenethyl, and other arylalkyl-isothiocyanates. Tumor development in the liver, lung, mammary gland, forestomach, and esophagus has been inhibited by feeding these compounds to animals. The anticarcinogenic action of isothiocyanates appears to be mediated by decreasing the activity of the P-450 isozymes involved in carcinogen activation and by inducing phase 2 xenobiotic metabolizing enzymes that detoxify and inactivate electrophilic intermediates generated from chemical carcinogens. These enzymes include glutathione transferases and NADPH-dependent quinone reductase. The induction of phase 2 enzymes appears to be the most likely mechanism of anticarcinogenic action of isothiocyanates.[12] Potent inducers of phase 2 enzymes such as sulforaphane (isolated from broccoli) and similar synthetic analogues block the formation of mammary tumors in rats treated with dimethylbenz[*a*]anthracene (DMBA).[12a] Since many of these agents are found in high levels in human diets and have thus proven to be safe to take, they are ideal candidates for the development of chemoprotective compounds. Indeed, epidemiological studies have suggested that individuals ingesting diets rich in cruciferous vegetables have a lower incidence of cancer.

Oltipraz

Oltipraz is a synthetic dithiolethione analogue, similar to some dithiolethiones found in cruciferous vegetables, that was developed in the late 1970s as an effective antischistosomal drug. During studies of its mechanism of action, Bueding et al.[13] observed increased levels of phase 2 enzymes and glutathione in tissues of rodents receiving the drug. Wattenberg and Bueding[14] first established the anticarcinogenic effects of oltipraz in inhibiting diethylnitrosamine-, benzo[*a*]pyrene-, or uracil mustard-induced carcinomas in the lungs and forestomachs of mice. Subsequent studies have shown the chemoprotective properties of oltipraz against various carcinogen-induced tumors of the breast, bladder, skin, trachea, and liver in rodents (reviewed in Refs. 15 and 16). This broad range of anticarcinogenic activity and relatively low toxicity has prompted the testing of the agent in phase I clinical trials, and only low grade (I/II) toxicities became apparent.[16] Phase II trials are under way, and the agent is intended for testing in some regions where aflatoxin ingestion is high, using the production of aflatoxin-N7-guanine adducts as a biomarker.[10]

Other Organosulfur Compounds

The organosulfur compounds include the isothiocyanates and dithiolethiones, discussed above, as well as diallyl sulfides present in garlic and onions and compounds endogenously formed in the body such as *N*-acetylcysteine and taurine (2-aminoethanesulfonate). These compounds have been shown to induce phase 2 enzymes glutathione S-transferase, NADPH-dependent quinone reductase, and UDP-glucuronosyl transferase in liver and colon tissue and to inhibit chemically induced tumors in the colon and other tissues of rodents (reviewed in Ref. 17). Diallyl sulfide has also been shown to inhibit P-450 2EI activity[18] and *N*-acetylcysteine to inhibit formation of DNA adducts in organs of rats treated with benzo[*a*]pyrene or exposed to cigarette smoke.[19] These activities may contribute to the anticarcinogenic actions of organosulfur compounds. *N*-Acetylcysteine, which also has antioxidant activity, has undergone phase I trials in patients in remission with oral, laryngeal, or lung cancer, and it has been shown to have a low frequency of side effects.[20]

Ellagic Acid

Ellagic acid is related to the coumarin class of lactones and is found in a variety of fruits and vegetables in the human diet. It has been shown to inhibit carcinogen-induced tumors in rodent skin, mammary gland, and forestomach (reviewed in Ref. 21). It appears to act by preventing formation of the activated forms of polycyclic aromatic hydrocarbons.

Dehydroepiandrosterone (DHEA)

DHEA is an adrenal steroid hormone precursor. Low serum levels have been found in women with breast cancer and in bladder cancers of both sexes (reviewed in Refs. 21 and 22). DHEA has also shown chemopreventive activity in animal models for skin, breast, colon, and lung carcinogenesis. Its anticarcinogenic effects appear to be due to its ability to inhibit P-450–mediated activation of carcinogens[21] and/or its ability to inhibit isoprenylation (and hence membrane targeting) of $p21^{ras}$ [22] (see Chap. 7). Other synthetic analogues of DHEA, such as 16-fluoro-DHEA, have been synthesized and may be better to use in human trials because they have less ability to be converted to testosterone or estrogen.[21]

Antiproliferative Agents

Retinoids/β-Carotene

Retinoids are synthetic or natural analogues of vitamin A, the fat-soluble vitamin first recognized in 1909[23] and first named in 1920.[24] It was noted early on that diets deficient in vitamin A led to keratinization of epithelium and squamous metaplasia (reviewed in Ref. 25). Wolbach and Howe,[26] in a classic paper published in 1925, described gastrointestinal, respiratory, and urogenital epithelial metaplasia in rats fed a vitamin A–deficient diet. In 1926, Fujumaki[27] observed that such a diet led to gastric carcinomas in rats. A paper in 1941 by Abels et al.[28] described an association of human cancer with vitamin A–deficient diets. The first demonstration that a vitamin A analogue could suppress premalignant epithelial lesions was by Lasnitzke,[29] who showed that such lesions could be reversed by the addition of retinyl acetate to the diet.

Retinoids play a key role in regulating differentiation and proliferation of a number of tissues, including normal epithelium and connective tissues as well as preneoplastic and neoplastic tissues. They act through a family of receptors (see below) found ubiquitously on cells. During animal development, retinoids have been shown to affect tissue determination, regional polarization of the embryo, and limb bud development. For this reason, retinoic acid has been thought to act as a "morphogen" in embryonic pattern formation (see Chap. 5).

The term *retinoid* has been used to include a large family of natural and synthetic compounds that encompasses vitamin A (retinol) and its esters, β-all-*trans* retinoic acid (tretinoin), 13-*cis*-retinoic acid (isotretinoin or Accutane), an aromatic ethyl ester derivative of retinoic acid (etretinate), the retinamides such as *N*-(4-hydroxyphenyl) retinamide (4HPR or Fenretinide), and retinoidal benzoic acid derivatives such as TTNPB.[25] Over 2000 retinoid analogues have been synthesized, and the structures of some of them are shown in Figure 15-3.

The activity of retinoids in reversion or slowing the growth of preneoplastic and neoplastic cells in vitro and in vivo has led to a number of clinical trials of agents in this class, with the goal of determining whether such agents could slow tumor progression or prevent the recurrence of cancers treated by other means. Clinical trials of this group of agents have included use in head and neck, lung, cervical, prostate, bladder, breast, and skin cancers, as well as in acute promyelocytic leukemia (APL).[30,31]

The clinical strategy for chemoprevention is threefold: (1) "to block or reverse carcinogenesis before the development of invasive cancer,"[30] (2) to prevent disease progression, and (3) to prevent the occurrence of second primary tumors. In some situations, chemopreventive agents are also used as adjuvants to chemotherapy or surgery. Clinical cancer chemoprevention is based on two concepts derived from experimental and epidemiologic studies of cancer. The first is that carcinogenesis is a multistep process beginning with premalignant changes that progress ultimately to invasive cancer. The second

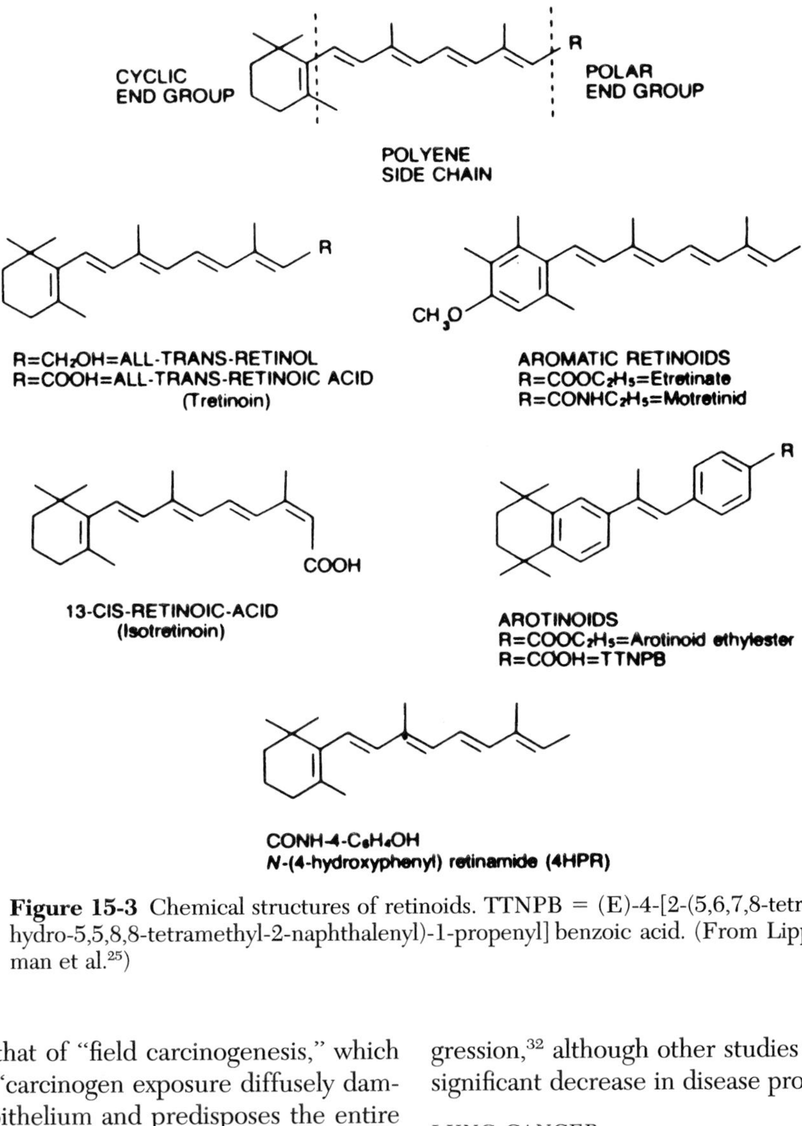

Figure 15-3 Chemical structures of retinoids. TTNPB = (E)-4-[2-(5,6,7,8-tetrahydro-5,5,8,8-tetramethyl-2-naphthalenyl)-1-propenyl] benzoic acid. (From Lippman et al.[25])

concept is that of "field carcinogenesis," which states that "carcinogen exposure diffusely damages the epithelium and predisposes the entire carcinogen-exposed field to the development of multiple independent cancers."[30] Exposure of the skin to sunlight, the colon to fatty acids, or the lung to cigarette smoke are examples of carcinogenic insults that can widely injure epithelial tissues and produce multiple primary tumors by field carcinogenesis.

HEAD AND NECK CANCER

One of the first successful clinical demonstrations of the usefulness of chemoprevention was in head and neck cancer. Clinical trials of isotretinoin in this disease have shown a significant decrease in the occurrence of second primary tumors but only minor effects on disease progression,[32] although other studies have shown a significant decrease in disease progression.[30,33]

LUNG CANCER

Positive results with systemic use of retinoids in preventing second primaries and disease progression in cancers of the head and neck and oral cancers has prompted trials of retinoids in lung cancer. The results have been mixed. One study employing etretinate in chronic smokers showed a decrease in bronchial squamous metaplasia compared to controls taking a placebo, while other studies with isotretinoin did not show a substantial difference in metaplasia between treated and control groups (reviewed in Ref. 30).

SKIN CANCER

A number of studies have suggested that retinoids can slow or prevent skin cancer progression. For example, isotretinoin has been reported to reverse preneoplastic skin lesions and to prevent invasive skin cancer in high-risk patients with xeroderma pigmentosum, and, in combination with interferon α-2a (IFN-α2a), to induce remissions in advanced squamous cell carcinomas of the skin (reviewed in Ref. 34).

CERVICAL CANCER

Oral isotretinoin (13-*cis*-retinoic acid) plus subcutaneously injected IFN-α2a has also been shown to produce a 58% (11 of 19) response rate in patients with stage IIB cervical cancer.[35] In another study, topical application of all-*trans*-retinoic acid reversed cervical intraepithelial neoplasia (CIN) type II but not more advanced dysplasia,[36] suggesting that, at least for early-stage cervical preneoplasia, disease progression can be slowed or halted.

ACUTE PROMYELOCYTIC LEUKEMIA (APL)

The use of all-*trans*-retinoic acid (tretinoin) in APL is not exactly chemoprevention because, in this instance, the retinoid is used as a primary or adjuvant therapy for the disease and is based on a specific genetic defect in retinoic acid receptors that renders this disease amenable to retinoid therapy. APL is characterized by a specific t(15;17) chromosomal translocation that fuses the retinoic acid receptor α(RARα) to a novel gene called PML, which produces a dominant-negative fusion protein, PML-RARα (see also Chap. 3). This fusion protein has abnormal intracellular localization and disrupts the assembly and/or structure of nuclear proteins that regulate myelocyte growth and differentiation.[37,38] The beneficial role of retinoic acid in promoting myeloid differentiation in APL may be related to its ability to restore the normal nuclear organization and structure of nucleoprotein complexes.[37,38]

All-trans-RA has been shown to induce complete remissions in a high percentage of patients with APL.[39] Unfortunately, resistance to RA develops rapidly, apparently from induction of its metabolism by a P-450 isozyme or a drug-induced increase in RA-binding proteins that compete for RA's binding to RARα.[40]

β-CAROTENE

β-carotene is one of the natural dietary sources of vitamin A (retinol). It is found in a variety of plant sources (including certain green and yellow vegetables) but is not synthesized by animals or humans. The other usual dietary source of vitamin A is the alcohol and aldehyde forms and their esters found in milk, eggs, and meat.[25] β-carotene is converted to retinol during intestinal absorption, from where it and preformed vitamin A from animal sources is transported to liver and fat tissue and stored. Retinol, from β-carotene or retinyl esters (the storage form of vitamin A which is converted to retinol upon mobilization from tissues), is transported in the blood as part of a complex with retinol-binding protein, transthyretin, and thyroxine.[25] Vitamin A has an important role in growth, reproduction, and epithelial differentiation.

β-carotene, vitamin A, and synthetic retinoids have attracted a lot of attention as possible chemopreventive agents. β-carotene is thought to function as an electron-scavenging antioxidant (see below), whereas retinol and retinoids are thought to enhance cellular differentiation by acting via their specific nuclear receptor proteins RAR and RXR, which regulate gene expression, cell proliferation, and tissue differentiation.[41]

A number of studies have been designed to test the ability of β-carotene to act as a chemopreventive agent for cancer. Only a few will be cited here. One large U.S.-based multicenter, two-armed, double-blind, randomized trial called CARET is designed to test whether oral administration of β-carotene (30 mg/day) plus retinyl palmitate (25,000 IU/day) can decrease the incidence of lung cancer in heavy smokers and asbestos workers.[41] This and other such studies are based on observations that retinoids have reversed cigarette smoking–induced preneoplastic bronchial lesions[42] and that retinyl palmitate increased the time to relapse or development of new primary tumors in patients with stage I lung cancer.[43] Another study found a decreased risk of developing lung cancer in *nonsmokers*, who had been consuming diets rich in β-carotene and raw fruits and vegetables.[44] Other studies have shown that oral ingestion of β-carotene induces regression of oral leukoplakia, a premalignant lesion for oral cancer (reviewed in Ref. 45). One study done in Finland has reported no decrease in the incidence of lung cancer in heavy smokers who were placed on β-carotene, alpha-tocopherol (vitamin E), or both and then followed for 5 to 8 years.[46] Moreover, this study also found that there was a small (8%) but statistically significant increase in mortality among the group taking β-carotene. The conclusions of this study are somewhat surprising (and controversial) because they are contradictory to a number of other studies. The reason that this is so may be that the Finnish study population were relatively heavy smokers (5 or more cigarettes a day) and had been smoking for many years (the study population was 50 to 69 years of age). Thus, one could argue that the lung tissue in these individuals had already been "initiated" and undergone several years of "promotion" at the time of entry into the study. Nevertheless, these results point out an important caveat for all chemoprevention studies, and this is that once tissues have undergone initiation and several stages of promotion-progression, it is unlikely that the tumorigenic process can be stopped—slowed perhaps, but not stopped.

Hormonal Chemoprevention

A large body of evidence indicates that hormones have an important role in the causation of cancer in women and men. For example, female hormones have been linked to the development of cancers of the breast, uterus, endometrium, and ovary. Male hormones are implicated in the development of prostate cancer. This type of hormone-induced carcinogenesis appears to result from the ability of hormones to stimulate cell division in hormone target organs. This may, in turn, lead to an increased probability of accumulating genetic errors over time. It is estimated that hormone-related cancers account for at least 20% of all male and 40% of all female cancers in the United States.[47] Thus, chemoprevention of cancer by the manipulation of hormone levels or by the use of "antihormones" is a legitimate strategy. Several hormonal chemopreventive agents have been shown to be effective as chemopreventive agents and others are under intense clinical study (Table 15-3).

ORAL CONTRACEPTIVES

The finding that sequential oral contraceptives (estrogen alone, followed by progesterone alone) and estrogen replacement therapy (at the earlier-used higher doses) increased the risk of endometrial cancer led to the "unopposed estrogen" hypothesis for the cause of this form of uterine cancer (reviewed in Ref. 47). In contrast, the now commonly used combination oral contraceptives (COCs), which contain lower amounts of estrogen plus progesterone, have markedly decreased the risk of endometrial and ovarian cancer. Epidemiologic studies show an 11.7% per year decrease in endometrial cancer and a 7.5% per year decrease in ovarian cancer with the use of COCs.[47]

Breast cancer is a different story, however. In breast tissue, both estrogen and progesterone stimulate cell proliferation. Thus, the combination of the two appears to have a greater stimulatory effect on breast cell division than estrogen alone, leading to the "estrogen-augmented by progesterone" theory of breast cancer etiology. The protective effects of late menarche, early menopause, and early childbearing fit this concept. Studies of breast cancer risk in COC users have not produced a clear answer. A 3.1% increase in breast cancer risk per year of COC use has been reported for women diagnosed under age 45, whereas no increased risk with COC use has been seen in women diagnosed over age 45.[47]

Table 15-3 Hormonal Chemopreventive Agents in Clinical Use

Chemopreventive Agent	Cancer Site	Mechanism of Action
Oral contraceptives	Endometrium	Antiestrogen
	Ovary	Inhibit ovulation
GnRH agonists	Breast, endometrium	Inhibit ovarian steroid hormone production
	Ovary	Inhibit ovulation
Progestogens (HRT)	Endometrium	Antiestrogen
Tamoxifen	Breast	Antiestrogen
Finasteride	Prostate	5α-reductase inhibitor

Abbreviations: GnRH, gonadotropin-releasing hormone; HRT, hormone replacement therapy.

Source: Reprinted with permission from Henderson et al., Hormonal chemoprevention of cancer in women. *Science* 259:633. Copyright 1993 by the American Association for the Advancement of Science.

GONADOTROPIN-RELEASING HORMONE ANALOGUES (GnRHAs)

GnRHAs reversibly inhibit ovulation and reduce production of ovarian steroid hormones. Clinical trials are under way to test whether such agents can act as contraceptives and also reduce the risk of breast, ovarian, and endometrial cancer.

HORMONE REPLACEMENT THERAPY (HRT)

HRT is being used to prevent the symptoms of menopause and osteoporosis associated with menopause. As noted above, the use of estrogen by itself is associated with an increased risk of endometrial cancer that is related to both dose and duration of therapy. This increased risk is calculated to be about 3.5-fold after 5 years of unopposed estrogen therapy.[47] However, the addition of progestogens to the regimen and lowering of the estrogen dose decreased the overall risk of endometrial cancer between 1973 and 1987 from 27.9% to 14.4%.[47]

The effect of HRT on breast cancer risk is less clear. Lowering the daily dose of from 1.25 mg to 0.625 mg of the commonly used conjugated equine estrogen, used by itself, causes the risk of breast cancer to go from 3.1% per each year of estrogen-alone replacement therapy to less than 2% per year of use.[48]

TAMOXIFEN

Tamoxifen has been a mainstay in the therapy of breast cancer for many years. It has been clearly established that in women who have had a primary in one breast, tamoxifen decreases the risk of developing a second tumor in the opposite breast.[47,49] These data, plus the well-documented epidemiologic evidence that the lifetime duration of exposure to estrogen (i.e., early menarche, late menopause, nulliparity) increases breast cancer risk, led to the concept that an antiestrogen such as tamoxifen could be a chemopreventive agent for women at high risk to develop breast cancer. As a result, large-scale clinical trials have been initiated in the United States and Europe to test this hypothesis. Such trials are not without controversy. After all, is it wise to give a hormonal agent to high-risk but otherwise healthy women for many years? The compelling argument to do so is the dramatically lower risk (35% decrease) for development of a cancer in the contralateral breast after a primary in one breast. Also, the side effects are relatively mild; they include hot flashes, nausea and vomiting, menstrual irregularities, and vaginal bleeding and discharge. The greatest objection to the long-term use of tamoxifen as a chemopreventive agent is the small but real increased risk of endometrial cancer. Even though it acts as an antiestrogen in breast tissue, tamoxifen has weak estrogen agonist action on the uterine endometrium. Thus, it can act as an unopposed estrogen in this tissue. Clearly, the risk-benefit ratio has to be considered in such clinical trials.

ANTIANDROGENS

Dihydrotestosterone (DHT) is a metabolic product derived from testosterone by the action of an enzyme called 5α-reductase. DHT stimulates prostatic cells to proliferate and its blood levels appear to correlate with prostate cancer risk; for example, DHT levels are lower in Japanese men, a low-risk group, compared to U.S. black males, a high-risk group.

The drug finasteride (Proscar) inhibits the enzyme testosterone 5α-reductase and thereby lowers the production of DHT and prevents hyperplasia of the prostatic stroma. Finasteride is already in phase III clinical trial as a chemopreventive agent for prostate cancer.[9] Based on what is known about the etiology of this disease, this approach should produce some positive results.

Anti-inflammatory Agents

A number of studies, but not all, suggest that chronic aspirin intake lowers the incidence of and mortality from colon cancer (reviewed in Ref. 50). Aspirin is one of a family of agents known as nonsteroidal anti-inflammatory drugs (NSAIDs); they inhibit the cyclooxygenase arm of the arachidonic acid cascade, which produces prostaglandins. Other members of the NSAID class include indomethacin, ibuprofen, piroxicam, and sulindac.

In rodent models of colonic carcinogenesis, NSAIDs have been shown to reduce the number of tumor-bearing animals and the number of tumors per animal.[50] Waddell and Loughry[51] first reported that sulindac reduced the size and number of rectal polyps in individuals with familial polyposis but that tumors recurred when treatment was stopped.[52] Additional studies have also shown a reversible regression of polyps in familial polyposis patients taking sulindac (reviewed in Ref. 50). While the mechanism of the antitumor effects of the NSAIDs is not clear, it may relate to inhibition of protaglandin-mediated stimulation of cell proliferation and/or to the effects of NSAIDs on reversing immune suppression.[50]

Ornithine Decarboxylase (ODC) Inhibitors

Hämäläinen[53] was the first to study in detail the levels of polyamines in human neoplasms. An increased polyamine content has since been found in a wide variety of human and animal neoplasms. Polyamines play an important role in cell proliferation and differentiation, and ODC is an essential enzyme in their biosynthetic pathway. ODC activity is increased in proliferating tissues and in tumors. Several ODC inhibitors can inhibit tumor formation in rodent models. One such inhibitor, D,L-α-difluoromethylornithine (DFMO), has been reported to inhibit chemical carcinogenesis in skin, tongue, liver, colon, breast, urinary bladder, kidney, and brain (reviewed in Ref. 54). Inhibitory effects of DFMO on the growth of bladder and kidney cancers in humans have also been reported.[55] Such data have led to clinical trials of DFMO as a chemopreventive agent for human cancer.[56]

Antioxidants

A number of chemopreventive agents discussed above also have the ability to scavenge oxygen radicals and to act as antioxidants. These include the organic sulfur–containing compounds, retinol, β-carotene, ellagic acid, oltipraz, and certain of the antiinflammatory agents.[9] Since oxygen radical damage to DNA is a well-recognized effect of a number of carcinogenic agents, including a number of chemical carcinogens and irradiation, the ability to block the production of oxygen radicals could potentially provide a major protective effect against carcinogenesis.

The damage produced by endogenously generated oxygen radicals has been suggested to be a major factor in aging and in such aging-related diseases as heart disease, senility, and cancer. Endogenous oxidants are produced continuously by normal cellular metabolism, and when not kept in check by cellular protective factors such as glutathione,[57] can result in extensive damage to proteins, lipids, and DNA. Oxidative damage to DNA, based on the urinary excretion of DNA adducts such as 8-hydroxyguanosine, is estimated to be about 10 "hits" per cell per day in humans.[58]

A number of vitamins and minerals have antioxidant actions. These include vitamins A, C, and E, carotenoids, and selenium. Vitamins E and C and carotenoids trap free radicals and reactive oxygen molecules. Micronutrient elements such as selenium, zinc, copper, iron, and manganese are essential cofactors for antioxidant enzymes. In addition, vitamin C can protect against cancer by inhibiting nitrosation of secondary amines and *N*-substituted amides to form nitrosamines and nitrosamides in acidic conditions such as those found in the stomach.[59] There is epidemiologic evidence that these nitroso compounds contribute to the etiology of cancers of the stomach, esophagus, and nasopharynx (reviewed in Ref. 59). Hence, there is good reason to believe that vitamin C may be an effective chemopreventive agent in those parts of the world where the incidence of these cancers is high, such as China, where such an intervention trial is under way.[60]

A number of clinical and epidemiologic studies have suggested a role for antioxidants in preventing human cancer. For instance, in the Iowa Women's Health Study, a high intake of vitamin E in the diet correlated with a decreased risk for colon cancer, especially in women under 65 years of age.[61] In contrast to these findings, a prospective study of breast cancer risk in over 89,000 women enrolled in the Nurses' Health Study showed no protective effect of diets rich in vitamins C or E and a modestly reduced risk for women with the lowest intake of vitamin A when vitamin A was added as a supplement to the diet.[62]

Protease Inhibitors

Certain protease inhibitors have been shown to suppress carcinogenesis in in vitro and in vivo assay systems. One of these, the Bowman-Birk inhibitor (BBI) derived from soybeans, has both antitrypsin and antichymotrypsin activity. BBI can block cell transformation of cells in culture exposed to chemical carcinogens and in several tissues (e.g., colon, liver, lung, esophagus, and oral cavity) in rodents exposed to various carcinogens (reviewed in Ref. 63). While the mechanism of anticarcinogenic action of BBI and similar protease inhibitors is not clear, several observed effects may be involved. For instance, protease inhibitors have been observed to inhibit the expression of the oncogenes c-*myc* and c-*fos* and to block carcinogen-induced gene amplification.[63]

There is circumstantial epidemiologic evidence for a role of soybean protease inhibitors in preventing human cancer. For example, populations who have a high intake of soybeans or soybean-derived products (e.g., Japanese and Seventh Day Adventists) have a lower risk of developing many cancers common in countries of the western world.[63]

REFERENCES

1. R. S. Sandler, C. M. Lyles, L. A. Peipins, C. A. McAuliffe, J. T. Woosley, and L. L. Kupper: Diet and risk of colorectal adenomas: Macronutrients, cholesterol, and fiber. *JNCI* 85:884, 1993.
2. R. Doll and R. Peto: The causes of cancer: Quantitative estimates of avoidable risks of cancer. *JNCI* 66:1191, 1981.
3. R. Weindruch, D. Albanes, and D. Kritchevsky: The role of calories and caloric restriction in carcinogenesis. *Hematol Oncol Clin North Am* 5:79, 1991.
4. D. F. Birt, H. J. Pinch, T. Barnett, A. Phan, and K. Dimitroff: Inhibition of skin tumor promotion by restriction of fat and carbohydrate calories in SENCAR mice. *Cancer Res* 53:27, 1993.
5. B. D. Roebuck, K. J. Baumgartner, and D. L. MacMillan: Caloric restriction and intervention in pancreatic carcinogenesis in the rat. *Cancer Res* 53:46, 1993.
6. A. B. Caragay: Cancer-preventive foods and ingredients. *Food Technol* 46:65, 1992.
7. C. Ip, D. J. Lisk, and J. A. Scimeca: Potential of food modification in cancer prevention. *Cancer Res* 54:1957s, 1994.
8. P. R. Taylor, B. Li, S. M. Dawsey, J. Li, C. S. Yang, W. Guo, W. J. Blot, and the Linxian Nutrition Intervention Trials Study Group: Prevention of Esophageal Cancer: The nutrition intervention trials in Linxian, China. *Cancer Res* 54:2029s, 1994.
9. G. J. Kelloff, C. W. Boone, V. E. Steele, J. A. Crowell, R. Lubet, and C. C. Sigman: Progress in cancer chemoprevention: Perspectives on agent selection and short-term clinical intervention trials. *Cancer Res* 54:2015s, 1994.
10. J. D. Groopman, G. N. Wogan, B. D. Roebuck, and T. W. Kensler: Molecular biomarkers for aflatoxins and their application to human cancer prevention. *Cancer Res* 54:1907s, 1994.
11. M. G. Simic: DNA markers of oxidative processes in vivo: Relevance to carcinogenesis and anticarcinogenesis. *Cancer Res* 54:1918s, 1994.
12. Y. Zhang and P. Talalay: Anticarcinogenic activities of organic isothiocyanates: Chemistry and mechanisms. *Cancer Res* 54:1976s, 1994.

12a. Y. Zhang, T. W. Kensler, C.-G. Cho, G. H. Posner, and P. Talalay: Anticarcinogenic activities of sulforaphane and structurally related synthetic norbornyl isothiocyanates. *Proc Natl Acad Sci USA* 91:3147, 1994.

13. E. Bueding, P. Dolan, and J. P. Leroy: The antischistosomal activity of oltipraz. *Res Commun Chem Pathol Pharmacol* 37:293, 1982.
14. L. W. Wattenberg and E. Bueding: Inhibitory effects of 5-(2-pyrazinyl)-4-methyl-1,2-dithiol-3-thione (oltipraz) on carcinogenesis induced by benzo[*a*]pyrene, diethyl-nitrosamine and uracil mustard. *Carcinogenesis (Lond).* 7:1379, 1986.
15. M. G. Bolton, A. Munoz, L. P. Jacobson, J. D. Groopman, Y. Y. Maxuitenko, B. D. Roebuck, and T. W. Kensler: Transient intervention with oltipraz protects against aflatoxin-induced hepatic tumorigenesis. *Cancer Res* 54:3499, 1993.
16. A. B. Benson: Oltipraz: A laboratory and clinical review. *J Cell Biochem* 17f:278, 1993.
17. B. S. Reddy, C. V. Rao, A. Rivenson, and G. Kelloff: Chemoprevention of colon carcinogenesis by organosulfur compounds. *Cancer Res* 53:3493, 1993.
18. C. S. Yang, T. J. Smith, and J. Hong: Cytochrome P-450 enzymes as targets for chemoprevention against chemical carcinogenesis and toxicity. *Cancer Res* 54:1982s, 1994.
19. A. Izzotti, F. D'Agostini, M. Bagnasco, L. Scatolini, A. Rovida, R. M. Balansky, C. F. Cesarone, and S. De Flora: Chemoprevention of carcinogen-DNA adducts and chronic degenerative diseases. *Cancer Res* 54:1994s, 1994.
20. N. De Vries and S. De Flora: *N*-acetyl-L-cysteine. *J Cell Biochem* 17F:270, 1993.
21. C. W. Boone, G. J. Kelloff, and W. E. Malone: Identification of candidate cancer chemopreventive agents and their evaluation in animal models and human clinical trials: A review. *Cancer Res* 50:2, 1990.
22. S. Schulz, R. C. Klann, S. Schönfeld, and J. W. Nyce: Mechanisms of cell growth inhibition and cell cycle arrest in human colonic adenocarcinoma cells by dehydroepiandrosterone: Role of isoprenoid biosynthesis. *Cancer Res* 52:1372, 1992.
23. W. Steppe: Versuche über Fütterung mit lipoidfreir Nahrung. *Biochem Z* 22:452, 1909.

24. J. C. Drummond: The nomenclature of the so-called accessory food factors (vitamins). *Biochem J* 14:660, 1920.
25. S. M. Lippman, J. F. Kessler, and F. L. Meyskens, Jr: Retinoids as preventive and therapeutic anticancer agents—Part I. *Cancer Treat Rep* 71:391, 1987.
26. S. B. Wolbach and P. R. Howe: Tissue changes following deprivation of fat soluble A vitamin. *J Exp Med* 42:753, 1925.
27. Y. Fujumaki: Formation of gastric carcinoma in albino rats fed on deficient diets. *J Cancer Res* 10:469, 1926.
28. J. C. Abels, A. T. Gorham, G. T. Pack, et al.: Metabolic studies in patients with cancer of the gastrointestinal tract: I. Plasma vitamin A levels in patients with malignant neoplastic disease, particularly of the gastrointestinal tract. *J Clin Invest* 20:749, 1941.
29. I. Lasnitzki: The influence of A hypervitaminosis on the effect of 20-methylcholanthrene on mouse prostate glands grown in vitro. *Br J Cancer* 9:434, 1955.
30. S. M. Lippman, S. E. Benner, and W. K. Hong: Retinoid chemoprevention studies in upper aerodigestivve tract and lung carcinogenesis. *Cancer Res* 54:2025s, 1994.
31. K. Slawin, D. Kadmon, S. H. Park, P. T. Scardino, M. Anzano, M. B. Sporn, and T. C. Thompson: Dietary fenretinide, a synthetic retinoid, decreases the tumor incidence and the tumor mass of *ras+myc*-induced carcinomas in the mouse prostate reconstitution model system. *Cancer Res* 53:4461, 1993.
32. S. E. Benner, T. F. Pajak, S. M. Lippman, C. Earley, and W. K. Hong: Prevention of second primary tumors with isotretinoin in patients with squamous cell carcinoma of the head and neck: Long-term follow-up. *JNCI* 86:140, 1994.
33. F. Chiesa, N. Tradati, M. Marazza, N. Rossi, P. Boracchi, L. Mariani, F. Formelli, R. Giardini, A. Costa, G. De Palo, and U. Veronesi: Fenretinide (4-HPR) in chemoprevention of oral leukoplakia. *J Cell Biochem* 17F:255, 1993.
34. S. M. Lippman, D. R. Parkinson, L. M. Itri, R. S. Weber, S. P. Schantz, D. M. Ota, M. A. Schusterman, I. H. Krakoff, J. U. Gutterman, and W. K. Hong: 13-cis retinoic acid and interferon α-2a: Effective combination therapy for advanced squamous cell carcinoma of the skin. *JNCI* 84:235, 1992.
35. S. M. Lippman, J. J. Kavanagh, M. Paredes-Espinoza, F. Delgadillo-Madrueño, P. Paredes-Casillas, W. K. Hong, E. Holdener, and I. H. Krakoff: 13-*cis*-retinoic acid plus interferon α-2a: Highly active systemic therapy for squamous cell carcinoma of the cervix. *JNCI* 84:241, 1992.
36. F. L. Meyskens, Jr., E. Surwit, T. E. Moon, J. M. Childers, J. R. Davis, R. T. Dorr, C. S. Johnson, and D. S. Alberts: Enhancement of regression of cervical intraepithelial neoplasia II (moderate dysplasia) with topically applied all-*trans*-retinoic acid: A randomized trial. *JNCI* 86:539, 1994.
37. J. A. Dyck, G. G. Maul, W. H. Miller, Jr., J. D. Chen, A. Kakizuka, and R. M. Evans: A novel macromolecular structure is a target of the promyelocyte-retinoic acid receptor oncoprotein. *Cell* 76:333, 1994.
38. K. Weis, S. Rambaud, C. Lavau, J. Jansen, T. Carvalho, M. Carmo-Fonseca, A. Lamond, and A. Dejean: Retinoic acid regulates aberrant nuclear localization of PML-RARα in acute promyelocytic leukemia cells. *Cell* 76:345, 1994.
39. R. P. Warrell, Jr., S. R. Frankel, W. H. Miller, Jr., D. A. Scheinberg, L. M. Itri, W. N. Hittelman, R. Vyas, M. Andreeff, A. Tafuri, A. Jakubowski, J. Gabrilove, M. S. Gordon, and E. Dmitrovsky: Differentiation therapy of acute promyelocytic leukemia with tretinoin (all-*trans*-retinoic acid). *N Engl J Med* 324:1385, 1991.
40. J. R. F. Muindi, S. R. Frankel, C. Huselton, F. DeGrazia, W. A. Garland, C. W. Young, and R. P. Warrell, Jr.: Clinical pharmacology of oral all-*trans* retinoic acid in patients with acute promyelocytic leukemia: *Cancer Res* 52:2138, 1992.
41. G. S. Omenn, G. Goodman, M. Thornquist, J. Grizzle, L. Rosenstock, S. Barnhart, J. Balmes, M. G. Cherniack, M. R. Cullen, A. Glass, J. Keogh, F. Meyskens, Jr., B. Valanis, and J. Williams, Jr.: The β-carotene and retinol efficacy trial (CARET) for chemoprevention of lung cancer in high risk populations: Smokers and asbestos-exposed workers. *Cancer Res* 54:2038s, 1994.
42. J. L. Misset, G. Mathe, G. Santelli, J. Gouveia, J. P. Homasson, M. C. Sudre, and H. Gaget: Regression of bronchial epidermoid metaplasia in heavy smokers with etretinate treatment. *Cancer Detect Prev* 9:167, 1986.
43. U. Pastorino, M. Infante, G. Chiesa, M. Maioli, M. Clerici, P. A. Belloni, M. Valente, and G. Ravasi: Lung cancer chemoprevention, in: U. Pastorino and W. K. Hong, eds.: *Chemoimmuno Prevention of Cancer.* New York: Thieme Medical Publishers, 1991, pp. 147–159.
44. S. T. Mayne, D. T. Janerich, P. Greenwald, S. Chorost, C. Tucci, M. B. Zaman, M. R. Melamed, M. Kiely, and M. F. McKneally: Dietary beta carotene and lung cancer risk in U.S. nonsmokers. *JNCI* 86:33, 1994.
45. H. S. Garewal: Beta-carotene and vitamin E in oral cancer prevention. *J Cell Biochem* 17F:262, 1993.
46. The Alpha-Tocopherol, Beta Carotene Cancer Prevention Study Group: The effect of vitamin E and beta carotene on the incidence of lung cancer and other cancers in male smokers. *N Engl J Med* 330:1029, 1994.
47. B. E. Henderson, R. K. Ross, and M. C. Pike: Hormonal chemoprevention of cancer in women. *Science* 259:633, 1993.
48. M. C. Pike, L. Bernstein, and D. V. Spicer: In J. E. Niederuber, ed.: *Current Therapy in Oncol-*

ogy. St. Louis: Dekker, Mosby-Year Book, 1993, pp. 292–303.
49. S. G. Nayfield, J. E. Karp, and L. G. Ford, F. A. Door, and B. S. Kramer: Potential role of tamoxifen in prevention of breast cancer. *JNCI* 83:1450, 1991.
50. L. J. Marnett: Aspirin and the potential role of prostaglandins in colon cancer. *Cancer Res* 52:5575, 1992.
51. W. R. Waddell and R. W. Loughry: Sulindac for polyposis of the colon. *J Surg Oncol* 24:83, 1983.
52. W. R. Waddell, G. F. Gasner, E. J. Cerise, and R. W. Loughry: Sulindac for polyposis of the colon. *Am J Surg* 157:175, 1989.
53. R. Hämäläinen: Über die quantitative Bestimmung des Spermins in Organismus und sein Verkommen in menschlichen Geweben und Körperflussigkeiten. *Acta Soc Med Fenn Duodecim* (Ser A) 23:97, 1947.
54. T. Kojima, T. Tanaka, T. Kawamori, A. Hara, and H. Mori: Chemopreventive effects of dietary D,L-α-difluoromethylornithine, and ornithine decarboxylase inhibitor, on initiation and postinitiation stages of diethylnitrosamine-induced rat hepatocarcinogenesis. *Cancer Res* 53:3903, 1993.
55. U. Dunzendorfer: The effect of α-difluoromethylornithine on tumor growth, acute phase reactants, β-2-microglobulins and hydroxyproline in kidney and bladder carcinomas. *Urol Invest* 36:128, 1981.
56. R. R. Love, P. P. Carbone, A. K. Verma, D. Gilmore, P. Carey, K. D. Tutsch, M. Pomplun, and G. Wilding: Randomized phase I chemoprevention dose-seeking study of α-difluoromethylornithine. *JNCI* 85:732, 1993.
57. A. Meister: Glutathione, ascorbate, and cellular protection. *Cancer Res* 54:1969s, 1994.
58. C. G. Fraga, P. A. Motchnik, M. K. Shigenaga, H. J. Helbock, R. A. Jacob, and B. N. Ames: Ascorbic acid protects against endogenous oxidative DNA damage in human sperm. *Proc Natl Acad Sci USA* 88:11003, 1991.
59. S. S. Mirvish: Experimental evidence for inhibition of *N*-nitroso compound formation as a factor in the negative correlation between vitamin C consumption and the incidence of certain cancers. *Cancer Res* 54:1948s, 1994.
60. P. R. Taylor, B. Li, S. M. Dawsey, J-Y. Li, C. S. Yang, W. Guo, W. J. Blot, and the Linxian Nutrition Trials Study Group: Prevention of esophageal cancer: The nutrition intervention trials in Linxian, China. *Cancer Res* 54:2029s, 1994.
61. R. M. Bostick, J. D. Potter, D. R. McKenzie, T. A. Sellers, L. H. Kushi, K. A. Steinmetz and A. R. Folsom: Reduced risk of colon with high intake of vitamin E: The Iowa Women's Health Study. *Cancer Res* 53:4230, 1993.
62. D. J. Hunter, J. E. Manson, G. A. Colditz, M. J. Stampfer, B. Rosner, C. H. Hennekens, F. E. Speizer, and W. C. Willett: A prospective study of the intake of vitamins C, E, and A and the risk of breast cancer. *N Engl J Med* 329:234, 1993.
63. A. R. Kennedy: Prevention of carcinogenesis by protease inhibitors. *Cancer Res* 54:1999s, 1994.

Index